Problems and Solutions of

Control Systems

with Essential Theory

Seventh Edition

AK JAIRATH ME, FIE, FIETE

Head, Department of Electrical Engineering
Former Chief Engineer and Deputy Director
Amity School of Engineering and Technology
Noida, India

CBS

CBS Publishers & Distributors Pvt Ltd

New Delhi • Bengaluru • Chennai • Kochi • Kolkata • Mumbai
Hyderabad • Jharkhand • Nagpur • Patna • Pune • Uttarakhand

Problems and Solutions of

Control Systems
with Essential Theory

Seventh Edition

ISBN: 978-81-948986-9-6

Copyright © Author and Publisher

Seventh Edition: 2021
First Edition: 1992; reprinted 1993
Second Edition: 1994; reprinted 1995, 1997, 1998
Third Edition: 1999; reprinted 2000, 2002
Fourth Edition: 2003; reprinted 2003, 2004, 2006, 2007
Fifth Edition: 2009
Sixth Edition: 2015

Published by Satish Kumar Jain and produced by Varun Jain for
CBS Publishers & Distributors Pvt Ltd
4819/XI Prahlad Street, 24 Ansari Road, Daryaganj, New Delhi 110 002, India.
Ph: 23289259, 23266861, 23266867   Website: www.cbspd.com
Fax: 011-23243014   e-mail: delhi@cbspd.com; cbspubs@airtelmail.in.
Corporate Office: 204 FIE, Industrial Area, Patparganj, Delhi 110 092
Ph: 4934 4934   Fax: 4934 4935   e-mail: publishing@cbspd.com; publicity@cbspd.com

Branches

- **Bengaluru:** Seema House 2975, 17th Cross, K.R. Road,
  Banasankari 2nd Stage, Bengaluru 560 070, Karnataka
  Ph: +91-80-26771678/79   Fax: +91-80-26771680   e-mail: bangalore@cbspd.com
- **Chennai:** 7, Subbaraya Street, Shenoy Nagar, Chennai 600 030, Tamil Nadu
  Ph: +91-44-26680620, 26681266   Fax: +91-44-42032115   e-mail: chennai@cbspd.com
- **Kochi:** 42/1325, 1326, Power House Road, Opp KSEB Power House,
  Ernakulam 682 018, Kochi, Kerala
  Ph: +91-484-4059061-65   Fax: +91-484-4059065   e-mail: kochi@cbspd.com
- **Kolkata:** 6/B, Ground Floor, Rameswar Shaw Road, Kolkata-700 014, West Bengal
  Ph: +91-33-22891126, 22891127, 22891128   e-mail: kolkata@cbspd.com
- **Mumbai:** PWD Shed, Gala no. 25/26, Ramchandra Bhatt Marg, Next to J.J. Hospital Gate No. 2
  Opp. Union Bank of India, Noorbaug, Mumbai-400009, Maharashtra, India
  Ph: +91-22-24902340/41   e-mail: mumbai@cbspd.com

Representatives

• **Hyderabad**	0-9885175004	• **Jharkhand**	0-9811541605	• **Nagpur**	0-9421945513
• **Patna**	0-9334159340	• **Pune**	0-9623451994	• **Uttarakhand**	0-9716462459

Printed at: Mudrak, Noida, UP, India

to
_____

my wife **Geeta** and son **Varun**
who are my source of encouragement and inspiration
and to
my dearest father late **Shri Ram Prasad Jairath** and
mother late **Shrimati Prem Rani Jairath**,
my elder brothers

Lt Col Ashok Jairath, Retd
and
Dr Subhash Jairath
who are responsible for
what I am today.

Preface to the Seventh Edition

It is a matter of immense happiness and satisfaction that in a span of 28 years, I have been able to publish seventh edition of this book. While teaching this subject in an engineering college and to the students preparing for GATE and IES, I was very pleased to discover that this book is very popular amongst the teachers and the students of control engineering. The book has again been carefully and thoroughly reviewed. An additional chapter on "Frequency response analysis" has been added to benefit the students.

I once again express my heartfelt gratitude to my publisher, CBS Publishers & Distributors in supporting my efforts.

AK Jairath

Prefaces to the Previous Editions

Preface to the Sixth Edition

It gives me immense pleasure to bring out the sixth edition of my book which over the last 22 years has become immensely popular with the faculty as well as the students who are pursuing undergraduate programs and also those students who are appearing in the professional examinations like GATE, IES, etc. The sixth edition has been thoroughly revised and restructured to ensure that it is student-friendly. Problems which already existed have been suitably clubbed under new chapters for easier understanding of the topics. New problems have been added in each chapter to make the concepts more clear and to give the students extra problem-solving practice. Chapter on feedback and its effect on system performance has been added. Nyquist stability criterion chapter has been fully revised and the explanation for each problem in this chapter will ensure better assimilation. Chapter 1 has been revised and the concept of drawing block diagrams for practical systems have been explained in a very simple way. Students must know how to recognize controllers, processes/plants and feedback elements of various types of control systems. I am sure that the explanation of the basics will lay the foundation for the control system subject in totality and make understanding of the other related topics easier.

It will be my endeavour to keep on improving the book in the coming years and the faculty and students will be seeing new editions of this book shortly. I plan to include design problems with case studies in my future editions. My aim is to ensure that this book is error-free. My students in Amity University have helped me a lot in pointing out typing/spelling errors and I owe a lot to them. I will be grateful to all the readers of this book to point out such mistakes in future also. I will acknowledge their efforts suitably.

I thank my publishers, CBS Publishers & Distributors, in supporting my efforts. Ours have been a very long association and hope this bond will last forever.

AK Jairath

Preface to the Fifth Edition

It gives me immense pleasure in bringing out the fifth edition of this book which has the patronage of the student community all over India. The main endeavour of this book is to make available a series of worked exercises to assist students to prevail over difficulties in understanding and correlating with the theory. The objective has been not to just present an assortment of problems with solutions, but by means of straightforward preparatory problems leading to intricate ones, introduce the students to the fundamentals of control engineering. In keeping with the aim, this edition has been compiled in a fashion to ensure the following:

- Step-by-step mathematical explanation for better understanding.
- Basics are explained in simple language.
- All possible variations to the principles of control engineering have been explained and illustrated by solving different problems.
- The problems and their solutions are listed in the order of difficulty and complexity.
- Step-by-step explanation will help the students in saving time which can be devoted elsewhere, fruitfully.
- Most of the problems have appeared in different examinations. This will help the students gain more confidence in tackling the examinations.
- Essential theory included under each chapter in the beginning is to ensure that students do not refer to any textbook for cross reference. It will also help students to immediately correlate to various graphical illustrations depicted in the theory portions.
- To keep the size and cost of the book within realistic limits, introductory material in the form of indispensable theory to each chapter contains only the vital features of the topics.

I once again thank CBS Publishers & Distributors and their publishing staff for providing help and support. I sincerely hope that this edition of the book will be as popular as the previous ones.

AK Jairath

Preface to the Fourth Edition

A decade has passed and with the extreme cooperation from the students and the publisher, I present the fourth revised edition. The book has been thoroughly revised, especially the chapter on "Root-locus". Basic essential theory on sensitivity has been added in the fourth chapter. In addition, important solved problems in each chapter have been incorporated.

I have no doubt that the fourth revised edition, with minimum errors will help the readers to assimilate the subject better.

AK Jairath

Preface to the Third Edition

It is a matter of happiness for me that within a short span of seven years this book has become immensely popular and the students have benefited after reading it. It has been my endeavour to improve the book constantly. Keeping this in view, I have revised the third edition by adding most essential theory which will help students to assimilate the subject better.

I have no doubt that the present revised edition will be liked by all the readers.

AK Jairath

Preface to the Second Edition

It gives me immense pleasure and satisfaction in bringing out this book in your hands. This book contains typical examples in control system engineering divided into various chapters. A brief giving out the basic theory has been added at the begining of each chapter which I am sure will help the students in understanding the solved problems. Problems have been solved with detailed explanation for better assimilation.

Effort has been made to keep printing mistakes to the minimum. Even then if the students are able to find any technical or printing mistakes, please bring the same to the notice of the author for rectification in the next edition. Also, any suggestions to improve this book are welcome. I thank CBS Publishers & Distributors for their cooperation and help.

AK Jairath

Contents

Preface to the Seventh Edition vii

Prefaces to the Previous Editions ix

1. CONTROL SYSTEMS: BASICS **1**

1.1 Introduction 1
1.2 Classification of Systems 1
1.3 Classification Based on the Parameters 2
1.4 Analysis of Control Systems 3
1.5 General Classification: Open and Closed-Loop Systems 3
1.6 Elements of Automatic or Feedback Control Systems 5
1.7 Requirements of Automatic Control Systems 6

2. MODELLING OF CONTROL SYSTEMS: ELECTRICAL SYSTEMS **20**

2.1 Introduction 20
2.2 Transfer Function 20

**3. MODELLING OF CONTROL SYSTEMS:
BLOCK DIAGRAM METHOD** **46**

3.1 Introduction 46
3.2 Block Diagram 46
3.3 Block Diagram of a Closed-Loop System 47

4. MODELLING OF CONTROL SYSTEMS: SIGNAL FLOW GRAPH **92**

4.1 Introduction 92
4.2 Signal Flow Graph 92
4.3 Signal-Flow Diagram Reduction 94
4.4 Mason's Gain Formula 95

5. MODELLING OF CONTROL SYSTEMS: PHYSICAL SYSTEMS **181**

5.1 Introduction 181

5.2 Basic Elements   181
5.3 Translatory Motion   181
5.4 Rotational Motion   183
5.5 Analogous System   185

6. TIME DOMAIN ANALYSIS   **227**

6.1 Introduction   227
6.2 Standard Test Signals   227
6.3 Static Accuracy   229
6.4 Computation of Steady State Errors   231
6.5 Transient Response: First Order System   233
6.6 Transient Response: Second Order System   237
6.7 Transient Response Specifications   242
6.8 Control Actions   246
6.9 Impulse Signal and Response   258
6.10 Sensitivity   259

7. CONTROL SYSTEMS: FEEDBACK CHARACTERISTICS   **327**

7.1 Introduction   327
7.2 Steady State Error/Accuracy   327
7.3 Sensitivity   329
7.4 Overall Gain   332
7.5 Time Constant   332
7.6 Stability   334
7.7 Disturbance   335

8. FREQUENCY RESPONSE ANALYSIS   **348**

8.1 General   348
8.2 Frequency Domain Specifications   349
8.3 Magnitude and Phase Angle Characteristics Plot   349
8.4 Frequency Response Specifications   350
8.5 Representation of Sinusoidal Transfer Function   351

9. POLAR PLOTS   **361**

9.1 General   361
9.2 Polar Plot Sketching Procedure   361

10. STABILITY   **373**

10.1 Introduction   373

10.2 Stability 373
10.3 Stability Study Methods 373

11. ROUTH'S STABILITY CRITERION 376

11.1 Introduction 376
11.2 Procedure for Application 376

12. NYQUIST STABILITY CRITERION 407

12.1 Introduction 407
12.2 Relationship 407
12.3 Encirclement 408
12.4 Enclosed 409
12.5 Stability Condition 410
12.6 Principle of Argument 411
12.7 Nyquist Contour and Nyquist Plot 414

13. BODE PLOT 483

13.1 Introduction 483
13.2 Method of Plotting 483

14. ROOT LOCUS 542

14.1 Introduction 542
14.2 Simple Control System 543
14.3 General Rules 543

15. STATE SPACE ANALYSIS 610

15.1 General 610
15.2 State 610
15.3 State Variables 610
15.4 State Vector 610
15.5 State Space 610
15.6 Controllability and Observability 611
15.7 Controllability 611
15.8 Observability 612

Control Systems: Basics

1.1 Introduction

Control system is a science which deals with systems, mechanisms, devices or collection of objects joined to have some form of interaction with a purpose. The arrangement of devices in a control system, their design and appearance vary with the objectives defined. Such systems may be manually controlled or totally independent of human involvement. With the advancement in science and changing requirements, the *human element* in a control system has almost been eliminated. The requirement today is of such automatic control systems which can serve the human needs.

A system is a basic unit consisting of an element or number of elements working in cohesion and desirably a single cohesive unit to perform desired controlled function. The main criterion is that this cohesive unit when actuated with any predetermined and predefined input signal, acts upon it and gives the desired output, after taking control over the disturbances, nonlinearities. Depending upon the nature or requirement, a system may be electrical, mechanical, pneumatic, hydraulic, digital, analog, etc.

1.2 Classification of Systems

There are several ways in which control systems can be classified. Control systems can be classified based on state, principle of superposition, nature of signal flow and also on input/output signal.

1.2.1 When classified on *State*, a system can be *Static* or *Dynamic*. In a *static* system the steady state values are reached instantly and remain for a long time, on application of input. *Dynamic* systems exhibit transients when subjected to input signal because of certain energy storage elements. A purely resistive circuit can be called a static system and an RLC circuit is an example of dynamic system. Based on the *principle of superposition*, the systems may be further classified as *Linear* and *Nonlinear*. When classified in terms of the *nature*, the system may be *Continuous or Discontinuous*. Based on the number of input/output signals, a system is further classified as *single-input-single-output* (SISO) or *multiple-input-multiple-output* (MIMO) systems.

1

1.3 Classification Based on the Parameters

Depending upon the parameters and the equations which describe the functioning of a system, a system may be classified as given below:

1.3.1 Distributed and Lumped Parameter Systems
Lumped parameter systems are described by simple differential equations whereas *Distributed* parameter systems make use of partial differential equations.

1.3.2 Random and Non-random Systems
Above mentioned systems may be *Random* or *Non-random*. In *Random* systems, describing function is based upon inputs whose characteristics are defined only in statistical terms. These are also called *Stochastic Systems*. If the input signal is *deterministic* or the parameter of the system can be described precisely all throughout, then such systems are called *Non-random* or *Deterministic systems*.

1.3.3 Discrete and Continuous Time Systems
Above mentioned systems can be further classified as *Discrete time* systems and *Continuous time* systems. A system in which all the variables are continuous with respect to time are called *Continuous time* systems. On the other hand, if few variables in a system are discrete with respect to time they are termed as *Discrete time* systems. *Discrete time* systems can be *Sample data* systems and *Digital* systems.

1.3.4 Linear and Nonlinear Systems
A *linear* system is one which satisfies the principle of superposition and principle of homogeneity. The output and input relationship can be represented on *x-y* plane by a straight line. Such systems can be easily analysed by simple mathematical tools. A *nonlinear* system on the other hand does not obey the principle of superposition and homogeneity. The nonlinearities in a system are generally caused by hysteresis, saturation effect in the components, frictional forces, etc. Any physical system when analysed in detail is nonlinear in nature. A linear system is difficult to realise as nonlinearities prop up. Such systems are approximately linearised so that application of linear mathematical equations can be made use of, though it may affect accuracy of results. This can be done if the deviation from the linearity is small and is not considered important enough for the problem under consideration.

1.3.5 Time-varying and Time-invariant Systems
Above mentioned systems may be *Time-varying* or *Time-invariant* in nature. If one or more parameters of a system vary with time, the system is termed as "*Time-varying*" and if all the system parameters are constant with respect to time, the system is called "*Time-invariant*" system.

1.4 Analysis of Control Systems

Analysis of control systems is followed by optimisation which means system improvement. Thereafter, the physical realisation of the proposed system is achieved. However, analysis of a control system is proceeded by:

(a) Specifications of the performance.

(b) Representation of the system and its subsystems in mathematical form.

(c) Deriving the mathematical equations by use of basic principles such as Voltage and Current laws in electrical systems, Newton's laws for mechanical systems, laws of Thermodynamics in fluid systems, by incorporating various forms of equation such as linear and nonlinear equations, differential, integral equations, etc.

(d) Simulation of the system by using analog, hybrid and computer simulation methods.

1.5 General Classification: Open and Closed-loop Systems

Control systems are analysed by forming a mathematical model and described by mathematical equations. These equations may be ordinary differential equations or difference equations. Hence, it would be correct to classify the systems based on their describing equations. Since this book is dealing with linear time-invariant systems, the readers need not be burdened with other definitions. In general, control system can be classified as either *open-loop* or *closed-loop*. The basic difference between these two types of control systems is the presence of feedback element in the closed-loop systems.

1.5.1 Open-loop Control System
This is also called a control system without feedback. This is the most simplest and economical type of control system. The block diagram representation of such a control system is depicted in Fig. 1.1. Such systems are cheap and easily assembled. However, it has many disadvantages in the sense that exact behaviour of the output is not available and the system does not change by itself to external disturbances. Human operator is required to control certain conditions to have desired output. A field controlled DC motor is an example of an open-loop control system. Such a control system is illustrated in Fig. 1.2. The cutting wheel shown is being driven at a constant speed. When a disturbing torque in the form of a piece of wood to be cut comes in contact with the cutting wheel, the speed of the cutting wheel reduces. Output cannot correct itself to the disturbing torque, such a system cannot correct for variations, either manually or automatically.

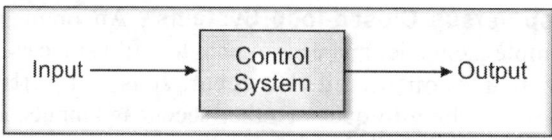

Fig. 1.1 Open-loop control system

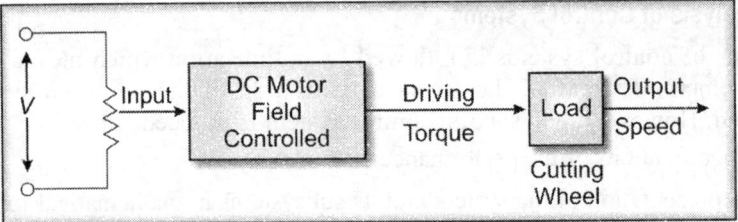

Fig. 1.2 A field-controlled DC motor

1.5.2 Closed-loop Control System Such systems are also termed control systems with feedback. These systems are costlier to build, difficult to assemble. However, major advantages associated with such systems are that they are self correcting to external disturbances and have faster response. Fig. 1.3 depicts a closed-loop control system. A part of terminal voltage KV_o has been fed back in the input so as to oppose the input voltage V_i. When the load is applied to the machine, the terminal voltage reduces. This in turn increases $A(V_i - KV_o)$ which increases V_g and on reaching steady-state, V_o is brought to the original value.

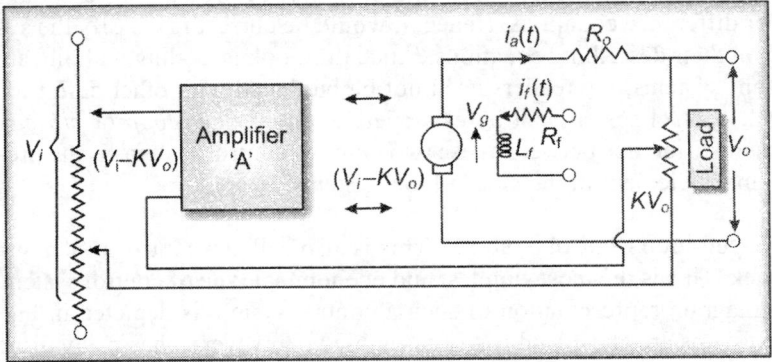

Fig. 1.3 Closed-loop control system

Consider the case when the load is removed. This increases KV_o and in turn reduces $A(V_i - KV_o)$ and V_g and as a result of reduction in excitation voltage V_g, the terminal voltage V_o comes back to original value. Students must note here that the feedback is negative. If the feedback is positive in nature, any decrease in the terminal voltage will decrease the excitation voltage and any increase in terminal voltage will increase the excitation voltage and the system will become unstable.

1.5.3 Open-loop versus Closed-loop Systems An open-loop system is simple, least complex, cheaper and economical to build since stability is not the major problem to be considered. However, it is not self correcting on encountering any disturbing torques or other secondary inputs. Input has to be manually corrected. Bandwidth is low and there is more effect of noise and disturbance. They also increase the effect of nonlinearity and distortion and hence, system is nonlinear. On the other hand feedback element present in a closed-loop system presents inherent capability to sense errors, disturbances and correct itself. Such systems are complex in design and costlier. Unless properly designed, a closed-loop system may inadvertently act as an oscillator due to overcorrecting tendency. Hence, stability presents a major problem to the control system designer of a closed-loop system.

Automatic control systems are closed-loop systems, which automatically respond to disturbances and variations without the presence or indulgence of human operator. Manual control systems are also closed-loop systems but in this, corrections are done by human operator. Automatic control systems are more accurate, expensive, complex in design, and with reduced effect of non-linearity and distortion. The error is continuously measured through feedback and self corrects at all instances of disturbances. Bandwidth of such systems is high.

1.6 Elements of Automatic or Feedback Control Systems

A general block diagram of an automatic feedback control system is shown in Fig. 1.4. Definitions of certain terms widely used in the feedback control systems are:

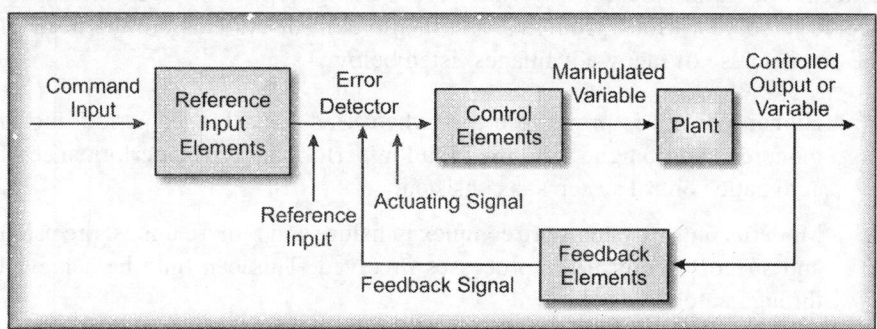

Fig. 1.4 Block diagram of an automatic feedback control system

✦ **Command Input:** Externally produced input independent of the feedback control system.

✦ **Reference Input Elements:** Elements which produce a signal proportional to the command. These are also called input transducers.

✦ **Reference Input:** It is the actual input signal proportional to the command.

✦ **Error Detector:** It is an element which sums or compares the signal obtained from feedback elements with the reference input signal.

✦ **Error or Actuating Signal:** It is the signal which is the difference between reference input and feedback signal. Since, the level of the signal is low to actuate the actuating device, it needs to be amplified.

✦ **Control Elements:** Elements which produce the desired output from the actuating signal. It in the form of amplifier. This amplitude signal actuates the actuating device.

✦ **Plant:** It is the device to be controlled.

✦ **Controlled Output:** It is the quantity which is required to be controlled at the desired level.

✦ **Feedback Elements:** Elements which help in providing the controlled output to be fed back to the error detector for comparison with the Reference Input Signal. Feedback elements are the transducers which convert the Controlled Output to a dimension which is generated by Reference Input Elements. Feedback increases accuracy, reduces sensitivity and effect of non-linearity and distortion, reduces time constant and increases bandwidth. Feedback controls the system dynamics by suitably adjusting roots of characteristic equation. However, it may cause instability even if open-loop system is stable. It decreases the gain of the system. Feedback can be negative as well as positive.

1.7 Requirements of Automatic Control Systems

Automatic control systems are needed today in all walks of life and the concepts of automatic control systems are being applied in ever increasing degree to solve various problems. Automatic control systems are the need of the day because of many advantages listed below:

(a) Human control is prone to errors when used for a longer period due to monotony and fatigue resulting in fall in performance. The performance of automatic control systems is consistent.

(b) Modern control systems are complex in nature and require utmost precision and speed to control the processes involved. This can only be achieved through automatic control.

(c) Employing automatic control reduces the labour element and wages in any industry, thereby indirectly increasing the profitability.

(d) Automatic control improves the quality of product being manufactured and ensures reduced scrap rate.

(e) Automatic control stops the operation of the control system immediately on occurrence of fault before an extensive damage occurs to the complete system. Hence, the maintenance and replacement cost is reduced.

Solved Problems

Problem 1.1 A servomotor having a linear torque-speed curve with T_s = Stall torque and No-load speed = N_o is shown in Fig. 1.5. Derive an expression for P_o (Shaft output power) vs. speed for $m = 1$. Also, calculate the following:

(a) The speed at which P_o is maximum

(b) The value of maximum P_o in terms of stall torque and no-load speed.

(c) What is the value of maximum power if $T_s = 0.03$ Nm and $N_o = 5000$ rpm and at what speed it will occur.

(d) Rotor moment of inertia is 4×10^{-7} kgm^2, find the maximum rotor acceleration at $N = 0$, 2500 rpm and 5000 rpm.

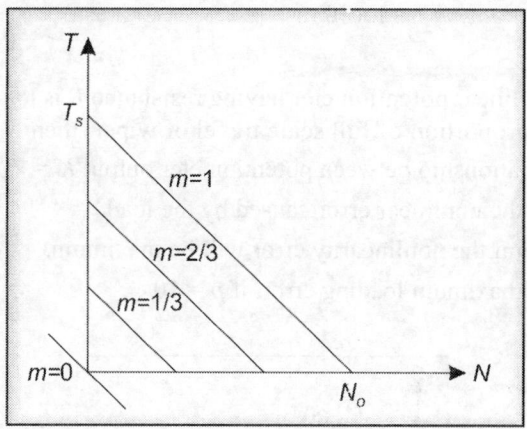

Fig. 1.5

Solution

The torque (T) and speed (N) relationship is

$$P_o = TN$$

From the linear relationship $T = T_s\left(1 - \dfrac{N}{N_o}\right)$

Substitution gives $P_o = T_s\left(1 - \dfrac{N}{N_o}\right)N$

(a) $\qquad \dfrac{dP_o}{dN} = T_s\left(1 - \dfrac{2N}{N_o}\right)$

Equating to zero, we get $T_s\left(1 - \dfrac{2N}{N_o}\right) = 0$ or $N = \dfrac{N_o}{2}$

(b) Therefore, $P_o = P_o(\text{max}) = T_s\left[\dfrac{N_o}{2} - \dfrac{1}{N_o}\left(\dfrac{N_o^2}{4}\right)\right] = \dfrac{1}{4}N_o T_s$

(c) $\quad P_o(\text{max}) = \dfrac{1}{4}\left(\dfrac{5000}{60/2\pi}\right)\left(\dfrac{0.03}{1}\right) = 3.93$ watts.

Speed at which maximum power will occur $N = \dfrac{5000}{2} = 2500$ rpm

(d) $T = J\alpha$ or $\alpha = \dfrac{T}{J} = \dfrac{0.03}{4 \times 10^{-7}} = 75000$ rad/sec^2

Problem 1.2 A linear potentiometer having resistance R is loaded by a resistor pR. If 'a' is the proportion of full scale travel of wiper, then

(a) Find the relationship between potentiometer output $K = V_o/V_i$ and a.

(b) Determine the nonlinear error caused by the load.

(c) At what point the nonlinearity error will be maximum.

(d) Determine maximum loading error, if $p = 10$

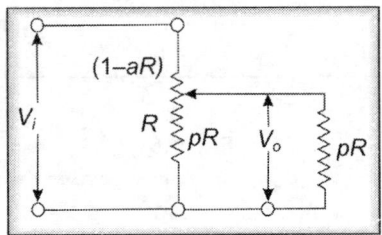

Fig. 1.6

Solution

(a) $\quad K = \dfrac{V_o}{V_i} = \dfrac{\dfrac{aR(pR)}{aR + pR}}{\dfrac{aR(pR)}{aR + pR} + (1 - a)R}$

or $\quad K = \dfrac{apR^2}{apR^2 + (1 - a)R^2(a - p)} = \dfrac{ap}{ap + a + p - a^2 - ap} = \dfrac{ap}{a - a^2 + p}$

(b) error(e) $= a - \dfrac{ap}{a - a^2 + p} = \dfrac{a^2(1 - a)}{a - a^2 + p} \simeq \dfrac{a^2(1 - a)}{p}$ for light loading.

(c) $\dfrac{de}{da} = \dfrac{d}{da}\dfrac{a^2(1 - a)}{p} = 0$

or $\quad a(2 - 3a) = 0,$ which gives $a = 0, 2/3.$

(d) $e = \dfrac{a^2(1-a)}{p}$

Putting $a = 2/3$ and $p = 10$, we get $e = \dfrac{\left(\dfrac{2}{3}\right)^2\left(1 - \dfrac{2}{3}\right)}{10} = \dfrac{4}{270} \simeq 1.5\%$

Problem 1.3 A tachometer has a gain of 0.05 V/rad/sec, Find

(a) The output voltage when the shaft speed is 40 degree/sec

(b) The output voltage when the shaft speed in 20 rad/sec

(c) The shaft speed in rad/sec and degree/sec when the output voltage is 1.8 V.

Solution Given, tachometer gain $K_T = 0.05$ V/rad/sec

(a) Output voltage $= K_T \times$ shaft speed in rad/sec $= \dfrac{0.05 \times 40 \times \pi}{180} = 0.035$ volt

(b) Output voltage $= 0.05 \times 20 = 1$ volt

(c) (i) Shaft speed (rad/sec) $= \dfrac{\text{output voltage}}{K_T} = \dfrac{1.8}{0.05} = 36$ rad/sec

(ii) Shaft speed (degree/sec) $= \dfrac{36 \times 180}{\pi} = 2061$ degree/sec

Problem 1.4 A two-turn 200 winding; 10 kW wire wound potentiometer has a reference of 20 V impressed across its terminals. Assume that it turns a full 720°. Find potentiometer constant or gain in (a) volts per degree, (b) volts per winding, (c) volts per radian, (d) volts per turn.

Solution

(a) Potentiometer gain in volts/degree $= \dfrac{\text{volt}}{\text{degree}} = \dfrac{20}{720} = \dfrac{1}{36} = 0.028$ volt/degree

(b) Potentiometer gain in volts/winding $= \dfrac{\text{volt}}{\text{No. of winding}} = \dfrac{20}{200} = 0.1$ volt/winding

(c) Potentiometer gain in volts/radian $= \dfrac{20}{720} \times \dfrac{180}{\pi} = 1.6$ volt/radian

(d) Potentiometer gain in volts/turn $= \dfrac{\text{volt}}{\text{No. of turns}} = \dfrac{10}{2} = 5$ volts/turn

Problem 1.5 (a) Find out the voltage resolution in percentage of a 1000 turns potentiometer, if the applied voltage is 20 V.

(b) Find out the angular resolution of 360 degree potentiometer having 7200 turns.

(c) Calculate the normal linearity of a potentiometer, if at the maximum permissible deviation point, the slider voltage is equal to 3.9 V, while the straight line gives a value of 4 V. The excitation level is 25 V.

Solution

(a) Voltage resolution $= \dfrac{\text{volt/turn}}{\text{volt}} \times 100 = \dfrac{20}{1000 \times 20} \times 100 = \dfrac{1}{10} = 0.1\%$

(b) Angular resolution $= \dfrac{\text{degree}}{\text{turns}} = \dfrac{360}{7200} = \dfrac{1}{20} = 0.05$ degree/turn

(c) Normal linearity (in %) $= \dfrac{\text{straight line value} - \text{slider voltage}}{\text{Excitation voltage}} = \dfrac{4 - 3.9}{25} \times 100$

$$= 0.4\%$$

Problem 1.6 Draw the block diagram representation of teacher–student learning system as open-loop and closed-loop system.

Solution The system components are

- Objective   : To exchange knowledge from teacher to students.
- Controller   : Teacher
- Process/Plant   : Student
- Control Inputs   : Desired knowledge to be imparted
- Outputs   : Knowledge acquired by the process
- Disturbances   : The entire environment other than academic environment
- Feedback   : Tests, Examinations, etc.

The open-loop system can be represented as:

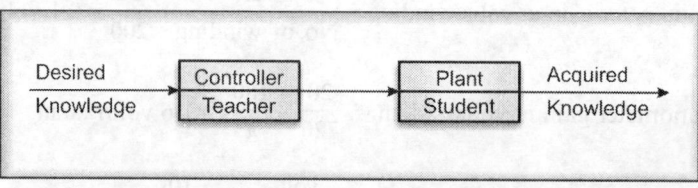

Fig. 1.7

The closed system caters for feedback also and can be represented as:

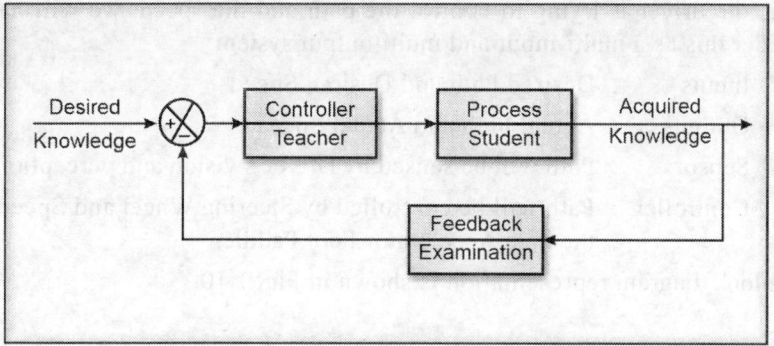

Fig. 1.8

Problem 1.7 Draw the block diagram representation of driver driving a car and trying to maintain the direction through steering wheel.

Solution
- Control Objective : To control direction of car
- Controller : Driver
- Process/Plant : Car
- Disturbances : Wind, Rain, Road surface, etc.
- Control Inputs : Desired direction of travel
- Output : Actual course of travel.

The block diagram representation is given in Fig. 1.9.

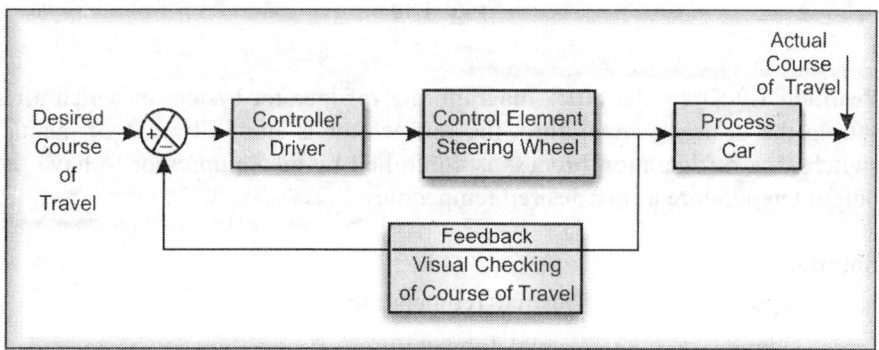

Fig. 1.9

Problem 1.8 Draw a block diagram of a system in which a driver is driving the car and is trying to control the direction and the speed.

Solution

Since, the driver is trying to control the path and the speed, we will have to consider this as a multi-input and multi-output system.

Inputs   : Desired Path and Desired Speed.

Outputs   : Actual Path and Actual Speed.

Sensors   : Path will be sensed by Driver's vision and perception, and

Controller : Path will be controlled by Steering Wheel and Speed with the help Accelerator Foot Paddle.

The block diagram representation is shown in Fig. 1.10.

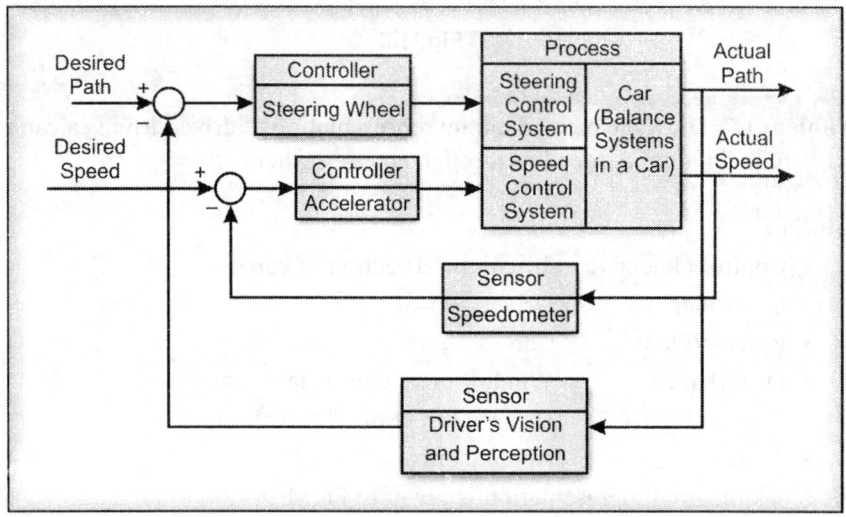

Fig. 1.10

Problem 1.9 Draw the block diagram of a refrigerator system in which after setting the desired temperature, the refrigerator is started by use of manual switch. The refrigeration process is controlled by the compressor to have the output temperature as per desired temperature.

Solution

Input   : Desired Temperature

Output   : Actual Temperature

Controller   : Manual Switch

Control Element : Compressor

Process   : Refrigerator

Sensor   : Temperature Sensor (Bi-metallic Strip)

The block diagram representation is shown in Fig. 1.11.

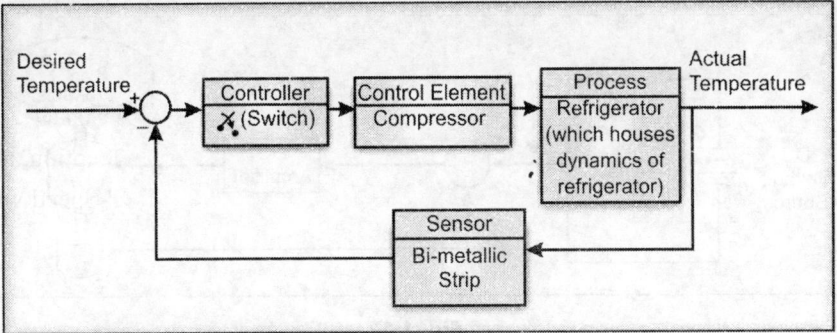

Fig. 1.11

Problem 1.10 It is desired to regulate the flow of water from the tank. The water into the tank is controlled by the valve manually and the measurements of the quantity of water is seen by the man by visual inspection of meter installed on the tank. Draw its block diagram representation.

Solution

Input	:	Desired Water Flow
Output	:	Actual Water Flow
Controller	:	Valve
Process	:	Tank
Sensor	:	Metre

The block diagram representation is shown in Fig. 1.12.

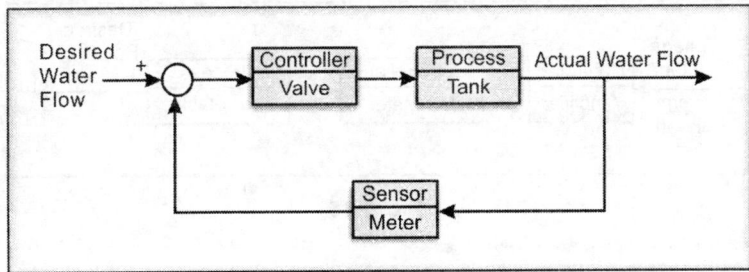

Fig. 1.12

Problem 1.11 It is desired to rotate the turntable with the help of a DC motor as shown in Fig. 1.13. The speed of the turntable is measured with the help of tachometer. Draw the block diagram of the closed-loop system is which the speed of the turntable is automatically controlled.

Solution

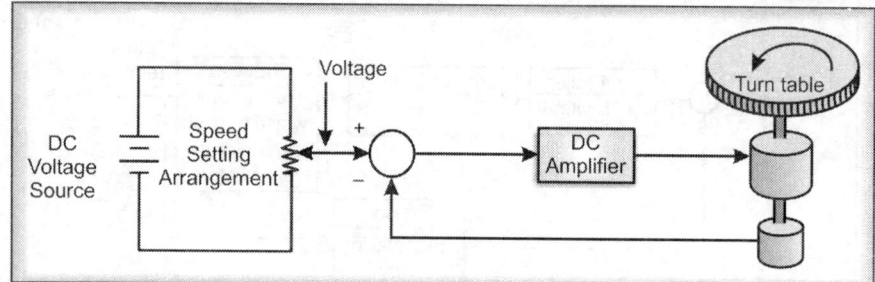

Fig. 1.13

The block diagram is shown in Fig. 1.14.

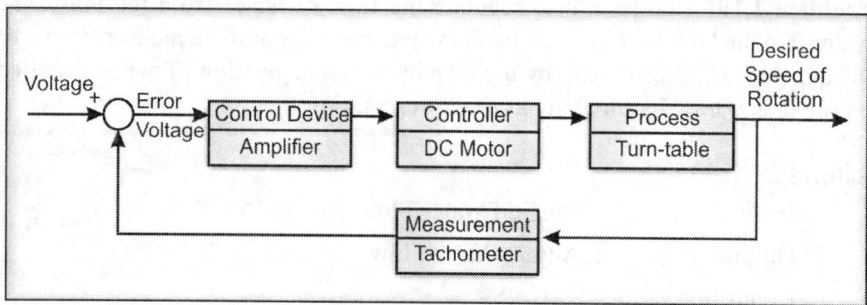

Fig. 1.14

The open-loop representation is shown in Fig. 1.15.

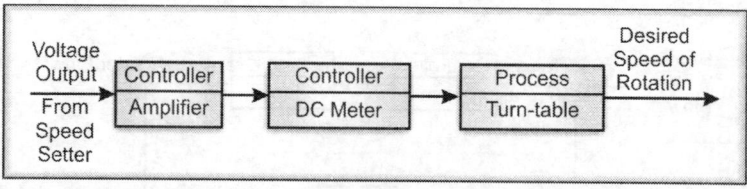

Fig. 1.15

Problem 1.12 It is desired to do air-conditioning of a room with the help of a window type air-conditioner. The air-conditioner is set to a desired temperature and put on by the manual switch which starts the compressor and the air-conditioning process controls the room temperature by sensing it through bimetallic strip.

Solution

The block diagram representation is shown in Fig. 1.16.

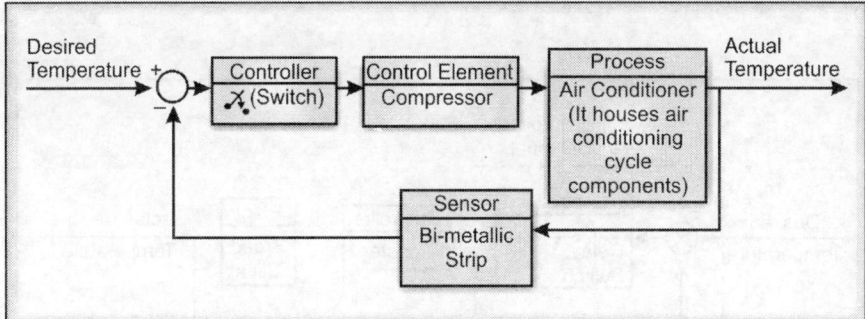

Fig. 1.16

Problem 1.13 It is desired to control the speed of rotating disc with the help of a DC Motor which obtains disc rotation speed proportional to the applied voltage. Tachometer measures the speed of the disc and converts it into output voltage proportional to the speed. The error voltage is amplified and applied to the motor.

Solution

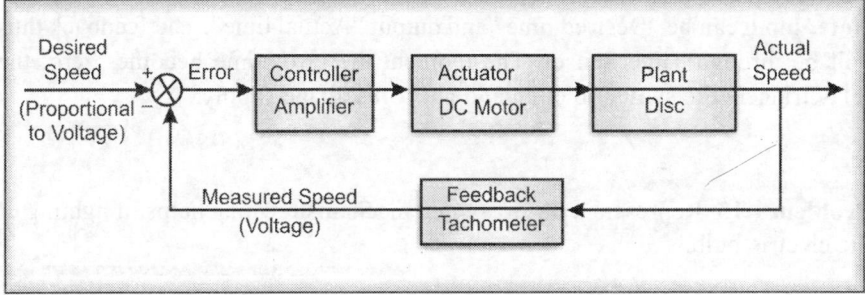

Fig. 1.17

Problem 1.14 Represent the automatic toaster system by block diagram method

(a) Bread is toasted based on time.

(b) Bread is toasted by the help of a temperature control.

Solution (a)

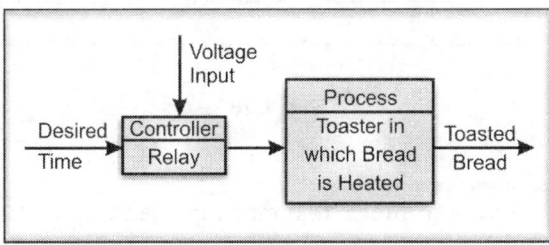

Fig. 1.18

(b)

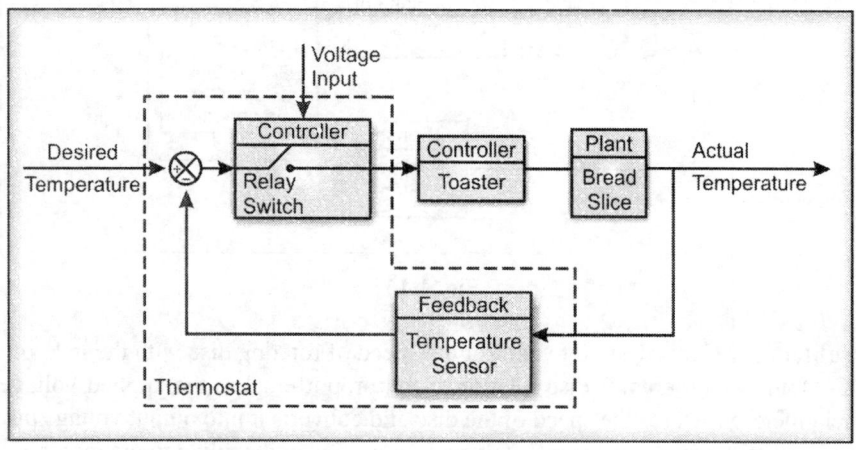

Fig. 1.19

Note: Input can be "Desired time" and output "Actual time". The feedback thus will be through Time Sensor. The moment the error time becomes zero, the relay triggers the switch to open and cuts off voltage supply.

Problem 1.15 Represent a on-off switch mechanism which helps in lighting of the electric bulb.

Solution

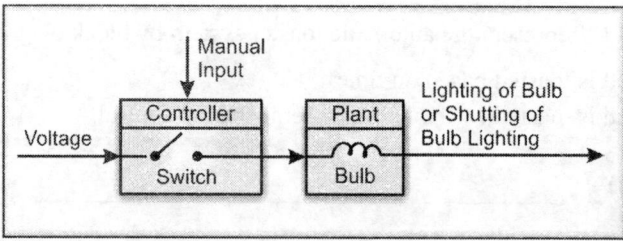

Fig. 1.20

Problem 1.16 Draw the block diagram representation of automatic air-conditioning control system.

Solution

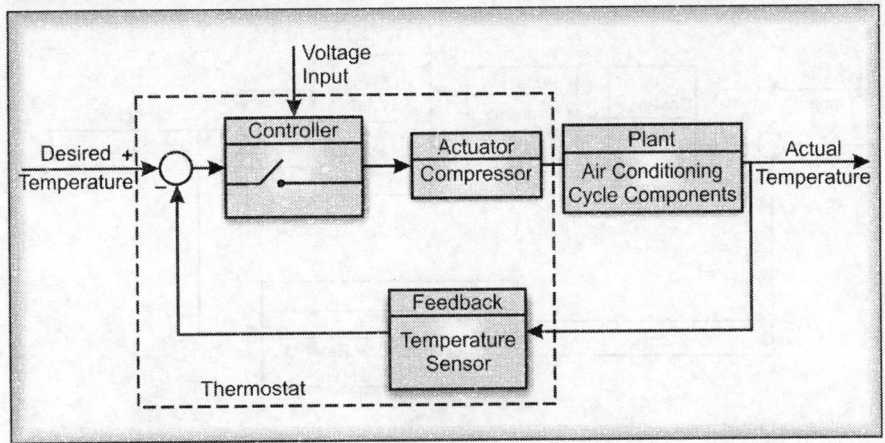

Fig. 1.21

Problem 1.17 Draw the block diagram representation of an automatic electric oven whose regulation is controlled through temperature setting.

Solution

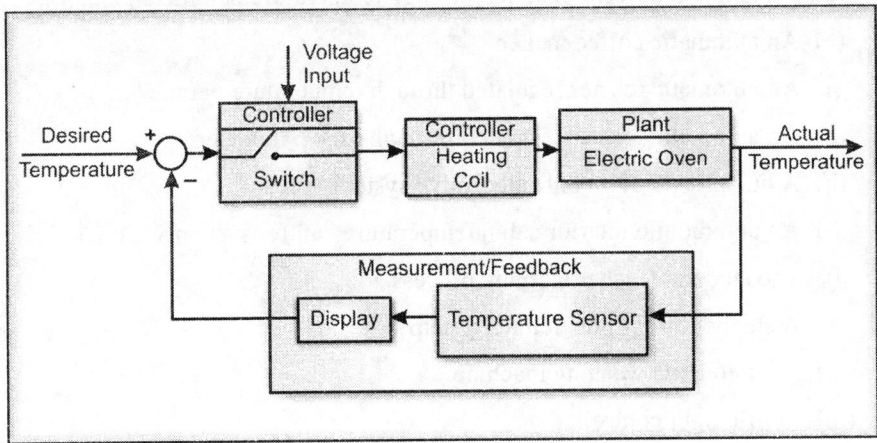

Fig. 1.22

Problem 1.18 Draw block diagram representation of hot and cold water mixing system.

Solution

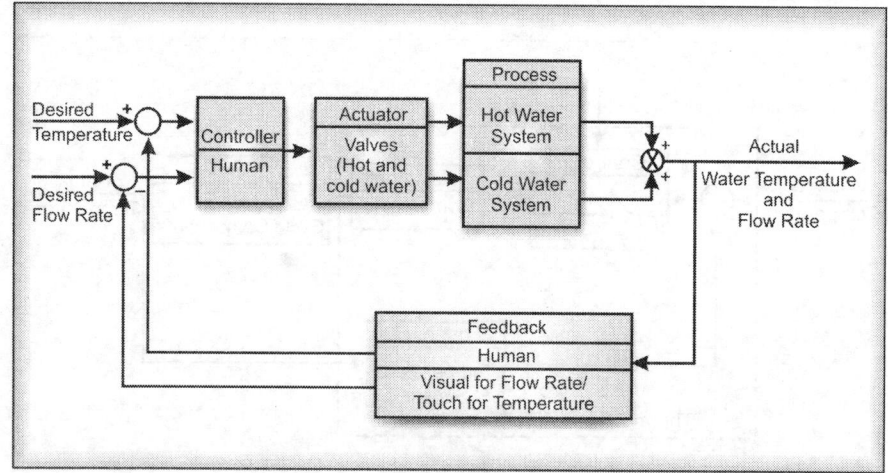

Fig. 1.23

Problem 1.19 Classify the following as Open or Closed-loop system.

(a) An electrical On-Off switch.

(b) A person driving a motor cycle.

(c) A deep freezer.

(d) An automatic oven regulated through temperature parameter.

(e) An automatic coffee maker.

(f) An automatic toaster regulated through temperature parameter.

(g) An automatic toaster regulated through time parameter.

(h) A human operator controlled valve system.

(i) An automobile interior cabin temperature control system.

(j) The student–teacher learning process.

(k) A stepper motor positioning system.

(l) An automatic washing machine.

(m) Traffic light controller.

(n) A room heater.

(o) Electric lift.

(p) Automatic dryer.

(q) Automatic dishwasher.

   (r) Household refrigerator.

   (s) Windscreen wiper

   (t) Missile

   (u) A man in a spaceship

   (v) Law of supply and demand

   (w) Room air-conditioner

   (x) Respiratory system

Solution

   (a) Open-loop

   (b) Closed-loop

   (c) Closed-loop

   (d) Closed-loop

   (e) Closed-loop

   (f) and (g) Closed-loop: However, if colour of the toast is the aim, then these are open-loop systems

   (h) Closed-loop

   (i) Closed-loop

   (j) If done through a feedback (examination), then it is a closed system

   (k) Open-loop

   (l) An open-loop system as the quality of cleanliness of clothes not ensured

   (m) Open-loop which can be converted into closed-loop, if the system senses the density of traffic

   (n) Open-loop as there is no provision of cut off

   (o) Closed-loop

   (p) Open-loop, as the amount of dryness is not sensed

   (q) Open-loop, as cleanliness of utensils is not ensured

   (r) Closed-loop

   (s) Open-loop

   (t) Closed-loop

   (u) Closed-loop

   (v) Closed-loop

   (w) Closed-loop

   (x) Closed-loop

Modelling of Control Systems: Electrical Systems

2.1 Introduction

The electrical circuit is made up of components like resistor, inductor and capacitor. Electrical circuits are analysed by use of Ohm's law, Kirchoff's law etc. A number of theorems like Thevenin's theorem, Norton's theorem, Superimposition theorem, etc. are available to solve them. Depending upon the requirements, any of them can be put to use. In control systems, we need to find the transfer function. It is obtained by writing differential equations which ultimately relates the input and output, following Kirchoff's law.

2.2 Transfer Function

Transfer function is defined as the ratio of the Laplace transform of the output signal to the Laplace transform of the input signal under the assumption that all initial conditions are zero.

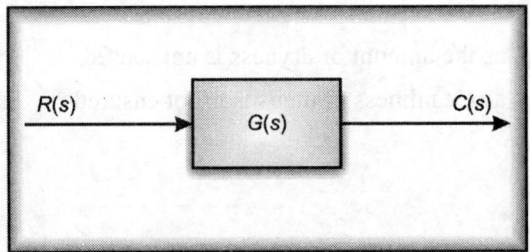

Fig. 2.1

$$\text{T.F.} = G(s) = \frac{C(s)}{R(s)}$$

2.3 The applicability and use of the concept of transfer function is:

  (a) It is limited to LTI differential equation systems.

  (b) It helps in writing the performance equations of the systems in s-domain.

  (c) It is property of the system per se. It is independent of nature and magnitude of the input.

  (d) The output response is analysed to ascertain the nature of the system for various input signals.

  (e) It does not provide information on the physical structure of the system.

  (f) It has no unit. It includes the units necessary to relate input and output of a system.

2.4 Table 2.1 gives the relationship of voltage and current under zero initial conditions for various components of the electrical circuit.

Table 2.1

Sr. No.	Component	Symbol	Current-voltage	Voltage-current
1.	Resistor	R	$i(t) = \dfrac{v(t)}{R}$	$v(t) = i(t)R$
2.	Inductor	L	$i(t) = \dfrac{1}{L}\int v(t)dt$	$v(t) = \dfrac{Ldi(t)}{dt}$
3.	Capacitor	C	$i(t) = C\dfrac{dv(t)}{dt}$	$v(t) = \dfrac{1}{C}\int i(t)dt$

2.5 In Laplace domain the relationship is tabulated in Table 2.2.

Table 2.2

Sr. No.	Component	Current-voltage	Voltage-current	Impedence $Z(s) = \dfrac{V(s)}{I(s)}$
1.	Resistor	$I(s) = \dfrac{V(s)}{R}$	$V(s) = RI(s)$	R
2.	Inductor	$I(s) = \dfrac{V(s)}{sL}$	$V(s) = sLI(s)$	sL
3.	Capacitor	$I(s) = sCV(s)$	$V(s) = \dfrac{I(s)}{sC}$	$\dfrac{1}{sC}$

$$\boxed{\textbf{SOLVED PROBLEMS}}$$

Problem 2.1 Derive the transfer function of the circuit as shown in Fig. 2.2

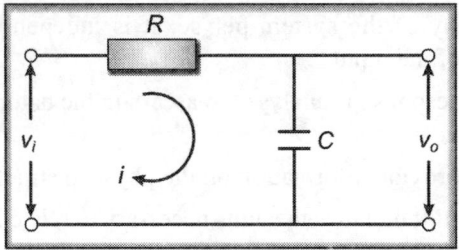

Fig. 2.2

Solution Writing the differential equations with the help of Kirchoff's law

$$v_i(t) = Ri(t) + v_o(t); \quad \text{and} \quad v_o(t) = \frac{1}{C}\int i(t)\, dt$$

Taking Laplace transform, we get

$$V_i(s) = RI(s) + V_o(s);$$

and

$$V_o(s) = \frac{1}{sC}I(s)$$

or

$$I(s) = sCV_o(s)$$

or

$$V_i(s) = sRCV_o(s) + V_o(s)$$

$$V_i(s) = V_o(s)(1 + sRC)$$

or

$$\frac{V_o(s)}{V_i(s)} = \frac{1}{1+sRC} = \frac{1}{1+sT} \quad \text{where,} \quad T = RC \qquad \textbf{Ans}$$

Problem 2.2 Derive the transfer function of the circuit as shown in Fig. 2.3.

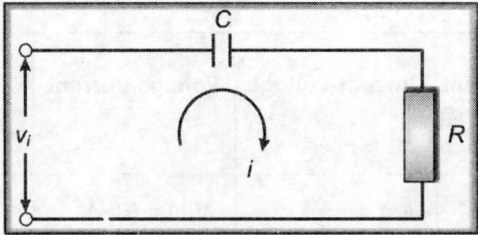

Fig. 2.3

Solution Writing the differential equation by Kirchoff's law, we get

$$v_i(t) = Ri(t) + \frac{1}{C}\int i(t)\, dt$$

Taking Laplace transform, we get

$$V_i(s) = RI(s) + \frac{1}{sC} I(s)$$

or $\quad \dfrac{I(s)}{V_i(s)} = \dfrac{1}{R + \dfrac{1}{sC}} = \dfrac{sC}{1 + sCR}$ **Ans**

Problem 2.3 Derive the transfer function of the circuit as shown in Fig. 2.4.

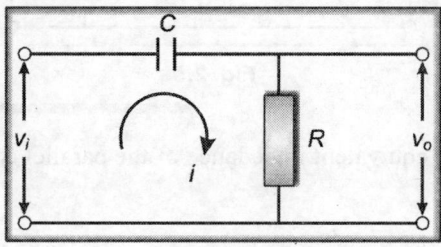

Fig. 2.4

Solution Writing the differential equations by Kirchoff's law, we get

$$v_i(t) = Ri(t) + \frac{1}{C}\int i(t)\, dt \tag{1}$$

$$v_o(t) = Ri(t) \tag{2}$$

Laplace transform gives

$$V_i(s) = RI(s) + \frac{I(s)}{sC} \tag{3}$$

and $\quad V_o(s) = RI(s)$

or $\quad I(s) = \dfrac{V_o(s)}{R}$ (4)

Substituting the value of $I(s)$ from Eqn. (4) in Eqn. (3), we get

$$V_i(s) = I(s)\left[R + \frac{1}{sC}\right]$$

or $\quad V_i(s) = \dfrac{V_o(s)}{R}\left[R + \dfrac{1}{sC}\right]$

or $\quad V_i(s) = V_o(s)\left[1 + \dfrac{1}{sCR}\right]$

or $\quad \dfrac{V_o(s)}{V_i(s)} = \dfrac{1}{1 + \dfrac{1}{sRC}} = \dfrac{sT}{1 + sT}$ where, $\quad T = RC$ **Ans**

Problem 2.4 Find the transfer function of the electrical network as shown in Fig. 2.5 in phase-lead form.

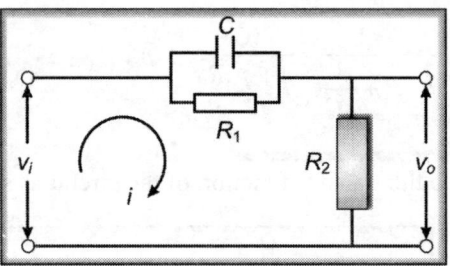

Fig. 2.5

Solution If Z_1 is the equivalent impedance of the parallel combination of R_1 and C, then

$$\frac{1}{Z_1} = \frac{1}{R_1} + \frac{j\omega C}{1}$$

$$\therefore \qquad Z_1 = \frac{R_1}{1 + j\omega CR_1} = \frac{R_1}{1 + sCR_1}, (j\omega = s)$$

Writing differential equations by Kirchoff's law, we get

$$v_i(t) = Z_1 i(t) + R_2 i(t)$$

and
$$v_o(t) = R_2 i(t)$$

Laplace transform of the above equations gives

$$V_i(s) = Z_1 I(s) + R_2 I(s) \quad \text{and} \quad V_o(s) = R_2 I(s)$$

$$\therefore \qquad I(s) = \frac{V_o(s)}{R_2}$$

or
$$V_i(s) = Z_1 \frac{V_o(s)}{R_2} + V_o(s)$$

or
$$V_i(s) = V_o(s) \left(\frac{Z_1}{R_2} + 1 \right)$$

Substituting the value of Z_1, we get

$$V_i(s) = V_o(s) \left(\frac{R_1}{1 + sCR_1} \cdot \frac{1}{R_2} + 1 \right)$$

$$\frac{V_o(s)}{V_i(s)} = \frac{R_2(1 + sCR_1)}{R_1 + R_2 + sCR_1 R_2} = \frac{R_2}{R_1 + R_2} \left[\frac{1 + sCR_1}{1 + \dfrac{sCR_1 R_2}{R_1 + R_2}} \right]$$

Putting $\dfrac{R_2}{R_1 + R_2} = a$ and $CR_1 = T$, we get

$$\frac{V_o(s)}{V_i(s)} = \frac{a(1+sT)}{1+asT}$$

Ans

Problem 2.5 Find the transfer function of the electrical network shown in Fig. 2.6.

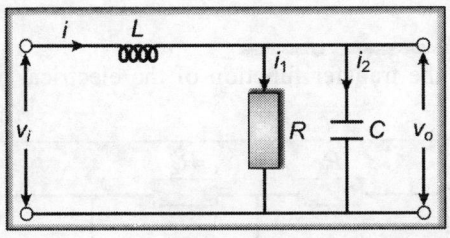

Fig. 2.6

Solution Assuming no external load

$\therefore \qquad\qquad i = i_1 + i_2$

Applying Kirchoff's law to the electrical network, we get

$$v_i(t) = \frac{Ldi(t)}{dt} + Ri_1(t)$$

and $\qquad Ri_1(t) = v_o(t)$

$$v_o(t) = \frac{1}{C}\int i_2(t)\, dt = Ri_1(t)$$

Laplace transform gives

$$V_i(s) = sLI(s) + V_o(s) \ \text{ and }$$

$$V_o(s) = \frac{I_2(s)}{sC} = RI_1(s)$$

or $\qquad V_i(s) = V_o(s) + sL\,(I_1(s) + I_2(s))$

or $\qquad V_i(s) = V_o(s) + sL\left(\dfrac{V_o(s)}{R} + sCV_o(s)\right)$

or $\qquad V_i(s) = V_o(s)\left[1 + \dfrac{sL}{R} + s^2 LC\right]$

or $\dfrac{V_o(s)}{V_i(s)} = \dfrac{1}{s^2 LC + \dfrac{sL}{R} + 1} = \dfrac{R}{s^2 LCR + sL + R} = \dfrac{R}{R\left(s^2 \dfrac{LCR}{R} + \dfrac{sL}{R} + \dfrac{R}{R}\right)}$

Putting $\dfrac{L}{R} = T_1$ and $CR = T_2$, we get

$$\dfrac{V_o(s)}{V_i(s)} = \dfrac{1}{T_1 T_2 s^2 + T_1 s + 1}$$ **Ans**

Problem 2.6 Find the transfer function of the electrical network as shown in Fig. 2.7

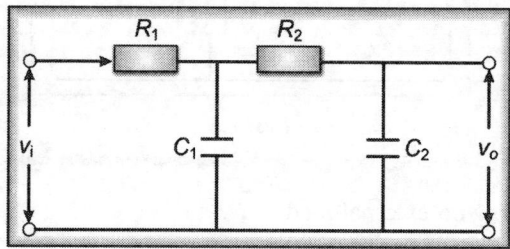

Fig. 2.7

Solution

Let $\quad R_1 = Z_1; \quad \dfrac{1}{sC_1} = Z_2; \quad R_2 = Z_3; \quad \dfrac{1}{sC_2} = Z_4$

Redrawing the Fig. 2.7 after substituting the values, we get Fig. 2.8

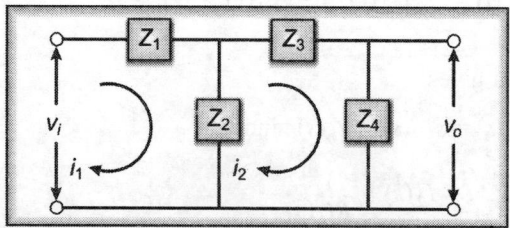

Fig. 2.8

Writing the differential equations with the help of Kirchoff's law and then taking the Laplace transform, we get

$$V_i(s) = (Z_1 + Z_2)I_1(s) - Z_2 I_2(s) \tag{1}$$

$$0 = Z_3 I_2(s) + Z_4 I_2(s) + Z_2 I_2(s) - Z_2 I_1(s) \tag{2}$$

$$V_o(s) = Z_4 I_2(s) \tag{3}$$

From equation (2)

$$I_1(s) = \frac{(Z_2 + Z_3 + Z_4)I_2(s)}{Z_2} \tag{4}$$

Substituting the value of $I_1(s)$ in equation (1), we get

$$V_i(s) = \frac{(Z_1 + Z_2)(Z_2 + Z_3 + Z_4)I_2(s) - Z_2^2 I_2(s)}{Z_2}$$

$$\frac{V_i(s)}{I_2(s)} = \frac{(Z_1 + Z_2)(Z_2 + Z_3 + Z_4) - Z_2^2}{Z_2}$$

But from equation (3)

$$I_2(s) = \frac{V_o(s)}{Z_4},$$

Therefore,

$$\frac{V_i(s)}{V_o(s)/Z_4} = \frac{(Z_1 + Z_2)(Z_2 + Z_3 + Z_4) - Z_2^2}{Z_2}$$

or

$$\frac{V_o(s)}{V_i(s)} = \frac{Z_2 Z_4}{Z_1 Z_2 + Z_1 Z_3 + Z_1 Z_4 + Z_2 Z_3 + Z_2 Z_4}$$

Substituting the values of Z_1, Z_2, Z_3, and Z_4, we get

$$\frac{V_o(s)}{V_i(s)} = \frac{\dfrac{1}{sC_1} \cdot \dfrac{1}{sC_2}}{\dfrac{R_1}{sC_1} + \dfrac{R_1 R_2}{1} + \dfrac{R_1}{sC_2} + \dfrac{R_2}{sC_1} + \dfrac{1}{s^2 C_1 C_2}}$$

$$= \frac{1}{sR_1 C_2 + s^2 R_1 R_2 C_1 C_2 + sR_1 C_1 + sR_2 C_2 + 1}$$

$$= \frac{1}{1 + s(R_1 C_1 + R_2 C_2 + R_1 C_2) + s^2 R_1 R_2 C_1 C_2} \qquad \textbf{Ans}$$

Problem 2.7 Derive the transfer function of the circuit as shown in Fig. 2.9. If $v_i = 8 \sin 10t$ volts, $R_1 = 50$ KΩ, $R_2 = 5$ KΩ, and $C = 1$ μF. Calculate output voltage in magnitude and in phase relative to input voltage.

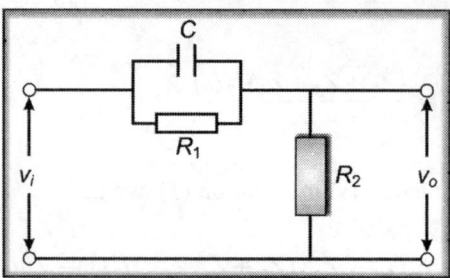

Fig. 2.9

Solution The transfer function of the circuit is derived in Problem 2.4 and is

$$G(s) = \frac{V_o(s)}{V_i(s)} = \frac{R_2(1 + sCR_1)}{R_1 + R_2 + sCR_1R_2}$$

$$\therefore \quad G(s) = \frac{5000(1 + s \times 1 \times 10^{-6} \times 50 \times 1000)}{50 \times 1000 + 5000 + s \times 1 \times 10^{-6} \times 50 \times 1000 \times 5000} = \frac{0.1(1 + 0.05\, s)}{1 + 0.0045\, s} \quad \textbf{Ans}$$

Put $\quad s = j\omega$; Therefore, $\quad G(j\omega) = \dfrac{0.1(1 + 0.05\, j\omega)}{1 + 0.0045\, j\omega}$

and $\qquad \angle G(j\omega) = \tan^{-1} 0.05\,\omega - \tan^{-1} 0.0045\,\omega$

Since $\qquad v_i(t) = 8\sin 10t, \quad$ therefore, $\quad \omega = 10$

$\therefore \qquad \angle G(j\omega) = \tan^{-1}(0.05 \times 10) - \tan^{-1}(0.0045 \times 10) = 23.989° \qquad \textbf{Ans}$

Also, $\quad M = |G(j\omega)| = \dfrac{0.1\sqrt{1 + (0.05\omega)^2}}{\sqrt{1 + (0.0045\omega)^2}}$

Putting $\omega = 10$, we get $M = \dfrac{0.1\sqrt{1 + (0.5)^2}}{\sqrt{1 + (0.045)^2}} = 0.1117$

Therefore, $\qquad V_o = 8 \times 0.1117 = 0.8936$ volts $\qquad\qquad$ **Ans**

Problem 2.8 If $C = 1\,\mu F$ in the circuit as shown in Fig. 2.10. What values of R_1 and R_2 will give $T = 0.6$ sec and $a = 0.1$.

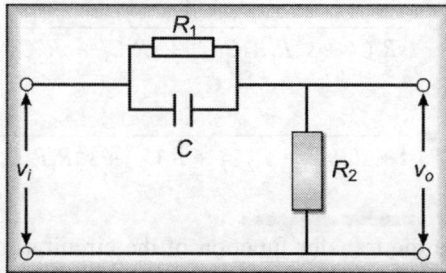

Fig. 2.10

Solution The transfer function of the network as shown in Fig. 2.10 has been found in Problem 2.4 and is of the form

$$\frac{V_o(s)}{V_i(s)} = \frac{a(1+sT)}{1+asT} \quad \text{where,} \quad T = CR_1 \quad \text{and} \quad a = \frac{R_2}{R_1 + R_2}$$

when $T = 0.6$ sec; $T = CR_1$

then $0.6 = 1 \times 10^{-6} \times R_1 \ (T = 0.6$ sec$)$

or $R_1 = \dfrac{0.6}{1 \times 10^{-6}} = 0.6$ MΩ **Ans**

also $a = \dfrac{R_2}{R_1 + R_2} \quad \text{or} \quad 0.1 = \dfrac{R_2}{0.6 + R_2}$

or $0.06 + 0.1R_2 = R_2 \quad \therefore \quad R_2 = \dfrac{0.06}{0.9} = 0.066$ MΩ **Ans**

Problem 2.9   Find the transfer function of the network as shown in Fig. 2.11. Plot its poles and zeros for $R = C = 1$.

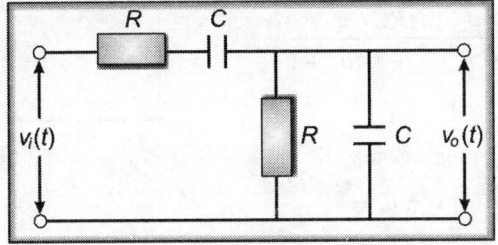

Fig. 2.11

Solution Assume current distribution as shown in Fig. 2.12.

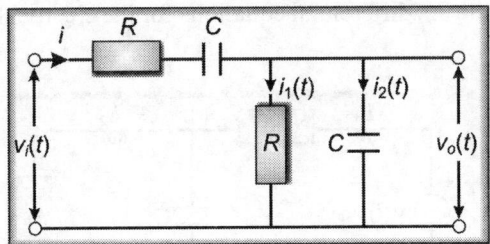

Fig. 2.12

Assuming no external load

$$i(t) = i_1(t) + i_2(t)$$

Applying Kirchoff's law to the electrical network, we get

$$v_i(t) = Ri(t) + \frac{1}{C}\int i(t)\,dt + Ri_1(t)$$

and $\qquad v_o(t) = \dfrac{1}{C}\int i_2(t)\,dt = Ri_1(t)$

Laplace transform gives

$$V_i(s) = RI(s) + \frac{I(s)}{sC} + V_o(s) \ \text{ and } \ V_o(s) = \frac{I_2(s)}{sC} = RI_1(s)$$

or $\qquad V_i(s) = (I_1(s) + I_2(s))\left[R + \dfrac{1}{sC}\right] + V_o(s)$

or $\qquad V_i(s) = V_o(s)\left[\dfrac{1}{R} + sC\right]\left[R + \dfrac{1}{sC}\right] + V_o(s)$

or $\qquad V_i(s) = V_o(s)\left[1 + \dfrac{1}{sRC} + sRC + 1 + 1\right]$

or $\qquad V_i(s) = V_o(s)\left[3 + \dfrac{1 + s^2R^2C^2}{sRC}\right]$

or $\qquad \dfrac{V_o(s)}{V_i(s)} = \dfrac{sRC}{s^2R^2C^2 + 3sRC + 1}$

But, $\qquad R = C = 1$

$$\frac{V_o(s)}{V_i(s)} = \frac{s}{s^2 + 3s + 1}$$

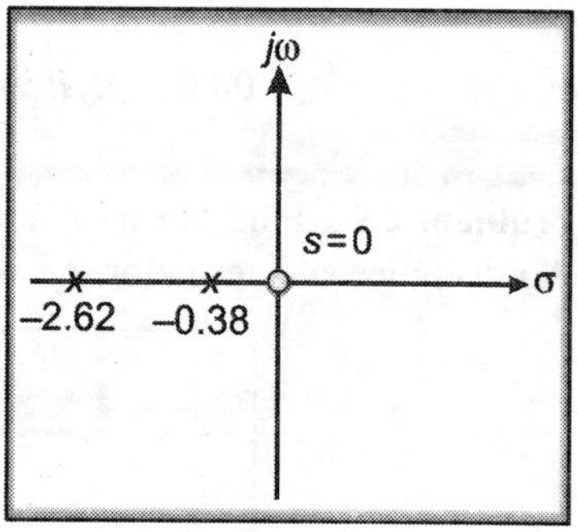

Fig. 2.13

Ans

There is one *zero* at origin and *poles* are at -2.62 and -0.38. The plot is shown in Fig. 2.16.

Problem 2.10 Write the differential equations for the electrical network as shown in Fig. 2.14.

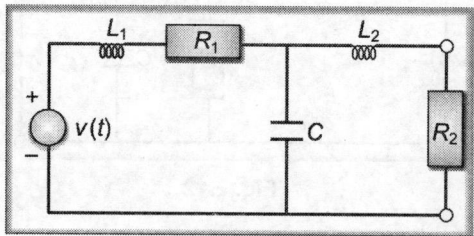

Fig. 2.14

Solution Assuming current distribution as shown in Fig. 2.15, the differential equations are obtained by the use of Kirchoff's law.

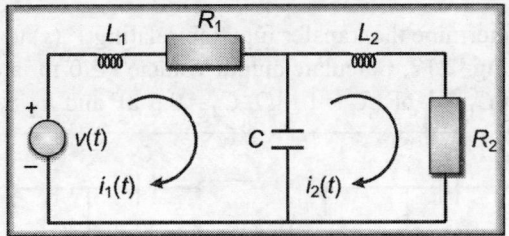

Fig. 2.15

$$v(t) = \frac{L_1 di_1(t)}{dt} + R_1 i_1(t) + \frac{1}{C}\int i_1(t)\,dt - \frac{1}{C}\int i_2(t)\,dt$$

and

$$0 = \frac{L_2 di_2(t)}{dt} + R_2 i_2(t) + \frac{1}{C}\int i_2(t)\,dt - \frac{1}{C}\int i_1(t)\,dt \qquad \textbf{Ans}$$

Problem 2.11 Determine the transfer function relation $V_o(s)$ to $V_i(s)$ for the network as shown in Fig. 2.16.

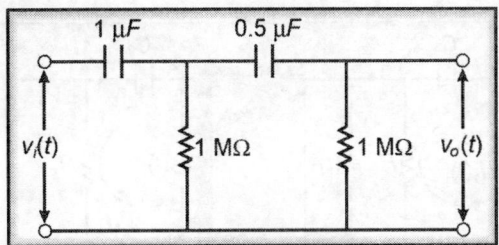

Fig. 2.16

Calculate output voltage, $t \geq 0$, for a unit step voltage input at $t = 0$.

Solution This is similar to Problem 2.6. The transfer function is

$$\frac{V_o}{V_i}(s) = \frac{Z_2 Z_4}{Z_1 Z_2 + Z_1 Z_3 + Z_1 Z_4 + Z_2 Z_3 + Z_2 Z_4}$$

where,

$$Z_1 = \frac{1}{s};\; Z_2 = 1;\; Z_3 = \frac{1}{0.5s};\; \text{and } Z_4 = 1$$

Substituting the values, we get

$$\frac{V_o}{V_i}(s) = \frac{s^2}{s^2 + 4s + 2} \qquad \textbf{Ans}$$

For a unit step voltage input

$$V_o(s) = \frac{s}{s^2 + 4s + 2} = \frac{A}{(s + 3.4)} + \frac{B}{(s + 0.59)} \quad \text{or } V_o(s) = \frac{1.2}{(s + 3.4)} + \frac{-0.21}{(s + 0.59)}$$

$$\therefore \qquad v_o(t) = 1.2 e^{-3.4t} - 0.21 e^{-0.59t} \qquad \textbf{Ans}$$

Problem 2.12   Determine the transfer function relating $V_o(s)$ to $V_i(s)$ for the network as shown in Fig. 2.17. Calculate output voltage $t \geq 0$ for a unit step voltage input at $t = 0$ when $C_1 = 1\ \mu F$, $R_1 = 1\ M\Omega$, $C_2 = 0.5\ \mu F$ and $R_2 = 1\ M\Omega$.

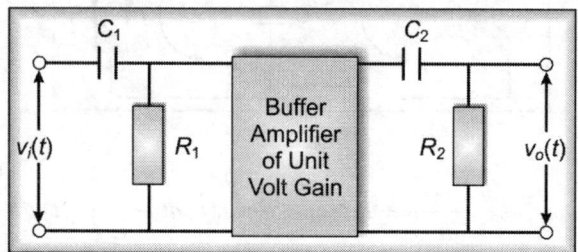

Fig. 2.17

Solution

Redrawing the circuit diagram as shown in Fig. 2.18 and applying Kirchoff's law, we get

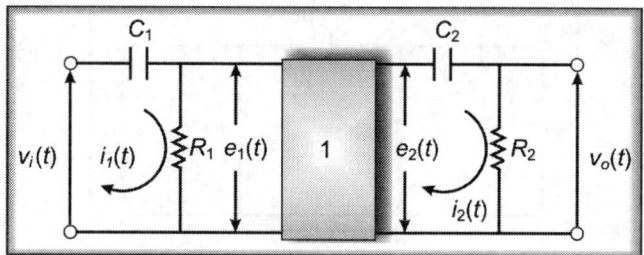

Fig. 2.18

$$V_i(s) = I_1(s)R_1 + \frac{I_1(s)}{sC_1}; \quad \text{and} \quad E_2(s) = I_2(s)R_2 + \frac{I_2(s)}{sC_2};$$

Also $E_1(s) = I_1(s)R_1$ and $V_o(s) = I_2(s)R_2$

Therefore,

$$\frac{V_o(s)}{V_i(s)} = \frac{I_2(s)R_2}{I_1(s)\left(R_1 + \dfrac{1}{sC_1}\right)} \quad \text{and} \quad \frac{I_2(s)}{I_1(s)} = \frac{E_2(s)\Big/\left(R_2 + \dfrac{1}{sC_2}\right)}{E_1(s)\,/\,R_1}$$

but $E_2(s) = E_1(s)$

Therefore, $$\frac{V_o(s)}{V_i(s)} = \frac{R_2\Big/\left(R_2 + \dfrac{1}{sC_2}\right)}{\left(R_1 + \dfrac{1}{sC_1}\right)\Big/R_1} = \frac{R_1 R_2}{\left(R_2 + \dfrac{1}{sC_2}\right)\left(1 + \dfrac{1}{sC_1}\right)}$$

or
$$\frac{V_o(s)}{V_i(s)} = \frac{1}{\left(1 + \dfrac{1}{sR_2C_2}\right)\left(1 + \dfrac{1}{sR_1C_1}\right)}$$

Substituting the values of R_1, R_2, C_1 and C_2, we get

$$\frac{V_o(s)}{V_i(s)} = \frac{s^2}{s^2 + 3s + 2} \qquad\qquad \textbf{Ans}$$

For unit step voltage input

$$V_o(s) = \frac{s}{s^2 + 3s + 2} = \frac{2}{(s+2)} - \frac{1}{(s+1)}$$

Therefore, $\qquad v_o(t) = 2e^{-2t} - e^{-t}$ $\qquad\qquad \textbf{Ans}$

Problem 2.13 Determine the transfer function of the electrical network as shown in Fig. 2.19

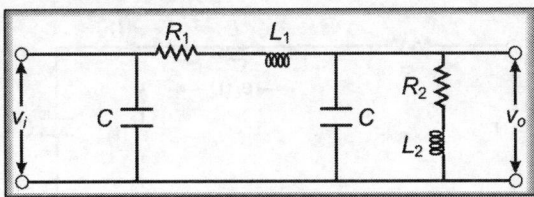

Fig. 2.19

Solution Assuming current distribution as shown in Fig. 2.20, the differential equations can be written as:

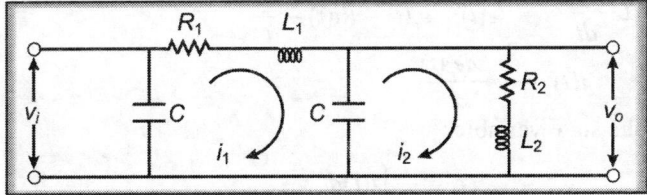

Fig. 2.20

$$v_i(t) = R_1 i_1(t) + L_1 \frac{di_1(t)}{dt} + \frac{1}{C}\int (i_1(t) - i_2(t))\, dt$$

$$0 = \frac{1}{C}\int (i_2(t) - i_1(t))\, dt + R_2 i_2(t) + L_2 \frac{di_2(t)}{dt}$$

and $\qquad v_o(t) = R_2 i_2(t) + L_2 \frac{di_2(t)}{dt}$

Assuming initial conditions to be *zero*, the Laplace transform of the above equations can be written as:

$$V_i(s) = R_1 I_1(s) + sL_1 I_1(s) + \frac{1}{sC}(I_1(s) - I_2(s)),$$

$$0 = \frac{1}{sC}(I_2(s) - I_1(s)) + R_2 I_2(s) + sL_2 I_2(s)$$

and $\quad V_o(s) = R_2 I_2(s) + sL_2 I_2(s)$ which gives

$$\frac{V_o(s)}{V_i(s)} = \frac{R_2 + sL_2}{s^3 CL_1 L_2 + s^2 C(R_1 L_2 + L_1 R_2) + s(L_1 + L_2 + CR_1 R_2) + (R_1 + R_2)}$$

Problem 2.14 A 10 V battery is switched on to a series circuit comprising a resistance of 3 Ω, inductance of 1 H and capacitance of 0.5 F. Write the state equations in matrix form, if the initial charges on the capacitor is 1 coulomb and initial current is zero.

Solution Circuit diagram of the problem is shown in Fig. 2.21

The loop equation of RLC circuit is:

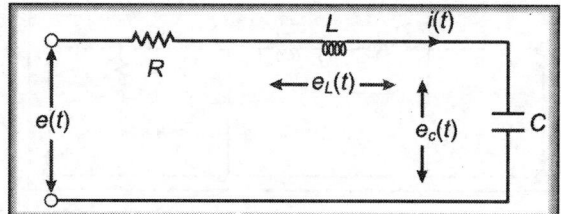

Fig. 2.21

$$e(t) = Ri(t) + \frac{L\,di(t)}{dt} + \frac{1}{C}\int i(t)dt + \frac{q}{C} \qquad (1)$$

or $\qquad L\dfrac{di(t)}{dt} = e(t) - e_c(t) - Ri(t) - \dfrac{q}{C} \qquad (2)$

and $\qquad i(t) = \dfrac{Cde_c(t)}{dt}$

Defining the state variables

$$x_1 = e_c(t) = \frac{1}{C}\int i(t)dt$$

and $\qquad x_2 = i(t)$

the following state equations are obtained

$$\dot{x}_1 = \frac{1}{C}x_2 \quad \text{and} \quad \dot{x}_2 = -\frac{1}{L}x_1 - \frac{R}{L}x_2 + \frac{e(t)}{L} - \frac{q}{LC}$$

In matrix form

$$\begin{bmatrix} \dot{x}_1 \\ \dot{x}_2 \end{bmatrix} = \begin{bmatrix} 0 & \dfrac{1}{C} \\ -\dfrac{1}{L} & -\dfrac{R}{L} \end{bmatrix} \begin{bmatrix} x_1 \\ x_2 \end{bmatrix} + \begin{bmatrix} 0 \\ \dfrac{1}{L} \end{bmatrix} \left[\left(e(t) - \dfrac{q}{C} \right) \right]$$

Substituting the values from the circuit, *i.e.* $R = 3\ \Omega$, $L = 1$ H, $C = 0.5$ F, $q = 1$ C and $e(t) = 10$

$$\begin{bmatrix} \dot{x}_1 \\ \dot{x}_2 \end{bmatrix} = \begin{bmatrix} 0 & 2 \\ -1 & -3 \end{bmatrix} \begin{bmatrix} x_1 \\ x_2 \end{bmatrix} + \begin{bmatrix} 0 \\ 1 \end{bmatrix} 8$$

or

$$\begin{bmatrix} \dot{x}_1 \\ \dot{x}_2 \end{bmatrix} = \begin{bmatrix} 0 & 2 \\ -1 & -3 \end{bmatrix} \begin{bmatrix} x_1 \\ x_2 \end{bmatrix} + \begin{bmatrix} 0 \\ 8 \end{bmatrix}$$

Problem 2.15 Find the transfer function of a system described by:

$$2\frac{dc(t)}{dt} + c(t) = r(t) * (t - 2)$$

where, $r(t)$ and $c(t)$ are the input and output respectively.

Solution Writing the given equation in Laplace form, we get

$$2sC(s) + C(s) = \frac{R(s)}{s^2} - \frac{2R(s)}{s}$$

$$C(s)(1 + 2s) = R(s)\left(\frac{1}{s^2} - \frac{2}{s}\right)$$

$$C(s)(1 + 2s) = R(s)\frac{(1 - 2s)}{s^2}$$

$$\therefore \qquad \frac{C(s)}{R(s)} = \frac{1 - 2s}{s^2(1 + 2s)} \qquad\qquad \textbf{Ans}$$

Problem 2.16 Derive the transfer function of the network as shown in Fig. 2.22.

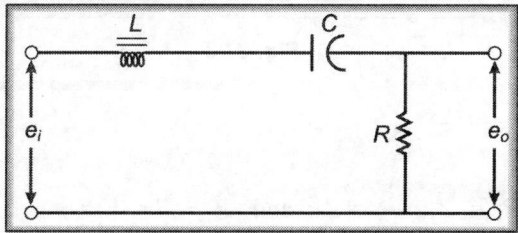

Fig. 2.22

Solution Writing Kirchoff's law equation, we get

$$e_i(t) = \frac{Ldi(t)}{dt} + \frac{1}{C}\int i(t)dt + Ri(t) + e_c(o)$$

where $i(t)$ is the current in the RLC loop and $e_c(o)$ = voltage across the capacitor initially. Assuming $e_c(o)$ = zero, *i.e.* zero initial conditions and taking Laplace, we get

$$E_i(s) = sLI(s) + \frac{I(s)}{sC} + RI(s) \qquad\qquad (1)$$

Also, $e_o(t) = Ri(t)$ and after taking Laplace, we get

$$E_o(s) = RI(s) \tag{2}$$

Rewriting Eqn. (1), we get

$$E_i(s) = I(s)\left[sL + \frac{1}{sC} + R\right] \tag{3}$$

But $I(s) = \dfrac{E_o(s)}{R}$ (from Eqn. (2))

Substituting value of $I(s)$ in Eqn. (3), we get

$$E_i(s) = \frac{E_o(s)}{R}\left[sL + \frac{1}{sC} + R\right]$$

$$E_i(s) = \frac{E_o(s)}{R}\left[\frac{s^2LC + 1 + sRC}{sC}\right]$$

$$\frac{E_o(s)}{E_i(s)} = \frac{sRC}{1 + sRC + s^2LC} \qquad \textbf{Ans}$$

Problem 2.17 Derive the transfer function of the network as shown in Fig. 2.23

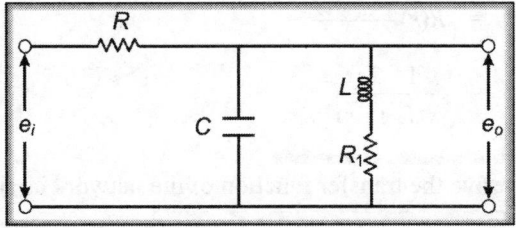

Fig. 2.23

Solution

Let $R = Z_1;\ \dfrac{1}{sC} = Z_2;$ and $R_1 + sL = Z_3$

Redrawing the Fig. 2.23 based on Z_1, Z_2, and Z_3, we get Fig. 2.24

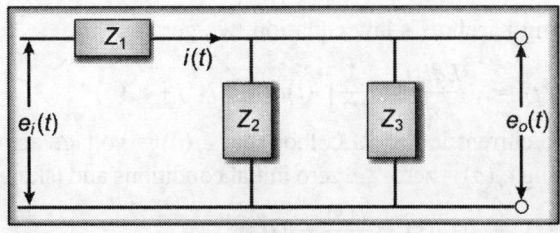

Fig. 2.24

Now, $$E_o(s) = \frac{I(s)Z_2Z_3}{Z_2 + Z_3} \qquad (1)$$

and $$I(s) = \frac{E_i(s)}{Z_1 + \dfrac{Z_2Z_3}{Z_2 + Z_3}} \qquad (2)$$

Substituting the value of $I(s)$ from Eqn. (2) in Eqn. (1), we get

$$E_o(s) = \frac{E_i(s)}{Z_1 + \dfrac{Z_2Z_3}{Z_2 + Z_3}} \times \frac{Z_2Z_3}{Z_2 + Z_3}$$

$$\frac{E_o(s)}{E_i(s)} = \frac{Z_2Z_3}{Z_1Z_2 + Z_2Z_3 + Z_3Z_1} \qquad (3)$$

Substituting the values of Z_1, Z_2, and Z_3, in Eqn. (3), we get

$$\frac{E_o(s)}{E_i(s)} = \frac{\dfrac{R_1 + sL}{sC}}{\dfrac{R}{sC} + \dfrac{R_1 + sL}{sC} + \dfrac{R(R_1 + sL)}{1}}$$

or $$\frac{E_o(s)}{E_i(s)} = \frac{R_1 + sL}{R + R_1 + sL + sRC(R_1 + sL)}$$

$$\frac{E_o(s)}{E_i(s)} = \frac{R_1 + sL}{R + R_1 + sL + sRR_1C + s^2RLC}$$

$$\frac{E_o(s)}{E_i(s)} = \frac{R_1 + sL}{R + R_1 + s(L + RR_1C) + s^2RLC} \qquad \textbf{Ans}$$

Problem 2.18 Find the transfer function of the electrical network as shown in Fig. 2.25

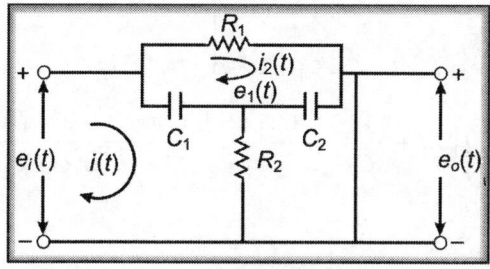

Fig. 2.25

Solution Applying Kirchoff's voltage law and the Laplace transform to the two loops, we get

$$-I_2 R_1 - \frac{1}{sC_2}I_2 - \frac{1}{sC_1}I_2 + \frac{1}{sC_1}I_1 = 0$$

or $\quad \frac{1}{sC_1}I_1 - \left(R_1 + \frac{1}{sC_1} + \frac{1}{sC_2}\right)I_2 = 0$ (1)

and $\quad -\frac{1}{sC_1}I_1 + \frac{1}{sC_1}I_2 - I_1 R_2 + E_i(s) = 0$

or $\quad \left(R_2 + \frac{1}{sC_1}\right)I_1 - \frac{1}{sC_1}I_2 = E_i(s)$ (2)

Also $\quad E_o(s) = I_1 R_2 + \frac{I_2}{sC_2}$ (3)

Using Cramer's rule

$$D_1 = \begin{vmatrix} \dfrac{1}{sC_1} & -\left(R_1 + \dfrac{1}{sC_1} + \dfrac{1}{sC_2}\right) \\ \dfrac{1}{sC_1} + R_2 & -\dfrac{1}{sC_1} \end{vmatrix}$$

$$= -\frac{1}{s^2 C_1^2} + \left(\frac{1}{sC_1} + R_2\right)\left(R_1 + \frac{1}{sC_1} + \frac{1}{sC_2}\right)$$

$$= \frac{1}{s^2 C_1 C_2} + \frac{R_1}{sC_1} + \frac{R_2}{sC_1} + \frac{R_2}{sC_2} + R_1 R_2$$

$$= \frac{1 + s(R_1 + R_2)C_2 + sR_2 C_1 + s^2 R_1 R_2 C_1 C_2}{s^2 C_1 C_2} s$$

$$D_2 = \begin{vmatrix} 0 & -\left(R_1 + \dfrac{1}{sC_1} + \dfrac{1}{sC_2}\right) \\ E_i(s) & -\dfrac{1}{sC_1} \end{vmatrix}$$

$$= E_i(s)\left(R_1 + \frac{1}{sC_1} + \frac{1}{sC_2}\right) = \frac{E_i(s)(sC_1 + sC_2 + s^2 R_1 C_1 C_2)}{s^2 C_1 C_2}$$

$$D_3 = \begin{vmatrix} \dfrac{1}{sC_1} & 0 \\ \dfrac{1}{sC_1} + R_2 & E_i(s) \end{vmatrix} = \frac{E_i}{sC_1}$$

Therefore,

$$I_1 = \frac{D_2}{D_1} = \frac{(sC_1 + sC_2 + s^2 R_1 C_1 C_2)E_i(s)}{1 + s(R_1 + R_2)C_2 + sR_2 C_1 + s^2 R_1 R_2 C_1 C_2}$$

and

$$I_2 = \frac{D_3}{D_1} = \frac{sC_2 E_i(s)}{1 + s(R_1 + R_2)C_2 + sR_2 C_1 + s^2 R_1 R_2 C_1 C_2}$$

Substituting the values of I_1 and I_2 in Eqn (3), we get

$$E_o(s) = \frac{E_i(s)}{1 + s(R_1 + R_2)C_2 + sR_2 C_1 + s^2 R_1 R_2 C_1 C_2} +$$

$$\frac{E_i(s)R_2(sC_1 + sC_2 + s^2 R_1 C_1 C_2)}{1 + s(R_1 + R_2)C_2 + sR_2 C_1 + s^2 R_1 R_2 C_1 C_2}$$

or

$$\frac{E_o(s)}{E_i(s)} = \frac{1 + s^2 R_1 R_2 C_1 C_2 + sR_2 C_1 + sR_2 C_2}{1 + s(R_1 + R_2)C_2 + sR_2 C_1 + s^2 R_1 R_2 C_1 C_2}$$

$$= \frac{1 + s^2 R_1 R_2 C_1 C_2 + sR_2(C_1 + C_2)}{1 + s^2 R_1 R_2 C_1 C_2 + s(R_1 C_2 + R_2 C_1 + R_2 C_2)} \qquad \textbf{Ans}$$

Problem 2.19 Find the transfer function of the electrical network as shown in Fig. 2.26.

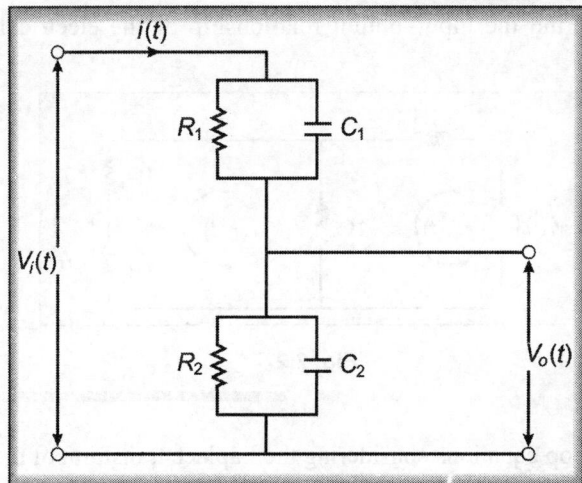

Fig. 2.26

Solution Considering the Laplace transform of the network

Let, $$Z_1 = \frac{R_1/sC_1}{R_1 + \dfrac{1}{sC_1}} = \frac{R_1}{1 + sR_1 C_1} (R_1 \text{ is in parallel with } C_1)$$

Similarly $$Z_2 = \frac{R_2}{1 + s R_2 C_2}$$

Also $\quad I(s) = \dfrac{V_i(s)}{Z_1 + Z_2}$

and $\quad V_o(s) = I(s)Z_2 = \dfrac{V_i(s) Z_2}{Z_1 + Z_2}$

Substituting values of Z_1 and Z_2, we get

$$\frac{V_o(s)}{V_i(s)} = \frac{\dfrac{R_2}{1 + sR_2C_2}}{\dfrac{R_1}{1 + sR_1C_1} + \dfrac{R_2}{1 + sR_2C_2}}$$

$$= \frac{R_2\,(1 + sR_1C_1)}{(R_1 + R_2)\left[1 + \dfrac{R_1R_2C_1 + R_1R_2C_2}{R_1 + R_2}\right]} = \frac{K(1 + sT_1)}{1 + sT_2} \qquad \textbf{Ans}$$

where,

$$T_1 = R_1C_1, \quad T_2 = \frac{R_1\,R_2\,C_1 + R_1R_2C_2}{R_1 + R_2} \text{ and } K = \frac{R_2}{R_1 + R_2}$$

Problem 2.20 Find the input–output relationship of the electrical network in Fig. 2.27

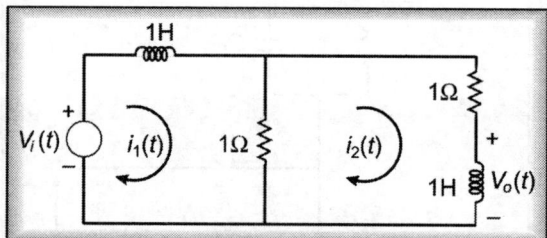

Fig. 2.27

Solution The loop equations considering the Laplace transform of the network is:

$$sI_1(s) + I_1(s) - I_2(s) = V_i(s)$$

or $\qquad (s + 1)I_1(s) - I_2(s) = V_i(s)$

and $\qquad -I_1(s) + sI_2(s) + I_2(s) + I_2(s) = 0 \qquad (1)$

or $\qquad -I_1(s) + (s + 2)\,I_2(s) = 0 \qquad (2)$

or $\qquad I_1(s) = (s + 2)I_2(s) \qquad (3)$

Substituting the value of $I_1(s)$ in Eqn. (1), we get

$$(s + 1)(s + 2)I_2(s) - I_2(s) = V_i(s)$$

or $\qquad \dfrac{I_2(s)}{V_i(s)} = \dfrac{1}{s^2 + 3s + 1}$ $\qquad\qquad\qquad$ (4)

Also $\qquad V_o(s) = sI_2(s)$

Therefore $\qquad V_o(s) = \dfrac{sV_i(s)}{s^2 + 3s + 1}$

or $\qquad \dfrac{V_o(s)}{V_i(s)} = \dfrac{s}{s^2 + 3s + 1}$ $\qquad\qquad\qquad$ **Ans**

Problem 2.21 Find the ratio $V_o(s)/V_i(s)$ for the electrical network as shown in Fig. 2.28

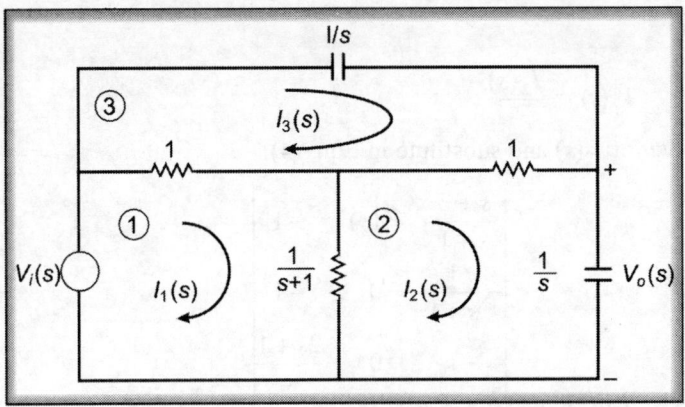

Fig. 2.28

Solution Laplace transform of the network of Fig. 2.28 gives the following circuit equations:

Loop 1

$$\left(1 + \frac{1}{s+1}\right)I_1(s) - I_3(s) - \frac{1}{s+1}I_2(s) = V_i(s)$$

$$\frac{s+2}{s+1}I_1(s) - \frac{1}{s+1}I_2(s) - I_3(s) = V_i(s) \qquad\qquad (1)$$

Loop 2

$$I_2(s) + \frac{I_2(s)}{s} + \frac{I_2(s)}{s+1} - \frac{I_1(s)}{s+1} - I_3(s) = 0$$

$$-\frac{1}{s+1}I_1(s) + \left(1 + \frac{1}{s} + \frac{1}{s+1}\right)I_2(s) - I_3(s) = 0$$

$$-\frac{1}{s+1}I_1(s) + \frac{s^2 + s + s + 1 + s}{s(s+1)}I_2(s) - I_3(s) = 0$$

$$-\frac{1}{s+1}I_1(s) + \frac{(s+1)^2 + s}{s(s+1)}I_2(s) - I_3(s) = 0 \tag{2}$$

Loop 3

$$-I_1(s) - I_2(s) + \frac{I_3(s)}{s} + I_3(s) + I_3(s) = 0$$

$$-I_1(s) - I_2(s) + \frac{2s+1}{s}I_3(s) = 0 \tag{3}$$

Also

$$V_o(s) = \frac{I_2(s)}{s} \tag{4}$$

We will solve for $I_2(s)$ and substitute in Eqn. (4).

$$I_2(s) = \frac{\begin{vmatrix} \dfrac{s+2}{s+1} & V_i(s) & -1 \\[2mm] -\dfrac{1}{s+1} & 0 & -1 \\[2mm] -1 & 0 & \dfrac{2s+1}{s} \end{vmatrix}}{\begin{vmatrix} \dfrac{s+2}{s+1} & -\dfrac{1}{s+1} & -1 \\[2mm] -\dfrac{1}{s+1} & \dfrac{(s+1)^2 + s}{s(s+1)} & -1 \\[2mm] -1 & -1 & \dfrac{2s+1}{s} \end{vmatrix}}$$

as

$$I_2(s) = V_i(s)\frac{s(s^2 + 3s + 1)}{2s^2 + 7s + 2} \tag{5}$$

Substituting the value of $I_2(s)$ from Eqn. (5) in Eqn. (4), we get

$$V_o(s) = V_i(s)\frac{s(s^2 + 3s + 1)}{s(2s^2 + 7s + 2)} = \frac{s^2 + 3s + 1}{2s^2 + 7s + 2}$$

Problem 2.22 Find the transfer function $V_o(s)/V_i(s)$ for the operational amplifier as shown in Fig. 2.29

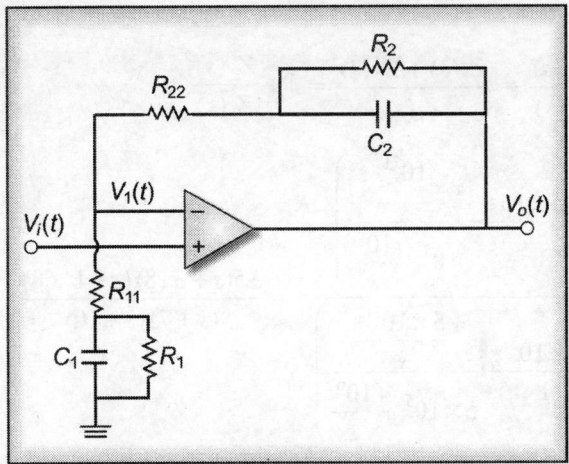

Fig. 2.29

$R_1 = 500\ \text{K}\Omega,\ C_1 = C_2 = 1\ \mu F,\ R_{22} = 500\ \text{K}\Omega,\ R_2 = 100\ \text{K}\Omega,\ R_{11} = 200\ \text{K}\Omega$

Solution Equivalent impedance circuit is shown in Fig. 2.30

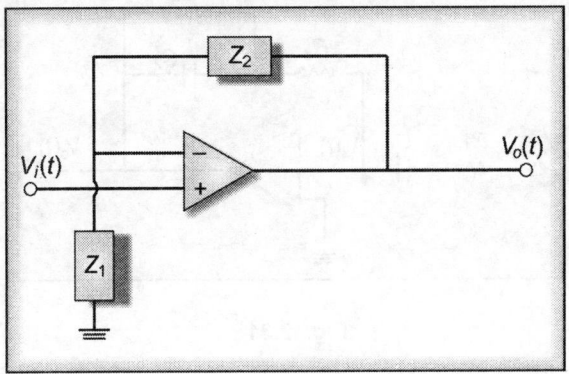

Fig. 2.30

$$Z_1(s) = R_{11} + \cfrac{1}{sC_1 + \cfrac{1}{R_1}} = 2 \times 10^5 + \cfrac{1}{s \times 10^{-6} + \cfrac{1}{5 \times 10^5}}$$

$$= 2 \times 10^5 + \cfrac{5 \times 10^{11}}{5 \times 10^5 + \cfrac{10^6}{s}}$$

$$Z_2(s) = 5 \times 10^5 + \cfrac{\cfrac{10^{11}}{s}}{10^5 + \cfrac{10^6}{s}}$$

Transfer function

$$\frac{V_o(s)}{V_i(s)} = \frac{Z_1(s) + Z_2(s)}{Z_1(s)} = 1 + \frac{Z_2(s)}{Z_1(s)}$$

$$= 1 + \frac{\left(\dfrac{5 \times 10^5}{1} + \dfrac{\dfrac{10^{11}}{s}}{10^5 + \dfrac{10^6}{s}}\right)}{\left(\dfrac{2 \times 10^5}{1} + \dfrac{\dfrac{5 \times 10^{11}}{s}}{5 \times 10^5 + \dfrac{10^6}{s}}\right)} = \frac{3.5(s + 3.18)(s + 11.68)}{(s + 7)(s + 10)}$$

Problem 2.23 Find the transfer function $V_o(s)/V_i(s)$ for the operational amplifier is shown in Fig. 2.31

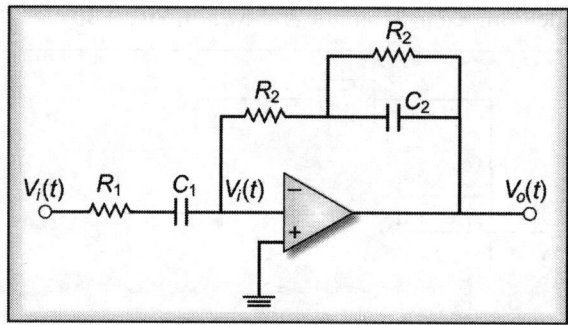

Fig. 2.31

$R_1 = 100 \ K\Omega, \ C_1 = 1\mu F, \ R_2 = 100 \ K\Omega, \ C_2 = 1\mu F.$

Solution Equivalent impedance circuit diagram is shown in Fig. 2.32

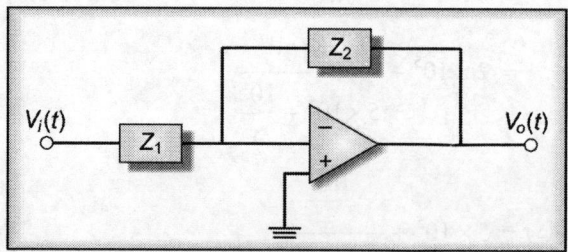

Fig. 2.32

$$Z_1(s) = R_1 + \frac{1}{sC_1} = 10^5 + \frac{1}{s \times 10^{-6}} = 10^5\left(\frac{s+10}{s}\right)$$

$$Z_2(s) = R_2 + \frac{1}{sC_2 + \dfrac{1}{R_2}} = 10^5 + \frac{1}{10^{-6}s + 10^{-5}} = 10^5\left(\frac{s+20}{s+10}\right)$$

$$\frac{V_o(s)}{V_i(s)} = -\frac{Z_2(s)}{Z_1(s)} = -\frac{10^5\left(\dfrac{s+20}{s+10}\right)}{10^5\left(\dfrac{s+10}{s}\right)} = -\frac{s(s+20)}{(s+10)^2}$$

Modelling of Control Systems: Block Diagram Method

3.1 Introduction

In order to carry out any analytical study on a control system, the first step is to evolve a mathematical model of the control system. This requires describing the system by a set of mathematical equations. Generally, we are interested in studying and knowing the behaviour of input and output of the system. For example, in analysing the electrical network, the interest is in knowing the output of the network for a specific input or a class of inputs. The internal behaviour can be studied by setting up differential equations of the network and then solving them. The terminal or input–output description is described by the transfer function.

3.2 Block Diagram

Control Systems are generally described by a set of differential equations. The inter-relationships of the equations are depicted pictorially with the help of block diagrams and signal flow graphs.

Block diagram is a pictorial representation of a control system. A simple open-loop control system having one input and one output is shown in Fig. 3.1. $R(s)$ is the signal into the block representing the input and $C(s)$ is the signal going out of the block representing the output signal. The block represents the transfer function $G(s)$. $R(s)$ and $C(s)$ are the Laplace transforms of the input and output signals.

The transfer function relating the input and output is thus given by:

$$G(s) = \frac{C(s)}{R(s)} \tag{3.1}$$

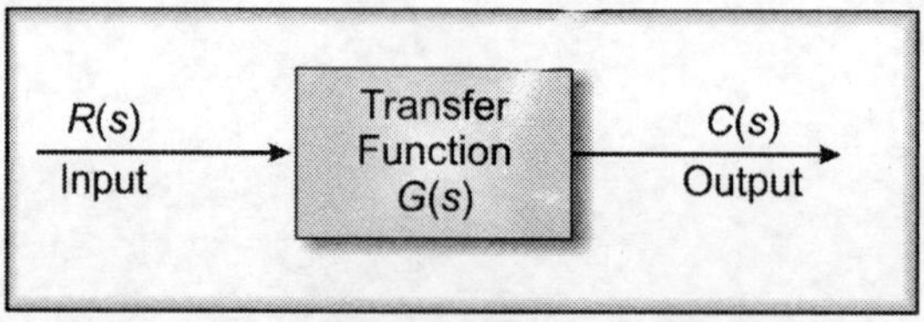

Fig. 3.1 Transfer function relationship

3.3 Block Diagram of a Closed-loop System

Let us consider a closed-loop system with a negative feedback represented by a block diagram as shown in Fig. 3.2.

The various notations used in the block diagram are explained below:

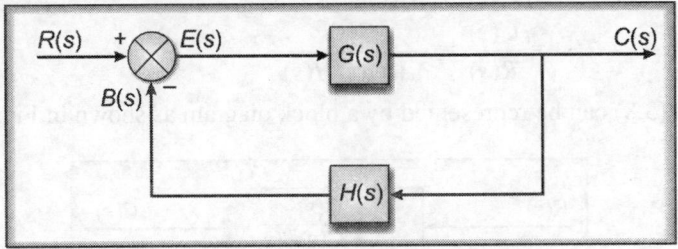

Fig. 3.2 Block diagram of a closed-loop system

$R(s)$: Input Signal

$C(s)$: Output Signal

$E(s)$: Actuating Signal

$B(s)$: Feedback Signal

$G(s)$: Forward Path Transfer Function

$H(s)$: Feedback Path Transfer Function

$T(s)$: Transfer Function of the Closed-loop system

From definition of transfer function, we can write

$$G(s) = \frac{C(s)}{E(s)},$$

or $\qquad C(s) = G(s)E(s)$ $\hfill$ (3.2)

$$T(s) = \frac{C(s)}{R(s)},$$ $\hfill$ (3.3)

and $\qquad H(s) = \frac{B(s)}{C(s)},$

or $\qquad B(s) = H(s)C(s)$ $\hfill$ (3.4)

From the block diagram, we can write the following equations

$$C(s) = G(s)E(s)$$ $\hfill$ (3.5)

$$E(s) = R(s) - B(s)$$ $\hfill$ (3.6)

or $\qquad E(s) = R(s) - H(s)C(s)$ $\hfill$ (3.7)

Substituting the value of $E(s)$ in equation (3.5), we

$$C(s) = G(s)\{R(s) - H(s)C((s)\}$$
$$C(s) + G(s)H(s)C(s) = G(s)R(s)$$
$$C(s)\{1 + G(s)H(s)\} = G(s)R(s)$$

or
$$T(s) = \frac{C(s)}{R(s)} = \frac{G(s)}{1 + G(s)H(s)} \qquad (3.8)$$

Equation (3.8) can be represented by a block diagram as shown in Fig. 3.3

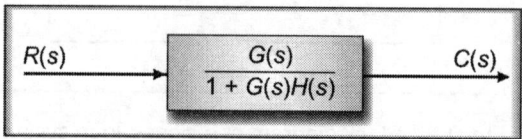

Fig. 3.3 Representation of equation (3.8)

The loop transfer function or open-loop transfer function of a closed-loop system is defined as:

$$\frac{B(s)}{E(s)} = G(s)H(s) \qquad (3.9)$$

This is derived as given below

$$\frac{B(s)}{E(s)} = \frac{B(s)}{C(s)} \times \frac{C(s)}{E(s)} = H(s)G(s) = G(s)H(s) \qquad (3.10)$$

In many cases, where $|G(s)H(s)| \gg 1$, the closed-loop transfer function

$$\frac{C(s)}{R(s)} = \frac{G(s)}{1 + G(s)H(s)}$$

can be approximated by

$$\frac{C(s)}{R(s)} \approx \frac{1}{H(s)} \qquad (3.11)$$

This is called the *approximate transfer function* and is the inverse of feedback transfer function. The significance of this concept lies in the fact that for cases where $|G(s)H(s)| \gg 1$, the closed-loop approximate transfer functions are independent of forward path transfer functions. It only depends upon the feedback transfer function $H(s)$. In the open-loop transfer function, the transfer function is dependent only on forward path transfer function $G(s)$, which is assembled with power elements. In an open-loop system, in order to achieve the desired objective, the power elements constituting $G(s)$, have to be selected very carefully. On the other hand in a closed-loop system, transfer function being dependent on $H(s)$, consists of measuring elements and hence, requires careful selection.

Any complex system can be conveniently represented by a block diagram and all complex representations of block diagrams can be reduced to a very simple block diagram by the use of block diagram reduction algebra. These rules have been illustrated in Table 3.1.

Table 3.1 Block Diagram Reduction Rules

Transformation	Original Block Diagram	Equivalent Block Diagram
1. Combining blocks in cascade	$R \to \boxed{G_1} \to \boxed{G_2} \xrightarrow{C}$	$R \to \boxed{G_1\ G_2} \xrightarrow{C}$ $C = (G_1\,G_2)R$
2. Combining blocks in parallel or eliminating a forward loop	$R \to \boxed{G_1}$, $\boxed{G_2}$, summing $\xrightarrow{C}$, $\pm$	$R \to \boxed{G_1 \pm G_2} \xrightarrow{C}$ $C = (G_1 \pm G_2)R$
3. Removing a block from a forward path	$R \to \boxed{G_1}$, $\boxed{G_2}$, summing $\xrightarrow{C}$, $\pm$ $C = (G_1 \pm G_2)R$	$R \to \boxed{G_1} \to \boxed{\dfrac{G_1}{G_2}} \to$ summing $\xrightarrow{C}$, $\pm$ $C = G_1\left(1 \pm \dfrac{G_1}{G_2}\right)R$ $C = (G_2 \pm G_1)R$
4. Eliminating a feedback loop	$R \to$ summing $\to \boxed{G_1} \xrightarrow{C}$, $\mp$, $\boxed{G_2}$	$R \to \boxed{\dfrac{G_1}{1 \mp G_1 G_2}} \xrightarrow{C}$ $C = \left(\dfrac{G_1}{1 \pm G_1 G_2}\right)R$
5. Removing a block from a feedback loop	$R \to$ summing $\to \boxed{G_1} \xrightarrow{C}$, $\mp$, $\boxed{G_2}$ $C = \dfrac{G_1}{1 \pm G_1 G_2}R$	$R \to \boxed{\dfrac{1}{G_2}} \to$ summing $\to \boxed{G_1 G_2} \xrightarrow{C}$, $\mp$ $C = \left(\dfrac{G_1 G_2}{1 \pm G_1 G_2}\right)\dfrac{1}{G_2} \times R$ $C = \dfrac{G_1}{1 \pm G_1 G_2}R$
6. Rearranging summing points	$R \to$ summing $\xrightarrow{R \pm X}$ summing $\xrightarrow{C}$ $X \uparrow \pm$, $\pm$ Y $C = R \pm X \pm Y$	$R \to$ summing $\xrightarrow{R \pm Y}$ summing $\xrightarrow{C}$ $Y \uparrow \pm$, $\pm$ X $C = R \pm X \pm Y$

Transformation	Original Block Diagram	Equivalent Block Diagram
7. Rearranging summing points	 $C = (R \pm X) \pm Y$ $= R + (\pm X \pm Y)$	 $C = R + (\pm X \pm Y)$
8. Moving a take off point ahead of a block	 $C = GR$	 $C = GR$
9. Moving a take off point beyond a block	 $C = GR$	 $C = GR$
10. Moving a summing point ahead of a block	 $C = RG \pm X$	 $C = G\left(R \pm \dfrac{1}{G}\right)X = RG \pm X$
11. Moving a summing point beyond a block	 $C = RG \pm XG = G(R \pm X)$	 $C = RG \pm GX = G(R \pm X)$
12. Moving a take off point ahead of a summing point	 $C = R \pm X$	 $C = R \pm X$
13. Moving a take off point beyond a summing point	 $C = R \pm X$	 $C = R \pm X$

Solved Problems

Problem 3.1 Obtain the transfer function for the block diagram as shown in Fig. 3.4.

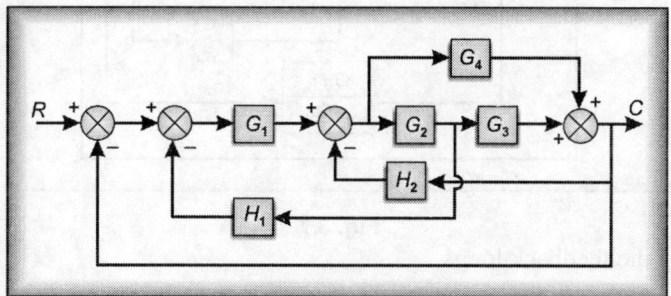

Fig. 3.4

Solution

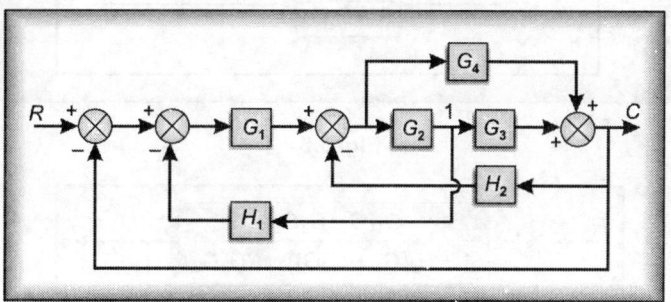

Fig. 3.5

Shifting take off point No. 1 ahead of G_2

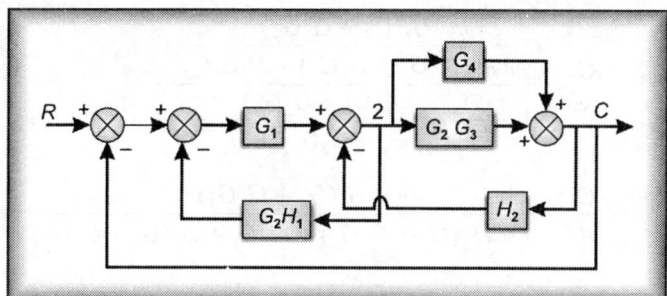

Fig. 3.6

Adding parallel paths G_4 and $G_2 G_3$ and shifting take off point No. 2 beyond block $G_2 G_3 + G_4$

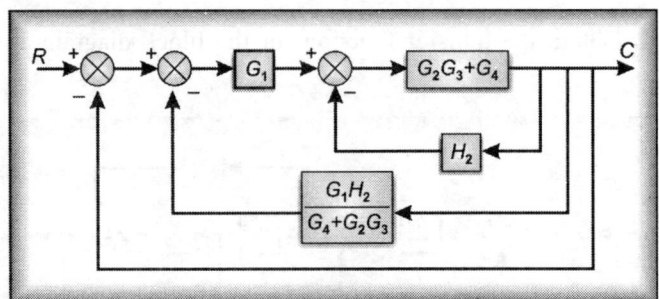

Fig. 3.7

Eliminating the feedback loops

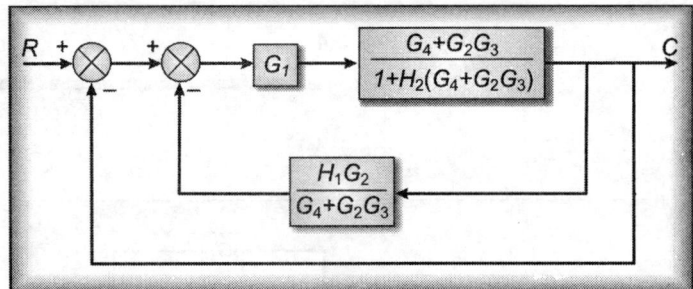

Fig. 3.8

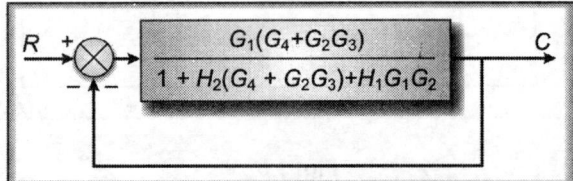

Fig. 3.9

$$\frac{C}{R} = \frac{\dfrac{G_1(G_4 + G_2 G_3)}{1 + H_2(G_4 + G_2 G_3) + H_1 G_1 G_2}}{1 + \dfrac{G_1(G_4 + G_2 G_3)}{1 + H_2(G_4 + G_2 G_3) + H_1 G_1 G_2}}$$

or

$$\frac{C}{R} = \frac{G_1(G_4 + G_2 G_3)}{1 + H_2(G_4 + G_2 G_3) + H_2 G_1 + G_1(G_4 + G_2 G_3)}$$

or

$$\frac{C}{R} = \frac{G_1(G_4 + G_2 G_3)}{1 + (G_4 + G_2 G_3)(H_2 + G_1) + H_1 G_1 G_2} \qquad \textbf{Ans}$$

Problem 3.2 Obtain the transfer function for the block diagram as shown in Fig. 3.10

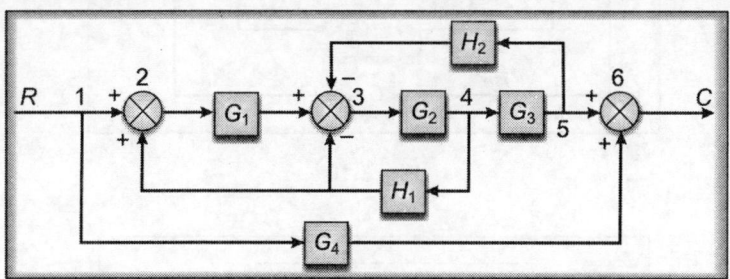

Fig. 3.10

Solution Moving take-off point No. 4 beyond block G_3

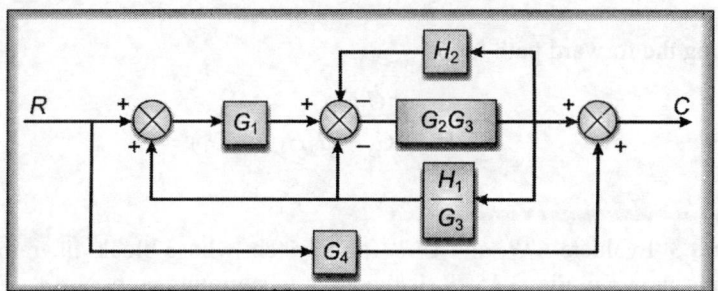

Fig. 3.11

Eliminating feedback path having transfer function H_2

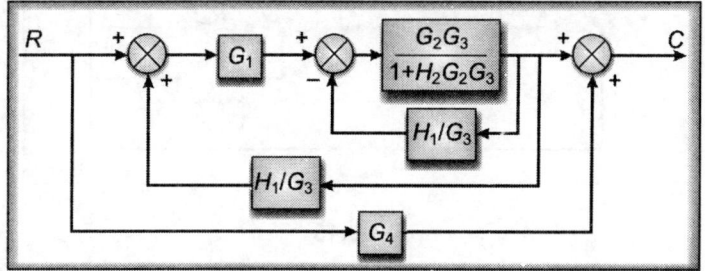

Fig. 3.12

Eliminating feedback loops one by one

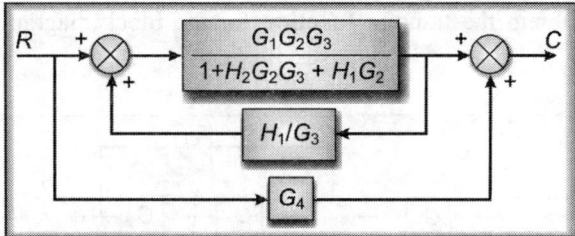

Fig. 3.13

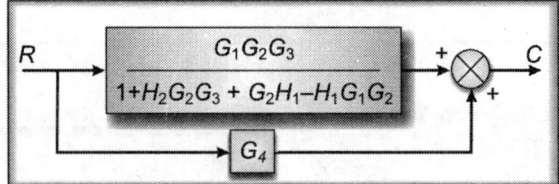

Fig. 3.14

Eliminating the forward path loops

$$\frac{C}{R} = G_4 + \frac{G_1G_2G_3}{1 + H_2G_2G_3 + G_2H_1(1 - G_1)} \qquad \textbf{Ans}$$

Problem 3.3 Evaluate C/R_1 and C/R_2 for a system whose block diagram representation is shown in Fig. 3.15. R_1 is the input to summing point No. 1.

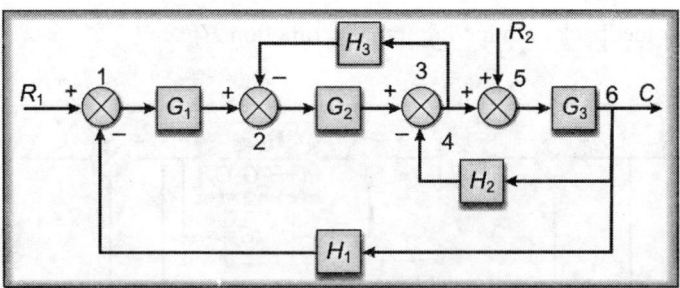

Fig. 3.15

Solution

Evaluation of C/R_1. Assume $R_2 = 0$. Therefore, summing point No. 5 can be removed. Shift take off point No. 4 beyond block G_3

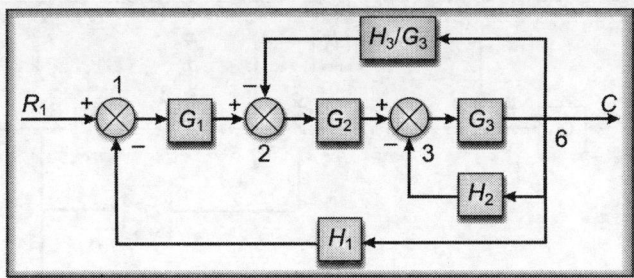

Fig. 3.16

Eliminate the feedback loop between points 3 and 6 and combine with block G_2

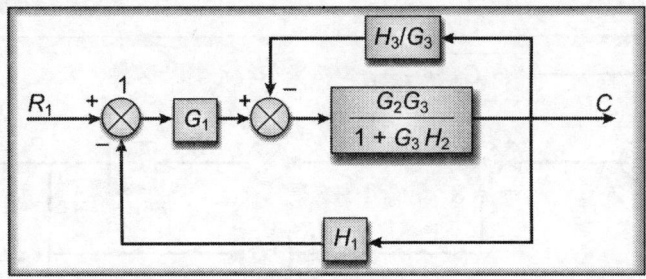

Fig. 3.17

Eliminating the feedback loop again and combine with block G_1

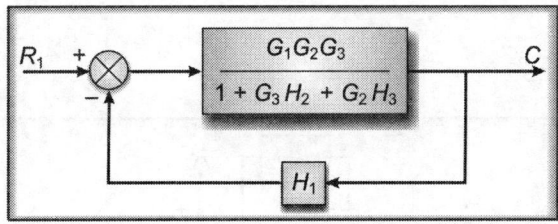

Fig. 3.18

$$\frac{C}{R_1} = \frac{G_1 G_2 G_3}{1 + G_2 H_3 + G_3 H_2 + G_1 G_2 G_3 H_1}$$ **Ans**

Evaluation of C/R_2

Assume $R_1 = 0$. Thus summing point No. 1 can be removed

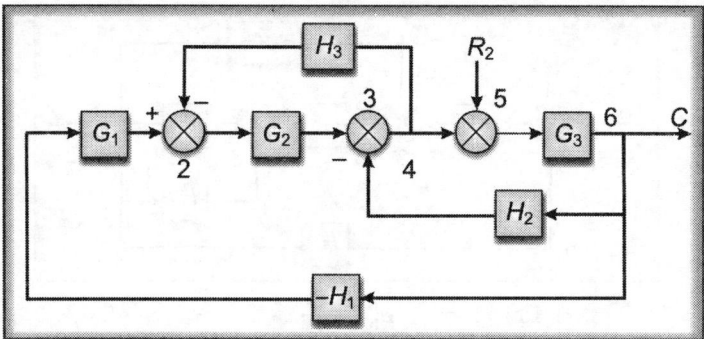

Fig. 3.19

Shifting the summing point No. 2 and rearranging beyond G_2

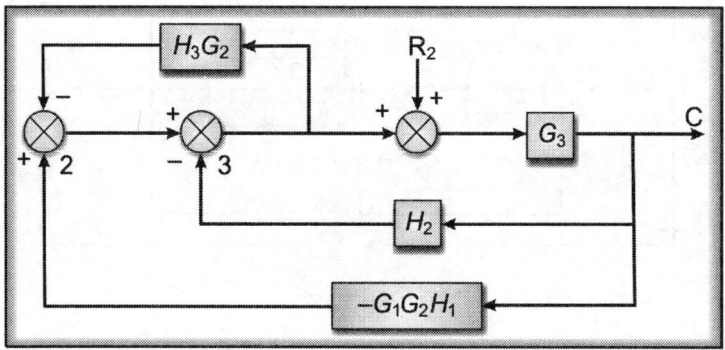

Fig. 3.20

Rearranging, we get

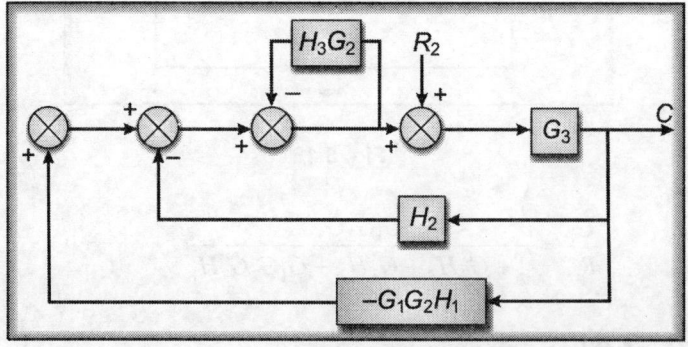

Fig. 3.21

Rearranging and eliminating the feedback loop

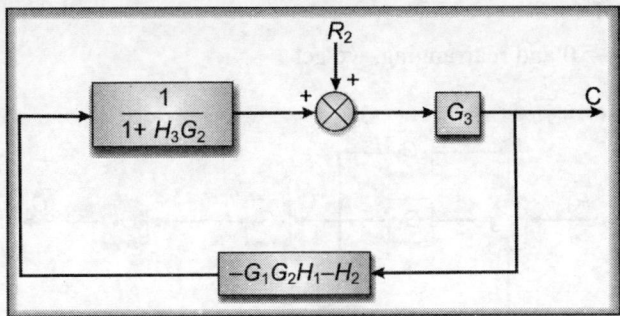

Fig. 3.22

Rearranging,

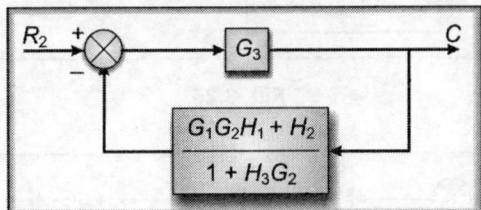

Fig. 3.23

Eliminating the feedback loop, we get

$$\frac{C}{R_2} = \frac{G_3(1+H_3G_2)}{1+H_3G_2+G_3(G_1G_2H_1+H_2)}$$ **Ans**

Problem 3.4 Evaluate C_1/R_1 and C_2/R_2 for a system whose block diagram representation is shown in Fig. 3.24.

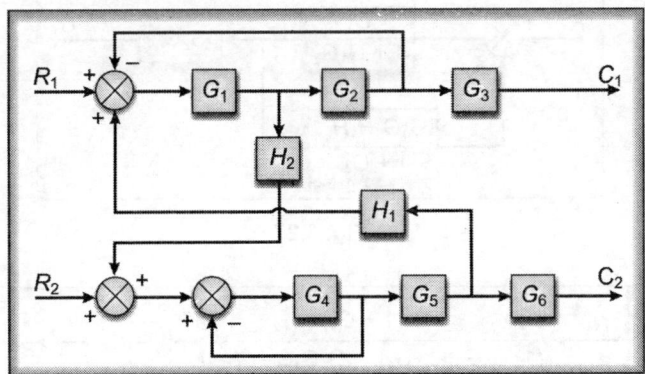

Fig. 3.24

Solution Assuming $R_2 = 0$

Evaluation of C_1/R_1

Assuming $C_2 = 0$ and rearranging, we get

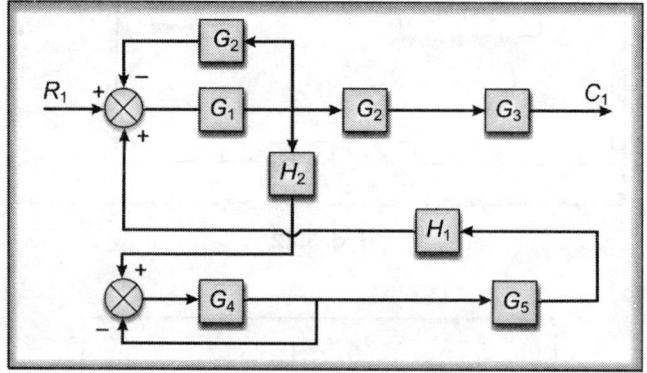

Fig. 3.25

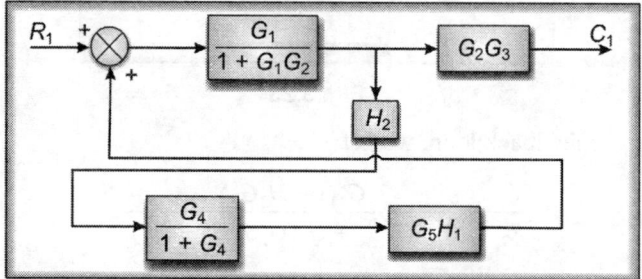

Fig. 3.26

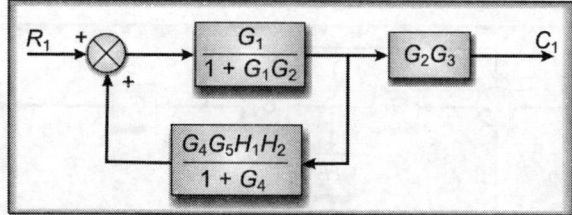

Fig. 3.27

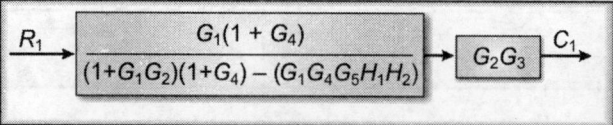

Fig. 3.28

$$\therefore \quad \frac{C_1}{R_1} = \frac{G_1 G_2 G_3 (1 + G_4)}{(1 + G_1 G_2)(1 + G_4) - G_1 G_4 G_5 H_1 H_2} \quad \textbf{Ans}$$

Evaluation of $\dfrac{C_2}{R_1}$

Assuming $C_1 = 0$ and rearranging, we get

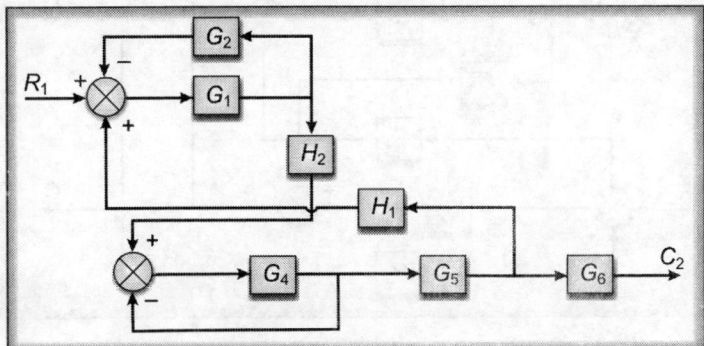

Fig. 3.29

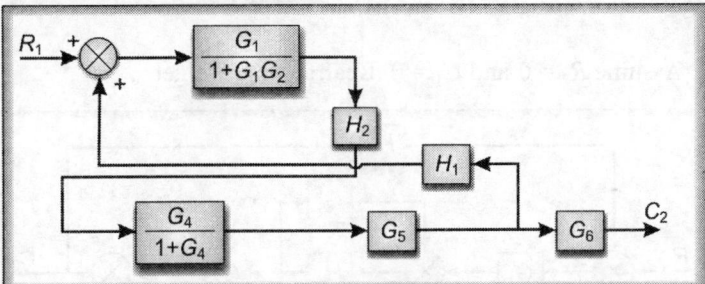

Fig. 3.30

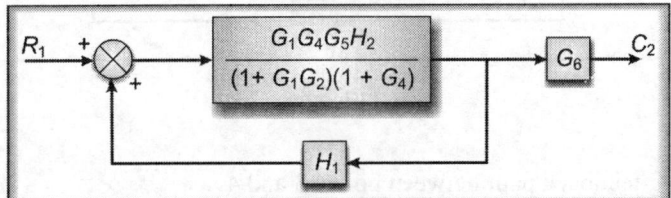

Fig. 3.31

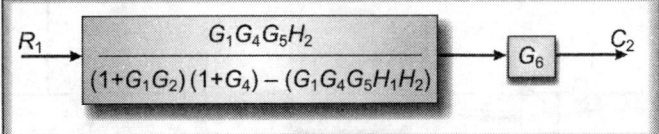

Fig. 3.32

$$\therefore \quad \frac{C_2}{R_1} = \frac{G_1 G_4 G_5 G_6 H_2}{(1 + G_1 G_2)(1 + G_4) - (G_1 G_4 G_5 H_1 H_2)} \quad \textbf{Ans}$$

Problem 3.5 Evaluate C_2/R_1 for the system whose block diagram representation is shown in Fig. 3.33

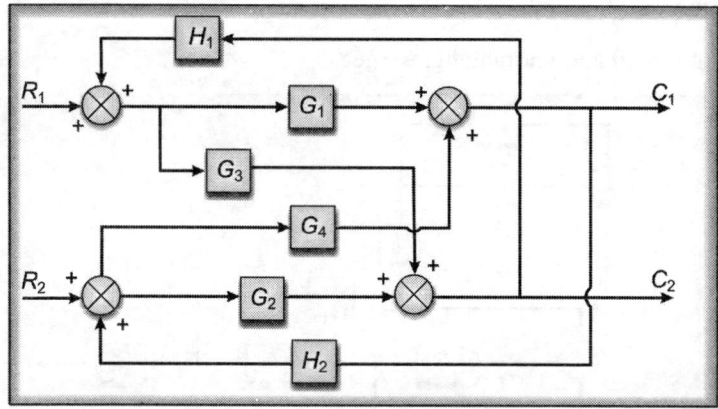

Fig. 3.33

Solution Assume $R_2 = 0$ and $C_1 = 0$. Rearranging, we get

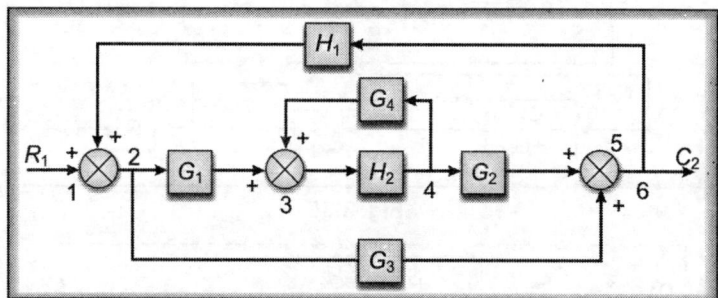

Fig. 3.34

Eliminating feedback path between points 3 and 4

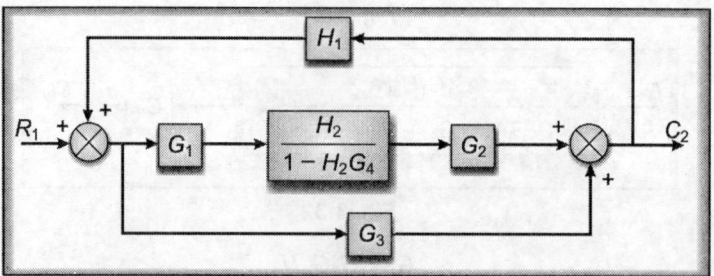

Fig. 3.35

Further simplifying

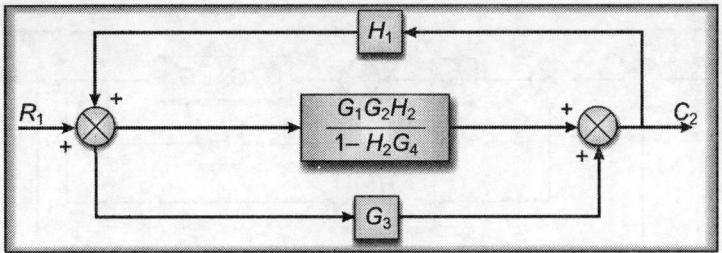

Fig. 3.36

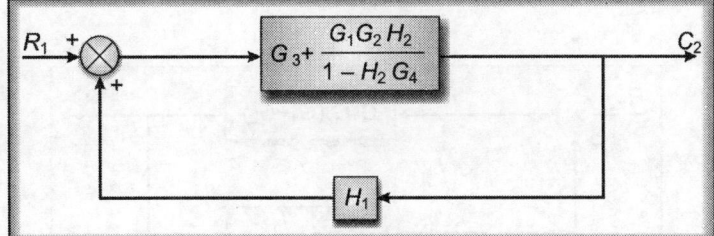

Fig. 3.37

$$\frac{C_2}{R_1} = \frac{G_3 + \dfrac{G_1 G_2 H_2}{1 - H_2 G_4}}{1 - \left(G_3 + \dfrac{G_1 G_2 H_2}{1 - H_2 G_4}\right) H_1}$$

$$= \frac{G_3 - H_2 G_3 G_4 + G_1 G_2 H_2}{(1 - H_2 G_4 - G_3 H_1 + H_1 H_2 G_4 G_3 - G_1 G_2 H_1 H_2)}$$

$$= \frac{G_3 (1 - H_2 G_4) + G_1 G_2 H_2}{(1 - G_4 H_2) + H_1 [G_3 (H_2 G_4 - 1) - G_1 G_2 H_2]} \qquad \textbf{Ans}$$

Problem 3.6 Find the transfer function for the system whose block diagram representation is shown in Fig. 3.38. *(Pune University)*

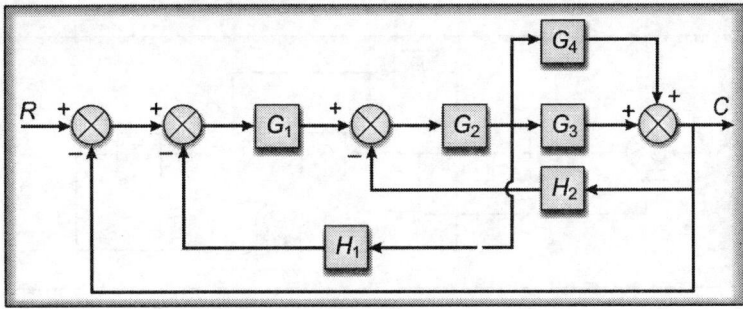

Fig. 3.38

Solution

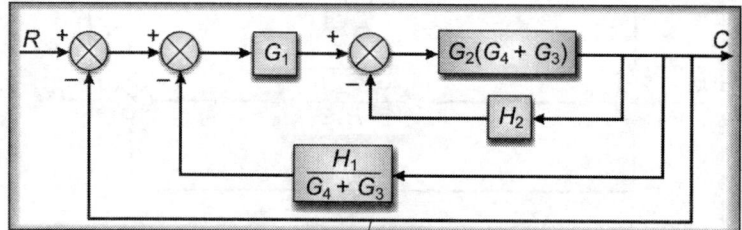

Fig. 3.39

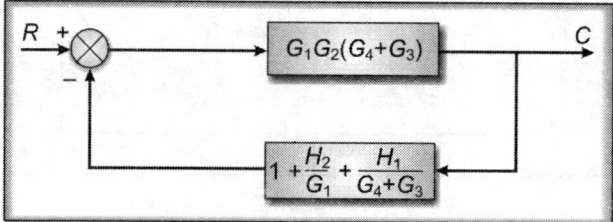

Fig. 3.40

$$\frac{C}{R} = \frac{G_1 G_2 (G_4 + G_3)}{1 + G_1 G_2 (G_4 + G_3) \left[1 + \dfrac{H_2}{G_1} + \dfrac{H_1}{G_4 + G_3} \right]}$$

$$= \frac{G_1 G_2 (G_4 + G_3)}{1 + H_2 G_2 (G_4 + G_3) + G_1 G_2 H_1 + G_1 G_2 (G_4 + G_3)} \quad \textbf{Ans}$$

Problem 3.7 Find transfer function for the block diagram as shown in Fig. 3.40

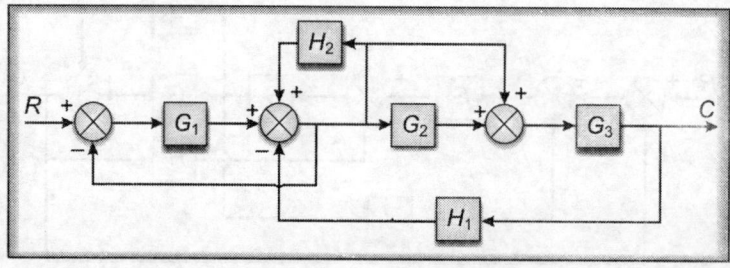

Fig. 3.41

Solution

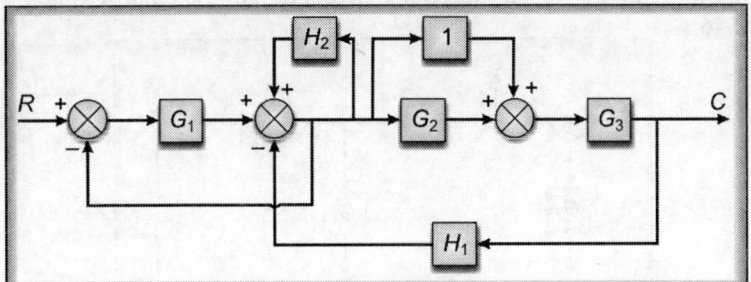

Fig. 3.42

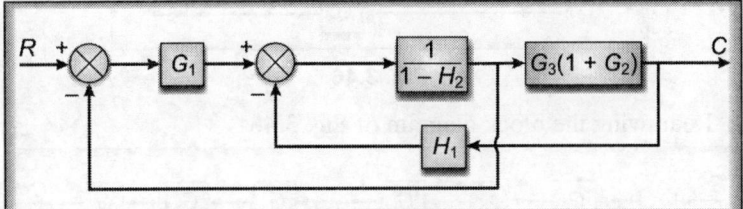

Fig. 3.43

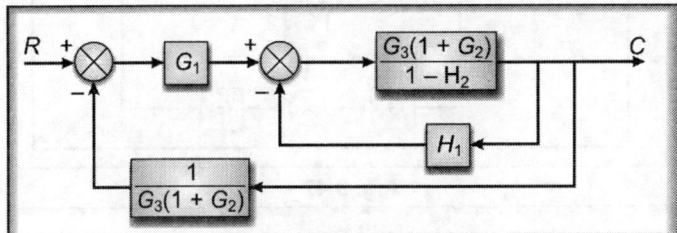

Fig. 3.44

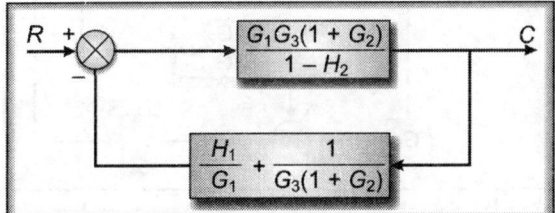

Fig. 3.45

$$\frac{C}{R} = \frac{\dfrac{G_1 G_3 (1 + G_2)}{1 - H_2}}{1 + \dfrac{G_1 G_3 (1 + G_2)}{1 - H_2} \left[\dfrac{H_1}{G_1} + \dfrac{1}{G_3 (1 + G_2)} \right]}$$

$$\therefore \quad \frac{C}{R} = \frac{G_1 G_3 (1 + G_2)}{(1 - H_2) + H_1 G_3 (1 + G_2) + G_1} \qquad \textbf{Ans}$$

Problem 3.8 Find the closed-loop transfer function of the control system shown in Fig. 3.46

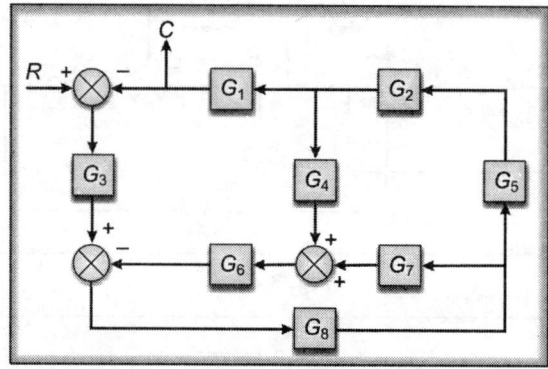

Fig. 3.46

Solution Redrawing the block diagram of Fig. 3.46

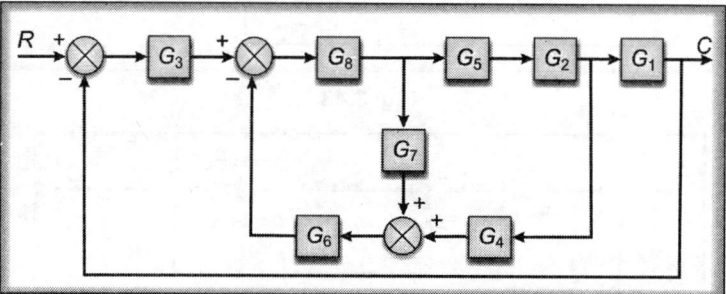

Fig. 3.47

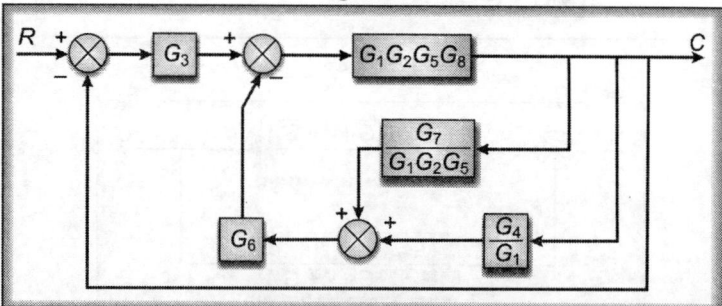

Fig. 3.48

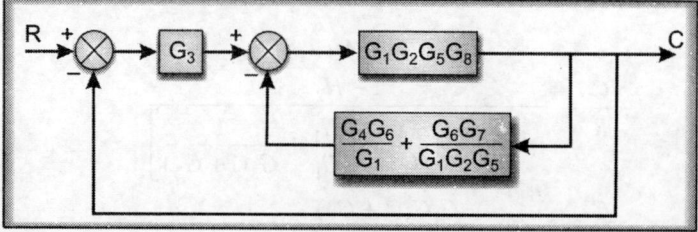

Fig. 3.49

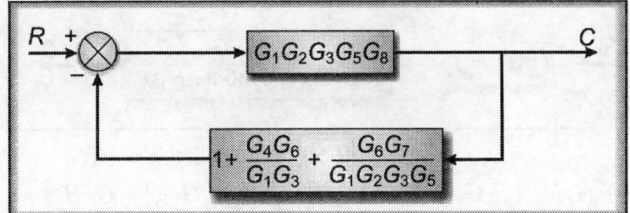

Fig. 3.50

$$\frac{C}{R} = \frac{G_1 G_2 G_3 G_5 G_8}{1 + G_1 G_2 G_3 G_5 G_8 \left(1 + \dfrac{G_4 G_6}{G_1 G_3} + \dfrac{G_6 G_7}{G_3 G_5 G_1 G_2}\right)}$$

$$\frac{C}{R} = \frac{G_1 G_2 G_3 G_5 G_8}{1 + G_1 G_2 G_3 G_5 G_8 + G_2 G_4 G_5 G_6 G_8 + G_6 G_7 G_8} \quad \textbf{Ans}$$

Problem 3.9 Find closed-loop transfer function of the control system as shown in Fig. 3.51

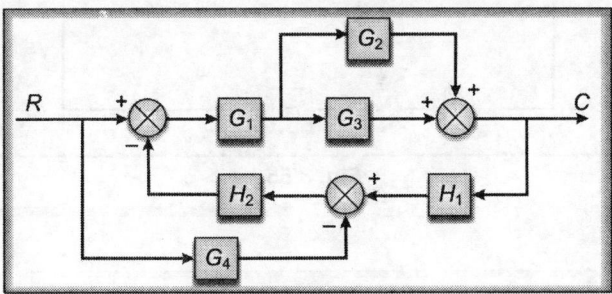

Fig. 3.51

Solution

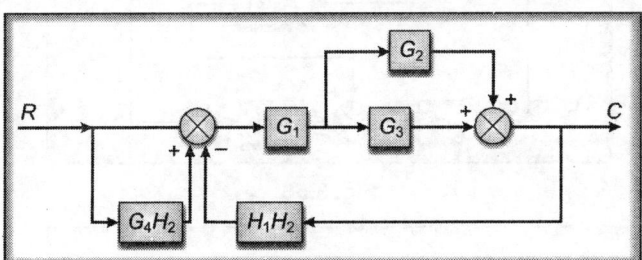

Fig. 3.52

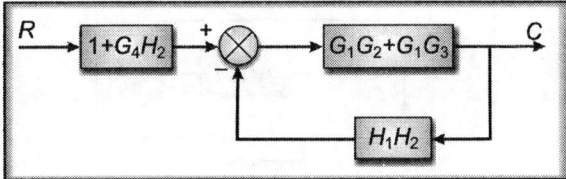

Fig. 3.53

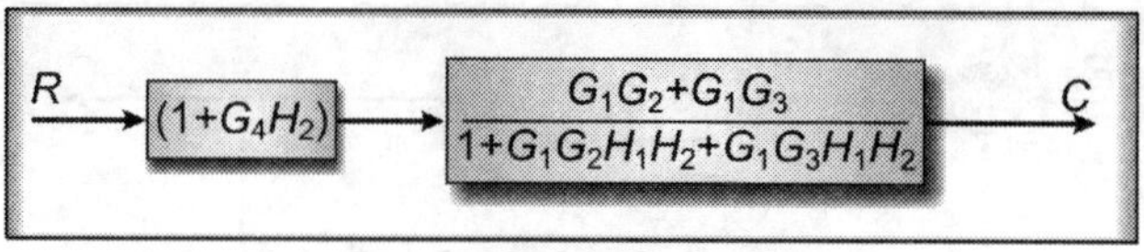

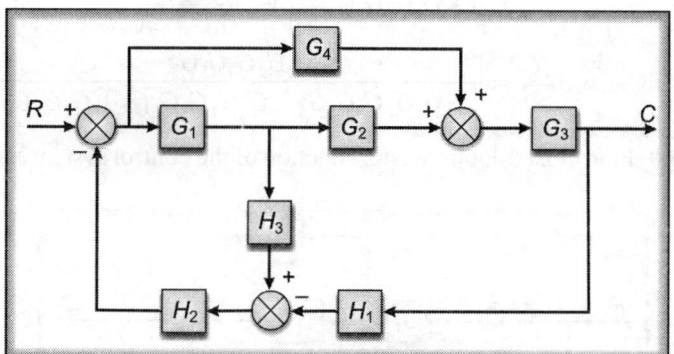

Fig. 3.54

$$\therefore \quad \frac{C}{R} = \frac{(1+G_4H_2)(G_1G_2+G_1G_3)}{1+G_1G_2H_1H_2+G_1G_3H_1H_2} = \frac{G_1(G_2+G_3)(1+G_4H_2)}{1+G_1H_1H_2(G_2+G_3)}$$ **Ans**

Problem 3.10 Find the closed-loop transfer function of the control system as shown in Fig. 3.55

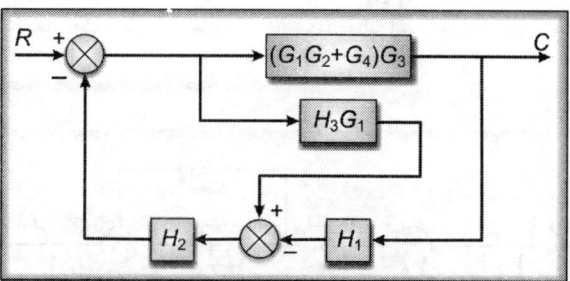

Fig. 3.55

Solution

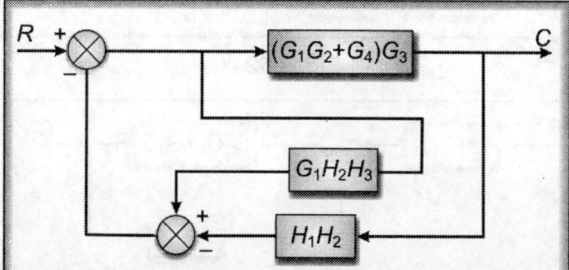

Fig. 3.56

Fig. 3.57

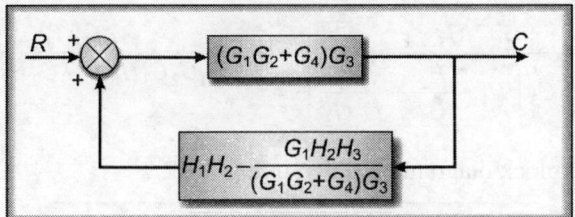

Fig. 3.58

$$\frac{C}{R} = \frac{G}{1 - GH} = \frac{(G_1 G_2 + G_4)G_3}{1 + G_1 H_2 (H_3 - G_2 G_3 H_1) - G_3 G_4 H_1 H_2} \quad \textbf{Ans}$$

Problem 3.11 Reduce the block diagram in Fig. 3.59 to its simplest possible form and hence obtain its closed-loop transfer function. (*Pune University*)

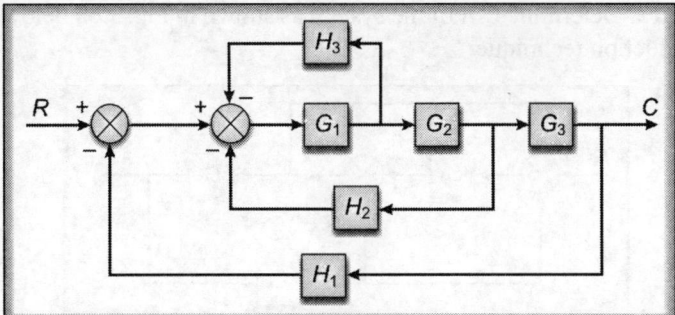

Fig. 3.59

Solution

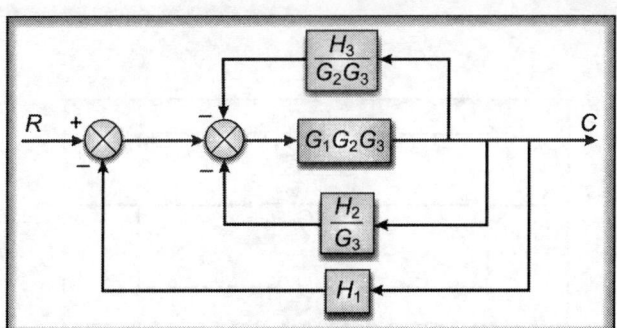

Fig. 3.60

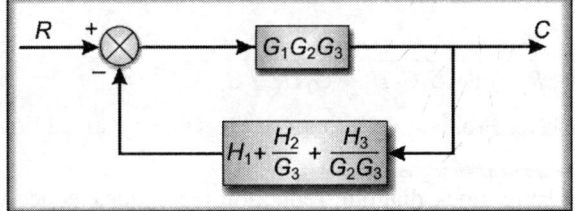

Fig. 3.61

$$\frac{C}{R} = \frac{G_1 G_2 G_3}{1 + G_1 G_2 G_3 \left(H_1 + \dfrac{H_2}{G_3} + \dfrac{H_3}{G_2 G_3} \right)} = \frac{G_1 G_2 G_3}{1 + G_1 G_2 G_3 H_1 + G_1 G_2 H_2 + G_1 H_3}$$

The simplest block diagram is shown in Fig. 3.62

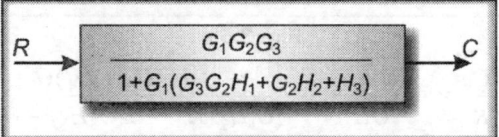

Fig. 3.62

Problem 3.12 Determine C/R of the system as shown in Fig. 3.63 below by block diagram reduction technique.

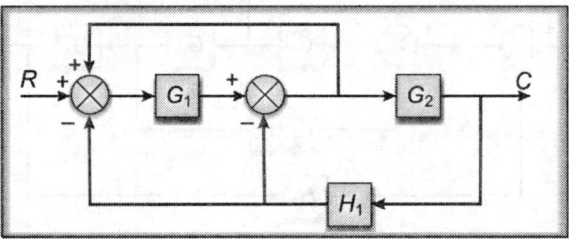

Fig. 3.63

Solution

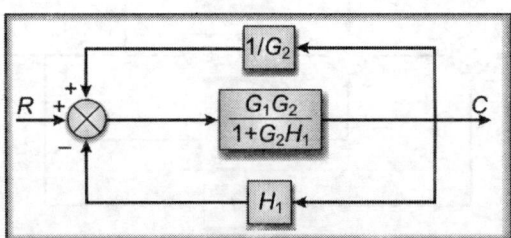

Fig. 3.64

$$\frac{C}{R} = \frac{G_1 G_2}{1 + G_1 G_2 H_1 + G_2 H_1 - G_1}$$ **Ans**

Note: Please refer Problem 4.23 to solve the same by signal flow graph.

Problem 3.13 Using block diagram reduction technique, reduce the block diagram as shown in Fig. 3.65 below and determine the overall transfer function.

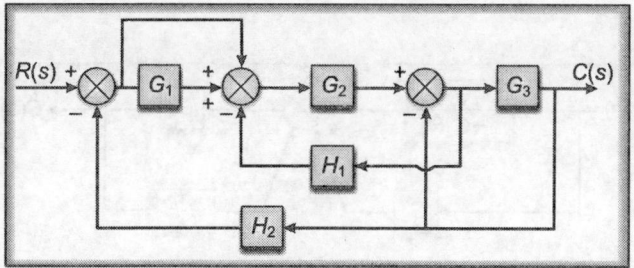

Fig. 3.65

Solution

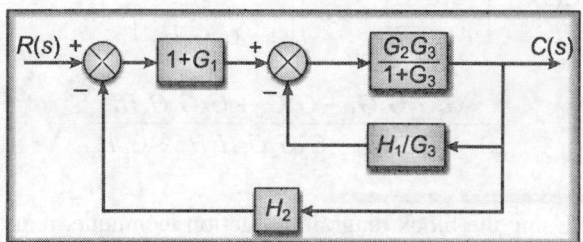

Fig. 3.66

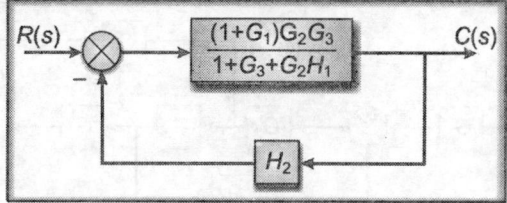

Fig. 3.67

$$\therefore \quad \frac{C(s)}{R(s)} = \frac{G}{1+GH} = \frac{G_1 G_2 G_3 + G_2 G_3}{1 + G_3 + G_2 H_1 + G_2 G_3 H_2 + G_1 G_2 G_3 H_2} \qquad \textbf{Ans}$$

Problem 3.14 Using the block diagram reduction technique, find the transfer function for the block diagram as shown in Fig. 3.68. (*AMIE*)

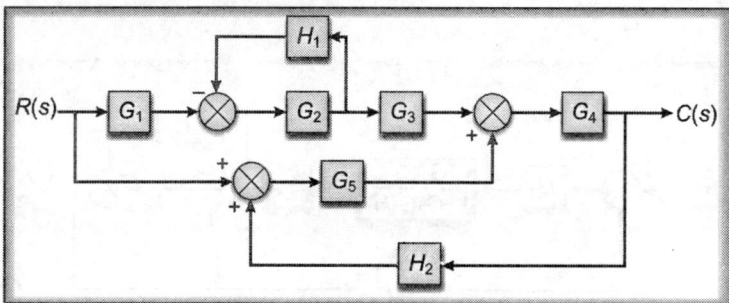

Fig. 3.68

Solution

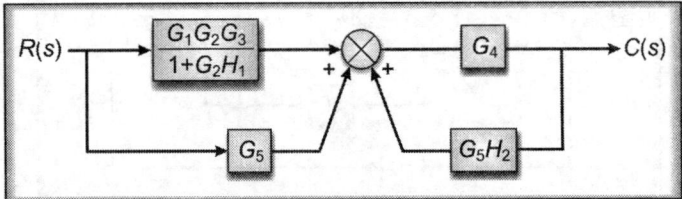

Fig. 3.69

$$\therefore \qquad \frac{C(s)}{R(s)} = \left(\frac{G_1 G_2 G_3}{1 + G_2 H_1} + G_5 \right) \left(\frac{G_4}{1 - G_4 G_5 H_2} \right)$$

$$= \frac{G_1 G_2 G_3 G_4 + G_4 G_5 + G_2 G_4 G_5 H_1}{1 - G_4 G_5 H_2 - G_2 G_4 G_5 H_1 H_2 + G_2 H_1} \qquad \textbf{Ans}$$

Problem 3.15 Using the block diagram reduction technique, reduce the control system as shown in Fig. 3.70 to simplest possible form and find the transfer function.

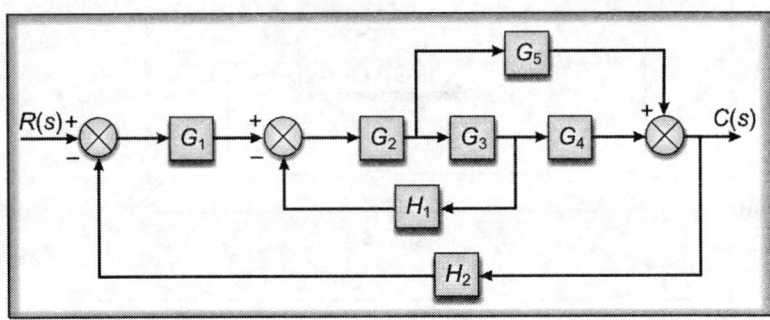

Fig. 3.70

Solution

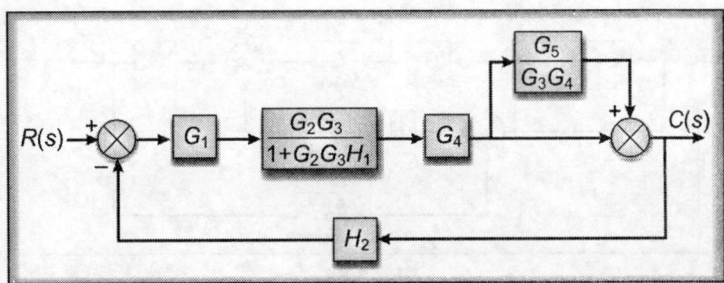

Fig. 3.71

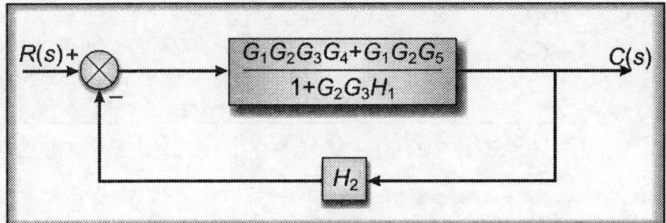

Fig. 3.72

$$\frac{C(s)}{R(s)} = \frac{G}{1+GH} = \frac{G_1G_2G_3G_4 + G_1G_2G_5}{1 + G_2G_3H_1 + G_1G_2G_3G_4H_2 + G_1G_2G_5H_2}$$

Ans

Problem 3.16 Determine the ratio $C(s)/R(s)$ for the multiple-loop system as shown in Fig. 3.73   (*Pune University*)

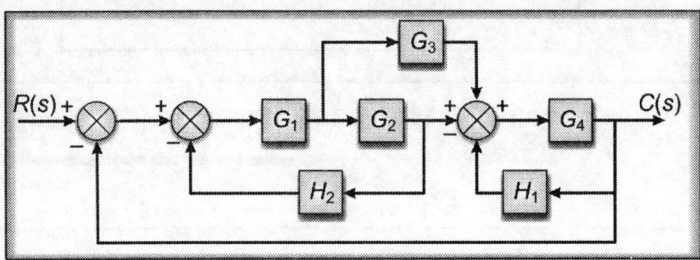

Fig. 3.73

Solution

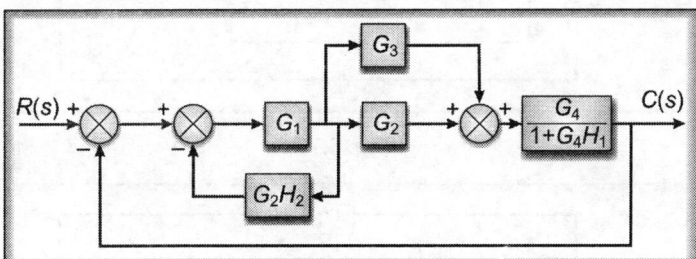

Fig. 3.74

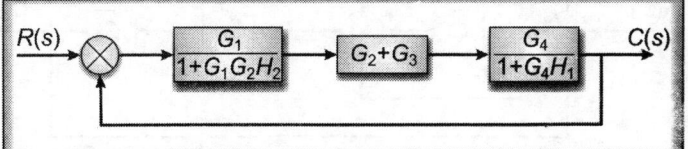

Fig. 3.75

$$\frac{C(s)}{R(s)} = \frac{G}{1+GH}$$

$$= \frac{G_1G_2G_4 + G_1G_3G_4}{1 + G_1G_2H_2 + G_4H_1 + G_1G_2G_4H_1H_2 + G_1G_2G_4 + G_1G_3G_4} \quad \textbf{Ans}$$

Problem 3.17 Find the closed-loop transfer function for the control system as shown in Fig. 3.76

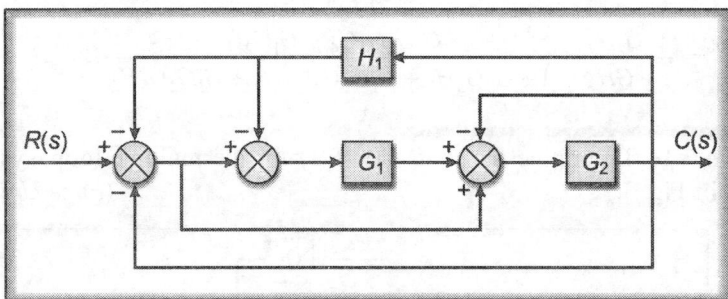

Fig. 3.76

Solution

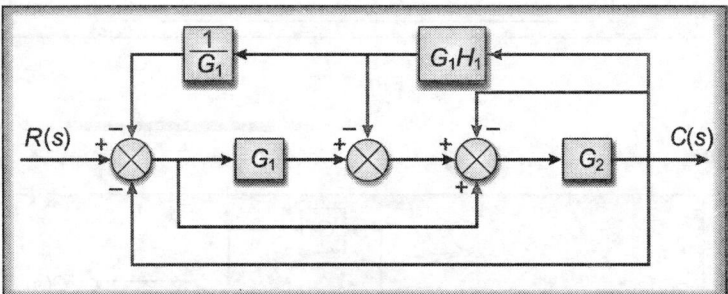

Fig. 3.77

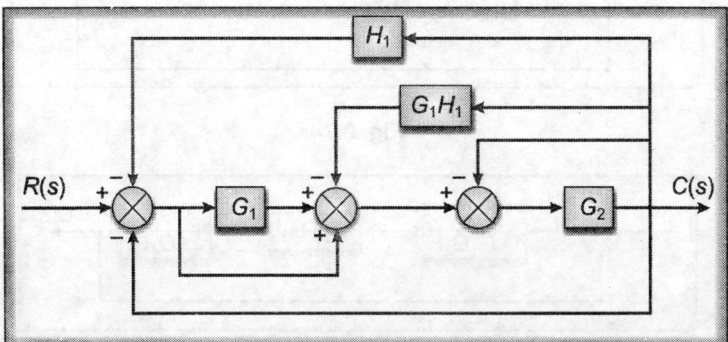

Fig. 3.78

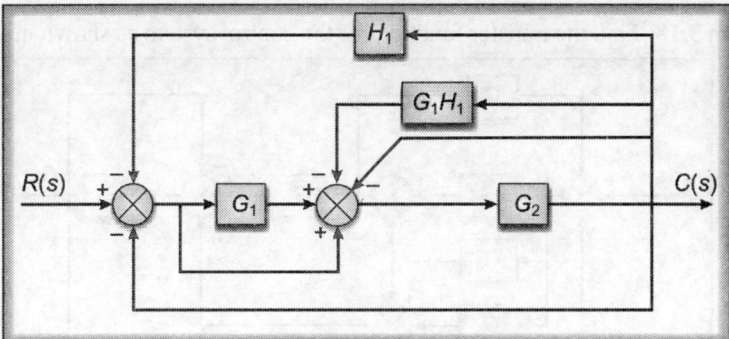

Fig. 3.79

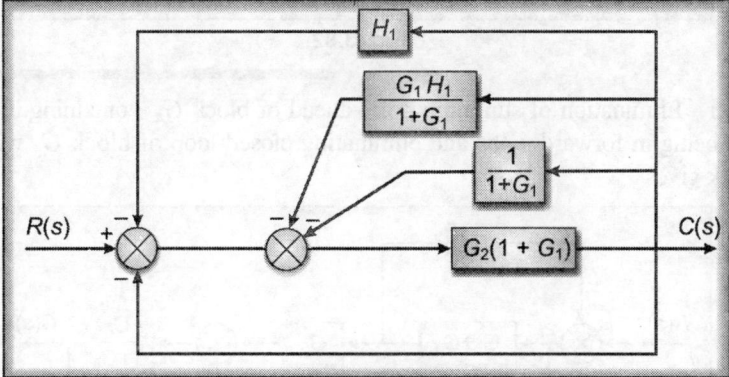

Fig. 3.80

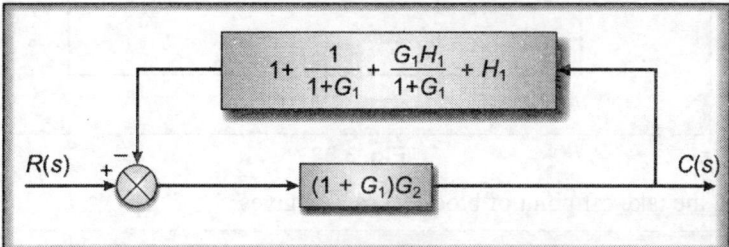

Fig. 3.81

$$\frac{C(s)}{R(s)} = \frac{(1+G_1)G_2}{1+[(1+G_1)G_2]\left[1+\dfrac{1}{1+G_1}+\dfrac{G_1H_1}{1+G_1}+H_1\right]}$$

$$= \frac{(1+G_1)G_2}{1+[(1+G_1)G_2+G_2+G_1G_2H_1+(1+G_1)G_2H_1]}$$

$$= \frac{(1+G_1)G_2}{1+[(G_2+G_1G_2H_1+(1+G_1)G_2+(1+G_1)G_2H_1]} \quad \textbf{Ans}$$

Problem 3.18 Find the transfer function for the control system as shown in Fig. 3.82

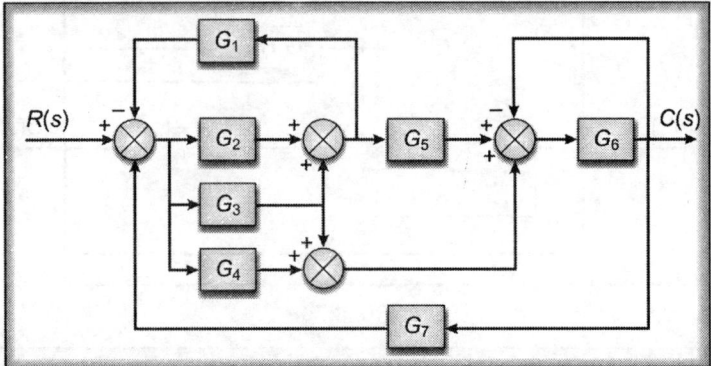

Fig. 3.82

Solution Elimination of summing point ahead of block G_4, containing blocks G_2 and G_3 being in forward paths and eliminating closed-loop of block G_6 with unity feedback gives

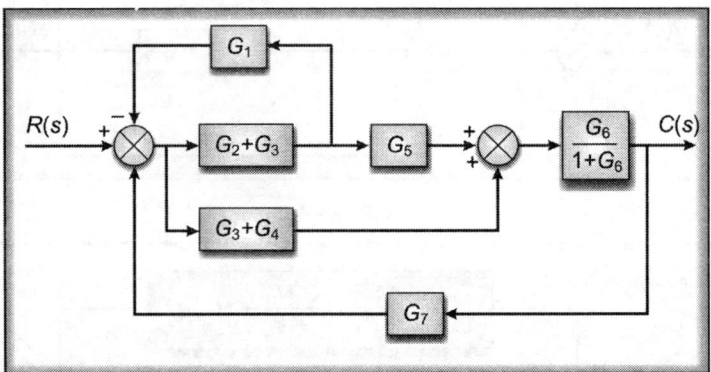

Fig. 3.83

Shifting the take off point of block $(G_3 + G_4)$ gives

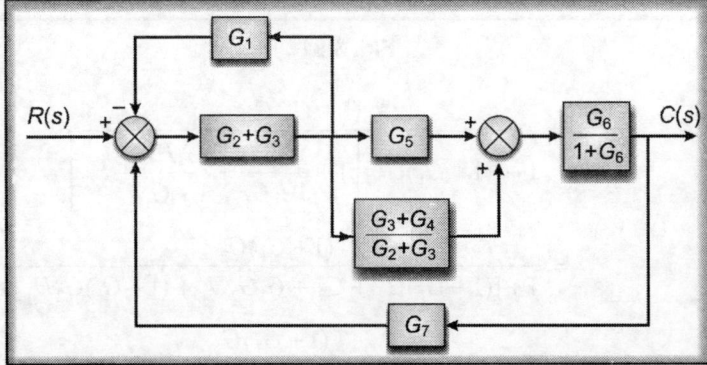

Fig. 3.84

Further reduction gives

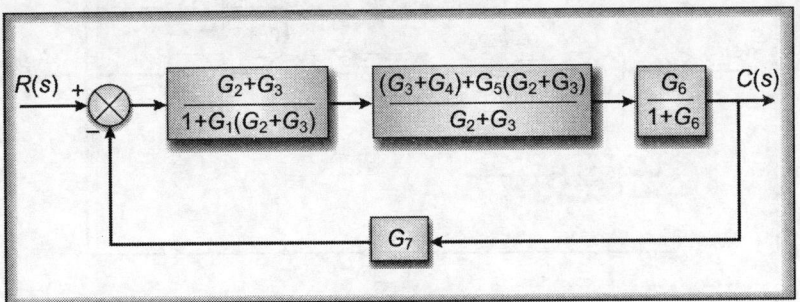

Fig. 3.85

$$\frac{C(s)}{R(s)} = \frac{G_3 G_6 + G_4 G_6 + G_2 G_5 G_6 + G_3 G_5 G_6}{1 + G_6 + G_1 G_2 + G_1 G_3 + G_1 G_2 G_6 + G_1 G_3 G_6 + G_3 G_6 G_7 + G_4 G_6 G_7}$$
$$+ G_2 G_5 G_6 G_7 + G_3 G_5 G_6 G_7$$

Ans

Problem 3.19 Determine $C(s)/R(s)$

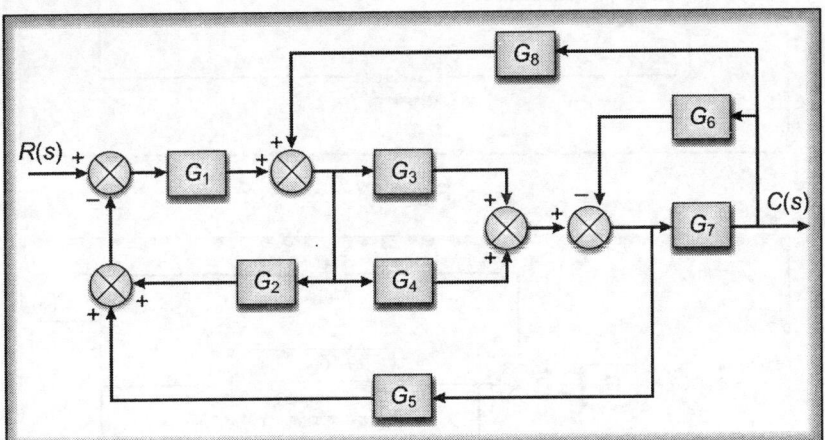

Fig. 3.86

Solution Combining blocks G_3 and G_4 and shifting take off point of block G_5 gives

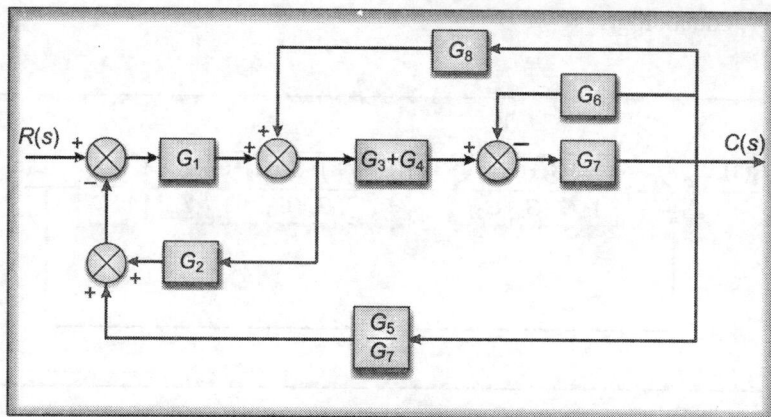

Fig. 3.87

Shifting block ($G_3 + G_4$) near block G_7 gives

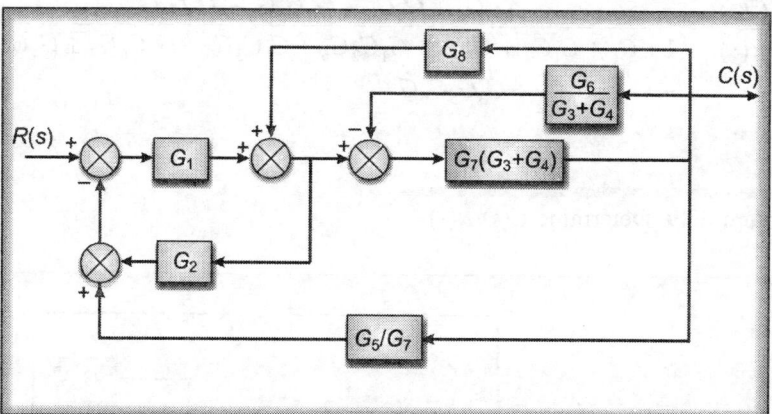

Fig. 3.88

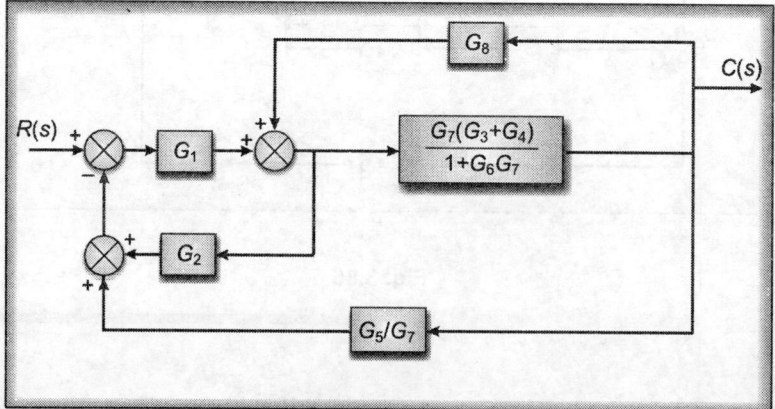

Fig. 3.89

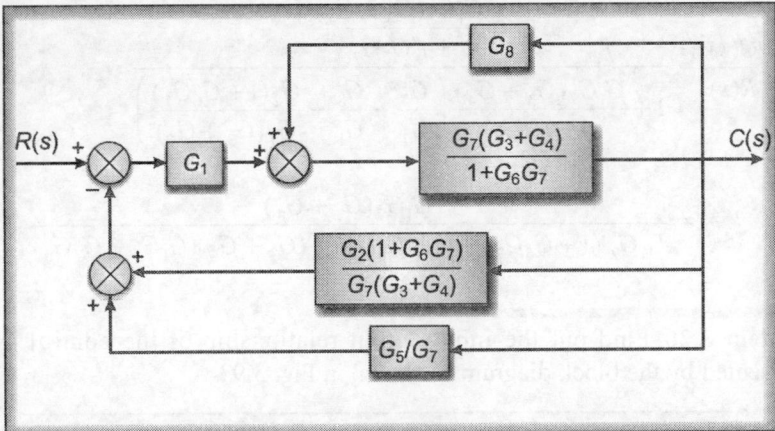

Fig. 3.90

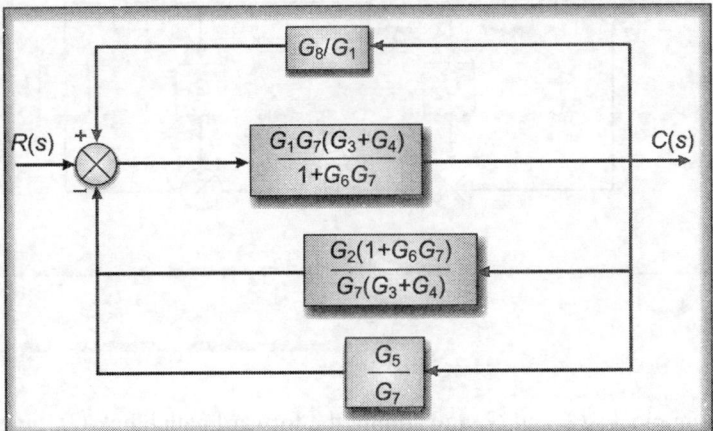

Fig. 3.91

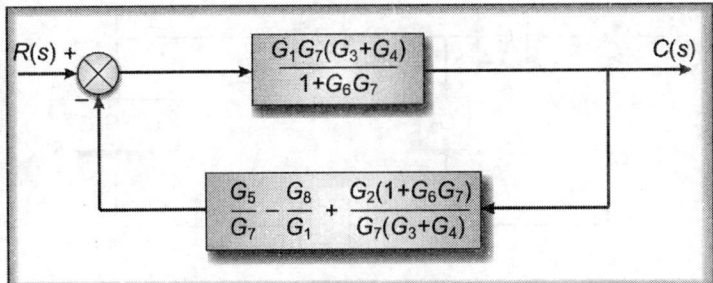

Fig. 3.92

$$\frac{C(s)}{R(s)} = \frac{\dfrac{G_1G_7(G_3+G_4)}{1+G_6G_7}}{1 + \dfrac{G_1G_7(G_3+G_4)}{1+G_6G_7}\left(\dfrac{G_5}{G_7} - \dfrac{G_8}{G_1} + \dfrac{G_2(1+G_6G_7)}{G_7(G_3+G_4)}\right)}$$

$$= \frac{G_1G_7(G_3+G_4)}{G_1G_2(1+G_6G_7) + (1+G_6G_7) + (G_3+G_4)(G_1G_5 - G_7G_8)} \qquad \textbf{Ans}$$

Problem 3.20 Find out the input-output relationship of the control system represented by the block diagram as shown in Fig. 3.93

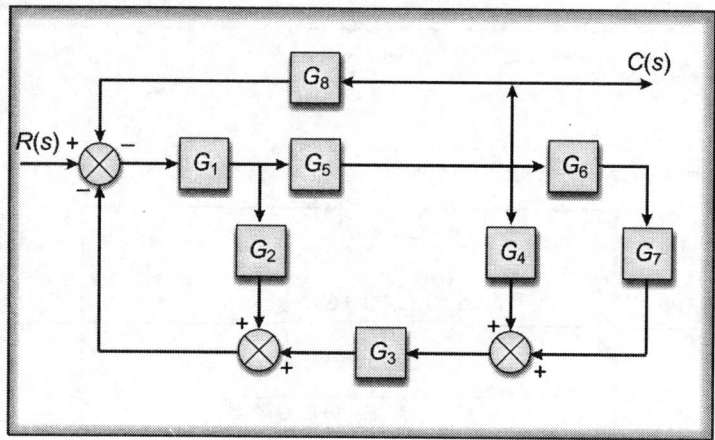

Fig. 3.93

Solution

Combining blocks G_6 and G_7 and adding the forward path block G_4, gives

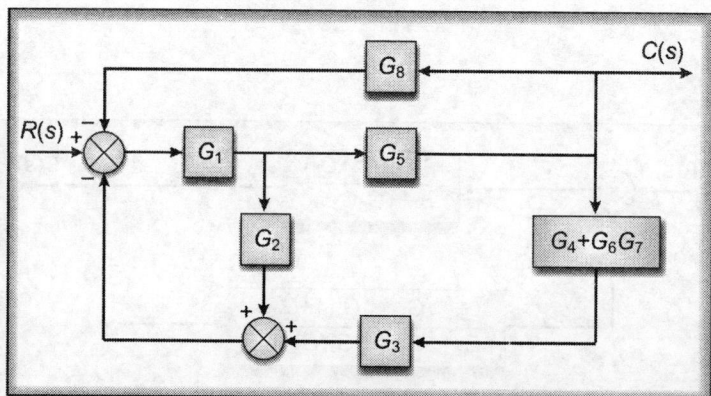

Fig. 3.94

Combining G_3 and $(G_4 + G_6G_7)$ gives

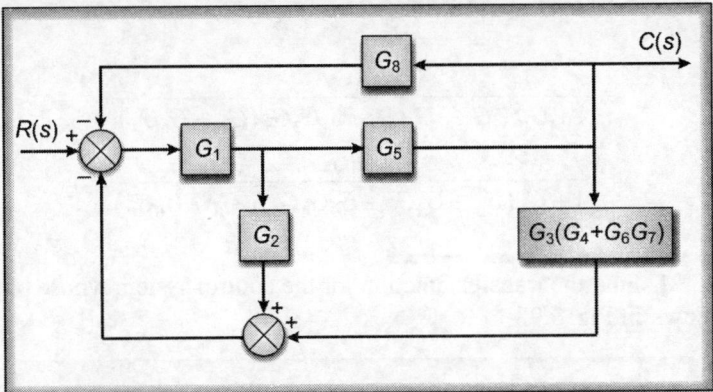

Fig. 3.95

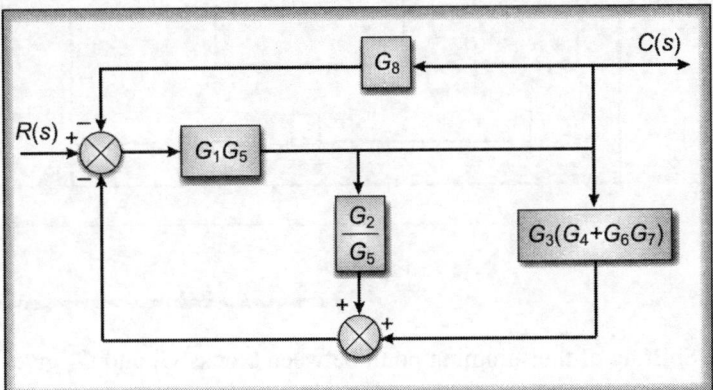

Fig. 3.96

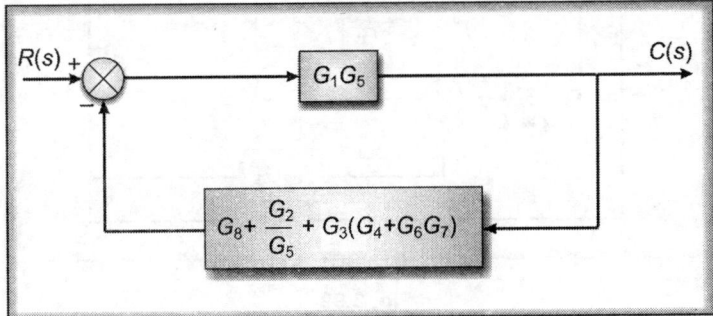

Fig. 3.97

$$\frac{C(s)}{R(s)} = \frac{G_1 G_5}{1 + G_1 G_5 \left[G_8 + \dfrac{G_2}{G_5} + G_3 (G_4 + G_6 G_7) \right]}$$

$$= \frac{G_1 G_5}{1 + G_1 G_5 G_8 + G_1 G_2 + G_1 G_3 G_5 (G_4 + G_6 G_7)}$$

$$= \frac{G_1 G_5}{1 + G_1 (G_2 + G_5 G_8 + G_3 G_4 G_5 + G_3 G_5 G_6 G_7)} \qquad \textbf{Ans}$$

Problem 3.21 Find the transfer function for the control system whose block diagram is shown in Fig. 3.98

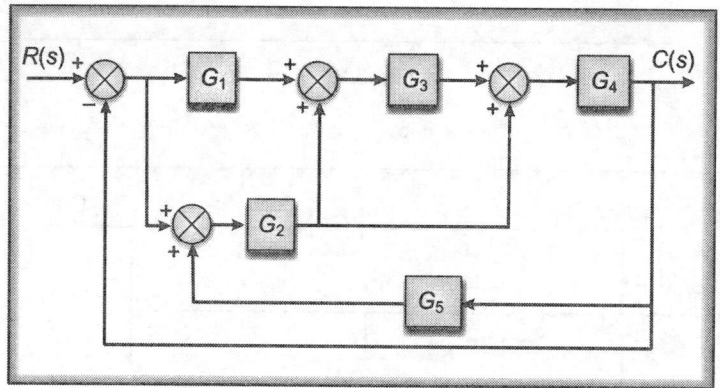

Fig. 3.98

Solution Shifting of the summing point between blocks G_1 and G_3 gives

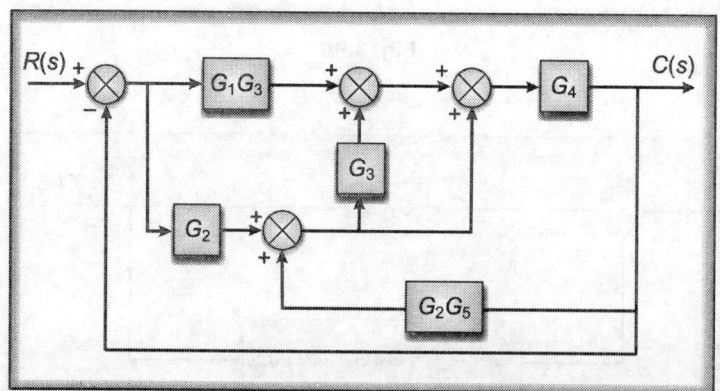

Fig. 3.99

Combining the parallel paths of G_3 and unity gives

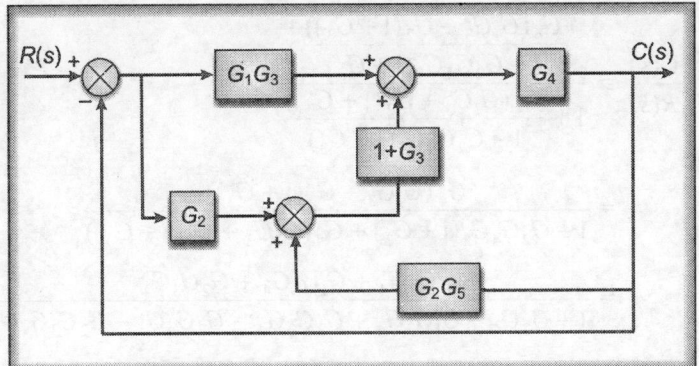

Fig. 3.100

Shifting the summing point between blocks G_2 and $(1 + G_3)$ gives

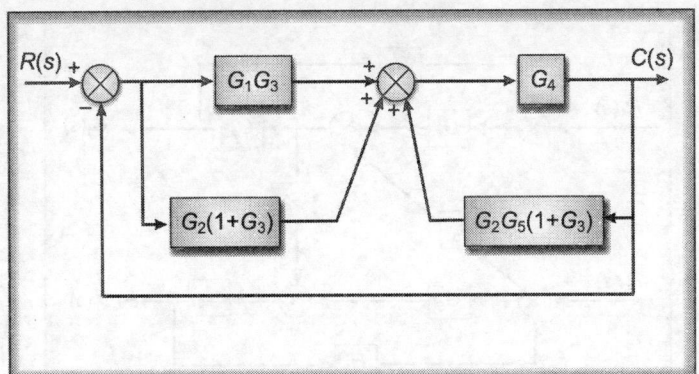

Fig. 3.101

Further reduction yields

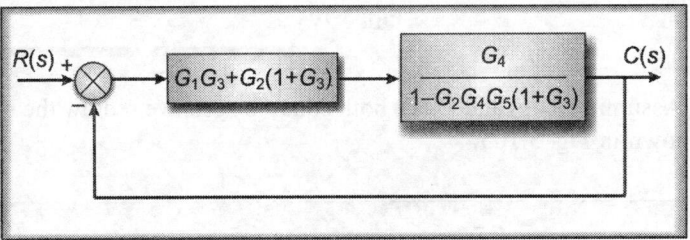

Fig. 3.102

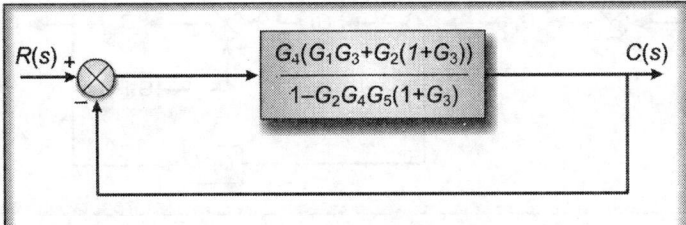

Fig. 3.103

$$\frac{C(s)}{R(s)} = \frac{\dfrac{G_4(G_1G_3 + G_2(1+G_3))}{1 - G_2G_4G_5(1+G_3)}}{1 + \dfrac{G_4(G_1G_3 + G_2(1+G_3))}{1 - G_2G_4G_5(1+G_3)}}$$

$$= \frac{G_4(G_1G_3 + G_2(1+G_3))}{1 - G_2G_4G_5(1+G_3) + G_4(G_1G_3 + G_2(1+G_3))}$$

$$= \frac{G_2G_4 + G_1G_3G_4 + G_2G_3G_4}{1 + G_2G_4 + G_1G_3G_4 + G_2G_3G_4 - G_2G_4G_5 - G_2G_3G_4G_5} \quad \textbf{Ans}$$

Problem 3.22 Find $C_2(s)/R_1(s)$

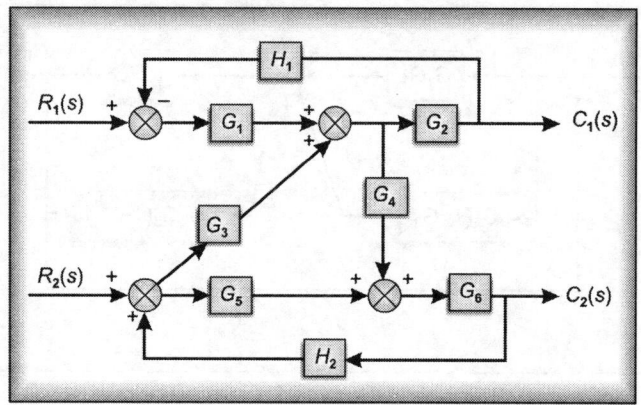

Fig. 3.104

Solution Assuming $R_2(s)$ and $C_1(s)$, both equal to zero, we redraw the block diagram as shown in Fig. 3.105.

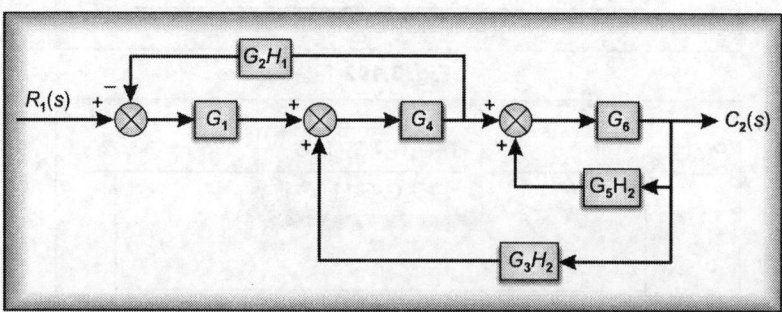

Fig. 3.105

Simplifying

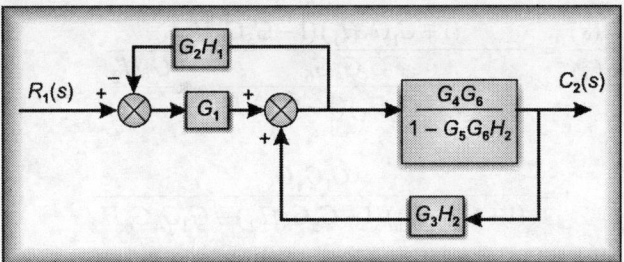

Fig. 3.106

Further simplification gives

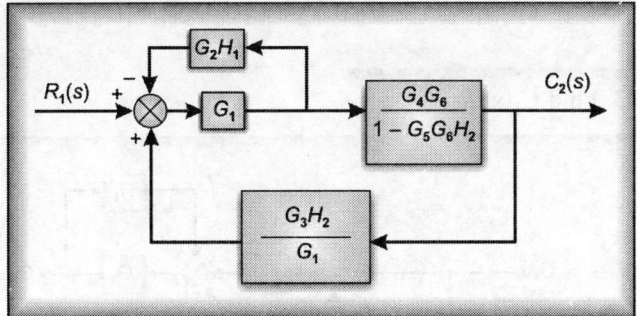

Fig. 3.107

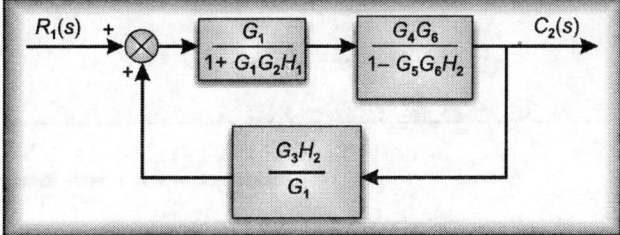

Fig. 3.108

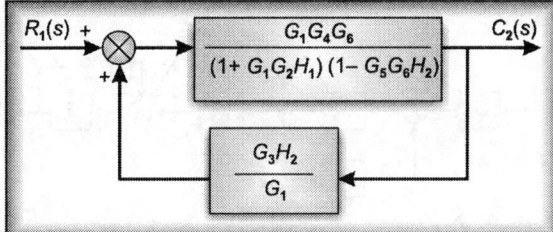

Fig. 3.109

$$\frac{C_2(s)}{R_1(s)} = \frac{\dfrac{G_1G_4G_6}{(1+G_1G_2H_1)(1-G_5G_6H_2)}}{1 - \dfrac{G_1G_4G_6}{(1+G_1G_2H_1)(1-G_5G_6H_2)} \times \dfrac{G_3H_2}{G_1}}$$

$$= \frac{G_1G_4G_6}{(1+G_1G_2H_1)(1-G_5G_6H_2) - G_3G_4G_6H_2}$$

$$= \frac{G_1G_4G_6}{1 - G_5G_6H_2 + G_1G_2H_1 - G_1G_2G_5G_6H_1H_2 - G_3G_4G_6H_2}$$

$$= \frac{G_1G_4G_6}{1 + G_1G_2H_1 - G_5G_6H_2 - G_1G_2G_5G_6H_1H_2 - G_3G_4G_6H_2} \quad \textbf{Ans}$$

Problem 3.23 Find $C_1(s)/R_2(s)$

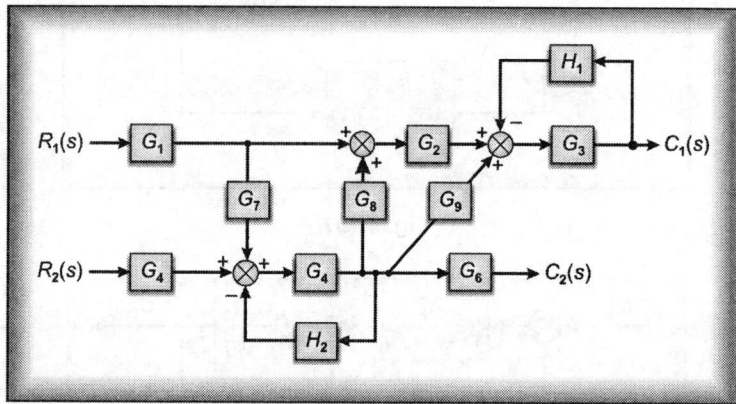

Fig. 3.110

Solution Assuming $R_1(s)$ and $C_2(s)$ equal to zero, we redraw the block diagram of Fig. 3.110. One must note that blocks G_1, G_7 and G_6 have no relevance due to the above mentioned assumptions.

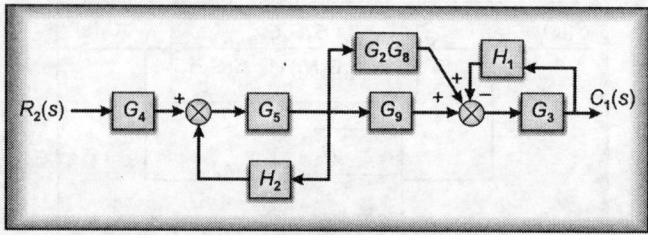

Fig. 3.111

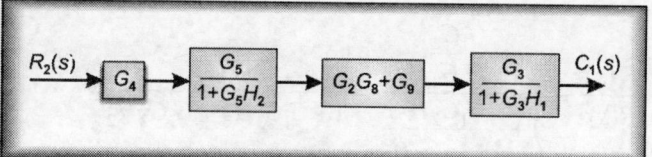

Fig. 3.112

$$\frac{C_1(s)}{R_2(s)} = \frac{G_3 G_4 G_5 (G_2 G_8 + G_9)}{(1 + G_3 H_1)(1 + G_5 H_2)}$$

$$\frac{C_1(s)}{R_2(s)} = \frac{G_2 G_3 G_4 G_5 G_8 + G_3 G_4 G_5 G_9}{1 + G_3 H_1 + G_5 H_2 + G_3 G_5 H_1 H_2}$$

Ans

Problem 3.24 Find $F_1(s)/R(s)$

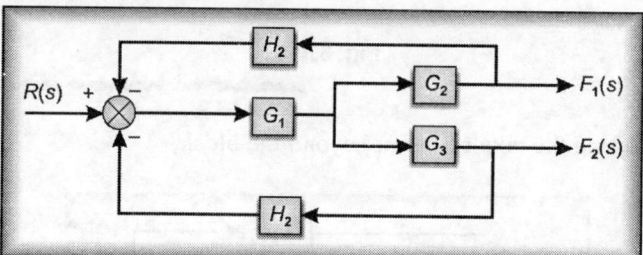

Fig. 3.113

Solution Assuming $F_2(s) = 0$, we redraw the block diagram

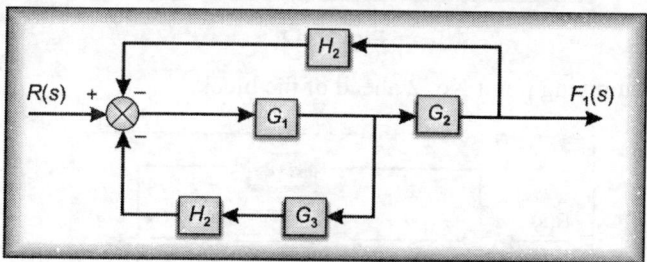

Fig. 3.114

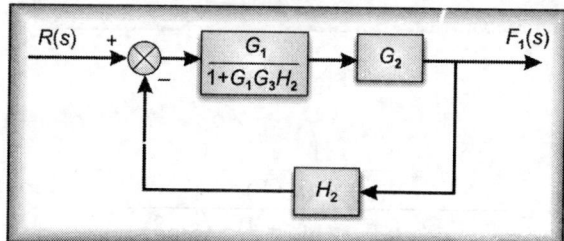

Fig. 3.115

$$\frac{F_1(s)}{R(s)} = \frac{\dfrac{G_1 G_2}{1 + G_1 G_3 H_2}}{1 + \dfrac{G_1 G_2 H_2}{1 + G_1 G_3 H_2}} = \frac{G_1 G_2}{1 + G_1 G_3 H_2 + G_1 G_2 H_2}$$ **Ans**

Problem 3.25 Find $C(s)/R(s)$

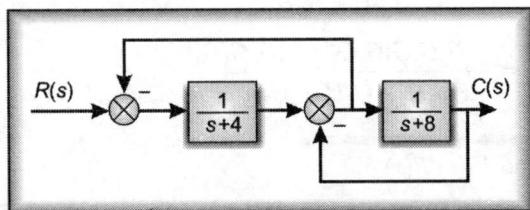

Fig. 3.116

Solution Shifting the take off point beyond the block,

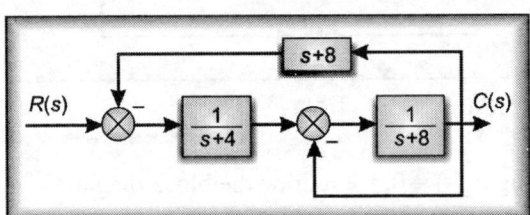

Fig. 3.117

Shifting the summing point No. 2 ahead of the block.

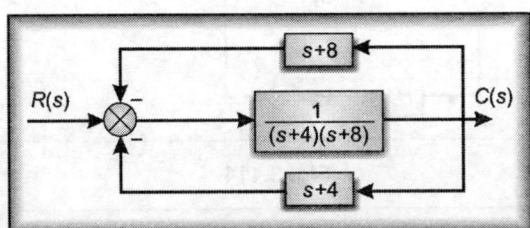

Fig. 3.118

$$\frac{C(s)}{R(s)} = \frac{\dfrac{1}{(s+4)(s+8)}}{1 + \dfrac{1}{(s+4)(s+8)}\left[\dfrac{(s+4)+(s+8)}{1}\right]}$$

$$\frac{C(s)}{R(s)} = \frac{\dfrac{1}{(s+4)(s+8)}}{1+\dfrac{1}{(s+4)(s+8)}(2s+12)}$$

$$= \frac{1}{(s+4)(s+8)+(2s+12)}$$

$$= \frac{1}{s^2+12s+32+2s+12}$$

$$= \frac{1}{s^2+14s+44}$$ **Ans**

Problem 3.26 Find $C(s)/R(s)$

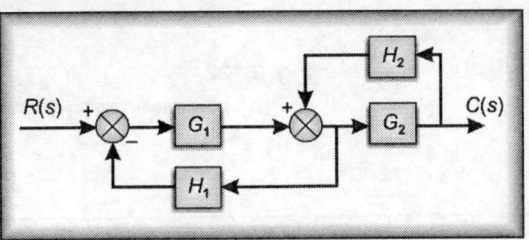

Fig. 3.119

Solution Shifting the take off point beyond block G_2 and summing point No. 2 of block G_1, we get

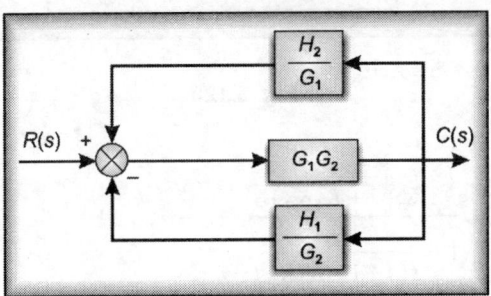

Fig. 3.120

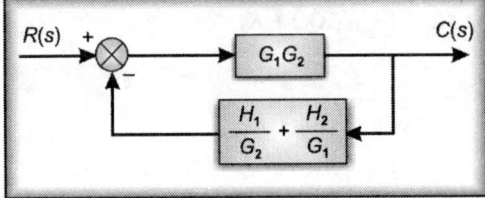

Fig. 3.121

$$\frac{C(s)}{R(s)} = \frac{G_1 G_2}{1 + G_1 G_2 \left[\dfrac{H_1}{G_2} + \dfrac{H_2}{G_1} \right]} = \frac{G_1 G_2}{1 + G_1 H_1 + G_2 H_2}$$ **Ans**

Problem 3.27 Find $C(s)/R(s)$

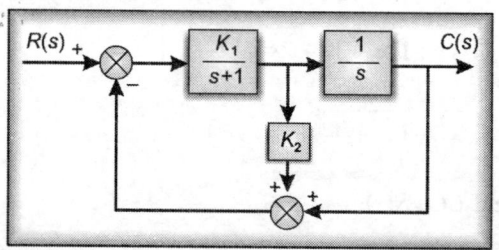

Fig. 3.122

Solution

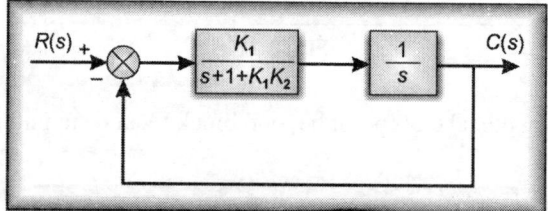

Fig. 3.123

$$\frac{C(s)}{R(s)} = \frac{\dfrac{K_1}{s(s + 1 + K_1 K_2)}}{1 + \dfrac{K_1}{s(s + 1 + K_1 K_2)}}$$

$$= \frac{K_1}{s(s + 1 + K_1 K_2) + K_1}$$

$$= \frac{K_1}{s^2 + s + s K_1 K_2 + K_1}$$

$$= \frac{K_1}{s^2 + s(1 + K_1 K_2) + K_1}$$ **Ans**

Problem 3.28 Simplify the block diagram as shown in Fig. 3.124.

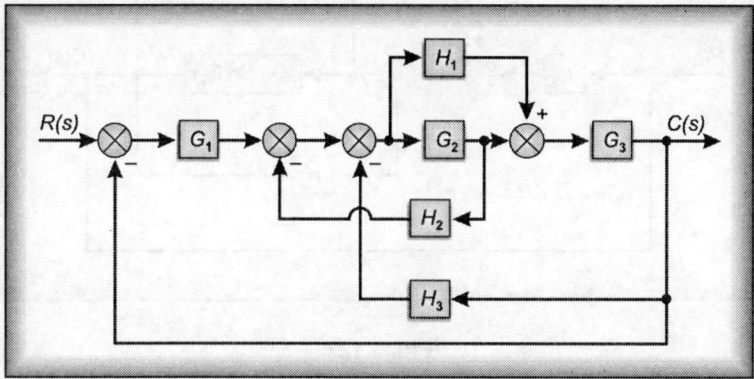

Fig. 3.124

Solution

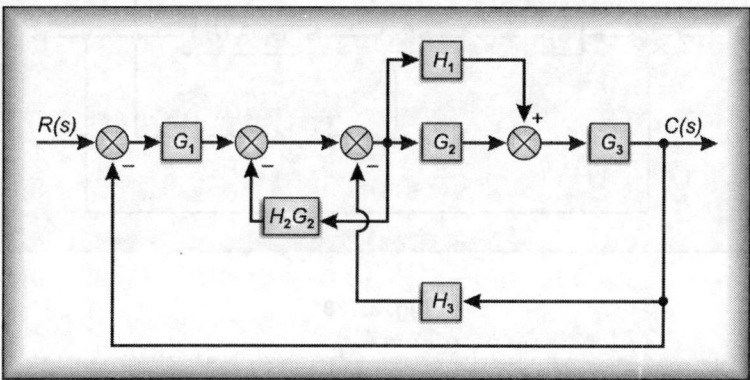

Fig. 3.125

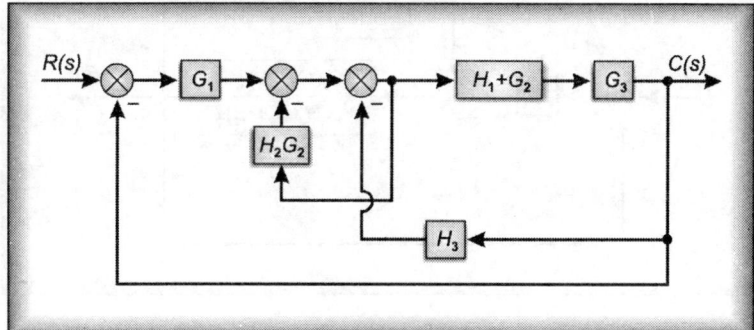

Fig. 3.126

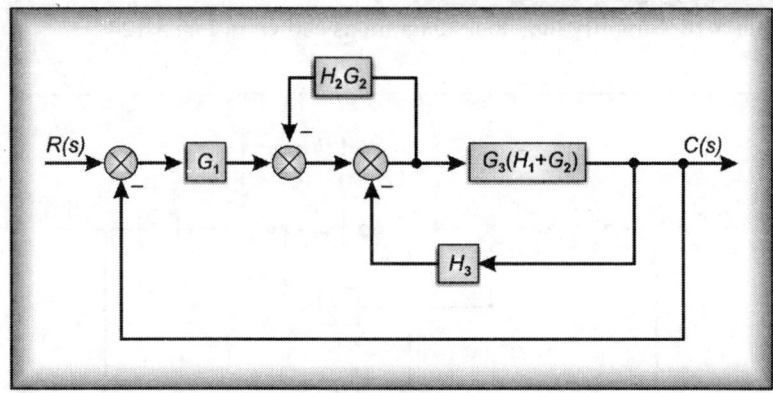

Fig. 3.127

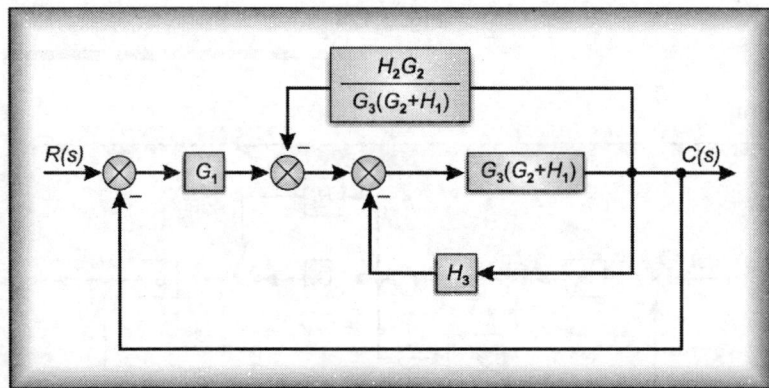

Fig. 3.128

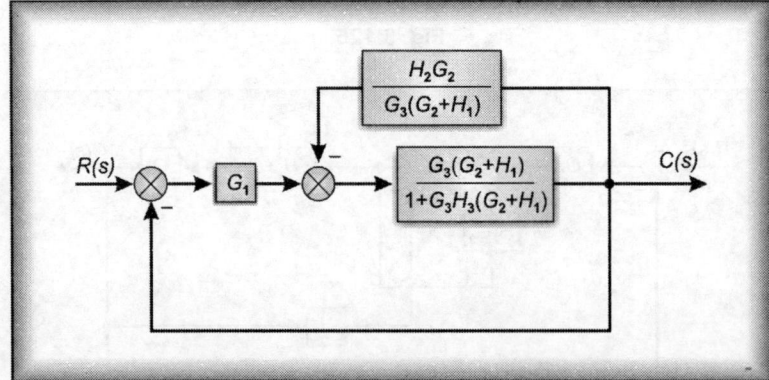

Fig. 3.129

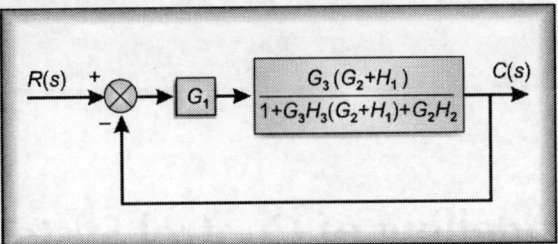

Fig. 3.130

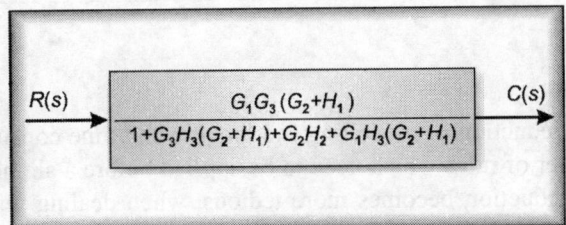

Fig. 3.131

4

Modelling of Control Systems: Signal Flow Graph

4.1 Introduction

Block diagram reduction of modelling is a lengthy and time consuming process. There are number of rules which need to be applied before a simplified model is obtained. The reduction becomes more tedious, when dealing with interwoven and very large systems depicted by block diagram representation. An alternate to block diagram reduction is signal flow graph which was advocated by SJ Mason.

4.2 Signal Flow Graph

Any linear control system can be described by a set of linear equations having the form

$$y_i = \sum_{j=1}^{j=n} a_{ij} y_j \quad \text{where, } i = 1, 2, \dots, n \text{ and } a_{ij} = \text{transmittance or gain from } y_j \text{ to } y_i$$

(4.1)

A signal flow graph represents a set of equations of this type written in the form of cause and effect and by means of *branches* and *nodes*. Each variable is represented by a *node* and *branches* connect the variables. Before proceeding further, let us understand commonly used terms in the signal flow diagrams. A signal flow diagram is shown in Fig. 4.1.

Node: Nodes are the variables or signal of the system, e.g. $x_1, x_2, x_3, x_4, x_5, x_6$.

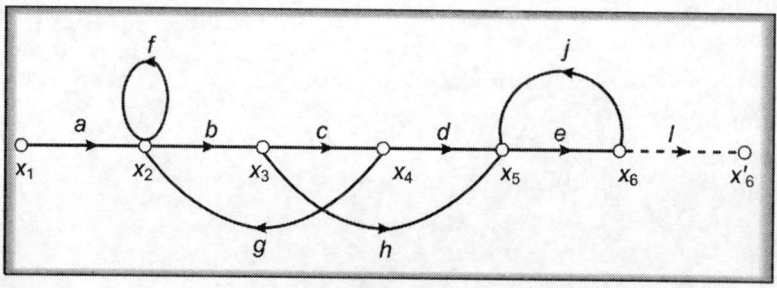

Fig. 4.1 Signal flow diagram

Branch: It is a line joining or connecting two nodes.

Arrow: It indicates the flow of signal, e.g. arrows shown on branch connecting nodes x_1 and x_2 indicate the direction of signal flow from node x_1 to node x_2. In terms of transfer function, it can be written as:

$$x_2 = ax_1,$$

where a is the transfer function and is generally written on the top of branch itself. It is also called *transmittance*.

Source Node: A source is a node having only outgoing branches such as x_1 in the preceding illustration.

Sink or Output Node: A sink is a node having incoming branches only. This condition may not always be met, as in the preceding illustration. In such cases an additional branch with unit transmittance as shown by the dotted line is introduced in order to meet the specified condition. In the preceding illustration, since x_6 node was not meeting the required condition, an outgoing branch from x_6 has been introduced terminating at x_6' having unit gain. Now x_6' meets the required condition. Therefore, x_6' is the sink node.

$$x_6' = 1 \times x_6 = x_6$$

Forward Path: Forward paths are paths which originate from a source node and terminate at a sink node and along which no node is encountered more than once. In the preceding illustration, there are two forward paths, namely

$$x_1 - x_2 - x_3 - x_4 - x_5 - x_6 - x_6' \text{ having gain 'abcde'.}$$
$$x_1 - x_2 - x_3 - x_5 - x_6 - x_6' \quad \text{having gain 'abhe'.}$$

Path Gain: It is the product of the gains of all branches along the path as explained in the preceding definition.

Loop: Loop is a closed path which originates from a node and terminates at the same node and along which no intermediate path is traversed twice, e.g.

$$x_2 - x_3 - x_4 - x_2 \text{ having gain 'bcg'.}$$
$$x_5 - x_6 - x_5 \quad \text{having gain 'ej'.}$$
$$x_2 - x_2 \quad \text{having gain 'f'.}$$

Self or Feedback Loop: It is a path originating from a node and terminating at the same node without encountering any other node. e.g. $x_2 - x_2$ is a self loop having transmittance 'f'.

Non-touching Loops: Non-touching loops are those loops having no path/ branches in common and in addition without any common node. In the preceding illustration

$$x_2 - x_3 - x_4 - x_2 \text{ and } x_5 - x_6 - x_5 \text{ and}$$

$x_2 - x_2$ and $x_5 - x_6 - x_5$, are the two possible combinations of non-touching loops.

4.3 Signal-flow Diagram Reduction

Complex signal-flow diagram of a system can be simplified by the use of signal-flow graph algebra as shown in Table 4.1.

Table 4.1   Signal-flow Graph Algebra

Sr. No.	Operation	Original	Equivalent equation or Diagram
1.	Addition		$x_3 = ax_1 + bx_2$ $x_2 = (a + b)x_1$
2.	Multiplication		 $x_2 = abx_1$
3.	Feedback loop elimination		$\dfrac{a/(1 \pm ab)}{}$ $x_1 \qquad x_2$ $x_2 = \dfrac{a}{1 \pm abx_1}$
4.	Self loop elimination		$\dfrac{a/(1 \pm ab)}{}$ $x_1 \qquad x_2$ $x_2 = \dfrac{a}{1 \pm bx_1}$
5.	Miscellaneous		 $x_4 = bcx_2 + acx_1$ $\quad = c(bx_2 + ax_1)$
6.	Miscellaneous		$\dfrac{ab/1 - bc}{}$ $x_1 \qquad x_3$

Let us prove the relationship given in Ser. No. 6 of Table 4.1. It will also help in proving the relationship given in Ser. No. 4 of Table 4.1.

$$x_3 = bx_2 \qquad (4.2)$$
$$x_2 = ax_2 + cx_3 \qquad (4.3)$$
or
$$x_3 = abx_1 + bcx_3 \qquad (4.4)$$

Redrawing the relationship of x_3, we get signal flow graph shown in Fig. 4.2. This gives

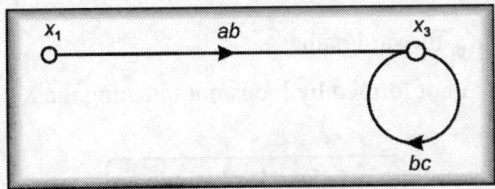

Fig. 4.2

$$x_3 = abx_1 + bcx_3 \qquad (4.5)$$
or
$$x_3 = abx_1 + b^2cx_2 \qquad (4.6)$$
$$= abx_1 + b^2c\,(ax_1 + cx_3) \qquad (4.7)$$
$$= abx_1 + b^2cax_1 + b^2c^2x_3 \qquad (4.8)$$
or
$$x_3(1 - b^2c^2) = abx_1(1 + bc) \qquad (4.9)$$
or
$$x_3 = \frac{abx_1(1+bc)}{1-b^2c^2} = \frac{abx_1(1+bc)}{(1-bc)(1+bc)} \qquad (4.10)$$

$$= \frac{ab}{1-bc}x_1 \qquad (4.11)$$

The relationship is shown in Fig. 4.3.

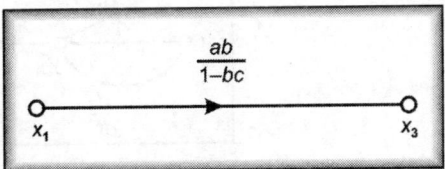

Fig. 4.3

4.4 Mason's Gain Formula

With the help of repeated applications of signal-flow graph algebra, it is possible to reduce complex and complicated graph to one source and sink nodes. This is time consuming and tedious. Mason was the first to spot the inherent property of the signal-flow graph and evolved a theorem which is popularly known as *Mason's gain formula*. This, by mere inspection of the signal-flow graph, gives the desired answer. The general expression for signal flow graph as given by Mason is:

$$T = \frac{1}{\Delta} \sum P_k \Delta_k \qquad\qquad (4.12)$$

where T = Overall transmittance of the system

$\Delta = 1-$ (sum of the gain of all individual loops) + (sum of the gain product of all possible combination of two non-touching loops) – (sum of the gain product of all possible combination of three non-touching loops) + (...) – (...) + ...

P_k = Gain of K_{th} forward path

Δ_k = Same as Δ but formed by loops not touching the K_{th} forward path

Solved Problems

Problem 4.1 Draw signal flow graph for the following equations:

(a) $y_2 = a_1 \dfrac{dy_1}{dt}$, (b) $y_3 = \dfrac{d^2 y_2}{dt^2} + \dfrac{dy_1}{dt} - y_1$, (c) $\dfrac{d^2 y}{dx^2} + \dfrac{2}{3}\dfrac{dy}{dx} + \dfrac{11}{2} y = x$

Solution

(a) $y_2 = a_1 \dfrac{dy_1}{dt}$

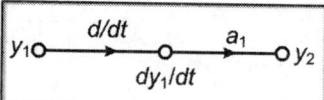

Fig. 4.4

(b) $y_3 = \dfrac{d^2 y_2}{dt^2} + \dfrac{dy_1}{dt} - y_1$

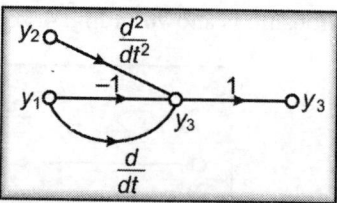

Fig. 4.5

(c) $\dfrac{dy^2}{dx^2} + \dfrac{2}{3}\dfrac{dy}{dx} + \dfrac{11}{2} y = x$

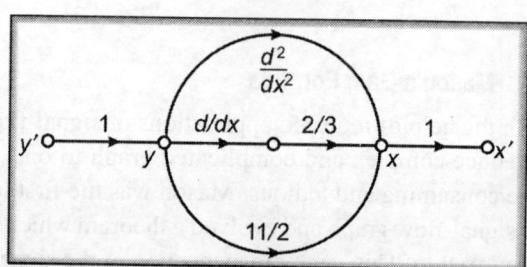

Fig. 4.6

Alternatively

$$\frac{d^2y}{dx^2} + \frac{2}{3}\frac{dy}{dx} + \frac{11}{2}y = x \quad \text{or} \quad s^2Y(s) + \frac{2}{3}sY(s) + \frac{11}{2}Y(s) = X(s)$$

$$\frac{Y(s)}{X(s)} = \frac{1}{s^2 + \frac{2}{3}s + \frac{11}{2}} = \frac{1/s^2}{1 + \frac{2}{3s} + \frac{11}{2s^2}} = \frac{1/s^2}{1 - \left(-\frac{2}{3s} - \frac{11}{2s^2}\right)}$$

Comparing with Mason's gain formula, we notice that there is:

(a) One forward path with gain of $1/s^2$

(b) Two feedback loops with gain of $-2/3s$, $-11/2s^2$

The signal flow graph is shown in Fig. 4.7. **Ans**

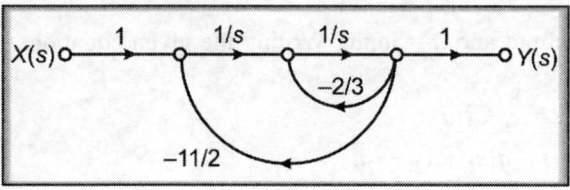

Fig. 4.7

Problem 4.2 Draw signal flow graph from the following equations:

$$x_2 = a_{21}x_1 + a_{23}x_3$$
$$x_3 = a_{31}x_1 + a_{32}x_2 + a_{33}x_3$$
$$x_4 = a_{42}x_2 + a_{43}x_3$$

Solution The signal flow diagram of three given equations is shown in Fig. 4.8

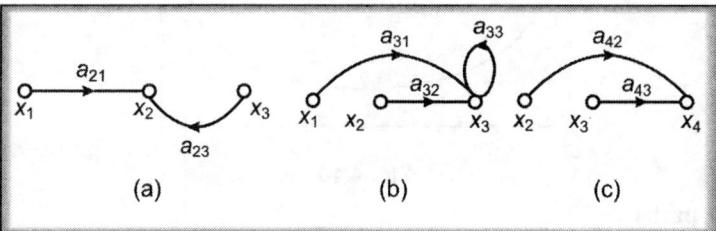

Fig. 4.8

Combining, the complete signal flow diagram is shown in Fig. 4.9

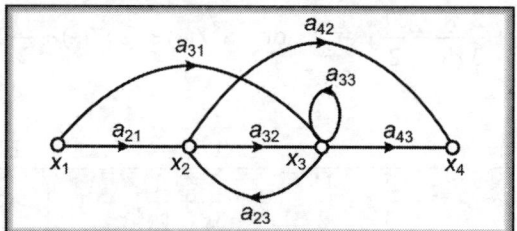

Fig. 4.9

Problem 4.3 A system is represented by the following set of equations. Construct the signal flow graph and find the closed-loop transfer function.

$$x = x_1 + t_3 u, \quad \dot{x}_1 = -q_1 x_1 + x_2 + t_2 u, \quad \dot{x}_2 = -q_2 x_1 + t_1 u$$

Solution

Consider, x = output and u = input. Writing the given equations in Laplace domain, we get

$$x = x_1 + t_3 u \tag{1}$$

$$sx_1 = -q_1 x_1 + x_2 + t_2 u$$

or $$x_1 = \frac{x_2}{s + q_1} + \frac{t_2}{s + q_1} u \tag{2}$$

$$sx_2 = -q_2 x_1 + t_1 u \quad \text{or} \quad x_2 = \frac{-q_2}{s} x_1 + \frac{t_1}{s} u$$

The signal flow graph is shown in Fig. 4.10

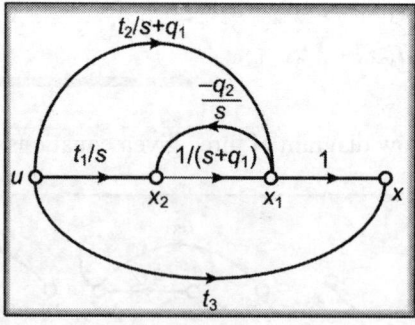

Fig. 4.10

Forward paths

$$P_1 = \frac{t_1}{s(s + q_1)}, \quad P_2 = \frac{t_2}{s + q_1}, \quad P_3 = t_3$$

Loops

$$L_1 = x_2 - x_1 - x_2 = -\frac{q_2}{s(s+q_1)}$$

$$\Delta = 1 + \frac{q_2}{s(s+q_1)}$$

$$\Delta_1 = \Delta_2 = 1$$

$$\Delta_3 = 1 + \frac{q_2}{s(s+q_1)} = \frac{s^2 + q_1 s + q_2}{s(s+q_1)}$$

$$T = \frac{P_1 \Delta_1 + P_2 \Delta_2 + P_3 \Delta_3}{\Delta}$$

$$= \frac{\dfrac{t_1}{s(s+q_1)} + \dfrac{t_2}{(s+q_1)} + \dfrac{t_3(s^2 + q_1 s + q_2)}{s(s+q_1)}}{1 + \dfrac{q_2}{s(s+q_1)}}$$

or $$T = \frac{t_1 + t_2 s + t_3(s^2 + q_1 s + q_2)}{s^2 + q_1 s + q_2} \qquad \textbf{Ans}$$

Problem 4.4 Draw the signal flow graph of the system of equations.

$$X_1 = a_{11}X_1 + a_{12}X_2 + a_{13}X_3 + b_1 u_1$$

$$X_2 = a_{21}X_1 + a_{22}X_2 + a_{23}X_3 + b_2 u_2$$

$$X_3 = a_{31}X_1 + a_{32}X_2 + a_{33}X_3$$

Solution The variables are X_1, X_2, X_3, u_1 and u_2. Choose five nodes representing the variables. Connect the various nodes choosing appropriate branch gain in accordance with the equations.

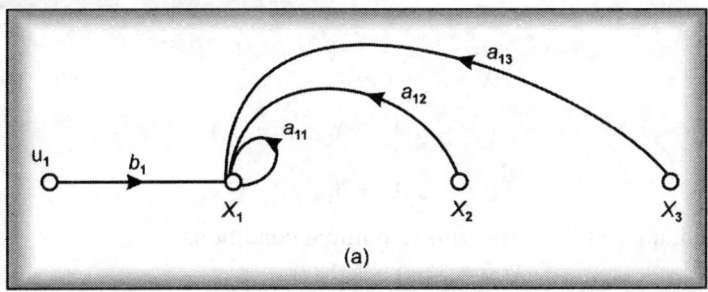

(a)

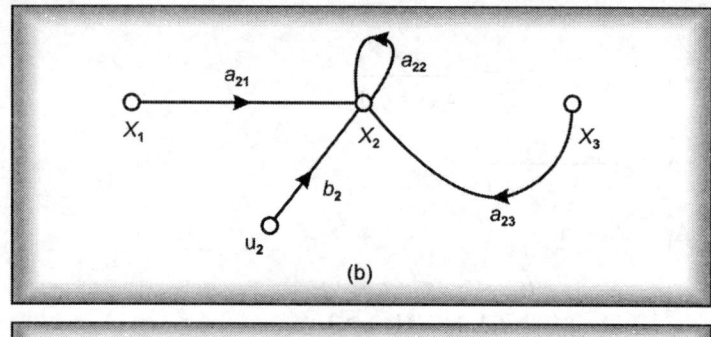

(b)

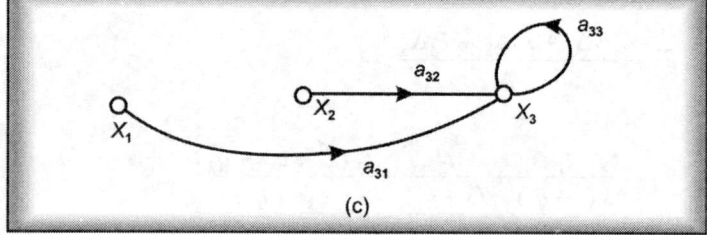

(c)

Fig. 4.11

The signal flow graph is shown in Fig. 4.12.

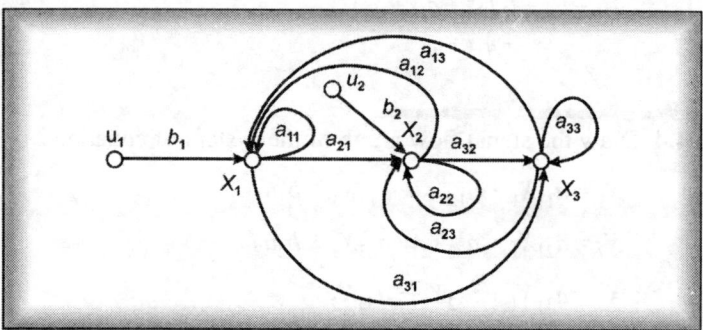

Fig. 4.12

Problem 4.5 For the system represented by the following equations find the transfer function $X(s)/U(s)$ using signal flow graph technique.

Solution

$$X = X_1 + b_3 u$$

$$\dot{X}_1 = -a_1 X_1 + X_2 + b_2 u$$

$$\dot{X}_2 = -a_2 X_1 + b_1 u$$

Taking Laplace transform with zero initial conditions

$$X(s) = X_1(s) + b_3 U(s)$$

$$sX_1(s) = -a_1X_1(s) + X_2(s) + b_2U(s)$$
$$sX_2(s) = -a_2X_1(s) + b_1U(s)$$

Rearranging the above equations, we get

$$X(s) = X_1(s) + b_3U(s)$$

$$X_1(s) = \frac{-a_1}{s}X_1(s) + \frac{1}{s}X_2(s) + \frac{b_2}{s}U(s)$$

$$X_2(s) = \frac{-a_2}{s}X_1(s) + \frac{b_1}{s}U(s)$$

The signal flow graph is shown in Fig. 4.13.

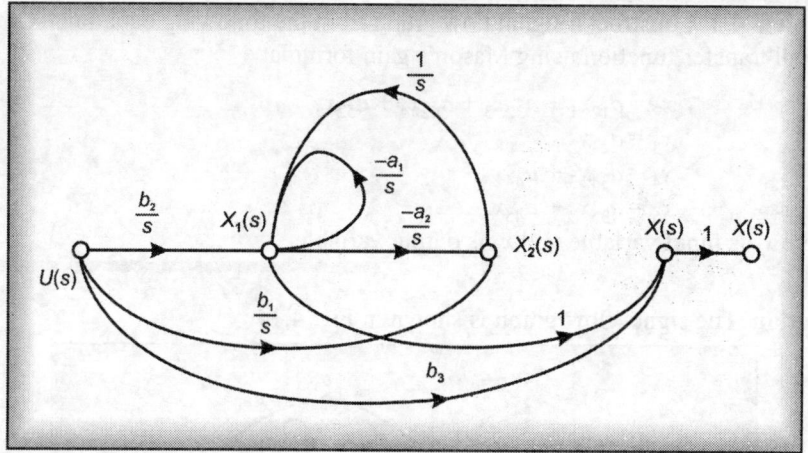

Fig. 4.13

There are three forward paths.

The gain of the forward paths are:

$$P_1 = b_3 \times 1 = b_3$$

$$P_2 = \frac{b_1}{s} \times \frac{1}{s} = \frac{b_1}{s^2}$$

$$P_3 = \frac{b_2}{s} \times 1 = \frac{b_2}{s}$$

There are two loops with loop gains:

$$L_1 = \frac{-a_1}{s} \quad \text{[self loop at node } X_1(s)]$$

$$L_2 = \frac{-a_2}{s^2} \quad [X_1(s) - X_2(s) - X_1(s)]$$

There is no combination of two loops which are non-touching.

$$\Delta = 1 + \frac{a_1}{s} + \frac{a_2}{s^2}$$

Forward path 1 does not touch loops L_1 and L_2. Therefore

$$\Delta_1 = 1 + \frac{a_1}{s} + \frac{a_2}{s^2}$$

Forward paths 2 and 3 touch the two loops. Hence, $\Delta_2 = 1$, $\Delta_3 = 1$.

The transfer function $\quad = \dfrac{X_{(s)}}{U_{(s)}} = \dfrac{P_1\Delta_1 + P_2\Delta_2 + P_3\Delta_3}{\Delta}$

$$= \frac{b_3(s^2 + a_1 s + a_2) + b_2 s + b_1}{s^2 + a_1 s + a_2} \qquad \textbf{Ans}$$

Problem 4.6 Construct a signal flow graph from the following equations. Obtain overall transfer function using Mason's gain formula

$$x_2 = a_{12}x_1 + a_{32}x_3 + a_{42}x_4 + a_{52}x_5$$
$$x_3 = a_{23}x_2$$
$$x_4 = a_{34}x_3 + a_{44}x_4$$
$$x_5 = a_{35}x_3 + a_{45}x_4$$

where x_1 is input variable and x_5 is output variable.

Solution The signal flow graph is shown in Fig. 4.14

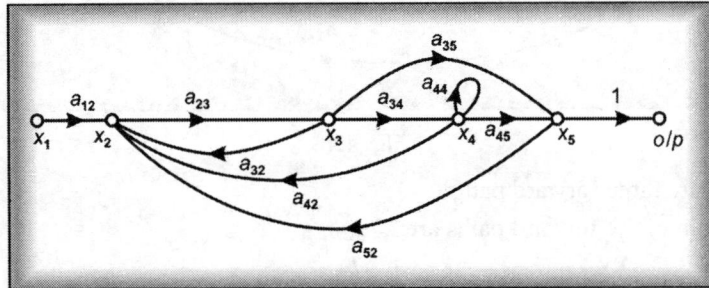

Fig. 4.14

Forward paths:

$P_1 = a_{12}\, a_{23}\, a_{34}\, a_{45}$

$P_2 = a_{12}\, a_{23}\, a_{35}$

Loops:

$L_1 = a_{23}\, a_{32}$

$L_2 = a_{44}$

$L_3 = a_{23}\, a_{34}\, a_{42}$

$L_4 = a_{23}\, a_{34} + a_{45}\, a_{52}$

$L_5 = a_{23}\, a_{35}\, a_{52}$

Two non-touching loops:

$L_1 L_2 = a_{23} a_{32} a_{44}$

$L_2 L_5 = a_{44} a_{23} a_{35} a_{52}$

Three non-touching loops $= 0$

Also, $\Delta_1 = 1$

$$\Delta_2 = 1 - a_{44} \qquad \text{[self loop } a_{44} \text{ does not touch path } P_1]$$

Mason's gain formula, $T = \dfrac{1}{\Delta} \Sigma P_K \Delta_K$

$$\Delta = 1 - \{L_1 + L_2 + L_3 + L_4 + L_5\} + \{L_1 L_2 + L_2 L_5\}$$

$\therefore \qquad T = \dfrac{a_{12}\, a_{23}\, a_{34}\, a_{45} + a_{12}\, a_{23}\, a_{35}\, (1 - a_{44})}{\begin{array}{c}1 - a_{23}\, a_{32} - a_{44} - a_{23}\, a_{34}\, a_{42} - a_{23}\, a_{34}\, a_{45}\, a_{52} \\ - a_{23}\, a_{35}\, a_{52} + a_{44}\, a_{23}\, a_{32} + a_{44}\, a_{23}\, a_{35}\, a_{52}\end{array}}$ **Ans**

Problem 4.7 Construct signal flow graph from the following equations and find the overall transfer function.

$$x_1 = t_{01} x_0$$
$$x_2 = t_{12} x_1 + t_{32} x_3 + t_{42} x_4$$
$$x_3 = t_{03} x_0 + t_{13} x_1 + t_{23} x_2$$
$$x_4 = t_{04} x_0 + t_{34} x_3 + t_{54} x_5$$
$$x_5 = t_{15} x_1 + t_{45} x_4 + t_{65} x_6$$
$$x_6 = t_{06} x_0 + t_{76} x_7 + t_{56} x_5$$
$$x_7 = t_{67} x_6 + t_{77} x_7$$
$$x_8 = t_{78} x_7$$

Solution: The signal flow graph is shown in Fig. 4.15

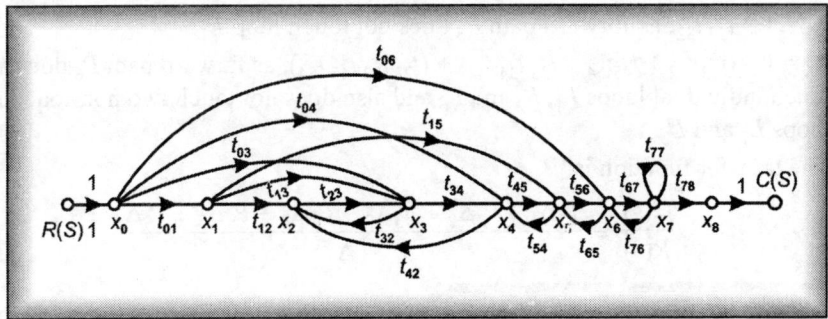

Fig. 4.15

Forward paths:

$P_1 = t_{01}\, t_{12}\, t_{23}\, t_{34}\, t_{45}\, t_{56}\, t_{67}\, t_{78} \qquad \Delta_1 = 1$

$P_2 = t_{01}\, t_{13}\, t_{34}\, t_{45}\, t_{56}\, t_{67}\, t_{78} \qquad \Delta_2 = 1$

$P_3 = t_{01}\, t_{15}\, t_{56}\, t_{67}\, t_{78} \qquad \Delta_3 = 1 - (t_{23}\, t_{32} + t_{23}\, t_{34}\, t_{42})$

$$P_4 = t_{03} t_{34} t_{45} t_{56} t_{67} t_{78}$$
$$P_5 = t_{04} t_{45} t_{56} t_{67} t_{78}$$
$$P_6 = t_{06} t_{67} t_{78}$$

$$\Delta_4 = 1$$
$$\Delta_5 = 1 - t_{23} t_{32}$$
$$\Delta_6 = 1 - (L_{12} + L_{21} + L_{61}) + (L_{11} L_{21})$$

Loops:

$$L_1 = t_{23} t_{32}$$
$$L_2 = t_{45} t_{54}$$
$$L_3 = t_{56} t_{65}$$
$$L_4 = t_{67} t_{76}$$
$$L_5 = t_{77}$$
$$L_6 = t_{23} t_{34} t_{42}$$

Two non-touching loops:

$$L_1 L_2 = t_{23} t_{32} t_{45} t_{54}$$
$$L_1 L_3 = t_{23} t_{32} t_{56} t_{65}$$
$$L_1 L_4 = t_{23} t_{32} t_{67} t_{76}$$
$$L_1 L_5 = t_{23} t_{32} t_{77}$$
$$L_2 L_5 = t_{45} t_{54} t_{77}$$
$$L_2 L_4 = t_{45} t_{54} t_{67} t_{76}$$
$$L_3 L_5 = t_{56} t_{65} t_{77}$$
$$L_6 L_3 = t_{23} t_{34} t_{42} t_{56} t_{65}$$
$$L_6 L_4 = t_{23} t_{34} t_{42} t_{67} t_{76}$$
$$L_6 L_5 = t_{23} t_{34} t_{42} t_{77}$$

Three non-touching loops:

$$L_1 L_2 L_5 = t_{23} t_{32} t_{45} t_{54} t_{77}$$
$$L_1 L_2 L_4 = t_{23} t_{32} t_{45} t_{54} t_{67} t_{76}$$
$$L_1 L_3 L_5 = t_{23} t_{32} t_{56} t_{65} t_{77}$$
$$L_6 L_3 L_5 = t_{23} t_{34} t_{42} t_{56} t_{65} t_{77}$$

Also,

$$\Delta = 1 - \{L_1 + L_2 + L_3 + L_4 + L_5 + L_6\}$$
$$+ \{L_1 L_2 + L_1 L_3 + L_1 L_4 + L_1 L_5 + L_2 L_5 + L_2 L_4 + L_3 L_5 + L_6 L_3 + L_6 L_4 + L_6 L_5\}$$
$$- \{L_1 L_3 L_5 + L_1 L_2 L_5 + L_1 L_3 L_5 + L_6 L_3 L_5\}$$

Calculation of Δ:

Since, there are six forward paths, there will be six Δs.

- $\Delta_1 = \Delta_2 = \Delta_4 = 1$; as forward paths, P_1, P_2 and P_4 touches all loops.
- $\Delta_3 = 1 - (t_{23} t_{32} + t_{23} t_{34} t_{42})$; as forward path, P_3 does not touch individual loops L_1 and L_6.
- $\Delta_5 = 1 - t_{23} t_{32}$; as forward path P_5 does not touch loop L_1.
- $\Delta_6 = 1 - (t_{23} t_{32} + t_{45} t_{54} + t_{23} t_{34} t_{42}) + (t_{23} t_{32} t_{45} t_{54})$; as farward path P_6 does not touch individual loops L_1, L_2 and L_6 and also does not touch two non-touching loops L_1 and L_2.

The transfer function is:

$$\frac{C(s)}{R(s)} = \frac{P_1 \Delta_1 + P_2 \Delta_2 + P_3 \Delta_3 + P_4 \Delta_4 + P_5 \Delta_5 + P_6 \Delta_6}{\Delta} \qquad \textbf{Ans}$$

Problem 4.8 Draw the signal flow graph of a system described by the following equations:

$$x_1 = a_{11} x_1 + a_{12} x_2 + r_1$$
$$x_2 = a_{21} x_1 + a_{22} x_2 + r_2$$

where, r_1 and r_2 are the inputs and x_1 and x_2 are the outputs.

Solution Draw the nodes and depict the equations as shown in Fig. 4.16 drawn, based on input-output relationship.

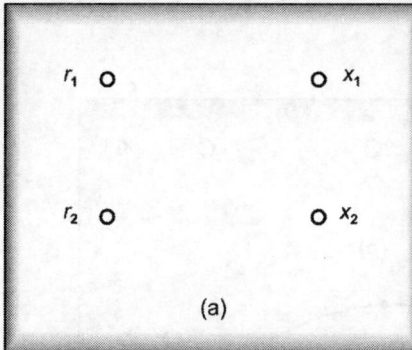

(a)

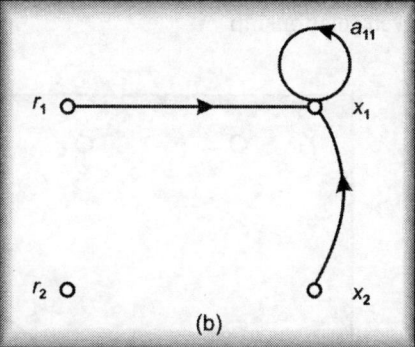

(b)

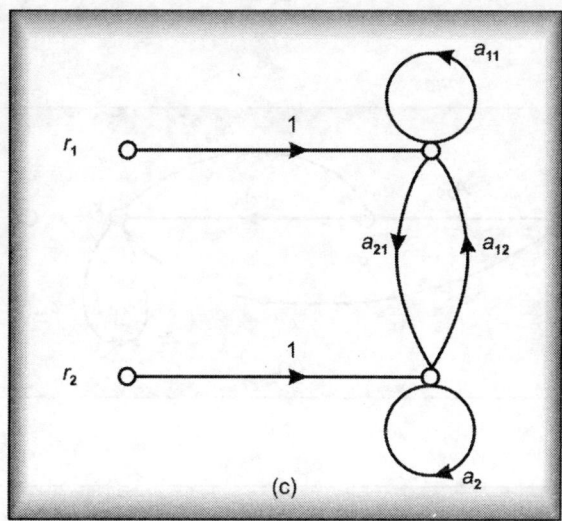

(c)

Fig. 4.16

Problem 4.9 A control system is described by the following relationship:

$$x_2 = A_{21}x_1 + A_{23}x_3$$
$$x_3 = A_{31}x_1 + A_{32}x_2 + A_{33}x_3$$
$$x_4 = A_{42}x_2 + A_{43}x_3$$

Draw signal flow graph by two methods.

Solution There are four variables x_1, x_2, x_3 and x_4 and hence we need four nodes.

First Method: Arrange the nodes from left to right, connect them as per the given relationship

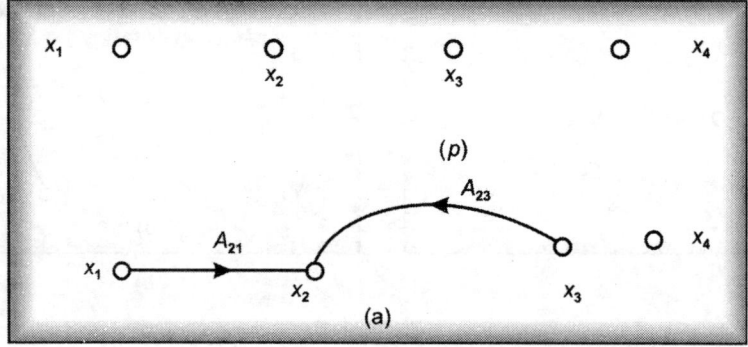

(p)

(a)

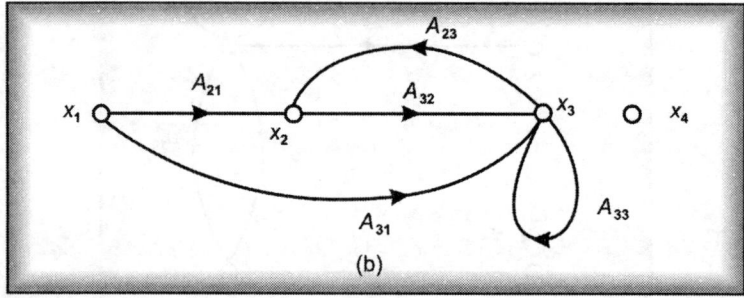

(b)

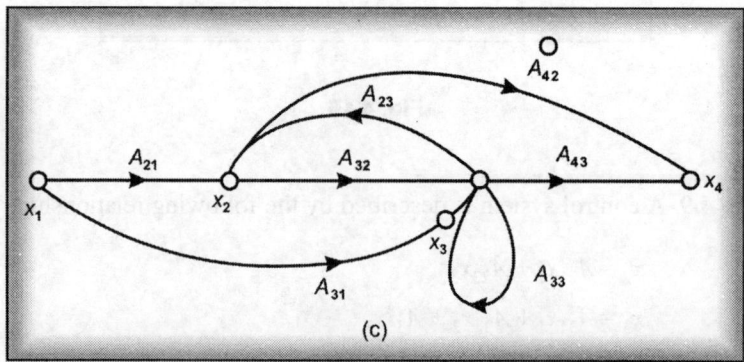

(c)

Fig. 4.17

Second Method: Arrange nodes as shown

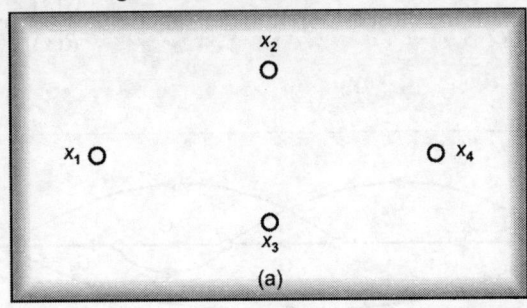

(a)

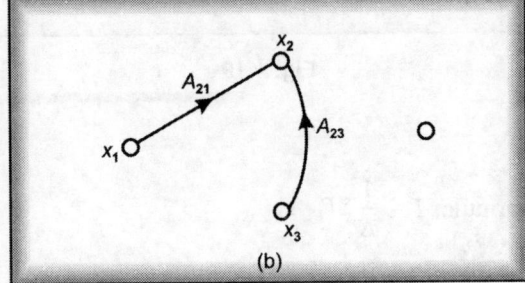

(b)

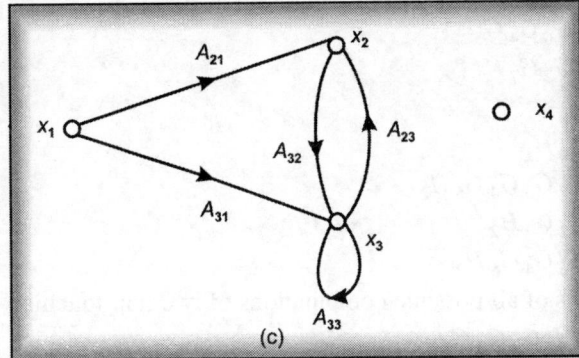

(c)

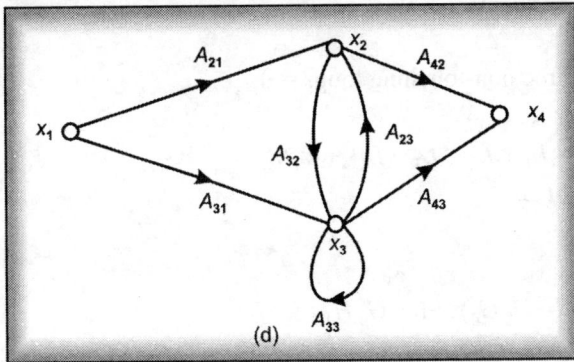

(d)

Fig. 4.18

Problem 4.10 Obtain the closed-loop transfer function, $\dfrac{C(s)}{R(s)}$ by using Mason's gain formula.

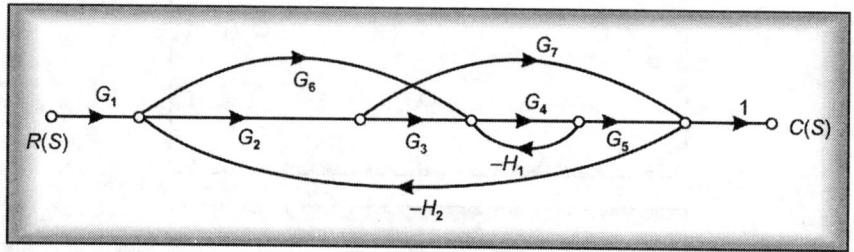

Fig. 4.19

Solution

Mason's gain formula, $T = \dfrac{1}{\Delta} \Sigma P_K \Delta_K$

Forward paths:

$P_1 = G_1 G_2 G_3 G_4 G_5$
$P_2 = G_1 G_6 G_4 G_5$
$P_3 = G_1 G_2 G_7$

Loops:

$L_1 = - G_4 H_1$
$L_2 = - G_2 G_3 G_4 G_5 H_2$
$L_3 = - G_2 G_7 H_2$
$L_4 = - G_6 G_4 G_5 H_2$

Gain products of all possible combinations of two non-touching loops:

$L_1 L_3 = H_1 H_2 G_2 G_4 G_7$

$\dfrac{C(s)}{R(s)} = \dfrac{P_1 \Delta_1 + P_2 \Delta_2 + P_3 \Delta_3}{\Delta}$

Number of three non-touching loops = 0

Also,

$\Delta = 1 - \{L_1 + L_2 + L_3 + L_4\} + L_1 L_3$

Calculation of Δ:

$\Delta_1 = 1$
$\Delta_2 = 1$
$\Delta_3 = 1 - (- H_1 G_4) = 1 + G_4 H_1$

$\therefore \quad T = \dfrac{G_1 G_2 G_3 G_4 G_5 + G_1 G_6 G_4 G_5 + G_1 G_2 G_7 [1 + G_4 H_1]}{1 + G_4 H_1 + G_2 G_3 G_4 G_5 H_2 + G_2 G_7 H_2 + G_6 G_4 G_5 H_2 + H_1 H_2 G_2 G_7} \cdot$ **Ans**

Problem 4.11 Obtain the closed-loop transfer function, $\dfrac{C(s)}{R(s)}$ by using Mason's gain formula.

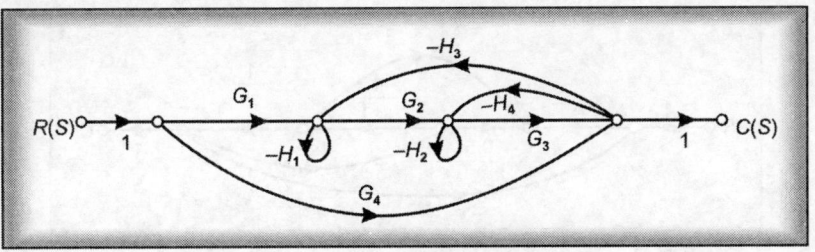

Fig. 4.20

Solution

Forward paths:

$$P_1 = G_1 G_2 G_3$$
$$P_2 = G_4$$

Loops:

$$L_1 = -H_1$$
$$L_2 = -H_2$$
$$L_3 = -G_3 H_4$$
$$L_4 = -G_2 G_3 H_3$$

Two non-touching loops:

$$L_1 L_2 = H_1 H_2$$
$$L_1 L_3 = H_1 H_4 G_3$$
$$\Delta = 1 - \{L_1 + L_2 + L_3 + L_4\} + \{L_1 L_2 + L_1 L_3\}$$

Calculation of Δ:

$$\Delta_1 = 1$$
$$\Delta_2 = 1 - (-H_1 H_2) + \{H_1 H_2\}$$

$$T = \frac{C(s)}{R(s)} = \frac{P_1 \Delta_1 + P_2 \Delta_2}{\Delta}$$

$$\therefore \quad T = \frac{G_1 G_2 G_3 + G_4(1 + H_1 + H_2 + H_1 H_2)}{(1 + H_1 + H_2 + G_3 H_4 + G_2 G_3 H_3 + H_3 H_2 + H_1 H_4 G_3)}. \qquad \textbf{Ans}$$

Problem 4.12 Using Mason's gain rule, obtain the overall transfer function of a control system represented by the signal flow graph shown below in Fig. 4.21.

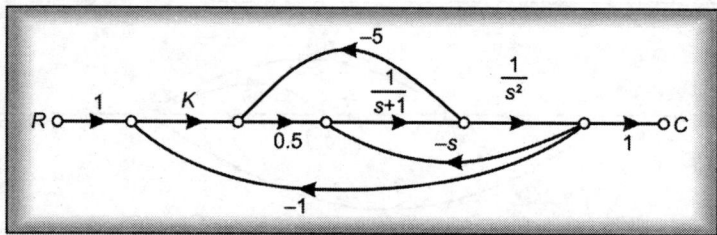

Fig. 4.21

Solution

Forward path:

$$P_1 = 1 \times K \times 0.5 \times \frac{1}{s+1} \cdot \frac{1}{s^2} = \frac{0.5K}{s^2(s+1)}; \qquad \Delta_1 = 1$$

Loops:

$$L_1 = 0.5 \times \frac{1}{s+1}(-5) = \frac{-2.5}{s+1}$$

$$L_2 = \frac{1}{s+1} \cdot \frac{1}{s^2}(-s) = \frac{-1}{s(s+1)}$$

$$L_3 = K(0.5) \cdot \frac{1}{s+1} \cdot \frac{1}{s^2}(-1) = \frac{-0.5K}{s^2(s+1)}$$

Two non-touching loops = 0

$$\Delta = 1 - \{L_1 + L_2 + L_3\}$$

$$\frac{C(s)}{R(s)} = \frac{P_1\Delta_1}{\Delta} = \frac{\dfrac{0.5K}{s^2(s+1)}}{1 - \left\{ \dfrac{-2.5}{s+1} - \dfrac{1}{s(s+1)} - \dfrac{0.5K}{s^2(s+1)} \right\}}$$

$$\therefore \frac{C(s)}{R(s)} = \frac{0.5K}{s^3 + 3.5s^2 + s + 0.5K}. \qquad \textbf{Ans}$$

Problem 4.13 Obtain the closed-loop transfer function for the control system as shown in Fig. 4.22

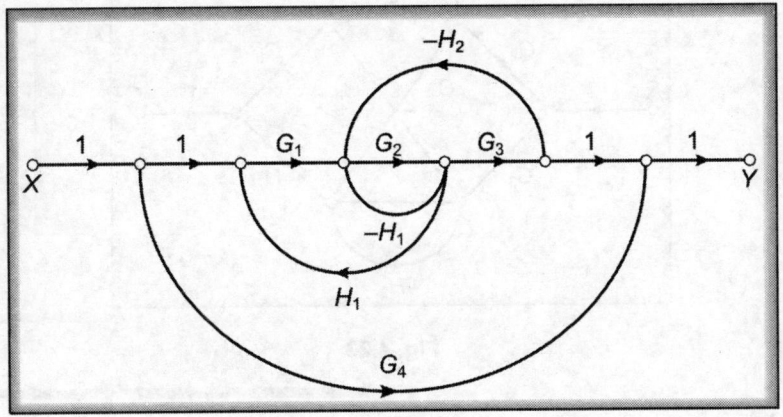

Fig. 4.22

Solution

Forward paths: There are two forward paths

$$P_1 = G_1 G_2 G_3 \quad \text{and} \quad P_2 = G_4$$

Loops: There are three feedback loops

$$L_1 = -G_2 H_1$$

$$L_2 = G_1 G_2 H_1$$

$$L_3 = -G_2 G_3 H_2$$

$$\Delta = 1 - (L_1 + L_2 + L_3)$$

$$= 1 + G_2 H_1 - G_1 G_2 H_1 + G_2 G_3 H_2$$

$$\Delta_1 = 1 \quad \text{and} \quad \Delta_2 = \Delta$$

$$\frac{Y}{X} = \frac{P_1 \Delta_1 + P_2 \Delta_2}{\Delta}$$

$$= \frac{G_1 G_2 G_3 + G_4 (1 + G_2 H_1 - G_1 G_2 H_1 + G_2 G_3 H_2)}{1 + G_2 H_1 - G_1 G_2 H_1 + G_2 G_3 H_2} \qquad \textbf{Ans}$$

Problem 4.14 Find C/R for the control system as shown in Fig. 4.23

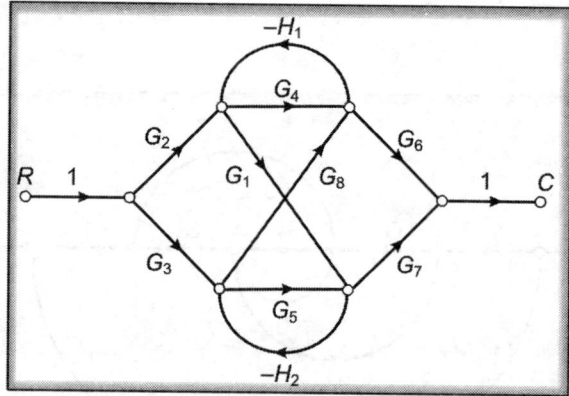

Fig. 4.23

Solution
Forward paths

$$P_1 = G_2G_4G_6$$
$$P_2 = G_3G_5G_7$$
$$P_3 = G_2G_1G_7$$
$$P_4 = G_3G_8G_6$$
$$P_5 = -G_2G_1H_2G_8G_6$$
$$P_6 = -G_3G_8H_1G_1G_7$$

Loops

$$L_1 = -G_4H_1, \quad L_2 = -G_5H_2, \quad L_3 = G_1H_2G_8H_1$$

Non-touching loops. There is one pair having gain product $= G_4H_1G_5H_2$

$$\Delta = 1 + G_4H_1 + G_5H_2 - G_1H_2G_8H_1 + G_4H_1G_5H_2$$
$$\Delta_1 = 1 + G_5H_2$$
$$\Delta_2 = 1 + G_4H_1$$
$$\Delta_3 = \Delta_4 = \Delta_5 = \Delta_6 = 1$$

$$\therefore \quad T = \frac{P_1\Delta_1 + P_2\Delta_2 + P_3\Delta_3 + P_4\Delta_4 + P_5\Delta_5 + P_6\Delta_6}{\Delta}$$

$$= \frac{\begin{aligned}&G_2G_4G_6(1 + G_5H_2) + G_3G_5G_7(1 + G_4H_1)\\ &+ G_2G_1G_7 + G_3G_8G_6 - G_2G_6G_8G_1H_2 - G_3G_7G_8G_1H_1\end{aligned}}{1 + G_4H_1 + G_5H_2 + G_4G_5H_1H_2 - G_1G_8H_1H_2} \quad \textbf{Ans}$$

Problem 4.15 Solve problem 3.1 by using Mason's gain formula.

Solution Converting the block diagram of Fig. 3.4 into signal flow graph, we get the network as shown in Fig. 4.24

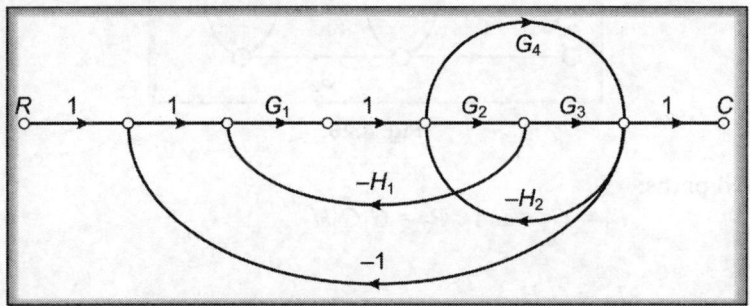

Fig. 4.24

Forward paths

$$P_1 = G_1G_2G_3 \quad \text{and} \quad P_2 = G_1G_4$$

Loops

$$L_1 = -G_1G_2H_1, \quad L_2 = -G_2G_3H_2$$
$$L_3 = -G_1G_2G_3, \quad L_4 = -G_4H_2$$
$$L_5 = -G_1G_4$$
$$\Delta = 1 + G_1G_2H_1 + G_2G_3H_3 + G_1G_2G_3 + G_4H_2 + G_1G_4$$
$$\Delta_1 = \Delta_2 = 1$$

$$T = \frac{G_1G_2G_3 + G_1G_4}{1 + G_1G_2H_1 + G_2G_3H_2 + G_1G_2G_3 + G_4H_2 + G_1G_4} \qquad \textbf{Ans}$$

Problem 4.16 Solve Problem 3.5 by using Mason's gain formula.

Solution Converting the block diagram of Fig. 3.33 into signal flow graph, the network diagram obtained is shown in Fig. 4.25.

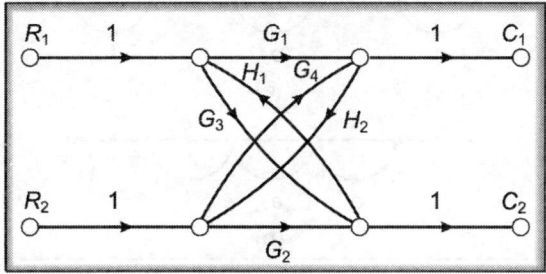

Fig. 4.25

This can be redrawn as shown in Fig. 4.26. Assuming $C_1 = 0$ and $R_2 = 0$

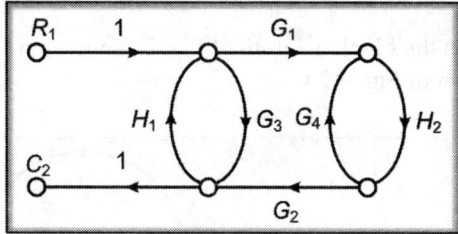

Fig. 4.26

Forward paths:

$$P_1 = G_3 \quad \text{and} \quad P_2 = G_1 G_2 H_2$$

Loops:

$$L_1 = G_3 H_1$$
$$L_2 = G_4 H_2$$
$$L_3 = G_1 H_2 G_2 H_1$$

Non-touching Loops: $G_3 G_4 H_1 H_2$

$$\Delta = 1 - (G_3 H_1 + G_4 H_2 + G_1 G_2 H_1 H_2) + G_3 G_4 H_1 H_2$$
$$\Delta_1 = 1 - G_4 H_2$$
$$\Delta_2 = 1$$
$$T = \frac{P_1 \Delta_1 + P_2 \Delta_2}{\Delta}$$

$\therefore$

$$T = \frac{G_3(1 - G_4 H_2) + G_1 G_2 H_2}{1 - G_3 H_1 - G_4 H_2 - G_1 G_2 H_1 H_2 + G_3 G_4 H_1 H_2}$$

or

$$T = \frac{G_3(1 - G_4 H_2) + G_1 G_2 H_2}{(1 - G_4 H_2) + H_1[G_3(G_4 H_2 - 1) - G_1 G_2 H_2]} \qquad \textbf{Ans}$$

Problem 4.17 Are the two systems given in Fig. 4.27 (a) and (b) equivalent? If not, prove.

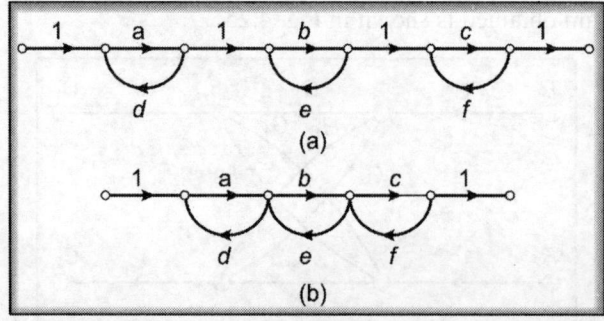

Fig. 4.27

Solution The systems are not equivalent as the transfer function for Fig. 4.27 (a) is:

$$T = \frac{abc}{1 - (ad + be + cf) + (adbe + adcf + becf) - (adbecf)}$$

and the transfer function for Fig. 4.27(b) is:

$$T = \frac{abc}{1 - (ad + be + cf) + (adcf)} \qquad \textbf{Ans}$$

Problem 4.18 Find the transfer function for the control system as shown in Fig. 4.28

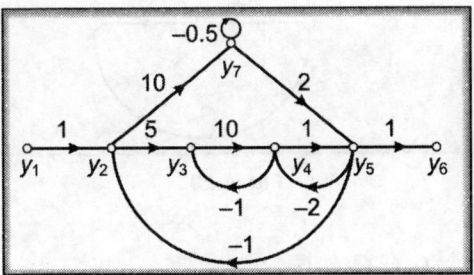

Fig. 4.28

Solution

Forward paths:

$$P_1 = y_1 y_2 y_3 y_4 y_5 y_6 = 1 \times 5 \times 10 \times 1 \times 1 = 50$$
$$P_2 = y_1 y_2 y_7 y_5 y_6 = 1 \times 10 \times 2 \times 1 = 20$$

Loops:

$$L_1 = y_2 y_3 y_4 y_5 y_2 = 5 \times 10 \times 1 \times -1 = -50$$
$$L_2 = y_3 y_4 y_3 = 10 \times -1 = -10$$
$$L_3 = y_4 y_5 y_4 = 1 \times -2 = -2$$
$$L_4 = y_7 y_7 = -0.5$$
$$L_5 = y_2 y_7 y_5 y_2 = 10 \times 2 \times -1 = -20$$

Non-touching loops:

(1) L_2 and $L_4 = -10 \times -0.5 = 5$
(2) L_3 and $L_4 = -2 \times -0.5 = 1$
(3) L_1 and $L_4 = -50 \times -0.5 = 25$

$$\Delta = 1 - (-50 - 10 - 2 - 0.5 - 20) + (5 + 1 + 25) = 114.5$$
$$\Delta_1 = 1 - (L_4) = 1 - (-0.5) = 1.5$$
$$\Delta_2 = 1 - (L_2) = 1 - (-10) = 11$$

$$T = \frac{P_1 \Delta_1 + P_2 \Delta_2}{\Delta} = \frac{50 \times 1.5 + 20 \times 11}{114.5} = 2.576 \qquad \textbf{Ans}$$

Problem 4.19 Solve Problem 3.8 by signal flow graph using Mason's gain formula.

Solution Converting the block diagram given in Fig. 3.46 into signal flow graph, the network obtained is shown in Fig. 4.29

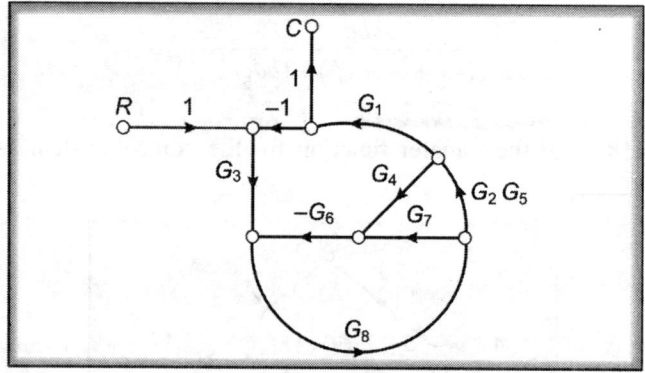

Fig. 4.29

Forward path:

$$P_1 = G_3 G_8 G_2 G_5 G_1$$

Loops:

$$L_1 = -G_3 G_8 G_2 G_5 G_1$$
$$L_2 = -G_6 G_8 G_7$$
$$L_3 = -G_6 G_8 G_2 G_5 G_4$$

$$\therefore \quad \Delta = 1 + G_1 G_2 G_3 G_5 G_8 + G_6 G_7 G_8 + G_2 G_4 G_5 G_6 G_8$$
$$\Delta_1 = 1$$

$$T = \frac{P_1 \Delta_1}{\Delta}$$

$$\therefore \quad T = \frac{G_1 G_2 G_3 G_5 G_8}{1 + G_1 G_2 G_3 G_5 G_8 + G_6 G_7 G_8 + G_2 G_4 G_5 G_6 G_8} \qquad \textbf{Ans}$$

Problem 4.20 Find the transfer function for the control system as shown in Fig. 4.30 by Mason's gain formula. Assume $N = 0$

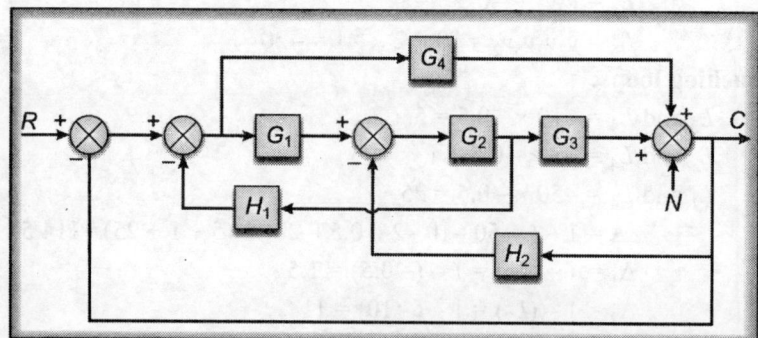

Fig. 4.30

Solution The signal flow graph for the system is shown in Fig. 4.31

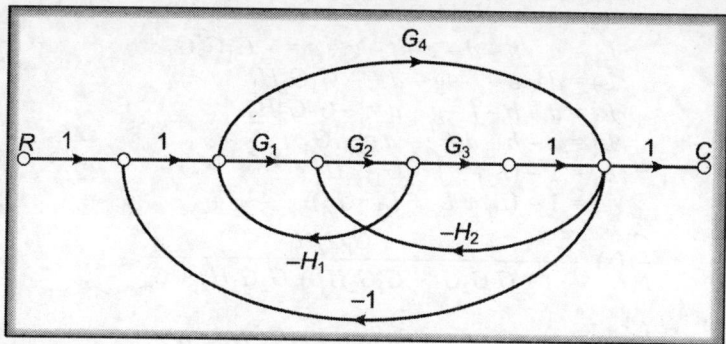

Fig. 4.31

Forward paths:

$$P_1 = G_1G_2G_3; \qquad P_2 = G_4$$

Loops:

$$L_1 = -G_1G_2H_1$$
$$L_2 = -G_1G_2G_3$$
$$L_3 = -G_2G_3H_2$$
$$L_4 = -G_4$$
$$L_5 = G_4H_2G_2H_1$$
$$\Delta = 1 + G_1G_2H_1 + G_1G_2G_3 + G_2G_3H_2 + G_4 - G_2G_4H_1H_2$$
$$\Delta_1 = \Delta_2 = 1$$

$$\therefore \qquad T = \frac{G_1G_2G_3 + G_4}{1 + G_1G_2H_1 + G_1G_2G_3 + G_2G_3H_2 + G_4 - G_2G_4H_1H_2} \qquad \textbf{Ans}$$

Problem 4.21 Solve Problem No. 4.20 and find out C/N, assuming $R = 0$

Solution The signal flow graph is shown in Fig. 4.32

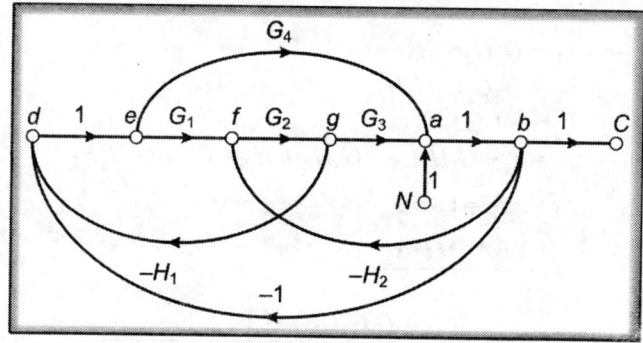

Fig. 4.32

Forward path: There is only one forward path
$$P_1 = 1 \times 1 \times 1 = 1 = N - a - b - C$$

Loops:

$$L_1 = a - b - d - e - f - g - a = - G_1G_2G_3$$
$$L_2 = d - e - f - g - d = - G_1G_2H_1$$
$$L_3 = a - b - f - g - a = - G_2G_3H_2$$
$$L_4 = a - b - d - e - a = - G_4$$
$$\Delta_1 = 1 - (L_2) = 1 + G_1G_2H_1$$
$$\Delta = 1 - (L_1 + L_2 + L_3 + L_4)$$

$$\frac{C}{N}(s) = \frac{1 + G_1G_2H_1}{1 + G_1G_2G_3 + G_1G_2H_1 + G_2G_3H_2 + G_4}$$

For $\dfrac{C}{N}(s) = 0$, the condition is $1 + G_1G_2H_1 = 0$ **Ans**

Problem 4.22 Solve Problem No. 3.10 by signal flow technique using Mason's gain formula.

Solution The block diagram of Fig. 3.55 is converted into signal flow graph as shown in Fig. 4.33.

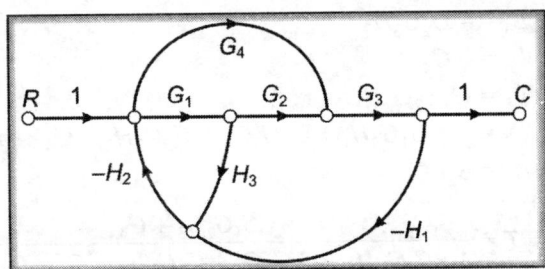

Fig. 4.33

Forward paths:
$$P_1 = G_1G_2G_3$$
$$P_2 = G_3G_4$$

Loops:

$$L_1 = G_3G_4H_1H_2$$
$$L_2 = G_1G_2G_3H_1H_2$$
$$L_3 = - G_1H_2H_3$$
$$\Delta = 1 + G_1H_2H_3 - G_3H_4H_1H_2 - G_1G_2G_3H_1H_2$$
$$\Delta_1 = \Delta_2 = 1$$

$$T = \frac{P_1\Delta_1 + P_2\Delta_2}{\Delta}$$

$$= \frac{G_1G_2G_3 + G_3G_4}{1 + G_1H_2(H_3 - G_2G_3H_1) - G_3G_4H_1H_2}$$ **Ans**

Problem 4.23 Find C/R for the following system using Mason's gain rule.

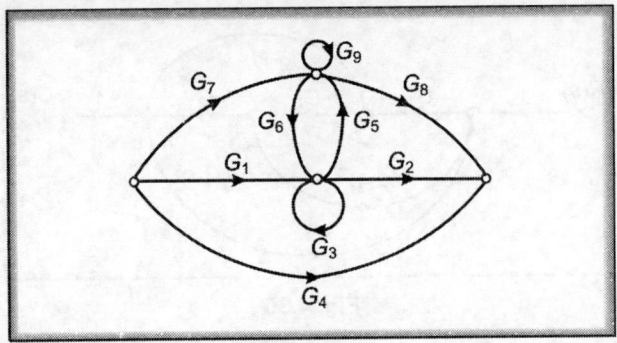

Fig. 4.34

Solution

Forward paths:

$$P_1 = G_1G_2$$

$$P_2 = G_4$$

$$P_3 = G_7G_8$$

$$P_4 = G_1G_5G_8$$

$$P_5 = G_7G_6G_2$$

Loops:

$$L_1 = G_9$$

$$L_2 = G_3$$

$$L_3 = G_5G_6$$

$$\Delta = 1 - (G_3 + G_9 + G_5G_6) + G_9G_3$$

$$\Delta_1 = 1 - G_9$$

$$\Delta_2 = 1 - (G_9 + G_3 + G_5G_6) + G_9G_3$$

$$\Delta_3 = 1 - G_3$$

$$\Delta_4 = 1$$

$$\Delta_5 = 1$$

$$T = \frac{P_1\Delta_1 + P_2\Delta_2 + P_3\Delta_3 + P_4\Delta_4 + P_5\Delta_5}{\Delta}$$

$$= \frac{\begin{array}{c} G_1G_2\,(1 - G_9) + G_4(1 - G_9 - G_3 - G_5G_6 + G_9G_3) \\ + G_7G_8(1 - G_3) + G_1G_5G_8 + G_7G_6G_2 \end{array}}{1 - G_3 - G_9 - G_5G_6 + G_9G_3} \qquad \textbf{Ans}$$

Problem 4.24   Find $C(s)/R(s)$ for a control system represented by signal flow graph as shown in Fig. 4.35. $C(s)$ is the output.

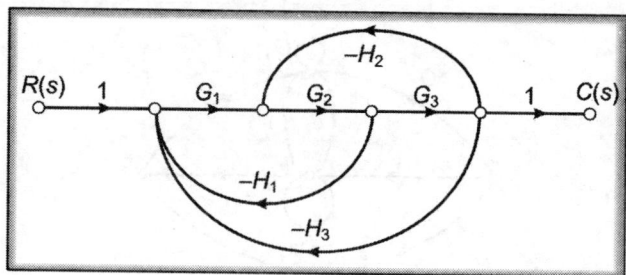

Fig. 4.35

Solution

Forward paths: Only one and the gain is $G_1 G_2 G_3$

Loops:

$$L_1 = - G_1 G_2 H_1$$
$$L_2 = - G_1 G_2 G_3 H_3$$
$$L_3 = - G_2 G_3 H_2$$

Non-touching loops: NIL

$$T = \frac{G_1 G_2 G_3}{1 + G_1 G_2 H_1 + G_1 G_2 G_3 H_3 + G_2 G_3 H_2}$$ **Ans**

Problem 4.25 Use Mason's gain formula for determining the overall transfer function of the control system as shown in Fig. 4.36

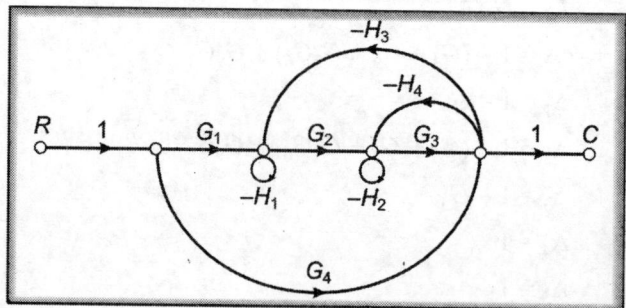

Fig. 4.36

Solution

Forward paths:

$P_1 = G_1 G_2 G_3$ and $P_2 = G_4;$ $\Delta_1 = 1$ and $\Delta_2 = (1 + H_1 + H_2 + H_1 H_2)$

Loops:

$$L_1 = -H_1; \quad L_2 = -H_2; \quad L_3 = -G_3 H_4; \quad L_4 = -G_2 G_3 H_3$$

Non-touching loops:

(1) $-H_1$ and $-H_2$ (2) $-H_1$ and $-G_3 H_4$

$$T = \frac{P_1\Delta_1 + P_2\Delta_2}{\Delta} = \frac{G_1 G_2 G_3 + G_4(1 + H_1 + H_2 + H_1 H_2)}{1 + H_1 + H_2 + G_3 G_4 + G_2 G_3 H_3 + H_1 H_2 + G_3 H_1 H_4} \qquad \textbf{Ans}$$

Problem 4.26 Solve Problem No. 3.12 by signal flow graph.

Solution The signal flow graph is shown in Fig. 4.37

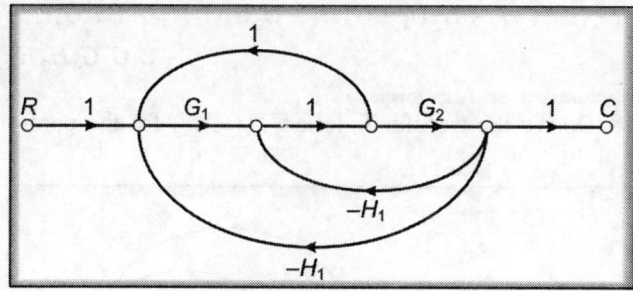

Fig. 4.37

Forward paths: $P_1 = a - b - c - d = G_1 G_2$

By inspection $\dfrac{C}{R} = \dfrac{G_1 G_2}{1 + G_1 G_2 H_1 + G_2 H_1 - G_1}$ **Ans**

Problem 4.27 Determine C/R using Mason's gain formula.

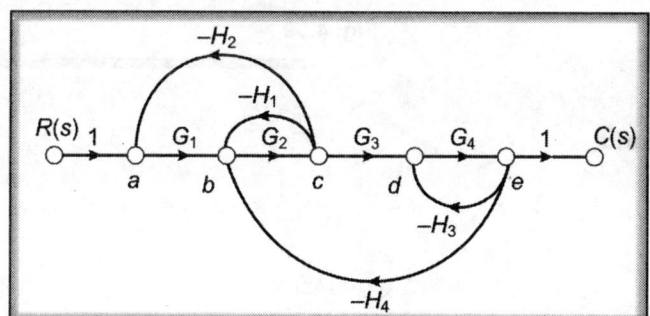

Fig. 4.38

Solution

Forward path:

There is only one forward path $R - a - b - c - d - e - C$, having gain $G_1 G_2 G_3 G_4$

Also, $\Delta_1 = 1$

Loops:

$$L_1 = a - b - c - a \qquad = -G_1 G_2 H_2$$
$$L_2 = b - c - b \qquad = -G_2 H_1$$
$$L_3 = d - e - d \qquad = -G_4 H_3$$
$$L_4 = b - c - d - e - b \ = -G_2 G_3 G_4 H_4$$

Non-touching loops:

1. $d - e - d$ and $b - c - b$, $\qquad (-G_4 H_3)(-G_2 H_1) = G_2 G_4 H_1 H_3$
2. $d - e - d$ and $a - b - c - a$, $\quad (-G_4 H_3)(-G_1 G_2 H_2) = G_1 G_2 G_4 H_2 H_3$

Therefore,

$$\frac{C}{R} = \frac{G_1 G_2 G_3 G_4}{1 + (G_1 G_2 H_2 + G_2 H_1 + G_4 H_3 + G_2 G_3 G_4 H_4) + (G_2 G_4 H_1 H_3 + G_1 G_2 G_4 H_2 H_3)} \qquad \textbf{Ans}$$

Problem 4.28 Derive the transfer function $Y(s)/X(s)$ for the control system as shown in Fig. 4.39

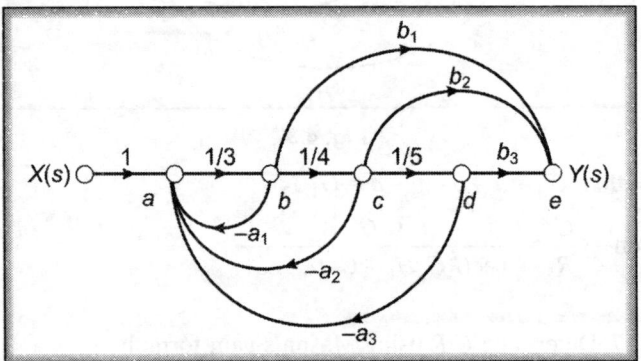

Fig. 4.39

Solution

Forward path:

$$P_1 = a - b - c - d - e = \frac{1}{3} \times \frac{1}{4} \times \frac{1}{5} \times b_3 = \frac{b_3}{60}; \qquad \Delta_1 = 1$$

$$P_2 = a - b - e = \frac{1}{3} \times b_1 = \frac{b_1}{3}; \qquad \Delta_2 = 1$$

$$P_3 = a - b - c - e = \frac{1}{3} \times \frac{1}{4} \times b_2 = \frac{b_2}{12}; \qquad \Delta_3 = 1$$

Loops:

$$L_1 = a - b - a = -\frac{a_1}{3}$$

$$L_2 = a - b - c - a = \frac{1}{3} \times \frac{1}{4} \times -a_2 = -\frac{a_2}{12}$$

$$L_3 = a - b - c - d - a = \frac{1}{3} \times \frac{1}{4} \times \frac{1}{5} \times -a_3 = -\frac{a_3}{60}$$

Therefore,

$$\frac{Y(s)}{X(s)} = \frac{\dfrac{b_3}{60} + \dfrac{b_1}{3} + \dfrac{b_2}{12}}{1 - \left(-\dfrac{a_1}{3} - \dfrac{a_2}{12} - \dfrac{a_3}{60}\right)} = \frac{b_3 + 20b_1 + 5b_2}{60 + 20a_1 + 5a_2 + a_3} = \frac{20b_1 + 5b_2 + b_3}{20a_1 + 5a_2 + a_3 + 60} \quad \textbf{Ans}$$

Problem 4.29 Using Mason's gain formula, find $\dfrac{C(s)}{R(s)}$ of the signal flow graph as shown in Fig. 4.40.

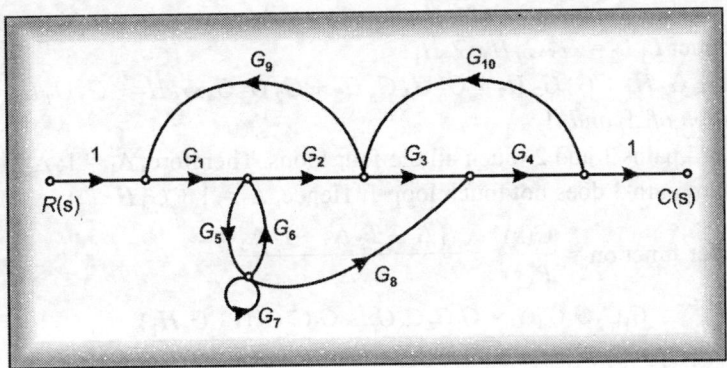

Fig. 4.40

Solution

$$\frac{C}{R} = \frac{P_1 \Delta_1 + P_2 \Delta_2}{\Delta}$$

$$= \frac{G_1 G_2 G_3 G_4 (1 - G_7) + G_1 G_5 G_8 G_4}{1 - [G_1 G_2 G_9 + G_3 G_4 G_{10} + G_1 G_5 G_8 G_4 G_{10} G_9 + G_5 G_6 + G_7]}$$
$$+ [G_1 G_2 G_9 G_7 + G_3 G_4 G_{10} G_5 G_6 + G_3 G_4 G_{10} G_7]$$

Problem 4.30 Obtain the transfer function of $\dfrac{C(s)}{R(s)}$ of the system whose signal flow graph is shown in Fig. 4.41

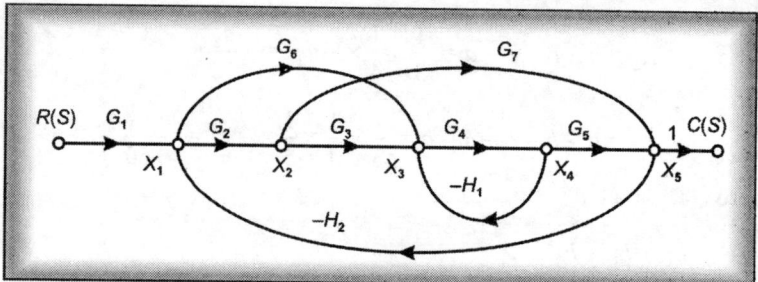

Fig. 4.41

Solution

Forward paths:

$$P_1 = G_1 G_2 G_3 G_4 G_5; \quad P_2 = G_1 G_6 G_4 G_5; \quad P_3 = G_1 G_2 G_7$$

Loops:

$$L_1 = -G_4 H_1; \ L_2 = -G_2 G_7 H_2; \ L_3 = -G_6 G_4 G_5 H_2; \ L_4 = -G_2 G_3 G_4 G_5 H_2$$

Non-touching loops:

There is one combination of loops L_1 and L_2 which are non-touching with loop gain product $L_1 L_2 = G_2 G_7 H_2 G_4 H_1$

$$\Delta = 1 + G_4 H_1 + G_2 G_7 H_2 + G_6 G_4 G_5 H_2 + G_2 G_3 G_4 G_5 H_2 + G_2 G_7 H_2 G_4 H_1$$

Calculation of Δ_1 and Δ_2:

Forward paths 1 and 2 touch all the four loops. Therefore, $\Delta_1 = 1; \Delta_2 = 1$.

Forward path 3 does not touch loop 1. Hence, $\Delta_3 = 1 + G_4 H_1$.

$$\text{Transfer function} = \frac{C(s)}{R(s)} = \frac{P_1 \Delta_1 + P_2 \Delta_2 + P_3 \Delta_3}{\Delta}$$

$$= \frac{G_1 G_2 G_3 G_4 G_5 + G_1 G_4 G_5 G_6 + G_1 G_2 G_7 (1 + G_4 H_1)}{1 + G_4 H_1 + G_2 G_7 H_2 + G_6 G_4 G_5 H_2 + G_2 G_3 G_4 G_5 H_2 + G_2 G_4 G_7 H_1 H_2}. \quad \textbf{Ans}$$

Problem 4.31 Find the gains $\dfrac{X_6}{X_1}, \dfrac{X_5}{X_2}, \dfrac{X_3}{X_1}$ for the signal flow graph as shown in Fig. 4.42.

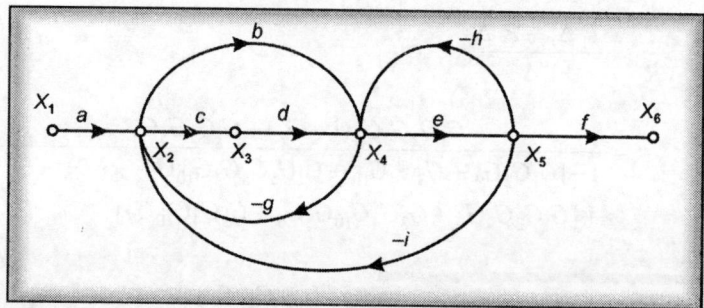

Fig. 4.42

Solution

(i) $\dfrac{X_6}{X_1}$

Forward paths:

$$P_1 = acdef \text{ and } P_2 = abcf$$

Loops:

$$L_1 = -cg; \quad L_2 = -eh; \quad L_3 = -cdei; \quad L_4 = -bei$$

Non-touching loops:

$$L_1 L_2 = cgeh$$

$$\Delta = 1 + cg + eh + cdei + bei + cgeh$$

Calculation of Δ_1 and Δ_2:

$\Delta_1 = \Delta_2 = 1$, as all loops touch forward paths P_1 and P_2.

$$\frac{X_6}{X_1} = \frac{P_1 \Delta_1 + P_2 \Delta_2}{\Delta} = \frac{cdef + abef}{1 + cg + eh + cdei + bei + cgeh}$$

(ii) $\dfrac{X_5}{X_2}$

The modified signal flow graph with X_5 as sink node and X_2 as the source node is shown in Fig. 4.43. The branches having transmittance a and f do not form the loops.

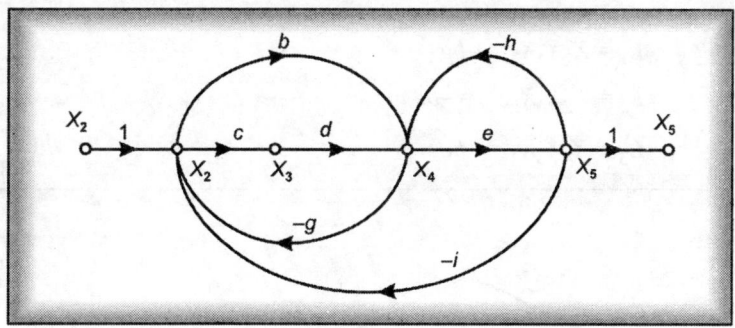

Fig. 4.43

$$\frac{X_5}{X_2} = \frac{P_1 \Delta_1 + P_2 \Delta_2}{\Delta}$$

$$= \frac{cde + be}{1 + cg + eh + cdei + bei + cgeh}$$

The signal flow graph as shown in Fig. 4.42 is redrawn with X_1 and X_3 as source and sink nodes, respectively.

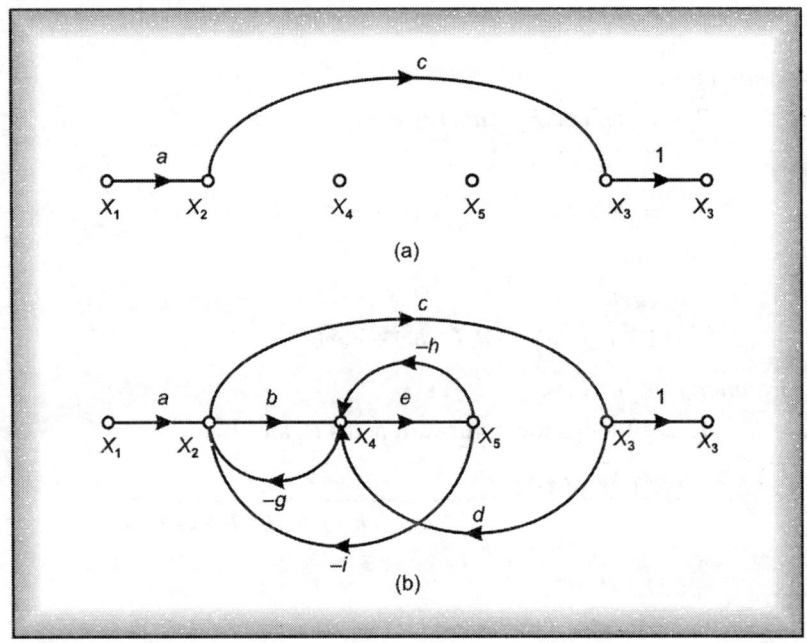

(a)

(b)

Fig. 4.44a and b

Forward paths:

$$P_1 = ac$$

Loops:

$$L_1 = X_2X_4X_2 = -bg$$
$$L_2 = X_2X_4X_5X_2 = -be$$
$$L_3 = X_4X_5X_4 = -he$$

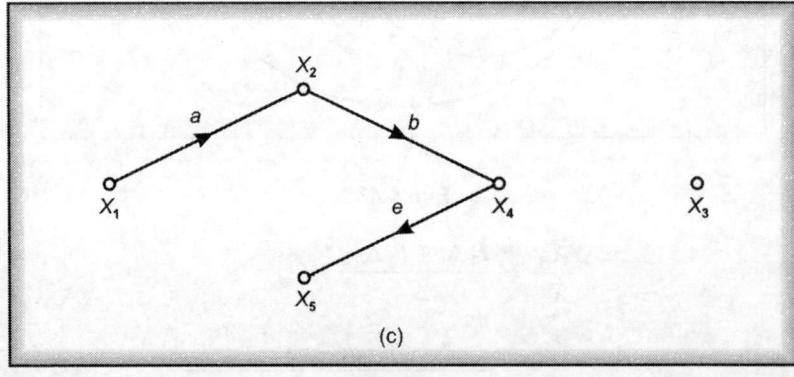

(c)

Fig. 4.44c

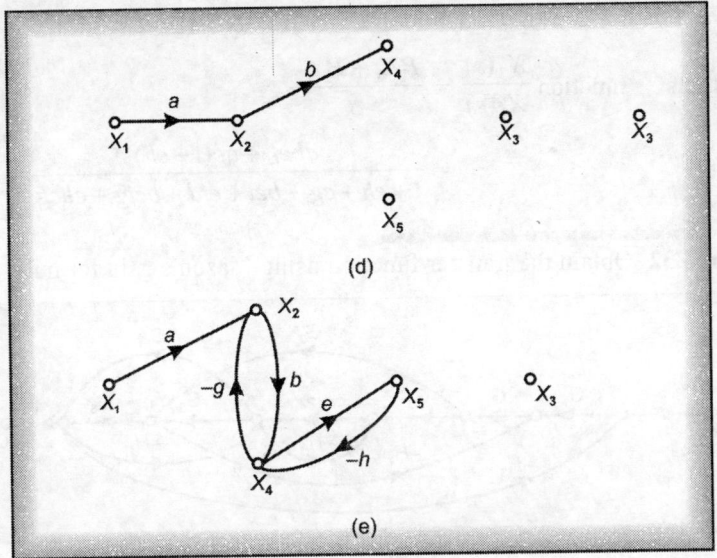

Fig. 4.44d and e

(iii) $\dfrac{X_3}{X_1}$

The signal flow graph is redrawn to obtain the clarity of the functional relation as shown in Fig. 4.44f.

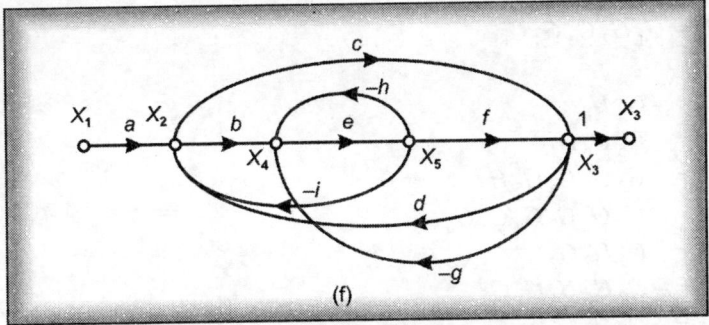

Fig. 4.44f

There are two forward paths. The gains of the forward paths are:

$$P_1 = abef$$
$$P_2 = ac$$

There are five loops with loop gains:

$L_1 = -eh$; $L_2 = -eg$; $L_3 = -bei$; $L_4 = efd$; $L_5 = -befg$

There is one combination of loops L_1 and L_2 which are non-touching with loop gain product $L_1 L_2 = ehcg$

$$\Delta = 1 + eh + cg + bei + efd + befg + ehcg$$

Forward path 1 touches all the five loops. Therefore $\Delta_1 = 1$.

Forward path 2 does not touch loop L_1. Hence, $\Delta_2 = 1 + eh$

The transfer function $\dfrac{X_3(s)}{X_1(s)} = \dfrac{P_1\Delta_1 + P_2\Delta_2}{\Delta}$

$$= \frac{abef + ac(1 + eh)}{1 + eh + cg + bei + efd + befg + ehcg}$$ **Ans**

Problem 4.32 Obtain the transfer function using Mason's gain formula.

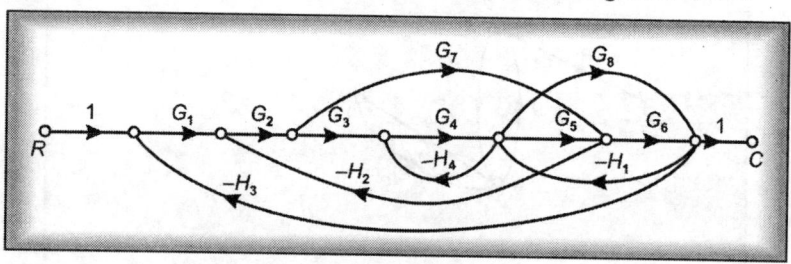

Fig. 4.45

Solution

Forward paths:

$P_1 = G_1 G_2 G_3 G_4 G_5 G_6$

$P_2 = G_1 G_2 G_7 G_6$

$P_3 = G_1 G_2 G_3 G_4 G_8$

Loops:

$L_1 = -G_4 H_4$

$L_2 = -G_5 G_6 H_1$

$L_3 = -G_2 G_3 G_4 G_5 H_2$

$L_4 = -G_1 G_2 G_3 G_4 G_5 G_6 H_3$

$L_5 = -G_7 H_2 G_2$

$L_6 = -G_7 G_6 H_3 G_1 G_2$

$L_7 = -G_8 H_1$

$L_8 = -G_8 H_3 G_1 G_2 G_3 G_4$

Calculation of Δ:

$\Delta_1 = 1$ (As all loops touch forward path P_1)

$\Delta_2 = 1 + G_4 H_4$ ($-G_4 H_4$ loop does not touch forward path P_2)

$\Delta_3 = 1$ (As all loops touch forward path P_3)

Two non-touching loops:

$L_1 L_5 = -G_4 H_4 (-G_7 H_2 G_2) = G_2 G_7 H_2 H_4$

$L_1 L_6 = -G_4 H_4 (-G_7 G_6 H_3 G_1 G_2) = G_1 G_2 G_4 G_6 G_7 H_3 H_4$

$L_5 L_7 = -G_8 H_1 (-G_2 G_7 H_2) = G_2 G_7 G_8 H_1 H_2$

Three non-touching loops = 0

$$\Delta = 1 - \{L_1 + L_2 + L_3 + L_4 + L_5 + L_6 + L_7 + L_8\} + \{L_1L_5 + L_1L_6 + L_5L_7\}$$

$$\boxed{\frac{C(s)}{R(s)} = \frac{1}{\Delta}\{P_1\Delta_1 + P_2\Delta_2 + P_3\Delta_3\}}$$

Ans

Problem 4.33 Find the transfer functions $\dfrac{C_1}{R_1}, \dfrac{C_2}{R_2}, \dfrac{C_1}{R_2}$ and $\dfrac{C_2}{R_2}$

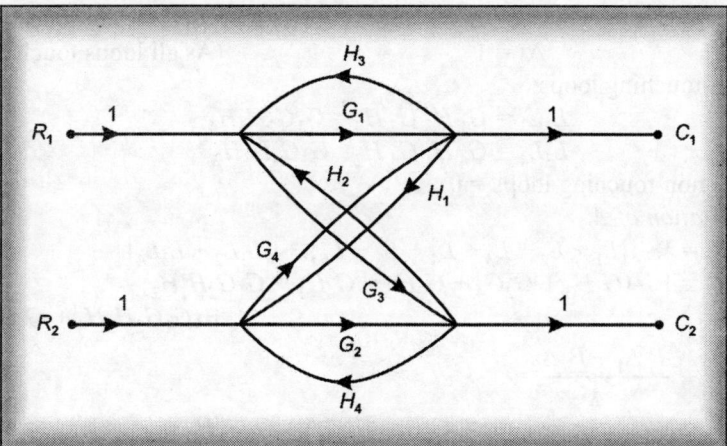

Fig. 4.46

Solution

(*i*) Let $R_2 = 0$, then we can find $\dfrac{C_{11}(s)}{R_1(s)}$

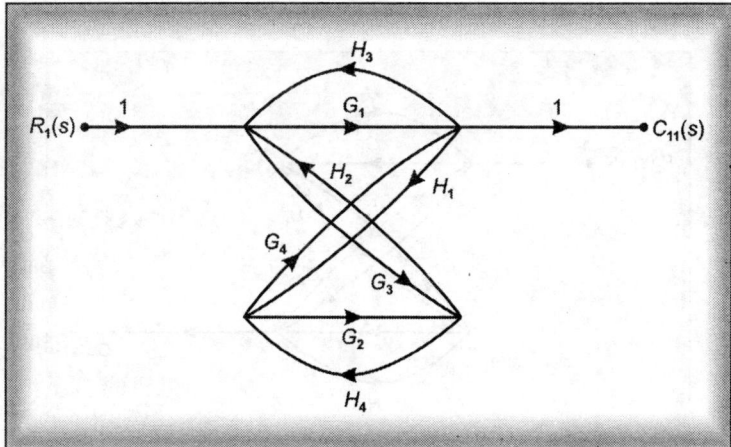

Fig. 4.47

Forward paths: $P_1 = G_1$
$$P_2 = G_3 H_4 G_4 = G_3 G_4 H_4$$
Loops: $L_1 = G_1 H_3$
$$L_2 = G_2 H_4$$
$$L_3 = G_3 H_2$$
$$L_4 = G_4 H_1$$
$$L_5 = G_1 H_1 G_2 H_2 = G_1 G_2 H_1 H_2$$
$$L_6 = H_3 G_3 H_4 G_4 = G_3 G_4 H_3 H_4$$

Calculation of Δ_1 and Δ_2:

$$\Delta_1 = 1 - G_2 H_4 \qquad \text{(As loop } L_2 \text{ does not touch path } P_1)$$
$$\Delta_2 = 1 \qquad\qquad\qquad \text{(As all loops touch path } P_2)$$

Two non-touching loops:

$$L_1 L_2 = G_1 H_3 G_2 H_4 = G_1 G_2 H_3 H_4$$
$$L_3 L_4 = G_4 H_1 G_3 H_2 = G_3 G_4 H_1 H_2$$

Three non-touching loops = 0

Calculation of Δ:

$$\Delta = 1 - \{L_1 + L_2 + L_3 + L_4 + L_5 + L_6\} + \{L_1 L_2 + L_3 L_4\}$$
$$= 1 - (G_1 H_3 + G_2 G_4 + G_3 H_2 + G_4 H_1 + G_1 G_2 H_1 H_2)$$
$$+ (G_1 G_2 H_3 H_4 + G_3 G_4 H_1 H_2)$$

$$\frac{C_{11}(s)}{R_1(s)} = \frac{P_1 \Delta_1 + P_2 \Delta_2}{\Delta}$$

$$= \frac{G_1 (1 - G_2 G_4) + G_3 G_4 H_4}{1 - (G_1 H_3 + G_2 H_4 + G_3 H_2 + G_4 H_1 + G_1 G_2 H_1 H_2 + G_3 G_4 H_3 H_4)}$$
$$+ (G_1 G_2 H_3 H_4 + G_3 G_4 H_1 H_2)$$

(*ii*) Let $R_2 = 0$, then we can find $\dfrac{C_{21}(s)}{R_1(s)}$

Forward paths: *Calculation of A_1 and A_2:*

$$P_1 = G_1 H_1 G_2 = G_1 G_2 H_1 \qquad \Delta_1 = 1$$
$$P_2 = G_2 \qquad\qquad\qquad\qquad \Delta_2 = 1 - G_4 H_1$$

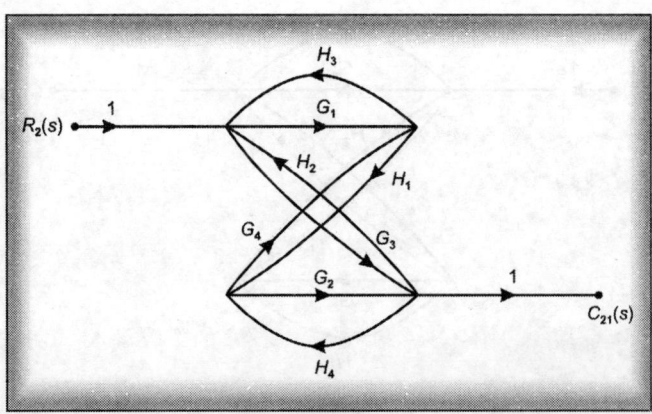

Fig. 4.48

Loops:

$$L_1 = G_1 H_3$$

$$L_2 = G_2 H_4$$

$$L_3 = G_3 H_2$$

$$L_4 = G_4 H_1$$

$$L_5 = G_1 H_1 G_2 H_2 = G_1 G_2 H_1 H_2$$

$$L_6 = H_3 G_3 H_4 G_4 = G_3 G_4 H_3 H_4$$

Calculation of Δ:

'Δ' remains same as in the first case

Calculation of Δ_1 *and* Δ_2

$$C_{21}(s) = \left\{ \frac{P_1 \Delta_1 + P_2 \Delta_2}{\Delta} \right\} R_1(s)$$

(iii) Let $R_1 = 0$, then we can find $\dfrac{C_{12}(s)}{R_2(s)}$

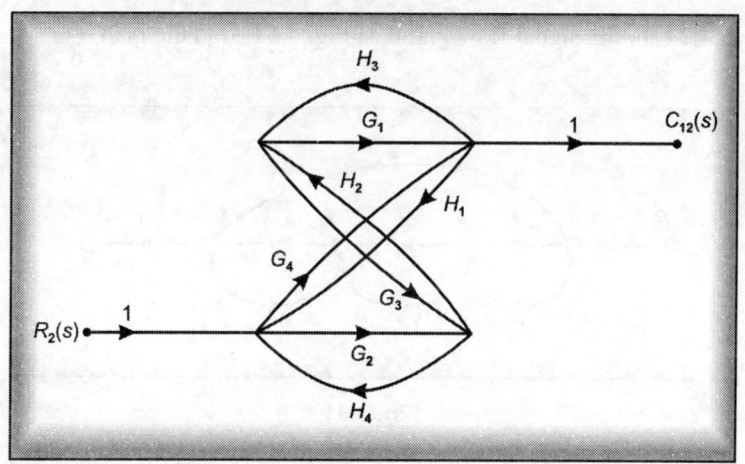

Fig. 4.49

Forward paths: *Calculation of* Δ_1 *and* Δ_2:

$$P_1 = G_4 \qquad\qquad \Delta_1 = 1 - G_3 H_2$$

$$P_2 = G_2 H_2 G_1 \qquad\qquad \Delta_2 = 1$$

'Δ' remains same as calculated earlier

$$C_{12}(s) = \left\{ \frac{P_1 \Delta_1 + P_2 \Delta_2}{\Delta} \right\} R_2(s)$$

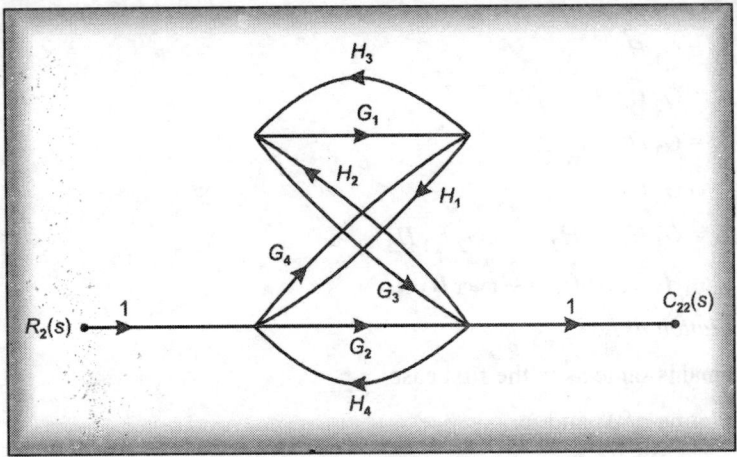

Fig. 4.50

Hence, $C_1(S) = C_{11}(S) + C_{12}(S)$

$C_2(S) = C_{21}(S) + C_{22}(S)$

Problem 4.34 In the signal flow graph of Fig. 4.51, find the gain $\dfrac{C}{R}$.

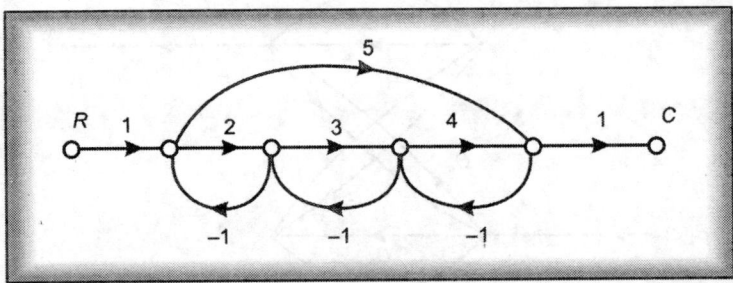

Fig. 4.51

Solution

Forward paths:

$$P_1 = 1 \times 2 \times 3 \times 4 \times 1 = 24$$
$$P_2 = 1 \times 5 = 5$$

Loops:

$$L_1 = 2 \times -1 = -2$$
$$L_2 = 3 \times -1 = -3$$
$$L_3 = 4 \times -1 = -4$$

Non-touching loops:

$$L_1L_3 = -2 \times -4 = 8$$

Therefore,

$$\Delta_2 = [1 - (-3)]$$

$$\therefore \quad \frac{C}{R} = \frac{24 \times 1 + 5[1 - (-3)]}{1 - (-2 - 3 - 4) + (-2 \times -4)}$$

$$= \frac{24 + 20}{1 + 9 + 8} = \frac{44}{18}.$$ **Ans**

Problem 4.35 The signal flow graph of a system is shown in Fig. 4.52. Find the transfer function $\dfrac{C(s)}{R(s)}$.

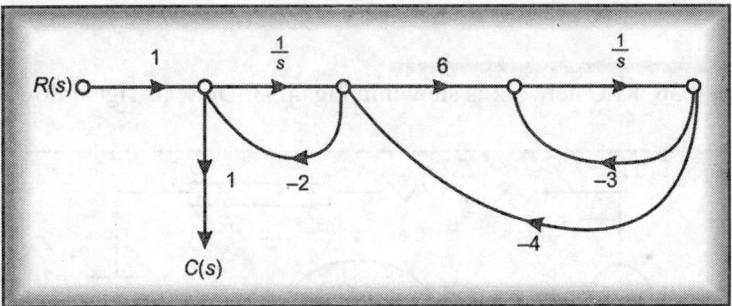

Fig. 4.52

Solution

Forward paths:

$$P_1 = 1 \times 1 = 1$$

Loops:

$$L_1 = \frac{1}{s} \times -2 = \frac{-2}{s}$$

$$L_2 = \frac{1}{s} \times -3 = \frac{-3}{s}$$

$$L_3 = 6 \times \frac{1}{s} \times -4 = \frac{-24}{s}$$

Non-touching loops:

$$L_1L_3 = \frac{-2}{s} \times \frac{-24}{s} = \frac{48}{s^2}$$

Calculation of Δ:

$$\Delta_1 = 1 - \left[\left(\frac{1}{s} \times -3\right) + \left(6 \times \frac{1}{s} \times -4\right)\right]$$

$$= 1 - \left[\frac{-3}{s} - \frac{24}{s}\right] = 1 + \frac{3}{s} + \frac{24}{s}$$

$$= \frac{s + 3 + 24}{s} = \frac{s + 27}{s}$$

$$\frac{C(s)}{R(s)} = \frac{P_1\Delta_1}{\Delta} = \frac{1 \times \left(\frac{s+27}{s}\right)}{1 - (L_1 + L_2 + L_3) + (L_1 L_2)} = \frac{\frac{s+27}{s}}{1 + \frac{29}{s} + \frac{48}{s^2}}$$

$$= \frac{\frac{s+27}{s}}{\frac{s^2 + 29s + 48}{s^2}} = \frac{s(s+27)}{s^2 + 29s + 48} \qquad \textbf{Ans}$$

Problem 4.36 RLC network is shown in Fig. 4.53. Draw its signal flow graph.

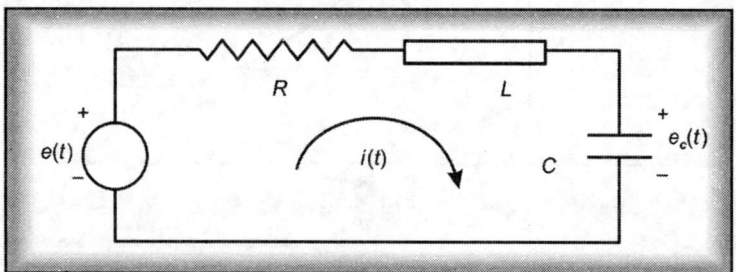

Fig. 4.53

Solution The governing differential equations are:

$$L\frac{di}{dt} + Ri + \frac{1}{C}\int idt = e(t) \qquad (1)$$

or $$L\frac{di}{dt} + Ri + e_c = e(t) \qquad (2)$$

$$C\frac{de_c}{dt} = i(t) \qquad (3)$$

Taking Laplace transform of eqns (1) and (2), and dividing eqn. (2) by L and eqn. 3 by C, we get

$$sI(s) - i(0^+) + \frac{R}{L}I(s) + \frac{1}{L}E_c(s) = \frac{1}{L}E(s) \qquad (4)$$

$$sI(s) + \frac{R}{L} I(s) = i(o^+) + \frac{1}{L} E_c(s) + \frac{1}{L} E(s) \tag{5}$$

$$I(s) = \frac{i(o^+)}{\left(s + \frac{R}{L}\right)} + \frac{-1}{L\left(s + \frac{R}{L}\right)} E_c(s) + \frac{1}{\left(s + \frac{R}{L}\right)} E(s) \tag{6}$$

and

$$sE_c(s) - e_c(0^+) = \frac{1}{C} I(s) \tag{7}$$

$$E_c(s) = \frac{e_c(0^+)}{s} + \frac{1}{sC} I(s)$$

Equations (4) and (5) are used for drawing the signal flow graph as shown in Fig. 4.54.

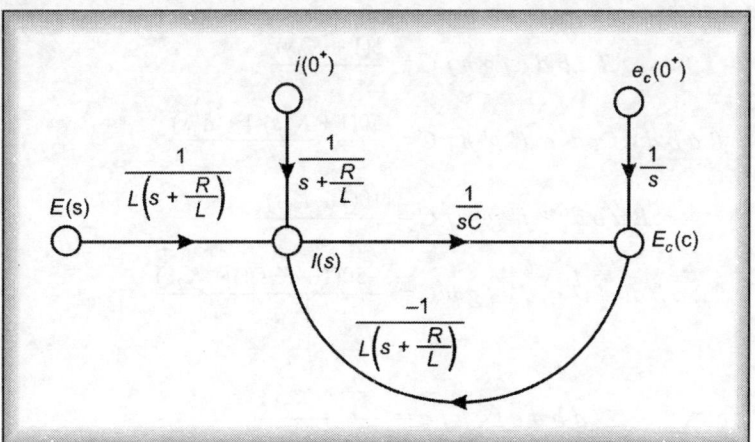

Fig. 4.54

Problem 4.37 Find C/R for the control system as shown in Fig. 4.55.

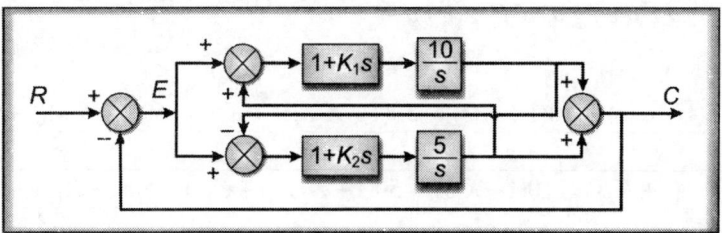

Fig. 4.55

Solution Evaluation of C/R

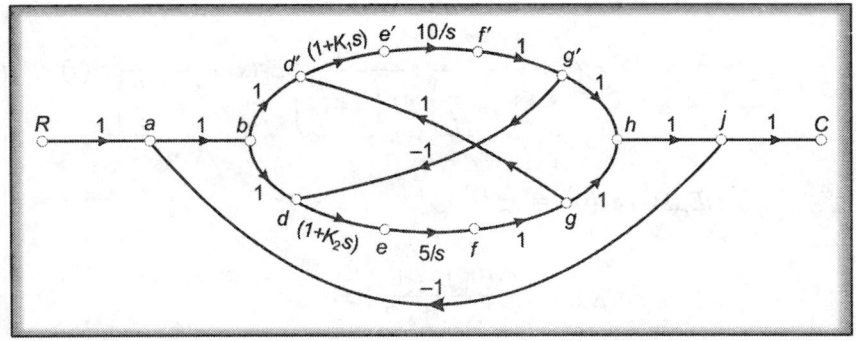

Fig. 4.56

Forward paths:

(1) $\qquad R\,a\,b\,d\,e\,f\,g\,h\,j\,C = \dfrac{5(1+K_2s)}{s}$

(2) $\qquad R\,a\,b\,d\,e\,f\,g\,d'\,e'\,f'\,g'\,h\,j\,C = \dfrac{50(1+K_2s)(1+K_1s)}{s^2}$

(3) $\qquad R\,a\,b\,d'\,e'\,f'\,g'\,h\,j\,C = \dfrac{10(1+K_1s)}{s}$

(4) $\qquad R\,a\,b\,d'\,e'\,f'\,g'\,d\,e\,f\,g\,h\,C = \dfrac{-50(1+K_1s)(1+K_2s)}{s^2}$

Loops:

(1) $\qquad a\,b\,d\,e\,f\,g\,h\,j\,a = -\dfrac{5(1+K_2s)}{s}$

(2) $\qquad a\,b\,d'\,e'\,f'\,g'\,h\,j\,a = -\dfrac{10(1+K_1s)}{s}$

(3) $\qquad a\,b\,d'\,e'\,f'\,g'\,d\,e\,f\,g\,h\,j\,a = \dfrac{50(1+K_1s)(1+K_2s)}{s^2}$

(4) $\qquad a\,b\,d\,e\,f\,g\,d'\,e'\,f'\,g'\,h\,j\,a = -\dfrac{50(1+K_1s)(1+K_2s)}{s^2}$

Also $\Delta_1 = \Delta_2 = \Delta_3 = \Delta_4 = 1$

$$\frac{C}{R} = \frac{\left[\dfrac{5(1+K_2s)}{s} + \dfrac{10(1+K_1s)}{s} + \dfrac{50(1+K_2s)(1+K_1s)}{s^2} - \dfrac{50(1+K_1s)(1+K_2s)}{s^2}\right]}{\left[1 + \dfrac{5(1+K_2s)}{s} + \dfrac{10(1+K_1s)}{s} + \dfrac{50(1+K_2s)(1+K_1s)}{s^2} - \dfrac{50(1+K_1s)(1+K_2s)}{s^2}\right]}$$

$$\frac{C}{R} = \frac{5(1+K_2s) + 10(1+K_1s)}{s + 5(1+K_2s) + 10(1+K_1s)} = \frac{5s(K_2 + 2K_1) + 15}{s(1 + 5K_2 + 10K_1) + 15} \qquad \textbf{Ans}$$

Problem 4.38 Draw the signal flow graph of the control system as shown in Fig. 4.57 and find the ratio $C(s)/R(s)$

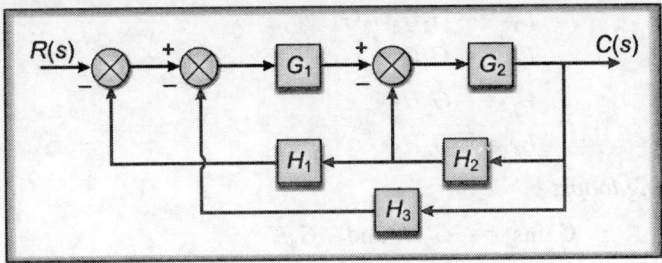

Fig. 4.57

Solution Signal flow graph is shown in Fig. 4.58

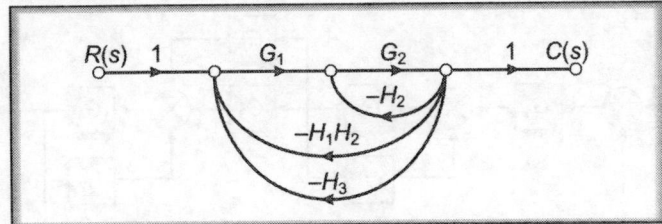

Fig. 4.58

By inspection, the ratio

$$\frac{C(s)}{R(s)} = \frac{G_1 G_2}{1 + G_1 G_2 H_3 + G_1 G_2 H_1 H_2 + G_2 H_2}$$ **Ans**

Problem 4.39 Draw a block diagram of the signal flow graph as shown in Fig. 4.59. Find the overall transfer function by block diagram reduction method. Verify the result by Mason's gain formula.

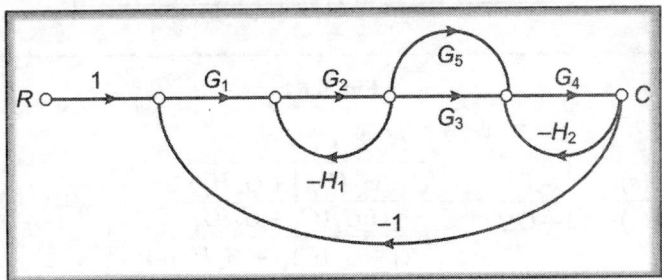

Fig. 4.59

Solution

Forward paths: $\quad P_1 = G_1 G_2 G_3 G_4 \Delta_1 = 1$

$\qquad\qquad\qquad P_2 = G_1 G_2 G_5 G_4 \Delta_2 = 1$

Loops: $\qquad\quad L_1 = -G_1 G_2 G_3 G_4$

$\qquad\qquad\qquad L_2 = -G_1 G_2 G_5 G_4$

$\qquad\qquad\qquad L_3 = -G_2 H_1$

$\qquad\qquad\qquad L_4 = -G_4 H_2$

Non-touching loops:

$\qquad\qquad\qquad$ Gains $= -G_2 H_1$ and $-G_4 H_2$

$$\frac{C}{R} = \frac{G_1 G_2 G_3 G_4 + G_1 G_2 G_4 G_5}{1 + G_1 G_2 G_3 G_4 + G_1 G_2 G_4 G_5 + G_2 H_1 + G_4 H_2 + G_2 G_4 H_1 H_2}$$

Block diagram of the given signal-flow graph is shown in Fig. 4.60

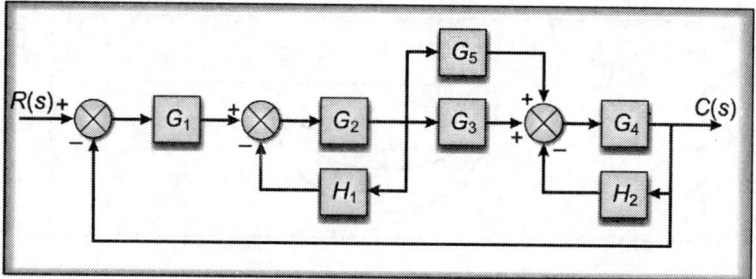

Fig. 4.60

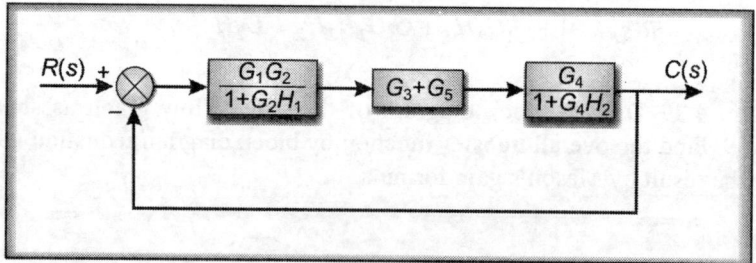

Fig. 4.61

$$\frac{C(s)}{R(s)} = \frac{G}{1 + GH} = \frac{\dfrac{G_1 G_2 (G_3 + G_5) G_4}{(1 + G_2 H_1)(1 + G_4 H_2)}}{1 + \dfrac{G_1 G_2 (G_3 + G_5) G_4}{(1 + G_2 H_1)(1 + G_4 H_2)}}$$

$$= \frac{G_1 G_2 G_3 G_4 + G_1 G_2 G_4 G_5}{1 + G_1 G_2 G_3 G_4 + G_1 G_2 G_4 G_5 + G_2 H_1 + G_4 H_2 + G_2 G_4 H_1 H_2}$$

Problem 4.40 Using signal flow graph and Mason's gain formula, obtain the overall gain of the control system depicted in Fig. 4.62

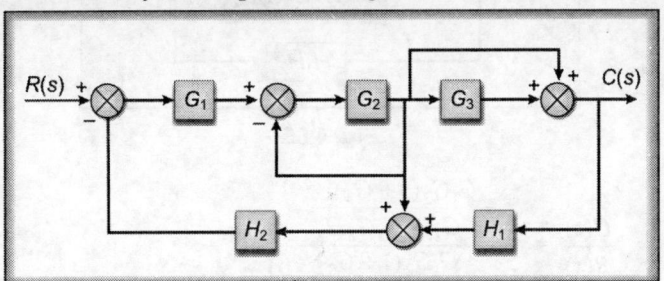

Fig. 4.62

Solution Signal flow graph of the given block diagram is shown in Fig. 4.63

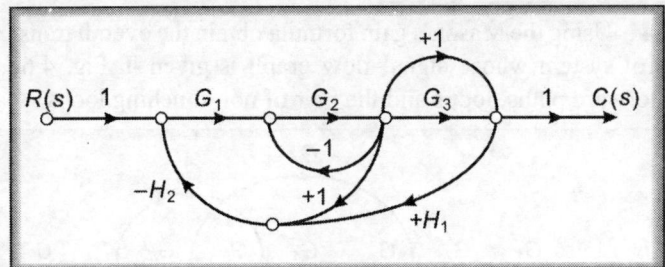

Fig. 4.63

Forward paths:
$$P_1 = G_1 G_2 G_3 \quad \text{and} \quad P_2 = G_1 G_2; \quad \Delta_1 = \Delta_2 = 1$$

Loops:
$$L_1 = - G_1 G_2 G_3 H_1 H_2$$
$$L_2 = - G_1 G_2 H_2$$
$$L_3 = - G_2 \text{ and}$$
$$L_4 = - G_1 G_2 H_1 H_2$$

Overall gain $= \dfrac{C(s)}{R(s)} = \dfrac{G_1 G_2 + G_1 G_2 G_3}{1 + G_2 + G_1 G_2 H_2 + G_1 G_2 H_1 H_2 + G_1 G_2 G_3 H_1 H_2}$ **Ans**

Block diagram reduction technique

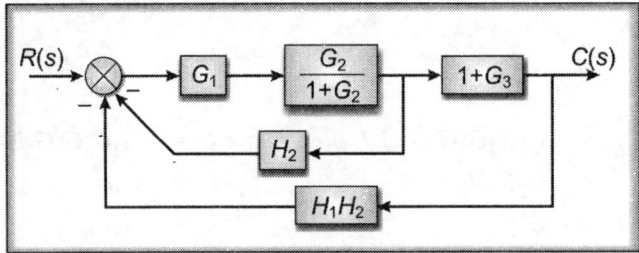

Fig. 4.64

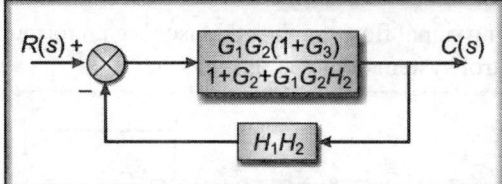

Fig. 4.65

$$\frac{C(s)}{R(s)} = \frac{\dfrac{G_1G_2(1+G_3)}{1+G_2+G_1G_2H_2}}{1+\dfrac{G_1G_2(1+G_3)}{1+G_2+G_1G_2H_2}H_1H_2}$$

$$= \frac{G_1G_2+G_1G_2G_3}{1+G_2+G_1G_2H_2+G_1G_2H_1H_2+G_1G_2G_3H_1H_2} \qquad \textbf{Ans}$$

Problem 4.41 Using the Mason's gain formula obtain the overall transfer function of the control system whose signal flow graph is given in Fig. 4.66. State the number of forward paths, loops and the pair of non-touching loops.

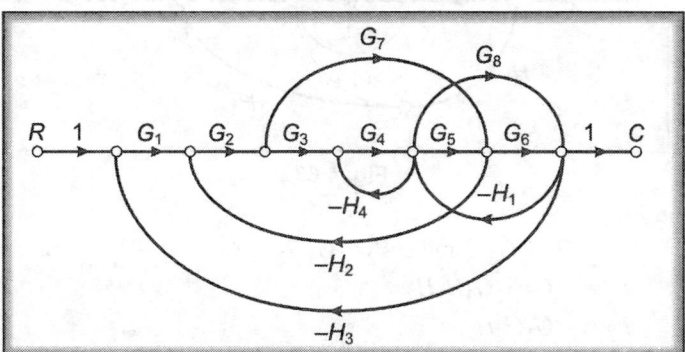

Fig. 4.66

Solution

Forward paths:

$$P_1 = G_1G_2G_3G_4G_5G_6; \quad P_2 = G_1G_2G_7G_6 \quad \text{and} \quad P_3 = G_1G_2G_3G_4G_8$$

$$\Delta_1 = 1 \qquad\qquad \Delta_2 = 1+G_4H_4 \qquad\qquad \Delta_3 = 1$$

Loops:

$$L_1 = -\,G_1G_2G_3G_4G_5G_6H_3 \qquad\qquad L_2 = -\,G_2G_3G_4G_5H_2$$

$$L_3 = -\,G_2G_7H_2 \qquad\qquad\qquad\qquad L_4 = -\,G_4H_4$$

$$L_5 = -\,G_8H_1 \qquad\qquad\qquad\qquad\quad L_6 = -\,G_5G_6H_1$$

$$L_7 = G_1G_2G_3G_4G_8H_3 \qquad\qquad\qquad L_8 = -\,G_1G_2G_7G_6H_3$$

Pair of non-touching loops:

   (1) $- G_4 H_4$ and $- G_2 G_7 H_2$

   (2) $- G_2 G_7 H_2$ and $- G_8 H_1$

Overall gain:

$$\frac{C(s)}{R(s)} = \frac{G_1 G_2 G_3 G_4 G_5 G_6 + G_1 G_2 G_6 G_7 (1 + G_4 H_4) + G_1 G_2 G_3 G_4 G_8}{\begin{aligned}&1 + (G_1 G_2 G_3 G_4 G_5 G_6 H_3 + G_2 G_3 G_4 G_5 H_2 + G_2 G_7 H_2 \\ &+ G_4 H_4 + G_8 H_1 + G_5 G_6 H_1 + G_1 G_2 G_3 G_4 G_8 H_3 + G_1 G_2 G_6 G_7 H_3) \\ &+ (G_4 H_4 G_2 G_7 H_2 + G_2 G_7 H_2 G_8 H_1)\end{aligned}} \quad \textbf{Ans}$$

Problem 4.42 For the control system as shown in Fig. 4.67, use block diagram reduction technique to obtain the equivalent transfer function from R to C and check the results using Mason's formula.

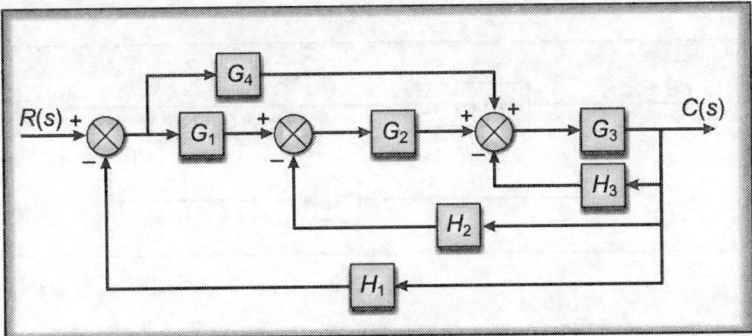

Fig. 4.67

Solution Eliminating the feedback H_3 and combining with G_3 to have a forward path only, we get

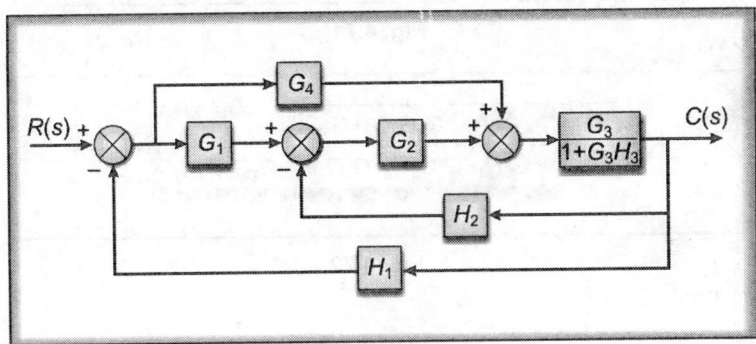

Fig. 4.68

Shifting the second summing point from left, towards the third summing point, and combining both of them, we get

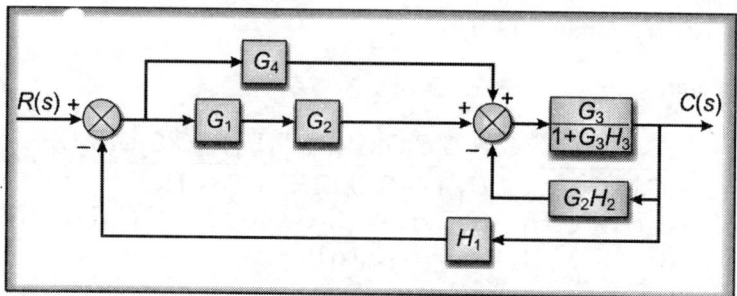

Fig. 4.69

Combining blocks G_1 and G_2 being in the same forward path, adding parallel path of G_4 with it and eliminating feedback element G_2H_2 and combining with $\dfrac{G_3}{1+G_3H_3}$ block, we get

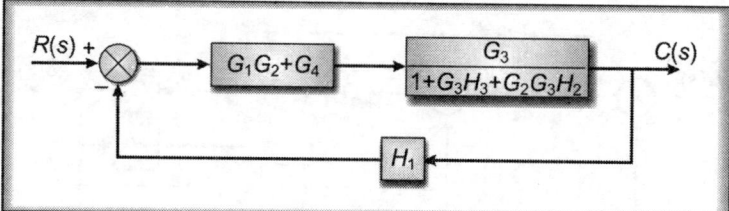

Fig. 4.70

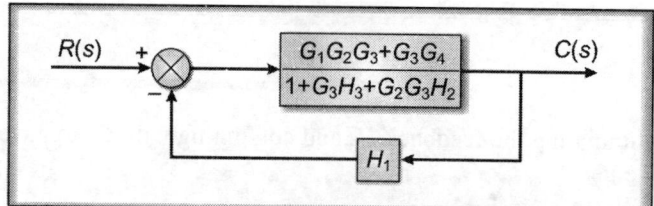

Fig. 4.71

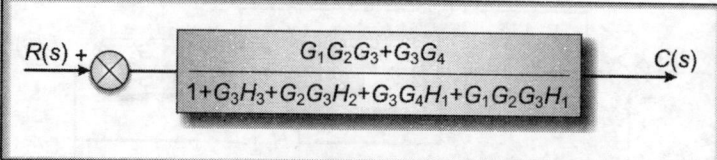

Fig. 4.72

Therefore,

$$\frac{C(s)}{R(s)} = \frac{G_1G_2G_3 + G_3G_4}{1 + G_3H_3 + G_2G_3H_2 + G_3G_4H_1 + G_1G_2G_3H_1}$$

Signal flow graph of the block diagram is shown in Fig. 4.73

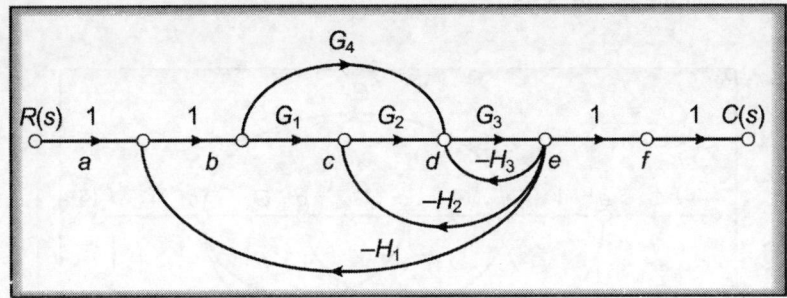

Fig. 4.73

Forward paths:
1. $F_1 = a - b - c - d - e - f,$ $G_1 G_2 G_3,$ $\Delta_1 = 1$
2. $F_2 = a - b - d - e - f,$ $G_4 G_3,$ $\Delta_2 = 1$

Loops:
1. $b - c - d - e - b,$ $- G_1 G_2 G_3 H_1$
2. $c - d - e - c,$ $- G_2 G_3 H_2$
3. $d - e - d,$ $- G_3 H_3$
4. $b - d - e - b,$ $- G_4 G_3 H_1$

There are no other loops. Therefore,

$$\Delta = 1 + G_3 H_3 + G_2 H_3 H_2 + G_3 G_4 H_1 + G_1 G_2 G_3 H_1$$

$$\frac{C(s)}{R(s)} = \frac{F_1 \Delta_1 + F_2 \Delta_2}{\Delta}$$

or

$$\frac{C(s)}{R(s)} = \frac{G_3 G_4 + G_1 G_2 G_3}{1 + G_3 H_3 + G_2 G_3 H_2 + G_3 G_4 H_1 + G_1 G_2 G_3 H_1} \qquad \textbf{Ans}$$

Problem 4.43 Draw the signal flow graph of the control system as shown in Fig. 4.74 and determine the transfer function by Mason's gain formula

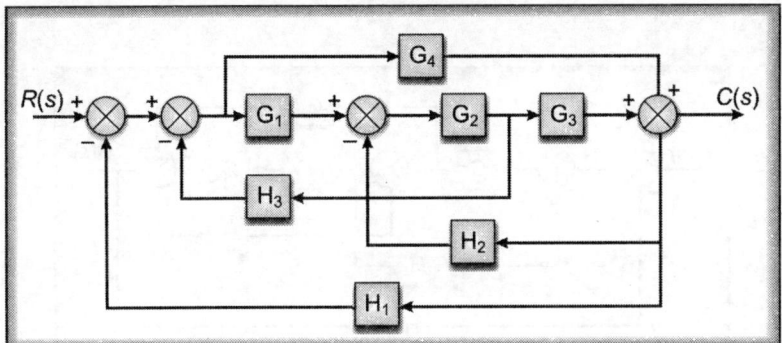

Fig. 4.74

Solution Signal flow graph of the block diagram can be drawn easily by inspection and as shown in Fig. 4.75

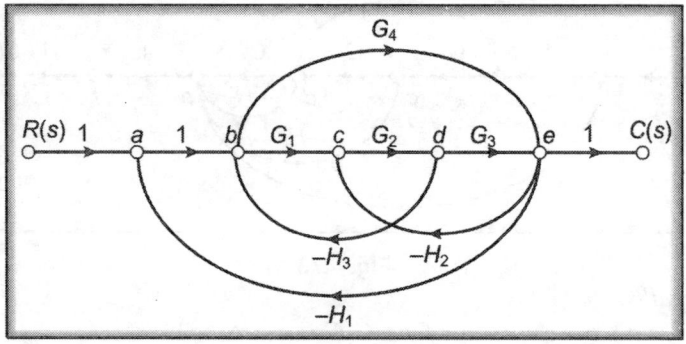

Fig. 4.75

There are two forward paths having gains $G_1G_2G_3$ and G_4 respectively.

Loops:

1. $b - c - d - e - a - b$ $G_1G_2G_3(-H_1) = -G_1G_2G_3H_1$
2. $b - c - d - b$ $G_1G_2(-H_3) = -G_1G_2H_3$
3. $c - d - e - c$ $G_2G_3(-H_2) = -G_2G_3H_2$
4. $b - e - a - b$ $G_4(-H_1) = -G_4H_1$
5. $c - d - b - e - c$ $G_2(-H_3)G_4(-H_2) = G_2G_4H_2H_3$

Non-touching loops: Nil

$$\therefore \quad \frac{C(s)}{R(s)} = \frac{G_1G_2G_3 + G_4}{1 + G_1G_2G_3H_1 + G_1G_2H_3 + G_2G_3H_2 + G_4H_1 - G_2G_4H_2H_3} \quad \textbf{Ans}$$

Block diagram reduction method

Taking the take off point of block G_4 after block G_1 and that of block H_3 before block G_2, we get.

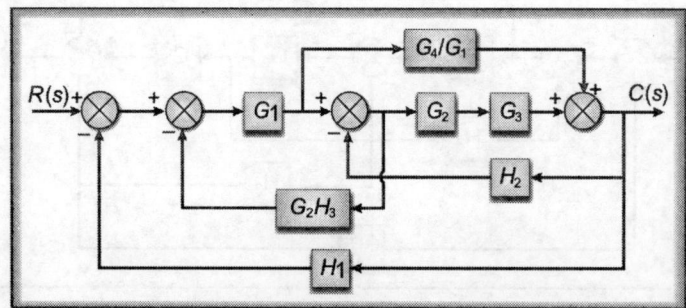

Fig. 4.76

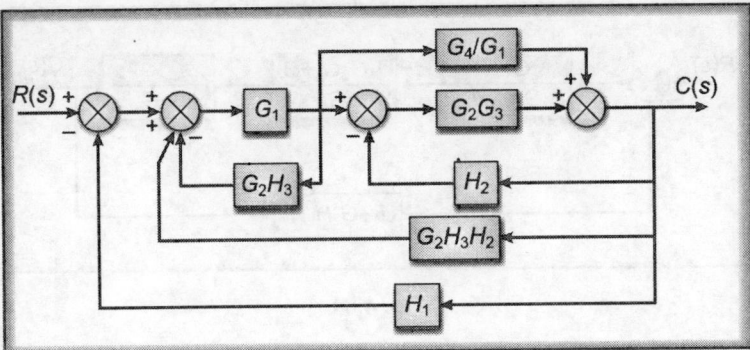

Fig. 4.77

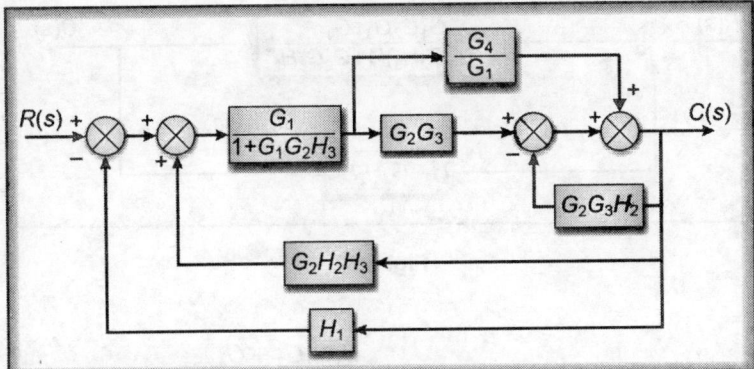

Fig. 4.78

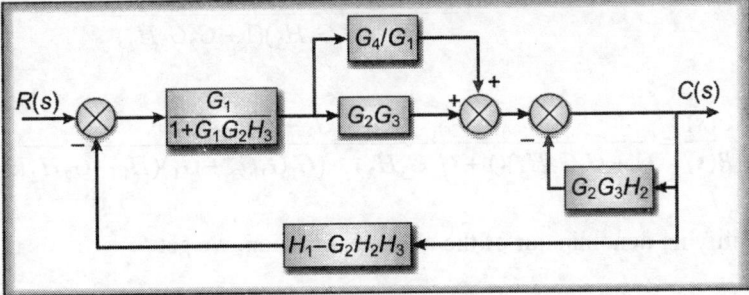

Fig. 4.79

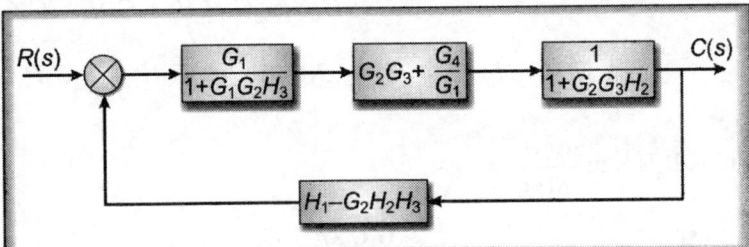

Fig. 4.80

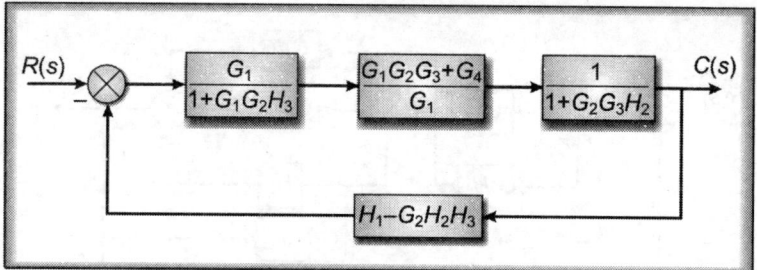

Fig. 4.81

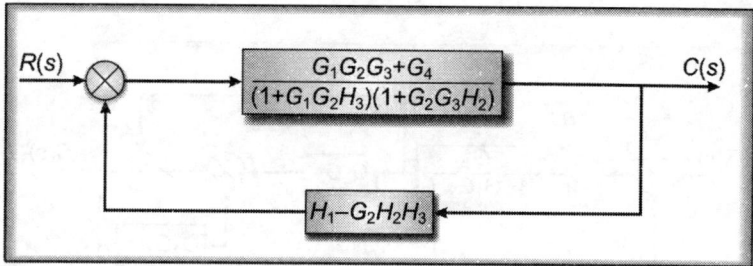

Fig. 4.82

$$\frac{C(s)}{R(s)} = \frac{G}{1 + GH} = \frac{\dfrac{G_1G_2G_3 + G_4}{(1 + G_1G_2H_3)(1 + G_2G_3H_2)}}{1 + \dfrac{(G_1G_2G_3 + G_4)(H_1 - G_2H_2H_3)}{(1 + G_1G_2H_3)(1 + G_2G_3H_2)}}$$

or $$\frac{C(s)}{R(s)} = \frac{G_1G_2G_3 + G_4}{(1 + G_1G_2H_3)(1 + G_2G_3H_2) + (G_1G_2G_3 + G_4)(H_1 - G_2H_2H_3)}$$

Simplifying denominator of the above expression, we get

$$\text{R.H.S.} = 1 + G_2G_3H_2 + G_1G_2H_3 + G_1G_2G_2G_3H_2H_3 + G_1G_2G_3H_1$$
$$- G_1G_2G_2G_3H_2H_3 + G_4H_1 - G_2G_4H_2H_3$$

$$= 1 + G_2G_3H_2 + G_1G_2H_3 + G_1G_2G_3H_1 + G_4H_1 - G_2G_4H_2H_3$$

Substituting R.H.S in $\dfrac{C(s)}{R(s)}$, we get,

$$\frac{C(s)}{R(s)} = \frac{G_1G_2G_3 + G_4}{1 + G_2G_3H_2 + G_1G_2H_3 + G_1G_2G_3H_1 + G_4H_1 - G_2G_4H_2H_3}$$

Problem 4.44 For the feedback control system as shown in Fig. 4.83

    (*a*) Find *C/R* using block diagram reduction method

    (*b*) Find *C/R* using Mason's gain formula

    (*c*) If $G_1 = 10$, $G_2 = 5$, $G_3 = 8$, $H_1 = 1$, $H_2 = 0.25$, $H_3 = 0.2$ and $R = 10.1$, find the input to the block G_2

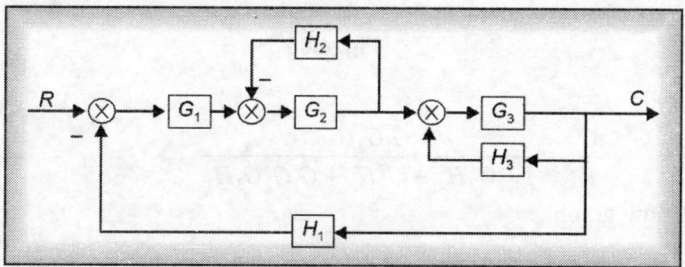

Fig. 4.83

Solution (*a*) Shifting block H_2 after/beyond block G_3 and eliminating feedback loop H_3 existing with forward path G_3 and shifting summing point number 2, from left before/ahead of block G_1 and then combining both the summing points, we get

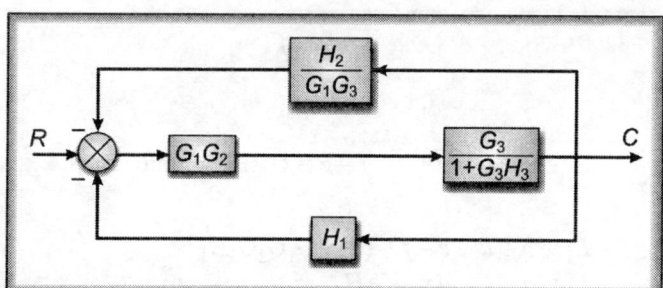

Fig. 4.84

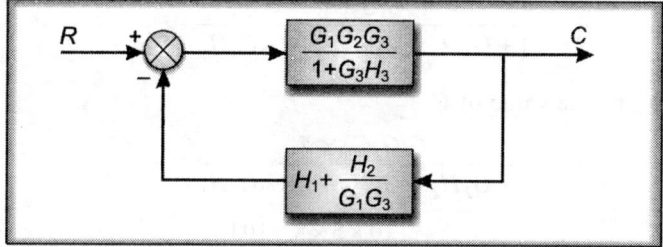

Fig. 4.85

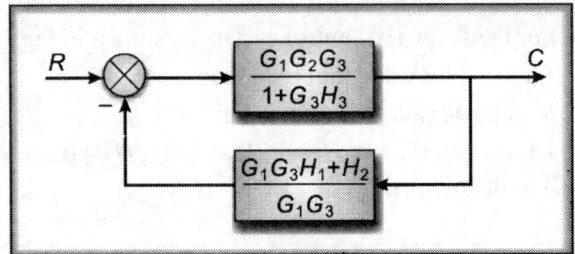

Fig. 4.86

Therefore,

$$\frac{C}{R} = \frac{G_1 G_2 G_3}{1 + G_2 H_2 + G_3 H_3 + G_1 G_2 G_3 H_1}$$ **Ans**

(b) Signal flow graph

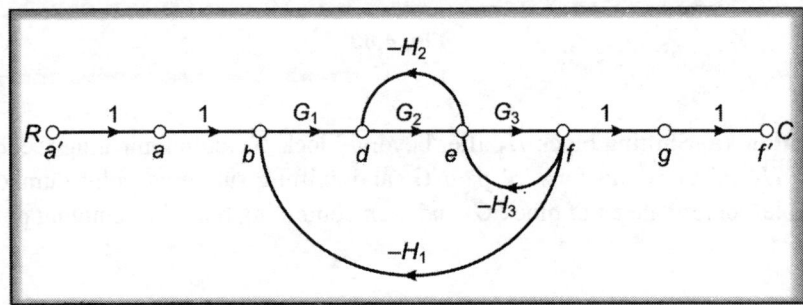

Fig. 4.87

Forward path:

$$P_1 = a - b - d - e - f - g = G_1 G_2 G_3 \quad \text{and} \quad \Delta_1 = 1$$

Loops:

$$L_1 = b - d - e - f - b = -G_1 G_2 G_3 H_1$$
$$L_2 = d - e - d = -G_2 H_2$$
$$L_3 = e - f - e = -G_3 H_3$$

Therefore,

$$\frac{C}{R} = \frac{G_1 G_2 G_3}{1 + G_2 H_2 + G_3 H_3 + G_1 G_2 G_3 H_1}$$ **Ans**

(c) Let us find the value of C

$$C = \frac{G_1 G_2 G_3 \times R}{1 + G_2 H_2 + G_3 H_3 + G_1 G_2 G_3 H_1}$$

$$= \frac{10 \times 5 \times 8 \times 10.1}{1 + (5 \times 0.25) + (8 \times 0.2) + (10 \times 5 \times 8 \times 1)}$$

$$= \frac{400 \times 10.1}{403.85} = 10.004$$

From signal flow graph

$$b = R \times 1 + C \times 1 \times (-H_1) = R - CH_1 \tag{1}$$

$$d = bG_1 + e(-H_2) = bG_1 - eH_2 \tag{2}$$

$$e = dG_2 + (C \times 1 \times -H_3) = dG_2 - CH_3 \tag{3}$$

Putting the values of b and e from Eqns (1) and (3) respectively in eqn. (2), we get,

$$d = bG_1 - eH_2$$

$$= (R - CH_1)G_1 - (dG_2 - CH_3) H_2$$

or $\qquad d = (R - CH_1)G_1 - dG_2H_2 + CH_2H_3$

or $\qquad d(1 + G_2H_2) = (R - CH_1)G_1 + CH_2H_3$

$$d = \frac{(R - CH_1)G_1 + CH_2H_3}{1 + G_2H_2}$$

Putting the values, we get

$$d = \frac{(10.1 - 10.004 \times 1)10 + 10.004 \times 0.25 \times 0.2}{1 + 5 \times 0.25} = 0.65$$

Therefore, input to the block $G_2 = 0.65$

Problem 4.45 Determine the transfer function of the block diagram as shown in Fig. 4.88 and verify the result by Mason's gain formula.

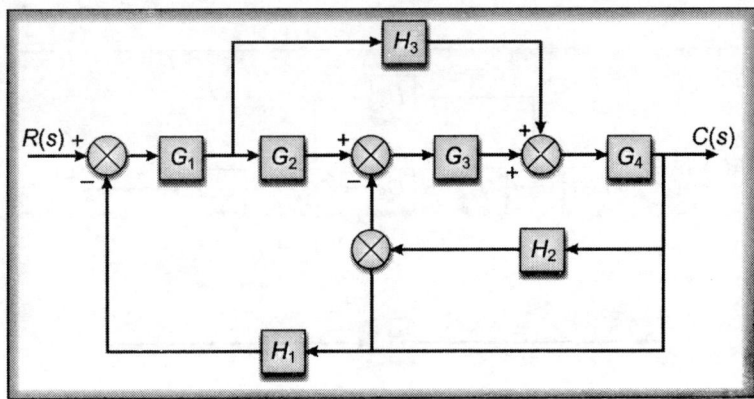

Fig. 4.88

Solution

Redrawing block diagram

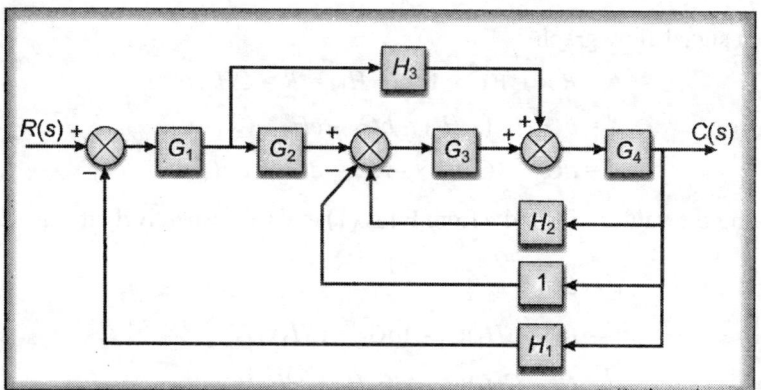

Fig. 4.89

Combining feedback H_2 and unity, we get

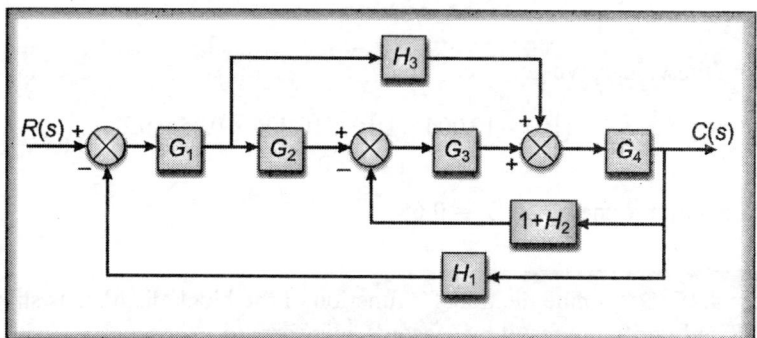

Fig. 4.90

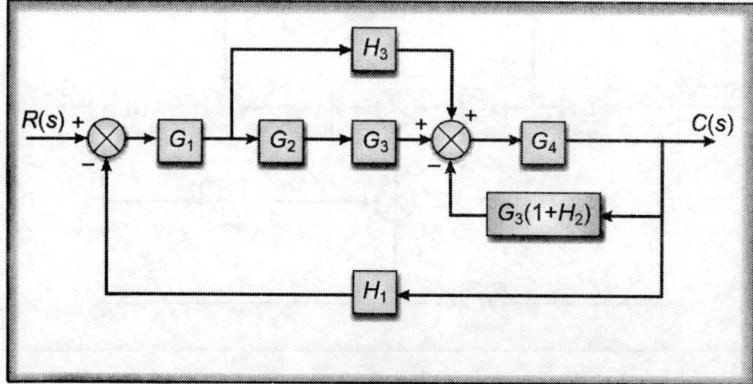

Fig. 4.91

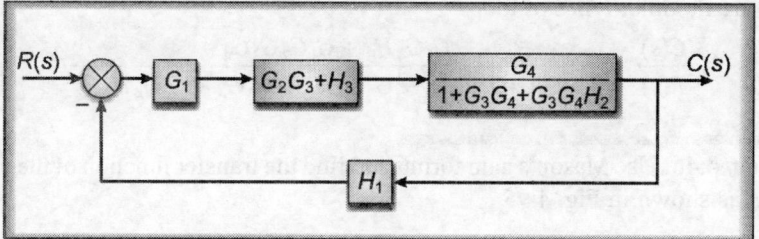

Fig. 4.92

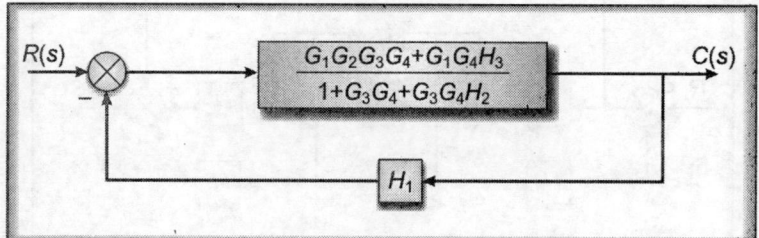

Fig. 4.93

$$\frac{C(s)}{R(s)} = \frac{G_1 G_2 G_3 G_4 + G_1 G_4 H_3}{1 + G_3 G_4 + G_3 G_4 H_2 + G_1 G_2 G_3 G_4 H_1 + G_1 G_4 H_1 H_3}$$ **Ans**

Signal flow graph:

Signal flow graph of the block diagram is given in Fig. 4.94

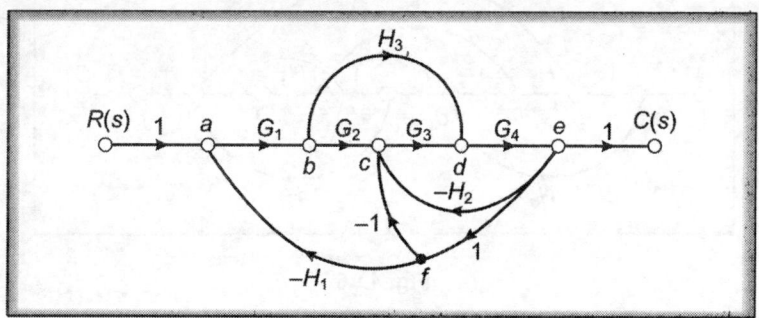

Fig. 4.94

Forward paths:

$$P_1 = a - b - c - d - e = G_1 G_2 G_3 G_4 \quad \text{and} \quad P_2 = a - b - d - e = G_1 G_4 H_3$$
$$\Delta_1 = 1 \qquad\qquad\qquad\qquad \text{and} \quad \Delta_2 = 1$$

Loops:

$$L_1 = a - b - c - d - e - a = -G_1 G_2 G_3 G_4 H_1; \quad L_2 = c - d - e - c = -G_3 G_4 H_2;$$
$$L_3 = c - d - e - f - c = -G_3 G_4; \quad \text{and} \quad L_4 = a - b - d - e - f - a = -G_1 H_3 G_4 H_1.$$

There are no other types of loops

$$\therefore \quad \frac{C(s)}{R(s)} = \frac{G_1 G_4 H_3 + G_1 G_2 G_3 G_4}{1 + G_1 G_2 G_3 G_4 H_1 + G_3 G_4 H_2 + G_3 G_4 + G_1 G_4 H_1 H_3} \qquad \textbf{Ans}$$

Problem 4.46 Use Mason's gain formula to find the transfer function of the control system as shown in Fig. 4.95

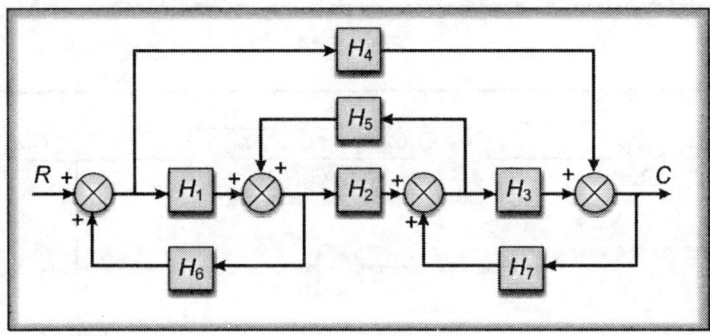

Fig. 4.95

Solution Signal flow graph is shown in Fig. 4.96

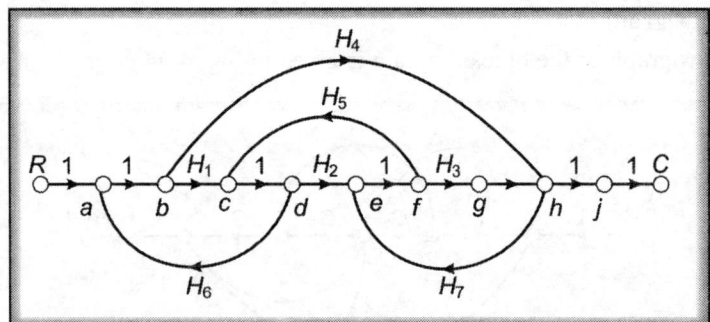

Fig. 4.96

Forward paths:

$$P_1 = H_1 H_2 H_3 = (a - b - c - d - e - f - g - h - j); \text{ and } \Delta_1 = 1$$
$$P_2 = H_4 = (a - b - h - j) \quad \text{and} \quad \Delta_2 = H_2 H_3 H_5$$

Note: Δ_2 is the path gain of the loop $c - d - e - f - g - c$, as it is not touching the forward path $a - b - h - j$ whose gain is H_4.

Loops:

$$L_1 = abcda = H_1 H_6; \qquad L_2 = efghe = H_3 H_7$$
$$L_3 = cdefgc = H_2 H_3 H_5; \quad L_4 = cdabhefgc = H_6 H_4 H_7 H_3 H_5 \text{ or } H_3 H_4 H_5 H_6 H_7$$

Non-touching loops:

$$abcda = H_1H_6 \quad \text{and} \quad efghe = H_3H_7$$

Their gain product is $H_1H_6H_3H_7$ or $H_1H_3H_6H_7$

$$\therefore \quad \frac{C}{R} = \frac{F_1\Delta_1 + F_2\Delta_2}{1 - (\text{sum of individaul loops}) + \left(\begin{array}{c}\text{sum of gain product of} \\ \text{non-touching loops}\end{array}\right)}$$

or

$$\frac{C}{R} = \frac{H_1H_2H_3 + H_4H_2H_3H_5}{1 - (H_1H_6 + H_3H_7 + H_2H_3H_5 + H_3H_4H_5H_6H_7) + (H_1H_3H_6H_7)} \quad \textbf{Ans}$$

Problem 4.47 Use the block diagram reduction method to obtain the overall transfer function of the control system is given in Fig. 4.97

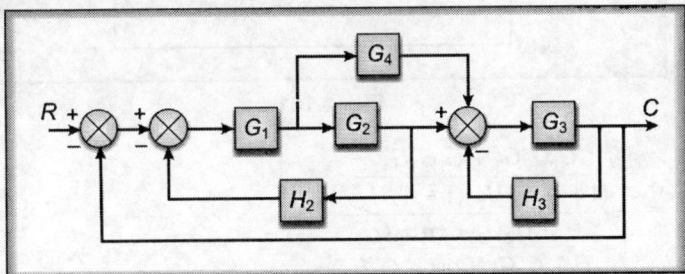

Fig. 4.97

Solution Shifting the take off point after the block G_2 to a point in between blocks G_1 and G_2 and eliminating feedback element H_3 by combining with block G_3, we get

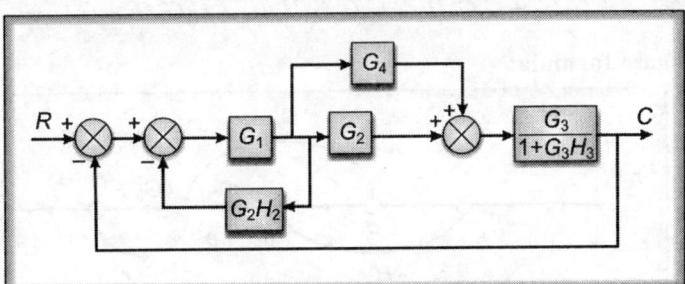

Fig. 4.98

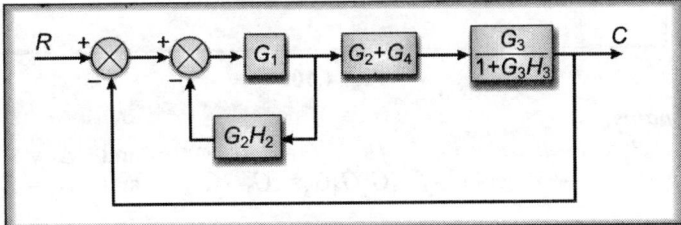

Fig. 4.99

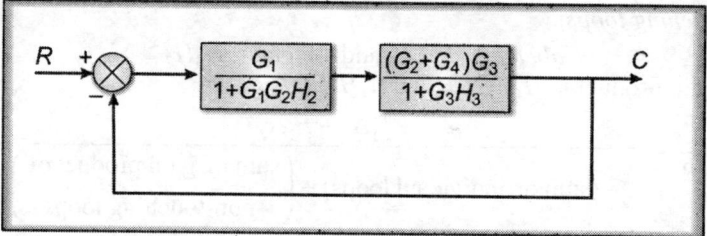

Fig. 4.100

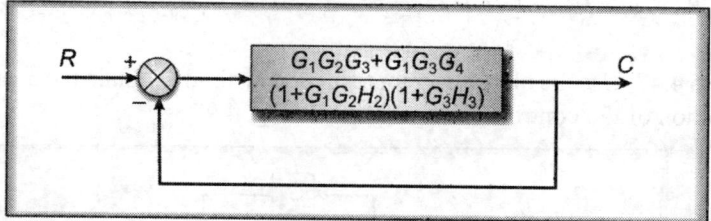

Fig. 4.101

$$\frac{C}{R} = \frac{\dfrac{G_1G_2G_3 + G_1G_3G_4}{(1 + G_1G_2H_2)(1 + G_3H_3)}}{1 + \dfrac{G_1G_2G_3 + G_1G_3G_4}{(1 + G_1G_2H_2)(1 + G_3H_3)}}$$

or $\quad \dfrac{C}{R} = \dfrac{G_1G_2G_3 + G_1G_3G_4}{(1 + G_1G_2H_2)(1 + G_3H_3) + (G_1G_2G_3 + G_1G_3G_4)}$

or $\quad \dfrac{C}{R} = \dfrac{G_1G_2G_3 + G_1G_3G_4}{1 + G_3H_3 + G_1G_2H_2 + G_1G_2G_3H_2H_3 + G_1G_2G_3 + G_1G_3G_4}$

Mason's gain formula:

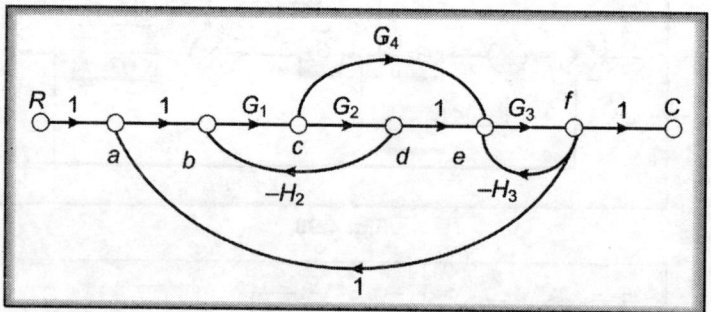

Fig. 4.102

Forward paths:

$$P_1 = a - b - c - d - e - f = G_1G_2G_3 \qquad \text{and} \quad \Delta_1 = 1$$
$$P_2 = a - b - c - e - f = G_1G_4G_3 = G_1G_3G_4 \qquad \text{and} \quad \Delta_2 = 1$$

Loops:

$$L_1 = a - b - c - d - e - f - a = -G_1G_2G_3$$

$$L_2 = b - c - d - b = -G_1G_2H_2$$
$$L_3 = e - f - e = -G_3H_3$$
$$L_4 = b - c - e - f - a - b = -G_1G_4G_3 = -G_1G_3G_4$$

Non-touching loops: $b - c - d - b$ and $e - f - e$ having gain product $(-G_1G_2H_2)$
$(-G_3H_3) = G_1G_2G_3H_2H_3$

$$\therefore \quad \frac{C}{R} = \frac{G_1G_2G_3 + G_1G_3G_4}{1 + (G_1G_2G_3 + G_1G_2H_2 + G_3H_3 + G_1G_3G_4) + G_1G_2G_3H_2H_3} \quad \textbf{Ans}$$

Problem 4.48 Reduce the block diagram is given in Fig. 4.103 and find overall transfer function and verify the result using Mason's gain formula.

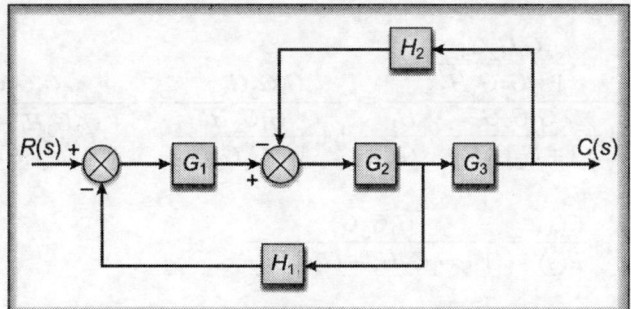

Fig. 4.103

Solution Shifting the take off point between blocks G_2 and G_3, after block G_3 and combining blocks G_2 and G_3 thereafter, being on the same forward path, we get

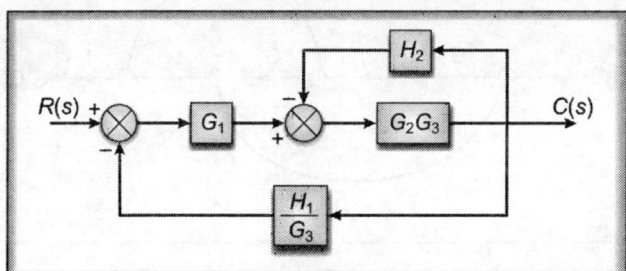

Fig. 4.104

Eliminating feedback element H_2 and combining with block G_2G_3, we get

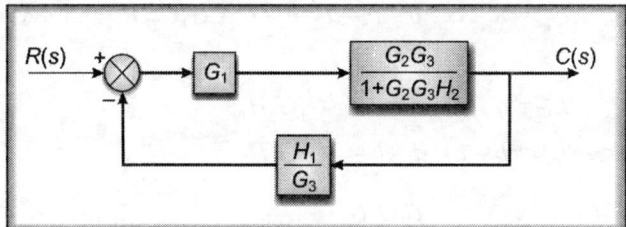

Fig. 4.105

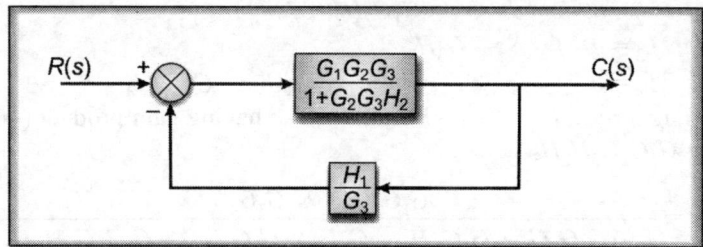

Fig. 4.106

Therefore,

$$\frac{C(s)}{R(s)} = \frac{\dfrac{G_1 G_2 G_3}{1 + G_2 G_3 H_2}}{1 + \dfrac{G_1 G_2 G_3}{1 + G_2 G_3 H_2} \times \dfrac{H_1}{G_3}} = \frac{G_1 G_2 G_3}{1 + \dfrac{G_1 G_2 H_1}{1 + G_2 G_3 H_2}} = \frac{G_1 G_2 G_3}{1 + G_2 G_3 H_2 + G_1 G_2 H_1}$$

or

$$\frac{C(s)}{R(s)} = \frac{G_1 G_2 G_3}{1 + G_1 G_2 H_1 + G_2 G_3 H_2}$$

Mason's gain formula:

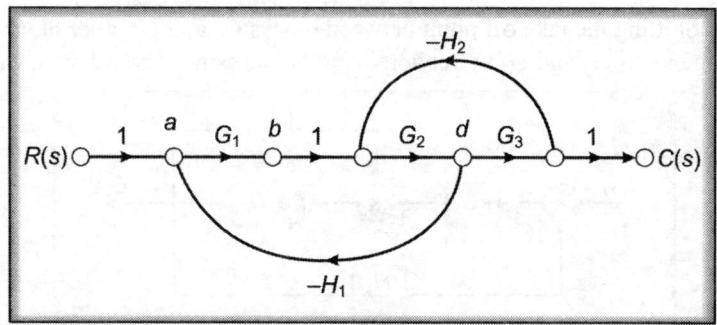

Fig. 4.107

Forward path (only one):

$$P_1 = R - a - b - c - d - e - C = G_1 G_2 G_3 \quad \text{and} \quad \Delta_1 = 1$$

Loops:

$$L_1 = a - b - c - a = -G_1 G_2 H_1$$
$$L_2 = b - c - d - b = -G_2 G_3 H_2$$

Therefore,

$$\frac{C(s)}{R(s)} = \frac{G_1 G_2 G_3}{1 + G_1 G_2 H_1 + G_2 G_3 H_2}$$

Ans

Problem 4.49 For the control system as shown in Fig. 4.108, draw the block diagram and evaluate $\theta_m(s)/E_f(s)$

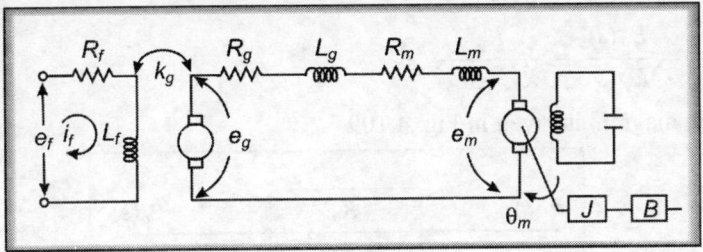

Fig. 4.108

Solution It is the configuration of a DC generator driving an armature controlled DC motor and is known as Ward-Leonard system. We will tackle this problem from basics and first develop transfer function and block diagram of a DC generator and then that of DC motor. Later on, we will combine both.

The equation relating the field voltage $e_f(t)$ and the field current $i_f(t)$ is:

$$e_f(t) = R_f i_f(t) + L_f \frac{d i_f(t)}{dt}$$

or $\qquad E_f(s) = I_f(s)(R_f + sL_f)$

or $\qquad \dfrac{I_f(s)}{E_f(s)} = \dfrac{1}{R_f + sL_f}$ \hfill (1)

The voltage induced in the armature, $e_g(t)$ is a function of the speed of rotation n and the flux developed by field ϕ. It can be expressed as:

$$e_g(t) = K_1 n\phi \hfill (2)$$

The flux ϕ depends upon the field current i_f and characteristics of the iron used in the field

$$\phi = K_2 i_f(t) \hfill (3)$$

Substituting the value of flux ϕ from eqn. (3) in eqn. (2), we get

$$e_g(t) = K_1 K_2 n i_f(t) \hfill (4)$$

or $\qquad e_g(t) = K_g i_f(t)$ \hfill (5)

or $\qquad E_g(s) = K_g I_f(s)$ \hfill (6)

where, $K_g = K_1 K_2 n$ is the generator constant having units of volt/amp.

or $\qquad I_f(s) = \dfrac{E_g(s)}{K_g}$ \hfill (7)

Substituting value of $I_f(s)$ in eqn. (1), we get

$$\frac{E_g(s)}{K_g E_f(s)} = \frac{1}{R_f + sL_f}$$

or $\qquad \frac{E_g(s)}{E_f(s)} = \frac{K_g}{R_f + sL_f}$ $\qquad\qquad$ (8)

The block diagram is given in Fig. 4.109

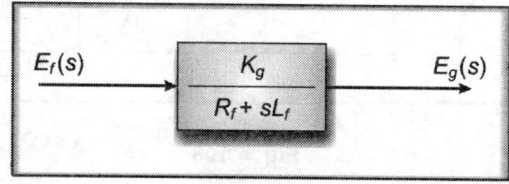

Fig. 4.109

Now, let us find out the transfer function of the motor which is connected to the generator. As the armature rotates, it develops an induced voltage $e_b(t)$ whose direction is opposite to $e_g(t)$. The induced voltage is proportional to the speed of rotation n and the flux created by the field current. Since, we are assuming that the field current is held constant, the flux must be constant. Therefore, the induced armature voltage is only dependent upon the speed of rotation and can be expressed as:

$$e_m(t) = K_m n = K_m \frac{d\theta_m(t)}{dt}$$ $\qquad\qquad$ (9)

$\therefore \qquad\qquad E_m(s) = sK_m \theta_m(s)$ $\qquad\qquad$ (10)

where, K_m is the voltage constant of the motor having unit volt/rad/sec.

or $\qquad \frac{E_m(s)}{\theta_m(s)} = sK_m$ $\qquad\qquad$ (11)

The block diagram representation is shown in Fig. 4.110

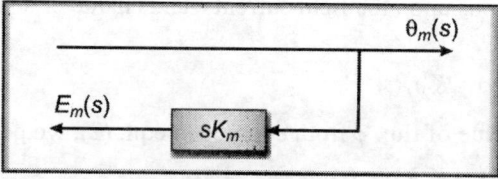

Fig. 4.110

The voltage equation of the armature circuit is:

$$e_g(t) = (R_g + R_m)(i_a(t)) + (L_g + L_m)\frac{di_a(t)}{dt} + e_m(t)$$ $\qquad\qquad$ (12)

Taking Laplace transfer of eqn. (12), we get

$$E_g(s) = (R_g + R_m)I_a(s) + (L_g + L_m)sI_a(s) + E_m(s)$$ $\qquad\qquad$ (13)

or $$\frac{I_a(s)}{E_g(s) - E_m(s)} = \frac{1}{(R_g + R_m) + s(L_g + L_m)}$$

or $$\frac{I_a(s)}{E_g(s) - E_m(s)} = \frac{1}{R + sL} \tag{14}$$

where, $R = R_g + R_m$, and

$L = L_g + L_m$

The block diagram representation of eqn. (14) is shown in Fig. 4.111

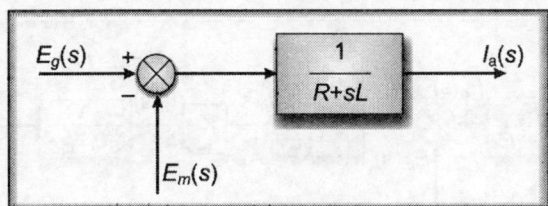

Fig. 4.111

The developed torque of the motor, T_L is a function of the flux developed by the field current, the armature current, and the length and number of conductors. Since, we are assuming that the field current is held constant, the expression for developed torque will be

$$T_L(t) = K_T i_a(t) \tag{15}$$

or $$\frac{T_L(s)}{I_a(s)} = K_T \tag{16}$$

The block diagram representation is shown in Fig. 4.112

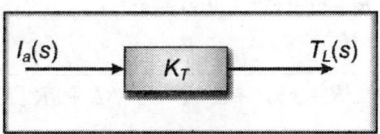

Fig. 4.112

And in terms of the mechanical load it drives, the expression for torque equation is

$$T_L(t) = \frac{J d^2 \theta_m(t)}{dt^2} + \frac{B d\theta_m(t)}{dt} \tag{17}$$

Taking Laplace transfer of equation (17), we get

$$T_L(s) = s^2 J \theta_m(s) + sB\theta_m(s)$$

or $$\frac{\theta_m(s)}{T_L(s)} = \frac{1}{s(sJ + B)} \tag{18}$$

The block diagram representation is shown in Fig. 4.113

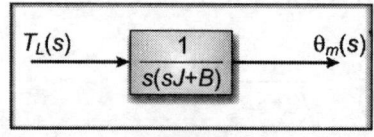

Fig. 4.113

Combining all the block diagrams, we get the complete block diagram as shown in Fig. 4.114

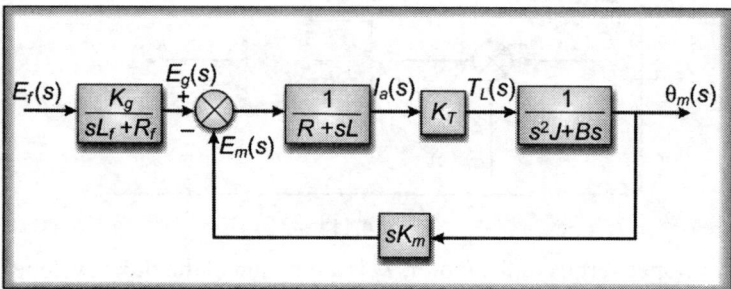

Fig. 4.114

$$\frac{\theta_m(s)}{E_g(s)} = \frac{\dfrac{K_T}{(R + sL)(s^2J + Bs)}}{1 + \dfrac{K_T}{(R + sL)(s^2J + Bs)} \times \dfrac{sK_m}{1}}$$

$$= \frac{K_T}{(R + sL)(s^2J + Bs) + sK_mK_T}$$

$$= \frac{K_T}{s^2JR + sBR + s^3JL + s^2BL + sK_mK_T}$$

$$= \frac{K_T}{s^3JL + s^2(JR + BL) + s(BR + K_mK_T)} \qquad (19)$$

Substituting

$$L = L_g + L_m$$

$$R = R_g + R_m$$

$$\gamma = \frac{B(R_g + R_m)}{K_mK_T} = \text{Damping factor}$$

$$T_a = \frac{L_g + L_m}{R_g + R_m} = \text{Armature time constant}$$

$$T_m = \frac{J(R_g + R_m)}{K_m K_T} = \text{Motor time constant}$$

we get,

$$\frac{\theta_m(s)}{E_g(s)} = \frac{\dfrac{1}{K_m}}{s[T_a T_m s^2 + (T_m + \gamma T_a)s + (\gamma + 1)]}$$

The block diagram of Fig. 4.114 reduces to as shown in Fig. 4.115

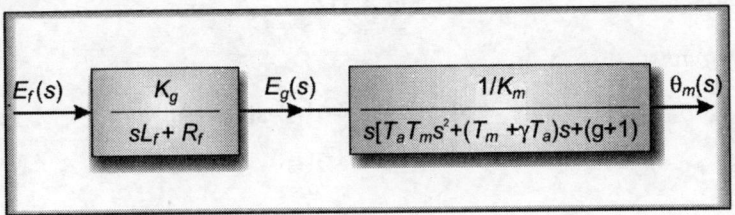

Fig. 4.115

Therefore, the transfer function

$$\frac{\theta_m(s)}{E_g(s)} = \frac{1}{sL_f + R_f}\left[\frac{K_g / K_m}{s[T_a T_m s^2 + (T_m + \gamma T_a)s + (\gamma + 1)]}\right]$$

Problem 4.50 Using Mason's gain formula, reduce the block diagram of Fig. 4.116 and determine $C(s)/R(s)$

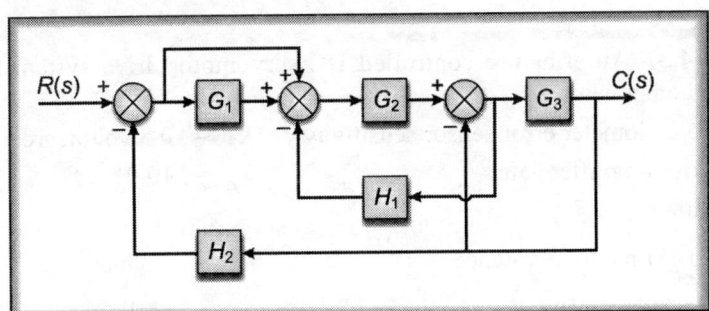

Fig. 4.116

Solution

Mason's gain formula: The signal flow graph is shown in Fig. 4.117. The forward paths and the loops with the gains are:

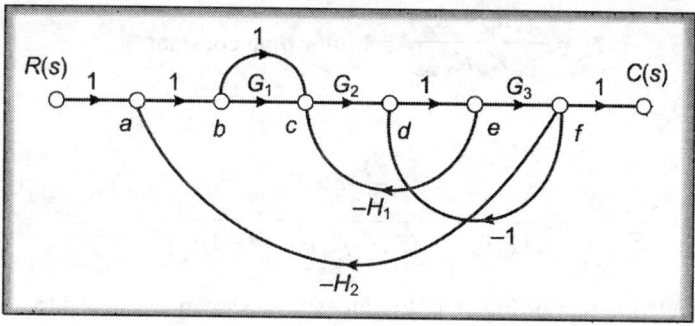

Fig. 4.117

Forward paths:

$$P_1 = a - b - c - d - e - f = G_1 G_2 G_3; \qquad \Delta_1 = 1$$
$$P_2 = a - b - c - d - e - f = G_2 G_3; \qquad \Delta_2 = 1$$

Loops:

$$L_1 = a - b - c - d - e - f - a = -G_1 G_2 G_3 H_2$$
$$L_2 = c - d - e - c = -G_2 H_1$$
$$L_3 = d - e - f - d = -G_3$$
$$L_4 = c - d - e - f - a - b - c = -G_2 G_3 H_2$$

Therefore, the transfer function is:

$$\frac{C(s)}{R(s)} = \frac{G_2 G_3 + G_1 G_2 G_3}{1 + G_3 + G_2 H_1 + G_1 G_2 G_3 H_2 + G_2 G_3 H_2} \qquad \textbf{Ans}$$

Problem 4.51 An armature controlled DC servomotor drive system has the following components

(a) Potentiometer error sensor sensitivity K_e = 0.5 volt/degree

(b) Error amplifier gain K_A = 10.0

(c) Motor

     (i) Armature resistance R_a = 5 ohms

     (ii) Inductance L_a = negligible

     (iii) Torque constant K_T = 1 Nm/A

     (iv) Inertia J_m = 2 kg/m^2

     (v) Friction D_m = negligible

(d) Gear ratio n = 1/10

(e) Load

    (i) Inertia J_L = 0.2 kgm^2

    (ii) Friction D_L = 0.2 Nm/(rad/sec)

Draw the schematic diagram of the system and find its overall transfer function.

Solution Schematic diagram is shown in Fig. 4.118

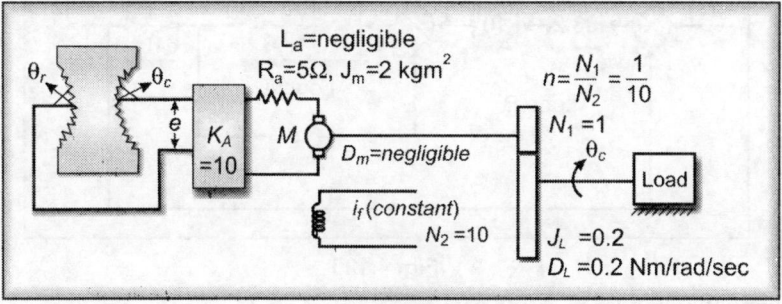

Fig. 4.118

The block diagram is shown in Fig. 4.119

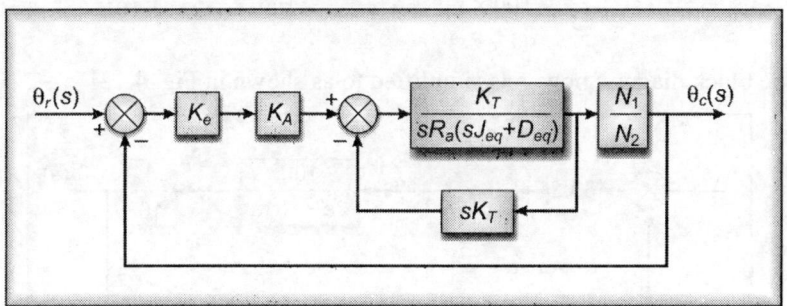

Fig. 4.119

Bringing all the given values to same units of measurements

$$K_e = 0.5 \text{ V/degree} = \frac{0.5 \times 180 \times 7}{22} = 33.2 \text{ V/rad}$$

$$J_m = 2 \text{ gmm}^2 = \frac{2}{1000} = 0.002 \text{ kgm}^2$$

Also, $$J_{eq} = J_m + J_L \left(\frac{N_1}{N_2}\right)^2 = 0.002 + \frac{0.2 \times 1}{10 \times 10} = 0.004 \text{ kgm}^2$$

$$D_{eq} = D_m + D_L \left(\frac{N_1}{N_2}\right)^2 = 0 + \frac{0.2 \times 1}{10 \times 10} = 0.002 \text{ Nm/rad/sec}$$

$$\frac{K_T}{sR_a(sJ_{eq} + D_{eq})} = \frac{1}{s \times 5(s \times 0.004 + 0.02)} = \frac{1000}{5s(4s + 2)} = \frac{100}{s(1 + 2s)}$$

$$sK_T = s \times 1 = s$$

Substituting all the values in the block diagram as shown in Fig. 4.119, we get block diagram as shown in Fig. 4.120

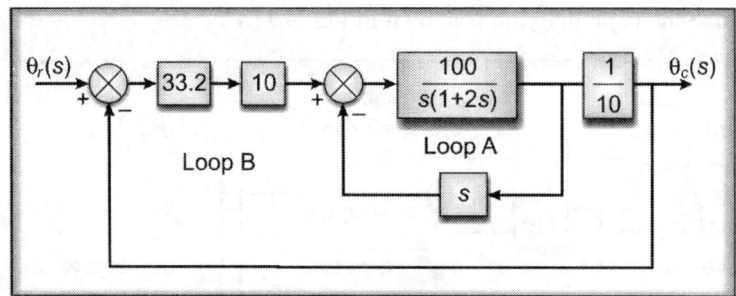

Fig. 4.120

Eliminating the Loop A, we get

$$\frac{\theta_c(s)}{\theta_r(s)} = \frac{\dfrac{100}{s(1 + 2s)}}{1 + \dfrac{100 \times s}{s(1 + 2s)}} = \frac{100}{s + 2s^2 + 100s} = \frac{100}{2s^2 + 101s}$$

The block diagram now gets simplified to as shown in Fig. 4.121

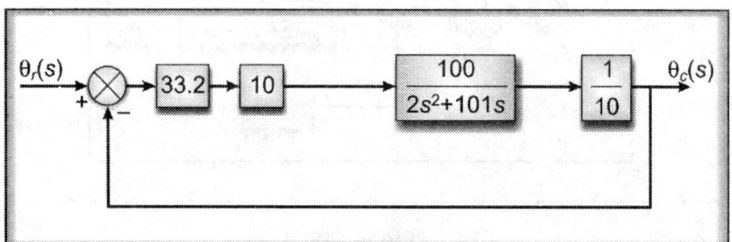

Fig. 4.121

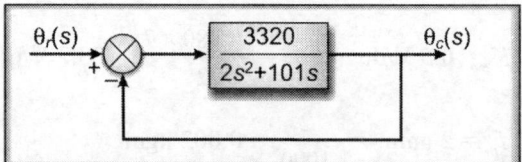

Fig. 4.122

Transfer function

$$\frac{\theta_c(s)}{\theta_r(s)} = \frac{\dfrac{3320}{2s^2 + 101s}}{1 + \dfrac{3320}{2s^2 + 101s}} = \frac{3320}{2s^2 + 101s + 3320} = \frac{1660}{s^2 + 50.5s + 1660}$$ **Ans**

Problem 4.52 Derive the expression for the transfer function of an AC servo motor and obtain the same in respect of a servo motor having following data

    (a) Starting torque   = 0.166 Nm
    (b) Moment of inertia, J = 1×10^{-5} kgm^2
    (c) Supply voltage   = 115 volts
    (d) No load speed   = 2904 rpm

Assume friction to be zero.

Solution

Construction. An AC servo motor is essentially a two-phase induction motor made up of two stator coils and a high resistance rotor. The two stator coils are called the control winding and reference winding. Control signal of variable voltage and polarity is applied to the control winding and the reference winding is supplied with a fixed signal that is phase-shifted by 90° relative to the control signal. Schematic diagram of such a construction is shown in Fig. 4.123

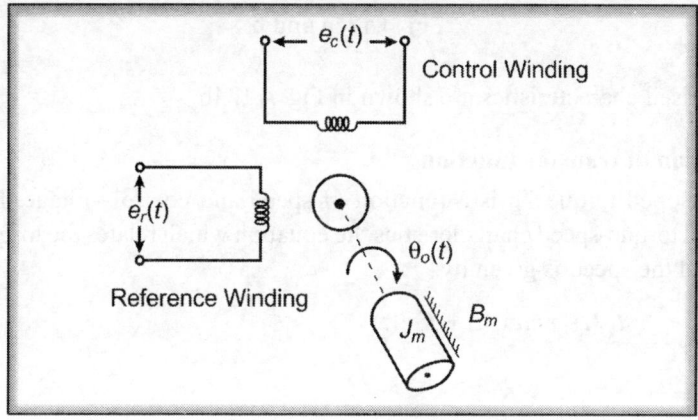

Fig. 4.123

Working

Let $e_r(t)$ = Reference field voltage

               J_m = Moment of inertia

               $e_c(t)$ = Control field voltage

               B_m = Friction

               T_D = Torque developed

               n = Speed

    As the control voltage $e_c(t)$ is varied, the torque T_D proportional to $e_r(t)$, $e_c(t)$ and the sine of the angle between them and speed n vary. Torque speed curves for varying values of control voltage are shown in Fig. 4.124(a). These curves are not straight lines. To develop a linear mathematical model to be represented by a set of linear differential equations, linearization of these curves is done.

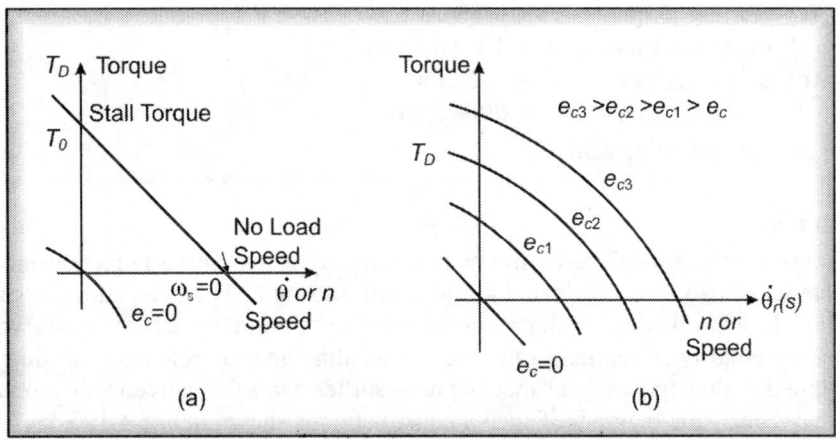

Fig. 4.124a and b

Linearised characteristics are shown in Fig. 4.124b

Calculation of transfer function

The developed torque T_D is a function of speed and control voltage. From the linearized torque-speed characteristics the equation which relates the torque of the motor and the speed is given by:

$$T_D(t) = m\omega_m(t) + Ke_c(t) \tag{1}$$

where,

(a) m = Slope of the linearized characteristics   (2)

$$= -\frac{T_0}{\omega_0} \text{ (slope being negative)}$$

(b) T_o = Stalling torque at speed equal to zero

$$= Ke_c(t) \tag{3}$$

Note: Stalling torque is proportional to $e_c(t)$

or $$K = \frac{T_0}{e_c(t)}, \text{ in Nm/volt} \tag{4}$$

(c) $$\omega_m = \frac{d\theta_m(t)}{dt} \tag{5}$$

Substituting the value in eqn. (1), we get

$$T_D(t) = \frac{md\theta_m(t)}{dt} + Ke_c(t) \tag{6}$$

In Laplace form

$$T_D(s) = ms\,\theta_m(s) + KE_c(s) \tag{7}$$

Writing torque developed in terms of moment of inertia and friction, we get

$$T_D(t) = \frac{J_m d^2\theta_m(t)}{dt^2} + \frac{B_m d\theta_m}{dt} \tag{8}$$

In Laplace form

$$T_D(s) = s^2 J_m \theta_m(s) + sB_m \theta_m(s) \tag{9}$$

Equating eqns (7) and (9), we get

$$ms\,\theta_m(s) + KE_c(s) = s^2 J_m \theta_m(s) + sB_m \theta_m(s) \tag{10}$$

Block diagram of eqn. (7) is given in Fig. 4.125

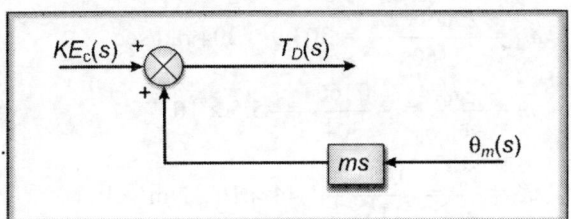

Fig. 4.125

Block diagram of eqn. (10) is given in Fig. 4.126

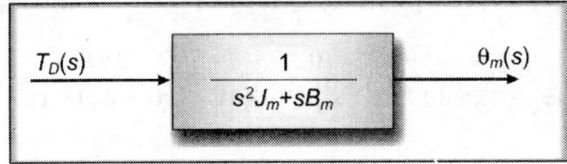

Fig. 4.126

$$KE_c(s) = \theta_m(s)(s^2 Jm + sB_m - ms)$$

Transfer function is

$$\frac{\theta_m(s)}{E_c(s)} = \frac{K}{s[sJ_m + (B_m - m)]} \tag{11}$$

Combining the two block diagrams, we get

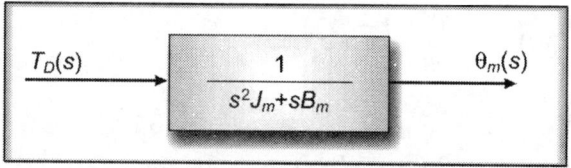

Fig. 4.127

Simplifying

$$\frac{\theta_m(s)}{KE_c(s)} = \frac{\dfrac{1}{s^2 J_m + sB_m}}{1 - \dfrac{1}{s^2 J_m + sB_m} \times sm}$$

or $\qquad \dfrac{\theta_m(s)}{KE_c(s)} = \dfrac{1}{s^2 J_m + sB_m - sm} = \dfrac{K}{s[sJ_m + (B_m - m)]}$ $\qquad\qquad$ (12)

Equations (11) and (12) are thus similar.

Solving the numerical

Converting no-load speed = 2904 rpm to rad/sec, we get

$$\omega_0 = \frac{2904 \times 2\pi}{60} = 303.9 = 304 \text{ rad/sec}$$

$$m = \frac{-T_0}{\omega_0} = -\frac{0.166}{304} = -5.5 \times 10^{-4}$$

$$K = \frac{T_0}{e_c} = \frac{0.166}{115} = 1.44 \times 10^{-3} \text{ Nm/volt}$$

Substituting the values in eqn. (12), we get

$$\frac{\theta_m(s)}{E_c(s)} = \frac{1.44 \times 10^{-3}}{s(s \times 1 \times 10^{-5} + 5.5 \times 10^{-4})}$$

$$= \frac{0.262 \times 10^{-3}}{s \times 10^{-4}(0.18 \times 10^{-1} s + 1)} = \frac{2.62}{s(1 + 0.018s)} \qquad \textbf{Ans}$$

Problem 4.53 Block diagram of the coupling between the signals of the turbo-prop engine is shown in Fig. 4.128. Draw the signal flow graph and find the following transfer functions:

(i) $\left.\dfrac{Y_1(s)}{R_1(s)}\right|_{R_2(s)=0}$ $\qquad$ (ii) $\left.\dfrac{Y_1(s)}{R_2(s)}\right|_{R_1(s)=0}$ $\qquad$ (iii) $\left.\dfrac{Y_2(s)}{R_1(s)}\right|_{R_2(s)=0}$ $\qquad$ (iv) $\left.\dfrac{Y_2(s)}{R_2(s)}\right|_{R_1(s)=0}$

Express the transfer function in matrix from $Y(s) = G(s)R(s)$

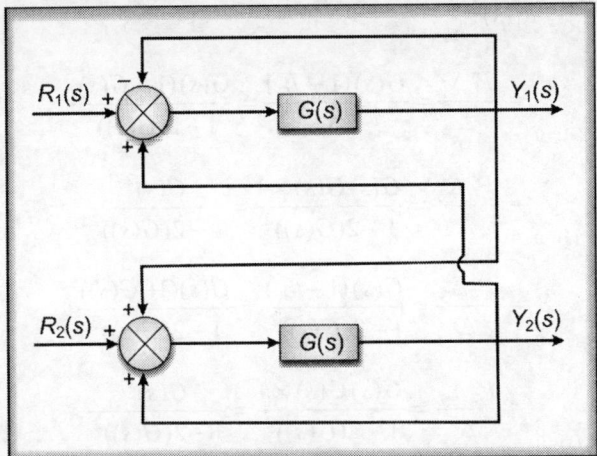

Fig. 4.128

Solution Equivalent signal flow graph is shown in Fig. 4.129

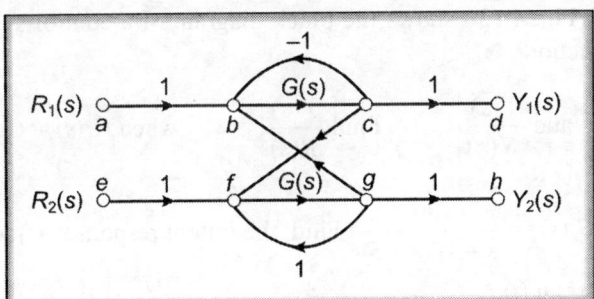

Fig. 4.129

Forward paths:

    $P_1 = abcd = G(s)$; $P_2 = abcfg = G(s)G(s)$; $P_3 = efgh = -G(s)$;
    $P_4 = efgbcd = G(s)G(s)$

Loops:

    $L_1 = efe = -G(s)$; $L_2 = efbce = G(s)G(s)$; $L_3 = bcb = -G(s)$;
    $L_4 = bcefb = G(s)G(s)$

Non-touching loops:

    efe and $bcd = -G(s)G(s)$
    $\Delta = 1 - (G(s) + G(s)G(s) - G(s) + G(s)G(s)) = 1 - 2(G(s))^2$

Transfer function $\dfrac{Y_1(s)}{R_1(s)}\Bigg|_{R_2(s)=0}$

$$\frac{Y_1(s)}{R_1(s)}\Bigg|_{R_2(s)=0} = \frac{P_1 \Delta_1}{\Delta} = \frac{G(s)(1-L_3)}{\Delta} = \frac{G(s)(1-G(s))}{1-2(G(s))^2}$$

$$\frac{Y_1(s)}{R_2(s)}\Bigg|_{R_1(s)=0} = \frac{P_2 \Delta_2}{\Delta} = \frac{G(s)G(s)\times 1}{1-2(G(s))^2} = \frac{G(s)^2}{1-2(G(s))^2}$$

$$\frac{Y_2(s)}{R_2(s)}\Bigg|_{R_1(s)=0} = \frac{P_3 \Delta_3}{\Delta} = \frac{G(s)(1-L_1)}{1-2(G(s))^2} = \frac{G(s)(1+G(s))}{1-2(G(s))^2}$$

$$\frac{Y_2(s)}{R_1(s)}\Bigg|_{R_2(s)=0} = \frac{P_4 \Delta_4}{\Delta} = \frac{G(s)G(s)\times 1}{1-2(G(s))^2} = \frac{G(s)^2}{1-2(G(s))^2}$$

Transfer function in matrix from

$$\frac{Y(s)}{R(s)} = \frac{1}{1-2(G(s))^2}\begin{bmatrix} G(s)[1-G(s)] & [G(s)]^2 \\ [G(s)]^2 & G(s)[1+G(s)] \end{bmatrix}$$

Problem 4.54 Fig. 4.130 shows the block diagram of a control system. Derive the transfer functions

(a) $\dfrac{Y(s)}{R(s)}\Bigg|_{N(s)=0}$ and $\dfrac{Y(s)}{N(s)}\Bigg|_{R(s)=0}$. Find $\dfrac{Y(s)}{R(s)}\Bigg|_{N(s)=0}$, when $G_c(s) = G_p(s)$

(b) If $G_p(s) = G_c(s) = \dfrac{100}{(s+1)(s+5)}$. Find the output response $y(t)$ when $N(s) = 0$

and $r(t) = u_s(t)$.

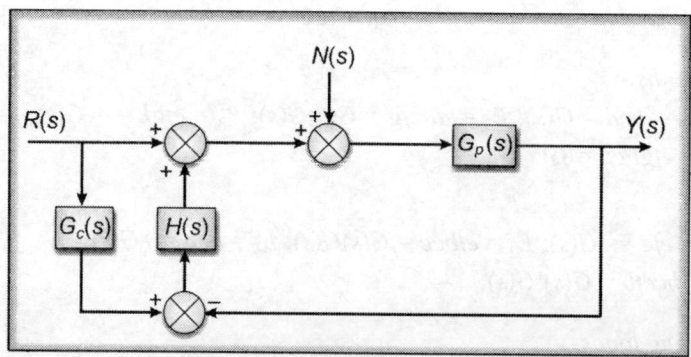

Fig. 4.130

Solution Equivalent signal flow graph is shown in Fig. 4.131

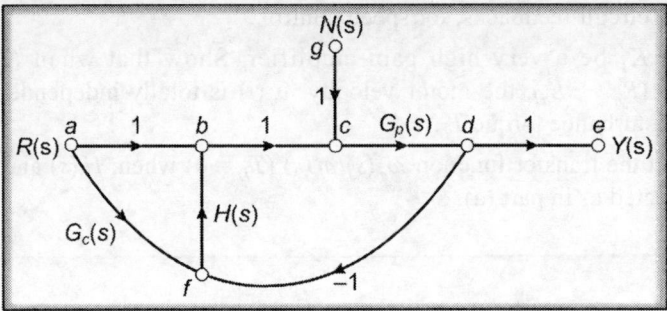

Fig. 4.131

Forward paths:
  $P_1 = abcde = G_p(s)$; $P_2 = afbcde = G_c(s)G_p(s) H(s)$; $P_3 = gcde = G_p(s)$
Loops:
$$L_1 = bcdfb = -G_p(s)H(s)$$
Non-touching loops: Nil
$$\Delta = 1 - (L_1) = 1 - (-G_p(s)H(s)) = 1 + G_p(s)H(s)$$

Transfer function:

(a) $\left.\dfrac{Y(s)}{R(s)}\right|_{N(s)=0} = \dfrac{P_1\Delta_1 + P_2\Delta_2}{\Delta} = \dfrac{G_p(s)\times 1 + G_c(s)G_p(s)H(s)}{1 + G_p(s)H(s)}$

$$= \dfrac{G_p(s)\times(1 + G_c(s)H(s))}{1 + G_p(s)H(s)}$$

$\left.\dfrac{Y(s)}{N(s)}\right|_{R(s)=0} = \dfrac{P_3\Delta_3}{\Delta} = \dfrac{G_p(s)\times 1}{1 + G_p(s)H(s)} = \dfrac{G_p(s)}{1 + G_p(s)H(s)}$

when $G_c(s) = G_p(s)\cdot\left.\dfrac{Y(s)}{R(s)}\right|_{N(s)=0} = \dfrac{G_p(s)[1 + G_p(s)H(s)]}{1 + G_p(s)H(s)} = G_p(s)$

(b) If $G_c(s) = G_p(s) = \dfrac{100}{(s+1)(s+5)}$

$\left.\dfrac{Y(s)}{R(s)}\right|_{N(s)=0} = G_p(s) = \dfrac{100}{(s+1)(s+5)}$

If $r(t) = u_s(t)$ then, $R(s) = \dfrac{1}{s}$

Therefore,

$$Y(s) = \dfrac{100}{s(s+1)(s+5)} = \dfrac{20}{s} - \dfrac{25}{s+1} + \dfrac{5}{s+5}$$

or $y(t) = (20 - 25e^{-t} + 5e^{-5t})\,u_s(t)$

Problem 4.55 The block diagram of Fig. 4.132 shows a DC-motor system, with voltage and circuit feedbacks, for speed control

(a) Let K_1 be a very high gain amplifier. Show that when $H_i(s)/H_c(s)$ $= - (R_a + sL_a)$, the motor velocity $\omega_m(t)$ is totally independent of load – disturbance torque T_L.

(b) Find the transfer function $\omega_m(s)/\omega_r(s)$ ($T_L = 0$) when; $H_i(s)$ and $H_e(s)$ are selected as in part (a).

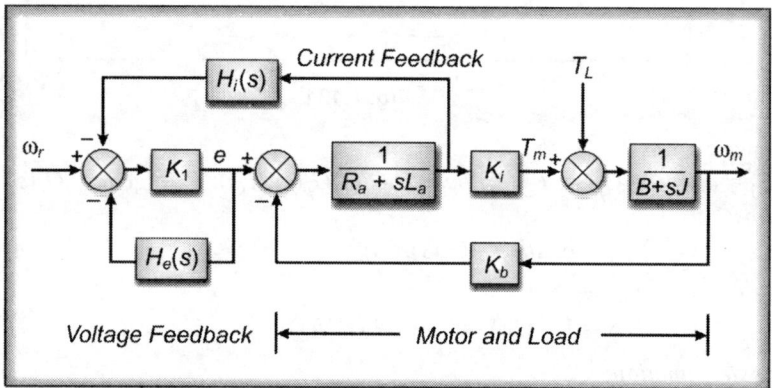

Fig. 4.132

Solution The equivalent signal flow graph is shown in Fig. 4.133

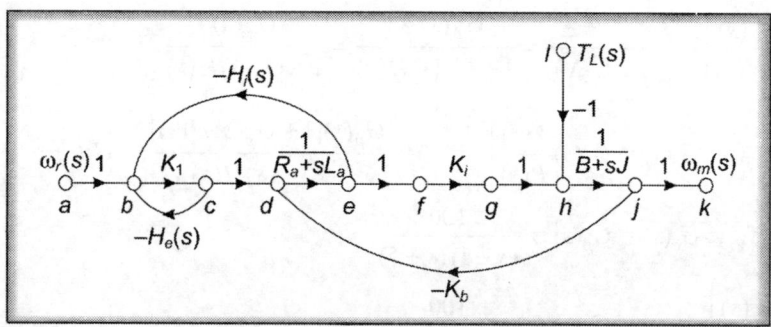

Fig. 4.133

Forward paths:

$$P_1 = lhjk = \frac{-1}{B + sJ}$$

$$P_2 = abcdefghjk = \frac{K_1 K_i}{(R_a + sL_a)(B + sJ)}$$

Loops:

$$L_1 = bcb = -K_1 H_e(s)$$

$$L_2 = bcdeb = \frac{-K_1 H_i(s)}{R_a + sL_a}$$

$$L_3 = defghjd = \frac{-K_b K_i}{(R_a + sL_a)(B + sJ)}$$

Non-touching loops:

(a) L_1 and $L_3 = bcb$ and $defghjb = \dfrac{K_1 K_i K_b H_e(s)}{(R_a + sL_a)(B + sJ)}$

$$\Delta = 1 - (L_1 + L_2 + L_3) + (L_1 L_3)$$

$$= 1 - \left[-K_1 H_e(s) - \frac{K_1 H_i(s)}{R_a + sL_a} - \frac{K_b K_i}{(R_a + sL_a)(B + sJ)} \right]$$

$$+ \left[\frac{-K_1 H_e(s) \times (-K_b K_i)}{(R_a + sL_a)(B + sJ)} \right]$$

$$= 1 + K_1 H_e(s) + \frac{K_1 H_i(s)}{R_a + sL_a} + \frac{K_b K_i}{(R_a + sL_a)(B + sJ)} + \frac{K_1 K_b K_i H_e(s)}{(R_a + sL_a)(B + sJ)}$$

$$\Delta_1 = 1 - (L_1 + L_2) = 1 - \left(-K_1 H_e(s) - \frac{K_1 H_i(s)}{R_a + sL_a} \right) = 1 + K_1 H_e(s) + \frac{K_1 H_i(s)}{R_a + sL_a}$$

$$\frac{\omega_m(s)}{T_L(s)} = \frac{P_1 \Delta_1}{\Delta} = \frac{-\dfrac{1}{B + sJ} \left(1 + K_1 H_e(s) + \dfrac{K_1 H_i(s)}{R_a + sL_a} \right)}{\Delta}$$

Since K_1 is a very high gain amplifier

$$\frac{\omega_m(s)}{T_L(s)} \simeq \frac{-\dfrac{K_1}{B + sJ} \left(H_e(s) + \dfrac{H_i(s)}{R_a + sL_a} \right)}{\Delta}$$

Equating $\dfrac{\omega_m(s)}{T_L(s)} = 0$, we get

$$H_e(s) + \frac{H_i(s)}{R_a + sL_a} = 0$$

or $\dfrac{H_i(s)}{H_e(s)} = -(R_a + sL_a)$

(b) $\dfrac{\omega_m(s)}{\omega_r(s)}\bigg|_{T_L=0} = \dfrac{P_2 \Delta_2}{\Delta}$

$$\Delta = 1 + K_1 H_e(s) + \frac{K_i K_b}{(R_a + sL_a)(B + sJ)} + \frac{K_1 H_i(s)}{R_a + sL_a} + \frac{K_1 K_i K_b H_e(s)}{(R_a + sL_a)(B + sJ)}$$

Putting $\dfrac{-H_i}{R_\infty + sL_a} = H_e(s)$, we get

$$\Delta = 1 + \frac{-K_iH_i(s)}{R_a + sL_a(B + sJ)} + \frac{K_iK_b}{(R_a + sL_a)} + \frac{K_iH_i(s)}{R_a + sL_a} + \frac{K_1K_iK_bH_e(s)}{(B + sL_a)(B + sJ)}$$

$$= 1 + \frac{K_iK_b}{(R_a + sL_a)(B + sJ)} + \frac{K_1K_iK_bH_e(s)}{(B + sL_a)(B + sJ)}$$

$$\frac{\omega_m(s)}{\omega_r(s)}\bigg|_{T_{L=0}} = \frac{\dfrac{K_1K_i}{(R_a + sL_a)(B + sJ)} \times 1}{1 + \dfrac{K_iK_b}{(R_a + sL_a)(B + sJ)} + \dfrac{K_1K_iK_bH_e(s)}{(R_a + sL_a)(B + sJ)}}$$

$$= \frac{K_1K_i}{(R_a + sL_a)(B + sJ) + K_iK_b + K_1K_iK_bH_e(s)}$$

$$\simeq \frac{K_1K_i}{K_1K_iK_bH_e(s)} \simeq \frac{1}{K_bH_e(s)}$$

Problem 4.56 The setup of the temperature control of an air-flow system is shown in Fig. 4.134. The hot water reservoir supplies the water that flows into the heat exchanger for heating the air, the temperature sensor senses the air temperature T_{AO} and sends it to be compared with the reference temperature T_r. The temperature error T_e is sent to the controller $u(t)$, which is an electrical signal, is converted to a pneumatic signal by a transducer. The output of the actuator controls the water flow rate through the three way value. Fig. 4.135 shows the block diagram of the system. The following parameters and variables are defined:

(a) dM_w is the flow rate of the heating fluid $= K_mu$

(b) $K_M = 0.054$ kg/sec/V

(c) T_w is the water temperature $= K_R dM_w$

(d) $K_R = 65°C/kg/sec$

(e) T_{AO} is the output air temperature

Heat-transfer equation between water and air

$$\tau_c \frac{dT_{AO}}{dt} = T_w - T_{AO}; \quad \tau_c = 10 \text{ seconds}$$

Temperature sensor equation

$$\tau_s \frac{dT_s}{dt} - T_{AO} - T_s; \quad \tau_s = 2 \text{ seconds}$$

(a) Draw the functional block diagram that includes all the transfer functions of the system

(b) Derive the transfer function $T_{AO}(s)/T_r(s)$, when $G_c(s) = 1$

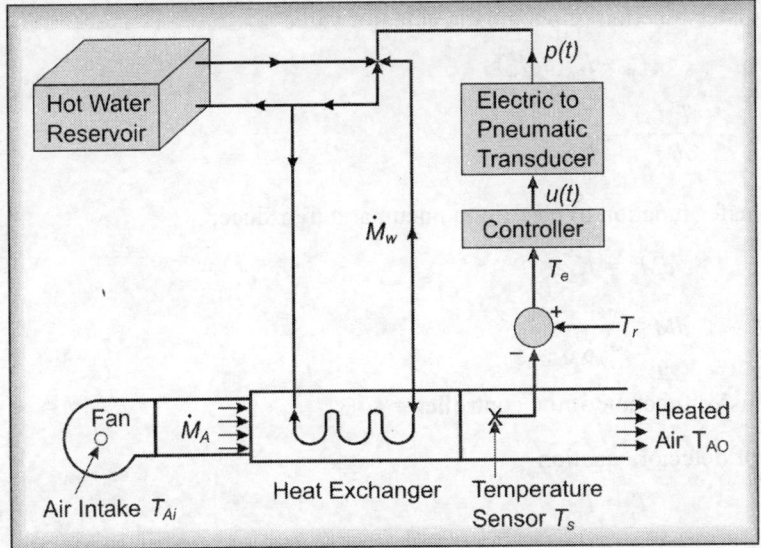

Fig. 4.134

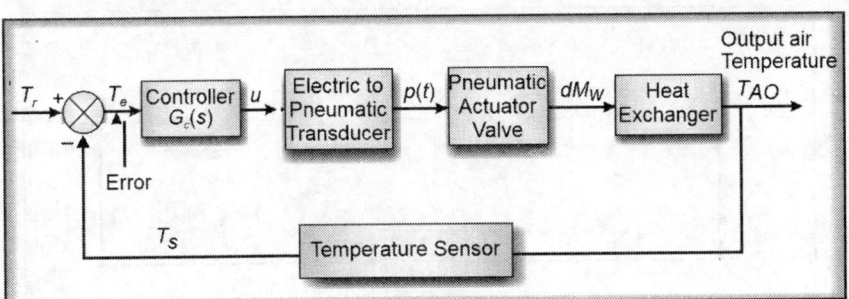

Fig. 4.135

Solution

Laplace transform of temperature sensor gives

$$\tau_s s T_s(s) = T_{AO}(s) - T_s(s)$$

or $\quad\dfrac{T_s(s)}{T_{AO}(s)} = \dfrac{1}{1 + s\tau_s}\quad$ (transfer function of temperature sensor)

Laplace transform of heat-transfer equation gives

$$\tau_c s T_{AO}(s) = T_w(s) - T_{AO}(s)$$

or $\quad\dfrac{T_{AO}(s)}{T_w(s)} = \dfrac{1}{1 + s\tau_c}\quad$ (transfer function of heat exchanges)

Transfer function of pneumatic actuator value

$$T_w = K_R(dM_w)$$

or $\qquad \dfrac{T_w(s)}{dM_w} = K_R$

Transfer function of electric to pneumatic transducer

$$dM_w = K_M u$$

or $\qquad \dfrac{dM_w}{u} = K_M$

Transfer function of the controller $= G_c(s)$

Error detector equation

$$T_e = T_r - T_s$$

(a) The functional block diagram (Fig. 4.136) is obtained by substituting the transfer functions and error detector equation in Fig. 4.136.

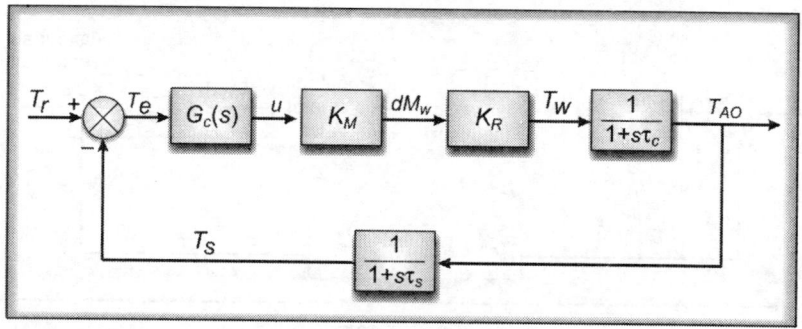

Fig. 4.136

(b) $\qquad \dfrac{T_{AO}(s)}{T_r(s)} = \dfrac{(G_c(s)K_M K_R)\dfrac{1}{1+s\tau_c}}{1+(G_c(s)K_M K_R)\dfrac{1}{(1+s\tau_c)(1+s\tau_s)}}$

$\qquad = \dfrac{(G_c(s)K_M K_R)(1+s\tau_s)}{(1+s\tau_c)(1+s\tau_s)(G_c(s)K_M K_R)}$

$\qquad = \dfrac{(1 \times 0.054 \times 65)(1+2s)}{(1+10s)(1+2s)+1 \times 0.054 \times 65}$

$\qquad = \dfrac{3.51(1+2s)}{20s^2 + 12s + 4.51}$

Problem 4.57 A linear analytical model of the automobile engine for the idle speed control system is shown in Fig. 4.137. The input of the system is the throttle position that controls the rate of airflow into the manifold (Fig. 4.138). Engine torque is developed from the manifold pressure due to air intake and the intake of the air-gas mixture into the cylinder. The engine variations are as follows:

$q_i(t)$ = amount of airflow across throttle into manifold

$dq_i(t)/dt$ = rate of airflow across throttle in any manifold

$q_m(t)$ = average air mass in manifold

$q_0(t)$ = amount of air leaving intake manifold through intake valves

$dq_0(t)/dt$ = rate of air leaving intake manifold through intake valves

$T(t)$ = engine torque

T_d = disturbance torque due to application of auto accessories = constant

$\omega(t)$ = engine speed

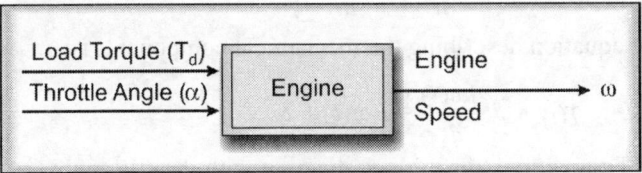

Fig. 4.137

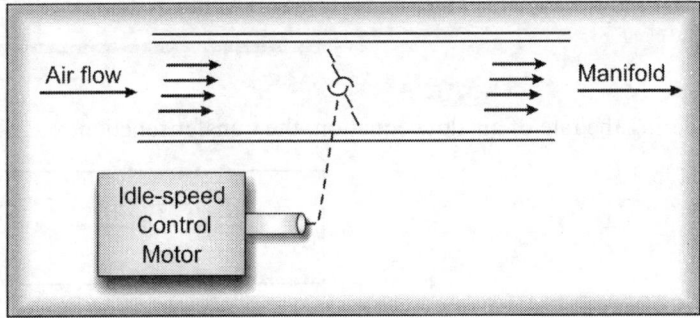

Fig. 4.138

$\alpha(t)$ = throttle position

τ_D = time delay in engine

J_e = inertia of engine

The following assumptions and mathematical relations between the engine variables are given:

1. The rate of airflow into the manifold is linearly dependent on the throttle position:

$$\frac{dq_i(t)}{dt} = K_1\alpha(t); \qquad K_1 = \text{proportional constant}$$

2. The rate of airflow leaving the manifold depends linearly on the air mass in the manifold and the engine speed:

$$\frac{dq_i(t)}{dt} = K_2 q_m(t) + K_3\omega(t); \qquad K_2, K_3 = \text{constants}$$

3. A pure time delay of τ_D seconds exist between the change in the manifold air mass and the engine torque:

$$T(t) = K_4 q_m(t - \tau_D); \qquad K_4 = \text{constant}$$

4. The engine drag is modeled by a viscous-friction torque where, B is the viscous-friction coefficient.

5. The average air mass $q_m(t)$ is determined from:

$$q_m(t) = \int\left[\frac{dq_i(t)}{dt} - \frac{dq_0(t)}{dt}\right]dt$$

6. The equation describing the mechanical components is:

$$T(t) = J\frac{d\omega(t)}{dt} + B\omega(t) + T_d$$

(a) Draw a functional block diagram with throttle angle (α) as the input and engine speed (ω) as the output and T_d as the disturbance input. Show the transfer function of each block.

(b) Find the transfer function ($\omega(s)/\alpha(s)$ of the system

---

Solution

Considering the rate of air flow equation, the transfer function is

$$\frac{\dot{q}_1}{\alpha} = K_1$$

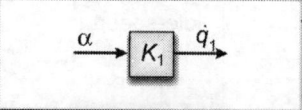

Fig. 4.139

Considering average air mass $q_m(t)$ relation, we get

$$\dot{q}_m(t) = \dot{q}_1(t) - \dot{q}_o(t)$$

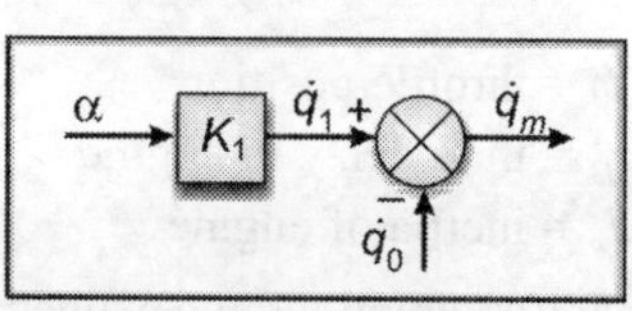

Fig. 4.140

Considering rate of air flow relation

$$\dot{q}_o(t) - K_3 \omega(t) = K_2 q_m(t)$$

Adding to Fig. 4.140 indicates the functional diagram as shown in Fig. 4.141

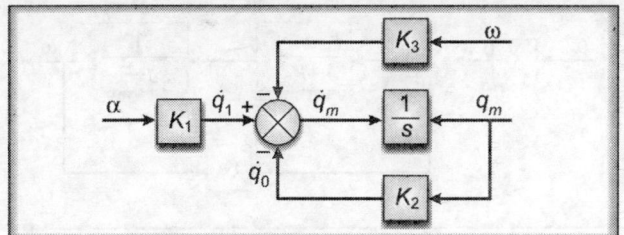

Fig. 4.141

Considering pure time delay of τ_o seconds

$$\frac{T}{q_m} = K_4 e^{-\tau_D s}$$

Adding to Fig. 4.141, the functional diagram is extended as shown in Fig. 4.142

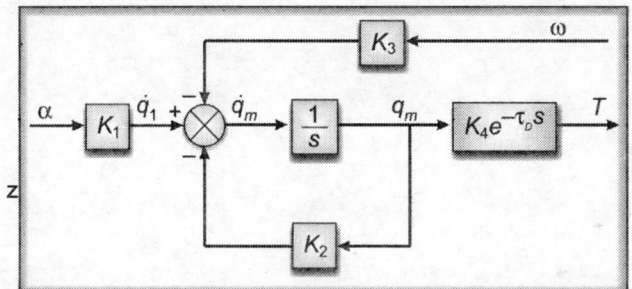

Fig. 4.142

Considering the equation describing mechanical components

$$T - B\omega - T_d = J\omega$$

or $$\frac{\dot{\omega}}{T - B\omega - T_d} = \frac{1}{J}$$

Adding to Fig. 4.142, the complete functional diagram is shown in Fig. 4.143

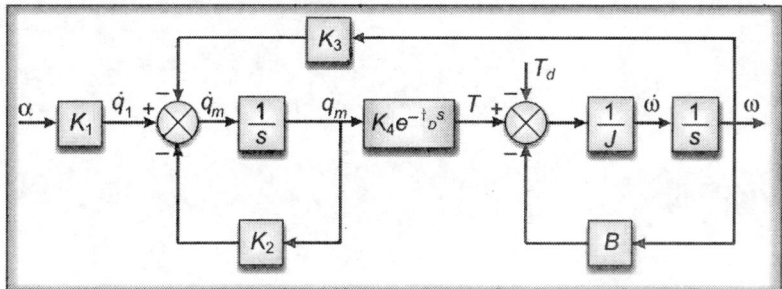

Fig. 4.143

Reducing the block diagram gives

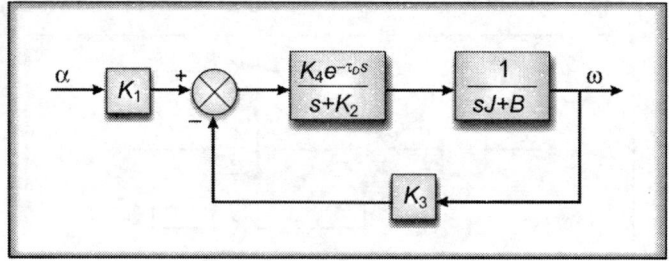

Fig. 4.144

Therefore the transfer function

$$\left.\frac{\omega(s)}{\alpha(s)}\right|_{\tau_D = 0} = \frac{K_1 K_4 e^{-\tau_D s}}{Js^2 + (JK_2 + B)s + K_2 B + K_3 K_4 e^{-\tau_0 s}}$$

Modelling of Control Systems: Physical Systems

5.1 Introduction

In the preceding chapters, we learnt the definition and computation of transfer function of electrical systems. In addition, we also learnt; how a control system is represented with the help of a block-diagram and signal-flow graph. In this chapter, analysis of mechanical systems have been explained.

5.2 Basic Elements

Most feedback control systems consist of electrical as well as mechanical equipment. Analysis of mechanical system makes use of three idealised elements, namely the *mass*, the *spring* and the *damper*. Equations for the mechanical systems are formulated from Newton's law of motion. The motion of mechanical elements can be *translatory*, *rotational* or combination of both.

5.3 Translatory Motion

The motion along a straight line is called the **translatory motion**. The variables which describe the translatory motion of mechanical systems are *velocity*, *acceleration* and *displacement*. The elements involved in the translatory motion are:

(a) Mass: Mass is the element symbolising inertia. Application of the force to the mass produces acceleration. Reaction force developed is represented in terms of the variables is shown below:

$$f_m(t) = ma = \frac{mdv}{dt} = M\frac{d^2x}{dt^2} = M\ddot{x} \tag{5.1}$$

where $\quad f_m(t)$ = force

$\qquad M$ = mass

$\qquad a$ = acceleration

$\qquad v$ = velocity and x = displacement.

Mass is represented with the help of two terminals as shown in Fig. 5.1. One of the terminals 'a' has the motion of mass and the other terminal 'b' is associated with the reference; with respect to which the variable of motion of terminal 'a' is measured. The reaction force $f_m(t)$ is a function of time and acts through mass M.

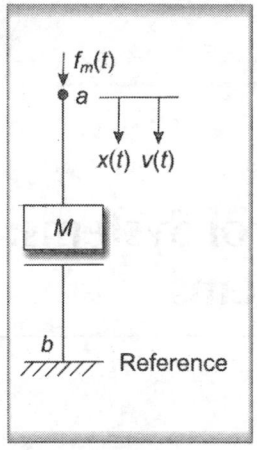

Fig. 5.1

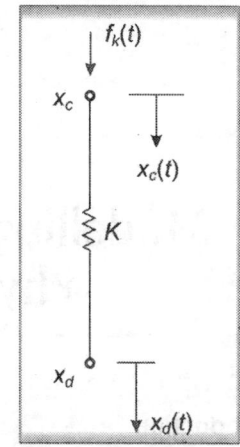

Fig. 5.2

(b) Spring: Spring is connected with the concept of elastic deformation of a body. If the spring is compressed, it tries to expand to its normal length and if stretched, the spring tries to contract. The reaction force $f_k(t)$ developed due compression or elongation of the spring is equal to the product of stiffness 'K' and the amount of deformation of the spring. The representation of the spring is shown in Fig. 5.2. If terminal 'c' has a position 'x_c' and d has 'x_d' measured from equilibrium position, then in accordance with Hooke's law

$$f_k(t) = K(x_c - x_d), \tag{5.2}$$

and if end d is stationary then

$$f_k(t) = Kx_c \tag{5.3}$$

(c) Friction. Friction is commonly experienced in mechanical systems. Generally, viscous friction predominates in comparison to coulomb friction force. It is also called *damping friction*. Presence of friction may not always be undesirable. Sometimes, viscous friction in the mechanical systems is intentionally introduced to improve the dynamic response of the system. Damping may be present unintentionally due to physical construction. The intentional incorporation in the system is done with the help of dashpot as shown in Fig. 5.3. It consists of a cylinder filled with oil and a piston which moves in the oil filled cylinder. Frictional force 'fv' is developed due to the movement of piston in the cylinder and is proportional to the difference in velocities of the two bodies. The network representation is shown in Fig. 5.4.

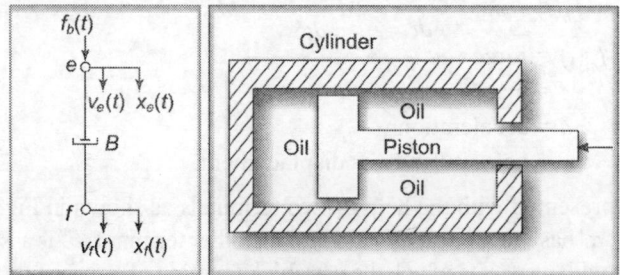

Fig. 5.3

$$f_b(t) = B(v_e(t) - v_f(t)) = fv$$

$$= B(\dot{x}_e(t) - \dot{x}_f(t)) = f\dot{x} \qquad (5.4)$$

5.4 Rotational Motion

The *movement of a body around its fixed axis is called the rotational motion*. Basic elements of rotational motion are *mass* (J), *spring stiffness* (K) and *rotational friction* (f). Variables of interest are the *torque* (T), *angular velocity* (θ) and *angular displacement* (θ). Network representation is shown in Fig. 5.4

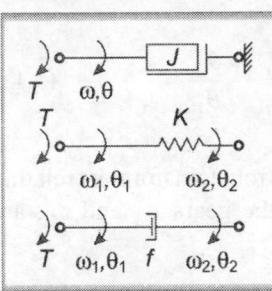

Fig 5.4

$$T = J\frac{d\omega}{dt} = J\frac{d^2\theta}{dt^2} = J\ddot{\theta} \qquad (5.5)$$

$$T = K(\theta_1 - \theta_2) \qquad (5.6)$$

$$T = f(\omega_1 - \omega_2) \qquad (5.7)$$

$$= f(\dot{\theta}_1 - \dot{\theta}_2) \qquad (5.8)$$

5.4.1 Gear Trains

Rotational systems have gear train as a component to transmit power from motor to load. The reduction in speed and the magnification of torque of the load driven by a high speed and low torque motor is achieved with the help of gear train. Figure 5.5 shows a motor driving a load through a geared system. Let T_1, θ_1 and N_1 and T_2, θ_2 and N_2 be the torque, angular displacement and number of teeth on the motor and load side, respectively. J_1 and f_1 are the moment of inertia of motor and gear 1 and J_2 and f_2 are the similar variables for gear 2 and load respectively. Assuming ideal gear conditions with no power loss, work done on the motor and load side is same.

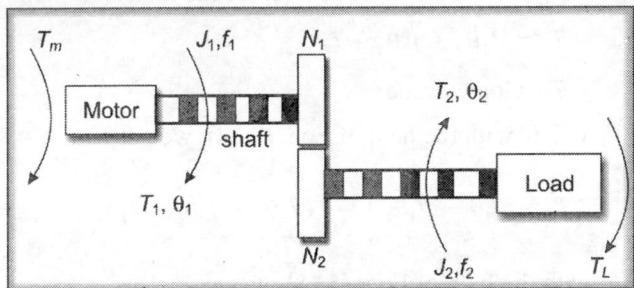

Fig. 5.5

$$T_1\theta_1 = T_2\theta_2 \qquad (5.9)$$

$$\frac{T_1}{T_2} = \frac{\theta_2}{\theta_1} \qquad (5.10)$$

Let r_1 and r_2 are the radius of gears 1 and 2 respectively. Assuming no slip taking place between the gear teeth, i.e. the gears are perfectly meshing, then the linear distance travelled by both the wheels is same.

Mathematically

$$r_1\theta_1 = r_2\theta_2 \quad \text{or} \quad \frac{r_1}{r_2} = \frac{\theta_2}{\theta_1} \qquad (5.11)$$

Also, the number of teeth on the gear wheel is proportional to the radius, i.e.

$$\frac{N_1}{N_2} = \frac{r_1}{r_2} \tag{5.12}$$

From Eqns (5.10), (5.11) and (5.12), we get

$$\frac{T_1}{T_2} = \frac{\theta_2}{\theta_1} = \frac{r_1}{r_2} = \frac{N_1}{N_2}, \quad \text{or} \quad \frac{T_1}{T_2} = \frac{N_1}{N_2} = \frac{\theta_2}{\theta_1} \tag{5.13}$$

The above relation is similar to the relation obtained from transformers relating voltages and currents. If angular velocities of both the gears ω_1 and ω_2, are considered, then

$$\frac{T_1}{T_2} = \frac{\theta_2}{\theta_1} = \frac{\omega_2}{\omega_1} = \frac{r_1}{r_2} = \frac{N_1}{N_2} \tag{5.14}$$

Equation (5.13), after differentiating twice gives

$$\frac{\ddot{\theta}_2}{\ddot{\theta}_1} = \frac{\dot{\theta}_2}{\dot{\theta}_1} = \frac{N_1}{N_2} \tag{5.15}$$

Thus, if $N_2 > N_1$, the gear train will reduce the speed and increase the torque. The differential equation of the motor side shaft is:

$$T_m = J_1\ddot{\theta}_1 + f_1\dot{\theta}_1 + T_1, \tag{5.16}$$

where, T_m = Torque developed by motor

Similarly, writing the differential equation for the load side shaft

$$T_2 = J_2\ddot{\theta}_2 + f_2\dot{\theta}_2 + T_L, \tag{5.17}$$

where, T_L = load torque.

Eliminating T_1 and T_2 with the help of Eqn. (5.14), we get

$$T_m = J_1\ddot{\theta}_1 + f_1\dot{\theta}_1 + \frac{N_1}{N_2}(J_2\ddot{\theta}_2 + f_2\dot{\theta}_2 + T_L) \tag{5.18}$$

Eliminating $\dot{\theta}_2$ with the help of Eqn. (5.18), we get

$$T_m = \left[J_1 + J_2\left(\frac{N_1}{N_2}\right)^2 \right]\ddot{\theta}_1 + \left[f_1 + f_2\left(\frac{N_1}{N_2}\right)^2 \right]\dot{\theta}_1 + T_L\left(\frac{N_1}{N_2}\right) \tag{5.19}$$

or we can write

$$T_m = J_{1eq}\ddot{\theta}_1 + f_{1eq}\dot{\theta}_1 + T_L\left(\frac{N_1}{N_2}\right) \tag{5.20}$$

where, $J_{1eq} = J_1 + J_2\left(\frac{N_1}{N_2}\right)^2 \tag{5.21}$

$$f_{1eq} = f_1 + f_2 \left(\frac{N_1}{N_2}\right)^2 \qquad (5.22)$$

and $\qquad T_L\left(\dfrac{N_1}{N_2}\right)$ = load torque referred to the motor side $\qquad (5.23)$

Torque equation referred to the load side is written as:

$$T_m\left(\frac{N_2}{N_1}\right) = J_{2eq}\ddot{\theta}_2 + f_{2eq}\dot{\theta}_2 + T_L \qquad (5.24)$$

where, $\qquad T_m\left(\dfrac{N_2}{N_1}\right)$ = motor torque referred to load side $\qquad (5.25)$

$$J_{2eq} = J_2 + J_1\left(\frac{N_2}{N_1}\right)^2 \qquad (5.26)$$

and $\qquad f_{2eq} = f_2 + f_1\left(\dfrac{N_2}{N_1}\right)^2 \qquad (5.27)$

5.5 Analogous System

As discussed earlier, the first step towards analysis of a control system is formulation of its mathematical model. The mathematical model is formed by application of one or more fundamental laws which relate to the physical nature of the system or its elements, e.g. *Kirchoff's laws*, and *Ohm's law* is used for electrical networks, mechanical systems are analysed by application of *Newton Law* and the *d'Alembert* principle, etc. The use of these basic laws help us in the formulation of differential equations. It has been seen that two different physical systems can be described by the same mathematical models, thereby originating the idea of *analogous systems*. The concept of analogous system is very useful in the study of complex systems like electrical, mechanical, hydraulic, etc. A non-electrical system is studied in terms of its electrical analog, because experiments can be very easily conducted on electrical systems and analysed. Any change in the parameters can be achieved very easily in the electric circuit and studied by altering the parameters, so as to accomplish the desired response. Afterwards, corresponding analogous mechanical quantities can be changed by the analogous amount to design the desired mechanical system.

However, a point of caution is that; this type of analogy cannot be extended to very complex and complicated systems, as the basis of analogy breaks down when applied. There are two possible analogies between mechanical and electrical systems utilised in the analysis of control systems.

5.5.1 Force-voltage Analogy
Consider a mechanical system is shown in Fig. 5.6. Mechanical network of the mechanical system under consideration is shown in Fig. 5.7. The nodal equation of the mechanical network at x is

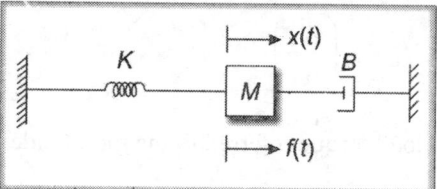

Fig. 5.6

$$f(t) = f_m(t) + f_B(t) + f_k(t) \tag{5.28}$$

$$f(t) = M\frac{d^2x}{dt^2} + B\frac{dx}{dt} + Kx \tag{5.29}$$

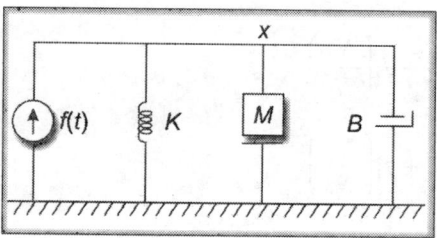

Fig. 5.7

Now consider an electrical network is shown in Fig. 5.8. The mesh equation by application of Kirchoff's law is:

$$e(t) = Ri(t) + \frac{Ldi(t)}{dt} + \frac{1}{C}\int i(t)dt \tag{5.30}$$

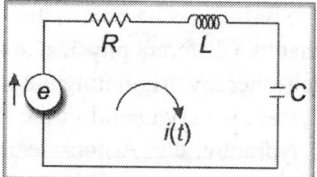

Fig. 5.8

Since $i(t) = \dfrac{dq}{dt}$,

Therefore $e(t) = \dfrac{Rdq}{dt} + \dfrac{Ld^2q}{dt^2} + \dfrac{q}{C}$ \hfill (5.31)

or $e(t) = \dfrac{Ld^2q}{dt^2} + \dfrac{Rdq}{dt} + \dfrac{q}{C}$ \hfill (5.32)

Comparing Eqns (5.29) and (5.32), we can infer that both these equations are identical in nature and hence, analogous. The terms which occupy corresponding positions in these two equations are called *analogous quantities*. Analogous quantities between mechanical and electrical systems based on force-voltage analogy are listed in Table 5.1.

Table 5.1 Analogous Quantities (Force-voltage Analogy)

Mechanical Systems		Electrical Systems
Translatory	Rotational	
Force (f)	Torque (T)	Voltage (e)
Mass (M)	Moment of Inertia (J)	Inductance (L)
Viscous Friction Coefficient (B)	Viscous Friction Coefficient (B)	Resistance (R)
Spring Stiffness (K)	Torsional Spring Stiffness (K)	Reciprocal of Capacitance $\left(\dfrac{1}{C}\right)$
Displacement (x)	Angular Displacement (θ)	Charge (q)
Velocity ($\dot{x}$)	Angular Velocity ($\dot{\theta}$)	Current (i)

5.5.2 Force-current Analogy

Consider an electrical network is shown in Fig. 5.9. Applying Kirchoff's law, we get

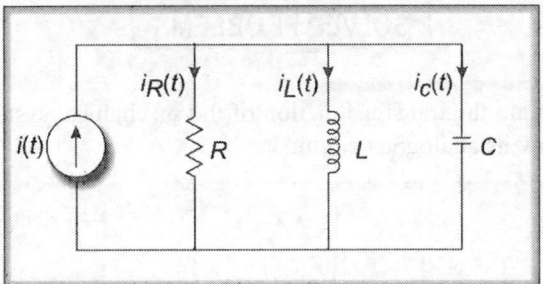

Fig. 5.9

We see that

$$i_R(t) + i_L(t) + i_C(t) = i(t) \tag{5.33}$$

$$\frac{e(t)}{R} + \frac{1}{L}\int e(t)dt + \frac{Cde(t)}{dt} = i(t) \tag{5.34}$$

But, $e(t) = \dfrac{d\Phi}{dt}$, where Φ = flux linkages

Therefore,

$$\frac{1}{R}\frac{d\Phi}{dt} + \frac{\Phi}{L} + \frac{Cd^2\Phi}{dt^2} = i(t) \tag{5.35}$$

Rearranging, we get

$$i(t) = \frac{Cd^2\Phi}{dt^2} + \frac{1}{R}\frac{d\Phi}{dt} + \frac{\Phi}{L} \tag{5.36}$$

Equations (5.29) and (5.36) are identical and hence, analogous. The analogous terms based on force-current analogy are listed in Table 5.2.

Table 5.2 Analogous Quantities (Force-current Analogy)

Mechanical Systems		Electrical Systems
Translatory	Rotational	
Force (f)	Torque (T)	Current (i)
Displacement (x)	Angular Displacement (θ)	Flux linkage (ϕ)
Velocity ($\dot{x}$)	Angular Velocity ($\dot{\theta}$)	Voltage (e)
Mass (M)	Moment of Inertia (J)	Capacitance (C)
Spring Stiffness (K)	Torsional Spring Stiffness (K)	Reciprocal of Inductance ($1/L$)
Viscous Friction (B) Coefficient	Viscous Friction Coefficient (B)	Reciprocal of Resistance ($1/R$)

SOLVED PROBLEMS

Problem 5.1 Obtain the transfer function of the mechanical system is shown in Fig. 5.10 and draw its analogous circuit.

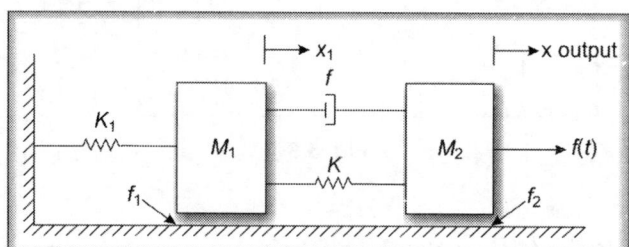

Fig. 5.10

Solution The network diagram of the above mechanical system is shown in Fig. 5.11

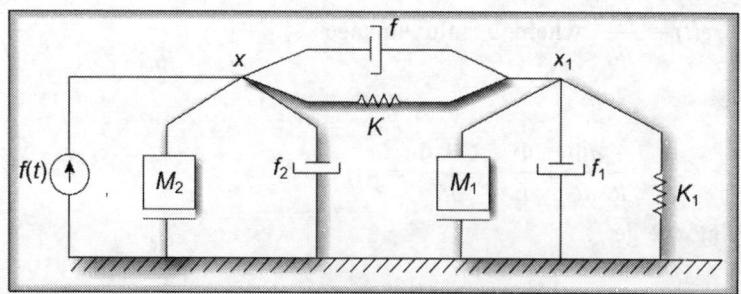

Fig. 5.11

Writing the nodal equations at each node, we get

Node x $M_2\ddot{x} + f_2\dot{x} + f(\dot{x} - \dot{x}_1) + K(x - x_1) = f(t)$

or $(M_2s^2 + f_2s + fs + K)X(s) - (fs + K)X_1(s) = F(s)$ $\qquad$ (1)

Node x_1 $M_1\ddot{x}_1 + f_1\dot{x}_1 + K_1x_1 = f(\dot{x} - \dot{x}_1) + K(x - x_1)$

or $(M_1s^2 + f_1s + fs + K + K_1)X_1(s) = (fs + K)X(s)$ $\qquad$ (2)

Substituting the value of $X_1(s)$ from Eq. (2) in Eqn.. (1), we get

$$(M_2s^2 + f_2s + fs + K)\,X(s) - \frac{(fs + K)(fs + K)\,X(s)}{(M_1s^2 + f_1s + fs + K + K_1)} = F(s)$$

or $$\{[M_2s^2 + f_2s + fs + K][M_1s^2 + f_1s + fs + K + K_1]$$
$$- [fs + K]^2\}\,X(s) = [M_1s^2 + f_1s + fs + K + K_1]F(s)$$

Transfer function $= \dfrac{X(s)}{F(s)} = \dfrac{M_1s^2 + f_1s + fs + K + K_1}{\begin{aligned}&\{M_1M_2s^4 + (M_1f_2 + M_2f_1 + M_1f + M_2f)s^3\\ &+ [M_2K_1 + K(M_1 + M_2) + f_1f_2 + f(f_1 + f_2)]s^2\\ &+ [K_1(f_1 + f_2) + K(f_1 + f_2)]s + KK_1\}\end{aligned}}$

Ans

Converting the nodal Eqns (1) and (2) into comparable electrical analogous equations, we get

$$L_2Di + R_2i + R(i - i_1) + \frac{1}{C}\int(i - i_1)\,dt = e(t) \qquad (3)$$

$$L_1Di_1 + R_1i_1 + \frac{1}{C_1}\int i\,dt = R(i - i_1) + \frac{1}{C}\int(i - i_1)\,dt \qquad (4)$$

Based on Eqns (3) and (4) the electrical analogous circuit based on force-voltage analogy is shown in Fig. 5.12

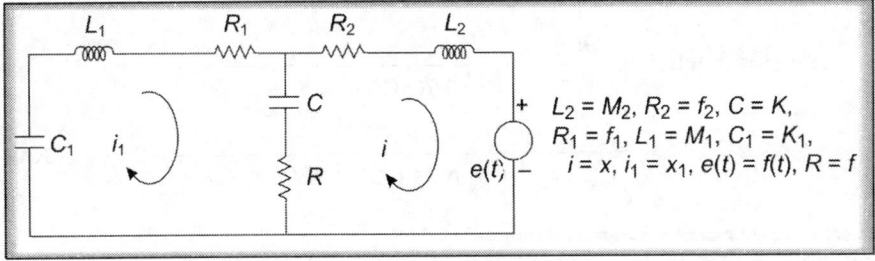

Fig. 5.12: Electrical analog circuit on force-voltage analogy

Problem 5.2 Obtain the transfer function of the mechanical network is shown in Fig. 5.13.

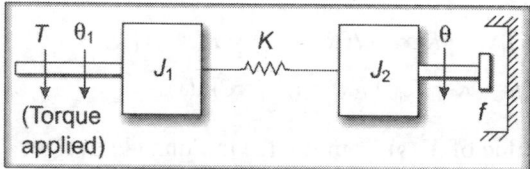

Fig. 5.13

Solution The network diagram for the above system is shown in Fig. 5.14.

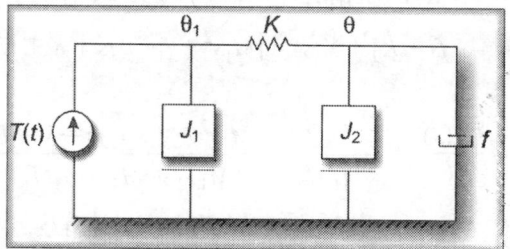

Fig. 5.14

The nodal equations are:

Node 'θ_1' $J_1\ddot{\theta}_1 + K(\theta_1 - \theta) = T(t)$

Node 'θ' $J_2\ddot{\theta} + f\dot{\theta} = K(\theta_1 - \theta)$

or $(J_1 s^2 + K)\theta_1(s) - K\theta(s) = T(s)$ (1)

and $(J_2 s^2 + fs + K)\theta(s) = K\theta_1(s)$ (2)

Substituting value of $\theta_1(s)$ from Eqn.. (2) in Eqn.. (1), we get

$$\left[\frac{(J_1 s^2 + K)(J_2 s^2 + fs + K)}{K} - \frac{K}{1} \right]\theta(s) = T(s)$$

∴   Transfer function $= \dfrac{\theta(s)}{T(s)} = \dfrac{K}{(J_1 s^2 + K)(J_2 s^2 + fs + K) - K^2}$

$$= \frac{K}{(J_1 J_2 s^4 + J_1 fs^3 + (KJ_1 + KJ_2)s^2 + Kfs}$$ **Ans**

Problem 5.3 Obtain the nodal equations for the mechanical system are shown in Fig. 5.15. Draw the analogous electric network also.

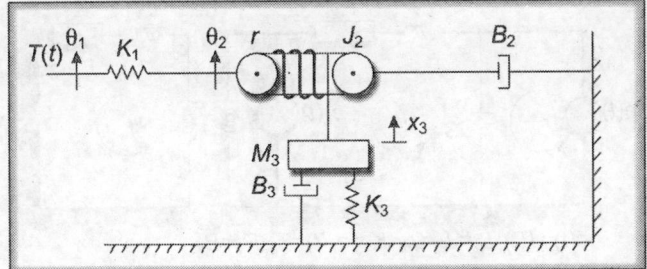

Fig. 5.15

Solution The network diagram for the above system is shown in Fig. 5.16

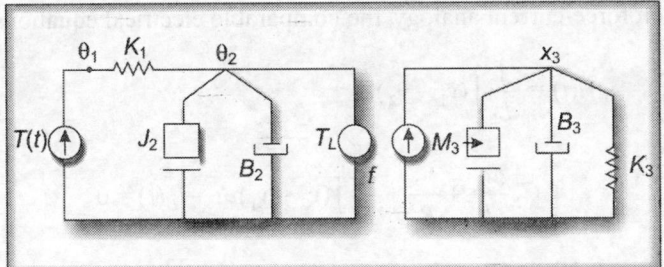

Fig. 5.16

The nodal equations are

$$T(t) = K_1(\theta_1 - \theta_2) \tag{1}$$

$$J_2\ddot{\theta}_2 + B_2\dot{\theta}_2 + K_1(\theta_2 - \theta_1) = -T_L \tag{2}$$

$$T_L = f \times r \tag{3}$$

$$T_L = f \times r = r[M_3\ddot{x}_3 + B_3\dot{x}_3 + K_3x_3] \tag{4}$$

$$x_3 = r\theta_2 \tag{5}$$

Converting the nodal equations into comparable electrical analogous equations, based on force-voltage analogy, we get

$$e(t) = \frac{1}{C_1}\int(i_1 - i_2)\,dt \tag{6}$$

$$L_2Di_2 + R_2i_2 + \frac{1}{C_1}\int(i_2 - i_1)\,dt + e_L(t) = 0 \tag{7}$$

$$e_L(t) = e_f(t) \times r \tag{8}$$

$$e_L(t) = r\left(L_3Di_3 + R_3i_3 + \frac{1}{C_3}\int i_3dt\right) \tag{9}$$

Electrical analogous circuit satisfying the above equations, based on force-voltage analogy is shown in Fig. 5.17

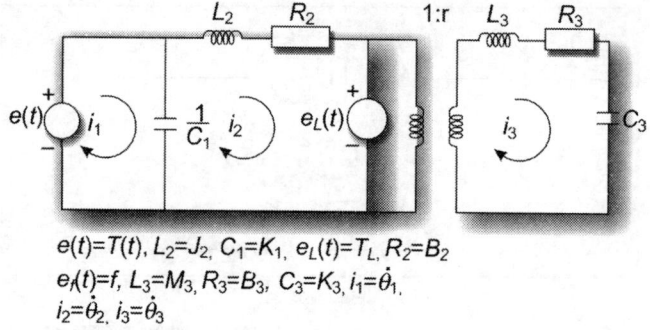

$e(t)=T(t)$, $L_2=J_2$, $C_1=K_1$, $e_L(t)=T_L$, $R_2=B_2$
$e_f(t)=f$, $L_3=M_3$, $R_3=B_3$, $C_3=K_3$, $i_1=\dot{\theta}_1$,
$i_2=\dot{\theta}_2$, $i_3=\dot{\theta}_3$

Fig. 5.17 Electrical analogous circuit based on force-voltage analogy

Based on force-current analogy, the comparable electrical equations are:

$$i(t) = \frac{1}{L_1}\int(v_1 - v_2)\,dt \tag{10}$$

$$C_2\frac{dv_2}{dt} + \frac{v_2}{R_2} + \frac{1}{L_1}\int(v_2 - v_1)\,dt + i_L(t) = 0 \tag{11}$$

$$i_L(t) = i_f(t) \times r \tag{12}$$

$$= r\left(C_3\frac{dv_3}{dt} + \frac{v_3}{R_3} + \frac{1}{L_3}\int v_3(dt)\right) \tag{13}$$

Electrical analogous circuit diagram based on force-current analogy is shown in Fig. 5.18

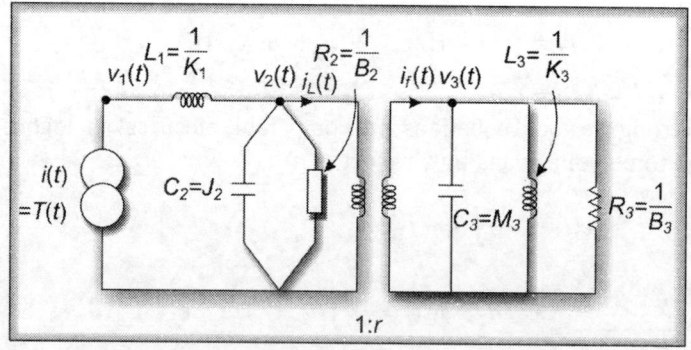

Fig. 5.18 Electrical analogous diagram based on force-current analogy

Problem 5.4 Write differential equations for the system are shown in Fig. 5.19. The force produced by the solenoid, when the coil is connected to voltage source is $f = K_i i$

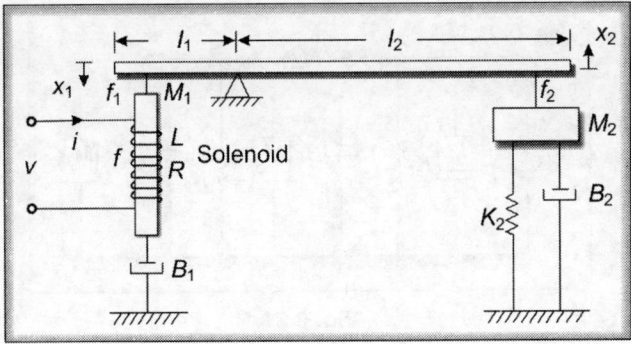

Fig. 5.19

Solution

$$v(t) = \frac{L\,di(t)}{dt} + Ri(t) \tag{1}$$

$$f(t) = K_i(t) = K_i = M_1\ddot{x}_1 + B_1\dot{x}_1 + f_1 \tag{2}$$

$$f_2(t) = f_1(t) \times \frac{l_1}{l_2} = M_2\ddot{x}_2 + B_2\dot{x}_2 + K_2 x_2$$

or

$$f_1(t) = \frac{l_2}{l_1}(M_2\ddot{x}_2 + B_2\dot{x}_2 + K_2 x_2)$$

or

$$f(t) = M_1\ddot{x}_1 + B_1\dot{x}_1 + \frac{l_2}{l_1}\left[M_2\ddot{x}_2 + B_2\dot{x}_2 + K_2 x_2\right]$$

Problem 5.5 Draw the mechanical network for the system in Fig. 5.20 and draw its analogous circuit.

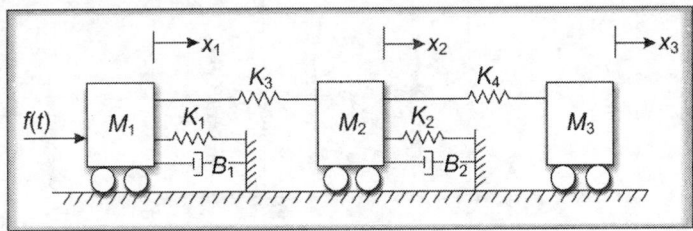

Fig. 5.20

Solution The network diagram for the above system is shown in Fig. 5.21

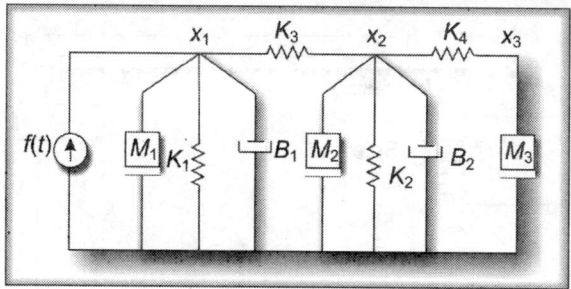

Fig. 5.21

Node 'x_1' $f(t) = M_1\ddot{x}_1 + B_1\dot{x}_1 + K_1x_1 + K_3x_1 - K_3x_2$

Node 'x_2' $K_3x_1 - K_3x_2 = M_2\ddot{x}_2 + B_2\dot{x}_2 + K_2x_2 + K_4x_2 - K_4x_3$

or $M_2\ddot{x}_2 + B\dot{x}_2 + (K_2 + K_3 + K_4)x_2 - K_3x_1 - K_4x_3 = 0$

Node 'x_3' $M_3\ddot{x}_3 = K_4x_2 - K_4x_3$

                 $M_3\ddot{x}_3 + K_4x_3 - K_4x_2 = 0$

Using force-current analogy

$$i(t) = C_1\frac{dv_1}{dt} + \frac{v_1}{R_i} + \frac{1}{L_1}\int v_1 dt + \frac{1}{L_3}\int v_1 dt - \frac{1}{L_3}\int v_2 dt$$

$$C_2\frac{dv_2}{dt} + \frac{v_2}{R_2} + \left(\frac{1}{L_2} + \frac{1}{L_3} + \frac{1}{L_4}\right)\int v_2 dt - \frac{1}{L_3}\int v_1 dt - \frac{1}{L_4}\int v_3 dt = 0$$

$$C_3\frac{dv_3}{dt} + \frac{1}{L_4}\int v_3 dt - \frac{1}{L_4}\int v_2 dt = 0$$

The analogous circuit is shown in Fig. 5.22

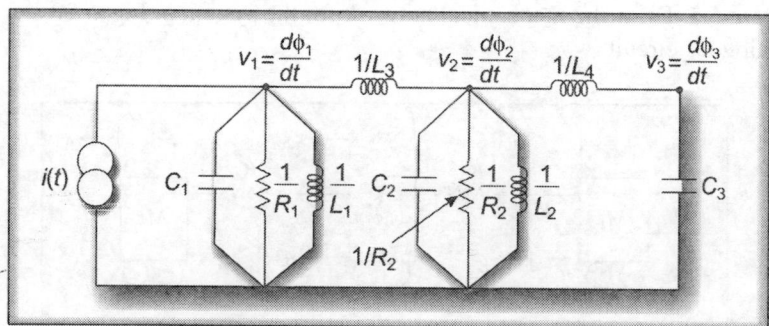

Fig. 5.22   Based on force-current analogy

Problem 5.6 A rack and pinion arrangement is shown in Fig. 5.23. Write the nodal equations and draw the electrical analog.

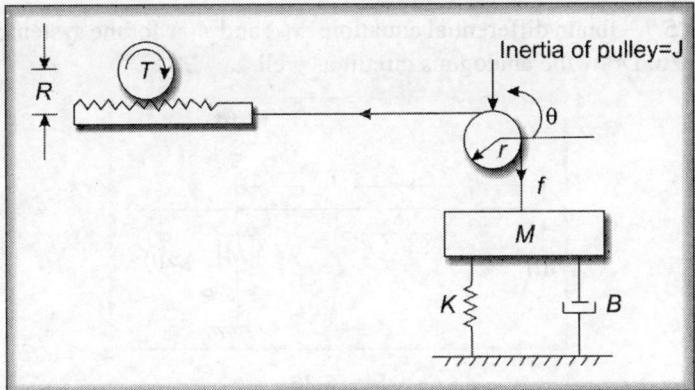

Fig. 5.23

Solution The tension will not be same on either side of the rope due to inertia of the pulley. The equations of performance are:

$$T = Rf'$$ (1)

$$f' = \frac{Js^2\theta}{r} + f$$ (2)

$$= \frac{Js^2x}{r^2} + f \quad \left(\because \theta = \frac{x}{r}\right)$$

$$f = M\ddot{x} + B\dot{x} + Kx$$ (3)

The network diagram is shown in Fig. 5.24

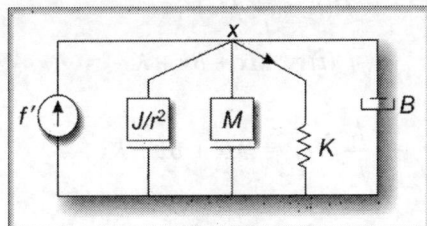

Fig. 5.24

The electrical analog based on force-current analogy is shown in Fig. 5.25

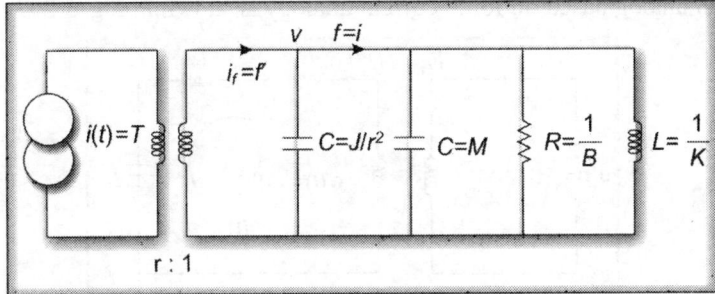

Fig. 5.25

Problem 5.7 Obtain differential equations $x(t)$ and $f(t)$ for the system are shown in Fig. 5.26. Draw the analogous circuit as well.

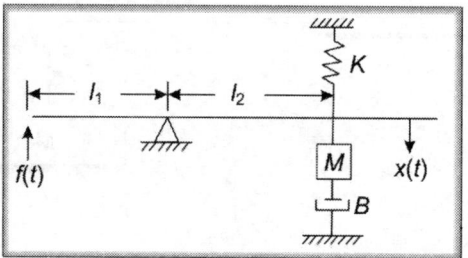

Fig. 5.26

Solution The network diagram is shown in Fig. 5.27. The nodal equations are:

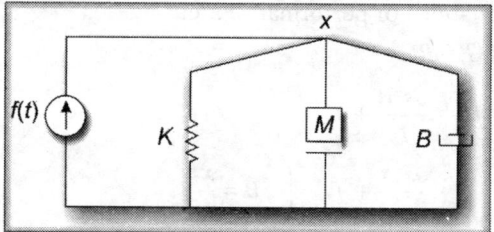

Fig. 5.27

$$f'(t) = \frac{l_1}{l_2} f(t)$$

and
$$f'(t) = M\ddot{x} + B\dot{x} + Kx$$

Equating, we get

$$\frac{l_1}{l_2} f(t) = M\ddot{x} + B\dot{x} + Kx$$

$$f(t) = \frac{l_2}{l_1} (M\ddot{x} + B\dot{x} + Kx)$$

Using force-voltage analogy, the analogous circuit is shown in Fig. 5.28. The electrical analog based on force-current analogy is shown in Fig. 5.29.

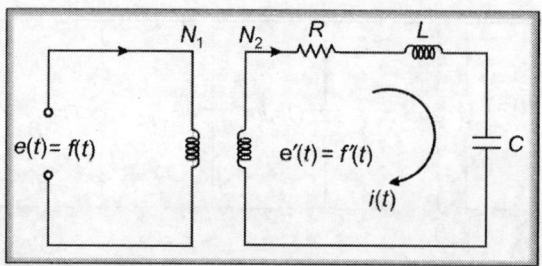

Fig. 5.28

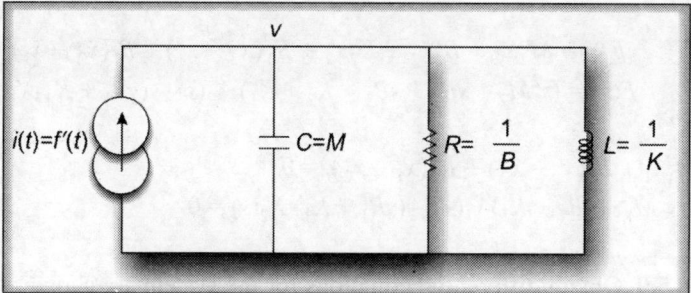

Fig. 5.29

Problem 5.8 Draw the mechanical network and write the differential equations for the system are shown in Fig. 5.30

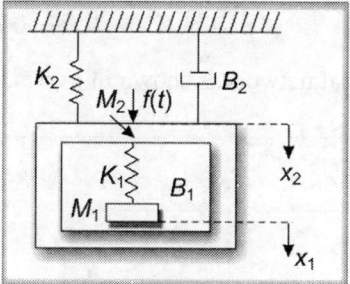

Fig. 5.30

Solution The mechanical network diagram is shown in Fig. 5.31

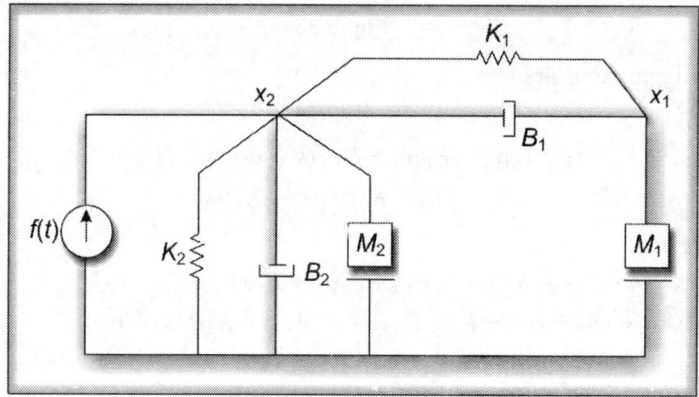

Fig. 5.31

Node 'x_2'

$$f(t) = M_2\ddot{x}_2 + B_2\dot{x}_2 + K_2 x_2 + K_1(x_2 - x_1) + B_1(\dot{x}_2 - \dot{x}_1)$$

or $\quad F(s) = (s^2 M_2 + sB_2 + sB_1 + K_2 + K_1) X_2(s) - (sB_1 + K_1)X_1(s)$

Node 'x_1'

$$M_1\ddot{x}_1 + B_1(\dot{x}_1 - \dot{x}_2) + K_1(x_1 - x_2) = 0$$

or $\quad (s^2 M_1 + sB_1 + K_1)X_1(s) - (sB_1 + K_1) X_2(s) = 0$

Problem 5.9 Obtain differential equations for the system are shown in Fig. 5.32 and draw the analogous electrical circuit.

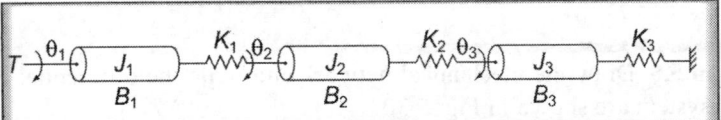

Fig. 5.32

Solution The mechanical network is shown in Fig. 5.33

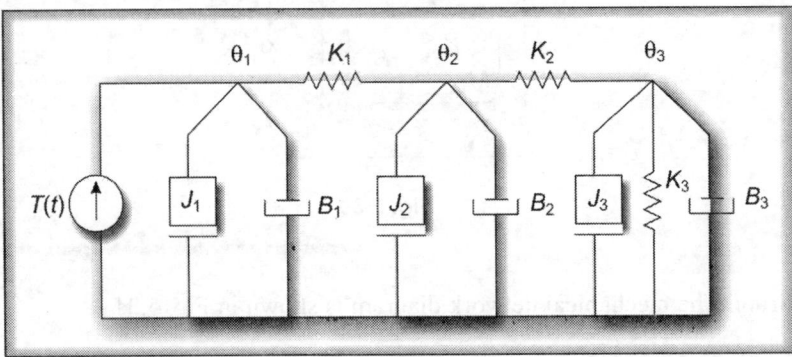

Fig. 5.33

The nodal equations are:

Node 'θ_1'

$$T(t) = J_1 \ddot{\theta}_1 + B_1 \dot{\theta}_1 + K_1 (\theta_1 - \theta_2)$$

or $\quad T(s) = (s^2 J_1 + sB_1 + K_1)\,\theta_1(s) - K_1\theta_2(s)$

Node 'θ_2'

$$K_1(\theta_1 - \theta_2) = K_2 (\theta_2 - \theta_3) + J_2\ddot{\theta}_2 + B_2\dot{\theta}_2$$

or $\quad [s^2 J_2 + sB_2 + (K_1 + K_2)]\,\theta_2(s) - K_1\theta_1(s) - K_2\theta_3(s) = 0$

Node 'θ_3'

$$K_2 (\theta_2 - \theta_3) = J_3\ddot{\theta}_3 + B_3\dot{\theta}_3 + K_3\theta_3$$

or $\quad [(s^2 J_3 + sB_3 + K_3 + K_2)\,\theta_3(s) = K_2\theta_2(s)$

Force-voltage analogy is shown in Fig. 5.34.

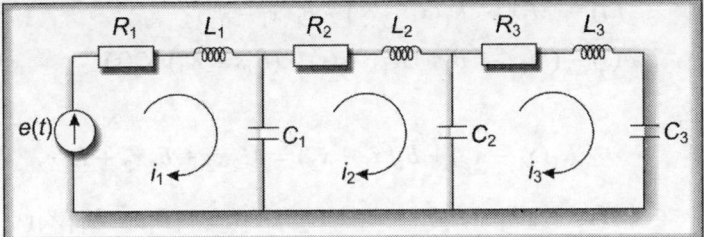

Fig. 5.34

The loop equations are:

$$e(t) = L_1 Di_1 + R_1 i_1 + \frac{1}{C_1}\int (i_1 - i_2)dt$$

$$0 = L_2 Di_2 + R_2 i_2 + \frac{1}{C_2}\int (i_2 - i_3)dt + \frac{1}{C_1}\int (i_2 - i_1)dt$$

$$0 = L_3 Di_3 + R_3 i_3 + \frac{1}{C_3}\int i_3\, dt + \frac{1}{C_2}\int (i_3 - i_2)dt$$

These equations are comparable with the nodal equations.

Problem 5.10 Obtain the differential equations describing the complete dynamics of the mechanical system are shown in Fig. 5.35

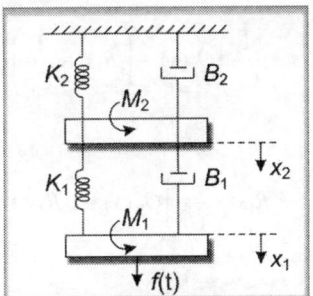

Fig. 5.35

Solution The mechanical network diagram is shown in Fig. 5.36

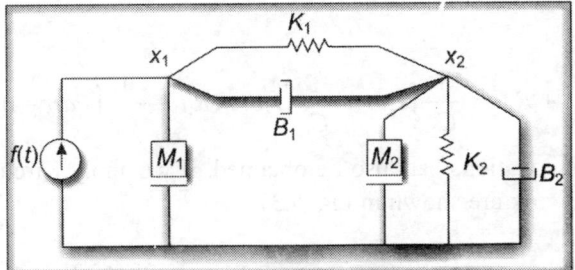

Fig. 5.36

Node 'x_1'

$$f(t) = M_1\ddot{x}_1 + K_1(x_1 - x_2) + B_1(\dot{x}_1 - \dot{x}_2)$$

or $\qquad F(s) = (M_1 s^2 + B_1 s + K_1)\, X_1(s) - (B_1 s + K_1)\, X_2(s)$

Node 'x_2'

$$K_1(x_1 - x_2) + B_1(\dot{x}_1 - \dot{x}_2) = M_2\ddot{x}_2 + B_2\dot{x}_2 + K_2 x_2$$

or $\qquad [M_2 s^2 + (B_1 + B_2)s + K_1 + K_2)]\, X_2(s) - [B_1 s + K_1]\, X_1(s) = 0$

The electrical analog based on force-voltage analogy is shown in Fig. 5.37. The electrical analog circuit is drawn with the help of electrical analog equations which are obtained from nodal equations in Laplace domain. The electrical analog equations are:

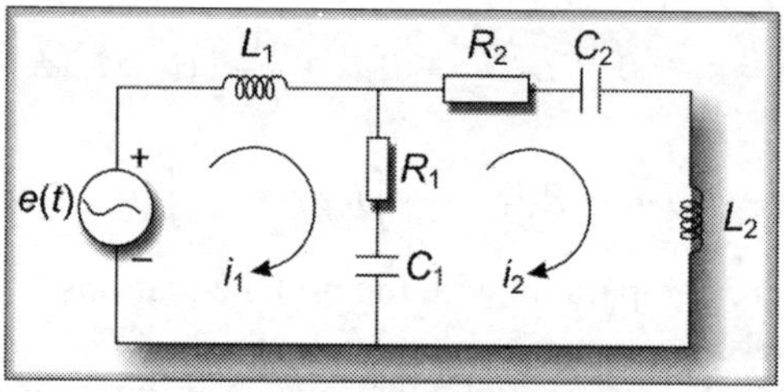

Fig. 5.37

$$E(s) = \left(L_1 s^2 + R_1 s + \frac{1}{C_1}\right) Q_1(s) - \left(R_1 s + \frac{1}{C_1}\right) Q_2(s)$$

and

$$\left(L_2 s^2 \frac{R_2}{C_2} s + R_1 s + \frac{1}{C_1}\right) Q_2(s) - \left(R_1 s + \frac{1}{C_1}\right) Q_1(s) = 0$$

where, Q = charge

Converting the above equations into differential equations form

$$e(t) = L_1 \frac{di_1}{dt} + \frac{1}{C_1}\int i_1 dt + R_1 i_1 - R_1 i_2 - \frac{1}{C_1}\int i_2 dt$$

and

$$L_2 \frac{di_2}{dt} + R_2 i_2 + R_1 i_2 + \frac{1}{C_2}\int i_2 dt + \frac{1}{C_1}\int i_2 dt - R_1 i_1 - \frac{1}{C_1}\int i_1 dt = 0$$

Note: Above equations can also be obtained, if we apply Kirchoff's laws to the electrical network are shown in Fig. 5.37

Problem 5.11 Obtain the transfer function of the system is shown in Fig. 5.38.

Solution The equations of performance for the system are:

$$B_1(\dot{x}_1 - \dot{x}_0) + K_1(x_1 - x_0) = K_2 x_0$$

$$(sB_1 + K_1) X_1(s) - (sB_1 + K_1) X_0(s) = K_2 X_0(s)$$

or $\quad X_0(s) [sB_1 + K_1 + K_2] = [sB_1 + K_1] X_1(s)$

$$\frac{X_0}{X_1}(s) = \frac{sB_1 + K_1}{sB_1 + K_1 + K_2} = \frac{K_1\left(1 + \dfrac{sB_1}{K_1}\right)}{K_1 + K_2\left(1 + \dfrac{sB_1}{K_1 + K_2}\right)}$$

$$= \frac{K_1}{K_1 + K_2}\left[\frac{\left(\dfrac{B_1}{K_1 + K_2} \times \dfrac{K_1 + K_2}{K_1}\right)s + 1}{\dfrac{B_1}{K_1 + K_2}s + 1}\right]$$

Put $\qquad a = \dfrac{K_1 + K_2}{K_1} \quad$ and $\quad T = \dfrac{B_1}{K_1 + K_2}$

$$\therefore \qquad \frac{X_o}{X_i}(s) = \frac{1}{a}\left(\frac{1 + aTs}{1 + Ts}\right)$$

Note: This corresponds to phase lead network.

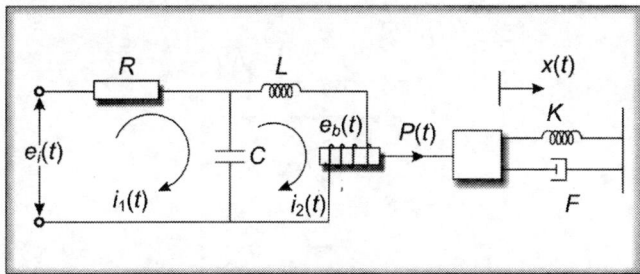

Fig. 5.38

Problem 5.12 Find the transfer function $X(s)/E_i(s)$ for the system is shown in Fig. 5.39. The following relations apply:

(a) force acting on mass M, $P(t) = K_2 i_2(t)$

(b) the back e.m.f. of coil $= \dfrac{K_1 dx(t)}{dt}$

where, K_1 and K_2 are constants.

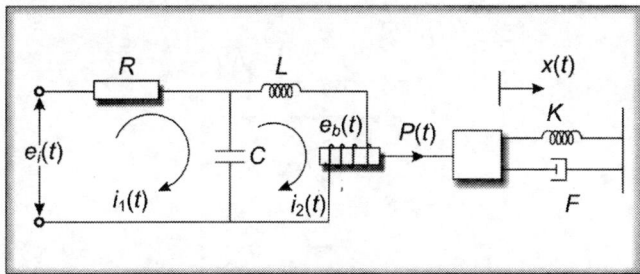

Fig. 5.39

Solution Writing the equations of performance for the given system in Laplace domain

$$E_i(s) = RI_1(s) + \frac{1}{sC}[I_1(s) - I_2(s)] \tag{1}$$

$$sLI_2(s) + \frac{1}{sC}[I_2(s) - I_1(s)] = sK_1 X(s) = -E_b(s) \left(\text{because } e_b(t) = K_1 \frac{dx}{dt} \right) \tag{2}$$

Substituting the value of $\frac{1}{sC}[I_2(s) - I_1(s)]$ in Eq. (2) from Eq. (1), we get

$$sLI_2(s) + RI_1(s) - E_i(s) = -sK_1 X(s) \tag{3}$$

Substituting for $I_1(s)$ from Eq. (2) in Eq. (1), we get

$$sLI_2(s) + sK_1 X(s) + sRC\left[sLI_2(s) + \frac{I_2(s)}{sC} + sK_1 X(s) \right] = E_i(s)$$

or $$(sL + R + s^2 RCL)\, I_2(s) + (s^2 K_1 RC + sK_1)\, X(s) = E_i(s) \tag{4}$$

also, $$P(t) = K_2 i_2(t) = M\ddot{x}(t) + F\dot{x}(t) + Kx(t)$$

or $$K_2 I_2(s) = (Ms^2 + Fs + K)\, X(s)$$

or $$I_2(s) = \left(\frac{Ms^2 + Fs + K}{K_2} \right) X(s) \tag{5}$$

Substituting this value of $I_2(s)$ in Eq. (4), we get

$$\left[(s^2 RLC + sL + R)\left(\frac{Ms^2 + Fs + K}{K_2} \right) + s^2 K_1 RC + sK_1 \right] X(s) = E_i(s)$$

$$\frac{X(s)}{E_i(s)} = \frac{K_2}{s^4 MCRL + s^3(FCLR + LM) + s^2(RM + KCRL + FL + K_1 K_2 CR)}{}$$
$$+ s(FR + KL + K_1 K_2) + RL$$

Problem 5.13 Obtain the nodal equations for the system are shown in Fig. 5.40 and draw its analogous electrical network.

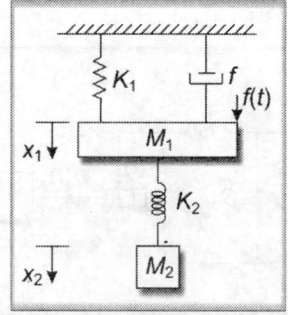

Fig. 5.40

Solution The mechanical network diagram for the given system is shown in Fig. 5.41

Node 'x_1'

$$f(t) = M_1\ddot{x}_1 + f\dot{x}_1 + K_1x_1 + K_2(x_1 - x_2)$$

or

$$F(s) = (s^2M_1 + sf + K_1 + K_2)\,X_1(s) - K_2X_2(s)$$

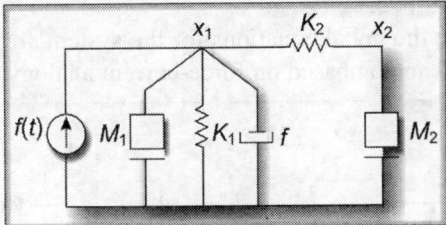

Fig. 5.41

Node 'x_2'

$$K_2(x_1 - x_2) = M_2\ddot{x}_2 \quad \text{or} \quad (s^2M_2 + K_2)\,X_2(s) - K_2X_1(s) = 0$$

Electrical analog based on force-voltage analogy is shown in Fig. 5.42

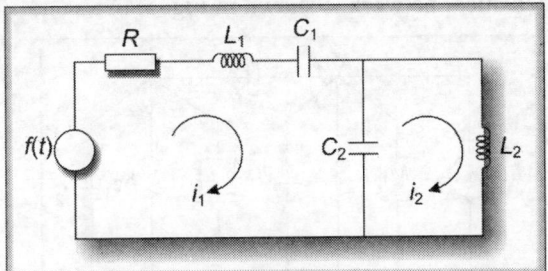

Fig. 5.42

The differential equations for the electrical analog are:

$$v(t) = Ri_1 + L_1\frac{di_1}{dt} + \frac{1}{C_1}\int i_1\,dt + \frac{1}{C_2}\int i_1\,dt - \frac{1}{C_2}\int i_2\,dt$$

or

$$v(t) = R\frac{dq_1}{dt} + L_1\frac{d^2q_1}{dt^2} + \frac{q_1}{C_1} + \frac{q_1}{C_2} - \frac{q_2}{C_2}$$

or

$$V(s) = \left(sR + s^2L_1 + \frac{1}{C_1} + \frac{1}{C_2}\right)Q_1(s) - \frac{1}{C_2}\times Q_2(s)$$

or

$$V(s) = \left(s^2L_1 + sR + \frac{1}{C_1} + \frac{1}{C_2}\right)Q_1(s) - \frac{Q_2(s)}{C_2}$$

and

$$L_2\frac{di_2}{dt} + \frac{1}{C_2}\int i_2\,dt - \frac{1}{C_2}\int i_1\,dt = 0$$

or

$$L_2\frac{d^2q_2}{dt} + \frac{q_2}{C_2} - \frac{q_1}{C_2} = 0$$

$$\left(s^2 L_2 + \frac{1}{C_2}\right)Q_2(s) - \frac{Q_1(s)}{C_2} = 0$$

Note: These equations are similar to the nodal equations.

Problem 5.14 Obtain the nodal equations for the system are shown in Fig. 5.43 and draw its electrical analog based on force-current analogy.

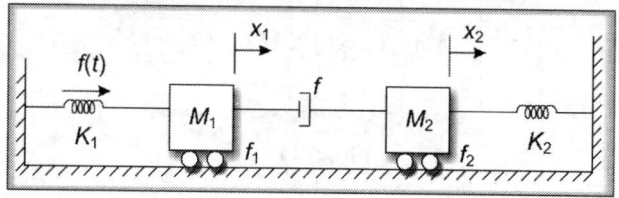

Fig. 5.43

Solution The mechanical network is shown in Fig. 5.44

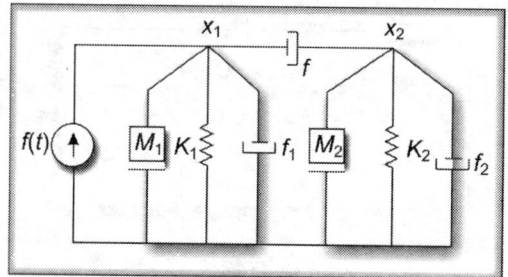

Fig. 5.44

Node 'x_1'

$$M_1\ddot{x}_1 + f_1\dot{x}_1 + f(\dot{x}_1 - \dot{x}_2) + K_1 x_1 = f(t)$$

Node 'x_2'

$$M_2\ddot{x}_2 + f_2\dot{x}_2 + K_2 x_2 = f(\dot{x}_1 - \dot{x}_2)$$

Electrical analog circuit based on force-current analogy is shown in Fig. 5.45

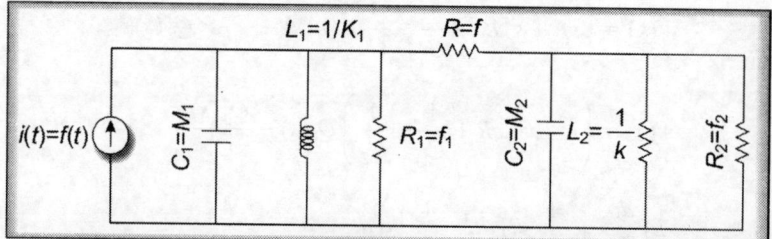

Fig. 5.45

Problem 5.15 Obtain the transfer function for the system is shown in Fig. 5.46

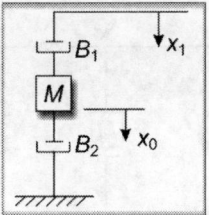

Fig. 5.46

Solution

$$B_1(\dot{x}_1 - \dot{x}_0) = M\ddot{x}_0 + B_2\dot{x}_0$$

$$B_1 s X_1(s) = (Ms^2 + B_2 s + B_1 s) X_0(s)$$

$$\frac{X_0(s)}{X_1(s)} = \frac{B_1 s}{M_1 s^2 + B_2 s + B_1 s} = \frac{B_1}{M_1 s + B_1 + B_2} \qquad \textbf{Ans}$$

Problem 5.16 Obtain transfer function for the system is shown in Fig. 5.47

Solution Writing equations for the given system based on Newton's law, we get

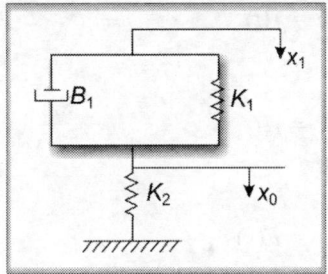

Fig. 5.47

$$K_1(x_1 - x_0) + B_1(\dot{x}_1 - \dot{x}_0) - K_2 x_0 = 0$$

$$[sB_1 + K_1] X_1(s) - [sB_1 + K_1 + K_2] X_0(s) = 0$$

$$\frac{X_0(s)}{X_1(s)} = \frac{sB_1 + K_1}{sB_1 + K_1 + K_2} \qquad \textbf{Ans}$$

Problem 5.17 Obtain transfer function of the system is shown in Fig. 5.48 and draw its electrical analog.

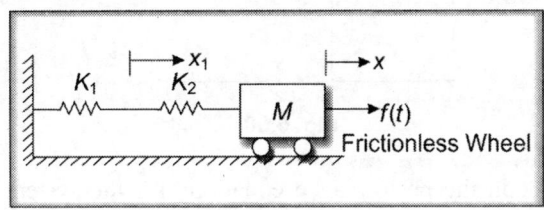

Fig. 5.48

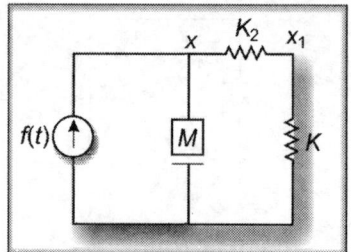

Fig. 5.49

Solution The mechanical network is shown in Fig. 5.49

Node 'x'; $M\ddot{x} + K_2(x - x_1) = f(t)$ (1)

Node 'x_1'; $K_1 x_1 = K_2 (x - x_1)$ (2)

Equation (1) can be rewritten as:

$\therefore$ $M\ddot{x} + K_2 x - K_2 x_1 = f(t)$ (3)

Substituting value of x_1 from Eq. (2) in Eq. (3), we get

$$M\ddot{x} + K_2 x - \frac{K_2 K_2}{K_1 + K_2} x = f(t)$$

$$M\ddot{x} + \frac{K_2 K_1}{K_1 + K_2} x = f(t)$$

or $\left(Ms^2 + \dfrac{K_1 K_2}{K_1 + K_2} \right) X(s) = F(s)$

$\therefore$ $$\frac{X(s)}{F(s)} = \frac{1}{Ms^2 + \dfrac{K_1 K_2}{K_1 + K_2}}$$

The electrical analog based on force-current analogy is shown in Fig. 5.50

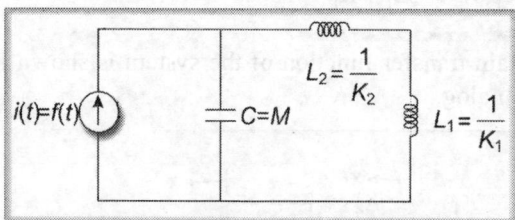

Fig. 5.50

Problem 5.18 Obtain the performance equations for the system are shown in Fig. 5.51 in Laplace domain.

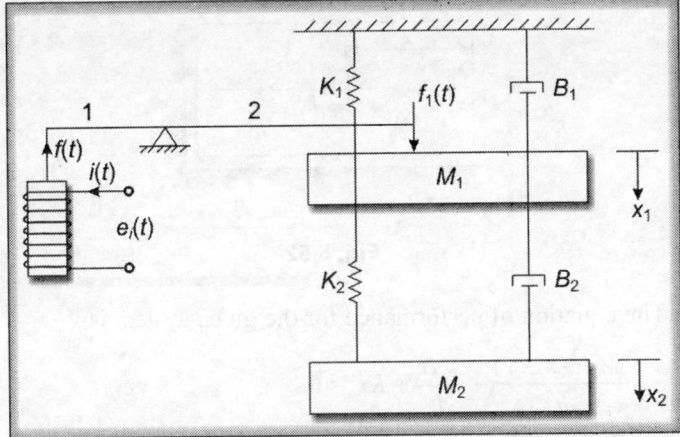

Fig. 5.51

Solution The equations in Laplace domain are:

$$E_i(s) = (sL + R)I(s) \tag{1}$$

$$F(s) = K_f I(s) = 2F_1(s) + M_1 s^2 \frac{X_1(s)}{2}$$

where, $K_f = $ constant

i.e. $$F(s) = 2F_1(s) + M_1 s^2 \frac{X_1(s)}{2}$$

or $$(F(s) - 2F_1(s)) = \frac{M_1 s^2 X_1(s)}{2} \tag{2}$$

The mechanical network is shown in Fig. 5.51a

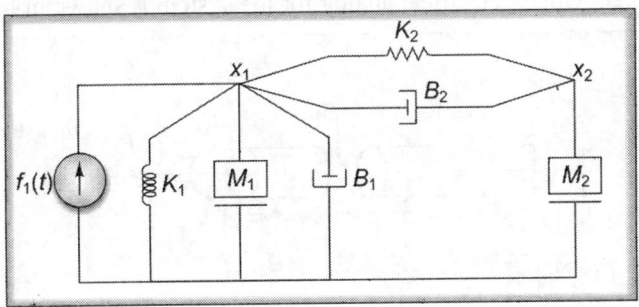

Fig. 5.51a

Also $$F_1(s) = (M_1 s^2 + B_1 s + K_1) X_1(s) + (X_1(s) - X_2(s)) (K_2 + B_2 s) \tag{3}$$

and $$M_2 s^2 X_2(s) + (B_2 s + K_2) (X_2(s) - X_1(s)) = 0 \tag{4}$$

Problem 5.19 Find out the transfer function for the mechanical accelerometer is shown in Fig. 5.52

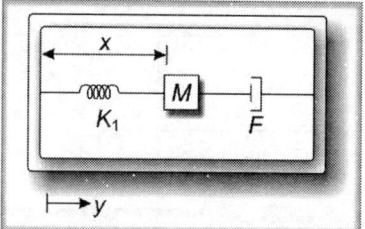

Fig. 5.52

Solution The equation of performance for the given system is:

$$\frac{Md^2(x-y)}{dt^2} + \frac{Fdx}{dt} + Kx = 0$$

$$\frac{Md^2x}{dt^2} + \frac{Fdx}{dt} + Kx = \frac{Md^2y}{dt^2}$$

In Laplace domain $(Ms^2 + Fs + K)\, X(s) = Ms^2 Y(s)$

$$\therefore \quad \frac{X(s)}{Y(s)} = \frac{Ms^2}{Ms^2 + Fs + K}$$

The transfer function between the input acceleration and output 'x' is found out as given below.

$$\frac{Md^2x}{dt^2} + \frac{Fdx}{dt} + Kx = \frac{Md^2y}{dt^2} = Ma$$

$$(Ms^2 + Fs + K)\, X(s) = MA(s)\;;\quad \frac{X(s)}{A(s)} = \frac{1}{Ms^2 + Fs + K} \qquad \textbf{Ans}$$

Problem 5.20 Obtain electrical analog for the system is shown in Fig. 5.53

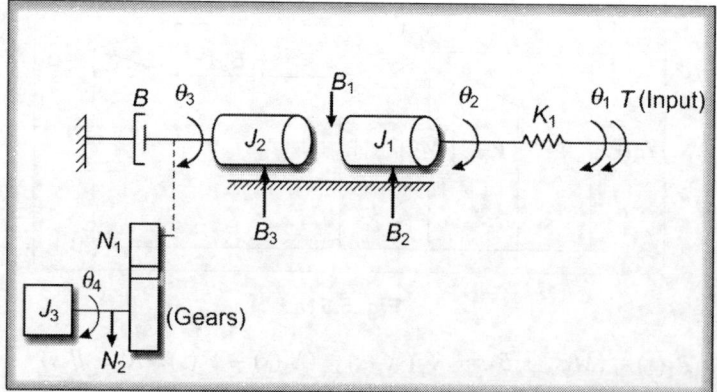

Fig. 5.53

Solution The mechanical network for the part of the system is shown in Figs 5.54 and 5.55

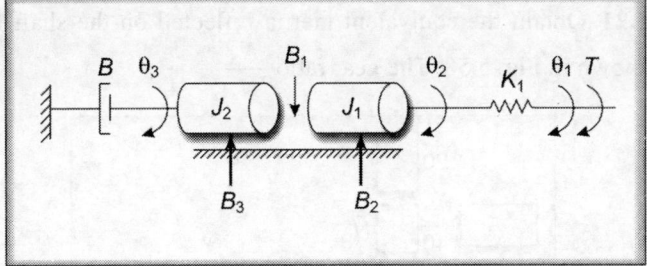

Fig. 5.54

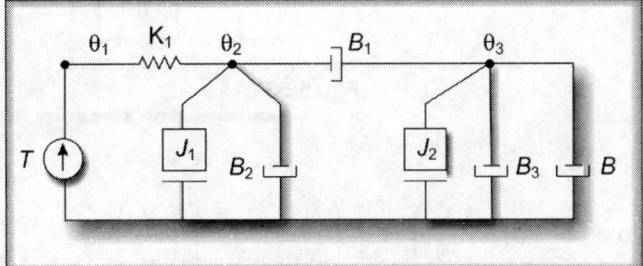

Fig. 5.55

The electrical analog is shown in Fig. 5.56

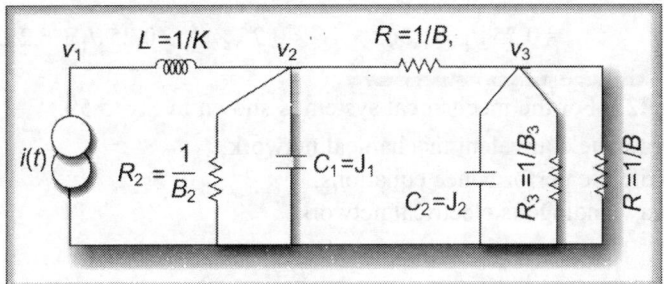

Fig. 5.56

Combining the remaining part of the system with the electrical analog obtained in Fig. 5.56, we get the electrical analog for the complete system is shown in Fig. 5.57.

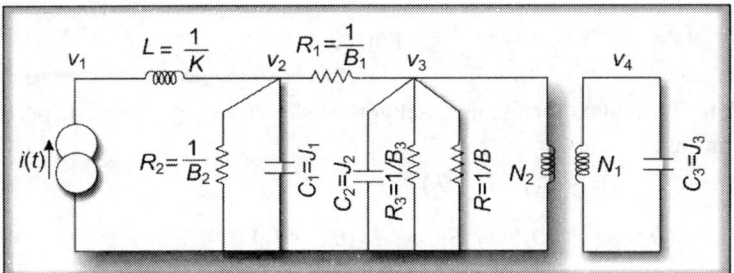

Fig. 5.57

Problem 5.21 Obtain the equivalent inertia reflected on the shaft D. The gear ratios are shown in Fig. 5.58. The gear ratio $\dfrac{N_3}{N_4} = \dfrac{5}{4}$.

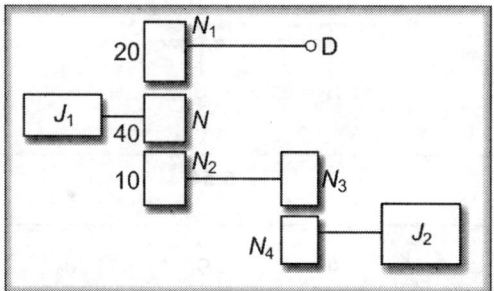

Fig. 5.58

Solution

$$\text{Equivalent inertia} = J_1 \times \left(\frac{N_1}{N}\right)^2 + J_2 \left(\frac{N_3}{N_4}\right)^2 \left(\frac{N}{N_2}\right)^2 \left(\frac{N_1}{N}\right)^2$$

$$= J_1 \times \left(\frac{20}{40}\right)^2 + J_2 \left(\frac{5}{4}\right)^2 \left(\frac{40}{10}\right)^2 \left(\frac{20}{40}\right)^2$$

$$= 0.25\, J_1 + 1.563 \times 16 \times 0.25 \times J_2 = 0.25\, J_1 + 6.252\, J_2 \quad \textbf{Ans}$$

Problem 5.22 For the mechanical system is shown in Fig. 5.59.
  (a) Draw the equivalent mechanical network.
  (b) Write the performance equations.
  (c) Draw analogous electrical network.

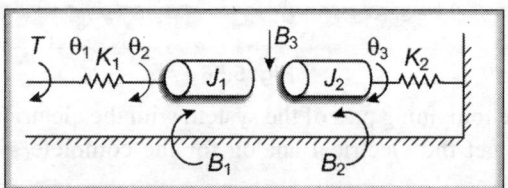

[*Pune University*]

Fig. 5.59

Solution Equivalent mechanical network is shown in Fig. 5.60. The performance equations are:

$$T(t) = K_1 (\theta_1 - \theta_2)$$

$$K_1(\theta_1 - \theta_2) = J_1\ddot{\theta}_2 + B_1\dot{\theta}_2 + B_3(\dot{\theta}_2 - \dot{\theta}_3)$$

$$B_3(\dot{\theta}_2 - \dot{\theta}_3) = J_2\ddot{\theta}_3 + B_2\dot{\theta}_3 + K_2\theta_3$$

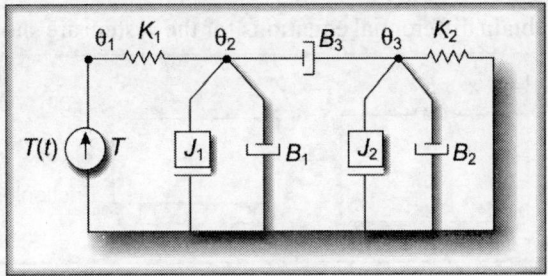

Fig. 5.60

Taking Laplace transform of the equations

$$T(s) = K_1(\theta_1(s) - \theta_2(s))$$
$$K_1(\theta_1(s) - \theta_2(s)) = (J_1s^2 + B_1s)\,\theta_2(s) + B_3s(\theta_2(s) - \theta_3(s))$$
$$B_3s(\theta_2(s) - \theta_3(s)) = (J_2s^2 + B_2s + K_2)\,\theta_3(s)$$

The electrical analog circuit is drawn with the help of electrical analog equations which are obtained from nodal equations in Laplace domain. The electrical analog equations based on force-voltage analogy are:

$$E(s) = \frac{1}{C_1}(Q_1(s) - Q_2(s))$$

$$\frac{1}{C_1}(Q_1(s) - Q_2(s)) = (L_1s^2 + R_1s)\,Q_2(s) + R_3s\,(Q_2(s) - Q_3(s))$$

$$R_3s\,(Q_2(s) - Q_3(s)) = \left(L_2s^2 + R_2s + \frac{1}{C_2}\right)Q_3(s)$$

where, Q is the charge. Converting the above equations into differential equations

$$e(t) = \frac{1}{C_1}\int (i_1 - i_2)dt$$

$$\frac{1}{C_1}\int (i_1 - i_2)dt = \frac{L_1 di_2}{dt} + R_1\,i_2 + R_3\,i_2 - R_3\,i_3$$

$$R_3 i_2 - R_3 i_3 = \frac{L_2 d\,i_3}{dt} + R_2\,i_3 + \frac{1}{C_2}\int i_3\,dt$$

Based on these equations electrical analog based on force-voltage analogy is shown in Fig. 5.61

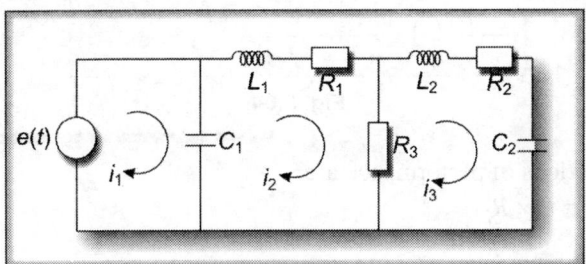

Fig. 5.61

Problem 5.23 Obtain differential equations for the system are shown in Fig. 5.62.

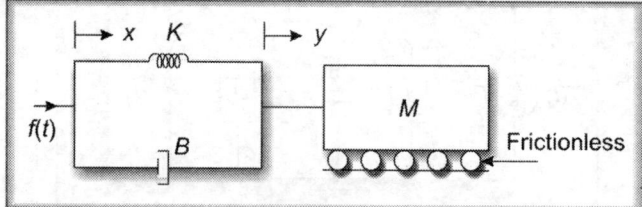

Fig. 5.62

Solution Mechanical network diagram is shown in Fig. 5.63. The equations of performance are:

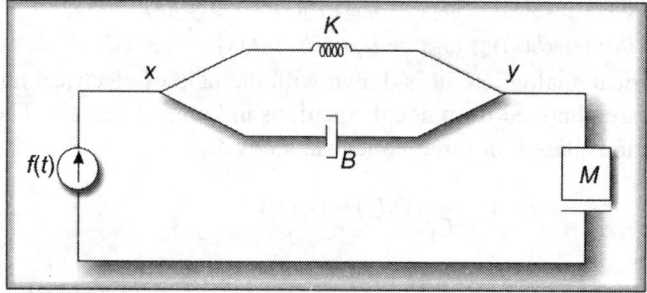

Fig. 5.63

$$f(t) = B(\dot{x} - \dot{y}) + K(x - y) = M\ddot{y}$$
$$F(s) = (Bs + K) X(s) - (Bs + K) Y(s) = Ms^2 Y(s)$$

Problem 5.24 Draw the mechanical network diagram of the system is shown in Fig. 5.64 and write the equations of performance.

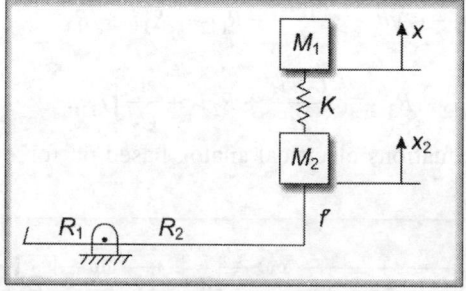

Fig. 5.64

Solution Equations of performance are:

$$f \times R_1 = f' \times R_2$$

or $\qquad f' = f \times \dfrac{R_1}{R_2}$; $f' = M_2\ddot{x}_2 + K(x_2 - x_1)$; and $K(x_2 - x_1) = M_1\ddot{x}_1$

Mechanical network diagram is shown in Fig. 5.65

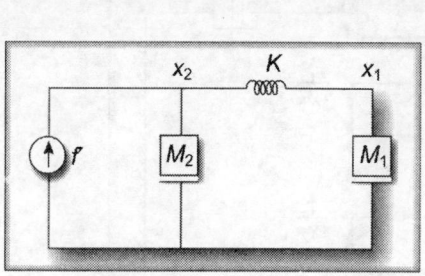

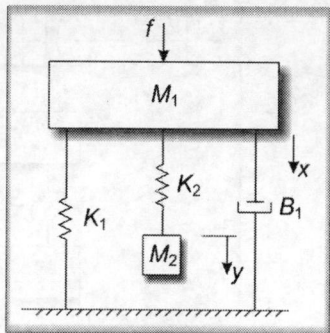

Fig. 5.65   Fig. 5.66

Problem 5.25 Figure 5.66 shows mechanical vibration absorber. Construct mechanical network diagram and obtain equations of performance.

Solution Mechanical network diagram is shown in Fig. 5.67. Equations of performance are:

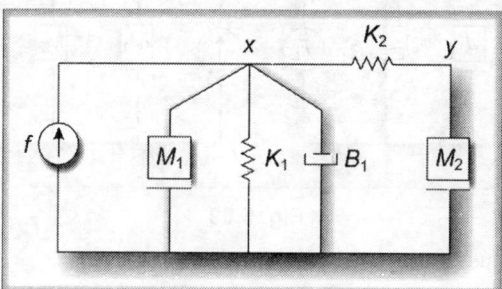

Fig. 5.67

$$f = M_1 \ddot{x} + B_1 \dot{x} + K_1 x + K_2 (x - y)$$

$$K_2 (x - y) = M_2 \ddot{y}$$

Problem 5.26 For the mechanical system is shown in Fig. 5.68, draw the schematic mechanical network and write the necessary differential (network) equations of performance of the system.

Solution The network diagram for the given mechanical network is shown in Fig. 5.69. The equations of performance are:

$$T(t) = J_1 \ddot{\theta} + B_1 \dot{\theta} + T_L \qquad (1)$$

$$T_L = f(t) \times r \qquad (2)$$

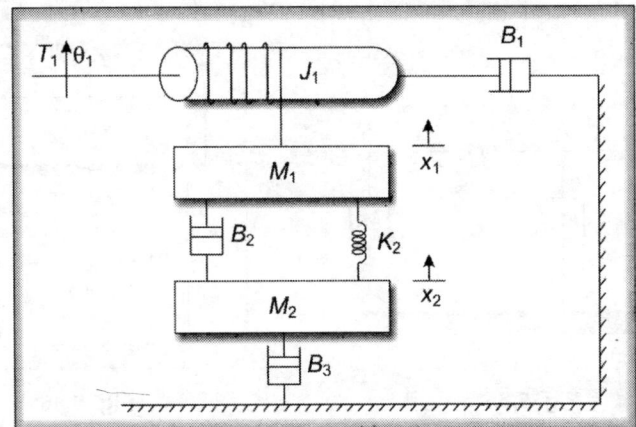

Fig. 5.68

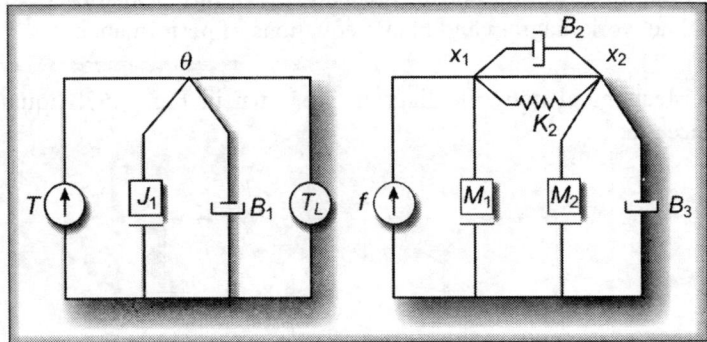

Fig. 5.69

where, r is the radius of the drum

$$x_1 = r\theta \tag{3}$$

$$f(t) = M_1\ddot{x}_1 + B_2(\dot{x}_1 - \dot{x}_2) + K_2(x_1 - x_2) \tag{4}$$

or

$$= M_1s^2x_1 + B_2s(x_1 - x_2) + K_2(x_1 - x_2) \tag{5}$$

$$B_2(\dot{x}_1 - \dot{x}_2) + K_2(x_1 - x_2) = M_2\ddot{x}_2 + B_3x_2 \tag{6}$$

or $\quad M_2s^2x_2 + B_3sx_2 + B_2s(x_2 - x_1) + K_2(x_2 - x_1) = 0 \tag{7}$

or

$$T = J_1 s^2 \theta + B_1 s\theta + T_L \tag{8}$$

$$= J_1s^2\theta + B_1s\theta + f(t) \times r \tag{9}$$

$$= J_1s^2\theta + B_1s\theta + r[M_1s^2x_1 + B_2s(x_1 - x_2) + K_2(x_1 - x_2)] \tag{10}$$

Problem 5.27 A process plant consists of two tanks of capacitance C_1 and C_2 respectively. If the flow into the top tank is Q_1, find the transfer function relating this flow, to the level in the bottom tank. Each tank has a resistance in its outlet pipe. The tanks be considered as non-interacting. Fig. 5.70 shows the process plant consisting of two tanks.

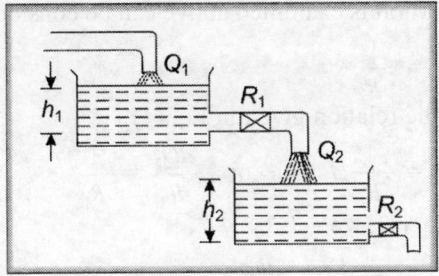

Fig. 5.70

Solution Let the resistances of the outlet pipes of the tanks be R_1 and R_2 respectively.

Top tank The rate of liquid stored in the top tank

$$= Q_1 - Q_2 = C_1 \frac{dh_1}{dt} \tag{1}$$

Differentiating and multiplying by R_1

$$R_1 \frac{dQ_1}{dt} - R_1 \frac{dQ_2}{dt} = R_1 C_1 \frac{d^2 h_1}{dt^2} \tag{2}$$

The relation between level of the liquid in a tank and outflow is $Q_2 = K\sqrt{h}$ where, K is the coefficient for the resistance. The plot is non-linear and is shown in Fig. 5.71. It must be linearised. The slope of the curve at any point is given by the gradient

$$\frac{dh}{dQ_2} = R \tag{3}$$

or

$$\frac{dQ_1}{dh} = \frac{1}{R} \tag{4}$$

Now $\qquad Q_2 = K\sqrt{h}$

Differentiating $\dfrac{dQ_2}{dh} = \dfrac{1}{2} \dfrac{K}{\sqrt{h}}$

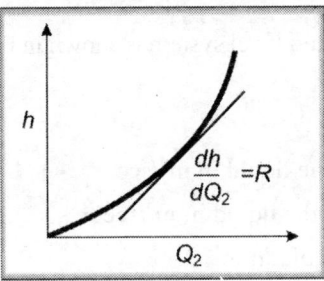

Fig. 5.71

Here, if a small portion is examined above can be considered constant and

$$\frac{1}{2}\frac{Q_2}{h} = \frac{1}{R} \tag{5}$$

Therefore, applying the relation given in Eq. (4)

$$Q_2(s) = \frac{1}{R_1}H_1(s) \quad \text{or} \quad \frac{dQ_2}{dh_1} = \frac{1}{R_1} \tag{6}$$

and substituting in Eq. (1), we get

$$R_1 C_1 \frac{d^2 h_1}{dt^2} = R_1 \frac{dQ_1}{dt} - \frac{dh_1}{dt} \tag{7}$$

Integrating $R_1 C_1 \dfrac{dh_1}{dt} = R_1 Q_1 - h_1$ $\tag{8}$

In Laplace domain $R_1 C_1 s H_1(s) = R_1 Q_1(s) - H_1(s)$

or $\qquad H_1(s) = \dfrac{R_1}{1 + R_1 C_1 s} Q_1(s)$ $\tag{9}$

Similarly, for the bottom tank

$$H_2(s) = \frac{R_2}{1 + R_2 C_2 s} Q_2(s) \tag{10}$$

Let $\qquad R_1 C_1 = T_1$, and $R_2 C_2 = T_2$

Then $\qquad H_1(s) = \dfrac{R_1}{1 + T_1 s} Q_1(s)$ $\tag{11}$

and $\qquad H_2(s) = \dfrac{R_2}{1 + T_2 s} Q_2(s)$ $\tag{12}$

From Eq. (6) $Q_2(s) = \dfrac{1}{R_1} H_1(s)$

Therefore, $\qquad H_2(s) = \dfrac{R_2}{R_1(1 + T_2 s)} H_1(s)$

Substituting the value of $H_{1(s)}$ from Eq. (11), we get

$$H_2(s) = \frac{R_1 R_2}{R_1 (1 + T_1 s)(1 + T_2 s)} Q_1(s)$$

or $\qquad H_2(s) = \dfrac{R_2}{(1 + T_1 s)(1 + T_2 s)} Q_1(s)$ **Ans**

Problem 5.28 For the liquid level system is shown in Fig. 5.72, obtain the governing mathematical model.

where

Q_i = inflow rate of the liquid in m^3/sec

Q_o = outflow rate of the liquid in m^3/sec

H_o = height of the liquid in m

Δ = change in respective values.

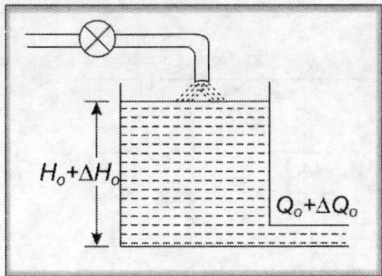

Fig. 5.72

Solution Figure 5.72 shows a tank supplying liquid through an outlet. Under steady conditions

$$Q_i = Q_o \qquad (1)$$

and H_o is the steady liquid level in the tank. Let ΔQ_i be the small increase in the liquid inflow rate which causes increase of ΔH_o in liquid level in the tank. The resultant increase of liquid outflow rate ΔQ_o is given as:

$$\Delta Q_o = \frac{\Delta H}{R} \qquad (2)$$

Liquid flow rate balance equation is:

(Rate of liquid in flow) – (rate of liquid outflow) = rate of liquid storage in the tank

or $$\Delta Q_i - \frac{\Delta H}{R} = C\frac{d(\Delta H)}{dt} \qquad (3)$$

where, C = capacitance of tank; R = resistance of outlet pipe.

or $$RC\frac{d(\Delta H)}{dt} \Delta H = R(\Delta Q_i) \qquad (4)$$

Problem 5.29 For the mechanical system is shown in Fig. 5.73, write the differential equations of performance with equivalent mechanical network.

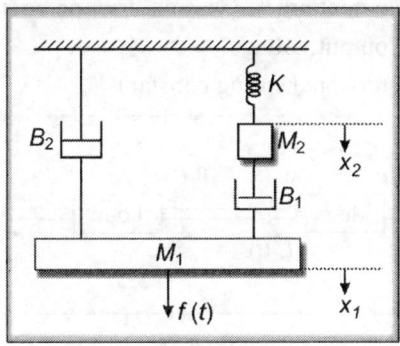

Fig. 5.73

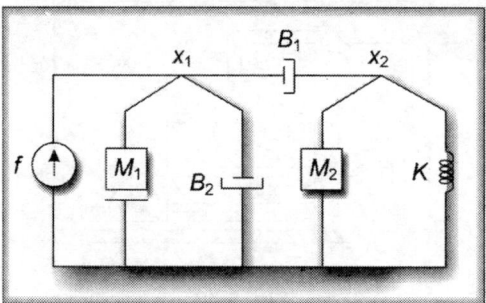

Fig. 5.74

Solution The equivalent mechanical network is shown in Fig. 5.74. Equations of performance are:

Node 'x_1'

$$f(t) = M_1\ddot{x}_1 + B_2\dot{x}_1 + B_1(\dot{x}_1 - \dot{x}_2)$$
$$= M_1s^2x_1 + B_2sx_1 + B_1sx_1 - B_1sx_2 = M_1s^2x_1 + B_2sx_1 + B_1s(x_1 - x_2)$$

Node 'x_2'

$$0 = M_2\ddot{x}_2 + Kx_2 + B_1(\dot{x}_2 - \dot{x}_1) = M_2s^2x_2 + Kx_2 + B_2s(x_2 - x_1)$$

Problem 5.30 An open-loop motor control system is shown in Fig 5.75. The potentiometer has a maximum range of 10 turns (20 π radians). Find the transfer function. The following parameters and variable are defined:

$\quad Q_m(t)$ = the motor displacement

$\quad Q_L(t)$ = the load displacement

$\quad T_m(t)$ = the motor torque

$\quad J_m$ = the motor inertia

$\quad B_m$ = the motor viscous-friction coefficient

$\quad B_p$ = the potentiometer viscous-friction coefficient

$\quad e_o(t)$ = the output voltage

$\quad K$ = the torsional spring constant

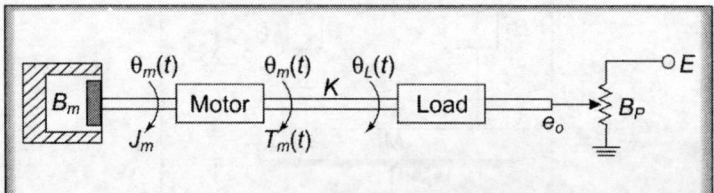

Fig. 5.75

Solution The equation describing the system dynamics are:

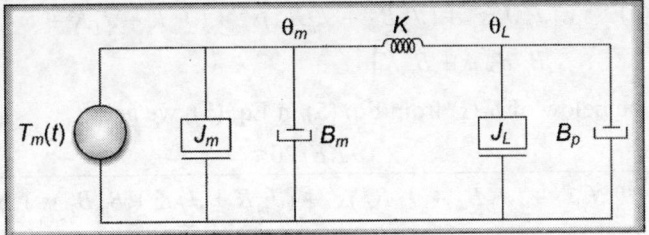

Fig. 5.76

$$J_m\ddot{\theta}_m + B_m\dot{\theta}_m + K(\theta_m(t) - \theta_L(t)) = T_m(t)$$

$$J_L\ddot{\theta}_L + B_P\dot{\theta}_L = K(\theta_m(t) - \theta_L(t))$$

Output equation is:

$$\frac{e_0(t)}{E(t)} = \frac{\theta_L(t)}{20\pi}$$

In Laplace transform from

$$s^2 J_m\theta_m(s) + sB_m\theta_m(s) + K\theta_m(s) - K\theta_L(s) = T_m(s) \qquad (1)$$

$$s^2 J_L\theta_L(s) + sB_P\theta_L(s) - K\theta_m(s) + K\theta_L(s) = 0 \qquad (2)$$

$$E_0(s) = \frac{\theta_L(s)E}{20\pi} \qquad (3)$$

From Eq. (2), we get

$$\theta_L(s)(s^2 J_L + sB_P + K) = K\theta_m(s)$$

or $\qquad \theta_m(s) = \dfrac{\theta_L(s)(s^2 J_L + sB_p + K)}{K} \qquad (4)$

From Eq. (1), we get

$$\theta_m(s)(s^2 J_m + sB_m + K) - K\theta_L(s) = T_m(s) \qquad (5)$$

Substituting the value of $\theta_m(s)$ from Eq. (4) in Eq. (5), gives

$$\frac{\theta_L(s)(s^2 J_L + sB_P + K)(s^2 J_m + sB_m + K)}{K} - K\theta_L(s) = T_m(s)$$

or $\qquad \dfrac{\theta_L(s)\left[(s^2 J_L + sB_P + K)(s^2 J_m + sB_m + K) - K^2\right]}{K} = T_m(s)$

or $\qquad \dfrac{\theta_L(s)}{T_m(s)} = \dfrac{K}{\left[(s^2 J_L + sB_P + K)(s^2 J_m + sB_m + K) - K^2\right]}$

or $\qquad \dfrac{\theta_L(s)}{T_m(s)} = \dfrac{K}{s\left[J_m J_L s^2 + (J_L B_m + J_m B_P)s^2 + (J_m K + J_L K + B_m B_P)s + B_m K\right]}$ (6)

Putting the below of $\theta_L(s)$ from Eq. (5) in Eq. (3), we get

$$\dfrac{E_0(s)}{T_m(s)} = \dfrac{KE/20\pi}{s\left[J_m J_L s^3 + (J_L B_m + J_m B_P)s^2 + (J_m K + J_L K + B_m B_P)s + B_m K)\right]} \quad \textbf{Ans.}$$

Problem 5.31 Find the transfer function $Y(s)/T_m(s)$ of the printwheel control system of a word processor whose simplified diagram is shown in Fig. 5.77.

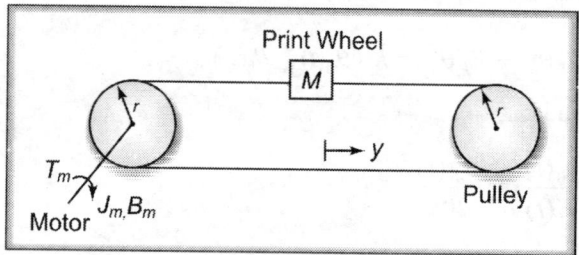

Fig. 5.77

Solution The equivalent inertia contributed by mass M that moves about the pulley of radius r (disregarding the inertia of pulley) that the motor sees is:

$$J = Mr^2$$

The mechanical diagram is shown in Fig. 5.78

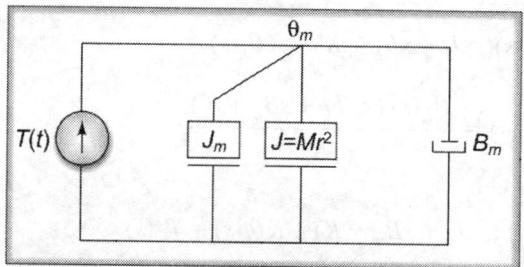

Fig. 5.78

The Torque equation at the motor shaft is:

$$T_m(t) = J_m \ddot{\theta}_m(t) + Mr^2 \theta_m(t) + B_m \theta_m(t) \qquad (1)$$

and the relation between the liner and rotational displacement is:

$$y(t) = r\theta_m(t) \qquad (2)$$

Laplace transform of Eqns (1) and (2) yields

$$T_m(s) = s^2 J_m \theta_m(s) + s^2 Mr^2 \theta_m(s) + sB_m \theta_m(s) \qquad (3)$$

and $\qquad\qquad Y(s) = s\theta_m(s) \qquad (4)$

From Eq. (3), we get

$$\frac{\theta_m(s)}{T_m(s)} = \frac{1}{s^2 J_m + s^2 M r^2 + s B_m}$$

or

$$\frac{\theta_m(s)}{T_m(s)} = \frac{1}{s\left[(J_m + M r^2)s + B_m\right]} \tag{5}$$

Substituting the value of $\theta_m(s)$ in Eq. (4) yields

$$\frac{Y(s)}{T_m(s)} = \frac{r}{s\left[(J_m + M r^2)s + B_m\right]} \qquad \textbf{Ans.}$$

Problem 5.32 For the mechanical system is shown in Fig. 5.79, write the equilibrium equations and draw an equivalent diagram based on force-voltage analogy.

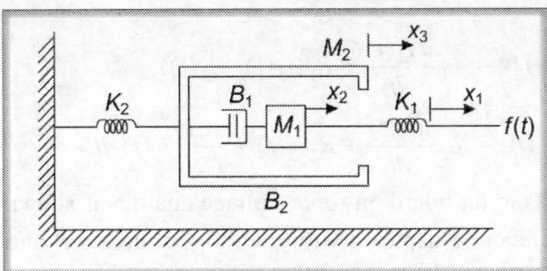

Fig. 5.79

Solution The mechanical diagram is shown in Fig. 5.80. The equilibrium equations are:

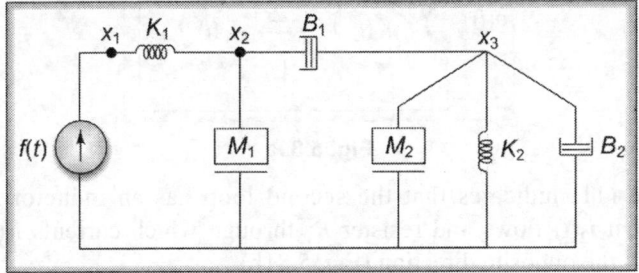

Fig. 5.80

$$f(t) = K_1(x_1 - x_2) \tag{1}$$

$$K_1(x_1 - x_2) = M_1 \ddot{x}_2 + B_1(\dot{x}_2 - \dot{x}_3) \tag{2}$$

$$B_1(\dot{x}_2 - \dot{x}_3) = M_2 \ddot{x}_3 + B_2 \dot{x}_3 + K_2 x_3 \tag{3}$$

Taking Laplace transform of the equations

$$F(s) = K_1(X_1(s) - X_2(s)) \tag{4}$$

$$K_1(X_1(s) - X_2(s)) = s^2 M_1 X_2(s) + s B_1 (X_2(s) - X_3(s)) \tag{5}$$

$$sB_1(X_2(s) - X_3(s)) = s^2M_2X_3(s) + sB_2X_3(s) + K_2X_3(s) \qquad (6)$$

Using $M_2 = L_2$; $M_1 = L_1$; $F(s) = E(s)$; $B_1 = R_1$; $B_2 = R_2$; $K_1 = 1/C_1$; $X = Q$ and $K_2 = 1/C_2$ as per force voltage analogy, we get

$$E(s) = \frac{1}{C_1}(Q_1(s) - Q_2(s)) \qquad (7)$$

$$\frac{1}{C_1}(Q_1(s) - Q_2(s)) = s^2L_1Q_2(s) + s\,R_1(Q_2(s) - Q_3(s)) \qquad (8)$$

$$sR_1(Q_2(s) - Q_3(s)) = s^2M_2Q_3(s) + sR_2Q_3(s) + \frac{1}{C_2}Q_3(s) \qquad (9)$$

Converting the above equations into differential equations

$$e(t) = \frac{1}{C_1}\int(i_1(t) - i_2(t))dt \qquad (10)$$

$$\frac{1}{C_1}\int(i_1(t) - i_2(t)) = L_1\frac{di_2(t)}{dt} + R_1(i_2(t) - i_3(t)) \qquad (11)$$

$$R_1(i_2(t) - i_3(t)) = L_2\frac{di_3(t)}{dt} + R_2 i_3(t) + \frac{1}{C_2}\int i_3(t)\,dt \qquad (12)$$

The electrical circuit board on force-voltage analogy is shown in Fig. 5.81

Equation (10) can be depicted with a voltage source $e(t)$ and a capacitor C_1; through which currents $i_1(t)$ and $i_2(t)$ flows, but in opposite direction (Fig. 5.81a). Hence, there are two loops.

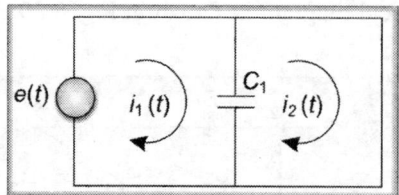

Fig. 5.81a

Equation (11) indicates that the second loop has an inductor L_1 through which current $i_2(t)$ flows and resister R_1 through which currents $i_2(t)$ and $i_3(t)$ flows; but in the opposite direction (Fig. 5.81b)

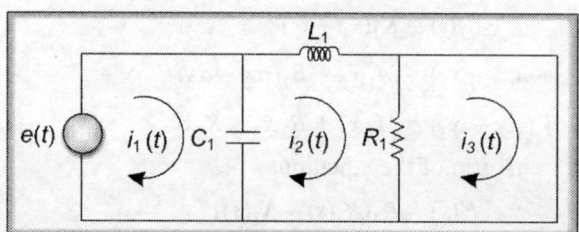

Fig. 5.81b

Eqn. (12) indicates that the third loop has an inductor L_2, resister R_2 and capacitor C_2 through which current $i_3(t)$ flows (Fig. 5.81c). The complete equivalent diagram based on force-voltage analogy is shown in Fig. 5.81c.

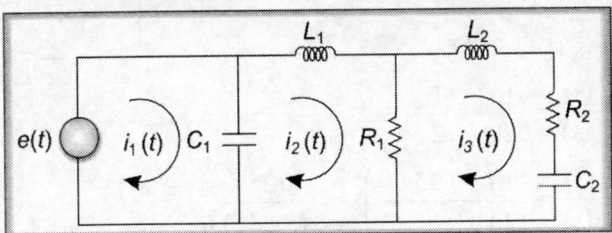

Fig. 5.81c

Problem 5.33 A load driven by a set up of motor and generator is shown in Fig. 5.82. If the generator output voltage is $e_g(t) = K_f i_f(t)$ where, $i_f(t)$ is the generator field current, find the transfer function, $G(s) = \theta_0(s)/E_i(s)$. For the generator, $K_f = 2\Omega$ and for the motor $K_t = 1$ Nm/A and $K_b = 1$Vs/rad

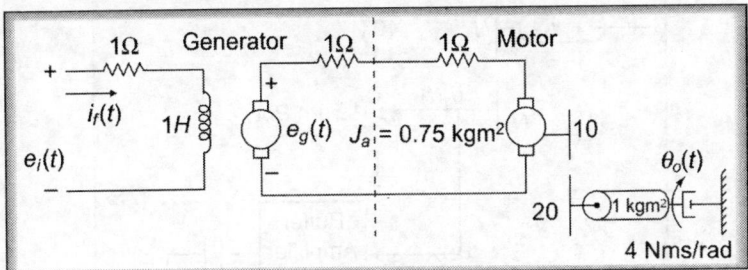

Fig. 5.82

Solution Generator

$$E_g(s) = K_f I_f(s) \text{ and } I_f(s) = \frac{E_i(s)}{R_f + sL_f}$$

Therefore, $\dfrac{E_g(s)}{E_i(s)} = \dfrac{K_f}{R_f + sL_f} = \dfrac{2}{1+s}$

Motor $\qquad R_a = $ Sum of both the resisters $= 1 + 1 = 2\Omega$

$$J_e = J_a + J_L \left(\frac{N_1}{N_2}\right)^2 = 0.75 + 1 \times \left(\frac{10}{20}\right)^2 = 1$$

$$B_e = B_L \left(\frac{N_1}{N_2}\right)^2 = 4 \times \left(\frac{10}{20}\right)^2 = 1$$

Therefore, $\dfrac{\theta_m(s)}{E_g(s)} = \dfrac{K_t/R_aJ_e}{s\left(s + \dfrac{1}{J_e}\left(B_e + \dfrac{K_t K_a}{R_a}\right)\right)} = \dfrac{\dfrac{1}{2\times1}}{s\left(s+1\left(1+\dfrac{1}{2\times1}\right)\right)} = \dfrac{0.5}{s(s+1.5)}$

We know that

$$\frac{\theta_0(s)}{\theta_m(s)} = \frac{10}{20} = \frac{1}{2}$$

Therefore, $\dfrac{\theta_0(s)}{E_g(s)} = \dfrac{1}{2}\left(\dfrac{0.5}{s(s+1.5)}\right) = \dfrac{0.25}{s(s+1.5)}$

Also $\dfrac{\theta_0(s)}{E_i(s)} = \dfrac{E_g(s)}{E_i(s)} \times \dfrac{\theta_0(s)}{E_g(s)} = \left(\dfrac{2}{s+1}\right)\left(\dfrac{0.25}{s(s+1.5)}\right) = \dfrac{0.5}{s(s+1)(s+1.5)}$

Problem 5.34 Find the transfer function $E_0(s)/T(s)$ for the systems is shown in Fig. 5.83

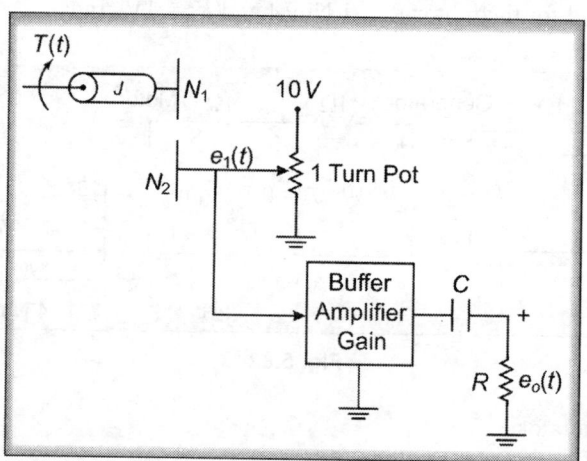

Fig. 5.83

Solution Writing the input-output equation of the mechanical system

$$T(s)\left(\frac{N_2}{N_1}\right) = J\left(\frac{N_2}{N_1}\right)s^2\theta_2(s) \tag{1}$$

Potentiometer

$$E_i(s) = \frac{10}{2\pi}\theta_2(s) \quad \text{or} \quad \theta_2(s) = \frac{\pi}{5}E_i(s) \tag{2}$$

Electrical circuit

$$E_0(s) = \frac{R}{R + \dfrac{1}{sC}}E_i(s) = \frac{s}{s+1/RC}E_i(s)$$

or $\qquad E_i(s) = \dfrac{s + 1/RC}{s} E_o(s)$ (3)

Substituting the value of $E_i(s)$ from Eq. (3) in Eq. (2), we get

$$\theta_2(s) = \frac{\pi}{5} E_i(s) = \left(\frac{\pi}{5}\left(\frac{s + 1/RC}{s}\right) E_o(s)\right)$$ (4)

Substituting in Eq. (1), we get

$$T(s)\frac{N_2}{N_1} = J\left(\frac{N_2}{N_1}\right)^2 s^2 \frac{\pi}{5}\left(\frac{s + 1/RC}{s}\right) E_o(s)$$

or $\qquad \dfrac{E_o(s)}{T(s)} = \dfrac{\dfrac{5}{J\pi}\left(\dfrac{N_1}{N_2}\right)}{s\left(s + \dfrac{1}{RC}\right)}$

Problem 5.35 Find the transfer function $E_0(s)/T(s)$ for the system is shown in Fig. 5.84

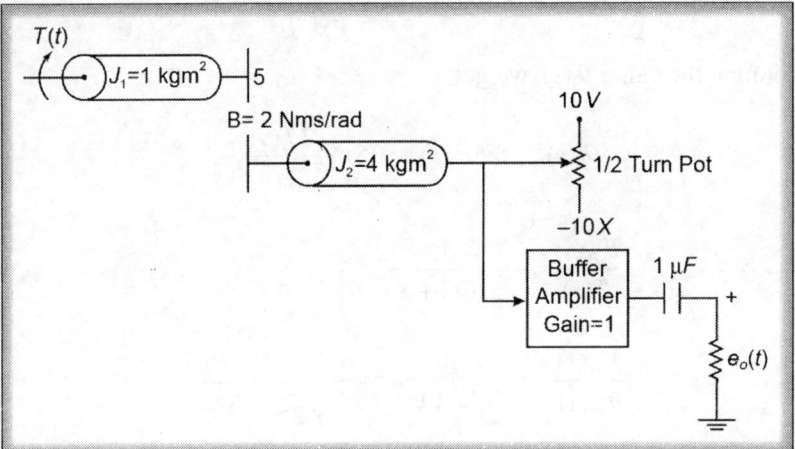

Fig. 5.84

Solution Reflecting mechanical impedances to the viscous damper side is shown in Fig. 5.85

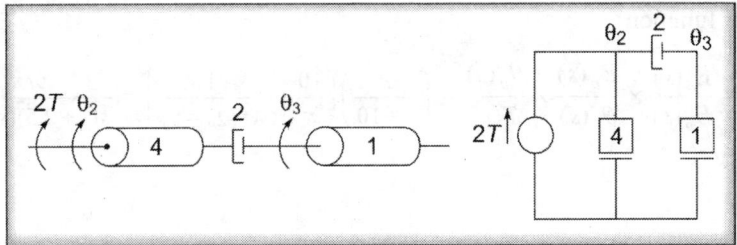

Fig. 5.85

Writing the equations

Mechanical System

$$4 s^2 \theta_2(s) + 2s(\theta_2(s) - \theta_3(s)) = 2T(s)$$

$$s^2 \theta_3(s) + 2s(\theta_3(s) - \theta_2(s)) = 0$$

or

$$(4s^2 + 2s)\, \theta_2(s) - 2s\theta_3(s) = 2T(s)$$

$$-2s\theta_2(s) + (s^2 + 2s)\theta_3(s) = 0$$

Solving for $\theta_3(s)$

$$\theta_3(s) = \frac{\begin{vmatrix} 4s^2 + 2s & T(s) \\ -2s & 0 \end{vmatrix}}{\begin{vmatrix} 4s^2 + 2s & -2s \\ -2s & s^2 + 2s \end{vmatrix}} = \frac{2\,T(s)}{s^2(2s+5)}$$

Potentiometer $\quad \theta_P(s) = 0.5\theta_3(s) \quad$ or $\quad \dfrac{\theta_P(s)}{\theta_3(s)} = 0.5$

Substituting the value $\theta_3(s)$, we get

$$\theta_P(s) = 0.5\theta_3(s) = 0.5\left(\frac{2\,T(s)}{s^2(2s+5)}\right)$$

or

$$\frac{\theta_P(s)}{T(s)} = \frac{1}{s^2(2s+5)}$$

Also

$$\frac{E_P(s)}{\theta_P(s)} = \frac{10}{(1/2)\,\text{turn}} = \frac{10}{\dfrac{1}{2} \times 2\pi} = \frac{10}{\pi}$$

Electrical circuit $\quad \dfrac{E_o(s)}{E_P(s)} = \dfrac{R}{R + \dfrac{1}{sc}} = \dfrac{100000}{100000 + \dfrac{1}{s \times 10^{-6}}} = \dfrac{s}{s+10}$

Transfer function

$$\frac{E_o(s)}{T(s)} = \frac{E_o(s)}{E_P(s)} \times \frac{E_P(s)}{\theta_p(s)} \times \frac{\theta_p(s)}{T(s)} = \left(\frac{s}{s+10}\right)\left(\frac{10}{\pi}\right)\left(\frac{1}{s^2(2s+5)}\right) = \frac{5/\pi}{s(s+2.5)(s+10)}$$

Time Domain Analysis

6.1 Introduction

It is a matter of interest to any control engineer to ascertain the performance of the control system and study its behaviour. Study and analysis of a control system with respect to time, is one of the methods employed to describe the performance of a feedback control system. Time response of a control system is evaluated since, time is used as an independent variable in most of the feedback control systems. Time response of a control system can be divided into two parts, *the transient* and the *steady state* response. All control systems exhibit transients before a steady state is attained, because of energy storing elements like mass, inertia, etc. which are always a part of a control system and cannot be avoided. These energy storing elements exhibit transients when subjected to input. This phenomenon is seen in control systems before a steady state is reached. Transients may be in the form of oscillations and may be sustained or decaying in nature. Steady state response is that part of the time response which is fixed when time approaches infinity. It is infact that part which remains when transients have died out. This points towards the accuracy of the control system. Time response is thus

$$c(t) = c_t(t) + c_{ss}(t) \qquad (6.1)$$

where $\qquad c(t)$ = time response,

$c_t(t)$ = transient response, and

$c_{ss}(t)$ = steady state response.

It would not be correct to subject the system to the random behaviour unlikely to experience during its operational life. A system is generally designed and analysed for a certain anticipated type of inputs generally experienced, namely the *step*, the *ramp* and the *parabolic* input signals. These are referred to as *standard test signals*.

6.2 Standard Test Signals

(a) *Step signal*: A step signal is a signal whose value changes from one level to another level in zero time. Graphically, the step signal is illustrated in Fig. 6.1a. Mathematically, the step signal is represented as given below:

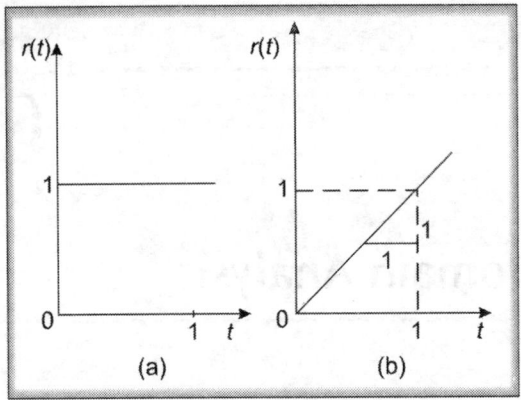
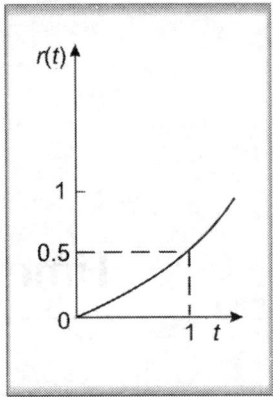

Fig. 6.1a and b   **Fig. 6.2**

$$r(t) = u(t), \tag{6.2}$$

where, $u(t) = 1; \quad t > 0$

              $= 0; \quad t < 0$

In the Laplace transform form $R(s) = \dfrac{1}{s}$ \hfill (6.3)

(b) *Ramp Signal*: A ramp signal is a signal which changes gradually with the time in a linear fashion. Graphical representation is shown in Fig. 6.1b. Mathematically, it can be represented as:

$$r(t) = t; \quad t > 0 \tag{6.4}$$

              $= 0; \quad t < 0$

In the Laplace transform form, it is written as $R(s) = \dfrac{1}{s^2}$ \hfill (6.5)

It is also referred to as *velocity input*.

(c) *Parabolic Signal*: It is also referred to as *acceleration input* and is graphically represented in Fig. 6.2 mathematically

$$r(t) = \frac{t^2}{2}; \quad t > 0 \tag{6.6}$$

              $= 0; \quad t < 0$

In the Laplace transform form $R(s) = \dfrac{1}{s^3}$ \hfill (6.7)

Note: Parabolic signal is integral of ramp signal and ramp signal is integral of step signal.

6.3 Static Accuracy

Accuracy is one of the most important criteria to describe the performance of control systems besides others, namely *stability, sensitivity* and *transient response.* The aim of the control system designers is to design a system having minimum error to certain anticipated class of inputs. This section deals with the methods, which will help us in determining the accuracy of the system.

6.3.1 Steady-state error
Steady-state error gives an idea to the system designer as to how accurate is the designed system. Steady-state error may be caused in the system due to the nature of input signal, type of system and non-linearities present in the system. A system may show steady-state error due to one type of input and the same may be absent due to other inputs. Consider a unity feedback system is shown in Fig. 6.3.

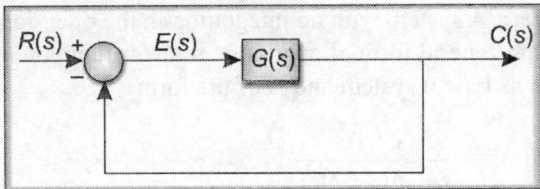

Fig. 6.3

The transfer function of the system as shown in Fig. 6.3 is:

$$\frac{C(s)}{R(s)} = \frac{G(s)}{1 + G(s)} \tag{6.8}$$

Also, $$C(s) = E(s) \, G(s) \tag{6.9}$$

$$\frac{E(s)G(s)}{R(s)} = \frac{G(s)}{1 + G(s)} \tag{6.10}$$

or $$E(s) = \frac{R(s)}{1 + G(s)} \tag{6.11}$$

Steady-state error is determined by application of final value theorem of Laplace transform. It is expressed as:

$$e_{(ss)} = \lim_{t \to \infty} e(t) \tag{6.12}$$

$$= \lim_{s \to 0} sE(s) \tag{6.13}$$

$$= \lim_{s \to 0} \frac{sR(s)}{1 + G(s)} \tag{6.14}$$

From Eqn. (6.13) it is seen that steady state depends upon the *input* and the *forward path transfer function G(s).*

6.3.2 Type of System

This is the way of designating a control system and is different from that of the order of a system. It is assumed that open-loop transfer function is of the form

$$G(s)H(s) = \frac{K(1 + sT_a)(1 + sT_b) + ... + (1 + sT_m)}{s^n(1 + sT_1)(1 + sT_2) + ... + (1 + sT_n)}$$ (6.15a)

where,

K = gain

s^n = represents pole of multiplicity n at the origin of the complex plane.

Type or *number* of system is governed by the number of integrations present in the open-loop transfer function and is thus, numerically equal to the number of poles of open-loop transfer function at origin of the complex plane.

(a) **Type 0 System:** A system with no integration in the open-loop transfer function or if in the general form of open-loop transfer function $n = 0$, the system is designated as type 0 system and is of the form

$$G(s)H(s) = \frac{K(s + 1)...}{(s + 2)(s + 3)...}$$ (6.15b)

(b) **Type 1 System:** A system with one integration in the open-loop transfer function or if in the general form of open-loop transfer function $n = 1$, then the system is designated as type 1 system and is of the form

$$G(s)H(s) = \frac{K(s + 1)...}{s(s + 2)(s + 3)...}$$ (6.16)

(c) **Type 2 System:** A system with two integration in the open-loop transfer function or if in the general form of open-loop transfer function $n = 2$, then the system is designated as type 2 system and is of the form

$$G(s)H(s) = \frac{K(s + 1)...}{s^2(s + 2)(s + 3)...}$$ (6.17)

6.3.3 Error Constants

Error constants or coefficients are the measure of steady-state errors and give an idea as to how steady-state errors can be reduced or totally eliminated.

(a) **Position error constant:** Position error constant is defined for a unit-step input.

In the Laplace form unit step input is $R(s) = 1/s$

$$e(ss) = \lim_{s \to 0} \frac{sR(s)}{1 + G(s)}$$

For unit step input

$$e(ss) = \lim_{s \to 0} \frac{s \times \frac{1}{s}}{1 + G(s)} = \lim_{s \to 0} \frac{1}{1 + G(s)} = \frac{1}{1 + G(o)} = \frac{1}{1 + K_p} \qquad (6.18)$$

where, $K_p = G(o)$ is defined as position error constant. $\qquad\qquad$ (6.19)

(b) *Velocity error constant*: Velocity error constant is defined for *unit ramp input,* i.e.

$R(s) = 1/s^2$

$$e(ss) = \lim_{s \to 0} \frac{sR(s)}{1 + G(s)} = \lim_{s \to 0} \frac{s \times \frac{1}{s^2}}{1 + G(s)} = \lim_{s \to 0} \frac{1}{s + sG(s)} = \lim_{s \to 0} \frac{1}{sG(s)} \qquad (6.20)$$

$$= \frac{1}{K_v} \text{ where, } K_v = \lim_{s \to 0} sG(s) \text{ is defined as velocity error constant.} \quad (6.21)$$

(c) *Acceleration error constant acceleration*: Acceleration error constant is defined for unit parabolic input, i.e.

$$R(s) = \frac{1}{s^3}$$

$$e(ss) = \lim_{s \to 0} \frac{sR(s)}{1 + G(s)} = \lim_{s \to 0} \frac{s \times \frac{1}{s^3}}{1 + G(s)} = \lim_{s \to 0} \frac{1}{s^2 + s^2 G(s)}$$

$$= \lim_{s \to 0} \frac{1}{s^2 G(s)} = \frac{1}{K_a} \qquad\qquad (6.22)$$

where, $K_a = \lim_{s \to 0} s^2 G(s)$ is defined as acceleration error constant. $\qquad$ (6.23)

Steady-state error constants describe steady-state errors when the inputs are three basic types—step, ramp and parabolic. For other types of inputs, no identification on steady-state error is available. Also, the manner in which steady-state errors will alter, is also not indicated.

6.4 Computation of Steady-state errors

(a) *Unit Step Input*: For a unit-step input the steady-state error in terms of position constant is given in Eqn. (6.18), i.e.

$$e(ss) = \frac{1}{1 + K_p} \quad \text{where, } K_p = \lim_{s \to 0} G(s)$$

Table 6.1 summarises the value of K_p as a function of type of system and corresponding steady-state error.

Table 6.1

Type of System	Position Constant	Steady-state error
0	K_p	$\dfrac{1}{1+K_p}$
1	∞	0
2	∞	0
.	.	.
.	.	.
.	.	.
n, where $n > 0$	∞	0

(b) Unit Ramp Input: For a unit-ramp input, the steady-state error in terms of velocity constant is given in Eqn. (6.21), i.e.

$$e(ss) = \frac{1}{K_v} \text{ where, } K_v = \lim_{s \to 0} sG(s)$$

Table 6.2 summarises the value of K_v as a function of type of system and corresponding steady-state error.

Table 6.2

Type of System	Velocity Constant	Steady-state error
0	0	∞
1	K_V	$\dfrac{1}{K_v}$
2	∞	0
.	.	.
.	.	.
.	.	.
n, where $n > 1$	∞	0

(c) Unit Parabolic Input: For a unit parabolic input, the steady-state error in terms of acceleration given in Eqn. (6.23), i.e.

$$e(ss) = \frac{1}{K_a} \quad \text{where,} \quad K_a = \lim_{s \to 0} s^2 G(s)$$

Table 6.3 summarises the value of K_a as a function of type of system and corresponding steady-state errors.

Table 6.3

Type of System	Acceleration Constant	Steady-state error
0	0	∞
1	0	∞
2	K_a	$\dfrac{1}{K_a}$
3	∞	0
.	.	.
.	.	.
.	.	.
n, where $n > 2$	∞	0

From the results summarised in Tables 6.1 to 6.3, students can compare the capabilities of various types of systems. It is very evident that error constants are either zero, finite or infinite. The magnitude of error constants will proportionally increase, if the inputs are greater than the unit value. Dimensionally, the error constants have the following units:

- Position error constant: no dimension;

- Velocity error constant: $\dfrac{1}{\sec}$; and

- Acceleration error constant: $\dfrac{1}{\sec^2}$

6.5 Transient Response: First Order System

Unit Step Input Let us consider a first order system whose block diagram is shown in Fig. 6.4.

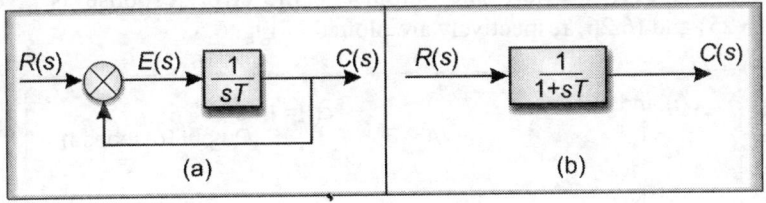

(a)   (b)

Fig. 6.4

The transfer function is:

$$\frac{C(s)}{R(s)} = \frac{1}{1+sT} \qquad (6.24)$$

For a unit-step input $\quad R(s) = \dfrac{1}{s}$

$$C(s) = \frac{1}{s(1+sT)} = \frac{A}{s} + \frac{B}{1+sT}$$

Solving for the A and B

$$A = sC(s)\Big|_{s=0} = s \times \frac{1}{s(1+sT)}\Big|_{s=0} = 1$$

$$B = (1+sT)\,C(s)\Big|_{s=-\frac{1}{T}} = (1+sT) \times \frac{1}{s(1+sT)}\Big|_{s=-\frac{1}{T}} = \frac{1}{s}\Big|_{s=-\frac{1}{T}} = -T$$

Therefore,

$$C(s) = \frac{1}{s} + \frac{-T}{(1+sT)} = \frac{1}{s} - \frac{T}{(1+sT)}$$

Output Response

Taking inverse Laplace transform, we get

$$c(t) = (1 - e^{-t/T}) \tag{6.25}$$

It can be seen that the output rises exponentially from zero value to unity.

Error Response

The error response is given by

$$e(t) = r(t) - c(t) = 1 - (1 - e^{-t/T}) = e^{-t/T} \tag{6.26}$$

Steady-state error: Steady-state error is given by

$$e(ss) = \lim_{t \to \infty} e(t) = \lim_{t \to \infty} e^{-t/T} = 0 \tag{6.27}$$

As the time approaches infinity, the output follows the input with zero steady-state error.

Graphical Representation The output and the error response as given by Eqns (6.25) and (6.26) respectively are plotted in Fig. 6.5

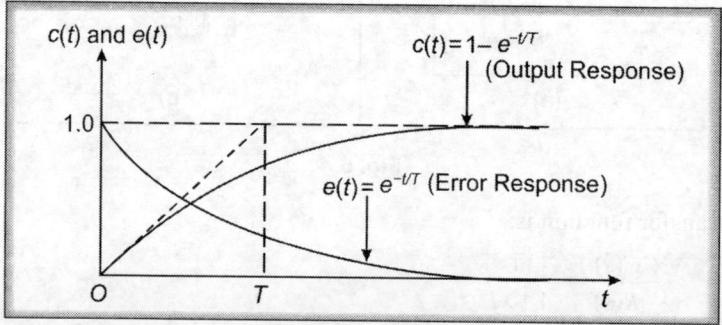

Fig. 6.5 First Order System: Unit step output and error response

Initial Slope of Curve: Initial slope of the curve is found out by putting $t = 0$ and differentiating the Eqn. (6.25).

$$\frac{dc}{dt}\bigg|_{t=0} = \frac{1}{T} e^{-t/T}\bigg|_{t=0} = \frac{1}{T} \tag{6.28}$$

Time Constant: T in Eqn. (6.28) is termed the *time constant* of the system. Time constant of a system is related to the speed of the response and gives an indication as to how quickly the system response reaches its final value. Smaller the time constant faster is the system response and larger its value, slower is the response indicating that the system response in sluggish. Table 6.4 lists the value of Eqn. (6.25).

Table 6.4

t/T	$e^{-t/T}$	$c(t)$
0.1	0.90	0.1
0.2	0.82	0.18
0.4	0.67	0.33
0.6	0.55	0.45
0.8	0.45	0.55
1.0	0.37	0.63
2.0	0.14	0.86
3.0	0.05	0.95
4.0	0.02	0.98

The tabulated data is plotted in Fig. 6.6

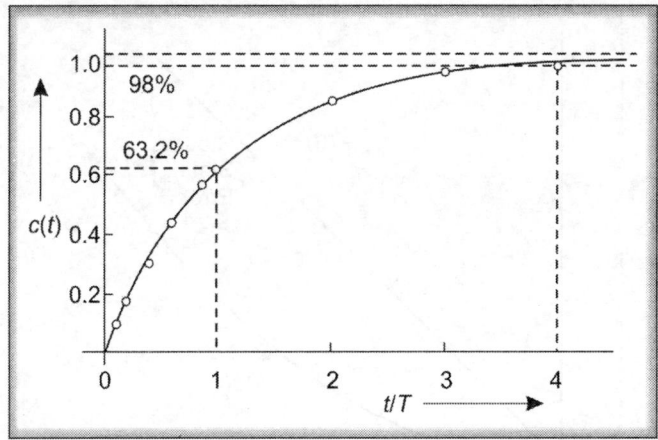

Fig. 6.6

Unit Ramp Input For a unit-ramp input

$$r(t) = t$$

i.e. $R(s) = \dfrac{1}{s^2}$ (In Laplace form)

Therefore, the transfer function is $\dfrac{C(s)}{R(s)} = \dfrac{1}{1 + sT}$

or $C(s) = \dfrac{1}{s^2(1 + sT)}$; $\dfrac{A}{s} + \dfrac{B}{s^2} + \dfrac{C}{1 + sT}$

Solving for A, B, and C, we get $A = -T$, $B = 1$ and $C = T^2$

Therefore,

$$C(s) = \frac{-T}{s} + \frac{1}{s^2} + \frac{T^2}{1 + sT} ; \text{ or}$$

$$C(s) = \frac{1}{s^2} + \left(-\frac{T}{s} + \frac{T^2}{1 + sT} \right)$$

Output Response

$$c(t) = t - T + Te^{-t/T} = t - T(1 - e^{-t/T}) \tag{6.29}$$

Error Response

$$e(t) = r(t) - c(t) = t - t + T(1 - e^{-t/T}) = T(1 - e^{-t/T}) \tag{6.30}$$

Steady-state error

$$e(ss) = \lim_{t \to \infty} e(t) = \lim_{t \to \infty} T(1 - e^{-1/T}) = T \tag{6.31}$$

The output response thus will track the unit ramp input by a steady-state error numerically equal to T. The response of such a system is plotted in Fig. 6.7

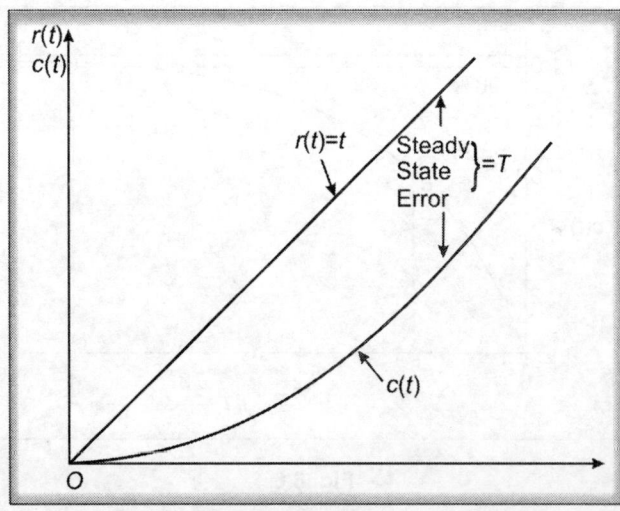

Fig. 6.7   First Order System: Unit ramp output
and error response

6.6 Transient Response: Second Order System

Let us consider a second order system whose block diagram is shown in Fig. 6.8

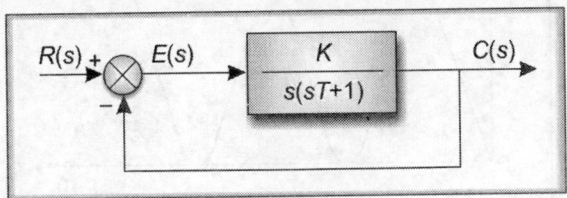

Fig. 6.8

The closed-loop transfer function of the system is given as:

$$\frac{C(s)}{R(s)} = \frac{K}{Ts^2 + s + K} = \frac{K/T}{s^2 + \frac{1}{T}s + \frac{K}{T}} \tag{6.32}$$

This can also be written in the following form:

$$\frac{C(s)}{R(s)} = \frac{\omega_n^2}{s^2 + 2\xi\omega_n s + \omega_n^2}; \quad \text{where,} \tag{6.33}$$

$$\xi = \text{damping ratio or factor} = \frac{1}{2\omega_n T}; \quad \omega_n = \text{undamped natural frequency} = \sqrt{\frac{K}{T}}$$

Characteristic Equation: The denominator polynomial $s^2 + 2\xi\omega_n s + \omega_n^2 = 0$ gives the roots of the equation. These roots are the poles of the transfer function. This equation is called the *characteristic equation*. The nature of roots of the characteristic equation gives us an indication of the time response. The exact solution of the output response is governed by the value of ξ and is discussed in detail below. Let us consider unit step input. Therefore

$$R(s) = \frac{1}{s}.$$ The Laplace transform of the output can be written as:

$$C(s) = \frac{\omega_n^2}{s(s^2 + 2\xi\omega_n s + \omega_n^2)} \tag{6.34}$$

$$= \frac{\omega_n^2}{s(s + \xi\omega_n + j\omega_n\sqrt{1 - \xi^2})(s + \xi\omega_n - j\omega_n\sqrt{1 - \xi^2})} \tag{6.35}$$

Case 1: **Damping Factor Equals Zero:** When $\xi = 0$, the output response Eqn. 6.34 is written as:

$$C(s) = \frac{\omega_n^2}{s(s^2 + \omega_n^2)} = \frac{1}{s} - \frac{s}{s^2 + \omega_n^2} \tag{6.36}$$

$$c(t) = 1 - \cos\omega_n t \tag{6.37}$$

The output response is plotted in Fig. 6.9 and is not damped but oscillatory in nature with angular frequency of ω_n rad/sec.

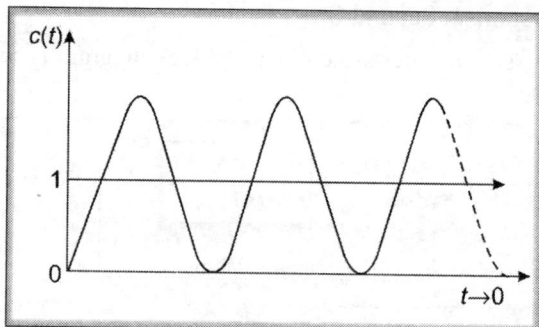

Fig. 6.9 Time response ($\xi = 0$), undamped case

Case II: **Damping Factor Equals Unity:** Where $\xi = 1$, the output response Eqn. (6.34) is written as:

$$C(s) = \frac{\omega_n^2}{s(s+\omega_n)^2} \tag{6.38}$$

$$C(s) = \frac{-\omega_n}{(s+\omega_n)^2} - \frac{1}{(s+\omega_n)} + \frac{1}{s} \quad \text{and} \tag{6.39}$$

$$c(t) = -\omega_n t e^{-\omega_n t} - e^{-\omega_n t} + 1 = 1 - e^{-\omega_n t}(\omega_n t + 1) \tag{6.40}$$

$$c_{ss} = \lim_{t \to \infty} c(t) = 1$$

The output response is plotted in Fig. 6.10. The response is termed critically damped and exhibits no overshoot. There are no oscillations but the response is slow.

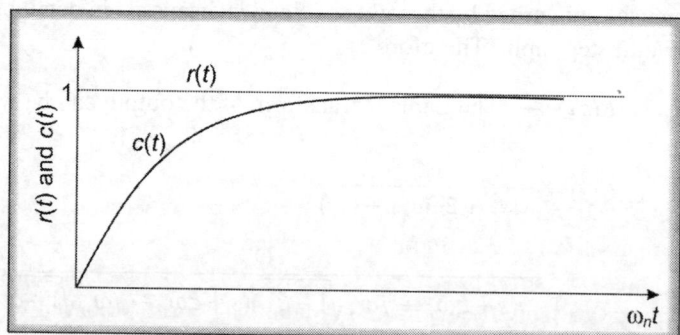

Fig. 6.10 Time response ($\xi = 1$), critically damped case

Case III: **Damping Factor Greater than Unity:** When $\xi > 1$, the roots are negative, real and unequal and the output response Eqn. (6.35) written in the partial fraction form is:

$$C(s) = \frac{A}{s} + \frac{B}{s + \xi\omega_n - \omega_n\sqrt{\xi^2 - 1}} + \frac{C}{s + \xi\omega_n + \omega_n\sqrt{\xi^2 - 1}} \tag{6.41}$$

Solving for *A*, *B*, *C*, we get

$$A = 1; \quad B = \frac{1}{\left[2(\xi^2 - \xi\sqrt{\xi^2 - 1} - 1)\right]}; \quad \text{and} \quad C = \frac{1}{\left[2(\xi^2 + \xi\sqrt{\xi^2 - 1} - 1)\right]},$$

Substituting the values of *A*, *B* and *C* in Eqn. (6.41) and then taking inverse Laplace transform, we get

$$C(t) = 1 - \frac{e^{-(\xi - \sqrt{\xi^2 - 1})\omega_n t}}{\left[2(\xi^2 - \xi\sqrt{\xi^2 - 1} - 1)\right]} + \frac{e^{-(\xi + \sqrt{\xi^2 - 1})\omega_n t}}{\left[2(\xi^2 + \xi\sqrt{\xi^2 - 1} - 1)\right]} \qquad (6.42)$$

The output response includes two exponentially decaying terms and does not exhibits overshoots and takes longer time to reach the final value as compared to $\xi = 1$. The response is considered as overdamped and is shown in Fig. 6.11

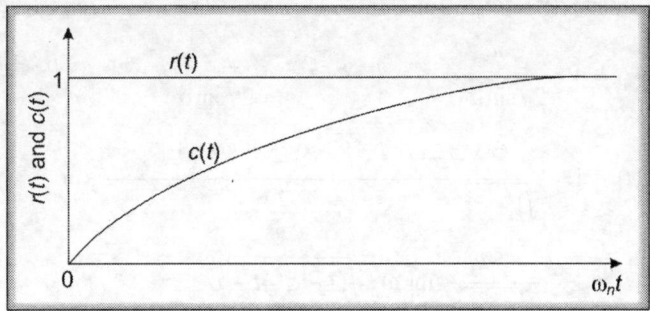

Fig. 6.11 Overdamped response

Case IV: **Damping Factor Less than Unity:** When $\xi < 1$, the output response Eqn. (6.34) for a unit-step is written as:

$$C(s) = \frac{A}{s} + \frac{B}{s + \xi\omega_n - j\omega_n\sqrt{1 - \xi^2}} + \frac{C}{s + \xi\omega_n + j\omega_n\sqrt{1 - \xi^2}} \qquad (6.43)$$

The location of the two conjugate poles is shown in Fig. 6.12

From Fig. 6.12, we get $\cos\alpha = -\xi$ $\qquad (6.44)$

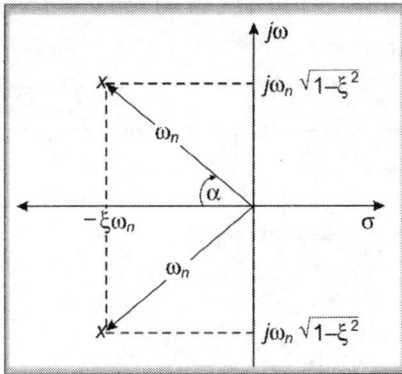

Fig. 6.12

$$\sin \alpha = \sqrt{1 - \xi^2} \qquad (6.45)$$

Using these relations constants A, B and C are found out as:

$$A = 1 \qquad (6.46)$$

$$B = \frac{e^{-j\alpha}}{2j\sin\alpha} \qquad (6.47)$$

$$C = -\frac{e^{j\alpha}}{2j\sin\alpha} \qquad (6.48)$$

Therefore Eqn. (6.43) can be written as:

$$C(s) = \frac{1}{s} + \frac{e^{-j\alpha}}{2j\sin\alpha(s + \xi\omega_n - j\omega_n\sqrt{1-\xi^2})} - \frac{e^{j\alpha}}{2j\sin\alpha(s + \xi\omega_n + j\omega_n\sqrt{1-\xi^2})} \qquad (6.49)$$

$$c(t) = 1 + \frac{e^{-j\alpha}}{2j\sin\alpha}e^{-(\xi\omega_n - j\omega_n\sqrt{1-\xi^2})t} - \frac{e^{j\alpha}}{2j\sin\alpha}e^{-(\xi\omega_n + j\omega_n\sqrt{1-\xi^2})t} \qquad (6.50)$$

$$c(t) = 1 + \frac{e^{-\xi\omega_n t}}{\sqrt{1-\xi^2}}\frac{e^{j(\omega_n t\sqrt{1-\xi^2} - \alpha)} - e^{-j(\omega_n t\sqrt{1-\xi^2} - \alpha)}}{2j}, \text{ or} \qquad (6.51)$$

$$c(t) = 1 - \frac{e^{-\xi\omega_n t}}{\sqrt{1-\xi^2}}\sin\left(\omega_n\sqrt{(1-\xi^2)}t - \alpha\right) \qquad (6.52)$$

From Eqns (6.44) and (6.45)

$$\frac{\sin\alpha}{\cos\alpha} = \frac{\sqrt{1-\xi^2}}{-\xi} \qquad (6.53)$$

$$\alpha = -\tan^{-1}\frac{\sqrt{1-\xi^2}}{\xi} \qquad (6.54)$$

Also, putting $\omega_d = \omega_n\sqrt{1-\xi^2}$, where ω_d is called a *damped natural frequency*, Eqn. (6.52) can be written as:

$$c(t) = 1 - \frac{e^{-\xi\omega_n t}}{\sqrt{1-\xi^2}}\sin\left(\omega_d t + \tan^{-1}\frac{\sqrt{1-\xi^2}}{\xi}\right), \text{ or} \qquad (6.55)$$

$$c(t) = 1 - \frac{e^{-\xi\omega_n t}}{\sqrt{1-\xi^2}}\sin(\omega_d t + \beta)$$

where,

$$\beta = \tan^{-1}\frac{\sqrt{1-\xi^2}}{\xi}$$

The output response of Eqn. (6.55) is plotted in Fig. 6.13 along with the step input.

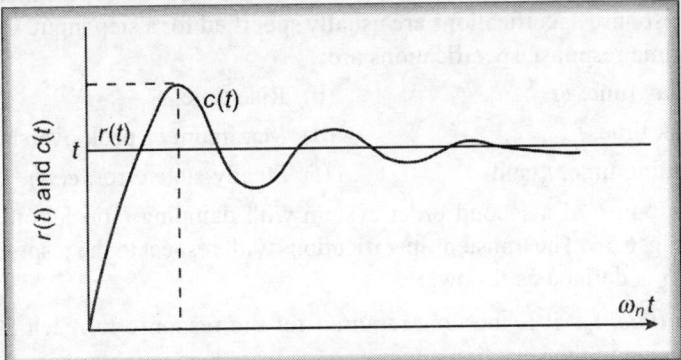

Fig. 6.13

The output response exhibits a number of overshoots/undershoots before settling down. The response is damped sinusoidal and is described as underdamped. Figure 6.14 shows the general pole locus for a second order system with fixed ω_n and variable damping ratio/factor. It can be seen that as ξ increases the poles sketch a circular locus of radius ω_n and move away from imaginary axis. The locus meets the negative real axis, and splits into two parts; one travels towards zero and other towards infinity.

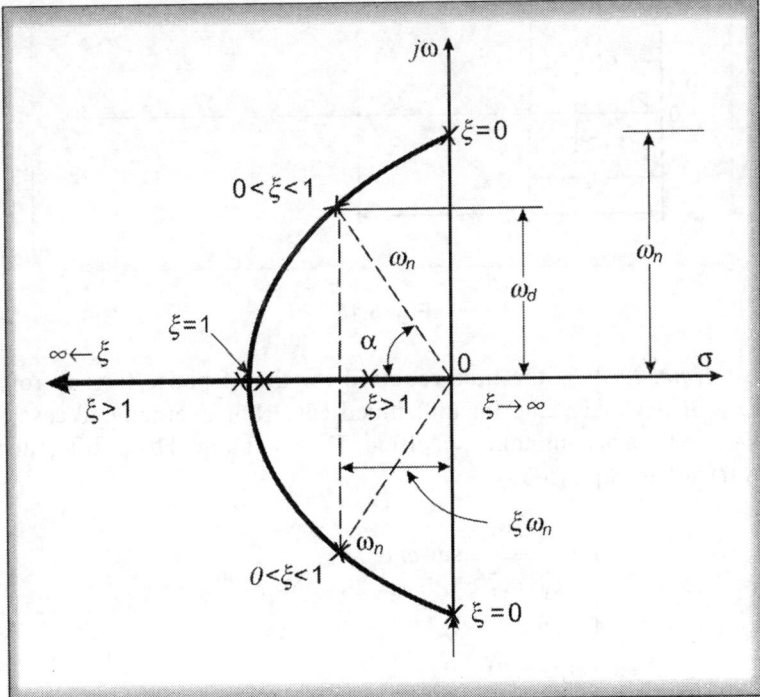

Fig. 6.14 Location of poles for a second-order system

6.7 Transient Response Specifications

Control systems are generaly designed with a damping ratio of less than one and transient response specifications are usually specified for a step input. Commonly specified time response specifications are:

(a) Delay time, t_d (b) Rise time, t_r

(c) Peak time, t_p (d) Maximum or peak overshoot, M_p

(e) Settling time, t_s and   (f) Steady-state error, $e(ss)$

Time response of a second order system with damping ratio less than one is shown in Fig. 6.15. The transient specifications with respect to the response shown in Fig. 6.15 is defined as follows:

(a) **Delay time, t_d:** It is the time required for the response to reach 50% of its final value in the very first attempt

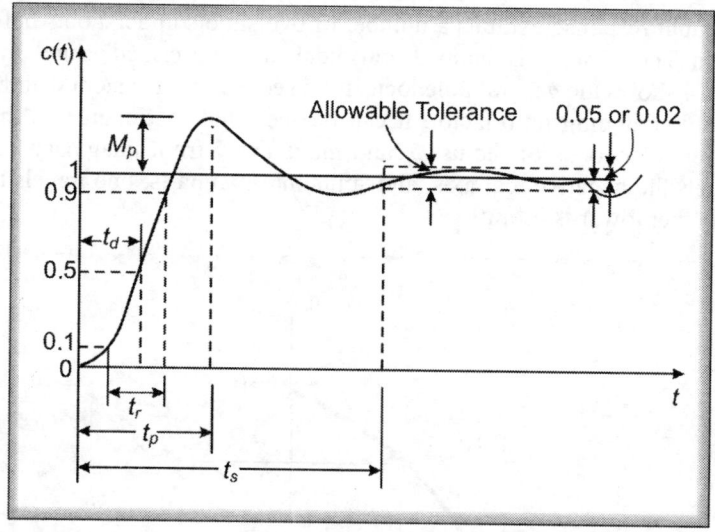

Fig. 6.15

(b) **Rise Time, t_r:** It is the time required for the response to rise from 0 to 100% of its final value for undamped case. For overdamped case, 10 to 90% rise time is commonly specified. This is obtained by putting the value of $c(t) = 1$ in Eqn. (6.55).

$$c(t) = 1 - \frac{e^{-\zeta\omega_n t}}{\sqrt{1-\xi^2}} \sin(\omega_d t + \beta)$$

Putting $c(t) = 1$ and $t = t_r$,

we get, $\sin(\omega_d t_r + \beta) = 0;$ (6.56)

or $\omega_d t_r + \beta = \pi,$

or $\qquad t_r = \dfrac{\pi - \beta}{\omega_d}$ $\qquad\qquad$ (6.57)

Smaller the value of t_r, faster is the response.

(c) Peak time, t_p: Peak time is the time required for the response to reach its maximum response or peak overshoot. This is obtained by differentiating Eqn. (6.55) with respect to time and equating it to zero.

$$\frac{dc(t)}{dt} = 0 + \frac{\xi\omega_n e^{-\xi\omega_n t_p}}{\sqrt{1-\xi^2}} \sin(\omega_d t_p + \beta) - \frac{e^{-\xi\omega_n t_p}}{\sqrt{1-\xi^2}} \times \omega_d \cos(\omega_d t_p + \beta) = 0 \quad (6.58)$$

Simplifying, we get

$$\xi\omega_n \sin(\omega_d t_p + \beta) = \omega_d \cos(\omega_d t_p + \beta)$$

$$\xi\omega_n \sin(\omega_d t_p + \beta) = \omega_n \sqrt{1-\xi^2} \cos(\omega_d t_p + \beta)$$

$$\xi \sin(\omega_d t_p + \beta) = \sqrt{1-\xi^2} \cos(\omega_d t_p + \beta) \qquad (6.59)$$

From Fig. 6.16

$$\xi = \cos \beta, \text{ and}$$

$$\sqrt{1-\xi^2} = \sin \beta \qquad\qquad (6.60)$$

Substituting the value of Eqn. (6.60) in Eqn. (6.59), we get

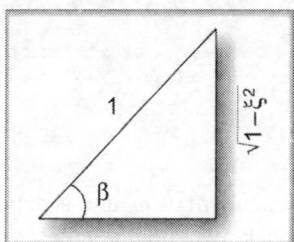

Fig. 6.16

$$\cos \beta \sin(\omega_d t_p + \beta) = \sin \beta \cos(\omega_d t_p + \beta) \qquad (6.61)$$

$$\cos \beta \sin(\omega_d t_p + \beta) - \sin \beta \cos(\omega_d t_p + \beta) = 0$$

$$\sin(\omega_d t_p + \beta - \beta) = 0$$

$$\sin(\omega_d t_p) = 0, \text{ or } \omega_d t_p = 0, \pi, 2\pi, 3\pi \ldots \qquad (6.62)$$

Since the peak time corresponds to overshoots, we get

$$\omega_d t_p = \pi, \text{ or} \qquad\qquad (6.63)$$

$$t_P = \frac{\pi}{\omega_d} \qquad\qquad (6.64)$$

Also, for first undershoot, $\qquad t_p = \dfrac{2\pi}{\omega_d}$, and

Second overshoot, $\qquad t_P = \dfrac{3\pi}{\omega_d} \qquad\qquad (6.65)$

(d) *Peak overshoot, M_p:* Peak overshoot is the maximum deviation of the response over the step input.

$$\text{Peak Overshoot} = \frac{c(t_p) - c(\infty)}{c(\infty)} \times 100\%$$

Peak overshoot is obtained as:

$$M_p = c(t_p) - 1$$

$$= 1 - \frac{e^{-\xi\omega_n t_p}}{\sqrt{1-\xi^2}}\sin(\omega_d t_p + \beta) - 1 = -\frac{e^{-\xi\omega_n t_p}}{\sqrt{1-\xi^2}}\sin(\omega_d t_p + \beta)$$

Putting the value of $\omega_d t_p = \pi$ from Eqn. (6.63), we get

$$M_p = -\frac{e^{-\xi\omega_n t_p}}{\sqrt{1-\xi^2}}\sin(\pi + \beta) = \frac{e^{-\xi\omega_n t_p}}{\sqrt{1-\xi^2}}\sin\beta$$

Putting, $t_p = \dfrac{\pi}{\omega_d}$ from Eqn. (6.64), we get

$$M_p = \frac{e^{\frac{-\xi\omega_n\pi}{\omega_d}}}{\sqrt{1-\xi^2}}\sin\beta$$

Putting, $\sin\beta = \sqrt{1-\xi^2}$, and $\omega_d = \omega_n\sqrt{1-\xi^2}$,

we get $M_p = \dfrac{e^{-\xi\omega_n\pi/\omega_n\sqrt{1-\xi^2}}}{\sqrt{1-\xi^2}} \times \sqrt{1-\xi^2} = e^{-\xi\pi/\sqrt{1-\xi^2}}$ \hfill (6.66)

or $\% M_p = 100\,e^{-\xi\pi/\sqrt{1-\xi^2}}$ \hfill (6.67)

Note: Peak overshoot is independent of ω_n and is dependent on ξ. As the value of damping ratio increases the peak overshoot decreases. The graph between M_p and ξ is plotted in Fig. 6.17.

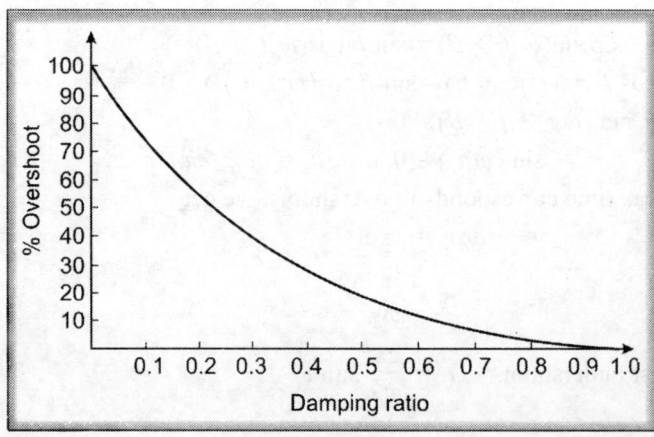

Fig. 6.17

(e) *Settling time* (t_s): Settling time is the time required for the response to reach and finally remain within a specified tolerance band (2 to 5%) of its final value. The response of Eqn. (6.55) consists of an exponentially decaying term $\dfrac{e^{-\xi\omega_n t}}{\sqrt{1-\xi^2}}$ sinusoidal oscillatory term $\sin(\omega_d t + \beta)$. The curves $1 \pm \dfrac{e^{-\xi\omega_n t}}{\sqrt{1-\xi^2}}$ are the envelope curves (Fig. 6.18).

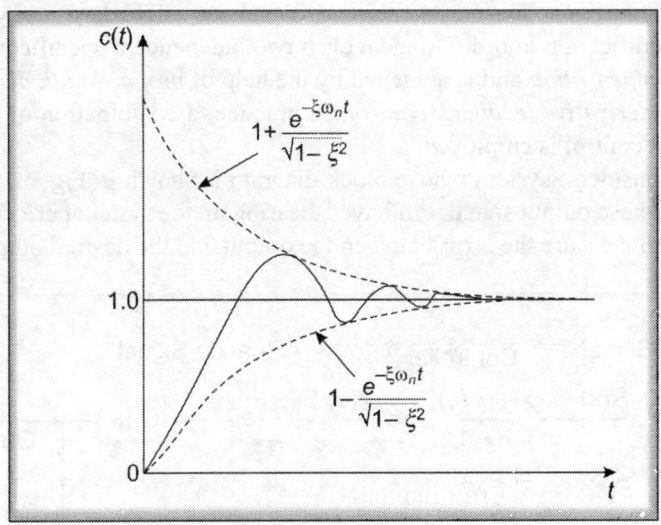

Fig. 6.18 Unit step response for $\xi < 1$

For the response to get into tolerance band of say 2%, the expression for setting time is obtained as:

$$1 + \frac{e^{-\xi\omega_n t_s}}{\sqrt{1-\xi^2}} = 1.02, \text{ or}$$

$$\frac{e^{-\xi\omega_n t_s}}{\sqrt{1-\xi^2}} = 0.02 \qquad (6.68)$$

For small values of ξ, we get

$$e^{-\xi\omega_n t_s} = 0.02$$

$$t_s = \frac{3.91}{\xi\omega_n} \approx \frac{4}{\xi\omega_n} \quad \text{or} \quad t_s = 4T, \quad \text{where} \quad T = \frac{1}{\xi\omega_n} \qquad (6.69)$$

Similarly for 4% tolerance band

$$t_s = 3T$$

(f) *Steady-state error*, e_{ss}:

$$e_{ss} = \lim_{t \to \infty} (r(t) - c(t)),$$

for a step input $\quad e_{ss} = \lim_{t \to \infty} (1 - c(t)) \qquad (6.70)$

Thus, for underdamped second order system, steady-state error to a unit-step input is zero.

6.8 Control Actions

While designing a system, the designer selects the resonable values for the *peak overshoot*, *rise time* and the *settling time*. The designer is never sure of the final design of the system as to whether it is good or not. For example, if the system has been designed for minimum overshoot, the rise time increases and on the other hand, if the rise time chosen is small, peak overshoot will be large. A system thus requires modification in order to meet even two independent specifications. This is called *compensation* and is achieved by the help of *proportional, derivative* or *integral* or *derivative feedback control*. In practice, a combination of derivative and integral control is employed.

Let us consider a system whose block diagram is shown in Fig. 6.19. It has a controller whose output signal will have a bearing on the system performance. Its purpose is to measure the error between the output and the desired output.

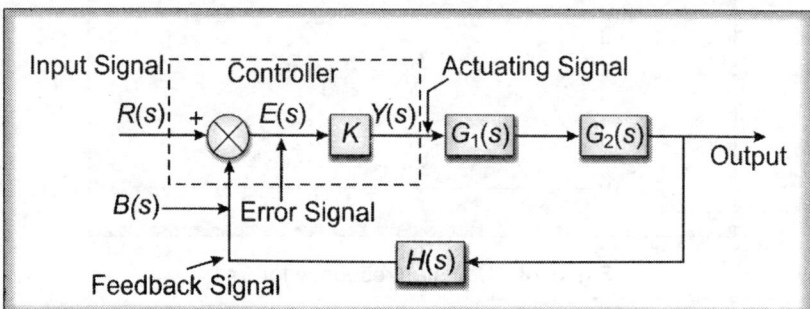

Fig. 6.19

The transfer function of the controller is:

$$K = \frac{Y(s)}{E(s)} \tag{6.71}$$

where

$$E(s) = R(s) - B(s) \tag{6.72}$$

or $\qquad E(s) = R(s) - H(s)C(s)$

This relationship is termed *control action* relationship. We will now discuss various control actions available to the control system engineer for improvement of the system performance.

6.8.1 Proportional Control Action
Let us consider a system whose block diagram is shown in Fig. 6.19. In this, the actuating signal is proportional to the error signal. Such type of system is called a control system. The relationship between the output of the controller $y(t)$ and the proportional actuating error signal $e(t)$ is

$$y(t) = Ke(t) \tag{6.73}$$

In Laplace transform form, it can be written as:

$$Y(s) = KE(s)$$

or
$$K = \frac{Y(s)}{E(s)} \qquad (6.74)$$

From Eqn. (6.73), one concludes that increase in gain K will increase the undamped frequency of oscillation ω_n, but will decrease the damping ratio ξ. the output response for various values of gain and damping factor is plotted in Fig. 6.20. For large values of K, which in turn reduces ω_n

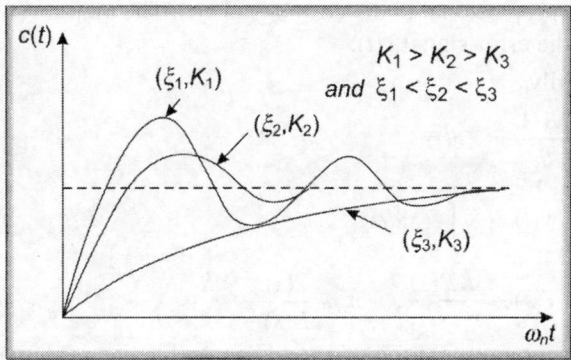

Fig. 6.20

(a) the output response to unit step input oscillates violently about its final steady-state value.

(b) Overshoot is large.

(c) Rise time and settling time reduce.

For low values of gain K, undamped frequency of oscillation reduces and the output response to unit step is overdamped (*refer to* Fig. 6.11)

The output response to a unit ramp input is shown in Fig. 6.21

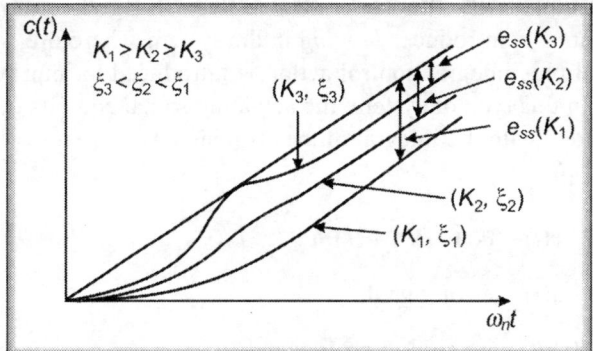

Fig. 6.21

It can be seen from the curves that steady-state error decreases as the gain is increased, but the response becomes more oscillatory and may also become unstable.

Therefore, while going in for proportional control action for improving system performance, a control system engineer will have to make a compromise in choosing the proper gain K, so that the steady-state error and the maximum overshoot of the output response are within the acceptable limits. This is a difficult proposition to achieve.

6.8.2 Integral Control Action In this the value of the controller output $y(t)$ is altered at a rate proportional to the error signal $e(t)$. The output $y(t)$ depends upon the integral of the error signal $e(t)$.

Mathematically,

$$\frac{dy(t)}{dt} = Ke(t) \tag{6.75}$$

or
$$y(t) = K \int_0^t e(t)dt \tag{6.76}$$

or
$$Y(s) = \frac{KE(s)}{s}, \quad \text{or} \quad \frac{Y(s)}{E(s)} = \frac{K}{s} \tag{6.77}$$

Block diagram representation of integral control action is shown in Fig. 6.22.

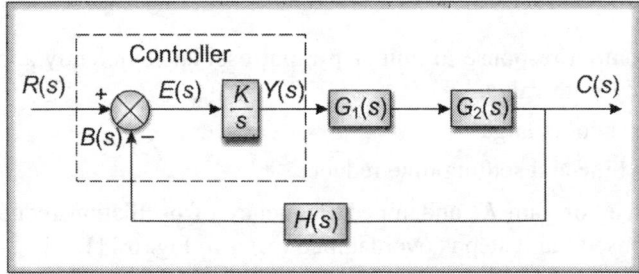

Fig. 6.22 Integral control

6.8.2.1 Proportional Plus Integral Control Action Integral control action itself is not sufficient, as it introduces *hunting* in the system. Therefore, a combination of proportional plus integral control action is introduced to improve the system performance. In this type of system, the actuating signal consists of proportional error signal added with the integral of the error signal.

Mathematically,

$$y(t) = e(t) + K \int_0^t e(t)\,dt \tag{6.78}$$

where, $e(t) =$ error signal;

and $\int_0^t e(t)dt =$ integral of error signal

or
$$Y(s) = E(s)\left[1 + \frac{K}{s}\right]$$
(6.79)

$$\frac{Y(s)}{E(s)} = \left(1 + \frac{K}{s}\right)$$
(6.80)

Block diagram representation of Eqn. (6.80) is shown in Fig. 6.23.

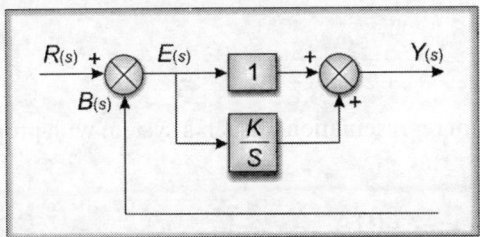

Fig. 6.23 Proportional plus integral control

For a second order unity feedback control system, employing proportional plus integral control action, the block diagram representation is shown in Fig. 6.24.

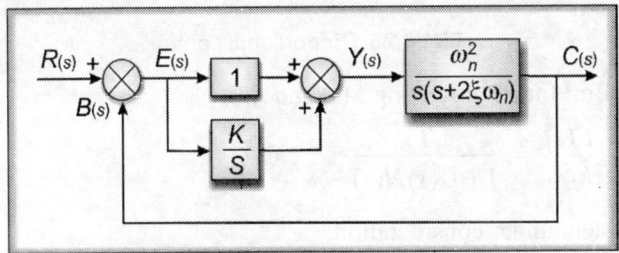

Fig. 6.24 Proportional plus integral control

The transfer function of such a system is given by:

$$\frac{C(s)}{R(s)} = \frac{(s + K)\omega_n^2}{s^3 + 2\xi\omega_n s^2 + \omega_n^2 s + K\omega_n^2}$$
(6.81)

The characteristics equation is given by:
$$s^3 + 2\xi\omega_n s^2 + \omega_n^2 + K\omega_n^2$$
(6.82)

s^3	1	ω_n^2
s^2	$2\xi\omega_n$	$K\omega_n^2$
s^1	$\dfrac{2\xi\omega_n^3 - K\omega_n^2}{2\xi\omega_n}$	0
s^0	$K\omega_n^2$	0

For the system to be stable

(a) $2\xi\omega_n > 0$, *i.e.* $\xi > 0$ and $\omega_n > 0$,

(b) $K\omega_n^2 > 0$, i.e. $K > 0$ and $\omega_n > 0$, and

(c) $2\xi\omega_n^2 - K\omega_n^2 > 0$, i.e $2\xi\omega_n > K$

Therefore, for the system to be stable $2\xi\omega_n > K$

Steady-state error

The steady-state error for a second-order system with unity feedback is obtained from the transfer function given by:

$$\frac{C(s)}{R(s)} = \frac{\omega_n^2}{s^2 + 2\xi\omega_n s + \omega_n^2} \tag{6.83}$$

The block diagram representation of such a system with proportional control is shown in Fig. 6.25.

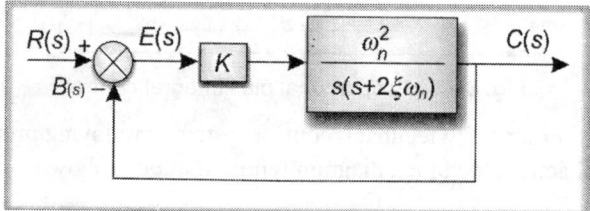

Fig. 6.25 Proportional control

The transform function relating $E(s)$ and $R(s)$

$$\frac{E(s)}{R(s)} = \frac{1}{1 + G(s)H(s)} \tag{6.84}$$

For the system under consideration

$$E(s) = \frac{R(s)}{1 + G(s)} = \frac{R(s)}{1 + \dfrac{K\omega_n^2}{s(s + 2\xi\omega_n)}}$$

or $\qquad E(s) = \dfrac{R(s)(s^2 + 2\xi\omega_n s)}{s^2 + 2\xi\omega_n s + K\omega_n^2} \tag{6.85}$

For unit step input

$$e_{(ss)} = \lim_{s \to 0} sE(s) = \lim_{s \to 0} \frac{1}{s}\left[\frac{s(s^2 + 2\xi\omega_n s)}{(s^2 + 2\xi\omega_n s + K\omega_n^2)}\right] = 0 \tag{6.86}$$

For unit ramp input

$$e_{(ss)} = \lim_{s \to 0} \frac{1}{s^2}\left[\frac{s(s^2 + 2\xi\omega_n s)}{(s^2 + 2\xi\omega_n s + K\omega_n^2)}\right] = \frac{2\xi\omega_n}{K\omega_n^2} = \frac{2\xi}{K\omega_n} \tag{6.87}$$

For unit parabolic input

$$e_{(ss)} = \lim_{s \to 0} \frac{1}{s^3}\left[\frac{s(s^2 + 2\xi\omega_n s)}{s^2 + 2\xi\omega_n s + K\omega_n^2}\right] = \infty \tag{6.88}$$

If the system is provided with integral control as shown in Fig. 6.22 generalised to Fig. 6.26, the transfer function relating $E(s)$ and $R(s)$ is given by:

$$E(s) = \frac{R(s)}{1 + G(s)H(s)} = \frac{R(s)}{1 + \dfrac{K\omega_n^2}{s^2(s + 2\xi\omega_n)}}$$

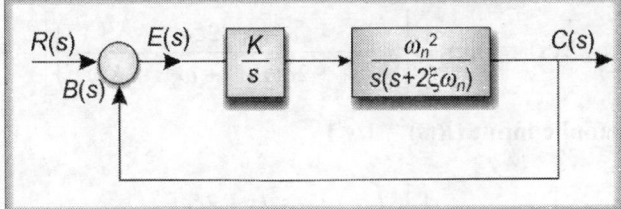

Fig. 6.26 Integral control

or
$$E(s) = \frac{s^2 R(s)(s + 2\xi\omega_n)}{s^3 + 2\xi\omega_n s^2 + K\omega_n^2} \tag{6.89}$$

For unit step input ($R(s) = 1/s$)

$$e(ss) = \lim_{s \to 0} sE(s) = \lim_{s \to 0} \frac{1}{s}\left[s\left(\frac{s^2(s + 2\xi\omega_n)}{(s^2 + 2\xi\omega_n s^2 + K\omega_n^2)}\right)\right] = 0 \tag{6.90}$$

For unit ramp input ($R(s) = 1/s^2$)

$$e(ss) = \lim_{s \to 0} sE(s) = \lim_{s \to 0} \frac{1}{s^2}\left[s^2\left(\frac{s^2(s + 2\xi\omega_n)}{(s^3 + 2\xi\omega_n s^2 + K\omega_n^2)}\right)\right] = 0 \tag{6.91}$$

For unit parabolic input ($R(s) = 1/s^3$)

$$e(ss) = \lim_{s \to 0} sE(s) = \lim_{s \to 0} \frac{1}{s^3}\left[s\left(\frac{s^2(s + 2\xi\omega_n)}{s^3 + 2\xi\omega_n^2 + K\omega_n^2}\right)\right] = \frac{2\xi}{K\omega_n} \tag{6.92}$$

A second-order system with unity feedback, if provided with proportional plus integral control action as shown in Fig. 6.24, the transfer function relating $E(s)$ with $R(s)$ is:

$$\frac{E(s)}{R(s)} = \frac{1}{1 + \dfrac{(s + K)\omega_n^2}{s^2(s + 2\xi\omega_n)}}$$

or
$$E(s) = \frac{s^2(s + 2\xi\omega_n)R(s)}{s^3 + 2\xi\omega_n s^2 + \omega_n^2 s + K\omega_n^2} \tag{6.93}$$

For unit step input ($R(s) = 1/s$)

$$e(ss) = \lim_{s \to 0} sE(s) = \lim_{s \to 0} \frac{1}{s}\left[s\left(\frac{s^2(s+2\xi\omega_n)}{(s^3+2\xi\omega_n s^2 + \omega_n^2 s + K\omega_n^2)}\right)\right] = 0 \qquad (6.94)$$

For unit ramp input ($R(s) = 1/s^2$)

$$e_{(ss)} = \lim_{s \to 0} sE(s) = \lim_{s \to 0} \frac{1}{s^2}\left[s\left(\frac{s^2(s+2\xi\omega_n)}{(s^3+2\xi\omega_n s^2 + \omega_n^2 s + K\omega_n^2)}\right)\right] = 0 \qquad (6.95)$$

For unit parabolic input ($R(s) = 1/s^3$)

$$e_{(ss)} = \lim_{s \to 0} sE(s) = \lim_{s \to 0} \frac{1}{s^3}\left[s\left(\frac{s^2(s+2\xi\omega_n)}{(s^3+2\xi\omega_n s^2 + \omega_n^2 s + K\omega_n^2)}\right)\right] = \frac{2\xi}{K\omega_n} \qquad (6.96)$$

Steady-state errors are tabulated in Table 6.5

Table 6.5

Input	Steady-state error		
	Proportional Control	Integral Control	Proportional Plus Integral Control
Unit Step	0	0	0
Unit Ramp	$\dfrac{2\xi}{K\omega_n}$	0	0
Unit Parabolic	∞	$\dfrac{2\xi}{K\omega_n}$	$\dfrac{2\xi}{K\omega_n}$

Summary

Proportional plus integral control increases the order and the type of system by one, respectively. Therefore, it improves steady-state performance. The steady-state error for step and ramp input is zero and for the parabolic input, it is finite, whereas, in the case of proportional control, it is infinite. In case of ramp input, the steady-state error is zero for integral control whereas, it is finite, in case of proportional control. Therefore, in order to eliminate error, incorporation of integral control action is a logical approach. The effect on the transient response is obtained by studying the open-loop transfer function of the system as shown in Fig. 6.24 employing proportional plus integral control.

$$G(s) = \frac{C(s)}{E(s)} = \frac{(s+K)\omega_n^2}{s^2(s+2\xi\omega_n)} \qquad (6.97)$$

The effect of such a control action is to add a zero at $s = -K$ and a pole at origin because of the factor $1/s$ in the term $1 + K/s$, i.e. $(s + K)/s$. It will cause peak overshoot to occur early in the output response and the peak overshoot will increase appreciably.

Moreover, the stability is reduced as the second order system becomes third order system and if the loop gain of the system is high, it will become more un-stable.

Therefore, integral control action improves the steady-state performance but makes the closed-loop system unstable, as the addition of pole-zero combination shifts the loci towards the right side in the s-plane. Readers will be more clear on these points after covering the chapters on Root locus and Nyquist stability criterion.

6.8.3 Derivative Control Action
A control system is said to have a derivative control action if the output $y(t)$ depends upon the rate of change of the error signal. Mathematically

$$y(t) = \frac{Kde(t)}{dt} \tag{6.98}$$

or $\qquad Y(s) = sKE(s)$

where, K is an adjustable constant and is called *variable of derivative control action*. Block diagram representation of a closed-loop system with unity feedback and derivative control action is shown in Fig. 6.27

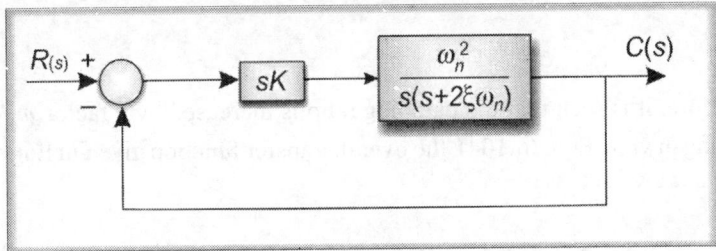

Fig. 6.27 Derivative control

6.8.3.1 Proportional Plus Derivative Control Action
In this type of control action, the actuating signal $y(t)$ depends upon the proportional derivative error signal.

$$y(t) = e(t) + \frac{Kde(t)}{dt} \tag{6.99}$$

or $\qquad Y(s) = E(s)(1+sK) \tag{6.100}$

The block diagram of such a control action is shown in Fig. 6.28.

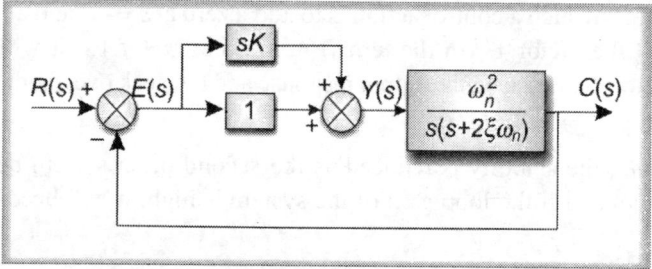

Fig. 6.28   Proportional plus derivative control

The overall transfer function is given by:

$$\frac{C(s)}{R(s)} = \frac{(1 + sK)\omega_n^2}{s^2 + (2\xi\omega_n + \omega_n^2 K)s + \omega_n^2} \qquad (6.101)$$

The characteristic equation is:

$$s^2 + (2\xi\omega_n + \omega_n^2 K)s + \omega_n^2 = 0 \qquad (6.102)$$

The characteristic equation of a second-order control system without using derivative control action is:

$$s^2 + 2\xi\omega_n s + \omega_n^2 = 0 \qquad (6.103)$$

Comparing Eqns (6.102) and (6.103), we get

$$\xi' = \frac{2\xi\omega_n + \omega_n^2 K}{2\omega_n}$$

or

$$\xi' = \xi + \frac{\omega_n K}{2} \qquad (6.104)$$

Therefore, it is seen that the damping ratio is increased by a factor $\omega_n K/2$

Keeping in view Eqn. (6.104), the overall transfer function given in Eqn. (6.101) can now be rewritten as:

$$\frac{C(s)}{R(s)} = \frac{\left(s + \frac{1}{K}\right)\omega_n^2 K}{s^2 + 2\xi'\omega_n s + \omega_n^2} \qquad (6.105)$$

Comparing the transfer function derived in Eqn. (6.105) and comparing with overall transfer function of a second-order control system without any control action as given in Eqn. (6.106) below

$$\frac{C(s)}{R(s)} = \frac{\omega_n^2}{s^2 + 2\xi\omega_n s + \omega_n^2} \qquad (6.106)$$

It is seen that

(a) there is no change in the natural frequency of oscillation ω_n; and

(b) the transfer function with derivative control action contains a zero at $s = -1/K$

The transfer function relating $E(s)$ and $R(s)$ is obtained from the expression given in Eqn. (6.84) and is:

$$\frac{E(s)}{R(s)} = \frac{s(s + 2\xi\omega_n)}{s^2 + (2\xi\omega_n + \omega_n^2 K)s + \omega_n^2} \tag{6.107}$$

For a unit step input ($R(s) = 1/s$)

$$e_{ss} = \lim_{s \to 0} sE(s) = \lim_{s \to 0} \frac{1}{s}\left[\frac{s \times s(s + 2\xi\omega_n)}{s^2 + (2\xi\omega_n + \omega_n^2 K)s + \omega_n^2}\right] = 0 \tag{6.108}$$

For a unit ramp input ($R(s) = 1/s^2$)

$$e_{ss} = \lim_{s \to 0} sE(s) = \lim_{s \to 0} \frac{1}{s^2}\left[\frac{s \times s(s + 2\xi\omega_n)}{s^2 + (2\xi\omega_n + \omega_n^2 K)s + \omega_n^2}\right] = \frac{2\xi}{\omega_n} \tag{6.109}$$

For a unit ramp input ($R(s) = 1/s^3$)

$$e_{ss} = \lim_{s \to 0} \frac{1}{s^3}\left[\frac{s \times s(s + 2\xi\omega_n)}{s^2 + ((2\xi\omega_n + \omega_n^2 K)s + \omega_n^2)}\right] = \infty \tag{6.110}$$

Summary

Comparing the expression derived in Eqns (6.108 to 6.110) with that of proportional control action tabulated in Table 6.5, it is seen that steady-state error remains unchanged even with derivative control action. The derivative control action has an impact on transient response. The increase in damping ratio reduces the peak overshoot. This can be visualised from Eqn. (6.67) and the graph is shown in Fig. 6.17. The increase in the coefficient of the term s as seen in Eqn. (6.102) when compared with Eqn. (6.103), also points towards the increase in the damping ratio. This results in slower system response with less overshoots. There is also reduction in settling time and rise time.

6.8.4 PID Control Action This type of control action employs proportional, integral and derivative control action together in a control system so as to derive the advantages of all the control actions into one. Mathematically

$$y(t) = e(t) + K_d\frac{de(t)}{dt} + K_i\int e(t)dt$$

$$Y(s) = E(s)\left(1 + sK_d + \frac{K_i}{s}\right) \tag{6.111}$$

The block diagram of a control system with unity feedback employing PID control action is shown in Fig. 6.29

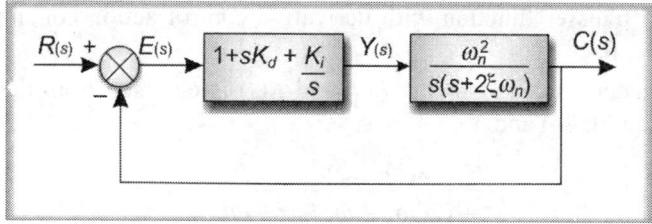

Fig. 6.29 PID Control

6.8.5 Derivative Feedback Control Action

Rate feedback and *tachometer feedback* control actions are other names of the derivative feedback control action. In this type of control action, the derivative of the output signal is feedback in such a way that the actuating signal $y(t)$ is the difference between it and the proportional error signal $e(t)$. Mathematically

$$y(t) = e(t) - K_t \frac{dc(t)}{dt} \tag{6.112}$$

where, K_t = feedback constant

or $\qquad Y(s) = E(s) - sK_t C(s) \tag{6.113}$

The block diagram representation of a second-order control system with unity feedback derivative feedback control action is shown in Fig. 6.30.

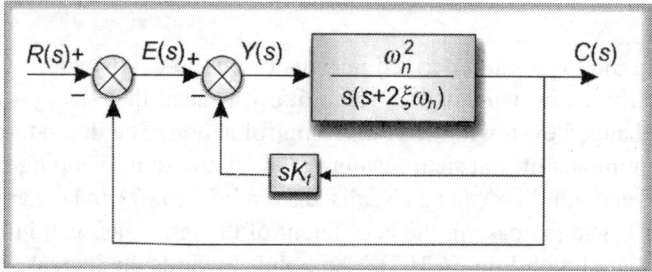

Fig. 6.30 Derivative feedback control

The transfer function of the system as shown in Fig. 6.30 is given by:

$$\frac{C(s)}{R(s)} = \frac{\omega_n^2}{s^2(2\xi\omega_n + \omega_n^2 K_t)s + \omega_n^2} \tag{6.114}$$

The characteristic equation is:

$$s^2 + (2\xi\omega_n + \omega_n^2 K_t)s + \omega_n^2 \tag{6.115}$$

Comparing with the characteristic equation of a second-order unity feedback system given in Eqn. (6.103), it is seen that the damping ratio has increased and is:

$$\xi' = \xi + \frac{\omega_n K_t}{2} \tag{6.116}$$

It should be noted that Eqns (6.115) and (6.106) are comparable. The transfer function relating $E(s)$ with $R(s)$ is obtained from the expression given in Eqn. (6.84) and is:

$$\frac{E(s)}{R(s)} = \frac{s^2(2\xi\omega_n + \omega_n^2 K_t)s}{s^2 + (2\xi\omega_n + \omega_n^2 K_t)s + \omega_n^2} \tag{6.117}$$

Now $$e_{ss} = \lim_{s \to 0} sE(s) \tag{6.118}$$

For a unit step input $(R(s) = 1/s)$

$$e_{ss} = \lim_{s \to 0} \frac{1}{s}\left[s\left(\frac{s^2 + (2\xi\omega_n + \omega_n^2 K_t)s}{s^2 + (2\xi\omega_n + \omega_n^2 K_t)s + \omega_n^2}\right)\right] = 0 \tag{6.119}$$

For a unit ramp input $(R(s) = 1/s^2)$

$$e_{ss} = \lim_{s \to 0} \frac{1}{s^2}\left[s\frac{s^2 + (2\xi\omega_n + \omega_n^2 K_t)s}{s^2 + (2\xi\omega_n + \omega_n^2 K_t)s + \omega_n^2}\right] = \frac{2\xi}{\omega_n} + K_t \tag{6.120}$$

When compared with Eqn. (6.109), which represents the steady-state error of a second order unity feedback system employing derivate control action, it is seen that steady-state error is increased by K_t, when the system is subjected to unit ramp input.

For a unit parabolic input $(R(s) = 1/s^3)$

$$e_{ss} = \lim_{s \to 0} \frac{1}{s^3}\left[s\left(\frac{s^2 + (2\xi\omega_n + \omega_n^2 K_t)s}{s^2 + (2\xi\omega_n + \omega_n^2 K_t)s + \omega_n^2}\right)\right] = \infty \tag{6.121}$$

6.8.5.1 Comparison with Derivative Control Action

(a) Both the control actions result in increase in the damping ratio. It results in reduced peak overshoot and settling time.

(b) The closed-loop transfer function of the derivative feedback control derived in Eqn. (6.114) does not have an additional zero added due to incorporation of derivative feedback control action, whereas the derivative control proportional action results in addition of a zero at $s = -1/K$ as seen from Eqn. (6.101). Therefore, the output responses will not be identical even if $K = K_t$.

(c) Since, the system remains type 1, the basic characteristics of steady-state error is not altered. However, for a unit ramp input the steady-state error increased by a term equal to K_t, in case of derivative feedback control action. This can be seen from Eqns (6.120) and (6.109).

6.9 Impulse Signal and Response

It is a signal which is applied suddenly for a very short duration of time. Fig. 6.31 shows a representation of impulse signal of magnitude R applied for a short duration δt. It is seen that at $t = 0$, the magnitude of the signal is R and exists for a very small time. Therefore, its value is zero. Mathematically

$$\delta(t) = 0 \quad \text{for} \quad t \neq 0$$

Also

$$\int\limits_{-t_1}^{+t_1} \delta(t)dt = 1 \quad as \quad t_1 \to 0$$

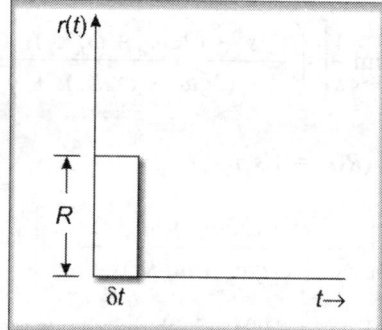

Fig. 6.31 Impulse signal

An impulse function $\delta(t)$ is the derivative of a step function. If the magnitude of an impulse function is unity, it is called unit impulse function

$$\delta(t) = \frac{d}{dt}u(t)$$

Taking Laplace of impulse function, we get

$$\mathcal{L}\,\delta(t) = \mathcal{L}\frac{d}{dt}u(t) = s \times \frac{1}{s} = 1 = R(s)$$

Impulse response of a system is given by:

$$C(s) = G(s)R(s) = G(s) \times 1 = G(s)$$

Therefore, $c(t) = \mathcal{L}^{-1}\,G(s) = g(t)$

The impulse response is obtained by taking inverse Laplace transform of the transfer function of the system and is shown by $g(t)$. It is also called *weighting function* of the system.

Let us now find out the time response of a first-order control system which is subjected to unit impulse input. Let the transfer function of the system is $1/1 + sT$

$$C(s) = G(s)R(s) = \frac{1}{1 + sT} \times R(s)$$

Since the input is unit impulse input, its Laplace transform is as found out above and hence

$$R(s) = 1$$

Therefore, $$C(s) = \frac{1}{1 + sT}$$

Taking inverse Laplace transform, we get

$$\mathcal{L}^{-1} C(s) = \mathcal{L}^{-1} \frac{1}{1 + sT}$$

or $$c(t) = \frac{1}{T} e^{-\frac{t}{T}}$$

when $t = 0$, $c(t) = \dfrac{1}{T}$ and when $t = \infty$, $c(t) = 0$

The output response is shown in Fig. 6.32

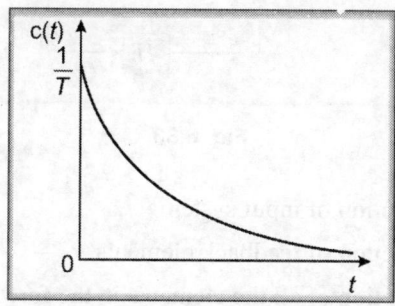

Fig. 6.32 Output response

6.10 Sensitivity

Performance of a control system depends upon a number of factors. We have already studied accuracy, transient and frequency response and seen how these affect the system performance. Few other factors are stability and sensitivity. Stability has been covered in Chapter 5 in detail and the subsequent chapters include various tools available to ascertain the stability of a control system.

6.10.1 Definition Sensitivity is defined as a ratio of variation in a system parameter to the variation in one of the system parameters. The system parameters may be gain, forward path elements, feedback path elements, impedance, friction, etc. Sensitivity is generally expressed in percentage. Sensitivity gives an idea of the performance of the system when subjected to parameter variation. This aids a control system designer to evolve an efficient and effective control system.

Mathematically, sensitivity can be expressed as:

$$\text{Sensitivity} = \frac{\% \text{ change in parameter } P}{\% \text{ change in } K}$$

or
$$S_K^P = \frac{\partial P/P}{\partial K/K}$$

where, S_K^P means sensitivity of variable P with respect to parameter K.

The definition is valid only for small variations. It should be kept in mind that sensitivity is function of frequency. Also, an ideal system should have least sensitivity, if not zero.

6.10.2 Concept

Let us consider a system whose block diagram is shown in Fig. 6.33

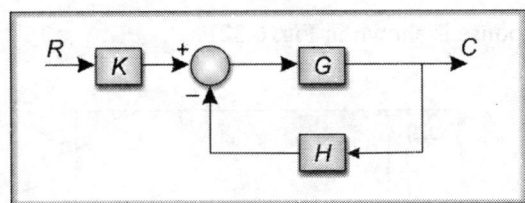

Fig. 6.33

where,

K = transfer function of input system

H = transfer function of feedback elements

G = transfer function of all the elements in the forward path of the feedback loop

C/R = overall transfer function of the system

$$T = \frac{C}{R} = \frac{KG}{1 + GH} \tag{6.122}$$

6.10.2.1 Sensitivity of Overall (closed-loop) Transfer Function with respect to G

$$S_G^T = \frac{dT|T}{dG|G} = \frac{G \, dT}{T \, dG} \tag{6.123}$$

Finding the expression dT/dG. This is achieved by differentiating T, the overall transfer function which is equal to $\dfrac{KG}{1 + GH}$

$$\frac{dT}{dG} = \frac{d}{dG}\left(\frac{KG}{1+GH}\right)$$

$$= K\left[\frac{(1+GH)\dfrac{dG}{dG} - G\dfrac{d}{dG}(1+GH)}{(1+GH)^2}\right]$$

$$= K\left[\frac{(1+GH)\times 1 - G(O+H)}{(1+GH)^2}\right]$$

$$= K\left[\frac{1+GH-GH}{(1+GH)^2}\right] = \frac{K}{(1+GH)^2} \qquad (6.124)$$

Substituting the value of $\dfrac{dT}{dG}$ from Eqn. (6.124) in Eqn. (6.123), we get

$$S_G^T = \frac{G}{T} \times \frac{K}{(1+GH)^2}$$

or
$$S_G^T = \frac{G}{\dfrac{KG}{1+GH}} \times \frac{K}{(1+GH)^2}$$

$$S_G^T = \frac{1}{1+GH} \qquad (6.125)$$

6.10.2.2 Sensitivity of Overall Transfer Function (Closed-loop) with respect to Feedback Elements H

This is given by:

$$S_G^T = \frac{dT/T}{dH/H} = \frac{H}{T}\frac{dT}{dH} \qquad (6.126)$$

Now
$$\frac{dT}{dH} = \frac{d}{dH}\left(\frac{KG}{1+GH}\right)$$

Differentiating $\dfrac{KG}{1+GH}$ with respect to H, we get

$$\frac{dT}{dH} = K\left[\frac{(1+GH)\dfrac{dG}{dH} - G\dfrac{d}{dH}(1+GH)}{(1+GH)^2}\right]$$

or
$$\frac{dT}{dH} = K\left[\frac{0 - G(G)}{(1+GH)^2}\right]$$

or
$$\frac{dT}{dH} = \frac{-KG^2}{(1+GH)^2} \qquad (6.127)$$

Substituting the value of $\dfrac{dT}{dH}$ from Eqn. (6.127) in Eqn. (6.126), we get

$$S_H^T = \frac{H}{T} \times \frac{-KG^2}{(1+GH)^2}$$

or

$$S_H^T = \frac{H}{\dfrac{KG}{1+GH}} \times - \frac{KG^2}{(1+GH)^2}$$

or

$$S_H^T = - \frac{GH}{1+GH} \tag{6.128}$$

6.10.2.3 Sensitivity of Overall Transfer Function (Closed-loop) with respect to K

This is given by:

$$S_K^T = \frac{dT/T}{dK/K} = \frac{K}{T}\frac{dT}{dK} \tag{6.129}$$

Now

$$\frac{dT}{dK} = \frac{d}{dK}\left[\frac{KG}{1+GH}\right]$$

$$= G\left[\frac{(1+GH)\dfrac{dK}{dK} - G\dfrac{d}{dK}(1+GH)}{(1+GH)^2}\right]$$

$$= G\left[\frac{1+GH-0}{(1+GH)^2}\right] = \frac{G}{1+GH} \tag{6.130}$$

Substituting the value of $\dfrac{dT}{dK}$ from Eqn. (6.130) in Eqn. (6.129), we get

$$S_K^T = \frac{K}{T} \times \frac{G}{1+GH}$$

or

$$S_K^T = \frac{K}{KG/1+GH} \times \frac{G}{1+GH}$$

or

$$S_K^T = 1 \tag{6.131}$$

6.10.2.4 Open-loop Control System

Let us now consider an open-loop system as shown in Fig. 6.34

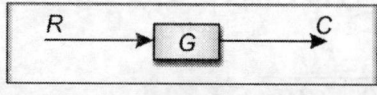

Fig. 6.34

Now,

$$T = \frac{C}{R} = G \tag{6.132}$$

Then
$$S_G^T = \frac{dT/T}{dG/G} = \frac{G}{T}\frac{dT}{dG}$$

$$\frac{dT}{dG} = \frac{dG}{dG} = 1$$

Therefore,

$$S_G^T = \frac{G}{T} \times 1 = \frac{G}{T} \tag{6.133}$$

Substituting the value of T from Eqn. (6.132) in Eqn. (6.133), we get

$$S_G^T = \frac{G}{G} = 1 \tag{6.134}$$

6.10.2.5 Interpretation of Results of Sensitivity

The results obtained of the sensitivity transfer function in Eqns (6.125), (6.128), (6.131) and (6.134) are very interesting. The interpretation is based on the fact that an ideal system should have minimum sensitivity function if not zero, with respect to any parameter of the control system.

If we compare the sensitivity function derived in Eqns (6.125) and (6.134), it can be seen that the sensitivity function of the overall transfer function for a closed-loop control system with respect to forward path transfer function reduces by a factor $1/(1 + GH)$ when compared with that of open-loop control system. Mathematically

$$\frac{S_G^T \text{ (closed-loop)}}{S_G^T \text{ (open-loop)}} = \frac{1/1+GH}{1} = \frac{1}{1+GH} \tag{6.135}$$

Now
$$S_H^T = \frac{-GH}{1+GH} \quad \text{[from Eqn. (6.100)]}$$

For cases where, GH >> 1, the sensitivity function reduces to

$$S_H^T \approx -1 \tag{6.136}$$

When the sensitivity of overall transfer function with respect to feedback element is compared with sensitivity of overall transfer function with respect to forward path element [Eqn. (6.135)], it is seen that it is divided by a factor $(1+ GH)$. Therefore, it is desirable that GH should be as large as possible, when sensitivity is taken as an important parameter to analyse the performance of a control system. Equations (6.128), (6.135) and (6.136) also indicate that input transfer function K and feedback element H are very critical. Any variation in their characteristics are directly seen in an overall system transfer function change. It is desirable that their characteristics should be stable when subjected to variation in temperature and over a period of time.

$$\boxed{\textbf{SOLVED PROBLEMS}}$$

Problem 6.1 Find the time response, initial value and final values of the following functions:

(a) $$F(s) = \frac{s(s+10)}{(s+2)(s+4)(s+6)}$$

(b) $$F(s) = \frac{12(s+1)}{s(s+2)^2(s+3)}$$

Solution

(a) $$F(s) = \frac{s(s+10)}{(s+2)(s+4)(s+6)} = \frac{-2}{(s+2)} + \frac{6}{(s+4)} + \frac{-3}{(s+6)}$$

$$f(t) = -2e^{-2t} + 6e^{-4t} - 3e^{-6t} \qquad \textbf{Ans}$$

Initial value $= \lim_{t \to 0} f(t) = -2 + 6 - 3 = 1$ **Ans**

Final value $= \lim_{s \to 0} sF(s) = \lim_{s \to 0} \dfrac{s^2(s+10)}{(s+2)(s+4)(s+6)} = 0$ **Ans**

(b) $$F(s) = \frac{12(s+1)}{s(s+2)^2(s+3)}$$

$$f = \frac{1}{s} + \frac{-9}{(s+2)} + \frac{6}{(s+2)^2} + \frac{8}{(s+3)}$$

$$f(t) = 1 - 9e^{-2t} + 6te^{-2t} + 8e^{-3t} \qquad \textbf{Ans}$$

Initial value $= \lim_{t \to 0} f(t) = 1 - 9 + 0 + 8 = 0$ **Ans**

Final value $= \lim_{s \to 0} sF(s) = \lim_{s \to 0} \dfrac{12(s+1)}{(s+2)^2(s+3)} = \dfrac{12 \times 1}{4 \times 3} = 1$ **Ans**

Problem 6.2 The unit step response of a system is given as:

$$c(t) = \frac{5}{2} + 5t - \frac{5}{2}e^{-2t}$$

Find the transfer function of the system.

Solution

$$c(s) = \frac{5}{2} + 5t - \frac{5}{2}e^{-2t} = \frac{5}{2s} + \frac{5}{s^2} + \frac{-5/2}{(s+2)} = \frac{5}{2}\left[\frac{1}{s} + \frac{2}{s^2} - \frac{1}{s+2}\right] = \frac{10(s+1)}{s^2(s+2)}$$

$$\therefore \text{ Transfer function} = \frac{C(s)}{R(s)} = \frac{\dfrac{10(s+1)}{s^2(s+2)}}{1/s} = \frac{10(s+1)}{s(s+2)} \qquad \textbf{Ans}$$

Problem 6.3 A servo system for the position control of a rotable mass is stabilised by viscous friction damping which is three-quarters of that is needed for critical damping. The undamped natural frequency of the system is 12 Hz. Derive an expression for the output of the system, if the input control is suddenly moved to a new position, being initially at rest. Hence, find the maximum overshoot.

Solution

The output response is given by:

$$\theta_0(t) = \theta_i \left[1 - \frac{e^{-\xi\omega_n t}}{\sqrt{1-\xi^2}} \sin(\omega_d t + \theta) \right]$$

Now $\qquad \xi = 0.75 \text{ (given)}$

$$\therefore \qquad \theta = \tan^{-1}\frac{\sqrt{1-\xi^2}}{\xi} = \tan^{-1}\frac{\sqrt{1-0.75^2}}{0.75} = 41.41°$$

But $\qquad 41.41° = \dfrac{41.4 \times \pi}{180} = 0.72$

$$\omega_d = \omega_n\sqrt{1-\xi^2}$$
$$f_n = 12 \text{ Hz} = 12 \text{ cycles/sec.}$$

or $\qquad \omega_n = 2\pi \times 12 = 75.4 \text{ rad/s}$

$\therefore \qquad \omega_d = 75.4\sqrt{1-0.75^2} = 50$

$$\frac{1}{\sqrt{1-\xi^2}} = \frac{1}{\sqrt{1-0.75^2}} = 1.5$$

$$\xi\omega_n = 0.75 \times 75.4 = 56.6$$

$\therefore \qquad \theta_0(t) = \theta_i(t)\,[1 - 1.5\,e^{-56.6t}\sin(50t + 0.72)] \qquad \textbf{Ans}$

Maximum overshoot

$$M_p = e^{-\pi\xi/\sqrt{1-\xi^2}} = e^{\left(-\frac{\pi \times 0.75}{\sqrt{1-0.75^2}}\right)} = 2.8\% \qquad \textbf{Ans}$$

Problem 6.4 An angular position of flywheel is controlled by an error actuated closed-loop automatic control system to follow the motion of input lever. The lever is maintained in sinusoidal oscillations through ±60° with an angular frequency $\omega = 2$ rad/sec. The inclusive moment of inertia of flywheel is 150 kg·m^2 and the stiffness of the control is 2400 Nm per radian of misalignment. Calculate the viscous frictional torque required to produce critical damping. Assuming critical damping, calculate the amplitude of swing of flywheel and the time lag between the flywheel and the control lever.

Solution

$$K(\theta_i - \theta_0) = (Js^2 + Fs)\theta_0$$

for critical damping

$$F = 2\sqrt{JK} = 2\sqrt{150 \times 2400} = 1200 \text{ Nm/rad/sec} \qquad \textbf{Ans}$$

$$\omega_n = \sqrt{\frac{K}{J}} = \sqrt{\frac{2400}{150}} = 4 \text{ rad/sec}$$

We know that

$$\frac{\theta_0}{\theta_i} = \frac{\omega_n^2}{s^2 + 2\xi\omega_n s + \omega_n^2}$$

∴ Torque equation is

$$(s^2 + 2\xi\omega_n s + \omega_n^2)\theta_0 = \omega_n^2 \theta_i$$

If $\theta_i = \theta \sin \omega t$ then the steady-state output is given by:

$$\frac{\theta}{\left\{\left[\frac{\omega^2}{\omega_n^2} - 1\right]^2 + \left[4\xi^2 \frac{\omega^2}{\omega_n^2}\right]\right\}^{1/2}}$$

Here $\theta = 60°$; $\omega = 2$ rad/sec; $\omega_n = 4$ rad/sec and $\xi = 1$

∴ Amplitude $= \dfrac{60}{\left\{\left[\dfrac{2 \times 2}{4 \times 4} - 1\right]^2 + \dfrac{4 \times 2 \times 2}{4 \times 4}\right\}^{1/2}} = 48°.$ \qquad **Ans**

The angle of lag is given by:

$$\tan \theta = \frac{2\omega\omega_n \xi}{\omega_n^2 - \omega^2} = \frac{2 \times 2 \times 4 \times 1}{16 - 4} = 53.13°$$

Therefore, time lag $= \dfrac{53.13 \times \pi}{2 \times 180} = 0.46$ sec. \qquad **Ans**

Problem 6.5 Figure 6.21 shows a mechanical system and the response when 10 N of force is applied to the system. Determine the values of M, F and K. The dimension 'x' is in meter.

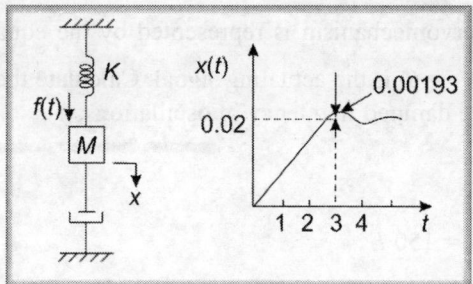

Fig. 6.35

Solution

The transfer function of the mechanical system is:

$$\frac{X(s)}{F(s)} = \frac{1}{Ms^2 + Fs + K}$$

$$F(s) = \frac{10}{s}$$

$\therefore$

$$X(s) = \frac{10}{s(Ms^2 + Fs + K)}$$

The steady-state value of 'x' is

$$x = \lim_{s \to 0} sX(s) = \frac{10}{K} = 0.02 \text{ (given)}$$

$\therefore$ $$K = \frac{10}{0.02} = 500$$ **Ans**

$$M_p = \frac{0.00193}{0.02} = 9.66\%$$

$$\frac{9.66}{100} = e^{-\pi\xi/\sqrt{1-\xi^2}} \quad \therefore \quad \xi = 0.6$$

$$t_p = \frac{\pi}{\omega_d} = \frac{\pi}{\omega_n\sqrt{1-0.6^2}} = 3$$

$\therefore$ $$\omega_n = 1.31 \text{ rad/sec.}$$

But $$\omega_n^2 = \frac{K}{M}$$

$\therefore$ $$M = \frac{K}{\omega_n^2} = \frac{500}{1.31 \times 1.31} = 291.81 \text{ kg}$$ **Ans**

Also $$\frac{F}{M} = 2\xi\omega_n$$

$\therefore$ $$F = 2 \times 0.6 \times 1.31 \times 291.81 = 458.73 \text{ N/m/sec}$$ **Ans**

Problem 6.6 A servomechanism is represented by the equation $\dfrac{d^2\theta}{dt^2} + 10\dfrac{d\theta}{dt}$ $= 150\,E$, where $E = (r - \theta)$ is the actuating signal. Calculate the value of damping ratio, undamped and damped frequency of oscillations.

Solution

$$\frac{d^2\theta}{dt^2} + 10\frac{d\theta}{dt} = 150\,E$$

or $\qquad \dfrac{d^2\theta}{dt^2} + 10\dfrac{d\theta}{dt} = 150\,(r - \theta)$

The equation in Laplace domain is:

$$s^2\,\theta(s) + 10\,s\theta(s) = 150\,(R(s) - \theta(s))$$

$$\frac{\theta(s)}{R(s)} = \frac{150}{s^2 + 10s + 150}$$

Comparing this with $\dfrac{\omega_n^2}{s^2 + 2\xi\omega_n s + \omega_n^2}$

$$\omega_n^2 = 150 \;\therefore\; \omega_n = 12.25 \text{ rad/sec} \hspace{3cm}\textbf{Ans}$$

$$2\xi\omega_n = 10 \;\therefore\; \xi = \frac{10}{2 \times 12.25} = 0.41 \hspace{2cm}\textbf{Ans}$$

$$\omega_d = \omega_n\sqrt{1 - \xi^2} = 12.25\sqrt{1 - 0.41^2} = 11.17 \text{ rad/sec} \hspace{1cm}\textbf{Ans}$$

Problem 6.7 Measurements conducted on a servomechanism show the system response to be $c(t) = 1 + 0.2\,e^{-60t} - 1.2\,e^{-10t}$, when subjected to a unit-step input. Obtain the expression for closed-loop transfer function, the damping ratio and undamped natural frequency of oscillations.

Solution

$$c(t) = 1 + 0.2e^{-60t} - 1.2\,e^{-10t}$$

$$C(s) = \frac{1}{s} + \frac{0.2}{s + 60} - \frac{1.2}{s + 10}$$

$$C(s) = \frac{600/s}{s^2 + 70s + 600} \hspace{4cm}\textbf{Ans}$$

Unit step input means $R(s) = \dfrac{1}{s}$

$$\therefore \qquad \frac{C(s)}{R(s)} = \frac{600}{s^2 + 70s + 600} \hspace{3cm}\textbf{Ans}$$

$$\omega_n^2 = 600 \;\therefore\; \omega_n = 24.49 \text{ rad/sec.} \hspace{2cm}\textbf{Ans}$$

$$2\xi\omega_n = 70 \quad \therefore \quad \xi = \frac{70}{2 \times 24.49} = 1.43 \qquad \textbf{Ans}$$

Problem 6.8 A unity feedback system is characterised by an open-loop transfer function $G(s) = K/s(s + 10)$. Determine the gain K so that the system will have a damping ratio of 0.5. For this value of K, determine settling time, peak overshoot and time to peak overshoot for a unit step input.

Solution

$$G(s) = \frac{K}{s(s+10)} \quad \text{and} \quad H(s) = 1$$

$$\therefore \qquad \frac{G(s)}{1 + G(s)H(s)} = \frac{K}{s^2 + 10s + K}$$

Comparing with standard second order transfer function

$$2\xi\omega_n = 10 \quad \therefore \quad \omega_n = \frac{10}{2\xi} = \frac{10}{2 \times 0.5} = 10$$

Also $\qquad\qquad K = \omega_n^2 \quad \therefore \quad K = 10 \times 10 = 100 \qquad$ **Ans**

$$t_s = \frac{4}{\xi\omega_n} = \frac{4}{0.5 \times 10} = 0.8 \text{ sec} \qquad \textbf{Ans}$$

$$M_p = e^{\frac{-\pi\xi}{\sqrt{1-\xi^2}}} = e^{\frac{-0.5\pi}{\sqrt{1-0.5^2}}} = 16.3\% \qquad \textbf{Ans}$$

$$T_p = \frac{\pi}{\omega_n\sqrt{1-\xi^2}} = \frac{\pi}{10\sqrt{1-0.25}} = 0.326 \text{ sec} \qquad \textbf{Ans}$$

Problem 6.9 The open loop transfer function of a unity feedback system is given by $G(s) = K/s(1 + sT)$ where, T and K are constants having positive values. By what factor the amplifier gain be reduced so that (a) the peak overshoot of unit step response of the system is reduced from 75% to 25%. (b) The damping ratio increases from 0.1 to 0.6.

Solution

$$G(s) = \frac{K}{s(1+sT)}$$

Let the value of damping ratio is ξ_1 when peak overshoot is 75% and ξ_2 when peak overshoot is 25%

$$M_p = e^{\frac{-\pi\xi}{\sqrt{1-\xi^2}}}$$

For $\qquad\qquad M_p = 75\%; \quad \xi = \xi_1 = 0.091$

and for $\qquad\qquad M_p = 25\%; \quad \xi = \xi_2 = 0.4037$

Transfer function $\quad = \dfrac{G(s)}{1 + G(s)H(s)} = \dfrac{K}{Ts^2 + s + K} = \dfrac{C(s)}{R(s)}$

or $\qquad \dfrac{C(s)}{R(s)} = \dfrac{K/T}{s^2 + \dfrac{1}{T}s + \dfrac{K}{T}}$

Therefore, $\qquad \omega_n = \sqrt{K/T}$ and $2\xi\omega_n = \dfrac{1}{T}$

Let the value of $K = K_1$ when $\xi = \xi_1$ and $K = K_2$ when $\xi = \xi_2$

Since $\qquad 2\xi\omega_n = \dfrac{1}{T}$

$$\xi = \dfrac{1}{2T\omega_n} = \dfrac{1}{2\sqrt{KT}}$$

$$\dfrac{\xi_1}{\xi_2} = \sqrt{\dfrac{K_2 T}{K_1 T}} = \sqrt{\dfrac{K_2}{K_1}}$$

or $\qquad \dfrac{K_2}{K_1} = \left(\dfrac{0.091}{0.4037}\right)^2 = 0.0508$

or $\qquad K_2 = 0.0508 K_1$

or the amplifier gain has to be reduced by a factor $\dfrac{1}{0.0508} = 20$ **Ans**

(b) Let $\xi = 0.1$, where gain is K_1, and $\xi = 0.6$ when gain is K_2

$\therefore \qquad \dfrac{0.1}{0.6} = \sqrt{\dfrac{K_2}{K_1}}$

or $\qquad K_2 = \dfrac{1}{36}K_1$ **Ans**

$\therefore$ the amplifier gain should be reduced by a factor 36.

Problem 6.10   A servomechanism is used to control the angular position (θ_o) of a mass through a common signal (θ_i). The moment of inertia of moving parts referred to the load shaft is 150 kg·m^2 and the motor torque constant at the load is 4×10^4 Nm/rad of error. The damping torque coefficient referred to the load shaft is 4×10^3 Nm/rad/sec.

(a) Find the step response of the servomechanism to a step input of one radian and the peak overshoot.

(b) Find the steady-state error when the command signal is a constant angular velocity of 1 rpm.

(c) Find the steady-state error which exists when a torque of 1200 Nm is applied to the load shaft.

Solution

See Fig. 6.36

(a) $J = 150$ kg·m^2, $B = 4 \times 10^3$ Nm/rad/sec and $K_T = 4 \times 10^4$ Nm/rad of error

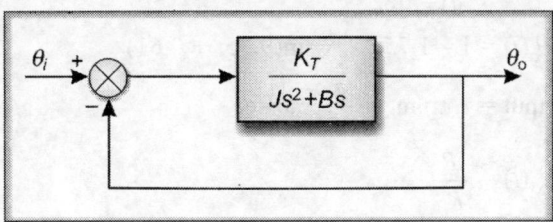

Fig. 6.36

$$G(s) = \frac{K_T}{Js^2 + Bs}$$

$$\frac{\theta_o(s)}{\theta_i(s)} = \frac{K_T}{Js^2 + Bs + K_T} = \frac{K_T/J}{s^2 + \dfrac{B}{J} + \dfrac{K_T}{J}}$$

$$\omega_n = \frac{K_T}{J} = \sqrt{\frac{4 \times 10^4}{150}} = 16.33 \text{ rad/sec}$$

Also $2\xi\omega_n = \dfrac{B}{J}$

∴ $\xi = \dfrac{B}{2J\omega_n} = \dfrac{4 \times 10^3}{2 \times 150 \times 16.33} = 0.82$

Peak time $= t_p = \dfrac{\pi}{\omega_d} = \dfrac{\pi}{16.33\sqrt{1 - 0.82^2}} = 0.336 \text{ sec}$

Peak overshoot $= M_p = e^{\dfrac{-0.82\,\pi}{\sqrt{1 - 0.82^2}}} = 1.11\%$

for under damped case, i.e. $\xi < 1$, the response is $= 1 - \dfrac{e^{-\xi\omega_n t}}{\sqrt{1 - \xi^2}} \sin(\omega_d t + \theta)$

$$\omega_d = \omega_n\sqrt{1 - \xi^2} = 16.33\sqrt{1 - 0.82^2} = 9.35 \text{ rad/sec}$$

$$\theta = \tan^{-1}\left(\frac{\sqrt{1 - 0.82^2}}{0.82}\right) = 34.92°$$

or $\theta = \dfrac{34.92 \times \pi}{180} = 0.61 \text{ radians}$

$$\xi\omega_n = 0.82 \times 16.33 = 13.4$$

and

$$\frac{1}{\sqrt{1-\xi^2}} = \frac{1}{\sqrt{1-0.82^2}} = 1.75$$

$\therefore$ $\theta_0(t) = 1 - 1.75e^{-13.4t} \sin(9.35t + 0.61)$ **Ans**

(b)   Input $= 1$ r.p.m. $= \dfrac{\pi}{30}$ rad/sec

$$e_{ss}(t) = \frac{R}{K_v}$$

$$K_v = \lim_{s\to 0} sG(s) = \lim_{s\to 0}\frac{sK_T}{Js^2 + Bs} = \frac{K_T}{B}$$

$\therefore$ $e_{ss}(t) = \dfrac{R \times B}{K_T} = \dfrac{\pi \times 4 \times 10^3}{30 \times 4 \times 10^4} = 0.011$ rad or 0.6 degree   **Ans**

(c) $T = 1200$ Nm   or $K_T = 4 \times 10^4$ Nm/rad of error.

$\therefore$ Error $= \dfrac{1200}{4 \times 10^4} = 0.03$ radians or 1.72 degrees.   **Ans**

Problem 6.11   In the position control system is shown in Fig. 6.37
(a) Sensitivity of error detector $K_p = 1$ volt/deg
(b) Transfer function of motor

$$\frac{\theta_m(s)}{V(s)} = \frac{K_m}{s(sT+1)}; \text{ where, } K_m = 10 \text{ rad/sec/volt}; T = 0.1 \text{ sec}$$

(c) Gear ratio $\dfrac{\dot\theta_c}{\dot\theta_L} = 1; \dfrac{\dot\theta_L}{\dot\theta_m} = \dfrac{1}{100}$

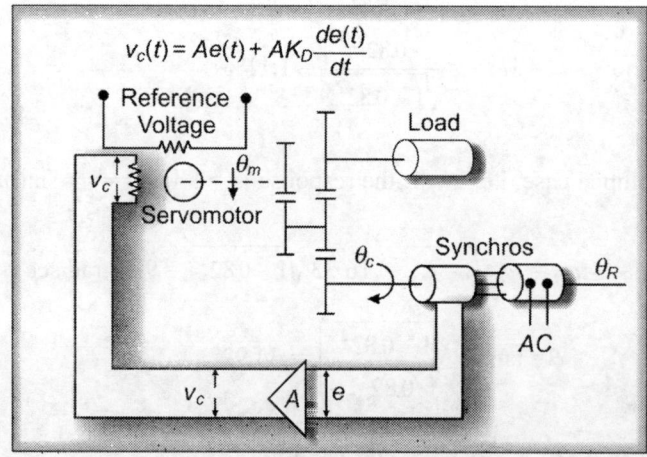

$$v_c(t) = Ae(t) + AK_D\frac{de(t)}{dt}$$

Reference
Voltage

Load

v_c

θ_m

Servomotor

θ_c

Synchros

θ_R

AC

v_c A e

Fig. 6.37

If the input shaft is driven at or speed of π rad/sec and is constant, determine the value of the amplifier gain such that the steady-state deviation between input and output positions is less than 3 degrees. If amplifier gain is 35, determine the damping ratio and setting time of the system.

(d) The amplifier is modified, the system dynamics by introducing an additional derivative term such that the output is given by

$$v_c(t) = Ae(t) + AK_D \frac{de(t)}{dt}$$

Determine the value of K_D so that the damping ratio is improved to 0.4. Does this improve the steady-state error as in part (c). Also, calculate the new settling time. Assume $A = 35$.

(e) The block diagram of the schematic diagram is shown in Fig. 6.38

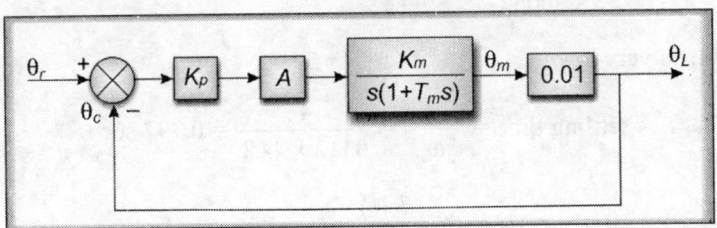

Fig. 6.38

(a) $$G(s) = \frac{K_p A K_m \times 0.01}{\pi \times s\,(1 + T_m s)}$$

$$K_P = 1 \text{ volt/degree} = \frac{180}{\pi} \text{ volt/radian}$$

$\therefore$ $$G(s) = \frac{180 \times A \times 10 \times 0.01}{\pi \times s(1 + 0.1s)}$$

$$e_{ss}(t) = 3° = \frac{3\pi}{180} = 0.05 \text{ radian}$$

But $$0.05 = \frac{R}{K_v}$$

where, $$R = \pi \text{ rad/sec}$$

and $$K_v = \lim_{s \to 0} sG(s) = \lim_{s \to 0} \frac{180 \times A \times 10 \times 0.01}{\pi\,(1 + 0.1s)} = \frac{18A}{\pi}$$

$\therefore$ $$0.05 = \frac{\pi}{18A/\pi}$$

$\therefore$ $$A = \frac{\pi^2}{18 \times 0.05} = 10.5 \qquad \textbf{Ans}$$

(b) If $$A = 35, \text{ then}$$

$$G(s) = \frac{180 \times 35 \times 10 \times 0.01}{\pi \times s\,(1 + 0.1s)} = \frac{200.6}{0.1s^2 + s}$$

or $\qquad G(s) = \dfrac{200.6}{s^2 + 10s}$

Transfer function with unity fedback

$$\frac{G(s)}{1 + G(s)H(s)} = \frac{2006}{s^2 + 10s + 2006}$$

Comparing with standard second order transfer function

$$2\xi\omega_n = 10 \text{ and } \omega_n = \sqrt{2006}$$

$\therefore \qquad \xi = \dfrac{10}{2 \times \sqrt{2006}} = \dfrac{10}{2 \times 44.8} = 0.112$ **Ans**

This value is very low,

$$t_s = \text{settling time} = \frac{4}{\xi\omega_n} = \frac{4}{0.112 \times 44.8} = 0.797 \text{ sec} \qquad \textbf{Ans}$$

(c) $\qquad v_c(t) = Ae(t) + AK_D\dfrac{de(t)}{dt}$

or $\qquad V_C(s) = E(s)\{A(1 + K_D s)\}$

Therefore, amplifier gain

$$= A(1 + K_D s) = 35(1 + K_D s)$$

$$G(s) = \frac{180 \times 35\,(1 + K_D s) \times 10 \times 0.01}{\pi \times s\,(1 + 0.1s)}$$

$$= \frac{2006\,(1 + K_D s)}{s^2 + 10s}$$

$$\frac{G(s)}{1 + G(s)H(s)} = \frac{2006\,(1 + K_D s)}{s^2 + 10s + 2006\,K_D s + 2006}$$

Comparing with the standard second order transfer function

$$\omega_n = \sqrt{2006} = 44.8 \text{ rad/sec}$$

$$2\xi\omega_n = 10 + 2006K_D$$

Given $\qquad \xi = 0.4$

$\therefore \qquad K_D = \dfrac{2 \times 0.4 \times 44.8 - 10}{2006} = 0.0128 \text{ sec}$

$$t_s = \frac{4}{\xi\omega_n} = \frac{4}{0.4 \times 44.8} = 0.223 \text{ sec} \qquad \textbf{Ans}$$

Steady-state error

$$G(s) = \frac{180 \times A\,(1+K_D s) \times 10 \times 0.01}{\pi \times s\,(1+0.1s)} = \frac{180\,A\,(1+K_D s)}{\pi \times s\,(1+0.1s)}$$

$$K_v = \lim_{s \to 0} sG(s) = \frac{18A}{\pi}$$

$$R = \pi$$

$$\therefore \qquad e_{ss}(t) = \frac{\pi \times \pi}{18A} < \frac{3\pi}{180} \text{ radians}$$

$$A > \frac{10\pi}{3} \text{ or } 10.5 \text{ (same as before)} \qquad \textbf{Ans.}$$

Problem 6.12 A DC position control system comprises a field control DC servomotor, potentiometer, an amplifier and a tachogenerator coupled to the motor shaft. A fraction K of the tachogenerator output is fedback to produce the stabilizing effect. The following values of the system are given:

Moment of inertia of motor $\qquad\qquad J_m = 3 \times 10^{-3} \text{ kgm}^2$

Moment of inertia of load $\qquad\qquad J_L = 5 \text{ kgm}^2$

Motor to load gear ratio, $\qquad\qquad \dfrac{\dot{\theta}_L}{\dot{\theta}_m} = \dfrac{1}{50} = N$

Motor to potentiometer gear ratio, $\qquad \dfrac{\dot{\theta}_L}{\dot{\theta}_C} = 1$

Motor torque constant, $\qquad\qquad K_T = 2 \text{ Nm/amp}$

Tachogenerator constant, $\qquad\qquad K_t = 0.2 \text{ volt/rad/sec}$

Sensitivity of error detector $\qquad\qquad K_p = 1 \text{ volt/rad}$

Amplifier gain $\qquad\qquad\qquad\qquad = A \text{ amps/volt}$

Motor field time constant and Motor and Load frictions are negligible.

   (a) Draw the block diagram.

   (b) Find the transfer function.

   (c) Obtain the amplifier gain required and the fraction of the tachogenerator voltage feedback to undamped natural frequency of 4 rad/sec and damping of system as 0.8.

Solution

(a) The block diagram showing connection of the hardware is shown in Fig. 6.39

(b)
$$J_{eq} = J_m + \left(\frac{N_1}{N_2}\right)^2 J_L = 3 \times 10^{-3} + \left(\frac{1}{50}\right)^2 \times \frac{5}{1} = 5 \times 10^{-3}$$

$$B_{eq} = 0$$

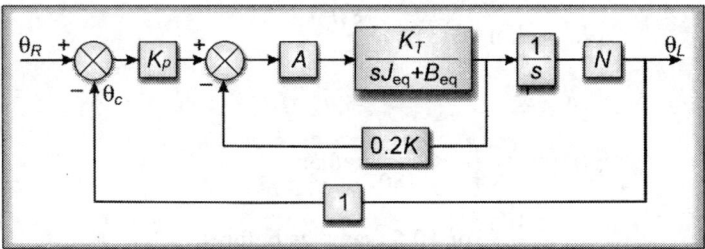

Fig. 6.39

Simplifying the inner loop

$$= \frac{\dfrac{AK_T}{sJ_{eq}}}{1 + \dfrac{0.2KAK_T}{sJ_{eq}}} = \frac{AK_T}{sJ_{eq} + 0.2KAK_T}$$

The block diagram reduces to

$$G(s) \;=\; \text{the forward path transfer function} = \frac{NK_p AK_T}{s(sJ_{eq} + 0.2KAK_T)}$$

$$H(s) \;=\; 1$$

$$\therefore \quad \frac{G(s)}{1 + G(s)H(s)} = \frac{\theta_L}{\theta_R} = \frac{\dfrac{NK_p AK_T}{s(sJ_{eq} + 0.2KAK_T)}}{1 + \dfrac{NK_p AK_T}{s(sJ_{eq} + 0.2KAK_T)}}$$

$$= \frac{NK_p AK_T}{s^2 J_{eq} + 0.2sKAK_T + NK_p AK_T}$$

$$= \frac{1 \times 1 \times A \times 2}{50[5 \times 10^{-3} \times s^2 + 0.2 \times K \times A \times 2 \times s + \dfrac{1}{50} \times 1 \times A \times 2]}$$

$$= \frac{0.04\,A}{5 \times 10^{-3} \times s^2 + 0.4\,KAs + 0.04\,A}$$

$$= \frac{8A}{s^2 + 80KAs + 8A} \qquad \textbf{Ans}$$

(c) Comparing with the standard second-order transfer function

$$8A = \omega_n^2$$

$$\therefore \qquad A = \frac{\omega_n^2}{8} = \frac{16}{8} = 2 \text{ amp/volt}$$

and $\qquad 80AK = 2\xi\omega_n$

$$\therefore \qquad K = \frac{2\xi\omega_n}{80A} = \frac{2 \times 0.8 \times 4}{80 \times 2} = 0.04 \qquad\qquad \textbf{Ans.}$$

Problem 6.13 A line diagram showing the various components of a servomechanism have been connected, as shown in Fig. 6.40. The following particulars refer to the system:

Sensitivity of Synchro, $\qquad K_s = 1$ volt/rad

Amplifier gain, $\qquad\qquad A = 30$ volt/volt

Motor torque constant, $\qquad K_T = 1 \times 10^{-5}$ Nm/volt

Load inertia, $\qquad\qquad\quad J_L = 1.5 \times 10^{-5}$ kgm^2

Viscous friction, $\qquad\qquad B_L = 1 \times 10^{-5}$ Nm/rad/sec

Tachometer constant, $\qquad K_t = 0.2$ volt/rad/sec

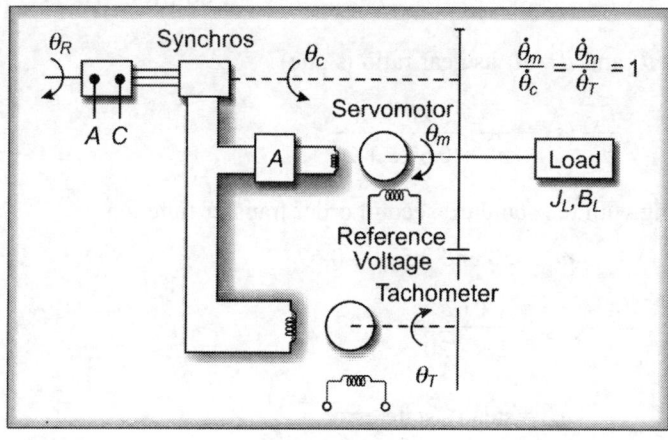

Fig. 6.40

(a) Estimate the value of damping ratio, if tachometer is disconnected. Find the steady-state error corresponding to input velocity of 1 rad/sec.

(b) What will be the value of damping ratio if tachometer is a part of the system?

(c) The amplifier is modified as

$v_c = Ae(t) + A\int e(t)dt$, and the tachometer is removed. Obtain the steady state behaviour and compare it with part (a).

Solution

The block diagram of the system is shown in Fig. 6.41

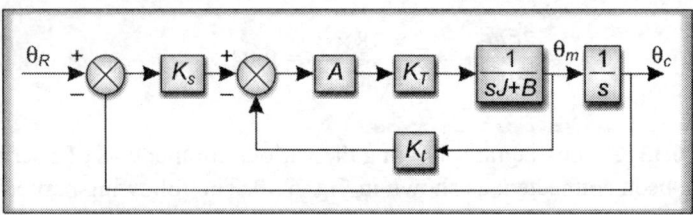

Fig. 6.41

(a) *When tachometer disconnected ($K_t = 0$)*

$$G(s) = \frac{K_s AK_T}{s(sJ + B)}$$

$$\frac{\theta_C}{\theta_R}(s) = \frac{\dfrac{K_s AK_T}{J}}{s^2 + s\dfrac{B}{J} + \dfrac{K_s AK_T}{J}} = \frac{\dfrac{1 \times 30 \times 1 \times 10^{-5}}{1.5 \times 10^{-5}}}{s^2 + s\dfrac{1 \times 10^{-5}}{1.5 \times 10^{-5}} + \dfrac{1 \times 30 \times 1 \times 10^{-5}}{1.5 \times 10^{-5}}}$$

(Note $J = J_L$ and $B = B_L$ as gear ratio is one)

$$\frac{\theta_C}{\theta_R}(s) = \frac{20}{s^2 + 0.67\,s + 20}$$

Comparing with the standard second order transfer function

$$\omega_n^2 = 20 \text{ and } 2\xi\omega_n = 0.67$$

$\therefore$
$$\xi = \frac{0.67}{2 \times \sqrt{20}} = 0.07 \qquad\qquad\qquad \textbf{Ans}$$

$$e_{ss} = \text{steady-state error} = \frac{1}{K_v}$$

$$K_v = \lim_{s \to 0} sG(s) = \frac{K_s AK_T}{B} = \frac{1 \times 30 \times 1 \times 10^{-5}}{1 \times 10^{-5}} = 30$$

$$R = 1 \text{ rad/sec}$$

$$e_{ss} = \frac{1}{30} = 0.033 \text{ rad} = 1.911 \text{ degrees} \qquad\qquad \textbf{Ans}$$

(b) *When tachometer is connected*

$$G(s) = \frac{K_s A K_T}{s(sJ + B + K_T K_t A)}$$

$$\frac{\theta_C}{\theta_R}(s) = \frac{K_s A K_T}{s(sJ + B + K_T K_t A) + K_s A K_T}$$

Comparing with the standard second order transfer function

$$\omega_n^2 = \frac{K_s A K_T}{J} = \frac{1 \times 30 \times 1 \times 10^{-5}}{1.5 \times 10^{-5}} = 20$$

$\therefore$ $\omega_n = 4.47$ rad/ sec

$$2\xi\omega_n = \frac{B + AK_t K_T}{J} = \frac{1 \times 10^{-5} + 30 \times 0.2 \times 1 \times 10^{-5}}{1.5 \times 10^{-5}}$$

$\therefore$ $2\xi\omega_n = 4.67$

or $$\xi = \frac{4.67}{2 \times \omega_n} = \frac{4.67}{2 \times 4.47} = 0.52$$ **Ans**

(c) *When amplifier modified and tachometer removed*

$$v_c(t) = Ae(t) + A\int e(t)\, dt$$

or $$V_c(s) = AE(s) + \frac{A}{s}E(s)$$

or $$\frac{V_c}{E}(s) = A\left(1 + \frac{1}{s}\right) = \frac{A(s+1)}{s}$$

$$G(s) = \frac{\dfrac{K_s K_T A (s+1)}{s}}{s^2 J + sB}$$

or $$G(s) = \frac{K_s K_T A}{s^2 (sJ + B)}$$

$$K_v = \lim_{s \to 0} sG(s) = \lim_{s \to 0} \frac{s K_s K_T A}{s^2 (sJ + B)}$$

$$= \lim_{s \to 0} \frac{K_s K_T A}{s(sJ + B)} = \infty$$

$\therefore$ $$e_{ss} = \frac{R}{K_v} = \frac{R}{\infty} = 0$$

Note: The error has become zero as the type of system has been increased to 2 by modifying the amplifier.

Problem 6.14 The control system, shown in Fig. 6.44, employs proportional plus error rate control. Determine the value of error rate constant K_e so that the damping ratio is 0.6. Determine the values of settling time, maximum overshoot and steady-state error, if the input is unit ramp. What will be the value of steady-state error without error rate control?

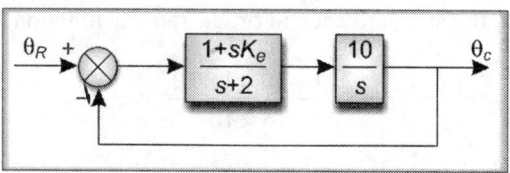

Fig. 6.42

·**Solution**

(a) With Error-rate Control

$$G(s) = \frac{10\,(1 + sK_e)}{s\,(s + 2)}$$

$$\frac{\theta_c}{\theta_R}(s) = \frac{10 + 10sK_e}{s^2 + s\,(2 + 10\,K_e) + 10}$$

Comparing with the standard second order transfer function

$$\omega_n = \sqrt{10} = 3.16 \text{ rad/sec}$$

$$2\xi\omega_n = 2 + 10K_e$$

$$K_e = \frac{2\,\xi\,\omega_n - 2}{10} = \frac{2 \times 0.6 \times 3.16 - 2}{10} = 0.18 \qquad \textbf{Ans}$$

$$t_s = \frac{4}{\xi\,\omega_n} = \frac{4}{0.6 \times 3.16} = 2.11 \text{ sec} \qquad \textbf{Ans}$$

$$M_p = e^{\frac{-0.6\pi}{\sqrt{1 - 0.6^2}}} = 9.49\% \qquad \textbf{Ans}$$

$$e_{ss} = \frac{R}{K_v} = \frac{1}{\lim\limits_{s \to 0} sG(s)} = \frac{1}{5} = 0.2 \text{ rad.} \qquad \textbf{Ans}$$

(b) Without Error-rate Control

$$G(s) = \frac{10}{s\,(s + 2)}$$

$$\frac{\theta_c}{\theta_R}(s) = \frac{10}{s^2 + 2s + 10}$$

∴ $$\omega_n = \sqrt{10} = 3.16 \text{ rad/sec}$$

and $$2\xi\omega_n = 2$$

$$\therefore \qquad \xi = \frac{2}{2 \times 3.16} = 0.32 \qquad \textbf{Ans.}$$

$$t_s = \frac{4}{\xi \omega_n} = \frac{4}{0.32 \times 3.16} = 4 \text{ sec} \qquad \textbf{Ans.}$$

$$M_p = e^{\frac{-0.32\,\pi}{\sqrt{1-0.32^2}}} = 35.1\% \qquad \textbf{Ans.}$$

Problem 6.15 The system, shown in Fig. 6.43, employs derivative of feedback in addition to unity feedback.

(a) Find the value of constant K_0 so that the damping ratio of the system is 0.7. With this calculated value of K_0 and unit ramp input, find the value of steady-state error.

(b) If $K_0 = 0$, calculate the value of damping ratio and undamped frequency of oscillations. What will be the steady-state error resulting from unit ramp input?

(c) The value of steady-state error is required to be 0.2 rad with derivative feedback and the damping ratio maintained at 0.7. How can this be achieved?

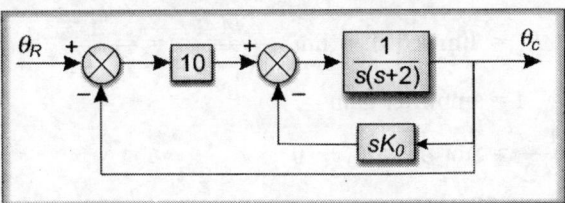

Fig. 6.43

Solution

(a)
$$\frac{\theta_c}{\theta_R}(s) = \frac{10}{s^2 + s(2 + K_0) + 10}$$

$$\therefore \qquad \omega_n = \sqrt{10} = 3.16 \text{ rad/sec}$$

$$2\xi\omega_n = 2 + K_0$$

$$K_0 = 2\xi\omega_n - 2 = 2 \times 0.7 \times 3.16 - 2 = 2.424 \qquad \textbf{Ans}$$

$$e_{ss} = \frac{R}{K_v} = \frac{1}{\lim_{s \to 0} sG(s)}$$

$$G(s) = \frac{10}{s(s + 2 + K_0)}$$

$$\therefore \qquad e_{ss} = \frac{10}{2 + K_0} = \frac{10}{2 + 2.424} = 2.26$$

and
$$K_v = \frac{1}{2.26} = 0.44 \text{ rad} \qquad \textbf{Ans}$$

(b) If $K_0 = 0$, then, $G(s) = \dfrac{10}{s(s+2)}$

$$\dfrac{\theta_c}{\theta_R}(s) = \dfrac{10}{s^2 + 2s + 10}$$

$\therefore$ $\omega_n = \sqrt{10} = 3.16$ rad/sec   **Ans**

Also $2\xi\omega_n = 2$

$\therefore$ $\xi = \dfrac{1}{3.16} = 0.316$ **Ans**

$$K_v = \lim_{s \to 0} sG(s) = \dfrac{10}{2} = 5$$

$\therefore$ $e_{ss} = \dfrac{R}{K_v} = \dfrac{1}{5} = 0.2$ rad   **Ans**

(c) $e_{ss} = 0.2$ rad

         $\xi = 0.7$

$$K_v = \dfrac{R}{e_{ss}} = \dfrac{1}{0.2} = 5$$

Also

$$K_v = \lim_{s \to 0} sG(s) = \lim_{s \to 0} \dfrac{sA}{s(K_0 + 2)} = \dfrac{A}{K_0 + 2}$$

where, A = amplifier gain

$\therefore$ $\dfrac{A}{K_0 + 2} = 5$ or $A = 5K_0 + 10$ (1)

Also $2\xi\omega_n = 2 + K_0$

where, $\xi = 0.7$ and $\omega_n = \sqrt{A}$

$\therefore$ $2 \times 0.7 \times \sqrt{A} = 2 + K_0$

or $1.96A = K_0^2 + 4K_0 + 4$ (2)

Putting the value of A from Eqn. (1) in Eqn. (2), we get

         $1.96(5K_0 + 10) = K_0^2 + 4K_0 + 4$

or $K_0^2 - 5.8K_0 - 15.6 = 0$

$\therefore$ $K_0 = 7.8$ and -2

Taking positive value of $K_0 = 7.8$ **Ans**

         $A = 5K_0 + 10 = 5 \times 7.8 + 10 = 49$ **Ans**

Therefore by keeping $K_0 = 7.8$ and $A = 49$, the desired result can be achieved.

Problem 6.16 The open-loop transfer function of a unity feedback system is:

$$G(s) = \dfrac{100}{s(s+10)}$$

(a) Find the static error constants and the steady-state error of the system when subjected to an input given by the polynomial

$$r(t) = P_0 + P_1 t + \frac{P_2}{2} t^2$$

(b) Find the dynamic error using the dynamic error coefficients

Solution

(a)
$$K_p = \lim_{s \to 0} G(s) = \lim_{s \to 0} \frac{100}{s\,(s+10)} = \infty \qquad \textbf{Ans}$$

$$K_v = \lim_{s \to} sG(s) = \lim_{s \to 0} \frac{100}{s+10} = 10 \qquad \textbf{Ans}$$

$$K_a = \lim_{s \to 0} s^2 G(s) = \lim_{s \to 0} \frac{100\,s}{s+10} = 0 \qquad \textbf{Ans}$$

$$r(t) = P_0 + P_1 t + \frac{P_2}{2} t^2$$

$$e_{ss} = \frac{R_1}{1 + K_p} + \frac{R_2}{K_v} + \frac{R_3}{K_a}$$

$$= \frac{P_0}{1 + \infty} + \frac{P_1}{10} + \frac{P_2}{0} = 0 + \frac{P_1}{10} + \infty = \infty \qquad \textbf{Ans}$$

(b)
$$G(s) = \frac{100}{s\,(s+10)} = \frac{10}{s\,(1 + 0.1\,s)}$$

$$\frac{E(s)}{R(s)} = \frac{1}{1 + G(s)} = \frac{1}{1 + \dfrac{10}{s(1 + 0.1\,s)}} = \frac{s + 0.1s^2}{10 + s + 0.1s^2}$$

$$\frac{E(s)}{R(s)} = \frac{s + 0.1s^2}{10 + s + 0.1s^2} = \frac{1}{10}s + \frac{0.9}{100}s^2 - \frac{1}{1000}s^3$$

The error constants are

$$C_0 = 0$$

$$C_1 = \frac{1}{10} = 0.1$$

$$C_2 = \frac{0.9}{100} = 0.009$$

$$C_3 = -\frac{1}{1000} = -0.001$$

$$r(t) = P_0 + P_1 t + \frac{P_2}{2} t^2$$

$$\dot{r}(t) = P_1 + P_2 t$$

$$\ddot{r}(t) = P_2$$

$$\dddot{r}(t) = 0$$

$$E(s) = 0.1sR(s) + 0.009\,s^2 R(s) - 0.001s^3 R(s)$$

In the time domain, the steady-state error is given by:

$$\lim_{t \to \infty} e(t) = (\, 0.1 \, \dot{r}(t) + 0.009 \, \ddot{r}(t) - 0.001 \, \dddot{r}(t))$$

$$= 0.1(P_1 + P_2 t) + 0.009 P_2$$

The steady-state error is then

$$e_{ss} = \lim_{t \to \infty} e(t) = \lim_{t \to \infty} \, [0.1[P_1 + P_2 \, t] + 0.009 P_2]$$

$$= \lim_{t \to \infty} \, [\, 0.1 P_1 + 0.009 P_2 + 0.1 P_2 t] \qquad \textbf{Ans}$$

Unless $P_2 = 0$, the steady-state error will be infinity.

Problem 6.17 Find the dynamic error co-efficients of the unity feedback system whose forward transfer function is $G(s) = 10/s \, (s + 1)$. Find the steady-state error to the polynomial input $r(t) = a_0 + a_1 t + a_2 t^2$

Solution

$$\frac{E(s)}{R(s)} = \frac{1}{1 + G(s)} \quad \text{for unity feedback system}$$

It can be expanded as a power series in s, by *Taylor's* series expansion leading to the following expression

$$\frac{1}{1 + G(s)} = C_0 + C_1 s + C_2 s^2 + C_3 s^3 + \dots \qquad (1)$$

where, C_0, C_1, C_2, etc. are defined as error coefficients.

If $\qquad G(s) = \dfrac{K_v}{s(1 + sT)}$

the $\qquad \dfrac{E}{R}(s) = \dfrac{s + s^2 T}{K_v + s + s^2 T}$

$$= \frac{1}{K_v} s + \frac{K_v T - 1}{K_v^2} s^2 - \frac{2 K_v T - 1}{K_v^3} s^3 + \dots \qquad (2)$$

By comparing (1) and (2), the error co-efficients are

$$C_0 = 0; \quad C_1 = \frac{1}{K_v}; \quad C_2 = \frac{K_v T - 1}{K_v^2}; \quad C_3 = -\frac{2 K_v T - 1}{K_v^3}$$

For $\qquad G(s) = \dfrac{10}{s \, (s + 1)}$

$$C_0 = 0 \qquad \textbf{Ans}$$

$$C_1 = \frac{1}{K_v} = \frac{1}{10} = 0.1 \qquad \textbf{Ans}$$

$$C_2 = \frac{K_v T - 1}{K_v^2} = \frac{10 \times 1 - 1}{100} = 0.09 \qquad \textbf{Ans}$$

$$C_3 = -\frac{2K_vT-1}{K_v^3} = -\frac{2\times10\times1-1}{1000} = -0.019 \qquad \textbf{Ans}$$

$$\therefore \qquad E(s) = 0.1\,sR(s) + 0.09\,s^2R(s) - 0.019s^3R(s)$$

In time domain, the steady-state error is given by:

$$\lim_{t\to\infty} e(t) = 0.1\,\dot{r}(t) + 0.09\,\ddot{r}(t) - 0.019\,\dddot{r}(t)$$

$$r(t) = a_0 + a_1\,t + a_2\,t^2$$

$$\dot{r}(t) = a_1 + 2a_2t$$

$$\ddot{r}(t) = 2a_2$$

$$\dddot{r}(t) = 0$$

then
$$\lim_{t\to\infty} e(t) = \lim_{t\to\infty}\ [0.1(a_1 + 2a_2t) + 0.09\times 2a_2]$$

$$= \lim_{t\to\infty}\ [0.1a_1 + 0.18a_2 + 0.2a_2t] \qquad \textbf{Ans}$$

Unless $a_2 = 0$, the steady-state error will become infinity.

Problem 6.18 The forward transfer function of a unity feedback type 1, second order system has a pole at -2. The nature of gain K is so adjusted that damping ratio is 0.4. The above system is subjected to input $r(t) = 1 + 4t$. Find the steady-state error.

Solution

$$G(s) = \frac{K}{s\,(s+2)}$$

$$\frac{C(s)}{R(s)} = \frac{K}{s^2 + 2s + K}$$

$$2\xi\omega_n = 2$$

$$\therefore \qquad \omega_n = \frac{1}{\xi} = \frac{1}{0.4} = 2.5$$

Also
$$K = \omega_n^2 = (2.5)^2 = 6.25$$

$$G(s) = \frac{6.25}{s\,(s+2)}$$

$$e_{ss} = \frac{1}{1+K_p} + \frac{4}{K_v}$$

$$K_p = \lim_{s\to0}\ G(s) = \infty$$

$$K_v = \lim_{s\to0}\ sG(s) = \frac{6.25}{2}$$

$$\therefore \qquad e_{ss} = \frac{1}{1+\infty} + \frac{4\times2}{6.25} = 1.28. \qquad \textbf{Ans}$$

Problem 6.19 A second order position control system has open loop transfer function $G(s) = 5.73A/s\,(1 + 0.1\,s)$, where A is the amplifier gain. Find the value of A so that steady-state error shall not exceed one degree when the input shaft rotates at 10 rpm.

Solution

Given

$$R = 10 \text{ r.p.m} = \frac{\pi}{3} \text{ rad/sec}$$

$$e_{ss} = 1^\circ = \frac{\pi}{180} \text{ radians}$$

$$K_v = \frac{R}{e_{ss}} = \frac{\pi/3}{\pi/180} = 60$$

Also

$$K_v = \lim_{s \to 0} sG(s) = \lim_{s \to 0} \frac{5.73A}{1 + 0.1\,s} = 5.73A$$

∴

$$5.73A = 60$$

or

$$A = \frac{60}{5.73} = 10.47 \qquad\qquad \textbf{Ans}$$

Problem 6.20 Give an example of Type 0, Type 1 and Type 2 systems

Solution

$$\text{Type } 0 = \frac{K_1(1 + sT)}{(1 + sT_2)(1 + sT_3)(1 + sT_1)} \qquad\qquad \textbf{Ans}$$

$$\text{Type } 1 = \frac{K_2(1 + sT)}{s(1 + sT_2)(1 + sT_1)} \qquad\qquad \textbf{Ans}$$

$$\text{Type } 2 = \frac{K_3(1 + sT)}{s^2(1 + sT_2)(1 + sT_1)} \qquad\qquad \textbf{Ans}$$

Problem 6.21 Find the steady-state response of a control system whose open-loop transfer function is given by:

$$G(s) = \frac{\theta_o}{\theta_e}(s) = \frac{K(s + 2)}{s(s^2 + 4s + 8)}$$

Consider step input having an error angle of θ_e. Assume K as the gain constant.

Solution

$$\theta_e(s) = \frac{\theta_o}{s}$$

$$\therefore \qquad \theta_o(s) = \frac{K(s+2)}{s(s^2 + 4s + 8)} \times \frac{\theta_e}{s}$$

or
$$s\theta_o(s) = \frac{K(s+2)\theta_e}{s(s^2 + 4s + 8)}$$

Applying the final value theorem,

$$\lim_{t \to \infty} \dot{\theta}_o(t) = \lim_{t \to \infty} \frac{d\theta_o(t)}{dt} = \lim_{s \to 0} \frac{K(s+2)\theta_e}{(s^2 + 4s + 8)} = \frac{K\theta_e}{4} \qquad \textbf{Ans.}$$

i.e. output velocity is proportional to the error angle.

Problem 6.22 Find the steady-state response to a step input to a control system whose open-loop transfer function is:

$$G(s) = \frac{\theta_o}{\theta_e}(s) = \frac{K_a(1 + sT_1)}{s^2(1 + sT_2)}$$

where K_a is the acceleration constant

Solution

$$\theta_e(s) = \frac{\theta_o}{s}$$

$$\therefore \qquad \theta_o(s) = \frac{K_a(1 + sT_1)}{s^2(1 + sT_2)}$$

$$s^2\theta_o(s) = \frac{K_a(1 + sT_1)}{(1 + sT_2)}$$

$$\lim_{t \to \infty} \ddot{\theta}_{a(t)} = \lim_{s \to 0} \frac{K_a(1 + sT_1)\theta_e}{1 + sT_2} = K_a\theta_e$$

i.e. output acceleration is proportional to the error angle.

Problem 6.23 A Type 1 control is used to track an input angle at a constant rate of 60 rpm. What is the required velocity constant, if the permissible tracking error is $2°$.

Solution

$$60 \text{ rpm} = \frac{60}{60} \times 360 = 360 \text{ deg/sec}$$

For a $2°$ lag,

$$K_v = \frac{360}{2} = 180 \text{ sec}^{-1} \qquad \textbf{Ans}$$

Problem 6.24 The transfer function of a control system is given by $G(s) = 1/(1 + sT)^2$. Show that if the input is a step displacement, the output will complete 98.26% of the step in $6T$ seconds for critical damping.

Solution

$$G(s) = \frac{1}{(1+sT)^2} = \frac{\dfrac{1}{T^2}}{s^2 + \dfrac{2}{T}s + \dfrac{1}{T^2}}$$

Comparing with the standard form

$$\omega_n = \frac{1}{T} \text{ and } \xi = 1$$

The standard form of solution of a second-order equation with critical damping to a unit step input is:

$$c(t) = 1 - e^{-\omega_n t}(1 + \omega_n t)$$

or

$$c(t) = 1 - e^{-t/T}\left(1 + \frac{t}{T}\right)$$

Putting $= t = 6T$

$$c(t) = 1 - e^{-6}(1 + 6) = 98.26\% \qquad\qquad\textbf{Ans}$$

Problem 6.25 A servo system for position control has the closed-loop transfer function $6/(s^2 + 2s + 6)$. Find the percentage overshoot; if the input is suddenly moved to a new position.

Solution

$$G(s) = \frac{6}{s^2 + 2s + 6}$$

Comparing with the standard form

$$\omega_n = \sqrt{6} \text{ and } 2\xi\omega_n = 2$$

∴

$$\xi = 0.408$$

$$M_p = e^{\dfrac{-\pi\xi}{\sqrt{1-\xi^2}}} = e^{\dfrac{-\pi\times 0.408}{\sqrt{1-0.408^2}}} = 24\% \text{ or } M_p = 0.24 \qquad\qquad\textbf{Ans}$$

Problem 6.26 A servo system for the positional control of a rotatable mass is stabilized by viscous-friction damping which three-quarters of that needed for critical damping. The undamped frequency of oscillation of the system is 12 Hz. Derive an expression for the output of the system if the input control is suddenly moved to a new position, the system being initially at rest. Hence, find the first overshoot.

Solution

$$\xi = 0.75$$
$$\omega_n = 12 \times 2\pi = 75.36 \text{ rad/sec.}$$

The solution is given by:

$$c(t) = \left[1 - \frac{e^{-\xi\omega_n t}}{\sqrt{1-\xi^2}} \sin(\omega_n\sqrt{1-\xi^2})t + \cos^{-1}\xi\right]$$

Substituting the values of ω_n and, we get

$$c(t) = [1 - 1.5e^{-56.6t}\sin(50t + 41.4)] \qquad \textbf{Ans}$$

$$\text{First overshoot} = e\frac{-\xi\pi}{\sqrt{1-\xi^2}} = e\frac{-0.75 \times 3.14}{\sqrt{1-0.75^2}} = 2.8\%$$

Problem 6.27 Figure 6.44 shows two generators cascaded together. The appropriate constants are:

$$L_1 = 100H, L_2 = 50H, R_1 = 200 \text{ ohm}, R_2 = 250 \text{ ohm},$$
$$K_1 = 250 \text{ V/A}, K_2 = 100 \text{ V/A}$$

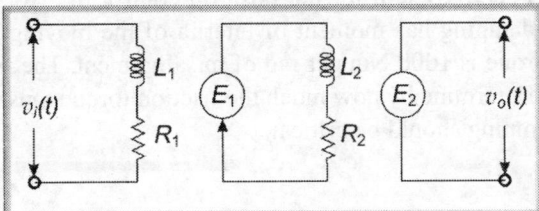

Fig. 6.44

Deduce a transfer function for the output/input voltage ratio and the output signal when a 10 V step is applied to the first generator. Generator back e.m.f. to be taken as zero

Solution

The transfer function is:

$$\frac{V_o}{V_i}(s) = \frac{K_1 K_2}{(sL_1 + R_1)(sL_2 + R_2)}$$

Substituting the values we get

$$\frac{V_o}{V_i}(s) = \frac{5}{(1 + 0.5s)(1 + 2s)} \qquad \textbf{Ans}$$

$$v_i(t) = 10$$

$$\therefore \qquad V_i(s) = \frac{10}{s}$$

$$\therefore \qquad V_o(s) = \frac{5 \times 10}{s\,(1 + 0.5s)\,(1 + 2s)}$$

or
$$V_o(s) = \frac{50}{s(1+0.5s)(1+2\ s)}$$

or
$$V_o(s) = \frac{50}{s(s+2)(s+0.5)} = \frac{A}{s} + \frac{B}{s+2} + \frac{C}{s+0.5}$$

$$A = \frac{50}{s(s+2)(s+0.5)}\bigg|_{s=0} \qquad \therefore\ A = 50$$

$$B = \frac{50}{s\ (1+0.5)}\bigg|_{s=-2} \qquad \therefore\ B = 16.67$$

$$C = \frac{50}{s\ (s+2)}\bigg|_{s=-0.5} \qquad \therefore\ C = -66.67$$

$$\therefore \qquad v_o(t) = 50\ [\ 1 + 0.33\ e^{-2t} - 1.33\ e^{-0.5t}] \qquad \textbf{Ans}$$

Problem 6.28 A servo system for the position control of a rotable mass using viscous-friction damping has moment of intertia of the moving parts 0.4 kg·m² and the motor torque is 1000 Nm per rad of misalignment. The friction torque is 30 N m/rad/sec. Determine by how much the friction torque must be increased in order that the damping should be critical.

Solution

The transfer function relating output and input is:

$$\frac{\theta_o}{\theta_i}(s) = \frac{1000}{0.4s^2 + 30s + 1000} = \frac{2500}{s^2 + 75s + 2500}$$

Comparing with standard second order function

$$\omega_n = \sqrt{2500} = 50, \text{ and } \xi = \frac{75}{2 \times 50} = 0.75$$

For critical damping the friction torque required is:

$$= \frac{30}{0.75} = 40 \text{ Nm/rad/sec} \qquad \textbf{Ans}$$

$$\therefore \qquad \text{Increase in friction} = 40 - 30.$$

Therefore, torque required is = 10 Nm/rad/sec $\qquad$ **Ans**

Problem 6.29 Determine the values of M, B and K from the response curve of the system are shown in Fig. 6.45

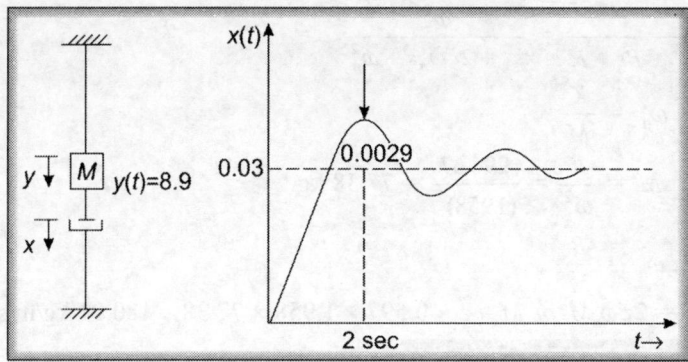

Fig. 6.45

Solution

The transfer function of the system is:

$$\frac{X(s)}{Y(s)} = \frac{1}{Ms^2 + Bs + K}$$

$$Y(t) = 8.9 \therefore Y(s) = \frac{8.9}{s}$$

$$\therefore \qquad X(s) = \frac{8.9}{s\,(Ms^2 + Bs + K)}$$

Also $x\,(\infty) = 0.03$ (from the response curve)

$$\therefore \qquad x\,(\infty) = \lim_{s \to 0} sX(s) = \lim_{s \to 0} \frac{8.9}{Ms^2 + Bs + K} = \frac{8.9}{K}$$

or $\qquad K = 296.67$ N/m (8.9/0.03) **Ans**

Also $\qquad t_p = 2 \text{ sec} = \dfrac{\pi}{\omega_d}$

or $\qquad \omega_d = \dfrac{\pi}{2} = 1.571$ rad/sec

Also $\qquad M_p = \dfrac{0.0029}{0.03}$ (for unit input) $= e^{\frac{-\xi\pi}{\sqrt{1-\zeta^2}}}$

or $\qquad \xi = 0.597$

Now $\qquad \omega_d = \omega_n \sqrt{1 - \xi^2}$

or $\qquad \omega_n = \dfrac{1.571}{\sqrt{1 - 0.597^2}} = 1.958$ rad/sec

Comparing the transfer function with standard second order transfer function

$$\frac{1}{Ms^2 + Bs + K} = \frac{\omega_n^2}{s^2 + 2\xi\omega_n s + \omega_n^2}$$

We get $\omega_n^2 = \dfrac{K}{M}$

or $\qquad M = \dfrac{K}{\omega_n^2} = \dfrac{296.67}{(1.958)^2} = 77.38$ kg $\qquad$ **Ans.**

and $\qquad 2\xi\omega_n = \dfrac{B}{M}$

or $\qquad B = 2\xi\omega_n M$ or $M = 2 \times 0.597 \times 1.958 \times 77.38 = 180.91$ kg/m/sec $\qquad$ **Ans.**

Problem 6.30 A servomechanism is represented by the equation:

$$\frac{d^2 y}{dt^2} + 4.8 \frac{dy}{dt} = 144E$$

Where $E = C - 0.5y$ is the actuating signal. Find the value of damping ratio, damped and undamped frequency of oscillations. Draw the block diagram of the system described by the above equation.

Solution

$$\frac{d^2 y}{dt^2} + 4.8 \frac{dy}{dt} = 144E$$

or $\qquad \dfrac{d^2 y}{dt^2} + 4.8 \dfrac{dy}{dt} = 144 (C - 0.5y)$

$\qquad s^2 Y(s) + 4.8 \, sY(s) = 144C(s) - 72 \, Y(s)$

or $\qquad \dfrac{Y(s)}{C(s)} = \dfrac{144}{s^2 + 4.8s + 72}$

The block diagram of the system is shown in Fig. 6.46

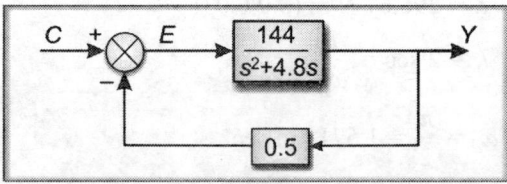

Fig. 6.46

Comparing $\dfrac{Y(s)}{R(s)}$ with standard second order transfer function, we get

$$\frac{144}{s^2 + 4.8s + 72} = \frac{\omega_n^2}{s^2 + 2\xi\omega_n s + \omega_n^2}$$

$$\omega_n = \sqrt{72} = 8.48 \text{ rad/sec} \qquad \text{Ans}$$

$$\xi = \frac{4.8}{2\omega_n} = \frac{4.8}{2 \times 8.48} = 0.281$$ **Ans**

$$\omega_d = \omega_n\sqrt{1-\xi^2} = 8.48\sqrt{1-0.281} = 8.14 \text{ rad/sec}$$ **Ans**

Problem 6.31 For a unity feedback system whose open-loop transfer function is:

$$G(s) = \frac{50}{(1+0.1\,s)\,(1+2\,s)},$$ find the position, velocity and acceleration error constants

Solution

$$G(s) = \frac{50}{(1+0.1\,s)\,(1+2\,s)}$$

(a) *Position error constant*

$$K_p = \lim_{s \to 0} G(s) = \lim_{s \to 0} \frac{50}{(1+0.1\,s)\,(1+2\,s)} = 50$$ **Ans**

(b) *Velocity error constant*

$$K_v = \lim_{s \to 0} sG(s) = \lim_{s \to 0} \frac{50s}{(1+0.1s)\,(1+2s)} = 0$$ **Ans**

(c) *Acceleration error constant*

$$K_a = \lim_{s \to 0} s^2 G(s) = \lim_{s \to 0} \frac{50\,s^2}{(1+0.1\,s)\,(1+2\,s)} = 0$$ **Ans**

Problem 6.32 A certain feedback system is described by the following transfer function:

$$G(s) = \frac{16}{s^2+4\,s+16}, \quad H(s) = Ks;$$

The damping ratio of the system is 0.8. Determine the overshoot of the system.

Solution

The transfer function of feedback system is $\dfrac{G(s)}{1+G(s)H(s)}$

Substituting the given values of $G(s)$ and $H(s)$; we get transfer function

$$\frac{C(s)}{R(s)} = \frac{16}{s^2+(4+16K)s+16}$$

Comparing with the standard second order transfer function, we get

$$\omega_n^2 = 16 \text{ or } \omega_n = 4$$

and

$$2\xi\omega_n = 4+16K$$

$$2 \times 0.8 \times 4 = 4+16K$$

$$K = 0.15$$

Therefore, the transfer function is:

$$\frac{C(s)}{R(s)} = \frac{16}{s^2 + 6.4s + 16}$$

Peak overshoot, $M_p = e^{\frac{-\pi \times 0.8}{\sqrt{1-0.8^2}}} = 0.015$ or 1.5%   **Ans**

Problem 6.33 Determine the error co-efficients and static error for unity and non-unity feedback system

$$G(s) = \frac{1}{s\,(s+1)\,(s+10)}$$

$$H(s) = (s+2)$$ *(Pune University)*

Solution

The transfer function of the given system is:

$$\frac{C(s)}{R(s)} = \frac{G(s)}{1 + G(s)H(s)}$$

Error co-efficients

(1) $K_p = \lim\limits_{s \to 0} G(s) = \lim\limits_{s \to 0} \dfrac{1}{s\,(s+1)(s+10)} = \infty$ **Ans**

(2) $K_v = \lim\limits_{s \to 0} sG(s) = \lim\limits_{s \to 0} \dfrac{1}{(s+1)(s+10)} = 0.1$ **Ans**

(3) $K_a = \lim\limits_{s \to 0} s^2 G(s) = \lim\limits_{s \to 0} \dfrac{s}{(s+1)(s+10)} = 0$ **Ans**

Static error

The given system is type 1 system as there is one pole at origin.

For a type 1 system, steady-state error for a unity feedback system is

(1) Unit step Input $= 0$ **Ans**

(2) Unit Ramp Input $= \dfrac{1}{K_v} = \dfrac{1}{0.1} = 10$ **Ans**

(3) Unit Parabolic Input $= \infty$ **Ans**

The error constants for a non-unity feedback system is obtained as given below.

$$G(s)\,H(s) = \frac{(s+2)}{s\,(s+1)(s+10)}$$

For a type 1 system

(1) $\quad K_p = \lim\limits_{s \to 0} G(s)H(s) = \lim\limits_{s \to 0} \dfrac{s+2}{s\,(s+1)(s+10)} = \infty$ **Ans**

(2) $\quad K_v = \lim\limits_{s \to 0} sG(s)H(s) = \lim\limits_{s \to 0} \dfrac{s+2}{(s+1)(s+10)} = \dfrac{1}{5} = 0.2$ **Ans**

(3) $\quad K_a = \lim\limits_{s \to 0} s^2 G(s)H(s) = \lim\limits_{s \to 0} \dfrac{s\,(s+2)}{(s+1)(s+10)} = 0$ **Ans**

Steady-state error for

(1) Unit step Input $\quad = 0$ **Ans**

(2) Unit Ramp Input $\quad = \dfrac{1}{0.2} = 5$ **Ans**

(3) Unit Parabolic Input $\quad = \infty$ **Ans**

Problem 6.34 A feedback control system is described as:

$$G(s) = \frac{50}{s\,(s+2)(s+5)}, \, , \, H(s) = \frac{1}{s}$$

For a unit step input, determine the steady-state error constants and errors.

(Pune University)

Solution

$$G(s)H(s) = \frac{50}{s^2\,(s+2)(s+5)}$$

$$K_p = \lim\limits_{s \to 0} G(s)H(s) = \lim\limits_{s \to 0} \frac{50}{s^2\,(s+2)(s+5)} = \infty \qquad \textbf{Ans}$$

$$K_v = \lim\limits_{s \to 0} sG(s)H(s) = \lim\limits_{s \to 0} \frac{50}{s\,(s+2)(s+5)} = \infty \qquad \textbf{Ans}$$

$$K_a = \lim\limits_{s \to 0} s^2 G(s)H(s) = \lim\limits_{s \to 0} \frac{50}{(s+2)(s+5)} = 5 \qquad \textbf{Ans}$$

$$1 + G(s)H(s) = \frac{s^2\,(s+2)(s+5)}{s^4 + 7s^3 + 10s^2 + 50}$$

$$\text{Steady-state error} = \lim\limits_{s \to 0} \frac{sR(s)}{1 + G(s)H(s)}$$

But $\qquad R(s) = \dfrac{1}{s}$

Therefore, steady-state error $= \lim\limits_{s \to 0} \dfrac{s^2\,(s+2)(s+5)}{s^4 + 7s^3 + 10s^2 + 50} = 0$ **Ans**

Problem 6.35 A certain feedback control system is described by the following transfer function:

$$G(s) = \frac{K}{s^2\,(s+20)(s+30)} \qquad H(s) = 1$$

Determine steady-state error coefficients and also determine the value of K to limit the error to 10 units due to input

$$r(t) = 1 + 10t + 20t^2 \qquad\qquad (Pune\ University)$$

Solution

(1) $K_p = \lim_{s\to 0} G(s) = \infty$ **Ans**

(2) $K_v = \lim_{s\to 0} sG(s) = \infty$ **Ans**

(3) $K_a = \lim_{s\to 0} s^2 G(s) = \dfrac{K}{600}$ **Ans**

Steady-state error

(1) Error due to unit step input $= \dfrac{1}{1 + K_p} = 0$

(2) Error due to $(r(t) = 10t)$ is $= \dfrac{10}{K_v} = 0$

(3) Error due to $r(t) = 20t^2$ is $= \dfrac{20}{K_a} = \dfrac{20 \times 600}{K} = \dfrac{12000}{K}$

Total steady-state error by principal of superposition

$$= 0 + 0 + \frac{12000}{K} = \frac{12000}{K}$$

Since, this has to be restricted to 10 units therefore

$$\frac{12000}{K} = 10 \text{ or } K = 1200 \qquad\qquad\qquad \textbf{Ans}$$

Problem 6.36 For the given closed-loop transfer function

$$\frac{C(s)}{R(s)} = \frac{\omega_n^2}{s^2 + 2\xi\omega_n s + \omega_n^2}$$

Obtain the following:
 (i) Controlled output $c(t)$ for $r(t) = u(t)$
 (ii) Location of poles on the s-plane as it changes from zero to infinity
 (iii) Percentage first peak overshoot and corresponding time constant
 (iv) Settling time for 2% tolerance band
 (v) Rise time   (*Pune University*)

Solution

(i) Assuming underdamped case

$$\frac{C(s)}{R(s)} = \frac{\omega_n^2}{(s + \xi\omega_n + j\omega_d)(s + \xi\omega_n - j\omega_d)}$$

where $\omega_d = \omega_n\sqrt{1-\xi^2}$

For a unit step input

$$C(s) = \frac{\omega_n^2}{s^2 + 2\xi\omega_n s + \omega_n^2} \times \frac{1}{s} = \frac{A}{s} + \frac{Bs + C}{s^2 + 2\xi\omega_n s + \omega_n^2}$$

$$C(s) = \frac{1}{s} - \frac{(s + 2\xi\omega_n)}{s^2 + 2\xi\omega_n s + \omega_n^2} = \frac{1}{s} - \frac{(s + \xi\omega_n)}{s^2 + 2\xi\omega_n s + \omega_n^2} - \frac{\xi\omega_n}{s^2 + 2\xi\omega_n s + \omega_n^2}$$

$$C(s) = \frac{1}{s} - \frac{(s + \xi\omega_n)}{(s + \xi\omega_n)^2 + \left(\omega_n\sqrt{1-\xi^2}\right)^2} - \frac{\xi\omega_n}{(s + \xi\omega_n)^2 + \left(\omega_n\sqrt{1-\xi^2}\right)^2}$$

Taking inverse laplace transform both sides:

$$c(t) = u(t) - e^{-\xi\omega_n t}\cos\left(\omega_n\sqrt{1-\xi^2}\right)t\,u(t) - \frac{\xi}{\sqrt{1-\xi^2}}\sin(\omega_n\sqrt{1-\xi^2})t\,u(t)$$

$$c(t) = u(t) - \frac{e^{-\xi\omega_n t}}{\sqrt{(1-\xi^2)}}\left\{\sqrt{(1-\xi^2)}\cos(\omega_n\sqrt{1-\xi^2})t + \xi\sin(\omega_n\sqrt{1-\xi^2})t\right\}u(t)$$

$$c(t) = u(t) - \frac{e^{-\xi\omega_n t}}{\sqrt{(1-\xi^2)}}\left\{\sin\Phi\cos(\omega_n\sqrt{1-\xi^2})t + \cos\Phi\sin(\omega_n\sqrt{1-\xi^2})t\right\}u(t)$$

$$c(t) = \left\{1 - \frac{e^{-\xi\omega_n t}}{\sqrt{(1-\xi^2)}}\sin\left[\omega_n\sqrt{1-\xi^2})t + \Phi\right]\right\}u(t) \quad \text{where } \Phi = \tan^{-1}\frac{\sqrt{(1-\xi^2)}}{\xi}$$

Ans

(ii)
$$s_1, s_2 = -\xi\omega_n \pm j\omega_n\sqrt{1-\xi^2} \ ; \ \text{for } 0 < \xi < 1$$
$$s_1, s_2 = -\omega_n \ ; \ \text{for } \xi = 1$$
$$s_1, s_2 = -\xi\omega_n \pm \omega_n\sqrt{\xi^2 - 1} \ ; \ \text{for } \xi > 1$$
$$s_1, s_2 = \pm j\omega_n \ ; \ \text{for } \xi = 0$$
$$s_1, s_2 = -\xi\omega_n \pm j\omega_n\sqrt{1-\xi^2} \ ; \ \text{for } \xi < 1$$

Locus of roots are shown in Fig. 6.47

(iii)
$$M_p = c(t_p) - 1$$
$$= -\frac{e^{-\xi\omega_n t_p}}{\sqrt{1-\xi^2}}\sin(\omega_d t_p + \beta)$$

where,
$$\omega_d = \omega_n\sqrt{1-\xi^2}$$

$$\beta = \tan^{-1}\frac{\sqrt{1-\xi^2}}{\xi}$$

$$c(t) = 1 - \frac{e^{-\xi\omega_n t}}{\sqrt{(1-\xi^2)}} \sin[\omega_d t + \Phi] \text{ where } \Phi = \tan^{-1}\frac{\sqrt{(1-\xi^2)}}{\xi} \text{ and } \omega_d = \omega_n\sqrt{1-\xi^2}$$

Also
$$t_p = \frac{\pi}{\omega_d}$$

or
$$\omega_d t_p = \pi$$

or
$$M_p = -\frac{e^{-\xi\omega_n t_p}}{\sqrt{1-\xi^2}} \sin(\pi + \beta)$$

or
$$M_p = \frac{e^{-\xi\omega_n t_p}}{\sqrt{1-\xi^2}} \sin \beta$$

or
$$M_p = e^{-\pi/\xi/\sqrt{1-\xi^2}}$$

$$\therefore \quad \left(\begin{array}{l} \sin\beta = \sqrt{1-\xi^2} \quad \text{and} \\ t_p = \dfrac{\pi}{\omega_n\sqrt{1-\xi^2}} \end{array} \right)$$

Therefore, percentage peak overshoot

$$= 100 \, e^{-\xi\pi/\sqrt{1-\xi^2}} \qquad \textbf{Ans}$$

(*iv*) The time response comprises a product of an exponentially decaying term $\dfrac{e^{-\xi\omega_n t}}{\sqrt{1-\xi^2}}$ and sinusoidal oscillating term $\sin(\omega_n t + \beta)$

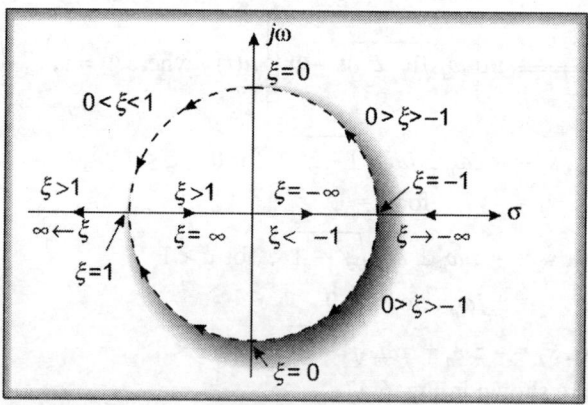

Fig. 6.47

For the time response to get into 2% tolerance band

$$1 + \frac{e^{-\xi\omega_n t_s}}{\sqrt{1-\xi^2}} = 1.02$$

or
$$\frac{e^{-\xi\omega_n t_s}}{\sqrt{1-\xi^2}} = 0.02$$

For lower values of ζ

$$e^{-\xi \omega_n t_s} = 0.02$$

or

$$t_s = \frac{3.91}{\xi \omega_n} \simeq \frac{4}{\xi \omega_n} = 4T$$

For 5% tolerance band

$$t_s = 3T$$

(iv)

$$c(t) = 1 - \frac{e^{-\xi \omega_n t_s}}{\sqrt{1 - \xi^2}} \sin (\omega_d t_r + \beta)$$

or $\quad \sin (\omega_d t_r + \beta) = 0$

or $\quad \omega_d t_r + \beta = \pi$

or $\quad t_r = \dfrac{\pi - \beta}{\omega_d}$ **Ans**

Problem 6.37 The closed-loop transfer function of a unity feedback control system is given by:

$$\frac{C(s)}{R(s)} = \frac{5}{s^2 + 4s + 5}$$

Determine (i) damping ratio, (ii) natural undamped resonance frequency, (iii) percentage peak overshoot, (iv) expression for error response

(Pune University)

Solution

Since $\quad \omega_n^2 = 5; \quad \therefore \quad \omega_n = 2.24$ rad/sec. **Ans**

and $\quad 2\xi \omega_n = 4$

or $\quad \xi = \dfrac{4}{2\omega_n} = \dfrac{4}{2 \times 2.24} = 0.9$ **Ans**

$$M_p = e^{\frac{-\xi \pi}{\sqrt{1 + \xi^2}}} = e^{\frac{-0.9 \times 3.14}{\sqrt{1 - 0.9^2}}} = 0.0015 \text{ or } 0.15\%$$ **Ans**

$$\omega_d = \omega_n \sqrt{1 - \xi^2} = 2.24 \sqrt{1 - 0.9^2} = 0.98$$

For a unit-step input

$$c(t) = 1 - \frac{e^{-0.9 \times 2.24 t}}{\sqrt{1 - 0.9^2}} \sin \left(0.98\, t + \tan^{-1} \frac{\sqrt{1 - 0.9^2}}{0.9} \right)$$

$$1 - e^{-4.6t} \sin(0.98\, t + 25.84°)$$

$$e(t) = r(t) - c(t) = e^{-4.6t} \sin (0.98 t + 25.84°)$$ **Ans**

Problem 6.38 The open loop transfer function of a feedback control system is given by:

$$G(s)H(s) = \frac{K(s+1)}{s(1+sT)(1+2s)}$$

Determine the error co-efficients and errors due to the unit positional input, unit ramp input and unit parabolic input; if $K = 10$ and $T = 4$ (*Pune University*)

Solution

Substituting the values of K and T,

$$G(s)\,H(s) = \frac{10(s+1)}{s(1+4s)(1+2s)}$$

(1) $K_p = \lim\limits_{s \to 0} G(s)\,H(s) = \infty$ **Ans**

(2) $K_v = \lim\limits_{s \to 0} s\,G(s)\,H(s) = 10$ **Ans**

(3) $K_a = \lim\limits_{s \to 0} s^2 G(s)\,H(s) = 0$ **Ans**

Error

(1) *Unit Step Input* $= \dfrac{1}{1+K_p} = 0$ **Ans**

(2) *Unit Ramp Input* $= \dfrac{1}{K_v} = \dfrac{1}{10} = 0.1$ **Ans**

(3) *Unit Parabolic Input* $= \dfrac{1}{K_a} = \infty$ **Ans**

Problem 6.39   Find all the time domain specifications for a unity feedback control system whose open-loop transfer function is given by:

$$G(s) = \frac{25}{s(s+6)}$$

(*BE Electrical Nov.* 1983, *Pune University*)

Solution

$$\frac{C(s)}{R(s)} = \frac{G(s)}{1+G(s)\,H(s)}$$

But $H(s) = 1$ and $G(s) = \dfrac{25}{s(s+6)}$.

Therefore, $\dfrac{C(s)}{R(s)} = \dfrac{25}{s^2 + 6s + 25}$

Therefore, $\omega_n^2 = 25$ or $\omega_n = 5$ rad/sec

$$2\xi\omega_n = 6 \text{ or } \xi = \frac{6}{2 \times 5} = 0.6$$

$$\omega_d = \omega_n \sqrt{1 - \xi^2} = 5\sqrt{1 - 0.6^2} = 4 \text{ rad/sec}$$

$$\beta = \tan^{-1} \frac{\sqrt{1 - \xi^2}}{\xi} = \tan^{-1} \frac{\sqrt{1 - 0.6^2}}{0.6}$$

$$= 53.13° = 0.92 \text{ rad}$$

(1) *Rise time*

$$t_r = \frac{\pi - \beta}{\omega_d} = \frac{3.14 - 0.92}{4} = 0.55 \text{ sec} \qquad \textbf{Ans}$$

(2) *Peak time*

$$t_p = \frac{\pi}{\omega_d} = \frac{3.14}{4} = 0.785 \text{ sec} \qquad \textbf{Ans}$$

(3) *Peak overshoot*

$$M_p = e^{\frac{-0.6 \times \pi}{0.8}} = 9.5\% \qquad \textbf{Ans}$$

(4) *Settling time (2%)*

$$t_s = \frac{4}{0.6 \times 5} = 1.33 \qquad \textbf{Ans}$$

Problem 6.40 Evaluate the error series for a unity feedback system having a forward path transfer function

$$G(s) = \frac{50}{s(s + 10)}$$

Estimate the steady-state error of the system for the input $r(t)$ given by:

$$r(t) = 1 + 2t + t^2 \qquad \textit{(B.E Electrical Pune University)}$$

Solution

$$G(s) = \frac{50}{s(s + 10)} = \frac{5}{s(1 + 0.1s)} = \frac{K}{s(1 + sT)}$$

$$\frac{E(s)}{R(s)} = \frac{1}{1 + G(s)}$$

$$\therefore \qquad \frac{E(s)}{R(s)} = \frac{s(1 + 0.1s)}{5 + s(1 + 0.1s)} = \frac{s(1 + sT)}{K + s(1 + sT)}$$

Rearranging

$$\frac{E(s)}{R(s)} = \frac{s + 0.1s^2}{5 + s + 0.1s^2} = \frac{s + s^2T}{K + s + s^2T}$$

and dividing

$$\frac{E(s)}{R(s)} = \frac{1}{K}s + \frac{KT - 1}{K^2}s^2 - \frac{2KT - 1}{K^3}s^3 + \dots$$

Here $K = 5$ and $T = 0.1$.

Therefore $\dfrac{E(s)}{R(s)} = \dfrac{s}{5} + \dfrac{0.5-1}{25}s^2 - \dfrac{1-1}{125}s^3 + \dots$

$$E(s) = \frac{1}{5}\, sR(s) - 0.02s^2R(s)$$

The error co-efficients

$$C_0 = 0;\; C_1 = \frac{1}{5}\, C_2 = -0.02$$

In the time domain, the steady-state error is given by:

$$\lim_{t \to \infty} e(t) = 0.2\,\dot{r}(t) - 0.02\ddot{r}(t)$$

the dynamic error co-efficients are:

$$K_1 = \infty$$

$$K_2 = \frac{1}{0.2} = 5$$

$$K_3 = \frac{1}{-0.02} = -50$$

$$r(t) = 1 + 2t + t^2$$

$$\dot{r}(t) = 2 + 2t$$

$$\ddot{r}(t) = 2$$

The steady-state error is then

$$\lim_{t \to \infty} e(t) = \lim_{t \to \infty} [0.2\,(2 + 2t) - 0.02 \times 2]$$

$$= \lim_{t \to \infty} [0.4 + 0.4t - 0.04] \qquad \textbf{Ans}$$

$$= \text{infinity (unless the co-efficient of } t \text{ is zero)}$$

Problem 6.41 Determine the position, velocity and acceleration error constants for a unity feedback control system whose open-loop transfer function given by:

$$G(s) = \frac{K}{s(s+4)(s+10)}$$

If $K = 400$, determine the steady-state error for a unit ramp input

(Pune University)

Solution

$$K_p = \lim_{s \to 0} G(s) = \infty$$

$$K_v = \lim_{s \to 0} s\, G(s) = \frac{K}{40}$$

$$K_a = \lim_{s \to 0} s^2\, G(s) = 0$$

Error for unit ramp input for $K = 400$ is:

$$= \frac{1}{K_v} = \frac{1}{K/40} = \frac{40}{400} = \frac{1}{10} \qquad \textbf{Ans}$$

Problem 6.42 A feedback system employing output rate damping is shown in Fig. 6.48. Find value of K_1 and K_2 so that close-loop system resembles a second order system with damping ratio equal to 0.5 and frequency of damped oscillations

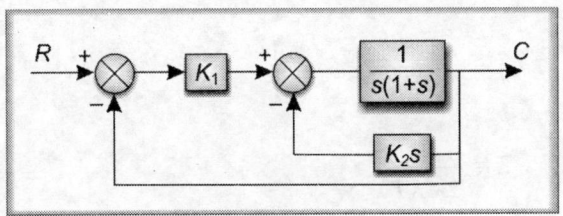

Fig. 6.48

9.5 rad/sec (b) with the above value of K_1 and K_2 find the percentage overshoot when input is step input. (c) What is the settling time for 2% tolerance?

Solution

(a) $$\frac{C}{R} = \frac{K_1}{s^2 + (1 + K_2)\, s + K_1}$$

Comparing with standard second order transfer function

$$\omega_n = \sqrt{K_1}, \text{ and } 2\xi\omega_n = 1 + K_2$$

Also $$\omega_d = \omega_n \sqrt{1 - \xi^2}$$

or $$\omega_n = \frac{\omega_d}{\sqrt{1 - \xi^2}} = \frac{9.5}{\sqrt{1 - 0.5^2}} = 10.97 \text{ rad}$$

∴ $$K_1 = \omega_n^2 = 10.97^2 = 120.34 \qquad \textbf{Ans}$$

and $$K_2 = 2\xi\omega_n - 1$$

$$= 2 \times 0.5 \times 10.97 - 1 = 9.97 \qquad \textbf{Ans}$$

(b) $$M_p = e^{\frac{-0.5 \times 3.14}{\sqrt{1 - 0.5^2}}} = 16.31\% \qquad \textbf{Ans}$$

(c) $$t_s = \frac{4}{\xi\omega_n} = \frac{4}{0.5 \times 10.97} = 0.729 \text{ sec} \qquad \textbf{Ans}$$

Problem 6.43 The open-loop transfer function of a unity feedback system is $G(s) = K/s(1 + Ts)$. (a) Find by what factor the gain K be reduced so that the overshoot is reduced from 60% to 15%. (b) Find by what factor the gain K should be reduced so that the damping ratio is increased from 0.1 to 0.6.

Solution

(b)
$$\frac{C(s)}{R(s)} = \frac{K/T}{s^2 + \dfrac{1}{T}s + \dfrac{K}{T}}$$

Comparing

$$\omega_n^2 = \frac{K}{T} \text{ and } 2\xi\omega_n = \frac{1}{T}$$

Therefore,

$$\xi = \frac{1}{2\sqrt{KT}}$$

Let
$\xi_1 = 0.1$ when gain is K_1, and
$\xi_2 = 0.6$ when gain is K_2

$\therefore$
$$0.1 = \frac{1}{2\sqrt{K_1 T}}, \text{ and}$$

$$0.6 = \frac{1}{2\sqrt{K_2 T}}$$

or
$$K_2 = \frac{K_1}{36}$$ **Ans**

Therefore the gain should be reduced by a factor 36

(a) Let ξ_1 be the damping ratio when the percentage overshoot is 60%

$\therefore$
$$0.60 = e^{-\xi_1 \times 3.14/\sqrt{1-\xi_1^2}} \text{ or } \xi_1 = 0.1604$$

Similarly

$$0.15 = e^{-\xi_1 \times 3.14/\sqrt{1-\xi_2^2}} \text{ or } \xi_2 = 0.52$$

or
$$\frac{\xi_1}{\xi_2} = \sqrt{\frac{K_2}{K_1}}$$

$$\frac{0.1604}{0.52} = \sqrt{\frac{K_2}{K_1}}$$

or
$$K_2 = \frac{K_1}{10.51}$$ **Ans**

Therefore, gain should be reduced by a factor 10.51.

Problem 6.44 For a closed-loop system whose block diagram is shown in Fig. 6.49, determine the values of K and T such that the maximum overshoot to the unit step input is 25% and time to peak is 2 sec.

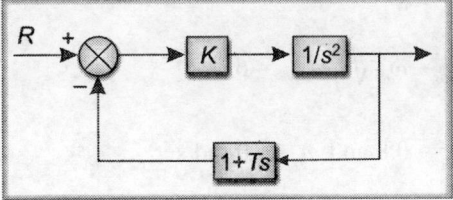

Fig. 6.49

Solution

$$\frac{C}{R} = \frac{K}{s^2 + TKs + K}$$

Comparing we get,

$$\omega_n^2 = K \text{ and } 2\xi\omega_n = TK$$

But

$$0.25 = e^{-\xi \times 3.14/\sqrt{1-\xi^2}}, \quad \therefore \quad \xi = 0.40$$

Also

$$t_p = \frac{3.14}{\omega_n \sqrt{1-\xi^2}}$$

or

$$2 = \frac{3.14}{\omega_n \sqrt{1-0.40^2}}$$

or

$$\omega_n = 1.71$$

Now

$$K = \omega_n^2 = 1.71^2 = 2.93 \qquad \textbf{Ans}$$

and

$$T = \frac{2\xi\omega_n}{K} = \frac{2 \times 0.40 \times 1.71}{2.93} = 0.467 \qquad \textbf{Ans}$$

Problem 6.45 Measurements conducted on a servomechanism show the error response to be

$$\frac{e}{r} = 1.66 \, e^{-8t} \sin(6t + 37°)$$

when the input is a sudden displacement r. Find the natural frequency of oscillations, damping ratio and damped frequency of oscillations.

Solution

The error response for a step input has the expression

$$e = r e^{-\xi\omega_n t} \sin(\omega_d t + \theta)$$

Comparing, we get

$$r = 1.66$$

$$\xi\omega_n = 8$$

$$\omega_d = \omega_n \sqrt{1-\xi^2} = 6$$

Solving, we get

$$\xi = 0.8 \text{ and } \omega_n = 10 \text{ rad/sec} \qquad \textbf{Ans}$$

Problem 6.46 In a steady-state AC position control system having unity feedback has $G(s) = 10K/s(1 + 0.1s)$. Find the minimum value of the amplifier gain K so that when input shaft rotates at half rev per sec, the steady-state position error is less than $0.2°$. With the above value of K, what will be the value of damping factor and natural frequency.

Solution

(a) $\dfrac{1}{2} rps = 3.14 \text{ rad/sec}$

$\therefore$ $R = 3.14 \text{ rad/sec}$

Also $0.2° = \dfrac{0.2 \times 3.14}{180} \text{ rad} = e_{ss}$

$$K_v = \frac{R}{e_{ss}} = \frac{3.14 \times 180}{0.2 \times 3.14} = 900$$

New $K_v = \lim_{s \to 0} sG(s) = \lim_{s \to 0} \dfrac{s10K}{s(1+0.1s)} = 900$

or $10K = 900$

$\therefore$ $K = 90$ \qquad\qquad\qquad\qquad\qquad **Ans**

(b) $\dfrac{C}{R} = \dfrac{100K}{s^2 + 10s + 100K}$

Comparing

$$\omega_n^2 = 100K = 100 \times 90$$

$\therefore$ $\omega_n = 94.87 \text{ rad/sec}$ \qquad\qquad **Ans**

and $\xi = \dfrac{10}{2 \times \omega_n} = \dfrac{10}{2 \times 94.87} = 0.053$ \qquad **Ans**

Problem 6.47 A unity feedback control system has the forward transfer function

$$G(s) = \frac{K_1 (2s + 1)}{s(4s + 1)(s + 1)^2}$$

The input $r(t) = 1 + 5t$ is applied to the system. It is desired that the steady-state value of the error should be equal to or less than 0.1 for the given input function. Determine the minimum value of K_1 to satisfy the requirement.

Solution

$$G(s) = \frac{K_1(2s+1)}{s(4s+1)(s+1)^2}$$

$$K_p = \lim_{s \to 0} G(s)\,H(s) = \infty$$

$$K_v = \lim_{s \to 0} sG(s)\,H(s) = K_1$$

$$e_{ss} = \frac{1}{1+K_p} + \frac{5}{K_v}$$

$$= \frac{1}{1+\infty} + \frac{5}{K_v} = \frac{5}{K_1}$$

Now $\qquad \dfrac{5}{K_1} \le 0.1 \quad \therefore \quad K_1 \ge 50$ **Ans**

Problem 6.48 The closed-loop poles of a system are shown in Fig. 6.50. Find the unit step response of the system and the settling time for 2% tolerance.

Solution

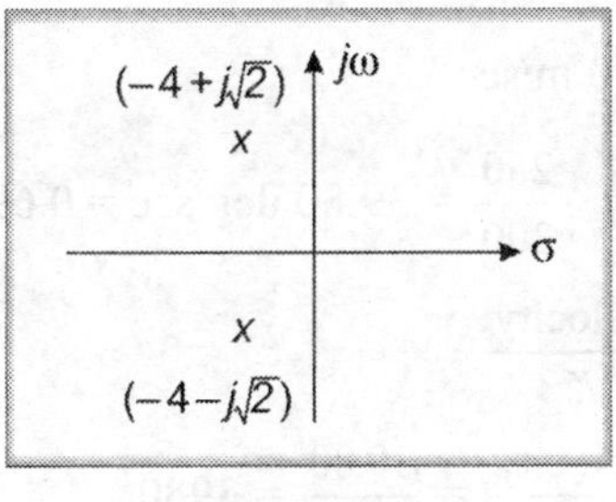

Fig. 6.50

$$-\xi\omega_n = -4$$

and $\qquad \omega_n\sqrt{1-\xi^2} = 2$

Solving, we get $\omega_n = 4.46$ rad/sec

$$\xi = 0.9$$

Time response is:

$$c(t) = 1 - \frac{e^{-\xi\omega_n t}}{\sqrt{1-\xi^2}} \sin\left[\left(\omega_n\sqrt{1-\xi^2}\right)t + \tan^{-1}\frac{\sqrt{1-\xi^2}}{\xi}\right]$$

$$= 1 - \frac{e^{-4t}}{0.44} \sin\left[2t + \tan^{-1}\frac{0.44}{0.9}\right]$$

The characteristic equation is:

$$(s+4+j\sqrt{2})(s+4-j\sqrt{2}) = s^2 + 8s + 20$$

$$\text{Settling time} = \frac{3.91}{\xi\omega_n} = \frac{3.91}{4} = 0.98 \text{ sec.}$$ **Ans**

Problem 6.49 A control system is designed to keep the antenna of a tracking radar pointed at a flying target. The system must be able to follow a target travelling in straight line with speed of 250 m/sec with maximum permissible error of 0.01°. The shortest distance from antenna to target is 300 m. Find the value of error constant K_v in order to satisfy the requirements.

Solution

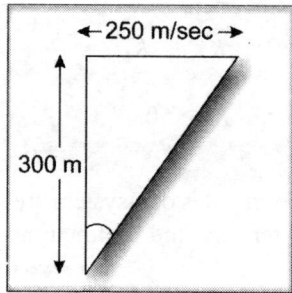

Fig. 6.51

$$\text{Speed} = 250 \text{ m/sec}$$

$$\text{Angular velocity} = \tan^{-1}\frac{250}{300} = 39.80 \text{ deg/sec} = 0.694 \text{ rad/sec}$$

Also

$$\text{error} = \frac{\text{velocity}}{K_v}$$

$\therefore$

$$K_v = \frac{\text{velocity}}{\text{error}} = \frac{39.80}{0.01} = 3980$$

or

$$\text{error} = 0.01° \text{ or } \frac{0.01 \times \pi}{180} \text{ radian}$$

$$K_v = \frac{0.694 \times 180}{0.01 \times \pi} = 3976.32$$ **Ans**

Problem 6.50 A turntable whose moment of inertia is 10 kg·m² is used in conjunction with a proportional controller having gain K in a unity feedback system. A torque of 55 Nm/rad of misalignment is developed by the controller. Damping factor produced by viscous friction is 0.25.

(a) Draw the signal flow graph of the system under consideration.

(b) Determine transfer function.

(c) If the constant velocity input is 0.05 rad/sec, determine the steady-state tracking error.

(d) If the input is ramp input suggest a modification to eliminate the error as in part 'c'.

Solution

(a) Signal flow graph is shown in Fig. 6.52

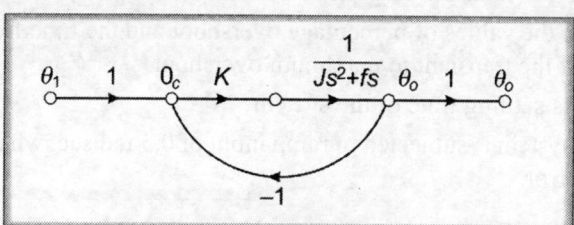

Fig. 6.52

(b)
$$\frac{\theta_o(s)}{\theta_i(s)} = \frac{K}{Js^2 + fs + K} = \frac{K/J}{s^2 + \dfrac{fs}{J} + \dfrac{K}{J}}$$

$$\omega_n^2 = K/J = 55/10 = 5.5 \text{ and } 2\xi\omega_n = f/J$$

$\therefore$ $$f/J = 2 \times 0.25 \times \sqrt{5.5} = 1.17$$

Therefore, $$\frac{\theta_o(s)}{\theta_i(s)} = \frac{5.5}{s^2 + 1.17s + 5.5}$$

(c) $$\theta_t(s) = \frac{0.05}{s^2} \text{ (given)}$$ **Ans**

$$e_{ss} = \lim_{s \to 0}\left[\frac{sR(s)}{1 + G(s)(Hs)}\right]$$

$$= \lim_{s \to 0}\left[s \times \frac{0.05}{s^2} \times \frac{s(s+1.17)}{s^2 + 1.17s + 5.5}\right]$$

$$= \frac{0.05 \times 1.17}{5.5} = 0.0106 \text{ rad}$$ **Ans**

(d) Error can be eliminated to a ramp input, if an integral controller is used. It makes the system type 2 (two) by introducing one more integration in the forward path. The transfer function of integral control is K_i/s and the transfer function controller becomes

$$= K + \frac{K_i}{s} = \frac{Ks + K_i}{s}$$ **Ans**

Problem 6.51 A second order servo system has unity feedback and an open-loop transfer function

$$G(s) = \frac{500}{s(s+15)}$$

(a) Draw a block diagram for the closed-loop system.

(b) What is the characteristic equation of the closed-loop?

(c) What are the numerical values of natural frequency (ω_n) and damping ratio (ξ)?

(d) Sketch the transient response for a unit-step input.

(e) Obtain the values of percentage overshoot and the time from the start of the transient to maximum overshoot.

(f) What is settling time of the system?

(g) If the system is subjected to ramp input of 0.5 rad/sec; what is the steady-state error.

Solution (a) Considering $R(s)$ as input and $C(s)$ as output, the block diagram is shown in Fig. 6.53

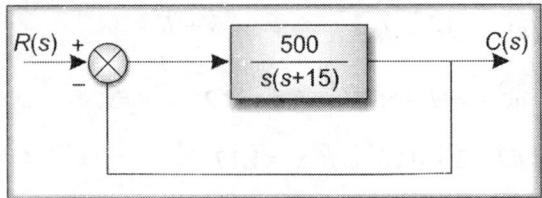

Fig. 6.53

(b) The transfer function is:

$$\frac{C(s)}{R(s)} = \frac{G(s)}{1 + G(s)H(s)} ; \quad H(s) = 1$$

Therefore,

$$\frac{C(s)}{R(s)} = \frac{\dfrac{500}{s(s+15)}}{1 + \dfrac{500}{s(s+15)} \times 1} = \frac{500}{s^2 + 15s + 500}$$

The characteristic equation is obtained by equating the denominator equal to zero. From the characteristic equation

$$\omega_n^2 = 500 \quad \therefore \ \omega_n = \sqrt{500} = 22.36 \text{ rad/sec}$$

and $$2\xi\omega_n = 15$$

or $$\xi = \frac{15}{2\omega_n} = \frac{15}{2 \times 22.36} = 0.34$$

(d) The transient response is shown in Fig. 6.15.

(e) $$\% \, M_p = 100 \times e^{\frac{-\xi}{\sqrt{1-\xi^2}}\pi}$$

Substituting the value of ξ found above, we get

$$\% \, M_p = 100 \times e^{-\dfrac{0.34}{\sqrt{1-0.34^2}}\pi} = 32.13\%$$

(f) Settling time $t_s = \dfrac{4}{\xi\omega_n} = \dfrac{4}{7.5} = 0.53$ sec.

(g)
$$\dfrac{E(s)}{R(s)} = \dfrac{1}{1+G(s)H(s)} = \dfrac{1}{1+\dfrac{500}{s(s+15)}\times 1}$$

$$\dfrac{E(s)}{R(s)} = \dfrac{s(s+15)}{s^2+15s+500}$$

$$R(s) = \dfrac{0.5}{s^2} \quad \text{(given)}$$

Steady-state error, $e_{ss} = \lim\limits_{s\to 0} sE(s)$

$$= \lim_{s\to 0} \dfrac{s\times R(s)\times s(s+15)}{s^2 \, 15s + 500}$$

$$= \lim_{s\to 0} \dfrac{0.5}{s^2}\left[\dfrac{s^2(s+15)}{s^2+15s+500}\right] = \dfrac{0.5\times 15}{500} = 0.015 \text{ rad/sec.} \qquad \textbf{Ans}$$

Problem 6.52 A unity feedback system is characterized by an open-loop transfer function $G(s) = K/s\,(s+10)$. Determine the gain K so that the system will have a damping ratio of 0.5. For this value of K, determine settling time for 5% tolerance band, peak overshoot and time to peak overshoot for a unit step-input.

Solution

The characteristic equation is given by $1 + GH = 0$;

i.e. $$1 + \dfrac{K}{s(s+10)}\times 1 = 0;$$

$s^2 + 10s + K = 0$. Comparing with characteristic equation of second order unity feedback closed-loop system, i.e. $s^2 + 2\xi\omega_n s + \omega_n^2 = 0$; we get

$$\omega_n = \sqrt{K}; \quad 2\xi\omega_n = 10 \text{ or } 2\times 0.5\times\sqrt{K} = 10 \text{ (given } \xi = 0.5)$$

Therefore,

gain $$K = \left(\dfrac{10}{2\times 0.5}\right)^2 = 100 \qquad\qquad \textbf{Ans}$$

$$\omega_n = \sqrt{100} = 10$$

Settling time $= t_s = \dfrac{3}{\xi\omega_n} = \dfrac{3}{0.5\times 10} = 0.6$ sec $\qquad\qquad \textbf{Ans}$

$$\text{\% peak overshoot} = M_p = 100 \times e^{\frac{-\xi\pi}{\sqrt{1-\xi^2}}} = 100 \times e^{\frac{-0.5\pi}{\sqrt{1-0.5^2}}} = 16.32\% \qquad \textbf{Ans}$$

$$\text{Time to peak overshoot, } t_p = \frac{\pi}{\omega_n\sqrt{1-\xi^2}} = \frac{\pi}{10\sqrt{1-0.5^2}} = 0.36 \text{ sec} \qquad \textbf{Ans}$$

Problem 6.53 A unity feedback control system has forward transfer function given by $G(s) = 8/s(s + 6)$. Find the output $c(t)$ when the system is subject to a step of 2 units.

Solution

The overall transfer function for closed-loop system with unity feedback is given by:

$$\frac{C(s)}{R(s)} = \frac{G}{1+GH} = \frac{\dfrac{8}{s(s+6)}}{1 + \dfrac{8}{s(s+6)} \times 1} = \frac{8}{s^2 + 6s + 8}$$

Comparing with $\dfrac{C(s)}{R(s)} = \dfrac{\omega_n^2}{s^2 + 2\xi\omega_n s + \omega_n^2}$,

we get $\qquad \omega_n = \sqrt{8} = 2.83 \text{ rad/sec}$

and $2\xi\omega_n = 6$, i.e. $\xi = \dfrac{6}{2\omega_n} = \dfrac{6}{2 \times 2.83} = 1.06$

Since $\xi > 1$, this is a case of overdamped system.

Now $\qquad \dfrac{C(s)}{R(s)} = \dfrac{8}{s^2 + 6s + 8} = \dfrac{8}{(s+2)(s+4)}$

Also $\qquad R(s) = \dfrac{2}{s}$ (given),

Therefore, $\qquad C(s) = \dfrac{8 \times 2}{s(s+2)(s+4)} = \dfrac{16}{s(s+2)(s+4)}$

or $\qquad C(s) = 16 \times \left[\dfrac{A}{s} + \dfrac{B}{s+2} + \dfrac{C}{s+4} \right]$

$$A = \dfrac{1}{(s+2)(s+4)}\bigg|_{s=0} = \dfrac{1}{8}$$

$$B = \dfrac{1}{s(s+4)}\bigg|_{s=-2} = -\dfrac{1}{4}$$

$$C = \dfrac{1}{s(s+2)}\bigg|_{s=-4} = \dfrac{1}{8}$$

Therefore,

$$C(s) = 16 \times \left[\frac{1/8}{s} + \frac{-1/4}{s+2} + \frac{1/8}{s+4} \right]$$

$$= \frac{2}{s} - \frac{4}{s+2} + \frac{2}{s+4}$$

$$c(t) = \mathcal{L}^{-1} \left[\frac{2}{s} - \frac{4}{s+2} + \frac{2}{s+4} \right]$$

or

$$c(t) = 2 - 4e^{-2t} + 2e^{-4t} \quad t \geq 0 \qquad \qquad \textbf{Ans}$$

Problem 6.54 A unity feedback servo-driven instrument has an open-loop transfer function $G(s) = 10/s(s + 2)$. Find

(a) the time domain response for a unit-step input
(b) the natural frequency of oscillation (ω_n) and damping ratio(ξ)
(c) maximum overshoot and the peak time
(d) steady-state error to an input $(1+ 4t)$
(e) In the above system, if two poles are introduced in the open-loop at $\pm j\sqrt{3}$, find the absolute stability of the closed-loop system.

Solution

The transfer function of the given system is $\dfrac{C(s)}{R(s)} = \dfrac{10}{s^2 + 2s + 10}$.

Here $H(s) = 1$ and $G(s) = \dfrac{10}{s(s+2)}$;

Therefore,

$$\frac{C(s)}{R(s)} = \frac{G(s)}{1 + G(s)H(s)}.$$

The characteristic equation is $s^2 + 2s + 10 = 0$.

Comparing with characteristic equation of second order unity feedback system, we get

$$\omega_n = \sqrt{10} = 3.16 \text{ rad/sec}$$

and

$$2\xi\omega_n = 2$$

or

$$\xi = \frac{1}{\omega_n} = \frac{1}{3.16} = 0.32$$

$$\omega_d = \omega_n \sqrt{1 - \xi^2} = 3.16 \sqrt{1 - 0.32} = 3 \text{ rad/sec}$$

$$\beta = \tan^{-1} \frac{\sqrt{1 - \xi^2}}{\xi} = \tan^{-1} \frac{\sqrt{1 - 0.32^2}}{0.32} = 71.34°$$

Now the output response is given by:

$$c(t) = 1 - \frac{e^{-\xi\omega_n t}}{\sqrt{1-\xi^2}} \sin(\omega_d t + \beta)$$

Substituting the values in the expression, we get

$$c(t) = 1 - \frac{e^{-0.32 \times 3.16t}}{\sqrt{1-0.32^2}} \sin(3t + 71.34°)$$

or $$c(t) = 1 - 1.05e^{-t} \sin(3t + 71.34°)$$ **Ans**

(b) The natural frequency (ω_n) and damping ratio (ξ) have been calculated in part (a) above as

$$\omega_n = 3.16 \text{ rad/sec.}$$

(c) Peak overshoot

$$\% M_P = 100 \times e^{-\frac{0.32}{\sqrt{1-\xi^2}}\pi} = 100 \times e^{-\frac{0.32}{\sqrt{1-0.32^2}} \times \pi} = 34.63\%$$

Peak time $$t_p = \frac{\pi}{\omega_n\sqrt{1-\xi^2}} = \frac{\pi}{3.16\sqrt{1-0.32^2}} = 1.04 \text{ sec.}$$

(d) Steady-state error

The input is $= 1 + 4t$. In Laplace transform for.

$$R(s) = \frac{1}{s} + \frac{4}{s^2} = \frac{s+4}{s^2}$$

Now $$E(s) = \frac{R(s)}{1 + G(s)H(s)} = \left[\frac{1}{1 + \dfrac{10}{s(s+2)} \times 1} \right] \times \frac{s+4}{s^2}$$

$$= \frac{s(s+4)(s+2)}{s^2(s^2+2s+10)}$$

$$e_{ss} = \lim_{s \to 0} sE(s) = \frac{s \times s(s+4)(s+2)}{s^2(s^2+2s+10)} = \frac{8}{10} = 0.8 \quad \textbf{Ans}$$

Now $$G(s) = \frac{10}{s(s+2)}$$

It is given that poles are added at $\pm j\sqrt{3}$. This means $G(s)$ will have another denominator term equal to (s^2+3). Therefore

$$G(s) = \frac{10}{s(s+2)(s^2+3)}$$

The characteristic equation is $1 + GH = 0$;

i.e. $$1 + \frac{10}{s(s+2)(s^2+3)} \times 1 = 0$$

or $\qquad s(s+2)(s^2+3)+10 = 0$

or $\qquad s^4 + 2s^3 + 3s^2 + 6s + 10 = 0$

Now, we have to find the absolute stability. This we will ascertain by Routh Hurwitz criterion. This is covered in detail in Chapter 8. Refer problems 8.2 and 8.4 also.

s^4	1	3	10
s^3	2	6	0
s^2	0	10	0

A zero has appeared in the first column of s^2 row. Therefore, put $s = 1/z$
Putting $s = 1/z$ in the characteristic equation, we get
$$10z^4 + 6z^3 + 3z^2 + 2z + 1 = 0$$

Developing the Routh's array, we get

	z^4	10	3	1
Sign change	z^3	6	2	0
Sign change	z^2	$-\dfrac{1}{3}$	1	0
	z^1	20	0	0
	z^0	1	0	0

There are two sign changes in the first column of the Routh's array. Hence, there are two roots lying on the right hand side of s-plane. Hence, the closed-loop system unstable.

Problem 6.55 For a negative feedback control system having forward path transfer function $G(s) = K/s(s+6)$ and $H(s) = 1$.

(a) Determine the value of gain K for the system to have damping ratio 0.832.

(b) For the value of gain K, determine the complete time response to an input $r(t) = 2u(t)$, where $u(t)$ is unit step input function.

Solution

(a) The transfer function

$$\frac{C(s)}{R(s)} = \frac{G(s)}{1+G(s)H(s)} = \frac{\dfrac{K}{s(s+6)}}{1+\dfrac{K}{s(s+6)}\times 1} = \frac{K}{s^2+6s+K}$$

Comparing with

$$\frac{C(s)}{R(s)} = \frac{\omega_n^2}{s^2 + 2\xi\omega_n s + \omega_n^2},$$

$$\omega_n^2 = K \text{ or } \omega_n = \sqrt{K}$$

$$2\xi\omega_n = 6$$

But $\xi = 0.832$ (given);

Therefore $2 \times 0.832 \times \sqrt{K}$ $= 6$

or $$K = \left(\frac{6}{2 \times 0.832}\right)^2 \text{ or } K = 13$$ **Ans**

(b) Now $K = 13$ and $R(s) = \dfrac{2}{s}$

The output response for unit step input $R(s) = \dfrac{1}{s}$ is given by:

$$c(t) = 1 - \frac{e^{-\xi\omega_n t}}{\sqrt{1 - \xi^2}} \sin(\omega_d t + \beta)$$

Let us find the values of ξ, ω_n, ω_d and β.
The transfer function with $K = 13$ is:

$$\frac{C(s)}{R(s)} = \frac{13}{s^2 + 6s + 13}$$

Therefore,

$$\omega_n = \sqrt{13} = 3.6 \text{ rad/sec.}$$

$$\xi = \frac{6}{2 \times \omega_n} = \frac{6}{2 \times 3.6} = 0.833$$

$$\omega_d = \omega_n\sqrt{1 - \xi^2} = 3.6\sqrt{1 - 0.833^2} = 2 \text{ rad/sec.}$$

$$\beta = \tan^{-1}\frac{\sqrt{1 - \xi^2}}{\xi} = \tan^{-1}\frac{\sqrt{1 - 0.833^2}}{0.833} = 33.6°$$

Therefore,

$$c(t) = 1 - \frac{e^{-0.833 \times 3.6t}}{\sqrt{1 - 0.8333^2}} (\sin(2t + 33.6°))$$

or $$c(t) = 1 - 1.81e^{-3t}\sin(2t + 33.6°)$$

For input $= 2\,u(t)$

$$c(t) = 2[1 - 1.81e^{-3t}\sin(2t + 33.6°)]$$

or $$c(t) = 2 - 3.62e^{-3t}\sin(2t + 33.6°)$$ **Ans**

Problem 6.56 Obtain the transfer function word for the system is shown in Fig. 6.54. Calculate the value of β to make the damping ratio ξ of the system equal to 0.5.

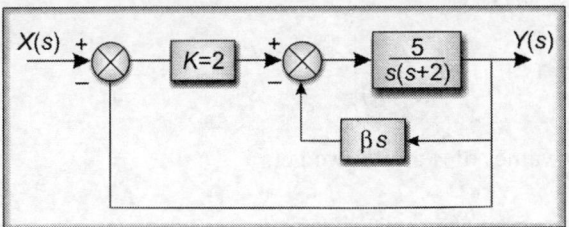

Fig. 6.54

Solution

The block diagram has a transfer function

$$\frac{Y(s)}{X(s)} = \frac{10}{s^2 + s(2 + 5\beta) + 10}$$

Comparing with the standard second order transfer function.

$$\frac{C(s)}{R(s)} = \frac{\omega_n^2}{s^2 + 2\xi\omega_n s + \omega_n^2}$$

We get $\quad 2\xi\omega_n = 2 + 5\beta$ and $\omega_n = \sqrt{10}$

But $\quad\quad\quad \xi = 0.5$ (given).

Therefore, $\quad 2 \times 0.5 \times \sqrt{10} = 2 + 5\beta$

or $\quad\quad\quad \beta = \dfrac{\sqrt{10} - 2}{5} = 0.232$ **Ans**

Problem 6.57 A first-order control system is defined by $G(s) = K/(s + a)$. If a step input is applied, the system response reaches 50% of the initial value in 10 seconds. How much time will be taken for the response to reach 99% of the steady-state value.

Solution

Consider that the system is represented by transfer function

$$\frac{C(s)}{R(s)} = \frac{K}{s + a}$$

$$C(s) = \frac{KR(s)}{s + a}, \text{ but } R(s) = \frac{1}{s} \text{ (given).}$$

Therefore, $\quad C(s) = \dfrac{K}{s(s + a)} = \dfrac{A}{s} + \dfrac{B}{s + a}$

$$A = sC(s)\big|_{s=0} = s \times \frac{K}{s(s+a)}\bigg|_{s=0} = \frac{K}{a}$$

$$B = (s+a) C(s)\big|_{s=-a}$$

$$= (s+a) \times \frac{K}{s(s+a)}\bigg|_{s=-a} = \frac{K}{s}\bigg|_{s=-a} = -\frac{K}{a}$$

Substituting the values of A and B, we get,

$$C(s) = \frac{K/a}{s} - \frac{K/a}{s+a}$$

Taking inverse Laplace transform, we get

$$c(t) = \frac{K}{a} - \frac{K}{a}e^{-at} = \frac{K}{a}(1-e^{-at})$$

Steady-state value

$$C_{ss} = \lim_{t \to \infty} c(t) = \lim_{t \to \infty}\left[\frac{K}{a}(1-e^{at})\right] = \frac{K}{a}$$

At $\qquad t = 10 \text{ sec}; \quad c(t) = 0.5 \times C_{ss}$ (given)

Substituting the values in the expression of $c(t)$, we get

$$0.5C_{ss} = \frac{K}{a}(1-e^{-10a})$$

or $\qquad 0.5 \times \dfrac{K}{a} = \dfrac{K}{a}(1-e^{-10a})$; because $C_{ss} = \dfrac{K}{a}$

or $\qquad e^{-10a} = 0.5$

or $\qquad a = 0.069$

Now, we have to find time t when

$$K = 0.069C_{ss}$$

Substituting the value $c(t) = 0.99C_{ss}$ in the expression of $c(t)$, we get

$$0.99C_{ss} = \frac{K}{a}(1-e^{-at})$$

but $\qquad C_{ss} = \dfrac{K}{a}$ and $a = 0.069$ (found earlier)

Therefore,

$$0.99 \times \frac{K}{a} = \frac{K}{a}(1 - e^{-0.069t})$$

or $\qquad 0.99 = 1 - e^{-0.069t}$

or $\qquad e^{-0.069t} = 1 - 0.99$

or $\qquad e^{-0.069t} = 0.01.$

or $\qquad t = 66.74$ sec. **Ans**

Problem 6.58 A circuit given in Fig. 6.55 has $e_i = -2$ volts, $R = 2M\Omega$, $C = 5\mu F$. The supply voltage $V_{cc} = 12$ V and input is applied at $t = 0$. Determine the output at $t = 30$ sec and plot the output for all time.

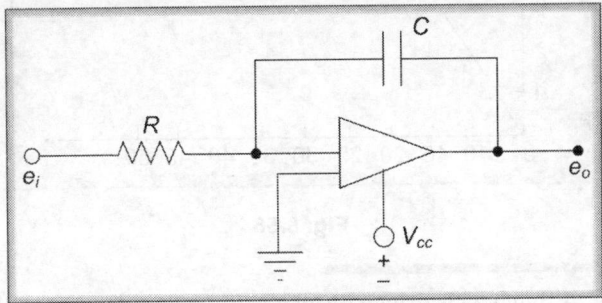

Fig. 6.55

Solution

$$e_0 = -\frac{1}{RC}\int e_i dt = -\frac{1}{2 \times 10^6 \times 5 \times 10^{-6}}\int(-2)dt = 0.2t$$

At $\quad t = 10$ **seconds;** $e_0 = 0.2 \times 10 = 2$ volts

At $\quad t = 15$ **seconds;** $e_0 = 0.2 \times 15 = 3$ volts

At $\quad t = 20$ **seconds;** $e_0 = 0.2 \times 20 = 4$ volts

At $\quad t = 25$ **seconds;** $e_0 = 0.2 \times 25 = 5$ volts

At $\quad t = 30$ **seconds;** $e_0 = 0.2 \times 30 = 6$ volts

At $\quad t = 45$ **seconds;** $e_0 = 0.2 \times 45 = 9$ volts

Since V_{cc} is given as 12 volts, the saturation voltage e_0 is 12 volts. Therefore, it will remain constant. The waveform is shown in the Fig. 6.56

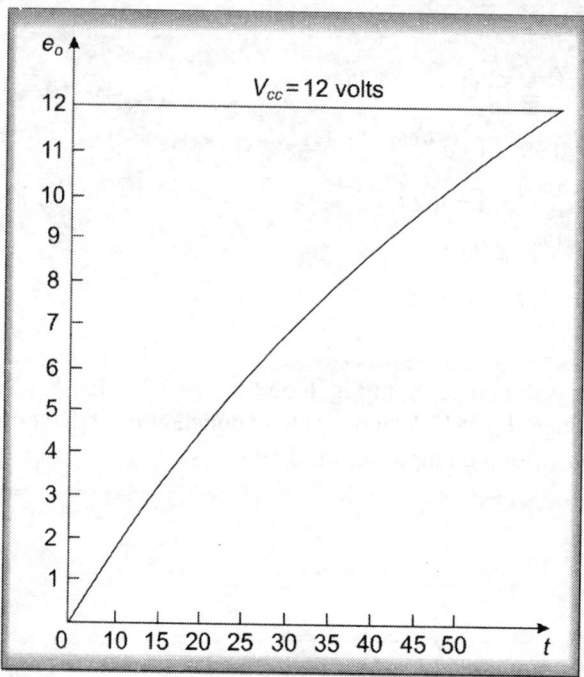

Fig. 6.56

Problem 6.59 A flywheel is driven by an electric motor. It is controlled automatically by a movement of handwheel and follows its movement. The effective moment of inertia of the flywheel is 150 kg·m². The torque developed by the motor is 2400 Nm/rad of misalignment between flywheel and handwheel. Viscous friction of the system is 600 Nm rad^{-1} sec. If the handwheel is suddenly moved through $\pi/3$ rad from rest, determine subsequent angular position of the flywheel.

Solution

Torque developed $= 2400\,(\theta_i(s) - \theta_0(s))$

This torque will be equal to $= 150\, s^2\, \theta_0(s) + 600\, s\, \theta_0(s)$

Equating

$$15\, s^2 \theta_o(s) + 600\, s\theta_o(s) = 2400(\theta_1(s) - \theta_0\,(s))$$

$$\frac{\theta_o(s)}{\theta_i(s)} = \frac{16}{s^2 + 4s + 16}$$

$$\theta_i(s) = \frac{\pi}{3s}$$

$$\bar{\theta}_o(s) = \frac{16\,\pi}{3\,s\,(s^2 + 4s + 16)} = \frac{16.75}{s(s^2 + 4s + 16)}$$

$$\theta_o(s) = \frac{1.05}{s} - \frac{1.05s + 4.2s}{(s+2)^2 + 12}$$

$$= 1.05 \left[\frac{1}{s} - \frac{s+4}{(s+2)^2 + 12} \right]$$

$$\theta_o(t) = 1.05 \{ 1 - e^{-2t} \cos 3.5\, t + 0.6 \sin 3.5t \}$$ **Ans**

Problem 6.60 Show that the steady-state error in the response to ramp input can be made to zero, if the closed-loop transfer function if given by:

$$\frac{C(s)}{R(s)} = \frac{a_{n-1}s + a_n}{s^n + a_1 s^{n-1} + \dots + a_{n-1}s + a_n}$$

Solution

$$\frac{C(s)}{R(s)} = \frac{a_{n-1}s + a_n}{s^n + a_1 s^{n-1} + \dots + a_{n-1}s + a_n}$$

$$\frac{E(s)}{R(s)} = \frac{R(s) - C(s)}{R(s)}$$

$$= \frac{s^n + a_1 s^{n-1} + \dots + a_{n-2}s^2}{s^n + a_1 s^{n-1} + \dots + a_{n-1}s + a_n}$$

For the unit ramp input $r(t) = t$

$$E(s) = \frac{s^n + a_1 s^{n-1} + \dots + a_{n-2}s^2}{s^n + a_1 s^{n-1} + \dots + a_{n-1}s + a_n} \times \frac{1}{s^2}$$

Therefore,

$$e_{ss} = \operatorname*{Lim}_{s \to 0} s\, E(s) = \operatorname*{Lim}_{s \to 0} \left(\frac{s^n + a_1 s^{n-1} + \dots a_{n-1}s^2}{s^n + a_1 s^{n-1} + \dots + a_{n-1}s + a_n} \right) \times \frac{1}{s} = 0$$

Problem 6.61 Consider a unity feedback control system with the closed-loop transfer function

$$\frac{C(s)}{R(s)} = \frac{Ks + b}{s^2 + as + b}$$

Determine the open-loop transfer function $G(s)$. Show that the steady-state error in the unit ramp response is given by:

$$e_{ss} = \frac{1}{K_v} = \frac{a - K}{b}$$

Solution

$$\frac{C(s)}{R(s)} = \frac{G(s)}{1 + G(s)H(s)} = \frac{Ks + b}{s^2 + as + b}$$

For $H(s) = 1$

$$\frac{G(s)}{1+G(s)} = \frac{Ks+b}{s^2 + as +b}$$

Hence $(s^2 + as + b) \, G(s) = (Ks + b) \, (1 + G(s))$

or $\qquad G(s) = \dfrac{Ks+b}{s(s+a-k)}$

The steady state error in the unit ramp response is

$$e_{ss} = \frac{1}{Kv} = \underset{s\to 0}{\text{Lim}} \frac{1}{sG(s)} = \underset{s\to 0}{\text{Lim}} \frac{s(s+a-k)}{s(K_s+b)} = \frac{a-k}{b} \qquad \textbf{Ans}$$

Problem 6.62 Consider the system shown in Fig. 6.57. Show that, if the disturbance is a ramp function, then the steady-state error due to this ramp disturbance may be eliminated only if two integrators precede the point where the disturbance enters.

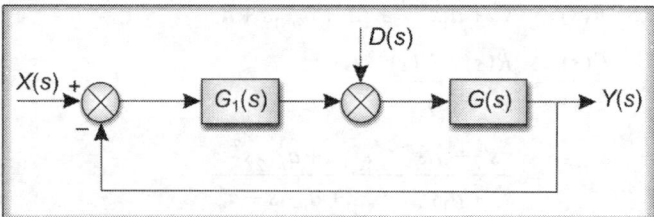

$$\text{Fig. 6.57}$$

Solution

$$\left.\frac{Y(s)}{D(s)}\right|_{X(s)=0} = \frac{G(s)}{1+G(s)\,G_1(s)}$$

For the ramp disturbance $\quad d(t) = at$

we have $\qquad D(s) = \dfrac{a}{s^2}$

$$Y(\infty) = \underset{s\to 0}{\text{Lim}} \, sY(s)$$

$$= \underset{s\to 0}{\text{Lim}} \frac{sG(s)}{1+G(s)\,G_1(s)} \times D(s)$$

$$= \underset{s\to 0}{\text{Lim}} \frac{sG(s)}{1+G(s)\,G_1(s)} \times \frac{a}{s^2}$$

$$= \underset{s\to 0}{\text{Lim}} \frac{a}{sG_1(s)}$$

If $G_1(s)$ has one integrator; then

$$C(\infty) = \underset{s \to 0}{\text{Lim}} \frac{as}{sG_1(s)} = \frac{a}{G_1(s)}$$

However, if $G_1(s)$ has two integrators; then

$$C(\infty) = \underset{s \to 0}{\text{Lim}} \frac{as^2}{sG_1(s)} = 0$$

Problem 6.63 Consider a unity feedback control system whose open-loop transfer function is:

$$G(s) = \frac{K}{s(Js + B)}$$

Discuss the effect that varying the values of K and B have on the steady-state error in unit ramp response.

Solution

$$G(s) = \frac{K}{s(Js + B)}$$

Closed-loop transfer function $= \dfrac{C(s)}{R(s)} = \dfrac{K}{Js^2 + Bs + K} = \dfrac{G(s)}{1 + G(s)H(s)}$

Here $H(s) = 1$

$$\frac{E(s)}{R(s)} = \frac{R(s) - C(s)}{R(s)} = \frac{Js^2 + Bs}{Js^2 + Bs + K}$$

or $\qquad E(s) = \dfrac{Js^2 + B(s)}{Js^2 + Bs + K} \times \dfrac{1}{s^2}$ (for ramp input)

$$e_{ss} = e(\infty) = \underset{s \to 0}{\text{Lim}}\ sE(s) = \underset{s \to 0}{\text{Lim}} \left(\frac{Js^2 + Bs}{Js^2 + Bs + K} \times \frac{s}{s^2} \right) = \frac{B}{K}$$

- Steady-state error e_{ss} can be reduced by increasing the gain K and decreasing the visous friction co-efficient B.
- However, the above will cause reduction in damping ratio (ξ) which in turn will make the transient response to be oscillatory.
- Since ξ is inversly proportional to the square root of gain K_b if K is doubted, e_{ss} will reduce to half its original value and ξ will decrease to 70.7% of its original valve.
- However, if B is decreased to half its original valve; both e_{ss} and ξ will reduce to the halves of there original values, respectively.

- Therefore, it is always advisable to increase the valve of K instead of decreasing the value of B.

The steady state error due to various valves of K is shown in Fig. 6.58.

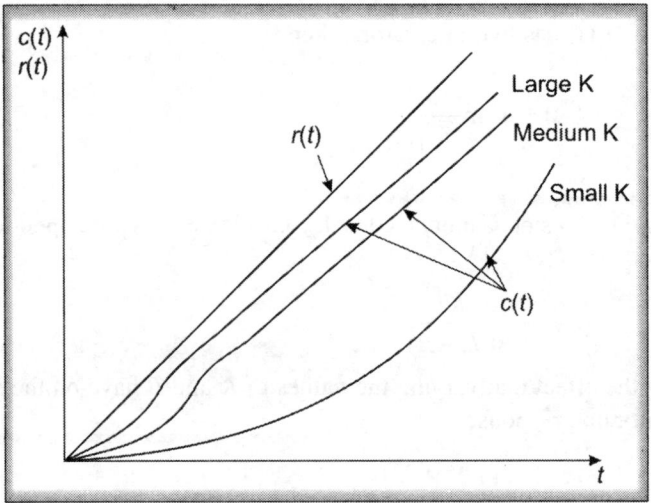

Fig. 6.58

Problem 6.64 A unity feedback control system employs a PD control as shown in Fig. 6.59. Calculate the value of K to make the system critically damped. If the input to the control system is ramp input, find the values of steady-state error, maximum overshoot and setting time (a) with control actions, (b) without control actions.

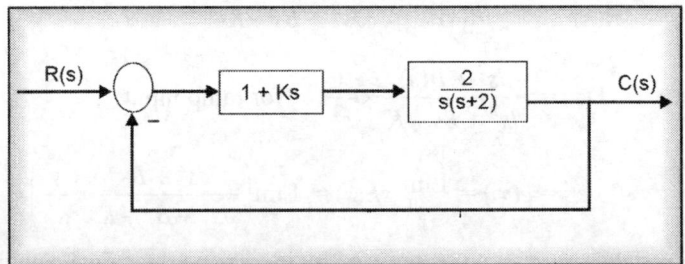

Fig. 6.59

Solution

(a) *With PD control action*

$$\frac{C(s)}{R(s)} = \frac{\dfrac{2(1+Ks)}{s(s+2)}}{1 + \dfrac{2(1+Ks)}{s(s+2)}} = \frac{2(1+Ks)}{s(s+2) + 2(1+Ks)} = \frac{2(1+Ks)}{s^2 + 2(1+K)s + 2}$$

The characteristic equation is:

$$s^2 + 2(1 + K)s + 2 = 0 \qquad (1)$$

Comparing with the standard form of characteristic equation, i.e.

$$s^2 + 2\xi\omega_n s + \omega_n^2 = 0 \qquad (2)$$

We get $\qquad \omega_n^2 = 2$ or $\omega_n = \sqrt{2} = 1.414$

and $\qquad 2\xi\omega_n = 2(1 + K)$

or $\qquad \xi\omega_n = 1 + K \qquad (3)$

For critically damped systems

$$\xi = 1$$

Therefore, equation (3) becomes

$$1 \times \omega_n = 1 + K$$

or $\qquad K = \omega_n - 1$

$$= 1.414 - 1$$

$$= 0.414$$

(b) *Without control action (K = 0)*

The characteristic equation (1) will become

$$s^2 + 2s + 2 = 0$$

and comparing with standard form of characterised equation (2), we get

$$\omega_n^2 = 2 \text{ or } \omega_n = \sqrt{2} = 1.414$$

and $\qquad 2\xi\omega_n = 2,$

or $\qquad \xi\omega_n = 1$

which gives $\qquad \xi = \dfrac{1}{\omega_n} = \dfrac{1}{0.414} = 0.707$

Steady-state error (e_{ss})

$$e_{ss} = \frac{2\xi}{\omega_n} = \frac{2 \times 0.707}{1.414} = 1$$

Setting time (t_s)

$$t_s = \frac{4}{\xi\omega_n} = \frac{4}{0.707 \times 1.414} = 4 \text{ seconds}$$

Maximum overshoot (M_p)

$$M_p = e^{\frac{-\pi\xi}{\sqrt{1\xi^2}}}$$

$$= e^{\frac{-\pi \times 0.707}{\sqrt{1 - 0.707^2}}} = 0.12 = 12\%$$

With control action

$$s(s + 2) + 2(1 + Ks) = 0$$

We had calculated $\qquad K = 0.414$

Therefore, the characteristic equation becomes

$$s(s + 2) + 2(1 + 0.414s) = 0$$

or $\qquad s^2 + 2.828s + 2 = 0$

Comparing with standard characteristic equation $s^2 + 2\xi\omega_n s + \omega_n^2 = 0$, gives

$$\omega_n^2 = 2 \text{ or } \omega_n = 1.414$$

Also $\qquad\qquad\qquad \xi = 1 \text{ (for critically damped system)}$

Steady-state error (e_{ss})

$$e_{ss} = \frac{2\xi}{\omega_n} = \frac{2 \times 1}{1.414} = 1.414 \text{ rad}$$

Settime time (t_s)

$$t_s = \frac{4}{\xi\omega_n} = \frac{4}{1 \times 1.414} = 2.828 \text{ sec}$$

Maximum overshoot (M_p)

$$M_p = e^{\frac{-\pi\xi}{\sqrt{1-\xi^2}}} = e^{\frac{-\pi \times 1}{\sqrt{1-1}}} = 0$$

Comparison

S. No.	Action	e_{ss} (rad)	t_s (secs)	M_p (%)
1	Without control action	1	4	12
2	Without control action	1.414	2.828	0

Control Systems: Feedback Characteristics

7.1 Introduction

Feedback is an essential element of an automatic control system. The feedback property of closed-loop system is a very useful property and all the control system designers use it to design control systems, because of the various advantages which accrue because of its use. Feedback allows the controlled output to be compared with the command input and if there is difference between the two, actuating error signal comes into existence which activates the controller to ensure that difference between the command input and controlled output disappears. The aim is that the control output should instantaneously reproduce the command input. However, in practical control systems, it may not happen and reason for the same is the presence of inertia, ageing of moving parts and energy storing elements which cannot be avoided while assembling a practical control system.

Ideally, the control law is that the command output must follow the command input instantaneously. However, when we consider the practical systems, control law for the feedback systems is modified to the extent that control output should follow the command input as closely as possible so that the time lag between them is as close to zero and in a very short time, control output corresponds to command input.

It also means that transients exhibited during early stages of the response should die out fast and the response should settle in the settling band as early as possible which means settling time should be as less as possible which further means that the response is faster.

Feedback has many advantages when compared with open-loop systems. These are discussed in succeeding paragraphs.

7.2 Steady-state Error/Accuracy

Let us consider an open-loop system as shown in Fig. 7.1

The error for a unity-step input is given by:

$$E(s) = R(s) - C(s) = R(s) - D_0(s)G(s)R(s)$$

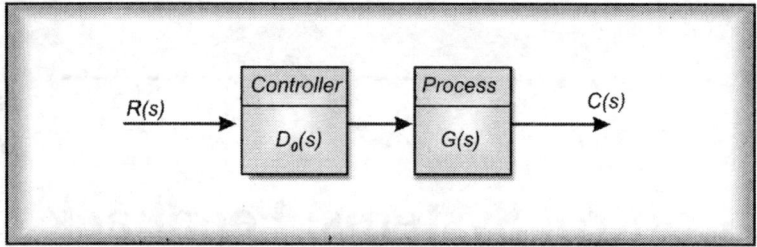

Fig. 7.1

$$= R(s)[1 - D_0(s)G(s)]$$

Therefore,

$$e_{ss} = \lim_{s \to 0} sE(s) = \lim_{s \to 0} sR(s)[1 - D_0(s)G(s)]$$

$$= \lim_{s \to 0} s \times \frac{1}{s} [1 - D_0(s)G(s)]$$

$$= \lim_{s \to 0} [1 - D_0(s)G(s)]$$

or $\quad\quad\quad e_{ss} = [1 - D_0(0)G(0)]$ $\quad\quad\quad\quad\quad\quad\quad\quad\quad\quad$ (7.1)

The error can be reduced to zero by adjusting $D_0(0)G(0)$ equal to one. However, due to parameter variations in $G(s)$ during working of the control system, the error may not become zero.

Now, let us consider a closed-loop system as shown in Fig. 7.2.

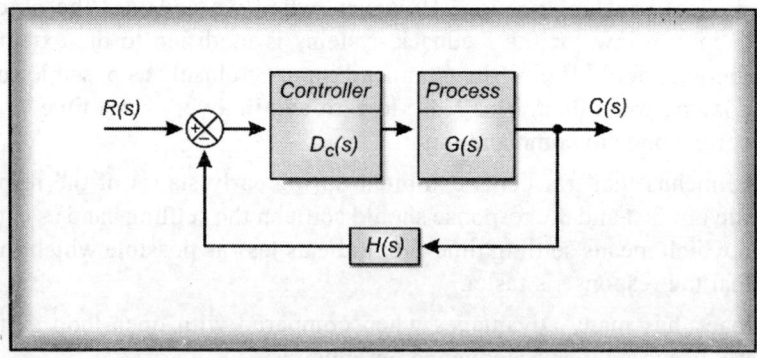

Fig. 7.2

$$C(s) = D_c(s)G(s)E(s)$$

$$E(s) = \frac{C(s)}{D_c(s)G(s)}$$

$$E(s) = R(s) - C(s)H(s)$$

or
$$C(s) = \frac{R(s) - E(s)}{H(s)}$$

$$E(s) = \frac{\frac{R(s) - E(s)}{H(s)}}{D_c(s)G(s)} = \frac{R(s) - E(s)}{H(s)D_c(s)G(s)}$$

$$E(s)[1 + D_c(s)G(s)H(s)] = R(s)$$

$$E(s) = \frac{R(s)}{1 + D_c(s)G(s)H(s)}$$

$$e_{ss} = \lim_{s \to 0} sE(s) = \lim_{s \to 0} s\left[\frac{1}{1 + D_c(s)\,G(s)H(s)}\right]$$

$$= \lim_{s \to 0} s\left[\frac{1}{1 + D_c(s)G(s)H(s)}\right] \times \frac{1}{s}$$

or
$$e_{ss} = \frac{1}{1 + D_c(0)G(0)H(0)} \tag{7.2}$$

In order to reduce the error, the *loop gain* $D_c(0)G(0)H(0)$ should be made reasonably large, *i.e.* $D_c(0)G(0)H(0) >>> 1$.

The inference is:

(a) The error in open-loop control system can be made equal to zero. However, practically due to parameter variations in $G(s)$ the error may not practically become zero.

(b) The error in closed-loop control system can be reduced considerably by making loop gain as large as possible.

7.3 Sensitivity

The variations in parameters of the system affect the performance of the control system. Uncertainties during the operation of a control system cannot be predicted in advance, however, to a great extent feedback incorporation in a control system negate the influence on its performance. Sensitivity is a measure of the effectiveness of the feedback in mitigating the effect of parameter variations/uncertainties in a closed-loop system in comparison to an open-loop control system. Parameter variations occur due to ageing, external disturbances, etc.

Generally, the sensitivity can be defined as effect of parameter variation of one quantity on the other. If the variation in T is caused by the variation in parameter K, then the sensitivity of the system parameter T to the parameter K is defined as:

$$S_K^T = \frac{\text{Percentage change in paramter } T}{\text{Percentage change in parameter } K} = \frac{\Delta T / T}{\Delta K / K}$$

Let us now extend the general definition to open-loop and closed-loop systems, where our interest is the output which may change or become sensitive to changes occuring in parameters of the control system.

Therefore, sensitivity of the system is defined as the ratio of the change in output of a closed-loop system to the change in the output in the open-loop system; input remaining the same. Let an open-loop and a closed-loop system are modelled as shown in Fig. 7.3.

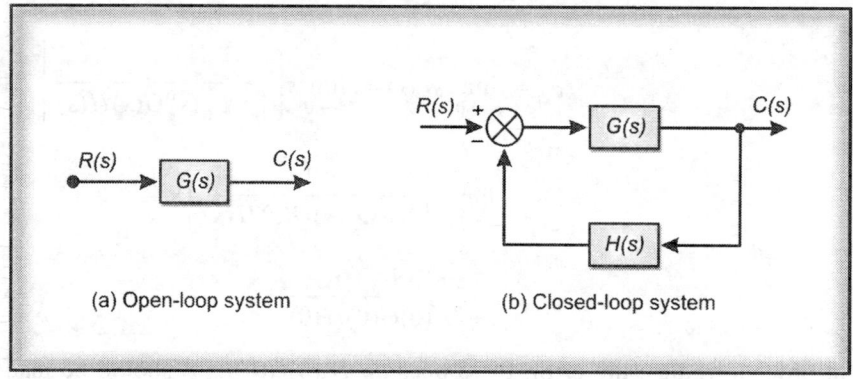

(a) Open-loop system   (b) Closed-loop system

Fig. 7.3

As seen from Fig. 7.3, the transfer function of the open-loop system is:

$$\frac{C(s)}{R(s)} = G(s)$$

and that of the closed-loop system is:

$$T(s) = \frac{C(s)}{R(s)} = \frac{G(s)}{1 + G(s)H(s)} \tag{7.3}$$

Then, the sensitivity (S_G^T) is given as:

$$S_G^T = \frac{\text{Percentage change in } T(s)}{\text{Percentage change in } G(s)} \tag{7.4}$$

$$= \frac{\Delta T(s) / T(s)}{\Delta G(s) / G(s)} = \frac{G(s)}{T(s)} \cdot \frac{\Delta T(s)}{\Delta G(s)} \tag{7.5}$$

Open-loop system
For the open-loop system

$$T(s) = G(s)$$

Therefore,

$$S_G^T = \frac{G(s)}{G(s)} \cdot \frac{\Delta G(s)}{\Delta G(s)} = 1 \text{ (unity)} \tag{7.6}$$

Closed-loop system

$$T(s) = \frac{C(s)}{R(s)} = \frac{G(s)}{1 + G(s)H(s)}$$

$$\frac{\Delta T(s)}{\Delta G(s)} = \frac{[(1 + G(s)H(s).1] - [G(s)\,0 + H(s)]}{[1 + G(s)H(s)]^2}$$

$$= \frac{(1 + G(s)H(s) - G(s)H(s)}{[1 + G(s)H(s)]^2}$$

$$= \frac{1}{[1 + G(s)H(s)]^2}$$

Now,

$$S_G^T = \frac{G(s)}{T(s)} \cdot \frac{\Delta T(s)}{\Delta G(s)}$$

$$= \frac{G(s)}{\dfrac{G(s)}{1 + G(s)H(s)}} \times \frac{1}{[1 + G(s)H(s)]^2}$$

or $\qquad S_G^T = \dfrac{1}{1 + G(s)H(s)}$ $\qquad\qquad$ (7.7a)

or $\qquad S_G^T = \dfrac{1}{1 + P(s)G(s)H(s)}$ $\qquad\qquad$ (7.7b)

where, $P(s)$ is controller transfer function.

Comparing Eqns (7.6) and (7.7a), it can be seen that the sensitivity is reduced by the factor $1 + G(s)H(s)$ in closed-loop system when compared with open-loop system.

Feedback Function H(s)

Mathematically, it is defined as:

$$S_H^T = \frac{\Delta T(s)/T(s)}{\Delta H(s)/H(s)} = \frac{H(s)}{T(s)} \cdot \frac{\Delta T(s)}{\Delta H(s)}$$

For a closed-loop system,

$$T(s) = \frac{G(s)}{1 + G(s)H(s)}$$

$$\frac{\Delta T(s)}{\Delta H(s)} = \frac{[(1 + G(s)H(s)] \cdot 0 - G(s)\,[0 + G(s)]}{[1 + G(s)H(s)]^2}$$

$$= \frac{(-G(s))^2}{(1 + G(s)H(s))^2}$$

Therefore,

$$S_H^T = \frac{H(s)}{T(s)} \cdot \frac{\Delta T(s)}{\Delta H(s)}$$

$$= \frac{H(s)}{\dfrac{G(s)}{1+G(s)H(s)}} \cdot \frac{-(G(s))^2}{1+G(s)H(s)}$$

$$= \frac{-G(s)H(s)}{1+G(s)H(s)} \tag{7.8}$$

Comparing, Eqns (7.7) and (7.8), it can be inferred that closed-loop system is more sensitive to parameter variations in feedback path parameters than the variations in forward path parameters.

7.4 Overall Gain

The overall gain of an open-loop control system is shown in Fig. 7.3(a).

$$G_0 = \text{overall gain} = \frac{C(s)}{R(s)} = G(s) \tag{7.9}$$

and that of a closed-loop system [Fig. 7.3(b)] is:

$$G_c = \frac{C(s)}{R(s)} = \frac{G(s)}{1+G(s)H(s)} \qquad \text{(for negative feedback)}$$

Therefore,

$$\frac{G_0}{G_c} = \frac{G(s)}{\dfrac{G(s)}{1+G(s)H(s)}} = 1 + G(s)H(s)$$

or

$$G_c = \frac{G_0}{1+G(s)H(s)} \tag{7.10}$$

Comparing Eqns (7.9) and (7.10), shows that overall gain of closed-loop system has reduced by the factor $1 + G(s)H(s)$.

7.5 Time Constant

Let us consider, open- and closed-loop systems as shown in Fig. 7.4.

Open-loop system (Fig. 7.4a)

$$\frac{C(s)}{R(s)} = \frac{K}{1+sT} \tag{7.11}$$

If $R(s) = 1/s$, then

$$C(s) = \frac{K}{1+sT} R(s) = \frac{K}{s(1+sT)}$$

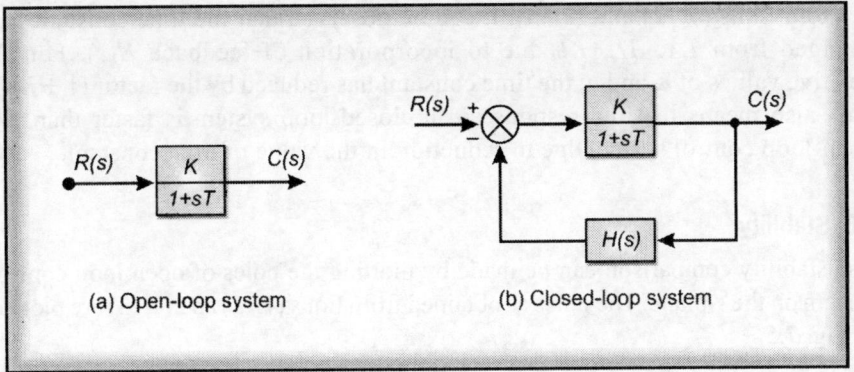

Fig. 7.4a and b

Hence,

$$c(t) = K\,[1 - e^{-s/T}] \tag{7.12}$$

Closed-loop systems (Fig. 7.4b)

Let $H(s) = h$,

then

$$\frac{C(s)}{R(s)} = \frac{G(s)}{1 + G(s)H(s)} = \frac{\dfrac{K}{1 + sT}}{1 + \dfrac{K}{1 + sT} \cdot h} = \frac{K}{1 + Kh + sT} \tag{7.13}$$

or

$$\frac{C(s)}{R(s)} = \frac{K/T}{s + \left(\dfrac{1 + Kh}{T}\right)}$$

If $R(s) = \dfrac{1}{s}$, then

$$
\begin{aligned}
c(t) &= L^{-1}\,\frac{K/T}{s\left[s + \dfrac{1 + Kh}{T}\right]} \\[2mm]
&= L^{-1}\left[\frac{A}{s} + \frac{B}{s + \dfrac{1 + Kh}{T}}\right] \\[2mm]
&= L^{-1}\left[\frac{K/1 + Kh}{1} + \frac{-K/1 + Kh}{s + \dfrac{1 + Kh}{T}}\right]
\end{aligned}
$$

or

$$c(t) = \frac{K}{1 + Kh} - \frac{K}{1 + Kh}\left[e^{\frac{t}{T/(1 + Kh)}}\right] \tag{7.14}$$

From Eqns (7.12) and (7.14), it can be observed that the time constant has changed from T to $T/1+Kh$ due to incorporation of feedback $H(s)$. For all positive valu~s of K and h; the time constant has reduced by the factor $(1 + Kh)$. This also means that the response of a closed-loop system is faster than the open-loop control system due to reduction in the value of time constant.

7.6 Stability

The stability comparison can be made by plotting the poles of open-loop control system on the s-plane. The poles as obtained from Eqns (7.11) and (7.13) are plotted in Fig. 7.5.

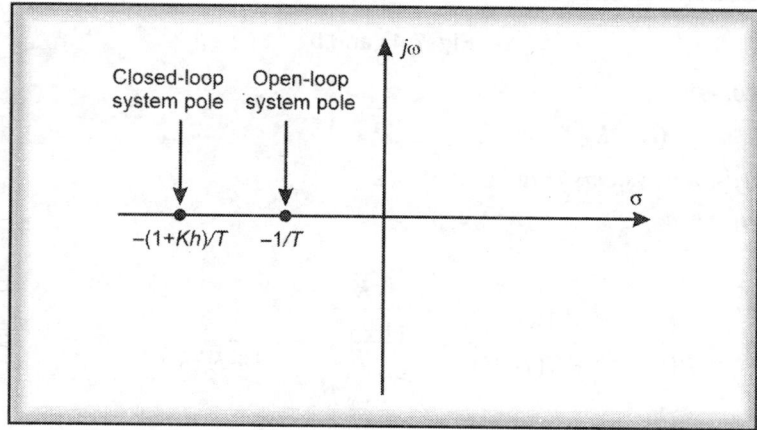

Fig. 7.5

This goes to show that stability of the control system is affected due to feedback as the poles location is shifted as shown in Fig. 7.5.

The dynamics of the system while considering the time response changes which means that feedback alters the dynamics of the open-loop control system by shifting the location of the poles.

Also, as seen in Eqn. (7.14) above, since, the time constant reduces, the dynamics/response of the system becomes faster. The transients die out fast and the steady state is reached earlier.

The relative stability is thus improved by the introduction of feedback and this is inferred from the fact that the feedback causes shifting of poles negatively away from the imaginary axis and also because of reduction in time constant, explained above.

Let us now consider the steady-state error of the closed-loop system is shown in Fig. 7.4 and also assume that the feedback is unity.

Then, the error is given as:

$$E(s) = \frac{R(s)}{1 + G(s)}$$

$$(7.15)$$

The steady state error for unity step input is:

$$e_{ss} = \lim_{s \to 0} sE(s)$$

$$= \lim_{s \to 0} s\frac{R(s)}{1+G(s)} = \lim_{s \to 0} \frac{1}{1+G(s)} \qquad \left[\because R(s) = \frac{1}{s}\right]$$

or
$$e_{ss} = \lim_{s \to 0} \frac{1}{1+\dfrac{K}{1+sT}} = \frac{1}{1+K} \qquad (7.16)$$

Therefore, if we make the gain K very large in comparison to 1, i.e. $Kp > 1$, then the value of the steady-state error can be made to zero; which means that the steady-state value of the output will be almost close to the input value.

However, variation of gain K to stabilise a closed-loop system and improve the dynamics and steady-state performance of the system, cut both ways. The designer plays with gain K to achieve best results. However, varying gain K to a very large value may lead to unstable response, i.e. the output response may become uncontrollable from the stability point of view with increase in time.

7.7 Disturbance

The desired operation of a control system is affected by the presence of non-linearities like ageing, saturation, backlash and other internal and external disturbances. The disturbances can occur in any part/path of the control system.

7.7.1 Forward Path/Input Disturbance
Let us consider a control system as shown in Fig. 7.6

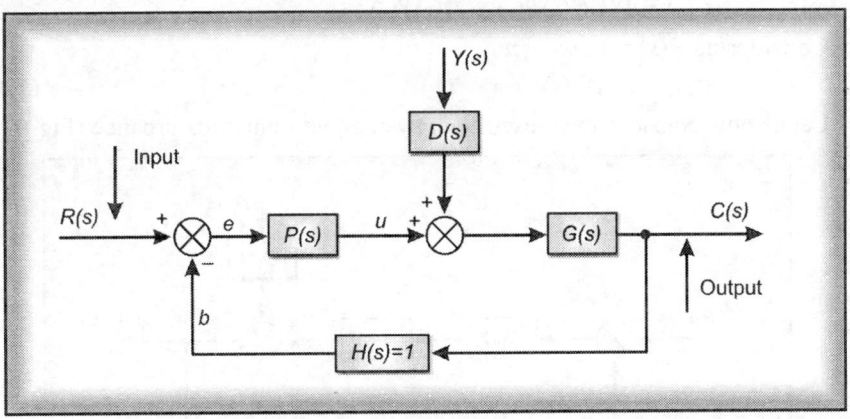

Fig. 7.6a

where, $P(s)$ = Controller transfer function

$D(s)$ = Plant transfer function due to disturbance input $Y(s)$

Assuming $R(s) = 0$

$$\frac{C(s)}{Y(s)} = \frac{G(s)}{1 + G(s)P(s)H(s)} D(s) = \frac{G(s)}{1 + P(s)G(s)} D(s)$$

or

$$\frac{C(s)}{Y(s)} = \frac{G(s)D(s)}{1 + P(s)G(s)} \quad [H(s) = 1] \tag{7.17}$$

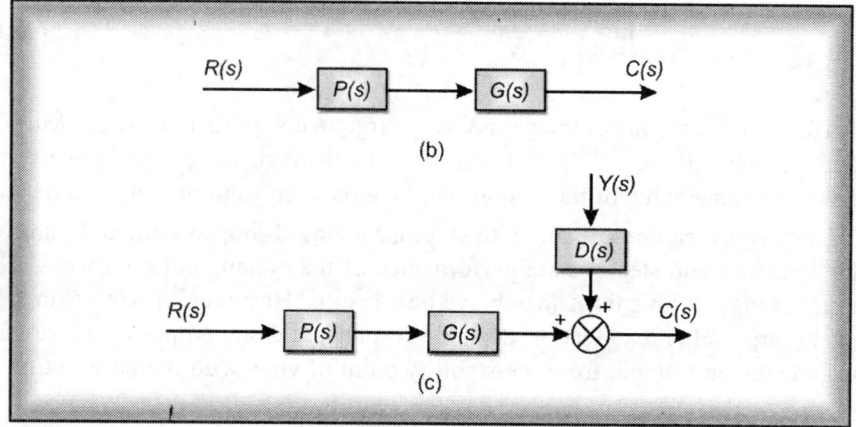

(b)

(c)

Fig. 7.6b and c

The transfer function of the open-loop system (Fig. 7.6b) is given as:

$$\frac{C(s)}{R(s)} = P(s)G(s) \tag{7.17a}$$

However, the transfer function changes because of disturbance in the forward path (Fig. 7.6b)

$$C(s) = P(s)G(s)R(s) + D(s)Y(s)$$

Considering $R(s) = 0$, we get

$$C(s) = D(s)Y(s) \tag{7.18}$$

Let us now consider the closed-loop system with input disturbance (Fig. 7.7).

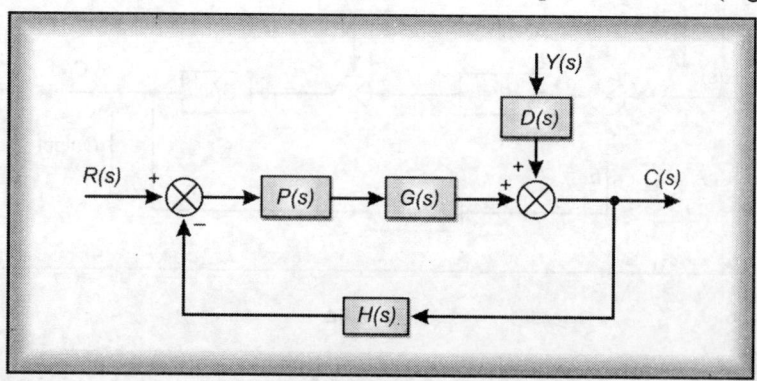

Fig. 7.7

For the control system is shown in Fig. 7.7

$$C(s) = \frac{P(s)G(s)}{1 + P(s)G(s)H(s)} R(s) + \frac{D(s)}{1 + P(s)G(s)H(s)} Y(s)$$

Considering $R(s) = 0$, we get

$$C(s) = \frac{D(s)}{1 + P(s)G(s)H(s)} Y(s) \qquad (7.19)$$

Comparing Eqns (7.18) and (7.19), clearly shows that feedback reduces input disturbance effect on $C(s)$ by the factor $[1 + P(s)G(s)H(s)]$.

If we consider $P(s)G(s)H(s) \gg 1$, then

$$C(s) = \frac{D(s)}{P(s)G(s)H(s)} Y(s) \qquad (7.20)$$

In order to achieve disturbance rejection

(a) we may increase $P(s)$, the controller gain, thus reducing $\dfrac{C(s)}{Y(s)}$. This also reduces sensitivity (Eqn. 7.7b).

(b) we can also make $D(s)$, the plant transfer function due to disturbance input $Y(s)$ as small as possible. This is not achievable as it has got incorporated in the system due to disturbance and is also a function of plant/process parameters.

7.7.2 Feedback Path   Let us consider a feedback control system is shown in Fig. 7.8.

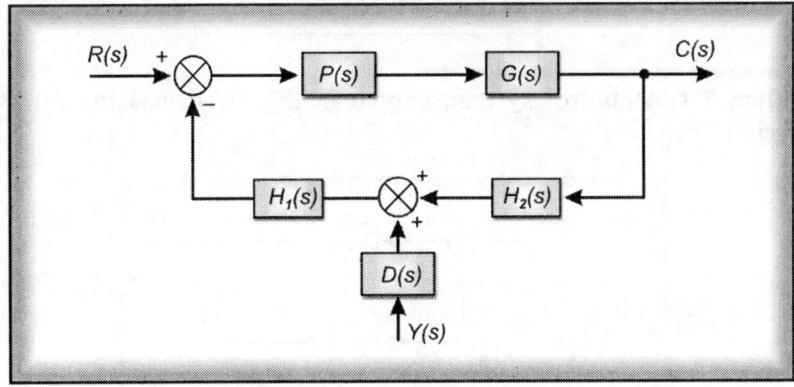

Fig. 7.8

Assuming $R(s) = 0$,

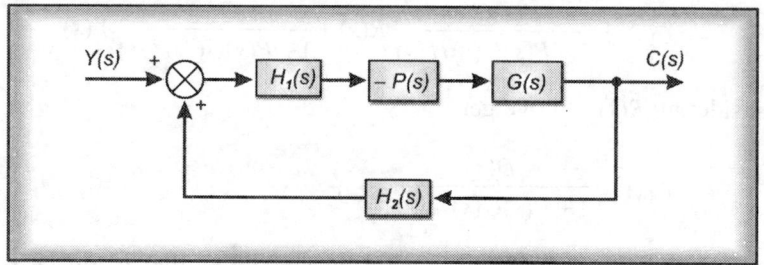

Fig. 7.9

we get

$$\frac{C(s)}{Y(s)} = \frac{-P(s)G(s)H_1(s)}{1 - P(s)G(s)\big(-H_1(s)\big)H_2(s)}$$

$$= \frac{-P(s)G(s)H(s)}{1 + P(s)G(s)H_1(s)H_2(s)}$$

For $P(s)G(s)H_1(s)H_2(s) \gg 1$

$$\frac{C(s)}{R(s)} = \frac{-P(s)G(s)H_1(s)}{P(s)G(s)H_1(s)H_2(s)} = -\frac{1}{H_2(s)}$$

Hence, disturbance rejection can be achieved by making

(a) $P(s)G(s)H_1(s)H_2(s) \gg 1$; and

(b) Properly designing $H_2(s)$.

Solved Problems

Problem 7.1 A control system, shown in Fig. 7.10, has the following characteristics:

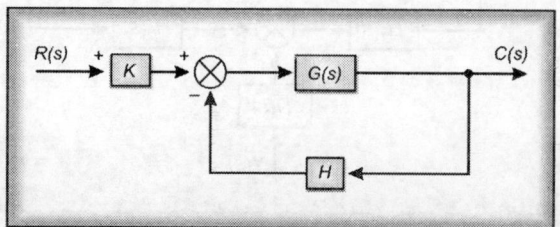

Fig. 7.10

$$K = 10 \text{ V/rad}, \quad H = 10 \text{ V/rad and } G(s) = \frac{100}{s(s+1)}$$

Determine the sensitivity of the system's transfer function T with respect to input transducer K, H and G respectively.

Solution

(a) With respect to K

$$S_K^T = 1$$

(b) With respect to H

$$S_H^T = \frac{-GH}{1+GH} = -\frac{\dfrac{100 \times 10}{s(s+1)}}{1 + \dfrac{100 \times 10}{s(s+1)}}$$

$$= -\frac{1000}{s(s+1) + 1000} = -\frac{1000}{s^2 + s + 1000}$$

(c) With respect to G

$$S_G^T = \frac{1}{1+GH} = \frac{1}{1 + \dfrac{100}{s(s+1)} \times 10} = \frac{s(s+1)}{s^2 + s + 1000}$$

Problem 7.2 In a position control system the forward path transfer function is $100/s(1+s)$ and feedback path transfer function is 10. Determine the sensitivity of T with respect to the feed forward feedback elements respectively in the vicinity of $\omega = 1$ rad/sec.

Solution

The following data have been given:

$$G = \frac{100}{s(1+s)}, \quad H = 10 \text{ and } \omega = 1 \text{ rad/sec.}$$

(a) $$S_G^T = \frac{1}{1+GH} = \frac{1}{1 + \dfrac{100}{s(s+1)} \times 10} = \frac{s(s+1)}{s^2 + s + 1000}$$

Substituting $s = j\omega$,

we get $$\left| S_G^T \right|_{j\omega} = \frac{j\omega(1 + j\omega)}{(j\omega)^2 + j\omega + 1000}$$

Putting $\omega = 1$ rad/sec,

we get, $S_G^T = \dfrac{j(j+1)}{j^2+j+1000} = \dfrac{-1+j}{j+999} = \dfrac{j-1}{j+999}$

Rationalising, we get

$$S_G^T = \dfrac{j-1}{j+999} \times \dfrac{j-999}{j-999}$$

$$= \dfrac{j^2-j-999j+999}{j^2-999^2}$$

$$= \dfrac{-1-j-999j+999}{j^2-999^2}$$

$$= \dfrac{-1000j+998}{-998002} = \dfrac{1000j-998}{998002}$$

$$|S_G^T| = \dfrac{\sqrt{1000^2+998^2}}{998002} = 0.0014 \quad \textbf{Ans}$$

(b) $S_H^T = \dfrac{-GH}{1+GH} = -\dfrac{\dfrac{100}{s(s+1)} \times 10}{1+\dfrac{100}{s(s+1)} \times 10} = \dfrac{-1000}{s^2+s+1000}$

Putting $\omega = 1$ rad/sec   and $s = j\omega$,

we get $\left|S_H^T\right|_{j\omega} = \dfrac{-1000}{999+j}$

Rationalising, we get

$$\left|S_H^T\right|_{j\omega} = \dfrac{-1000(999-j)}{(999)^2-j^2} = -\dfrac{999000-j1000}{999^2+1} = \dfrac{j1000-999000}{998002}$$

$$= \dfrac{\sqrt{1000^2+999000^2}}{998002} \simeq 1 \quad \textbf{Ans}$$

Problem 7.3 Consider a unity feedback control system whose open-loop transfer function is:

$$G(s) = \dfrac{K}{s(Js+B)}$$

Discuss the effect, that varying the values of K and B have on the steady-state error in unit ramp response.

Solution

$$G(s) = \frac{K}{s(Js + B)}$$

Closed-loop transfer function $= \dfrac{C(s)}{R(s)} = \dfrac{K}{Js^2 + Bs + K} = \dfrac{G(s)}{1 + G(s)H(s)}$

Here, $H(s) = 1$

$$\frac{E(s)}{R(s)} = \frac{R(s) - C(s)}{R(s)} = \frac{Js^2 + Bs}{Js^2 + Bs + K}$$

or $\qquad E(s) = \dfrac{Js^2 + B(s)}{Js^2 + Bs + K} \times \dfrac{1}{s^2}$ (for ramp input)

$$e_{ss} = e(\infty) = \lim_{s \to 0} sE(s) = \lim_{s \to 0}\left(\frac{Js^2 + Bs}{Js^2 + Bs + K} \times \frac{s}{s^2}\right) \times \frac{B}{K}$$

- Steady-state error e_{ss} can be reduced by increasing the gain K and decreasing the viscous friction coefficient B.
- However, the above will cause reduction in damping ratio (ξ) which in turn will make the transient response to be oscillatory.
- Since, ξ is inversely proportional to the square root of gain K, if K is doubled, e_{ss} will reduce to half its original value and ξ will decrease to 70.7% of its original valve.
- However, if B is decreased to half its original valve; both e_{ss} and ξ will reduce to the halves of there original values, respectively.
- Therefore, it is always advisable to increase the valve of K instead of decreasing the value of B.

The steady-state error due to various values of K is shown in Fig. 7.11.

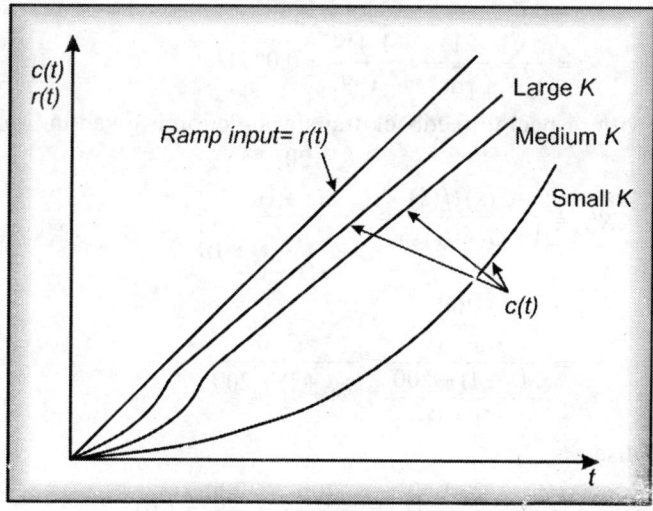

Fig. 7.11

Problem 7.4 A position control system is modelled as shown in Fig. 7.12. Find the sensitivity of the feedback control system transfer function with respect to forward path transfer function and feedback transfer function.

$$H = 10 \text{ and } \omega = 1 \text{ rad/sec.}$$

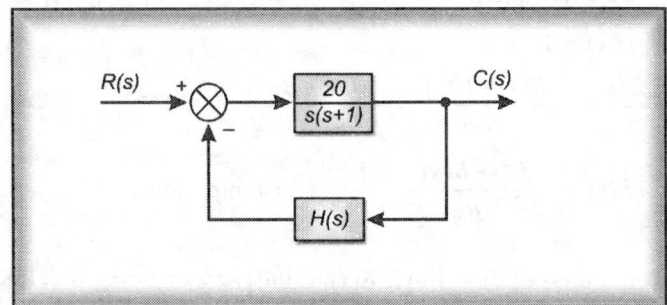

Fig. 7.12

Solution

Sensitivity with respect to forward path transfer function is given as:

$$S_G^T = \frac{1}{1 + G(s)H(s)}$$

$$= \frac{1}{1 + \dfrac{20}{s(s+1)} \times 10} = \frac{s(s+1)}{s(s+1) + 200}$$

Therefore,

$$\left| S_G^T(j\omega) \right|_{j\omega=j1} = \frac{j\omega(j\omega+1)}{j\omega(j\omega+1) + 200} = \frac{-1+j}{j+200}$$

$$= \frac{\sqrt{1^2 + 1^2}}{\sqrt{1^2 + 199^2}} = \frac{1.414}{199} = 0.0071$$

Sensitivity with respect to feedback transfer function is given in Eqn. (7.8).

$$S_H^T = \frac{-G(s)H(s)}{1 + G(s)H(s)} = \frac{-\dfrac{20}{s(s+1)} \times 10}{1 + \dfrac{20}{s(s+1)} \times 10}$$

$$= \frac{\dfrac{-200}{s(s+1)}}{\dfrac{s(s+1) + 200}{s(s+1)}} = \frac{-200}{s(s+1) + 200}$$

for $\omega = 1$ rad/sec

$$\left| S_H^T(j\omega) \right|_{\omega=1} = \frac{200}{j(j+1) + 200} = \frac{200}{j+199} = \frac{200}{199} = 1.01.$$

Problem 7.5 A control system is represented as shown in Fig. 7.13.

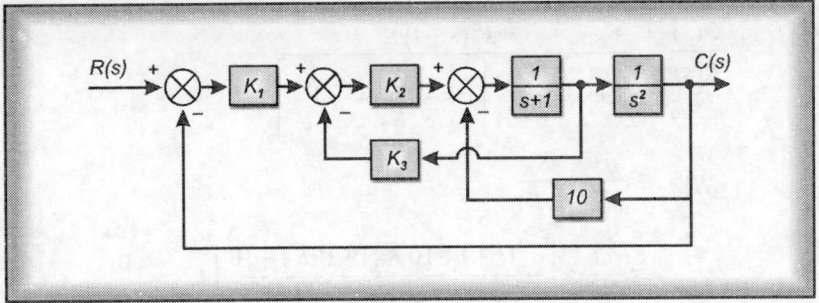

Fig. 7.13

Find:

(a) Overall transfer function $T(s) = \dfrac{C(s)}{R(s)}$

(b) Sensitivity of $T(s)$ with respect to K_2. Assume $K_1 = K_3 = 10$

(c) If $K_2 = 5$, find the sensitivity, when $\omega = 0$.

Solution

Simplifying, we get

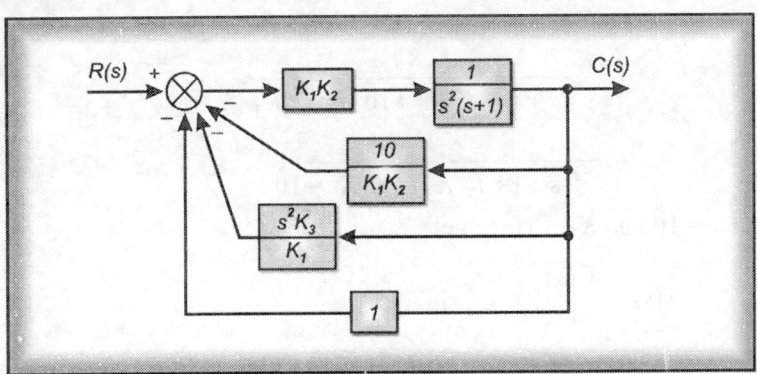

Fig. 7.14

Further simplification gives

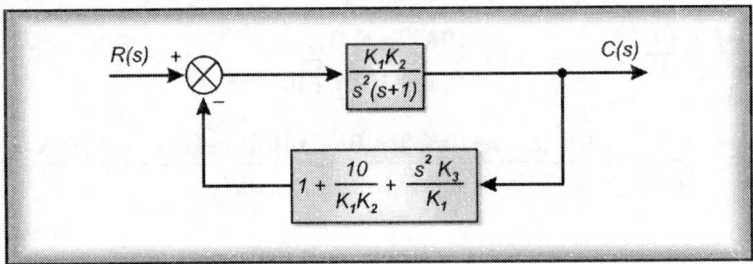

Fig. 7.15

Therefore,

$$\frac{C(s)}{R(s)} = \frac{\dfrac{K_1 K_2}{s^2(s+1)}}{1 + \dfrac{K_1 K_2}{s^2(s+1)}\left[1 + \dfrac{10}{K_1 K_2} + \dfrac{s^2 K_3}{K_1}\right]}$$

$$S_{K_3}^T = \frac{\partial T}{\partial K_3} \times \frac{K_3}{T}$$

$$= \frac{\partial T}{\partial K_3} \times \left[\frac{s^2(s+1+10\,K_3) + 10K_3 + 10}{10}\right]$$

$$= \frac{\dfrac{K_1 K_2}{s^2(s+1)}}{1 + \left[\dfrac{K_1 K_2}{s^2(s+1)} + \dfrac{10}{s^2(s+1)} + \dfrac{K_2 K_3}{(s+1)}\right]}$$

$$= \frac{\dfrac{K_1 K_2}{s^2(s+1)}}{\dfrac{s^2(s+1) + K_1 K_2 + 10 + s^2 K_2 K_3}{s^2(s+1)}}$$

$$= \frac{K_1 K_2}{s^2(s+1) + K_1 K_2 + 10 + s^2 K_2 K_3}$$

$$= \frac{K_1 K_2}{s^2(s+1+K_2 K_3) + K_1 K_2 + 10}$$

Putting $K_1 = 10$ and $K_3 = 1$, we get

$$T(s) = \frac{C(s)}{R(s)}$$

$$= \frac{10K_2}{s^2(s+1+10K_2) + 10K_2 + 10}$$

The sensitivity to variations in K_2 is given as:

$$\frac{\partial T}{\partial K_2} = \frac{10K_2}{s^2(s+1+10K_2) + 10K_2 + 10}$$

$$\frac{\partial T}{\partial K_2} = \frac{[(s^2(s+1+10K_2) + 10K_2 + 10)10] - [10s^2 + 10]10K_2}{[s^2(s+1+10K_2) + 10K_2 + 10]^2}$$

$$S_{K_2}^T = \frac{\partial T}{\partial K_2} \times \frac{K_2}{T}$$

$$= \frac{10[s^2(s+1+10K_2)+10K_2+10]-10K_2[10s^2+10]}{[s^2(s+1+10K_2)+10K_2+10]^2}$$

$$\times \frac{\dfrac{K_2}{10K_2}}{s^2(s+1+10K_2)+10K_2+10}$$

$$= \frac{10[s^2(s+1+10K_2)+10K_2+10]-10K_2[10s^2+10]}{10[s^2(s+1+10K_2)+10K_2+10]}$$

$$= \frac{s^2(s+1+10K_2)+10K_2+10-K_2(10s^2+10)}{s^2(s+1+10K_2)+10K_2+10}$$

$$= \frac{s^2(s+1+10K_2)+10K_2+10-10s^2K_2-10K_2}{s^2(s+1+10K_2)+10K_2+10}$$

$$= \frac{s^2(s+1+10K_2)+10-10s^2K_2}{s^2(s+1+10K_2)+10K_2+10}$$

$$\left| S_{K_2}^T(j\omega) \right|_{\omega=0} = \frac{10}{10K_2+10}$$

Putting $K_2 = 1$, we get

$$\left| S_{K_2}^T(j\omega) \right|_{\substack{\omega=0 \\ K_2=1}} = \frac{10}{10+10}$$

$$= \frac{10}{100} = 0.1.$$

(c) $\left| S_{K_2}^T(j\omega) \right|_{\substack{\omega=0 \\ K_2=5}} = \dfrac{10}{10\times 5+10}$

$$= \frac{10}{60} = \frac{1}{6}$$

Problem 7.6 A control system is represented by a signal flow diagram as shown in Fig. 7.16. Find:

(a) Transfer function $\dfrac{C(s)}{R(s)} = T(s)$

(b) Sensitivity $S_{G_4}^T$

(c) Check, if sensitivity depends upon $G_2(s)$ or $G_1(s)$.

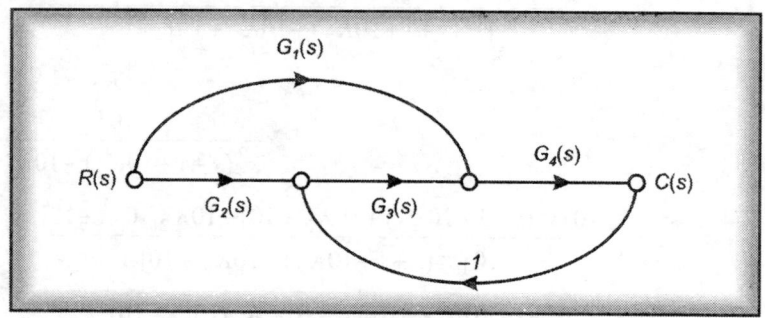

Fig. 7.16

Solution

(a)
$$P_1 = G_2(s)\ G_3(s)\ G_4(s) \qquad \Delta_1 = 1$$
$$P_2 = G_1(s)\ G_4(s) \qquad \Delta_2 = 1$$
$$L_1 = -G_3(s)\ G_4(s)$$

$$T = \frac{C(s)}{R(s)}$$

$$= \frac{G_2(s)\ G_3(s)\ G_4(s) + G_1(s)\ G_4(s)}{1 + G_3(s)\ G_4(s)} = \frac{N(s)}{D(s)}$$

(b)
$$S_{G4(s)}^{T(s)} = S_{G4(s)}^{N(s)} - S_{G4(s)}^{D(s)}$$

$$S_{G4(s)}^{N(s)} = \frac{\partial N(s)}{\partial G_4(s)} \times \frac{G_4(s)}{N(s)}$$

$$= \frac{\left(G_2(s)\ G_3(s) + G_1(s)\right)}{1} \times \frac{G_4(s)}{G_2(s)\ G_3(s)\ G_4(s) + G_1(s)\ G_4(s)}$$

$$= \frac{G_4(s)\ [G_2(s)\ G_3(s) + G_1(s)]}{G_2(s)\ G_3(s)\ G_4(s) + G_1(s)\ G_4(s)}$$

$$= \frac{G_2(s)\ G_3(s)\ G_4(s) + G_1(s)\ G_4(s)}{G_2(s)\ G_3(s)\ G_4(s) + G_1(s)\ G_4(s)} = 1$$

$$S_{G4(s)}^{D(s)} = \frac{\partial D(s)}{\partial G_4(s)} \times \frac{G_4(s)}{D(s)}$$

$$= \frac{G_3(s)}{1} \times \frac{G_4(s)}{1 + G_3(s)\ G_4(s)}$$

$$= \frac{G_3(s)\ G_4(s)}{1 + G_3(s)\ G_4(s)}$$

Therefore,

$$S_{G_4(s)}^{T(s)} = 1 - \frac{G_3(s)\,G_4(s)}{1 + G_3(s)\,G_4(s)}$$

$$= \frac{1}{1 + G_3(s)\,G_4(s)}$$

(c) From the sensitivity $S_{G_4(s)}^{T(s)} = \dfrac{1}{1 + G_3(s)\,G_4(s)}$ found in part 'b' above, it can

be seen that $S_{G_4(s)}^{T(s)} = \dfrac{1}{1 + G_3(s)\,G_4(s)}$ does not have any terms $G_2(s)$ or $G_1(s)$,

hence, sensitivity $S_{G_4(s)}^{T(s)}$ does not depend either on $G_2(s)$ or on $G_1(s)$.

Frequency Response Analysis

8.1 General

In Chapter 4, various standard signals which are generally employed to study the behaviour of a control system were discussed. In this chapter, we will discuss the system performance when the input is sinusoidal. The frequency response study is carried out by subjecting the input of the system to a range of frequencies and then recording the output relating the amplitude and phase angle. Frequency response analysis is a complimentary method to the step and ramp input testing. This result is applicable to linearized systems and does not take into account the harmonics. It deals only with steady state and measurements are taken when transients have disappeared. Therefore, the frequency response test is generally not carried out for the systems with large time constants.

Let us consider a unity feedback system having input

$$r(t) = X \sin \omega t \tag{8.1}$$

The steady-state output may be written as:

$$c(t) = Y \sin(\omega t + \phi) \tag{8.2}$$

The frequency response is thus independent of amplitude and phase of the input signal. Let us consider a unity feedback control system having closed-loop transfer function

$$\frac{C(s)}{R(s)} = \frac{G(s)}{1 + G(s)} \tag{8.3}$$

By putting $s = j\omega$, we get

$$\frac{C(j\omega)}{R(j\omega)} = \frac{G(j\omega)}{1 + G(j\omega)} = M(j\omega) \tag{8.4}$$

or $$M(j\omega) = M(j\omega) \angle \phi_m(j\omega)$$

where $$M(j\omega) = \left| \frac{G(j\omega)}{1 + G(j\omega)} \right|, \tag{8.5}$$

and $$\Phi_m(j\omega) = \angle G(j\omega) - \angle 1 + G(j\omega) \tag{8.6}$$

8.2 Frequency Domain Specifications

Let us consider a second order unity feedback control system having transfer function

$$\frac{C(j\omega)}{R(j\omega)} = \frac{\omega_n^2}{(j\omega)^2 + 2\xi\omega_n(j\omega) + \omega_n^2} \tag{8.7}$$

$$= \frac{1}{1 + j2\left(\dfrac{\omega}{\omega_n}\right)\xi - \left(\dfrac{\omega}{\omega_n}\right)^2} \tag{8.8}$$

Let $$u = \frac{\omega}{\omega_n}$$

then
$$\frac{C(j\omega)}{R(j\omega)} = \frac{1}{1 + j2u\xi - u^2} \tag{8.9}$$

$$= \frac{1}{\left[(1 - u^2) + j(2u\xi)\right]} \tag{8.10}$$

and
$$M = |M(j\omega)| = \frac{1}{\left[(1 - u^2)^2 + (2u\xi)^2\right]^{1/2}} \tag{8.11}$$

$$\angle M(j\omega) = \Phi = -\tan^{-1}\frac{2u\xi}{1 - u^2}$$

Now
$$c(t) = Y\sin(\omega t + \Phi)$$

Considering sinusoidal input of unity magnitude

$$c(t) = \frac{1}{\left[(1 - u^2)^2 + (2u\xi)^2\right]^{1/2}}\sin\left(\omega t - \tan^{-1}\frac{2u\xi}{1 - u^2}\right) \tag{8.12}$$

8.3 Magnitude and Phase Angle Characteristics Plot

Magnitude and phase angle characteristics plot can be plotted by finding out their values for varying values of $u = \omega/\omega_n$

If $\quad u = \infty$ then $M = 0$ and $\phi = -\pi$ $\qquad$ (8.13)

$\quad u = 0$ then $M = 1$ and $\phi = 0$ $\qquad$ (8.14)

and, if $\quad u = 0$ then $M = 1/2\xi$ and $\phi = -\pi/2$ $\qquad$ (8.15)

Peak value: This occurs at a frequency called **resonant frequency** ($u_r = \omega_j/\omega_n$). This is obtained by differentiating Eqn. (8.10) with respect to u and equating to zero.

$$\left[\frac{dM}{du}\right]_{u-u_r} = \frac{\left[-4u_r(1 - u_r^2) + 8\xi^2 u_r\right]}{\left[2[(1 - u_r^2)^2 + (2\xi u_r)^2]\right]^{3/2}} = 0 \tag{8.16}$$

This gives
$$4u_r^3 - 4u_r + 8\xi^2 u_r = 0,$$

or
$$u_r^2 - 1 + 2\xi^2 = 0, \tag{8.17}$$

or
$$u_r = \sqrt{1 - 2\xi^2} \tag{8.18}$$

Since
$$u_r = \frac{\omega_r}{\omega_n}, \text{ we also get} \tag{8.19}$$

$$\omega_r = \omega_n \sqrt{1 - 2\xi^2} \tag{8.20}$$

Putting the value of u_r from Eqn. (8.19) in Eqn. (8.11), we get

$$M = M_r = \frac{1}{2\xi\sqrt{1 - \xi^2}} \tag{8.21}$$

Similarly, by putting the value in Eqn. (8.11), we get

$$\Phi_r = -\tan^{-1} = \frac{\sqrt{1 - 2\xi^2}}{\xi} \tag{8.22}$$

The magnitude and phase angle characteristics plot is shown in Fig. 8.1

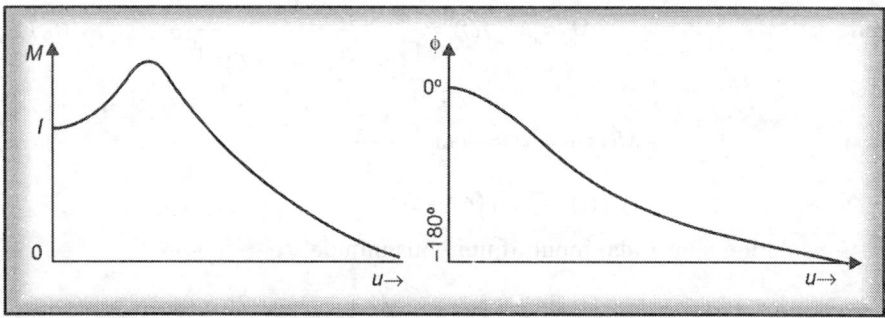

Fig. 8.1

Also, from Eqns (8.20) and (8.22), one infers that the values of M_t and ϕ are dependent on ξ.

Case I: $\xi = 0$. When ξ approaches zero, M_r approaches infinity and ω approaches ω_n.

Case II: $\xi = 1/\sqrt{2}$ or 0.707. When ξ is equal to 0.707, $M_r = 1$.

Case III: $\xi > 1/\sqrt{2}$ or 0.707. When $\xi > 1/\sqrt{2}$ or 0.707, we see from Eqn. (8.20) that ω_r becomes imaginary. This means that for values $\xi > 1/\sqrt{2}$ or 0.707, there is no resonance peak.

8.4 Frequency Response Specifications

A typical magnification curve of a feedback control system is shown in Fig. 8.2.

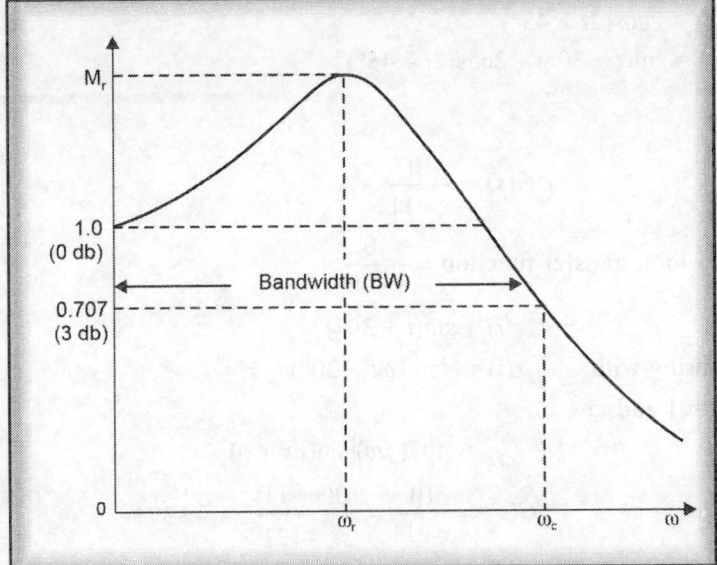

Fig. 8.2

(a) **Bandwidth (BW):** The bandwidth (BW) is the frequency at which the magnitude drops to 70.7% of its zero frequency value or 3 db level lower than the 0 db value. Bandwidth gives measure of the transient response properties.

(b) **Resonant peak (M_r):** Resonance peak is the peak value of the magnitude.

(c) **Resonant frequency (ω_r):** Resonant frequency is the frequency at which resonant peak occurs.

8.5 Representation of Sinusoidal Transfer Function

Frequency response is a function of frequency and is characterized by its magnitude and phase angle. Two methods used for the graphical representation of frequency response and polar diagram/plot and Bode diagram/plot. These have been explained in detail in subsequent chapters.

Solved Problems

Problem 8.1 Consider the unity feedback control system with the open-loop transfer function

$$G(s) = \frac{10}{s+11}$$

Obtain the steady-state output when it is subjected to each of the following:

(a) $r(t) = \sin(t + 30°)$

(b) $r(t) = 2\cos(2t + 45°)$

(c) $r(t) = \sin(t + 30°) - 2\cos(2t - 45°)$

Solution

(a)
$$G(s) = \frac{10}{s+11}$$

Closed-loop transfer function $= \dfrac{10}{s+11}$

$$r(t) = \sin(t + 30°)$$

Comparing with $\qquad r(t) = A\sin(\omega t + 30°)$

we get $A = 1$ and $\omega = 1$

$$C_{ss} = A|G(j\omega)|\sin(\omega t + \phi)$$

Now
$$G(s) = \frac{10}{s+11} = \frac{10(s-11)}{s^2 - 121}$$

Putting
$$s = j\omega = j \times 1 = j,$$

we get
$$G(j\omega) = \frac{10(j-11)}{j^2 - 121} = 0.902 - j0.082$$

or
$$|G(j\omega)| = \sqrt{(0.902)^2 + (0.082)^2} = 0.905$$

$$\theta = \tan^{-1}\left(-\frac{0.082}{0.902}\right) = -5.19°$$

Therefore
$$\phi = 30° - 5.19° = 24.81°$$

Putting the values of A, ω, ϕ and $|G(j\omega)|$ in C_{ss}; we get

$$C_{ss} = A|G(j\omega)|\sin(\omega t + \phi)$$
$$= 1 \times 0.905\sin(1 \times t + 24.81°)$$
$$= 0.905\sin(t + 24.81°)$$

(b)
$$r(t) = 2\cos(2t - 45°)$$

Comparing with $\qquad r(t) = A\cos(\omega t - 45°),$

we get $\qquad A = 2$ and $\omega = 2$

Putting $\qquad s = j\omega = 2j$

$$G(j\omega) = \frac{10(2j - 11)}{(2j)^2 - 121} = 0.88 - j0.16$$

$$= \sqrt{(0.88)^2 + (0.16)^2} = 0.894$$

$$\theta = \tan^{-1}\left(-\frac{0.16}{0.88}\right) = -10.3°$$

$$\phi = -45° - 10.3° = -55.3°$$

Putting the values of A, ω, ϕ and $G(j\omega)$,

we get
$$C_{ss} = 2 \times 0.894\cos(2t - 55.3°)$$
$$= 1.788\cos(2t - 55.3°)$$

(c)
$$r(t) = \sin(t + 30°) - 2\cos(2t - 45°)$$

The response will be the sum of the response calculated in parts (a) and (b) above

or
$$C_{ss} = 0.905\sin(t + 24.81°) - 1.788\cos(2t - 55.3°)$$

Problem 8.2 Consider the system whose closed-loop transfer function is:

$$\frac{C(s)}{R(s)} = \frac{K(1+T_2s)}{(1+T_1s)}$$

Obtain the steady-state output of the system when it is subjected to input $r(t) = R\sin\omega$.

Solution

$$\frac{C(s)}{R(s)} = \frac{K(T_2s+1)}{(T_1s+1)}$$

or
$$\frac{C(s)}{KR(s)} = \frac{(T_2s+1)}{(T_1s+1)} = \frac{(T_2s+1)(T_1s-1)}{(T_1^2s^2-1)}$$

Putting $s = j\omega$, we get

$$\frac{C(j\omega)}{KR(j\omega)} = \frac{(T_2j\omega+1)(T_1j\omega-1)}{1+T_1^2\omega^2} = \frac{1+T_1T_2\omega^2}{1+\omega^2T_1^2} + j\frac{T_2\omega-T_1\omega}{1+\omega^2T_1^2}$$

$$\frac{1}{K}\left|\frac{C(j\omega)}{R(j\omega)}\right| = \sqrt{\frac{(1+T_1T_2\omega^2)^2 + (T_2\omega-T_1\omega)^2}{(1+\omega^2T_1^2)^2}}$$

$$= \sqrt{\frac{1+T_1^2T_2^2\omega^4 + 2T_1T_2\omega^2 + T_2^2\omega^2 + T_1^2\omega^2 - 2T_1T_2\omega^2}{(1+T_1^2\omega^2)^2}}$$

$$= \sqrt{\frac{1+T_1^2\omega^2}{(1+T_1^2\omega^2)^2} + \frac{T_2^2\omega^2(1+T_1^2\omega^2)}{(1+T_1^2\omega^2)^2}}$$

$$= \sqrt{\frac{1}{1+T_1^2\omega^2} + \frac{T_2^2\omega^2}{1+T_1^2\omega^2}} = \sqrt{\frac{1+T_2^2\omega^2}{1+T_1^2\omega^2}}$$

or
$$\left|\frac{C(j\omega)}{R(j\omega)}\right| = K\sqrt{\frac{1+T_2^2\omega^2}{1+T_1^2\omega^2}}$$

or
$$|G(j\omega)| = \left|\frac{C(j\omega)}{R(j\omega)}\right| = RK\sqrt{\frac{1+T_2^2\omega^2}{1+T_1^2\omega^2}}$$

$$\phi = \tan^{-1}(\omega T_2) - \tan^{-1}(\omega T_1)$$
$$C_{ss} = A|G(j\omega)|\sin(\omega t + \phi)$$

Substituting the values, we get

$$C_{ss} = RK\sqrt{\frac{1+T_2^2\omega^2}{1+T_1^2\omega^2}}\sin(\omega t + \tan^{-1}(\omega T_2) - \tan^{-1}(\omega T_1))$$

Problem 8.3 Consider the system as shown in Fig. 8.3.

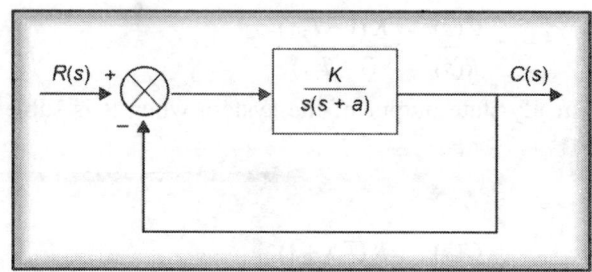

Fig. 8.3

(a) Find the value of K and a to satisfy the frequency domain specification $M_r = 10.4$ and $\omega_r = 11.55$ rad/sec.

(b) For the values of K and a determined in part (a), calculate the settling line and bandwidth of the system.

Solution

$$\frac{G(s)}{R(s)} = \frac{G(s)}{1+G(s)H(s)} = \frac{K}{s^2 + as + K}$$

Comparing with the standard second order transfer function

$$\omega_n = \sqrt{K} \text{ and } 2\xi\omega_n = a \text{ or } \xi = \frac{a}{2\sqrt{K}}$$

Now
$$M_r = \frac{1}{2\xi\sqrt{1-\xi^2}} \text{ or } 1.04 = \frac{1}{2\xi\sqrt{1-\xi^2}}$$

∴
$$\xi = 0.77 \text{ or } 0.632$$

If $\xi > 0.707$, there is no resonant peak, therefore $\xi = 0.77$ is neglected.

Also
$$\omega_r = \omega_n\sqrt{1-2\xi^2}$$

or
$$11.55 = \omega_n\sqrt{1-2\times0.632^2} \text{ or } \omega_n = 25.75 \text{ rad/sec}$$

Now
$$\omega_n = \sqrt{K} \quad \therefore 25.75 = \sqrt{K} \text{ or } K = 663 \qquad \textbf{Ans.}$$

Also
$$a = 2\xi\omega_n = 2\times0.632\times25.75 = 32.55$$

(b) Settling line $\qquad t_s = \dfrac{4}{\xi \omega_n}$ for 2% tolerance

$$= \dfrac{4}{0.632 \times 25.75} = 0.246 \text{ sec.} \qquad \textbf{Ans.}$$

(c) Bandwidth (BW) $= \omega_n [(1 - 2\xi^2) + \sqrt{(2 - 4\xi^2 + 4\xi^4)}]^{1/2}$

$$= 25.75[1 - 2 \times 0.62^2 + \sqrt{2 - 4 \times 0.632^2 + 4 \times 0.632^4}]^{1/2}$$

$$= 28.45 \text{ rad/sec.} \qquad \textbf{Ans.}$$

Problem 8.4 The closed-loop transfer function of a feedback system is given by:

$$T(s) = \dfrac{1000}{(s + 22.5)\,(s^2 + 2.45s + 44.4)}$$

(a) Determine the resonance peak M_r and resonant frequency ω_r of the system by drawing the frequency response curve.
(b) What should be the values of damping ratio and undamped natural frequency of oscillations of an equivalent second-order system which will produce the same M_r and ω_r as determined in part (a).
(c) Determine the bandwidth of the equivalent second-order system.

Solution

$$T(j\omega) = \dfrac{1000}{(j\omega + 22.5)\,[(j\omega)^2 + 2.45\,j\omega + 44.4]}$$

$$= \dfrac{1}{(1 + 0.044\,j\omega)\,(1 - 0.0225\omega^2 + j0.055\omega)}$$

$$M = \dfrac{1}{\sqrt{1 + (0.044\omega)^2}\,\sqrt{(1 - 0.0225\omega^2)^2 + (0.055\omega)^2}}$$

Sr. No.	ω (rad/sec)	M
1	0	1.00
2	1	1.02
3	2	1.087
4	4	1.46
5	5	1.907
6	6	2.588
7	6.5	2.632
8	6.7	2.68
9	7	2.45
10	10	0.68

Frequency response curve is shown in Fig. 8.4.

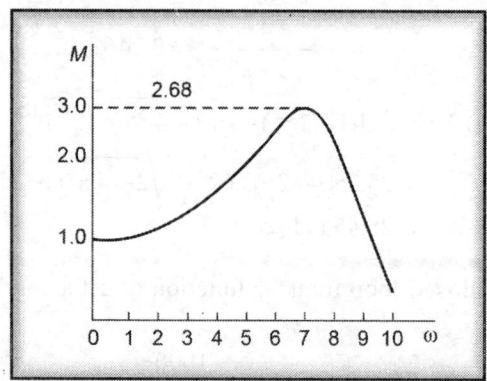

Fig. 8.4

From the curve it is seen that

$$\omega_r = 6.7 \text{ rad/sec and } M_r = 2.68$$

(b) Now, $$M_r = \frac{1}{2\xi\sqrt{1-\xi^2}} \text{ or } 2.68 = \frac{1}{2\xi\sqrt{1-\xi^2}}$$

or $$\xi = 0.190, 0.968$$ **Ans.**

($\xi = 0.968$ is neglected because for $\xi > 0.707$, there is no resonant peak)

Also $$\omega_r = \omega_n\sqrt{1-\xi^2}$$

or $$6.7 = \omega_n\sqrt{1-2\times0.190^2} \text{ or } \omega_n = 6.82 \text{ rad/sec.}$$ **Ans.**

(c) $$BW = \omega_n[1-2\xi^2 + \sqrt{2-4\xi^2 + 4\xi^4}]^{1/2}$$

Putting $$\xi = 0.190 \text{ and } \omega_n = 6.82,$$

we get $$BW = 10.32$$ **Ans.**

Problem 8.5 The closed-loop frequency response $|M(j\omega)|$ versus frequency of compound order prototype system is shown in Fig. 8.5. Sketch the

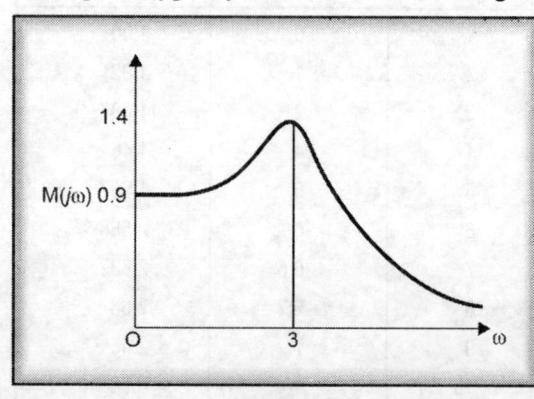

Fig. 8.5

corresponding unit step response of the system; indicate the values of the maximum overshoot, peak time and the steady-state error due to unit step response.

Solution

From Fig. 8.5, $M_r = 1.4$ and $\omega_r = 3$ rad/sec

$$M_r = \frac{1}{2\xi\sqrt{1-\xi^2}}$$

$$1.4 = \frac{1}{2\xi\sqrt{1-\xi^2}}$$

or

$$\xi = 0.387$$

Maximum overshoot $= e^{-\pi\xi/\sqrt{1-\xi^2}}$

$$= e^{-\pi\times0.387/\sqrt{1-0.387^2}} = 0.2675 = 26.75\%$$

$$\omega_r = \omega_n\sqrt{1-2\xi^2}$$

$$3 = \omega_n\sqrt{1-2\times0.387^2}$$

$$\omega_n = 3.586 \text{ rad/sec}$$

$$t_{\text{peak}} = \frac{\pi}{\omega_n\sqrt{1-\xi^2}} = \frac{\pi}{3.586\sqrt{1-0.387^2}} = 0.95 \text{ sec}$$

Also, at
$$\omega = 0, |M| = 0.9$$

Therefore, the steady-state value of the unit step response is 0.9.

Hence, the error $e_{ss} = 1 - 0.9 = 0.1$

Unit step response is depicted in Fig. 8.6.

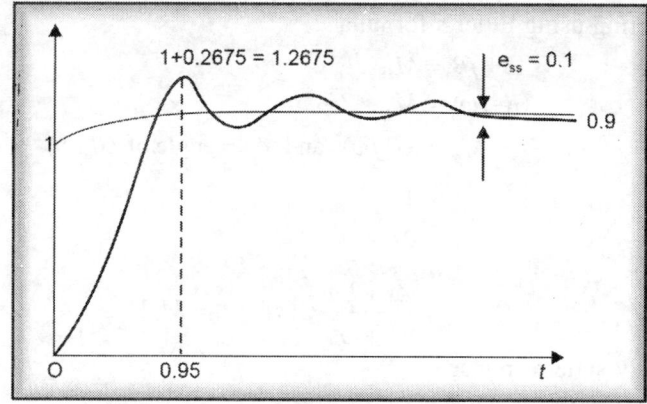

Fig. 8.6

Problem 8.6 A control system having transfer function $G(s)$ is subjected to input $r(t) = A \cos \omega t + B \sin \omega t$. Find the expression for frequency response $c(t)$.

Solution

$$\frac{C(s)}{R(s)} = G(s)$$

$$r(t) = A \cos \omega t + B \sin \omega t$$

$$= \sqrt{A^2 + B^2} \cos \omega t - \tan^{-1}(B/A)$$

$$R(s) = \frac{As}{s^2 + \omega^2} + \frac{B\omega}{s^2 + \omega^2} = \frac{As + B\omega}{s^2 + \omega^2}$$

$\therefore$

$$C(s) = \frac{As + B\omega}{s^2 + \omega^2} G(s)$$

$$= \frac{As + B\omega}{(s + j\omega)(s - j\omega)} G(s)$$

$$= \frac{A_1}{s + j\omega} + \frac{A_2}{s - j\omega} + \text{terms due to } G(s)$$

$$A_1 = \frac{As + B\omega}{s - j\omega} G(s) \Big|_{s = -j\omega}$$

$$= \frac{-Aj\omega + B\omega}{-2j\omega} G(-j\omega)$$

$$= \frac{A}{2} - \frac{B}{2j} G(-j\omega)$$

$$= \frac{1}{2}(A + jB)G(-j\omega)$$

Representing using Euler's formula

$$A + jB = M_i e^{-j\phi_i}$$

$$G(-j\omega) = M_g e^{-j\phi_g}$$

where

$$M_g = |G(j\omega)| \text{ and } \phi_g = \text{angle of } G(j\omega)$$

$\therefore$

$$A_1 = \frac{1}{2} M_i M_g e^{-j(\phi_i + \phi_g)}$$

Similarly

$$A_2 = \frac{As + B\omega}{s^2 + \omega^2} G(s) \Big|_{s = j\omega} = M_i M_g e^{j(\phi_i + \phi_g)}$$

The steady-state response

$$C_{ss}(s) = \frac{A_1}{s + j\omega} + \frac{A_2}{s - j\omega}$$

Note: The steady-state response is due to poles of the input waveform and hence is contributed by first two terms of $C(s)$ expressed earlier.

Substituting the values of A_1 and A_2, we get

$$C_{ss}(s) = \frac{\dfrac{M_i M_g}{2} e^{-j(\phi_i + \phi_g)}}{s + j\omega} + \frac{\dfrac{M_i M_g}{2} e^{+j(\phi_i + \phi_g)}}{s - j\omega}$$

inverse Laplace transform gives

$$c(t) = M_i M_g \left(\frac{e^{-j(\omega t + \phi_i + \phi_g)} + e^{+j(\omega t + \phi_i + \phi_g)}}{2} \right)$$

$$= M_i M_g \cos(\omega t + \phi)$$

where

$$\phi = \phi_i + \phi_g$$

Problem 8.7 Given $G(s) = \omega_n^2 / (s^2 + 2\xi\omega_n s + \omega_n^2)$, show that $|G(j\omega_n)| = 1/2\xi$.

Solution

$$G(j\omega) = \frac{\omega_n^2}{(j\omega)^2 + 2\xi\omega_n(j\omega) + \omega_n^2}$$

$$= \frac{1}{\left(j\dfrac{\omega}{\omega_n}\right)^2 + 2\xi\left(j\dfrac{\omega}{\omega_n}\right) + 1}$$

$$|G(j\omega)| = \frac{1}{\left[\left(1 - \left(\dfrac{\omega}{\omega_n}\right)^2\right)^2 + \left(2\dfrac{\omega}{\omega_n}\xi\right)^2\right]^{1/2}}$$

Put

$$\omega = \omega_n$$

$$|G(j\omega_n)| = \frac{1}{[(1-1)^2 + (2 \times 1 \times \xi)^2]^{1/2}} = \frac{1}{[4\xi^2]^{1/2}} = \frac{1}{2\xi}$$

Problem 8.8 A first order system described by $1.5 dx/dt + x = y$ is subjected to sinusoidal input $y(t) = 3\sin(\omega_o t)$ mV. Assuming that the output of the system goes into the filter that cut offs all input signals smaller in amplitude than 0.01 mV, determine the critical frequency ω_e such that for all $\omega_o > \omega_e$ the filter will not observe the input signal.

Solution

For the equation $1.5\dfrac{dx}{dt} + x = y = 3\sin(\omega_o t)$

the solution is

$$\frac{3}{\sqrt{1 + (1.5\omega_o)^2}} \sin(\omega_o t + \phi)$$

where
$$\phi = \tan^{-1}(1.5\omega_o)$$

$$A = \frac{3}{\sqrt{1+(1.5\omega_o)^2}}$$

$\omega_o = \omega_e$ can be found by setting $A = 0.01$

$$0.01 = \frac{3}{\sqrt{1+(1.5\omega_o)^2}}$$

$$= \frac{3}{\sqrt{1+2.25\omega_o^2}}$$

$$\sqrt{1+2.25\omega_o^2} = 300$$

$$\omega_o^2 = \frac{300^2-1}{2.25} = \frac{(300+1)(300-1)}{2.25} = \frac{301\times299}{2.25}$$

or
$$\omega_o = \sqrt{\frac{301\times299}{2.25}}$$

Problem 8.9 A system is represented by the differential equation
$$A\ddot{x}(t) + B\dot{x}(t) + C\ddot{x}(t) = f(t)$$
Determine the sinusoidal response of the equation.

Solution

If amplitude is unity and angular frequency is ω, then
$$\phi(t) = e^{j\omega t} \tag{8.1}$$
then
$$x(t) = Ke^{j(\omega t + \phi)} \tag{8.2}$$
where K is the relative amplitude and ϕ is the phase angle with respect to the input vector.

Differentiating Eqn. (8.2), we get

$$\frac{dx}{dt} = j\omega Ke^{j(\omega t+\phi)} \tag{8.3}$$

and
$$\frac{d^2x}{dt^2} = (j\omega)^2 Ke^{j(\omega t+\phi)} \tag{8.4}$$

Substituting in A
$$A(j\omega)^2 Ke^{j(\omega t+\phi)} + Bj\omega Ke^{j(\omega t+\phi)} + CKe^{j(\omega t+\phi)} = e^{j\omega t}$$
or
$$Ke^{j\phi}[A(j\omega)^2 + B(j\omega) + C] = 1$$

or
$$Ke^{j\phi} = \frac{1}{A(j\omega)^2 + B(j\omega) + C} \tag{8.5}$$

The Eqn. (8.5) depicts the relative amplitude K and phase shift ϕ with respect to the sinusoidal input.

Polar Plots

9.1 General

The *polar plot* of a sinusoidal transfer function $G(j\omega)$ is a plot of $|G(j\omega)|$ versus phase angle of $G(j\omega)$ on polar coordinates as 'ω' is varied from zero to infinity. It is also called *Nyquist plot*. The polar plot is plotted on a complex plane. As ω is varied, magnitude and phase angle change and if magnitude is plotted for varying phase angles, the locus obtained is the polar plot. Polar plots are easier to construct and at any desired frequency and ready information on magnitude and phase angle can be obtained. However, calculations are tedious and if a system is modified by adding a pole or zero, the polar plot has to be constructed again.

9.2 Polar Plot Sketching Procedure

Let us assume that the transfer function

$$G(s) = \frac{C(s)}{R(s)} = \frac{(a + j\omega b)(c + j\omega d)}{(e + j\omega f)(g + j\omega h)}$$

Then,

$$M = \left| \frac{(a + j\omega b)(c + j\omega d)}{(e + j\omega f)(g + j\omega h)} \right| = \frac{\sqrt{a^2 + (\omega b)^2} \sqrt{c^2 + (\omega d)^2}}{\sqrt{e^2 + (\omega f)^2} \sqrt{g^2 + (\omega h)^2}}$$

and

$$\phi = \tan^{-1}\frac{\omega b}{a} + \tan^{-1}\frac{\omega d}{c} - \tan^{-1}\frac{\omega f}{e} - \tan^{-1}\frac{\omega h}{g}$$

Value of 'M'

The value of M can be easily found out by varying ω from 0 to ∞ and the same can be tabulated.

Value of ϕ

The following must be remembered to arrive at the values of tan functions. The value of tan function depends upon which quadrant, i.e. the function $(a + jb)$ is located.

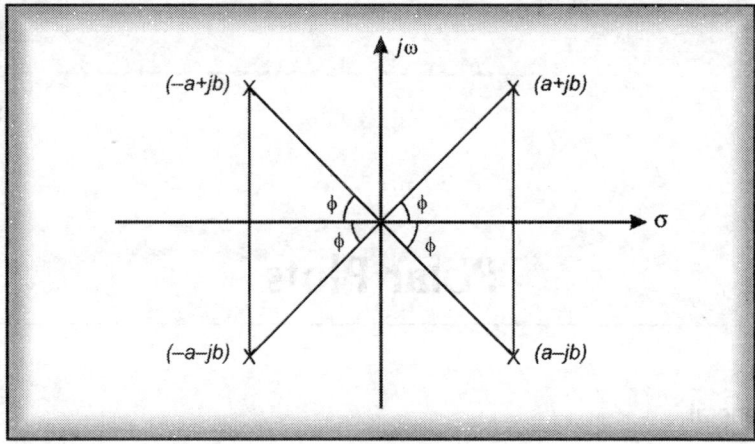

Fig. 9.1

(a) $\tan^{-1} \dfrac{b}{a}$ θ; ϕ will be positive (measured CW)

    or $\phi = 360° - \theta$ (negative when measured CCW)

(b) $\tan^{-1} \dfrac{b}{-a}$ θ; $\phi = 180 + \theta$ (negative when measured CCW)

(c) $\tan \dfrac{-b}{-a}$; $\phi = 180 - \theta$ (negative when measured CCW)

(d) $\tan^{-1} \dfrac{-b}{a} = \phi$ (negative when measured CCW)

Polar plots for a few typical transfer functions are shown in Table 9.1.

SOLVED PROBLEMS

Problem 9.1 Sketch the polar plots of $1/s$ and $1/s^2$.

Solution

(a) Polar plot of $1/s$

$$G(s) = \frac{1}{s}$$

Putting $s = j\omega$, we get

$$G(j\omega) = \frac{1}{j\omega}$$

If $\omega = 0$, then $|G(j\omega)| = \infty$

If $\omega = \infty$, then $|G(j\omega)| = 0$

(b) Polar plot of $1/s^2$

$$G(s) = \frac{1}{s^2}$$

Putting $s = j\omega$, we get

$$G(j\omega) = \frac{1}{j^2\omega^2} = -\frac{1}{\omega^2}$$

If $\omega = 0$ then $|G(j\omega)| = \infty$

If $\omega = \infty$, then $|G(j\omega)| = 0$

Table 9.1: Typical Polar Plots

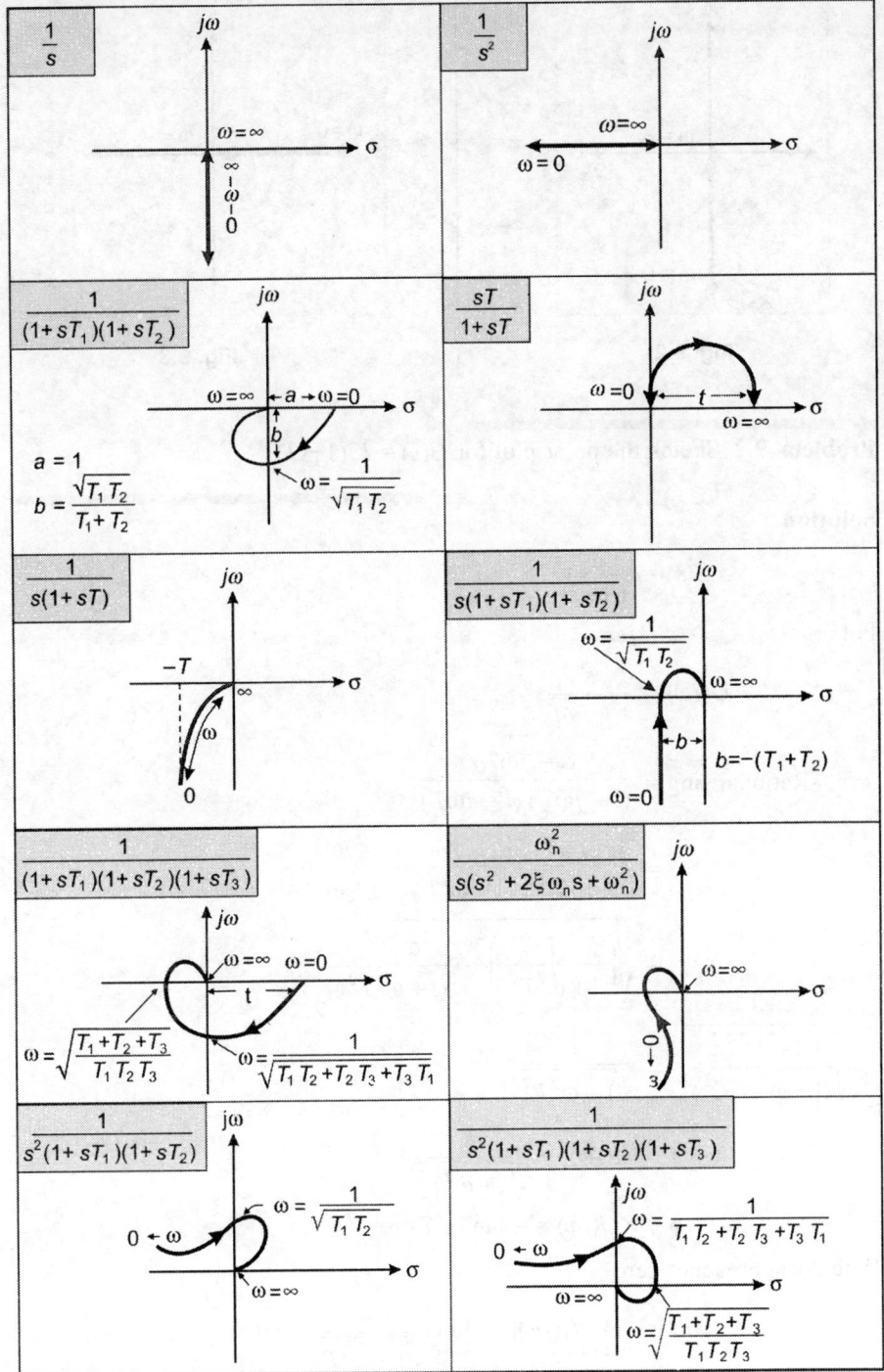

Polar plot is shown in Fig. 9.2

Polar plot is shown in Fig. 9.3

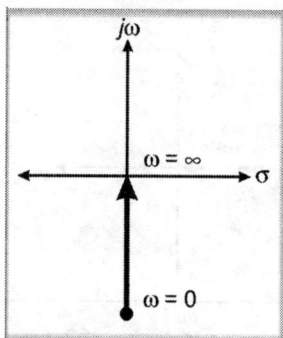

Fig. 9.2

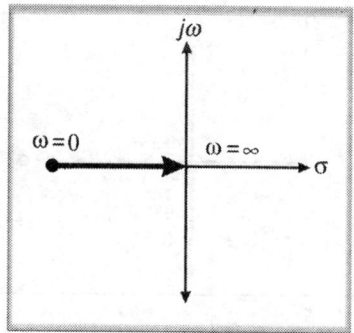

Fig. 9.3

Problem 9.2 Sketch the polar plot for $G(s) = K/(1+sT)$

Solution

$$G(s) = \frac{K}{1 + sT}$$

Put $\qquad s = j\omega$

$$G(j\omega) = \frac{K}{1 + j\omega T}$$

Rationalising $= \dfrac{(1 - j\omega T)}{(1 + j\omega T)(1 - j\omega T)}$

$$= \frac{1 - j\omega T}{1 + \omega^2 T^2} = \frac{1}{1 + \omega^2 T^2} - \frac{j\omega T}{1 + \omega^2 T^2}$$

$$M = \sqrt{\left(\frac{1}{1 + \omega^2 T^2}\right)^2 + \frac{\omega^2 T^2}{(1 + \omega^2 T^2)}}$$

$$\sqrt{\frac{1 + \omega^2 T^2}{(1 + \omega^2 + T^2)^2}} = \frac{1}{\sqrt{1 + \omega^2 T^2}}$$

$$M = |G(j\omega)| = \frac{K}{\sqrt{1 + \omega^2 T^2}}$$

$$\phi = \angle G(j\omega) = -\tan^{-1}\omega T$$

When ω approaches zero

$$M = \lim_{\omega \to 0} |G(j\omega)| = \lim_{\omega \to 0} \frac{K}{\sqrt{1 + \omega^2 T^2}} = 0$$

and $$\Phi = \lim_{\omega \to 0} \angle G(j\omega) = \lim_{\omega \to 0} (-\tan^{-1}\omega T) = 0$$

when ω approaches infinity

$$M = \lim_{\omega \to \infty} |G(j\omega)| = \lim_{\omega \to \infty} \frac{K}{\sqrt{1+\omega^2 T^2}} = K,$$

and $$\Phi = \lim_{\omega \to \infty} \angle G(j\omega) = \lim_{\omega \to \infty} (-\tan^{-1}\omega T) = -90°$$

Polar plot is shown in Fig. 9.4

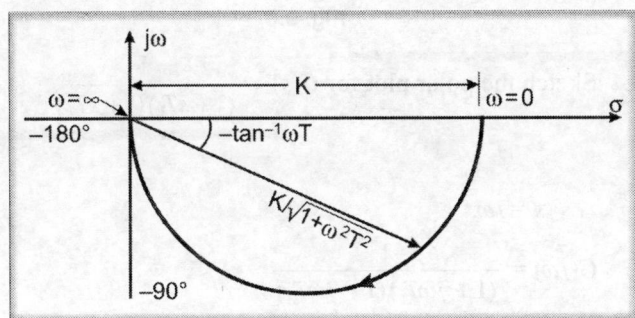

Fig. 9.4

Problem 9.3 Sketch the polar plot for $G(s) = \dfrac{sT}{1+sT}$

Solution

$$G(s) = \frac{sT}{1+sT}$$

Putting $s = j\omega,$

we get $$G(j\omega) = \frac{j\omega T}{1+j\omega T}$$

Now, $$M = \frac{\omega T}{\sqrt{1+\omega^2 T^2}}$$

and $$\Phi = 90° - \tan^{-1}\omega T,$$

when ω approaches zero

$$M = 0 \text{ and } \Phi = 90°,$$

when ω approaches infinity

$$M = 1 \text{ and } \Phi = 0$$

Polar plot is shown in Fig. 9.5

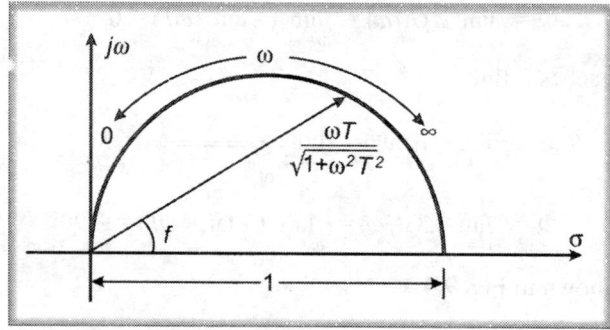

Fig. 9.5

Problem 9.4 Sketch the polar plot for $G(s) = \dfrac{1}{(1 + sT_1)(1 + sT_2)}$

Solution

Putting $\qquad s = j\omega,$

we get $\qquad G(j\omega) = \dfrac{1}{(1 + j\omega T_1)(1 + j\omega T_2)}$

$$M = \dfrac{1}{\sqrt{1 + \omega^2 T_1^2}\,\sqrt{1 + \omega^2 T_2^2}}$$

and $\qquad \Phi = -\tan^{-1}\omega T_1 - \tan^{-1}\omega T_2,$

when ω approaches zero

$$M = 1 \quad \text{and} \quad \Phi = 0,$$

when ω approaches infinity

$$M = 0 \quad \text{and} \quad \Phi = -90° - 90° = -180°$$

Now, $\qquad G(j\omega) = \dfrac{1}{(1 + j\omega T_1)(1 + j\omega T_2)}$

Separating $G(j\omega)$ into real and imaginary parts, we get

$$G(j\omega) = \dfrac{1 - \omega^2 T_1 T_2}{(1 + \omega^2 T_1^2)(1 + \omega^2 T_2^2)} - j\dfrac{\omega T_1 + \omega T_2}{(1 + \omega^2 T_1^2)(1 + \omega^2 T_2^2)}$$

Equating real part to zero, we get $1 - \omega^2 T_1 T_2 = 0$, i.e. $\omega = \dfrac{1}{\sqrt{T_1 T_2}}$. This is the value at which the polar plot crosses the imaginary axis.

Putting the value of ω in M, we get

$$M = \dfrac{1}{\sqrt{1 + \dfrac{T_1^2}{T_1 T_2}}\,\sqrt{1 + \dfrac{T_2^2}{T_1 T_2}}} \quad \text{or} \quad M = \dfrac{\sqrt{T_1 T_2}}{T_1 + T_2}$$

The polar plot is shown in Fig. 9.6

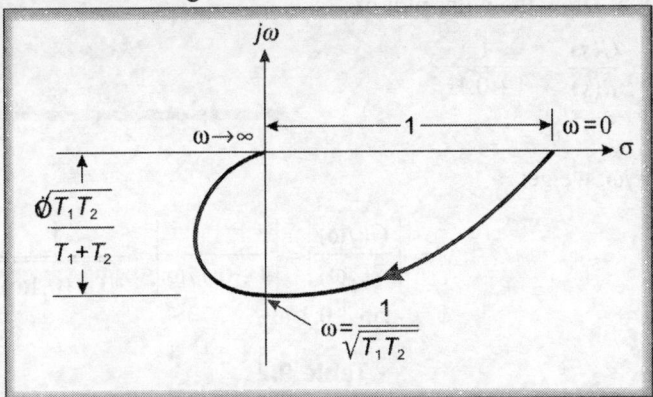

Fig. 9.6

Problem 9.5 Sketch the polar plot of $G(s) = \dfrac{1}{s(1 + sT)}$

Solution

Putting $\quad s = j\omega$, we get

$$G(j\omega) = \frac{1}{j\omega(1 + j\omega T)}$$

Now, $\quad M = \dfrac{1}{\omega\sqrt{1 + \omega^2 T^2}} \quad$ and $\quad \Phi = -90° - \tan^{-1} \omega T$

When ω approaches zero, $M = \infty$ and $\Phi = -90°$, and when ω approaches infinity $M = 0$ and $\Phi = -90° - 90° = -180°$. Polar plot is shown in Fig. 9.7.

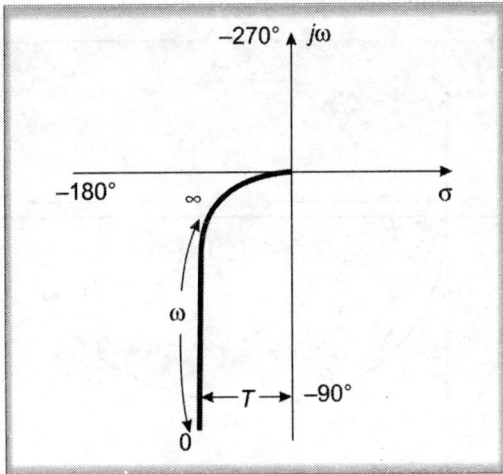

Fig. 9.7

Problem 9.6 Draw the polar plot of

$$\frac{C(s)}{R(s)} = \frac{1}{1+0.1s}$$

Solution

Putting $s = j\omega$, we get

$$M = \left|\frac{C(j\omega)}{R(j\omega)}\right| = \left|\frac{1}{1+0.1j\omega}\right| = \frac{1}{\sqrt{1+(0.1\omega)^2}}$$

$$\phi = -\tan^{-1} 0.1\,\omega$$

Table 9.2

S. No.	ω (rad/sec)	$M = \dfrac{1}{\sqrt{1+(0.1\omega)^2}}$	$\phi = -\tan 0.1\,\omega$ (degree)
1.	0	1	0
2.	1	0.99	−5.71
3.	2	0.98	−11.13
4.	4	0.93	−21.80
5.	5	0.89	−26.60
6.	6	0.86	−30.90
7.	10	0.71	−45.00
8.	20	0.45	−63.40
9.	40	0.24	−76.00
10.	50	0.196	−78.70
11.	75	—	−82.40
12.	100	0.099	−84.29
13.	∞	0	−90

The polar plot is shown in Fig. 9.8.

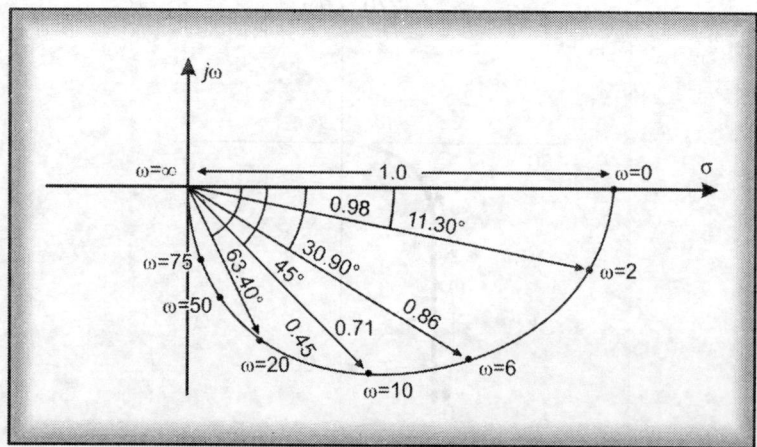

Fig. 9.8

Problem 9.7 Draw the polar plot of

$$G(s) = \frac{10}{s(s+2)(s+10)}$$

Solution

$$G(s) = \frac{10}{s(s+2)(s+10)} = \frac{0.5}{s(1+0.5s)(1+0.1s)}$$

$$M = \frac{0.5}{\omega\sqrt{1+(0.5\omega)^2}\sqrt{1+(0.1\omega)^2}} = \frac{0.5}{\omega\sqrt{1+0.25\omega^2}\sqrt{1+0.01\omega^2}}$$

$$\phi = -90° - \tan^{-1} 0.25\,\omega - \tan^{-1} 0.01\,\omega$$

Table 9.3

S. No.	ω (rad/sec)	$M = \dfrac{0.5}{\omega\sqrt{1+0.25\omega^2}\sqrt{1+0.01\omega^2}}$	$\phi = -90° - \tan^{-1} 0.25\,\omega - \tan^{-1} 0.01\,\omega$ (degree)
1.	0	∞	−90
2.	1	0.40	−104.61
3.	3	0.09	−128.57
4.	10	0.022	−163.90°
5.	100	0.00009	−222.71°
6.	∞	0	−270°

The polar plot is shown in Fig. 9.9.

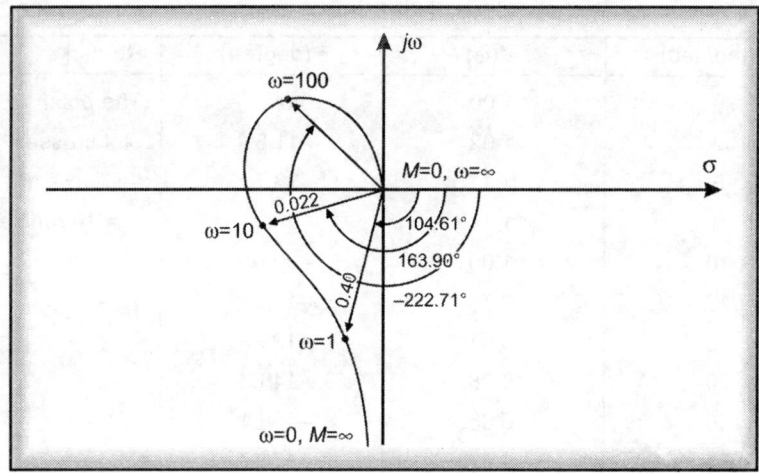

Fig. 9.9.

Problem 9.8 Draw the polar plot of control system having

$$G(s) = \frac{100}{s^2 + 10s + 100}$$

Solution

Put

$$s = j\omega$$

$$G(j\omega) = \frac{100}{(j\omega)^2 + 10\,j\omega + 100} = \frac{100 + j(0)}{-\omega^2 + 10\,j\omega + 100}$$

$$\therefore \qquad M = \left| \frac{100 + j(0)}{(100 - \omega^2) + j(10\omega)} \right| = \frac{\sqrt{100^2}}{\sqrt{(100 - \omega^2)^2 + (10\omega)^2}}$$

$$= \frac{100}{\sqrt{(100 - \omega^2)^2 + (10\omega)^2}}$$

$$\phi = \angle \frac{100 + j(0)}{(100 - \omega^2) + j(10\omega)}$$

$$= \tan^{-1}\left(\frac{0}{100}\right) - \tan^{-1}\left(\frac{10\omega}{100 - \omega^2}\right)$$

Various values of M and ϕ for ω ranging from 0 to ∞ are given in Table 9.4

Table 9.4

ω (rad/sec)	$M(\omega)$	ϕ (degree)	Remarks
0	1.00	0.0	The polar
2	1.02	−11.8	plot crosses
5	1.11	−33.7	'−jω' axis at
8	1.14	−65.8	ω = 10 rad/sec
10	1.00	−90.0	
12	0.78	−110.1	
15	0.51	−129.8	
20	0.28	−146.3	
40	0.06	−165.1	
70	0.02	−171.7	
∞	0.00	−180.0	

With the help of the data given in Table 9.4, the same is plotted on the complex plane with varying ω (Fig. 9.10)

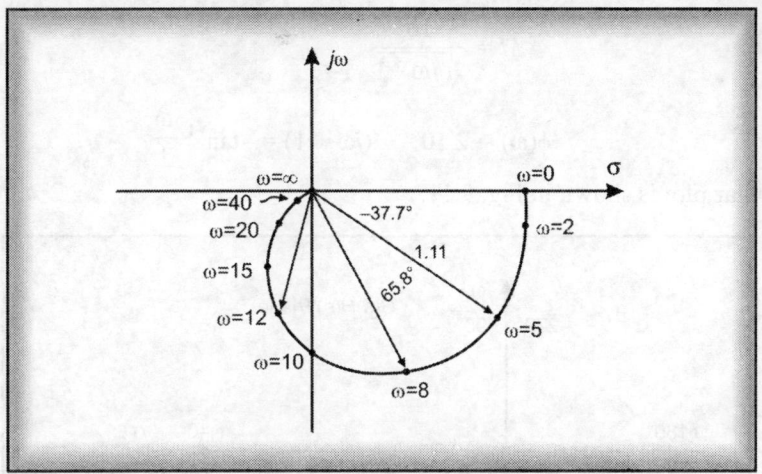

Fig. 9.10

Note: The polar plot intersects the imaginary axis at a frequency equal to the natural frequency of the system = ω_n = 10 rad/sec.

Problem 9.9 Draw the polar plot for the transfer function

$$G(s) = \frac{10}{(s+1)}$$

Solution

Put $\qquad\qquad s = j\omega$

Table 9.5

ω (rad/sec)	M(ω)	φ (degree)	Remarks
0	1.00	0.0	The polar plot
0.2	9.8	−11	never crosses
0.4	9.3	−21	the '−jω' axis
0.6	8.6	−31	
0.8	7.8	−39	
2.0	4.5	−63	
3.0	3.2	−72	
4.0	2.5	−76	
5.0	1.9	−79	
10	0.99	−84	
∞	0.00	−90	

$$G(j\omega) = \frac{10}{j\omega + 1}$$

$$M = \frac{10}{\sqrt{j\omega + 1}}$$

$$\phi(\omega) = \angle 10 - \angle(j\omega + 1) = -\tan^{-1}\frac{\omega}{1}$$

The polar plot is shown in Fig. 9.11.

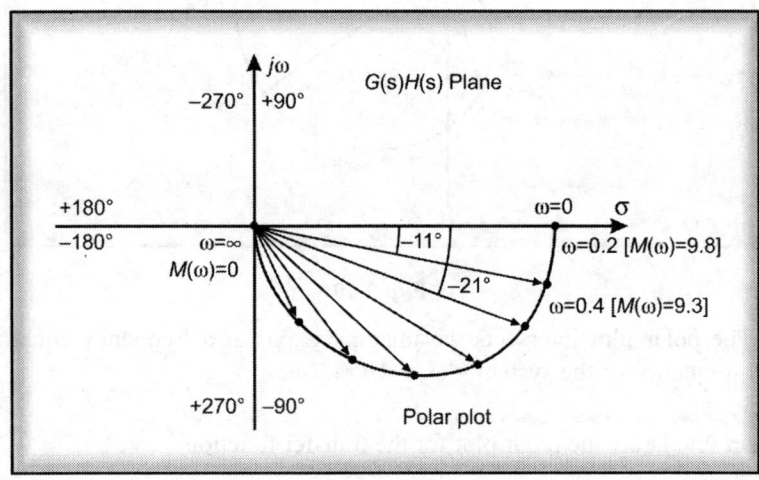

Fig. 9.11

Stability

10.1 Introduction

Stability is a very important performance criteria. It is always the concern of the system designer that the control system is stable when the system is subjected to command signals, variations in power supply and parameters of feedback loop and due to disturbances. There are various methods of defining the stability of a linear time-invariant system. One most frequently used definition of stability is "*A system is said to be stable, if the output of the system is bounded for any bounded input.*" This implies that if the system is subjected to bounded input, it should generate a bounded output response, for the system to be termed as stable, and if the response produced is an unbounded response, the system is unstable.

10.2 Stability

The mere statement that the system is stable has no meaning. It merely indicates the absolute stability, i.e. whether a system is stable or unstable. The main question "as to how much stable is the system" is very important. Therefore, the degree of stability has to be determined. The degree of stability is a measure of relative stability. In order to obtain answer to this question, we study the following relation:

$$\frac{C(s)}{R(s)} = \frac{G(s)}{1 + G(s)H(s)} \tag{10.1}$$

Stability is thus determined by studying the denominator of the equation by equating it to zero (called the characteristic equation).

$$1 + G(s)H(s) = 0 \tag{10.2}$$

10.3 Stability Study Methods

10.3.1 In linear systems, stability is independent of the system input. Thus, the above equation determines system stability. There are various methods available to evaluate the system stability. All methods, in one way or the other, evaluate the characteristic equation. The system stability can be determined by finding out the roots of the characteristic equation and its location. The system is considered to be stable if the roots of the characteristic equation are located in left half of the *s*-plane. This causes the output response due to bounded input to decrease to zero as the time approaches infinity.

10.3.2 If the roots of the characteristic equation are located on the imaginary axis but with no roots in the right hand side of the *s*-plane, then the system is unstable and the responses will be either undamped sinusoidal oscillatory or will increase in magnitude, respectively as time increases. The stability defined above leads to two methods by which system stability can be ascertained:

(a) The first method finds out the region of parameters of the system which will ensure that the roots of the characteristic equation have negative real parts. Routh-Hurwitz and Nyquist criteria utilize this approach to evaluate system stability. These criteria have been explained in Chapters 11 and 12, respectively and illustrated with number of solved problems.

(b) The second method calculates the exact roots to find out system stability. This is done by finding out the direct solution. This is extremely tedious. The other method is the Root Locus method and is explained in Chapter 14.

10.3.3 Another method available for stability studies is Bode diagram. However, the result is conclusive, if the transfer function is of the minimum phase type. Plotting of Bode diagram is explained in Chapter 13.

10.3.4 The above methods are applicable only for the linear systems. Thus, it is seen that output response depends upon the types of roots. These have been plotted in Table 10.1.

Regions of stability obtained from the preceding discussions is marked on the s-plane (Fig. 10.1) The degree of stability or relative stability not only depends upon the type of roots but also on its location on the left half of *s*-plane. If two different systems have one real root '*a*' and '*b*' respectively, then the system whose location of real root is farther away from the imaginary axis will be relatively more stable. Same is true for the complex conjugate roots with negative real parts.

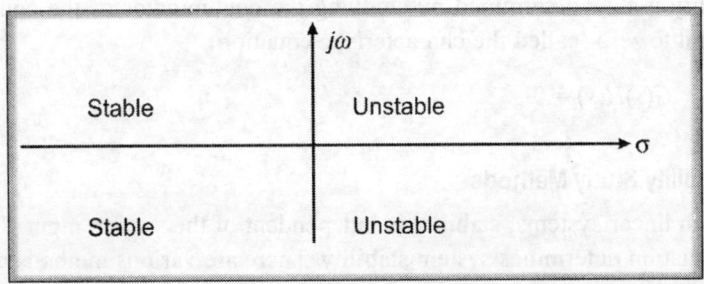

Fig. 10.1

Table 10.1 Responses due to Various Types of Roots

Location of roots	Response

Routh's Stability Criterion

11.1 Introduction

Routh's stability criterion can be readily applied to the characteristic equation to find out the presence of the roots without having to solve the actual equation. As per Routh's stability criterion the *necessary conditions* for a control system to be stable are:

- None of the coefficients of the characteristic equation should be missing or zero.
- All the coefficients should be real and should have the same sign.

The *sufficient condition* for a system to be stable is that each term of the first column of Routh's array be positive and should have the same sign.

11.2 Procedure for Application

Routh's array for the characteristic equation

$$a_0 s^n + a_1 s^{n-1} + a_2 s^{n-2} + \ldots + a_{n-1} s + a_n = 0$$

where, $n = 7$ is formed as given below:

s^7	a_0	a_2	a_4	a_6
s^6	a_1	a_3	a_5	a_7
s^5	b_1	b_3	b_5	×
s^4	c_1	c_3	$c_5(a_7)$	×
s^3	d_1	d_3	×	×
s^2	e_1	$e_2(c_5)$	×	×
s^1	f_1	×	×	×
s^0	$g_1(e_2)$	×	×	×

where,

$$b_1 = \frac{\begin{matrix} a_0 & a_2 \\ a_1 & a_3 \end{matrix}}{a_1} = \frac{a_1 a_2 - a_0 a_3}{a_1}$$

$$b_3 = \frac{\begin{matrix} a_0 & a_4 \\ a_1 & a_5 \end{matrix}}{a_1} = \frac{a_1 a_4 - a_0 a_5}{a_1}$$

$$b_5 = \frac{\begin{matrix} a_0 & a_6 \\ a_1 & a_7 \end{matrix}}{a_1} = \frac{a_1 a_6 - a_0 a_7}{a_1}$$

$$c_1 = \frac{\begin{matrix} a_1 & a_3 \\ b_1 & b_3 \end{matrix}}{b_1} = \frac{b_1 a_3 - a_1 b_3}{b_1}$$

$$c_3 = \frac{\begin{matrix} a_1 & a_5 \\ b_1 & b_5 \end{matrix}}{b_1} = \frac{b_1 a_5 - a_1 b_5}{b_1}$$

$$c_5 = \frac{\begin{matrix} a_1 & a_7 \\ b_1 & 0 \end{matrix}}{b_1} = \frac{b_1 a_7 - a_1 0}{b_1} = a_7$$

$$d_1 = \frac{\begin{matrix} b_1 & b_3 \\ c_1 & c_3 \end{matrix}}{c_1} = \frac{c_1 b_3 - b_1 c_3}{c_1}$$

$$d_3 = \frac{\begin{matrix} b_1 & b_5 \\ c_1 & c_5 \end{matrix}}{c_1} = \frac{c_1 b_5 - b_1 c_5}{c_1}$$

and so on.

The study of the array reveals that successive rows have one term fewer than the preceding row, and hence, the array is triangular.

The following are the limitations of Routh's stability criterion:

- It is valid only if the characteristic equation is algebraic.
- If any coefficient of the characteristic equation is complex or contains power of 'e', this criterion cannot be applied.
- It gives us information as to how many roots are lying in the RHS of the s-plane. Values of the roots are not available. Also, it cannot distinguish between real and complex roots.

SOLVED PROBLEMS

Problem 11.1 The open-loop transfer function of a feedback control system is given by:

$$G(s)H(s) = \frac{K(s+1)}{s(1+Ts)(1+2s)}$$

The parameters K and T may be represented in a plane with K as the horizontal axis and T as vertical axis. Determine the region in which the closed-loop system is stable.

Solution

The characteristic equation is given by:

$$1 + G(s)H(s) = 2Ts^3 + (2+T)s^2 + (K+1)s + K = 0$$

The Routh's array is

s^3	$2T$	$K+1$
s^2	$2+T$	K
s^1	$K+1-\dfrac{2KT}{2+T}$	$\times$
s^0	K	$\times$

For stability (by considering terms of first column one by one)

$$2T > 0 \quad \text{and} \quad 2+T > 0 \quad \therefore \quad T > 0$$

Also, $\qquad K > 0,$

$$K+1-\frac{2KT}{2+T} > 0 \quad \text{gives} \quad K < \frac{2+T}{T-2} \quad \text{and} \quad T < \frac{2(1+K)}{K-1}$$

Therefore, for stability, we have

$$K > 0, T > 0 \quad \text{and} \quad K > 1 \quad \text{and} \quad T > 2$$

The region of stability is shown in Fig. 11.1

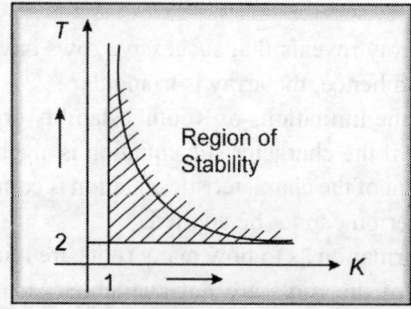

Fig. 11.1

Problem 11.2 Using Routh's stability criterion, ascertain stability for each of the following cases:

(a) $s^6 + 2s^5 + 8s^4 + 12s^3 + 20s^2 + 16s + 16 = 0$

(b) $s^6 + s^5 - 2s^4 - 3s^3 - 7s^2 - 4s - 4 = 0$

(c) $s^6 + 3s^5 + 4s^4 + 6s^3 + 5s^2 + 3s + 2 = 0$ (*Pune University*)

Solution

(a) $s^6 + 2s^5 + 8s^4 + 12s^3 + 20s^2 + 16s + 16 = 0$

The Routh's array is

s^6	1	8	20	16
s^5	2	12	16	×
	1	6	8	×
s^4	2	12	16	×
	1	6	8	×
s^3	0	0	×	×
	1	3	×	×
s^2	3	8	×	×
s^1	1/3	×	×	×
s^0	8	×	×	×

Subsidiary equation of s^4 row
$$s^4 + 6s^2 + 8 = 0$$

Derivative of subsidiary equation
$$4s^3 + 12s = 0$$
or $s^3 + 3s = 0$
(form row s^3 with its coefficients).

The s^3 row has become a zero row because the elements of s^5 and s^4 rows are same. The zero row was replaced by writing the subsidiary equation of the row above the s^3 row and then the first derivative of the subsidiary equation was taken as shown with the Routh's array. The s^3 row was then formed by the coefficients of s^3 row.

The first column of the Routh's array shows no sign change. Therefore, no roots are on the R.H.S. of the s-plane. However, presence of a zero row indicates the presence of symmetrically located roots in the s-plane. In order to find the roots, we proceed as given below:

$$s^6 + 2s^5 + 8s^4 + 12s^3 + 20s^2 + 16s + 16 = (s^4 + 6s^2 + 8)(s^2 + 2s + 2) = 0$$

Therefore, $(s^4 + 6s^2 + 8) = 0$ or $(s^2 + 4)(s^2 + 2) = 0$

gives $s = \pm 2j$; $s = \pm\sqrt{2}j$ and $s^2 + 2s + 2 = 0$, gives $s = -1 \pm j1$

These roots are shown in Fig. 11.2.

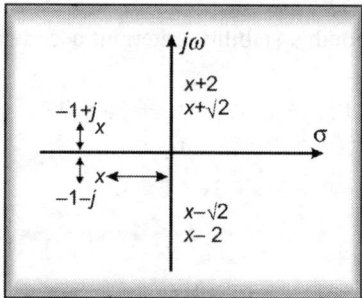

Fig. 11.2

It can be seen that there are two pairs of imaginary roots and one pair of complex conjugate root having negative real parts. Therefore, the system is limitedly stable.

(b) $s^6 + s^5 - 2s^4 - 3s^3 - 7s^2 - 4s - 4 = 0$

The Routh's array is

s^6	1	-2	-7	-4
s^5	1	-3	-4	×
s^4	1	-3	-4	×
s^3	0	0	×	×
	2	-3	×	×
s^2	-1.5	-4	×	×
s^1	-25/3	×	×	×
s^0	-4	×	×	×

Sign change

Subsidiary equation of s^4 row
$$s^4 - 3s^2 - 4 = 0$$

Derivative of subsidiary equation
$$4s^3 - 6s = 0$$
or $2s^3 - 3s = 0$
(form row s^3 with its coefficients).

There is one sign change in the first column of the Routh's array between s^3 and s^2 rows. Therefore, one root is lying on the R.H.S. of the s-plane. The system is thus unstable. Roots can be ascertained as given below:

$$s^6 + s^5 - 2s^4 - 3s^3 - 7s^2 - 4s - 4 = 0$$
or $$(s^4 - 3s^2 - 4)(s^2 + s + 1) = 0$$
$$\Rightarrow \quad s^4 - 3s^2 - 4 = 0 \quad \text{or} \quad (s^2 + 1)(s^2 - 4) = 0$$
$$\therefore \qquad\qquad\qquad\qquad s = \pm j \text{ and } s = \pm 2$$

and $$(s^2 + s + 1) = 0, \text{ given, } s = \frac{-1}{2} \pm j\frac{\sqrt{3}}{2}$$

Therefore, the roots are $+j, -j, +2, -2, -\dfrac{1}{2} \pm j\dfrac{\sqrt{3}}{2}$

(c) $s^6 + 3s^5 + 4s^4 + 5s^3 + 5s^2 + 3s + 2 = 0$

The Routh's array is

s^6	1	4	5	2
s^5	1	2	1	×
s^4	1	2	1	×
s^3	0	0	×	×
	1	1	×	×
s^2	1	1	×	×
s^1	0	×	×	×
	2	×	×	×
s^0	1	×	×	×

Subsidiary equation of s^4 row

$$s^4 + 2s^2 + 1 = 0$$

Derivative of subsidiary equation
$s^3 + s = 0$ (form s^3 row with its coefficients).

Subsidiary equation of s^2 row

$$s^2 + s = 0$$

Derivative of subsidiary equation
$2s = 0$ (form s^1 row with its coefficient).

There are two rows which become zero and there is no sign change in the first column of the Routh's array. The original characteristic equation can be written as:

$$(s^4 + 2s^2 + 1)(s^2 + 3s + 2) = 0$$
$$(s^4 + 2s^2 + 1) = 0 \quad \text{or} \quad (s^2 + 1)^2 = 0$$

gives $\qquad s = \pm j$ and $s = \pm j$

and $\qquad (s^2 + 3s + 2) = 0 \quad \text{or} \quad (s + 1)(s + 2) = 0$

gives $\qquad s = -1$ and -2

Because of the multiplicity of the pair of the imaginary root, the system is unstable.

Problem 11.3 The characteristic equation of a servo system is given by:

$$b_0 s^4 + b_1 s^3 + b_2 s^2 + b_3 s + b_4 = 0$$

Determine the conditions for stability.

Solution

The Routh's array is

s^4	b_0	b_2	b_4
s^3	b_1	b_3	×
s^2	$\dfrac{b_1 b_2 - b_0 b_3}{b_1} = a$	$\dfrac{b_1 b_4}{b_1}$	×
s^1	$\dfrac{\dfrac{b_3(b_1 b_2 - b_0 b_3) - b_1^2 b_4}{b_1}}{a}$	×	×
s^0	b_4	×	×

The conditions for stability are:

$$b_1 > 0, \quad b_0 > 0, \quad (b_1 b_2 - b_0 b_3) > 0$$
$$(b_1 b_2 b_3 - b_0 b_3^2 - b_1^2 b_4) > 0 \quad \text{and} \quad b_4 > 0$$

Problem 11.4 Determine the stability of the following cases, which represent characteristic equations of two different control systems:

(a) $3s^4 + 10s^3 + 5s^2 + 5s + 2 = 0$

(b) $s^5 + s^4 + 2s^3 + 2s^2 + 3s + 5 = 0$

Solution

(a) $3s^4 + 10s^3 + 5s^2 + 5s + 2 = 0$

The Routh's array is

s^4	3	5	2
s^3	10	5	×
s^2	3.5	2	×
s^1	$\dfrac{-2.5}{3.5}$	×	×
s^0	2	×	×

Sign change (between s^3 and s^2), Sign change (between s^2 and s^1).

There are two sign changes in the first column of the Routh's array. Hence, two roots lie on the R.H.S. of the s-plane. Therefore, the system is unstable.

(b) $s^5 + s^4 + 2s^3 + 2s^2 + 3s + 5 = 0$

The Routh's array is

s^5	1	2	3
s^4	1	2	5
s^3	$0(d*)$	-2	×
s^2	$\dfrac{2d + 2}{d}$	5	×
s^1	$\dfrac{-4d - 4 - 5d^2}{2d + 2} \simeq -2(\text{as } d \to 0)$		
s^0	5		

* Since, in s^3 row, first column has encountered a zero (0), it is replaced by d

Sign change (between s^2 and s^1), Sign change.

The first term of the fifth (s^1) row has value of -2 as $d \to 0$. Thus, there are two sign changes in the first column making the system unstable.

Alternatively,

Put $\qquad s = \dfrac{1}{z}$ in the characteristic equation which gives

$$5z^5 + 3z^4 + 2z^3 + 2z^2 + z + 1 = 0$$

The Routh's array is

z^5	5	2	1
z^4	3	2	1
z^3	−1.33	−0.67	
z^2	0.5	1	
z^1	2		
z^0	1		

Sign change
Sign change

There are two sign changes in the first column indicating the system to be unstable.

Problem 11.5 The characteristic equation of a feedback control system is $s^3 + 3Ks^2 + (K + 2)s + 4 = 0$. Determine the range of K for which the system is stable.

Solution

The Routh's array is

s^3	1	$K + 2$
s^2	$3K$	4
s^1	$\dfrac{3K^2 + 6K - 4}{3K}$	
s^0	4	

For stability $\qquad 3K > 0$, i.e. $K > 0$

and $\qquad 3K^2 + 6K - 4 > 0$, i.e. $K > -1 \pm 1.53$

The range of K is thus $\infty > K > 0.53$

Problem 11.6 The characteristic equation of a feedback control system is:

$$s^4 + 20Ks^3 + 5s^2 + 10s + 15 = 0$$

Find the range of K for which system is stable.

Solution

The Routh's array is

s^4	1	5	15
s^3	$20K$	10	×
s^2	$5 - \dfrac{1}{2K}$	15	×
s^1	$10 - \dfrac{600K^2}{10K - 1}$	×	×
s^0	15	×	×

The system will be stable, if

$$20K > 0, \text{ i.e. } K > 0$$

$$5 - \frac{1}{2K} > 0 \quad \text{or} \quad 5 > \frac{1}{2K} \quad \text{or} \quad K > \frac{1}{10}$$

Also,

$$10 - \frac{600K^2}{10K - 1} > 0$$

which means

$$-600K^2 + 100K - 10 > 0$$

gives complex roots $\dfrac{1}{12} \pm j\dfrac{1}{10}$, hence, the feedback control system is unstable.

Problem 11.7 Find the conditions for stability for the systems whose characteristic equations are given below. The case where stability is suggested for real values of K, determine the values of K which will cause sustained oscillations. Find the frequency of oscillations.

(a) $\quad s^4 + 20s^3 + 224s^2 + 1240s + 2400 + K = 0$

(b) $\quad s^3 + (K + 0.5)s^2 + 4Ks + 50 = 0$

Solution

(a) $\qquad s^4 + 20s^3 + 224s^2 + 1240s + 2400 + K = 0$

The Routh's array is

s^4	1	224	$2400 + K$
s^3	20	1240	×
s^2	162	$2400 + K$	×
s^1	$1240 - \dfrac{(2400 + K)20}{162}$	×	×
s^0	$2400 + K$	×	×

For stability

$$2400 + K > 0 \quad \text{or} \quad K > -2400$$

Also,

$$1240 - \frac{(2400 + K)20}{162} > 0 \quad \text{or} \quad K < 7644$$

The range for K for stability is
$$-2400 < K < 7644$$

If $K = 7644$, then the first column is
$$(1, 20, 162, 0, 10044)$$

Thus, row s^1 becomes a zero row. The *subsidiary* equation for the row above this zero row, i.e. s^2 is
$$162s^2 + 10044 = 0$$

$\therefore$ $\qquad\qquad s = \pm j7.9$

The presence of an imaginary pair of roots gives oscillatory response with frequency of oscillations 7.9 rad/sec.

(b) $\qquad s^3 + (K + 0.5) s^2 + 4Ks + 50 = 0$

The Routh's array is

s^3	1	$4K$
s^2	$(K + 0.5)$	50
s^1	$4K - \dfrac{50}{K + 0.5}$	
s^0	50	

For stability
$$K + 0.5 > 0 \quad \text{or} \quad K > -0.5,$$

$$4K - \frac{50}{K + 0.5} > 0$$

or $\qquad 4K(K + 0.5) > 50$

or $\qquad K(K + 0.5) > 12.5$

or $\qquad K^2 + \dfrac{K}{2} - 12.5 > 0$

or $\qquad K > -3.8 \quad \text{and} \quad K > 3.3$

Therefore, the condition for stability is:
$$K > -3.8, \ K > -0.5 \ \text{and} \ K > 3.3$$

Thus, if $K > 3.3$, then all conditions are satisfied.

If we put $K = 3.3$ then s^1 row is zero row. The *subsidiary* equation for s^2 row for $K = 3.3$ is:
$$3.8s^2 + 50 = 0$$

$\therefore$ $\qquad\qquad s = \pm j\, 3.63$

Thus, frequency of oscillation is 3.63 rad/sec.

Problem 11.8 The open-loop transfer function of a unity feedback control system is given as:

$$G(s)H(s) = \frac{K}{s(1+Ts)}$$

It is desired that all the roots of the characteristics equation must lie in the region to the left of the line $s = -a$. Determine the values of K and T required so that there are no roots to right of the line $s = -a$.

Solution

The characteristic equation is:

$$1 + G(s)H(s) = 0$$

or
$$1 + \frac{K}{s(1+Ts)} = 0$$

or
$$s(1+Ts) + K = 0$$

or
$$Ts^2 + s + K = 0$$

or
$$s^2 + \frac{1}{T}s + \frac{K}{T} = 0$$

Replace s by $(s-a)$

Thus,
$$(s-a)^2 + \frac{1}{T}(s-a) + \frac{K}{T} = 0$$

or
$$s^2 + \left(-2a + \frac{1}{T}\right)s + \left(a^2 + \frac{K}{T} - \frac{a}{T}\right) = 0$$

The Routh's array is

s^2	1	$a^2 + \dfrac{K}{T} - \dfrac{a}{T}$
s^1	$\left(-2a + \dfrac{1}{T}\right)$	0
s^0	$a^2 + \dfrac{K}{T} - \dfrac{a}{T}$	

If no root is to lie on the line $s = -a$, then the first column of the Routh's array should have no sign change,

i.e.
$$\left(-2a + \frac{1}{T}\right) > 0 \quad \text{or} \quad \frac{1}{T} > 2a \quad \text{or} \quad aT < \frac{1}{2}$$

Also,
$$a^2 + \frac{K}{T} - \frac{a}{T} > 0$$

or
$$\frac{K}{T} > \left(\frac{a}{T} - a^2\right)$$

or $\qquad K > (a - a^2 T)$

or $\qquad K > a(1 - aT)$

∴ if $\qquad aT = 0$ then $K > a$

and if $\qquad aT = \dfrac{1}{2}$ then $K > \dfrac{a}{2}$

Problem 11.9 The open-loop transfer function of a unity feedback system is given by:

$$G(s) = \frac{K}{s(s+3)(s^2 + s + 1)}$$

Determine the values of K that will cause sustained oscillations in the closed-loop system. Also, find the oscillation frequency.

Solution

The characteristic equation is:

$$1 + G(s) = 1 + \frac{K}{s(s+3)(s^2 + s + 1)} = 0$$

or $\qquad s(s+3)(s^2 + s + 1) + K = 0$

or $\qquad s^4 + 4s^3 + 4s^2 + 3s + K = 0$

The Routh's array is

s^4	1	4	K
s^3	4	3	0
s^2	$\dfrac{13}{4}$	K	
s^1	$\dfrac{\left(\dfrac{39}{4} - 4K\right)}{\dfrac{13}{4}}$		
s^0	K		

The condition for system stability is:

$$K > 0 \quad \text{and} \quad \left(\frac{39}{4} - 4K\right) > 0$$

Therefore for stability, K should lie in the range $\quad \dfrac{39}{16} > K > 0$

When $K = -39/16$, there will be a zero at the first entry in the fourth row (s^1 row). This will indicate presence of symmetrical roots, which is shown below, will be pure imaginary.

$\therefore$ $\qquad$ $K = \dfrac{39}{16}$ will cause sustained oscillations.

The *subsidiary* equation of third row for $K = 39/16$, is:

$$\frac{13}{4}s^2 + \frac{39}{16} = 0$$

$\therefore$ $\qquad$ $s = \pm j0.75$

Therefore, the frequency of sustained oscillations is 0.75 rad/sec

---

Problem 11.10 Determine the values of K and b, so that the system whose open-loop transfer function is:

$$G(s) = \frac{K(s+1)}{s^3 + bs^2 + 3s + 1}$$

oscillates at a frequency of oscillations of 2 rad/sec. Assume unity feedback.

Solution

The characteristic equation is:

$$s^3 + bs^2 + 3s + 1 + K(s+1) = 0$$

or

$$s^3 + bs^2 + (K+3)s + (K+1) = 0$$

The Routh's array is

s^3	1	$K+3$
s^2	b	$K+1$
s^1	$(K+3) - \dfrac{(K+1)}{b}$	
s^0	$K+1$	

The system will have sustained oscillations if row s^1 is zero

i.e. $\qquad \dfrac{b(K+3) - (K+1)}{b} = 0$

i.e. $\qquad b = \dfrac{K+1}{K+3}$ $\hfill$ (1)

The *subsidiary* equation of row s^2 is:

$$bs^2 + (K+1) = 0$$

$\therefore$ $\qquad$ $s^2 = -4$ or $(j\omega)^2 = (j2)^2$ $\hfill$ (because $\omega = 2$ rad/sec)

putting $\quad b = \dfrac{K+1}{4}$, in equation (1), we get

$$\frac{K+1}{4} = \frac{K+1}{K+3}$$

or $\qquad K + 3 = 4$

or $\qquad K = 1$ **Ans.**

or $\qquad b = \dfrac{K+1}{4} = \dfrac{1+1}{4} = \dfrac{1}{2} = 0.5$ **Ans.**

Problem 11.11 The open-loop transfer function of a control system is:

$$G(s)H(s) = \frac{K(s+2)}{s(s+3)\,(s+5)}$$

If $K = 8$, check all the roots of the characteristic equation of the control system have damping factor greater than 0.5.

Solution

$1 + G(s)H(s) = 0$ is the characteristic equation.

Therefore,

$$1 + \frac{K(s+2)}{s(s+3)(s+5)} = 0$$

$$s(s+3)(s+5) + K(s+2) = 0$$

$$s^3 + 8s^2 + 15s + Ks + 2K = 0$$

$$s^3 + 8s^2 + (15+K)s + 2K = 0$$

$$(s+1)(s^2 + 7s + 16) = 0$$

The oscillations will be caused due to the quadratic factor $s^2 + 7s + 16$. The factor $(s+1)$ will cause the response e^{-t} to die out. Therefore, considering $(s^2 + 7s + 16)$ and comparing with $s^2 + 2\xi\omega_n s + \omega_n^2 = 0$, we get

$$\omega_n^2 = 16 \ \therefore \ \omega_n = 4$$

and $\qquad 2\xi\omega_n = 7$

$\therefore \qquad \xi = \dfrac{7}{2\omega_n} = \dfrac{7}{2 \times 4} = \dfrac{7}{8} = 0.874$

This is greater than 0.5.

Problem 11.12 Determine whether the largest time constant of the characteristic equation given below is less than, greater than or equal to 1.0 sec

$$s^3 + 5s^2 + 8s + 6 = 0$$

Solution

$$s^3 + 5s^2 + 8s + 6 = 0$$

Its roots are

$$(s+3)(s^2 + 2s + 2) = 0$$

The roots are:

$$s = -3 \quad \text{and} \quad s = -1 + j1$$

for $\qquad s = -3$

$$T = \frac{1}{3} = 0.33 \text{ sec} \qquad\qquad\qquad (\text{as } \xi\omega_n = 1)$$

for $\qquad s = -1 \pm j1$

See Fig. 11.3, where the roots have been plotted

$$\xi = \cos\beta = \frac{1}{\sqrt{1+1}} = \frac{1}{\sqrt{2}}$$

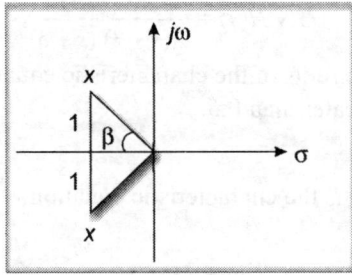

Fig. 11.3

Also,

$$\omega_n^2 = 2$$

$\therefore \qquad \omega_n = \sqrt{2}$

Now $\qquad \xi\omega_n = \dfrac{1}{\sqrt{2}} \times \sqrt{2} = 1$

$$T = \frac{1}{\xi\omega_n} = \frac{1}{1} = 1 \text{ sec}$$

$\therefore$ Largest time constant is equal to 1 sec.

Problem 11.13 A feedback control system has an open-loop transfer function of:

$$G(s)H(s) = \frac{Ke^{-s}}{s(s^2 + 2s + 1)}$$

Determine the maximum value of K for the closed-loop stability.

Solution

For low frequencies

$$e^{-s} = (1 - s)$$

$$G(s)H(s) = \frac{K(1-s)}{s(s^2 + 2s + 1)}$$

$$1 + G(s)H(s) = 1 + \frac{K(1-s)}{s(s^2 + 2s + 1)}$$

$$\therefore \qquad s(s^2 + 2s + 1) + K(1 - s) = 0$$

$$s^3 + 2s^2 + s + K - Ks = 0$$

$$s^3 + 2s^2 + s(1 - K) + K = 0$$

The Routh's array is

s^3	1	$1 - K$
s^2	2	K
s^1	$\dfrac{2(1 - K) - K}{2}$	
s^0	K	

For stability $K > 0$ and

$$2(1 - K) - K > 0$$

or $\qquad\qquad 2 - 3K > 0 \quad$ or $\quad K < \dfrac{2}{3}$

Hence, the restriction on K is $0 < K < \dfrac{2}{3}$ **Ans**

Problem 11.14 Decide the stability of the system, whose characteristic equation is given by:

$$s^5 + 2s^4 + 5s^3 + 10s^2 + 4s + 8 = 0$$

(Pune University)

Solution

The Routh's array is

s^5	1	5	4
s^4	2	10	8
	1	5	4
s^3	0	0	×
	2	5	×
s^2	$\dfrac{5}{2}$	4	×
s^1	$\dfrac{9}{5}$	×	×
s^0	4	×	×

Subsidiary equation of s^4 row

$$s^4 + 5s^2 + 4 = 0$$

Derivative of subsidiary equation

$$4s^3 + 10s$$

or $\quad 2s^3 + 5s = 0$

(form s^3 row with its coefficients).

s^3 row was a zero row and hence, it was formed by writing the *subsidiary* function of s^4 and then taking its derivative. There is no sign change in the first column of the new array so formed. However, there is a zero row.

The *subsidiary* equation formed of s^4 row is one of the factors of the characteristic equation. The other factor is $(s + 2)$

$$s^5 + 2s^4 + 5s^3 + 10s^2 + 4s + 8 = 0$$
$$\Rightarrow (s^4 + 5s^2 + 4)(s + 2)$$

Roots of
$$s^4 + 5s^2 + 4 \Rightarrow s = \pm j\,2.14, \pm j\,0.66$$
$$s + 2 \Rightarrow s = -2$$

Imaginary roots make system limitedly stable and the response, continuous oscillatory. Therefore, the system is limitedly stable.

Problem 11.15 Determine the range of K such that feedback control system having characteristic equation:

$$s(s^2 + s + 1)(s + 4) + K = 0, \text{ will be stable.}\qquad(Pune\ University)$$

Solution

The characteristic equation is:
$$s^4 + 5s^3 + 5s^2 + 4s + K = 0$$

The Routh's array is

s^4	1	5	K
s^3	5	4	
s^2	4.2	K	
s^1	$4 - 1.2\,K$		
s^0	K		

For stability
$$K > 0$$
and $$4 - 1.2K > 0, \quad \text{i.e.} \quad K < 3.33$$
Therefore, the range of K for stability is $0 < K < 3.33$

Problem 11.16 The open-loop transfer function of a unity feedback system is given by:

$$G(s) = \frac{K}{s(1 + 0.4s)(1 + 0.25s)}$$

Find the restriction on K so that the closed-loop system is absolutely stable.

(Pune University)

Solution

The characteristic equation is:

$$1 + G(s) = 1 + \frac{K}{s(1 + 0.4s)(1 + 0.25s)} = 0$$

or
$$s(1 + 0.4s)(1 + 0.25s) + K = 0$$

or
$$s^3 + 6.5s^2 + 10s + 10K = 0$$

The Routh's array is

s^3	1	10
s^2	6.5	10K
s^1	$\dfrac{65 - 10K}{6.5}$	
s^0	10K	

For absolute stability

$10K > 0$ or $K > 0$ and $65 - 10K > 0$ or $K < 6.5$

Therefore, the system will be absolutely stable, if the range of K is $0 < K < 6.5$

Ans

Problem 11.17 The closed-loop transfer function of a system is given by:

$$\frac{C(s)}{R(s)} = \frac{50}{s(1 + sT)(1 + 0.5s) + 50}$$

Find the value of T such that the system is driven onto the verge of instability and the resulting frequency of oscillation.

Solution

The characteristic equation is: $Ts^3 + (1 + 2T)s^2 + 2s + 100 > 0$

The Routh's array is

s^3	T	2
s^2	$1 + 2T$	100
s^1	$\dfrac{2(1 + 2T) - 100T}{1 + 2T}$	0
s^0	100	

The restriction on T is:
$$T > 0;$$

$$T > -\frac{1}{2};$$

and $\qquad 2 - 96T > 0 \quad$ or $\quad T < \dfrac{1}{48}$

The system will become unstable if $T = \dfrac{1}{48}$

If we substitute $T = 1/48$, we will get a zero row in the Routh's array and since the zero row occurs on the odd power of s, the s^1 row will be a zero row. The *subsidiary* equation of s^2 row is:

$$1.042s^2 + 100 = 0$$

which gives $\qquad s = \pm j9.8$

Therefore, the angular frequency of oscillation is 9.8 rad/sec.

Problem 11.18 The transfer function of a closed-loop system with unity feedback system is:

$$\frac{K(s + 2)\,(s + 1)}{(s + 0.1)\,(s - 1)}. \text{ Comment on stability.}$$

Solution

The characteristic equation is:

$$1 + G(s) = 1 + \frac{K(s + 2)(s + 1)}{(s + 0.1)(s - 1)} = 0$$

or $\qquad (K + 1)s^2 + (3K - 0.9)s + (2K - 0.1) = 0$

The Routh's array is

s^2	$K + 1$	$2K - 0.1$
s^1	$3K - 0.9$	
s^0	$2K - 0.1$	

Comments

1. For stability first column should be greater than unity,

i.e. (a) $K > -1$

(b) $K > \dfrac{0.9}{3}$, i.e. $K > 0.3$

(c) $K > \dfrac{0.1}{2}$, i.e. $K > 0.05$

There will be no sign change in the first column of Routh's array, if the condition $K > 0.3$ is satisfied and this condition satisfies all the remaining conditions.

Problem 11.19 Applying Routh's criterion, find range of K for stability for a system whose characteristic equation is given by: $s^3 + 3Ks^2 + (K+2)s + 4 = 0$.

Solution

The Routh's array is

s^3	1	$K+2$
s^2	$3K$	4
s^1	$\dfrac{3K(K+2)-4}{3K}$	
s^0	4	

Condition for stability

(a) $\qquad\qquad 3K > 0$ or $K > 0$

(b) $\qquad\qquad \dfrac{3K(K+2)-4}{3K} > 0$ or $K > 0.527$

Therefore, range of K is $\infty > K > 0.527$

Problem 11.20 A unity feedback control system has an open-loop transfer function $G(s) = \dfrac{K(s+13)}{s(s+3)(s+7)}$. Using Routh's stability criterion, find the range of K for the system to be stable. If $K = 1$, check if all the poles of the closed-loop transfer function have damping factor greater than 0.5. Assume unity feedback.

Solution

Characteristic equation:

$$1 + G(s) = 1 + \frac{K(s+13)}{s(s+3)(s+7)} = 0$$

or $s^3 + 10s^2 + (21+K)s + 13K = 0$

The Routh's array is

s^3	1	$21+K$
s^2	10	$13K$
s^1	$\dfrac{210-3K}{10}$	
s^0	$13K$	

Condition for stability

(a) $13K > 0$ or $K > 0$

(b) $\dfrac{210 - 3K}{10} > 0$ or $K < 70$

Therefore, range is $0 < K < 70$

When $K = 1$, the characteristic equation is:
$$s^3 + 10s^2 + 22s + 13 = 0$$
or $(s + 7.2)(s^2 + 2.8s + 1.84) = 0$

The roots are:
$$s = -7.2, -1.746, -1.0535$$

All roots are negative and real and hence, lie on the negative real axis.

Now, the damping factor $= |\cos \beta|$

Since, all roots are negative and real; $\beta = 180°$. Therefore, damping ratio $= 1$. Hence, for all roots damping ratio is greater than 0.5.

Problem 11.21   Test the stability of this control system characterized by the characteristic equation:
$$s^4 + 2s^3 + 3s^2 + 4s + 5 = 0 \qquad\qquad (AMIE)$$

Solution

The Routh's array is   *Without Error rate Control*

	s^4	1	3	5
	s^3	2	4	
Sign change	s^2	1	5	
Sign change	s^1	-6		
	s^0	5		

$G(s) = *$

$* = *$

System is unstable as there are two sign changes in the first column of Routh's array. It indicates two roots lie on the R.H.S. of the s-plane.

Problem 11.22   A unity feedback control system is described by the following characteristic equation:
$$s^4 + 4s^3 + 7s^2 + 16s + 12 = 0$$

Test its stability and find the roots on the imaginary axis. $AMIE)$

Solution

The Routh's array is

s^4	1	7	12
s^3	4	16	×
s^2	3	12	×
s^1	0	×	×
	2	×	×
s^0	12	×	×

Subsidiary equation of s^2 row
$$3s^2 + 12 = 0$$
or $\quad s^2 + 4 = 0$

Derivative of subsidiary equation
$$2s = 0$$
(form s^1 row with its coefficient).

The s^1 row becomes a zero row. s^1 row is, therefore, formed by the procedure explained earlier (Problem 11.2 (a)). First column of the Routh's array shows no sign change. But the presence of zero row indicates, presence of symmetrically located roots in the s-plane. Roots are found as given below:

$$s^4 + 4s^3 + 7s^2 + 16s + 12 = (s^2 + 4)(s^2 + 4s + 3)$$
$$s^2 + 4 \quad \text{gives} \quad \pm j2$$
$$s^2 + 4s + 3 \quad \text{gives} \quad -3, -1$$

Problem 11.23 Determine the range of values of K ($K > 0$) such that the characteristic equation $s^3 + 3(K + 1)s^2 + (7K + 5)s + (4K + 7) = 0$ has roots more negative than $s = -1$. *(AMIE)*

Solution

Substitute $s = z - 1$

The characteristic equation becomes

$$z^3 + 3Kz^2 + (K + 2)z + 4 = 0$$

z^3	1	$K + 2$
z^2	$3K$	4
z^1	$\dfrac{3K(K + 2) - 4}{3K}$	×
z^0	4	×

For stability

$$3K > 0 \Rightarrow K > 0,$$

and

$$\frac{3K(K + 2) - 4}{3K} > 0$$

or $\qquad 3K^2 + 6K - 4 > 0$

or $\qquad K = -2.53, \ 0.53$

Hence, the range of K is $\quad \infty > K > 0.53$ **Ans**

Problem 11.24 Comment on the stability of the following system using the Routh's criteria $s^4 + 6s^3 + 21s^2 + 36s + 20 = 0$ (*AMIE*)

Solution

s^4	1	21	20
s^3	6	36	×
	1	6	×
s^2	15	20	×
s^1	$\dfrac{14}{3}$	×	×
s^0	4	×	×

There is no sign change in the first column of the Routh's array and hence, the system under consideration is stable.

Problem 11.25 Using Routh's criterion, check the stability of the following control systems:

(a) $2s^4 + s^3 + 2s^2 + s + 20 = 0$

(b) $s^5 + s^4 + 2s^3 + s^2 + 4s + 2 = 0$

Comment on the location of roots.   (*AMIE*)

Solution

(a) $2s^4 + s^3 + 2s^2 + s + 20 = 0$

The Routh's array is

s^4	2	2	20
s^3	1	1	×
s^2	0	20	×
Sign change	K	20	×
Sign change s^1	$\dfrac{K-20}{K}$	×	×
s^0	20	×	×

There is a zero in s^2 row. It is replaced by $K \to 0$. First column of s^1 row has the term $1 - 20/K$ which is a negative term as $K \to 0$. It means, there are two sign changes which imply that two roots lie on R.H.S. of the s-plane. Therefore, the system is unstable.

(b) $s^5 + s^4 + 2s^3 + s^2 + 4s + 2 = 0$

The Routh's array is

s^5	1	2	4
s^4	1	1	2
s^3	1	2	×
s^2	–1	2	×
s^1	4	×	×
s^0	2	×	×

Sign change → (from s^3 to s^2)
Sign change → (from s^2 to s^1)

There are two sign changes in the first column of the Routh's array and hence, two roots of the characteristic equation lie on the R.H.S. of the s-plane. Therefore, the system is unstable.

Problem 11.26 The characteristic equation for a certain feedback control system is given by $s^4 + a_1 s^3 + a_2 s^2 + a_3 s + K = 0$. If the numerical values of $a_1 = 22$, $a_2 = 10$ and $a_3 = 2$, find K which corresponds to the stable system. *(AMIE)*

Solution

The characteristic equation is:
$$s^4 + 22s^3 + 10s^2 + 2s + K = 0$$

The Routh's array is

s^4	1	10	K
s^3	22	2	×
s^2	$\dfrac{218}{22}$	K	×
s^1	$2 - \dfrac{484K}{218}$	×	×
s^0	K	×	×

For stability

$$2 - \frac{484K}{218} > 0$$

or
$$\frac{484K}{218} < 2$$

or
$$K < \frac{2 \times 218}{484}$$

or
$$K < 0.9$$

and $\qquad\qquad K > 0$

∴ the range of K for stability is:

$$0 < K < 0.9$$

$K = 0.9$ will cause sustained oscillations in the closed-loop system.

Problem 11.27 Using Routh-Hurwitz's criterion, check whether systems represented by the following characteristic equations are stable or not. Comment on the location of roots. Determine the frequency of sustained oscillations, if any

(a) $\qquad s^3 + 20s^2 + 9s + 100 = 0$

(b) $\qquad s^4 + 2s^3 + 6s^2 + 8s + 8 = 0$ $\qquad\qquad\qquad\qquad$ *(AMIE)*

Solution

(a) $\qquad s^3 + 20s^2 + 9s + 100 = 0$

The Routh's array is

s^3	1	9
s^2	20	100
s^1	4	×
s^0	100	×

Since, there is no sign change in the first column of Routh's array, no roots of the characteristic equation are located on the R.H.S. of the s-plane. Therefore, the system is stable.

(b) $\qquad s^4 + 2s^3 + 6s^2 + 8s + 8 = 0$

The Routh's array is

s^4	1	6	8		*Subsidiary equation of s^2 row*
s^3	2·	8	×		$s^2 + 4 = 0$
	1	4	×		*Derivative of subsidiary equation*
s^2	2	8	×		$2s = 0$
	1	4	×		(form s^1 row with its coefficient).
s^1	0	×	×		
	2	×	×		
s^0	4	×	×		

s^1 row becomes a zero row. Problem was overcome by forming the auxiliary polynomial from the coefficients of s^2 row. Derivative of the polynomial was taken which was $2s = 0$ and s^1 row was replaced by the coefficient of s which is 2.

There is no sign change in the first column of the Routh's array.

Now,
$$s^4 + 2s^3 + 6s^2 + 8s + 8 = (s^2 + 4)(s^2 + 2s + 2)$$

Roots are:
$$s^2 + 4 = 0 \Rightarrow s = \pm j2$$
and
$$s^2 + 2s + 2 = 0 \Rightarrow s = -1 \pm j1$$

No roots lie on the R.H.S. of the s-plane.

Frequency of sustained oscillation is 2 rad/sec. Therefore, the system is limitedly stable.

Problem 11.28 The open-loop transfer function of a unity feedback system is given by:

$$\frac{K}{(s+2)(s+4)(s^2+6s+25)}$$

By applying the Routh's criterion, discuss the stability of the closed-loop system as a function of K. Determine the value of K which will cause sustained oscillations in the closed-loop system. What are the corresponding oscillation frequencies? *(AMIE)*

Solution

Characteristic equation is:
$$1 + G(s)H(s) = 0$$
or
$$1 + \frac{K}{(s+2)(s+4)(s^2+6s+25)} = 0$$
$$s^4 + 12s^3 + 69s^2 + 198s + (200 + K) = 0$$

The Routh's array is

s^4	1	69	$200 + K$
s^3	12	198	$\times$
s^2	52.5	$200 + K$	$\times$
s^1	$198 - \dfrac{12(200 + K)}{52.5}$	$\times$	$\times$
s^0	$200 + K$	$\times$	$\times$

System will be stable, if
$$200 + K > 0 \quad \text{or} \quad K > -200$$

and $198 - \dfrac{12(200 + K)}{52.5} > 0$ or $K < 666.25$

Oscillations will occur when $K = 666.25$

or $\qquad\qquad 52.5s^2 + (200 + K) = 0$

or $\qquad\qquad 52.5s^2 + (200 + 666.25) = 0$

or $\qquad\qquad\qquad s = \pm j\, 4.06$

Hence, the frequency of sustained oscillation is 4.06 rad/sec.

Problem 11.29   A unity feedback system has the following open-loop transfer function

$$G(s) = \dfrac{Ks(3s + 1)}{s^2 + 2s + 3}$$

Discuss the stability of the closed-loop system in terms of parameter K.

Solution

$$G(s) = \dfrac{Ks\,(1 + 3s)}{s^2 + 2s + 3}$$

$$1 + G(s)H(s) = 1 + \dfrac{Ks(1 + 3s)}{s^2 + 2s + 3} = 0$$

or $\qquad s^2 + 2s + 3 + Ks + 3Ks^2 = 0$

or $\qquad (1 + 3K)s^2 + (2 + K)s + 3 = 0$

The Routh's array is

s^2	$1 + 3K$	3
s^1	$2 + K$	×
s^0	3	×

For stability

$$1 + 3K > 0 \Rightarrow K > -\dfrac{1}{3}$$

and $\qquad\qquad\qquad 2 + K > 0 \Rightarrow K > -2$

for system to be stable K should be greater than $-\dfrac{1}{3}$.

Problem 11.30   The characteristic equation of the system is $s^4 + Ks^3 + 2s^2 + s + 3 = 0$. For what values of K is the equilibrium state of the system asymptotically stable.

Solution

The Routh's array is

s^4	1	2	3
s^3	K	1	×
s^2	$\dfrac{2K-1}{K}$	3	×
s^1	$1-\dfrac{3K^2}{2K-1}$	×	×
s^0	3	×	×

For the system to be asymptotically stable, all the elements of the first column of Routh's array must be positive. Therefore,

$$K>0 \; ; \quad \frac{2K-1}{K}>0 \; ; \quad 1-\frac{3K^2}{2K-1}>0$$

- The first two conditions, i.e. $K>0$ and $\dfrac{2K-1}{K}>0$ give the condition $K>\dfrac{1}{2}$
  and this condition satisfies the first two conditions.

- The third condition of $1-\dfrac{3K^2}{2K-1}>0$ gives the equation

$$K>\frac{-2\pm\sqrt{-8}}{-6}$$

- The equation yields two imaginary roots. It can be seen that the value of the polynomial will always remain negative for any real K.

- From the above, it implies that the three conditions for K cannot be altered simultaneously.

- Therefore, K cannot be assigned any value to make the system under consideration to be stable.

Problem 11.31 A control system is modelled in Fig. 11.4. If $K=2$, find out how many times the gain may be increased before on stability access.

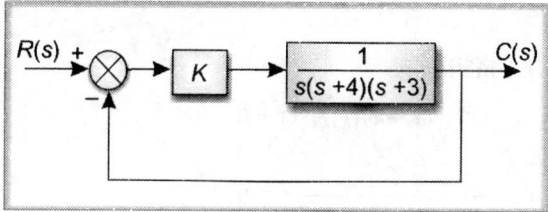

Fig. 11.4

Solution

$$\frac{C(s)}{R(s)} = \frac{G(s)}{1 + G(s)H(s)} = \frac{K}{s(s+4)(s+3) + K}$$

The characteristic equation is:

$$s(s+4)(s+3) + K = 0$$
$$s^3 + 7s^2 + 12s + K = 0$$

The Routh's array is

s^3	1	12
s^2	7	K
s^1	$\dfrac{84-K}{7}$	0
s^0	K	$\times$

For the system to be stable, the conditions are $84 - K > 0$ or $K < 84$; and which

gives the condition $0 < K < 84$. If $K = 2$; then the margin $= \dfrac{84}{2} = 42$ **Ans**

Problem 11.32   A position control system is shown in Fig. 11.5.

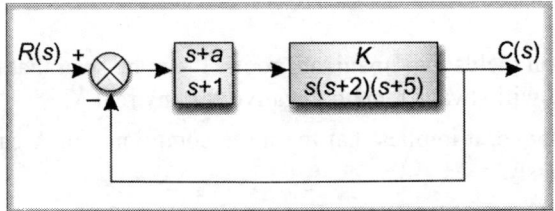

Fig. 11.5

K and a are the parameters of the system. Determine the range of K and a for which the system is stable.

Solution

The characteristic equation is:

$$1 + G(s)H(s) = 0$$

$$1 + \frac{K}{s(s+1)(s+2)(s+5)} = 0$$

or $s^4 + 8s^3 + 17s^2 + (K+10)s + Ka = 0$

The Routh's array is

s^4	1	17	Ka
s^3	8	$K+10$	×
s^2	$\dfrac{126-K}{8}$	Ka	×
s^1	$\dfrac{\dfrac{(126-K)}{8}(K+10)-8Ka}{\dfrac{126-K}{8}}$	×	×
s^0	Ka	×	×

The conditions of stability are:

- $126 < K$
- $(K+10)(126-K)-64Ka > 0$
- $Ka > 0$

Figure 10.6 depicts the shaded area satisfying the above three conditions where the system is stable.

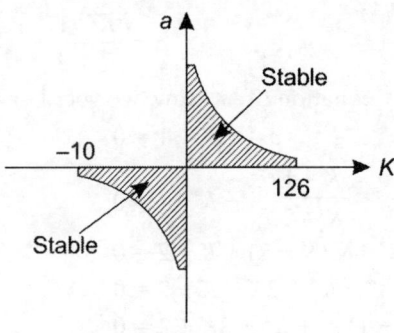

Fig. 11.6

Problem 11.33 The feedback system, shown below, oscillates at 2 rad/sec. Find the value of K and a.

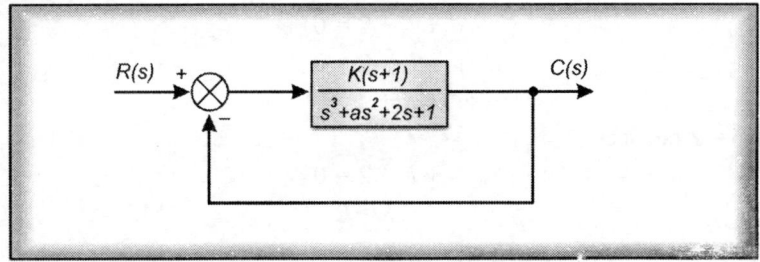

Fig. 11.7

Solution

The characteristic equation is:

$$1 + G(s)H(s) = 1 + \frac{K(s+1)}{s^3 + as^2 + 2s + 1} = 0$$

or $\quad s^3 + as^2 + 2s + 1 + K(s + 1) = 0$

or $\quad s^3 + as^2 + (2 + K) s + K + 1 = 0$

Forming the Routh's array, we get

s^3	1	$K+2$
s^2	a	$K+1$
s^1	$\dfrac{aK + 2a - K - 1}{a}$	×
s^0	$K+1$	

For marginally stable system 's^1' row element has to be zero.

$$\frac{aK + 2a - K - 1}{a} = 0$$

or $$a = \frac{K+1}{K+2}$$

Forming subsidiary equation of 's^2' row, we get

$$as^2 + K + 1 = 0$$

or $$\frac{K+1}{K+2} s^2 + K + 1 = 0$$

or $\quad (K + 1) s^2 + K (K + 2) + K + 2 = 0$

or $\quad (K + 1) s^2 + K^2 + 2K + K + 2 = 0$

or $\quad (K + 1) s^2 + K^2 + 3K + 2 = 0$

or $$s^2 + \frac{K^2 + 3K + 2}{K+1} = 0$$

or $$s^2 + \frac{(K+2)(K+1)}{K+1} = 0$$

or $\quad s^2 + K + 2 = 0$

or $\quad (j\omega)^2 + K + 2 = 0$

or $\quad -\omega^2 + K + 2 = 0$

Put $\omega = 2$ rad/sec

or $\quad -4 + K + 2 = 0$

or $\quad K = 2$

Therefore, $\quad a = \dfrac{K+1}{K+2} = \dfrac{2+1}{2+2} = \dfrac{3}{4} = 0.75$

Nyquist Stability Criterion

12.1 Introduction

Let us consider a feedback control system having closed-loop transfer function

$$\frac{C(s)}{R(s)} = \frac{G(s)}{1 + G(s)H(s)} = M(s) \tag{12.1}$$

where the characteristic equation is:

$$1 + G(s)H(s) = 0$$

and its poles give an indication of the stability of the control system. The Nyquist stability criterion tells us whether any roots of the characteristic equation lie on the RHS of the s-plane.

12.2 Relationship

Let the open-loop transfer function is:

$$G(s)H(s) = \frac{K(s + z_1)(s + z_2)\ldots\ldots(s + z_m)}{(s + p_1)(s + p_2)\ldots\ldots(s + p_n)} \tag{12.2}$$

where the numerator gives open-loop zeros; and the denominator gives open-loop poles.

Therefore, if $H(s) = 1$; then

$$M(s) = \frac{\dfrac{K(s + z_1)(s + z_2)\ldots\ldots(s + z_m)}{(s + p_1)(s + p_2)\ldots\ldots(s + p_n)}}{1 + \dfrac{K(s + z_1)(s + z_2)\ldots\ldots(s + z_m)}{(s + p_1)(s + p_2)\ldots\ldots(s + p_n)}} \tag{12.3}$$

$$= \frac{\dfrac{K(s + z_1)(s + z_2)\ldots\ldots(s + z_m)}{(s + p_1)(s + p_2)\ldots\ldots(s + p_n)}}{\dfrac{(s + p_1)(s + p_2)\ldots\ldots(s + p_n) + K(s + z_1)(s + z_n)\ldots\ldots(s + z_m)}{(s + p_1)(s + p_2)\ldots\ldots(s + p_n)}} \tag{12.4}$$

$$M(s) = \frac{G(s)}{1 + G(s)H(s)} = \frac{K(s + z_1)(s + z_2)\ldots\ldots(s + z_m)}{(s + p_1)(s + p_2)\ldots\ldots(s + p_n) + K(s + z_1)(s + z_2)} \tag{12.5}$$

From Eqn. (12.5) the denominator is:

$$1 + G(s)H(s) = \frac{(s+p_1)(s+p_2)\ldots.(s+p_n)+K(s+z_1)(s+z_2)\ldots.(s+z_m)}{(s+p_1)(s+p_2)\ldots.(s+p_3)} \qquad (12.6)$$

Comparing Eqns (12.2) and (12.6), we gather the following relationship

> poles of open-loop transfer function $[G(s)\,H(s)]$
> $\quad$ = poles of characteristic equation $[1 + G(s)H(s)]$ $\qquad$ (12.7a)

Equation (12.4) is

$M(s)$ = Transfer function of closed-loop system

$$= \frac{K(s+z_1)(s+z_2)\ldots(s+z_m)}{(s+p_1)(s+p_2)\ldots(s+p_n)+K(s+z_1)(s+z_2)\ldots(s+z_{m)}} \qquad (12.7b)$$

$$= \frac{K(s+z_1)(s+z_2)\ldots(s+z_m)}{(s+z_1')(s+z_2')\ldots(s+z_m')} \qquad (12.8)$$

$$= \frac{G(s)}{1+G(s)H(s)} \qquad (12.9)$$

From Eqns (11.6) and (11.9), we get the relationship

> poles of closed-loop transfer function
> $\quad$ = zeros of the characteristic equation $\qquad$ (12.10)

For the closed-loop system to be stable, Nyquist criterion indicates stability by finding out how many poles of the closed-loop transfer function or zeros or the characteristic equation are located on the LHS of s-plane. There is no need to factorise the characteristic equation.

12.3 Encirclement

If an area or a point is found inside a closed contour/path in a complex plane, then that area or point is said to be encircled. In the Fig. 12.1, a closed path τ_s along with two points s_1 and s_2 are shown. Hence, it can be stated that the closed path τ_s is encircling the origin and point s_1 clockwise.

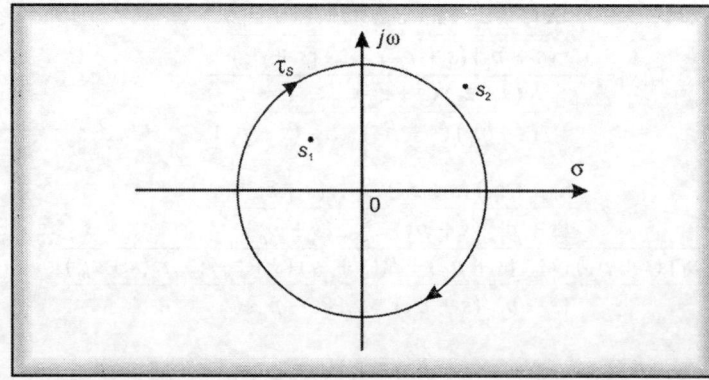

Fig. 12.1 s-plane

It can also be noted that the number of encirclements of origin and point s_1 is once, i.e. one time in CW direction. In short, it can be written as:

$$N = 1$$

[CW encirclement has been taken as positive]

However, point s_2 has not been encircled by closed path τ_s.

Applying, the same logic to Figs 12.2 and 12.3, it can be stated that t_{s1} encircles the origin two times in CCW direction and t_{s2} does not encircle the origin at all. Therefore in first case $N = -2$ and in the second case $N = 0$.

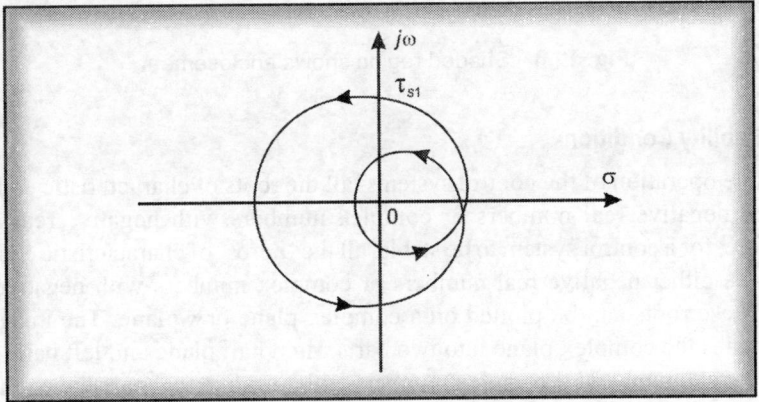

Fig. 12.2 Encirclement of origin by τ_{s_1}

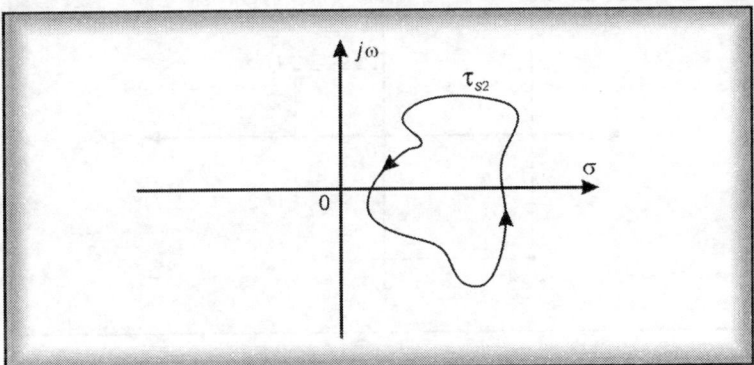

Fig. 12.3 Encirclement of origin by τ_{s_2}

12.4 Enclosed

If an area is encircled by a contour or path in CCW direction, then that area is said to be enclosed in Fig. 12.4.

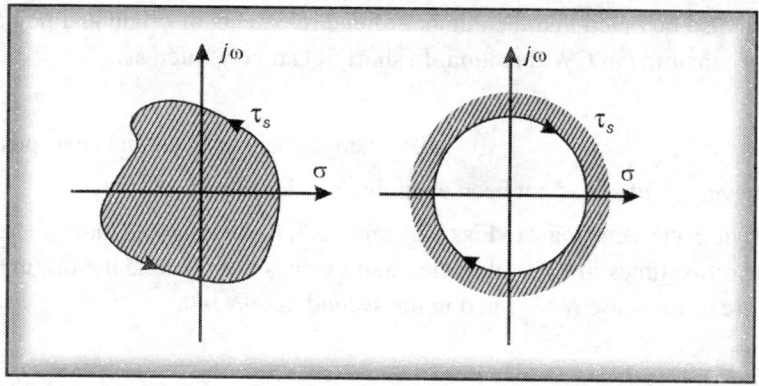

Fig. 12.4   Shaded region shows enclosement

12.5 Stability Condition

For stable operation of the control systems, all the roots of characteristic equation must be negative real numbers or complex numbers with negative real parts. Therefore, for a control system to be stable, all the 'Zeros' of characteristic equation should be either negative real numbers or complex numbers with negative real parts. These roots can be plotted on a complex plane or s-plane. The imaginary axis divides the complex plane into two parts: right half plane and left half plane. Negative real numbers or complex numbers with negative real parts lie on the left of s-plane as shown in Fig. 12.5.

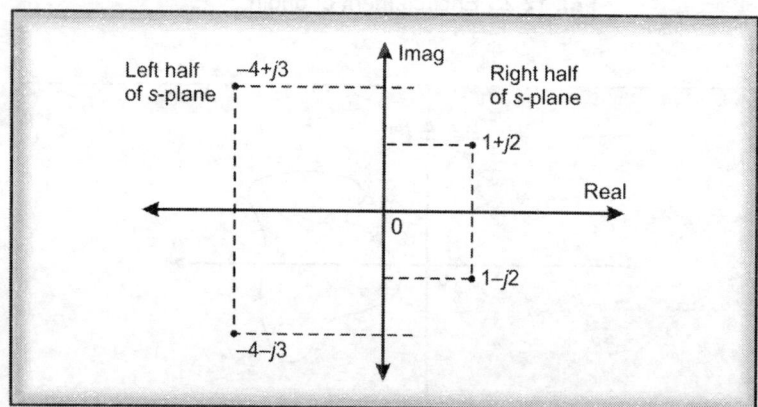

Fig. 12.5   Complex plane

Therefore, the roots which are positive real numbers or complex numbers with positive real parts lie on the right half of s-plane. The condition for stability can be stated as **"for a system to be stable all the zeros of characteristic equation should lie on the left half of s-plane."** Therefore, to ascertain control system stability, we have to search for 'Zeros' on the right half of s-plane, which leads the control system to instability. However, it may not be practical to look for every point on s-plane to which half of s-plane is located. A shortcut method is available which makes use of polar plot to find out presence of zeros and is called Nyquist stability criterion.

12.6 Principle of Argument

To find out stability on the polar plot, it is first necessary to correlate the region of instability on the s-plane along with the region of instability on the polar plot, or $1 + GH$ plane. The $1 + GH$ plane is generally the name given to the plane where $1 + G(s)H(s)$ is plotted in complex coordinates with s replaced by $j\omega$. Likewise, the plot of $G(s)H(s)$ with s replaced by $j\omega$ is often termed GH plane. This terminology will be often used in the remaining part of this chapter.

The Nyquist criterion is based on the Cauchy's principle of argument of complex variable theory. It states that if **$F(s)$ is single valued function that has finite number of poles and zeros in the s-plane and τ_s is contour in the s-plane which does not pass through the poles and zeros of the $F(s)$, the corresponding closed path τ_F in the F-plane will encircle the origin as many times as the difference between zeros of $F(s)$ and poles of $F(s)$ located in the area enclosed by the contour τ_s.** Consider $[F(s) = 1 + G(s)H(s)]$ be a single valued rational function which is **analytic** everywhere in a specified region except at a finite number of points in s-plane. (A function $F(s)$ is said to be analytic if the function and all its derivatives exist). The points where the function and its derivatives do not exist are called **singular points**.

Let τ_s be a closed path chosen in s-plane as shown in Fig. 12.6 such that the function $F(s)$ is analytic at all points on it. For each point on τ_s represented on s-plane, there is a corresponding mapping point in $F(s)$ plane. Thus, when mapping is done on $F(s)$ plane, the contour τ_s mapped by the function $F(s)$ plane is also a closed path as shown in Fig. 12.7. The direction of traverse of τ_F in $F(s)$ plane may be CW or CCW, depending upon the function $F(s)$.

The Cauchy's principle of argument states that: **The mapping made on $F(s)$ plane will encircle its origin as many number of times as the difference between the number of zeros and poles of $F(s)$ enclosed by the s-plane locus τ_s in the s-plane.**

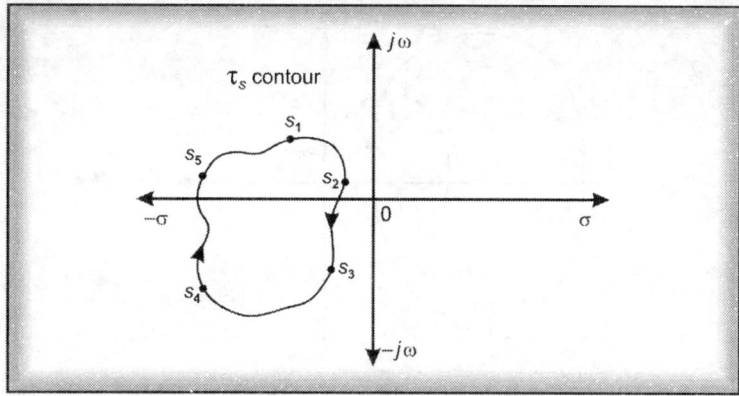

Fig. 12.6 τ_s contour in s-plane

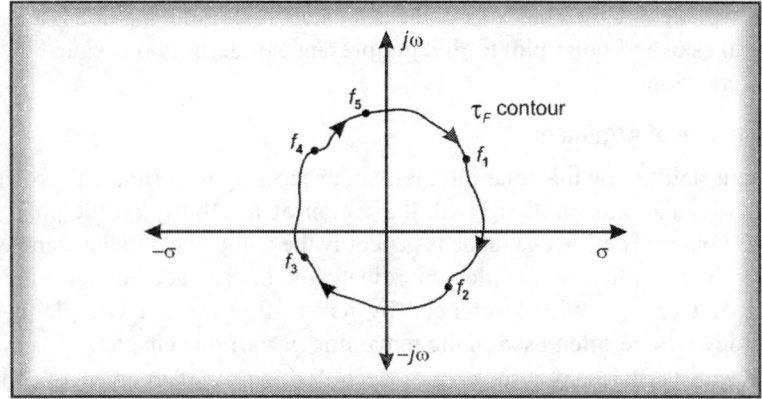

Fig. 12.7 $F(s) = 1 + G(s)\,H(s)$ plane

Thus, $N = Z - P$

where N: number of encirclements made by $F(s)$ plane plot τ_F of its origin.

$Z =$ Number of zeros of $F(s)$ enclosed by τ_s in s-plane.

$P =$ Number of poles of $F(s)$ enclosed by τ_s in s-plane.

Let us consider a function $F(s)$

$$\frac{K(s+1)(s+2)}{s(s+3)(s+4)(s+6+j2)}$$

Therefore,

$$\text{Zeros} = -1, -2$$

$$\text{Poles} = 0, -3, -4, -6 \pm j2$$

The zeros ($\bullet$) and poles ($\times$) are plotted in Fig. 12.8, also shows two closed paths τ_{s_1} and τ_{s_2}.

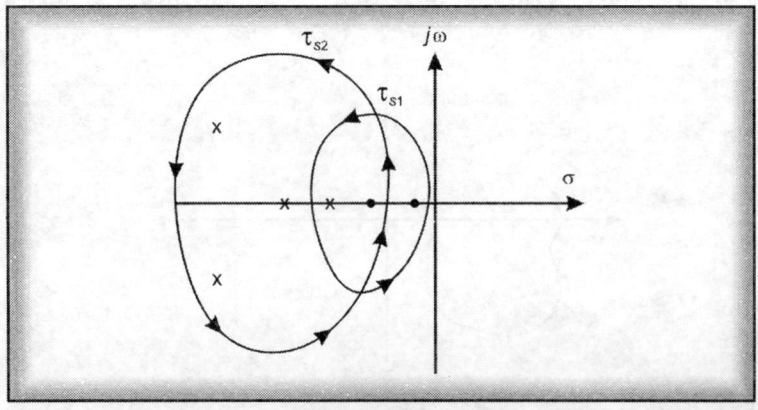

Fig. 12.8

The closed path τ_{s_1} encircles two zeros and one pole. Therefore, $Z = 2$ and $P = 1$. Substituting in the equation

$$N = Z - P$$
$$= 2 - 1 = 1$$

If we map this into $F(s)$ plane, the closed path τ_{F_1} will encircle the origin once in CW direction as shown in Fig. 12.9.

The closed path τ_{s_2} encircles four poles and one zero. Therefore, $Z = 1$ and $P = 4$. Substitution gives

$$N = Z - P = 1 - 4 = -3$$

(three encirclements of origin in CCW direction as shown in Fig. 12.10)

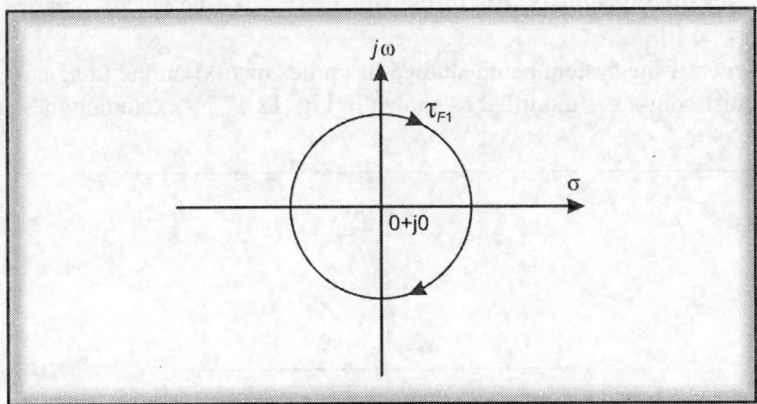

Fig. 12.9

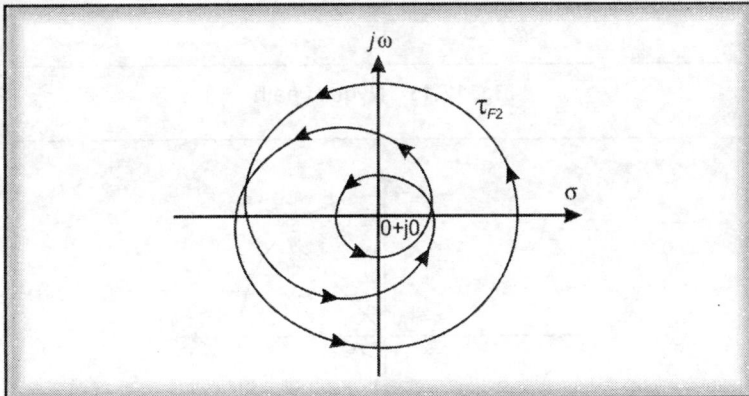

Fig. 12.10

Note: The mapping on $F(s)$ plane will encircle its origin as many number of times as the difference between the number of zeros and poles of $F(s)$ enclosed by the s-plane locus.

12.7 Nyquist Contour and Nyquist Plot

The principle of argument can be used to ascertain the stability of control systems. We have seen that if the zeros of characteristic function lie on the right half of s-plane, it will lead to system instability. In order to encircle the entire right half of s-plane, we select a closed path as shown in Fig. 12.11 such that all the zeros lying on the right half of s-plane will lie inside this path. This path in s-plane is known as **Nyquist path.** Nyquist path is generally taken in CW direction. The Nyquist path consists of the following sub-paths.

(a) Positive imaginary axis of the s-plane ($s = 0 + j\omega$, where ω varies from 0 to ∞^+)

(b) Semi circle of infinite radius enclosing the right hand side of s-plane

(c) Negative imaginary axis of the s-plane ($s = 0 - j\omega$ where ω varies from ∞^- to 0).

However, if the system being studied has poles of $F(s)$ on the imaginary axis, the Nyquist contour is modified as shown in Fig. 12.12 by excluding these poles from the path.

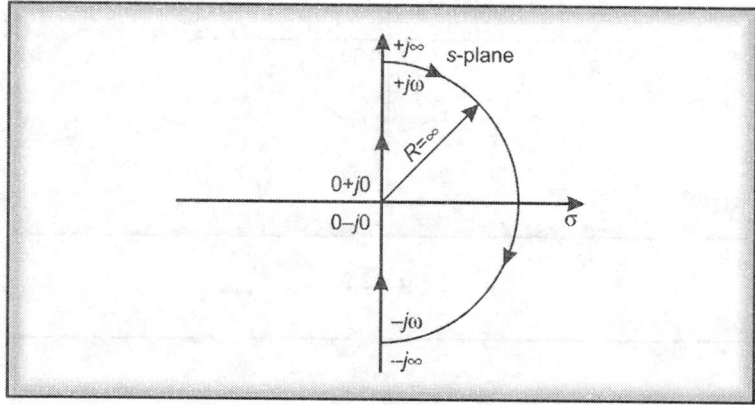

Fig. 12.11 Nyquist path

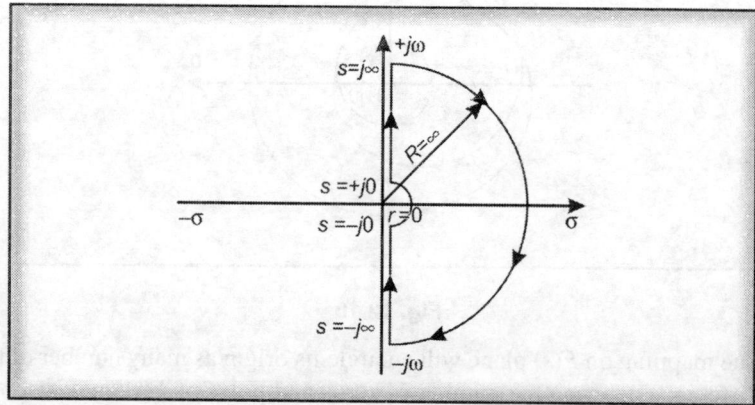

Fig. 12.12 Nyquist path

Corresponding to Nyquist contour, a plot can be mapped on $F(s) = 1 + G(s)H(s)$ plane as shown in Fig. 12.13 and the number of encirclements made by this $F(s)$ plot about its origin are counted.

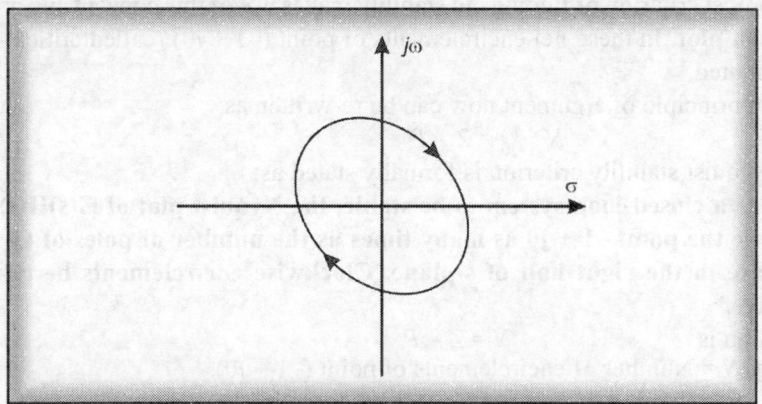

Fig. 12.13 $1 + G(s)H(s)$ plane

From the principle of argument
$$N = Z - P,$$
where, N = number of encirclements made by $F(s)$ plane plot of the point $(0 + j0)$,

Z and P are the zeros and poles lying on right half of s-plane

For the system to be stable: $Z = 0$ which mean $N = -P$.

Apart from this, the Nyquist contour can also be mapped on $G(s)H(s)$ plane (open-loop transfer function of a closed-loop system is shown in Fig. 12.14).

Now consider,

$F(s) = 1 + G(s)H(s)$ for which the origin $(0 + j0)$ is shown in Fig. 12.13

Therefore, $\qquad G(s)H(s) = F(s) - 1$
$$= (0 + j0) - 1 = (-1 + j0)$$

Note: The origin $(0 + j0)$ gets shifted to $(-1 + j0)$ in $G(s)H(s)$ plane.

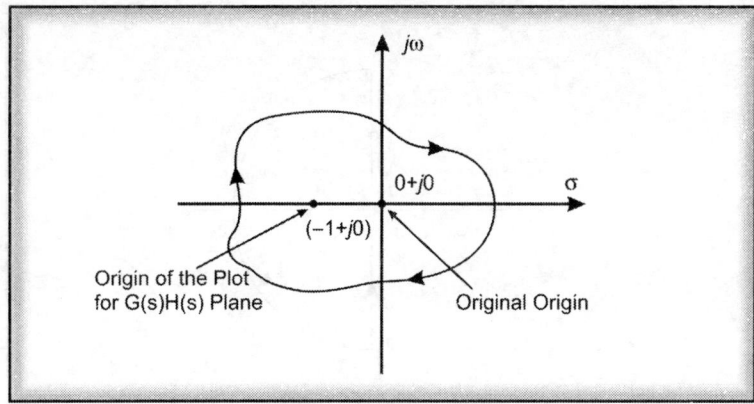

Fig. 12.14 $G(s)H(s)$ plane

Thus, a contour on $1 + G(s) H(s)$ plane can be easily mapped to a path on $G(s)H(s)$ plane or open loop transfer function plane. This path will be identical to that of $1 + G(s)H(s)$ path except that the origin gets shifted to the $(-1 + j0)$ from $(0 + j0)$.

Nyquist criterion of finding out stability makes use of this concept and is called **Nyquist plot.** In these net encirclements of point $(-1 + j0)$ (called critical point) are counted.

The principle of argument now can be re-written as:
$$N_{-1+j0} = Z - P$$
The Nyquist stability criterion is formally stated as:

"For a closed-loop system to be stable, the Nyquist plot of G(s)H(s) must encircle the point $-1 + j0$ as many times as the number of poles of G(s)H(s) that are in the right half of s-plane. Clockwise encirclements be taken as positive."

Equation is $\qquad\qquad N = Z - P$
where, N = Number of encirclements of point $(-1 + j0)$.

P = Number of poles of $G(s)H(s)$ that are on the right half of s-plane.

For stability of the closed-loop, Z should be zero, i.e. $N = -P$.

$$\boxed{\textbf{SOLVED PROBLEMS}}$$

Problem 12.1 Draw the Nyquist plot for

$$G(s)H(s) = \frac{K}{s(s+a)} \quad K > 0; \ a > 0$$

Solution

Poles = $0, -a$. The poles are marked on Fig. 12.15.

Since, there are no poles on the imaginary axis, the Nyquist path is shown in Fig. 11.15. The Nyquist path on $G(s)H(s)$ plane bypasses the pole $s = 0$ by the path defined by *'efa'* having radius $r \to 0$.

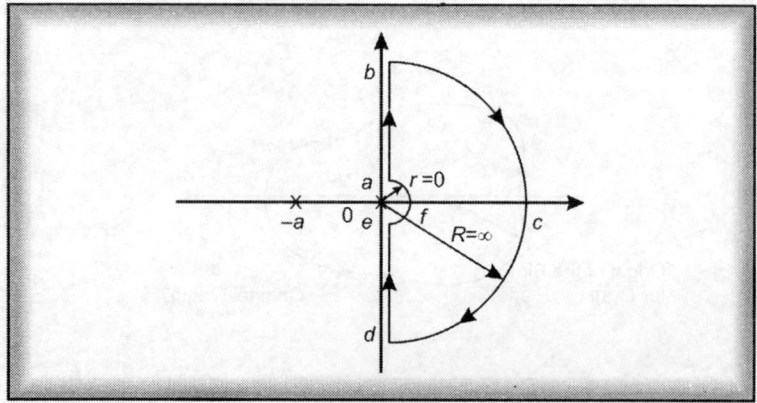

Fig. 12.15 Nyquist path of $G(s)H(s)$

$$M = \left| \frac{K}{s(s+a)} \right| = \left| \frac{K/a}{s\left(1+\dfrac{1}{a}s\right)} \right| = \frac{K/a}{\omega\sqrt{1+\left(\dfrac{\omega}{a}\right)^2}}$$

$$\phi = -90° - \tan^{-1}\frac{\omega}{a}$$

Nyquist plot of path 'ab'

Here $s = j\omega$ varies from $j0^+$ to $j\infty^+$. This corresponds to sketching of polar plot.

Table 12.1

Sr. No.	ω	M	ϕ
1	0	∞	$-90°$
2	∞	0	$-180°$

The Nyquist plot for Nyquist path 'ab' is shown in Fig. 12.16a.

Nyquist plot for Nyquist path 'bcd'

In this $s = j\omega$ varies from $+j\infty$ to $-j\infty$ and ϕ varies from $+90°$ to $-90°$ in direction. Putting $s = Re^{j\phi}$ (equation of circle in exponential form), we get

$$G(s)H(s) = \frac{K/a}{s\left(1+\dfrac{s}{a}\right)} = \frac{K/a}{Re^{j\phi}\left(1+\dfrac{Re^{j\phi}}{a}\right)}$$

Applying the limit $R \to \infty$, we get

$$\lim_{R \to \infty} G(j\omega)H(j\omega) = \lim_{R \to \infty} \frac{K/a}{Re^{j\phi}\left(1+\dfrac{Re^{j\phi}}{a}\right)}$$

$$= \lim_{R \to \infty} \frac{K}{R^2\,e^{j2\phi}}$$

$$= \lim_{R \to \infty} \frac{K}{R^2}\,e^{-j2\phi} \qquad (\phi \text{ varies from } +90° \text{ to } -90°)$$

$$= 0\ \angle{-2\phi}$$

$$= 0\ \angle{-[2\,(+90°) \text{ to } 2(-90°)]}$$

$$= 0\ \angle{-180° \text{ to } +180°}$$

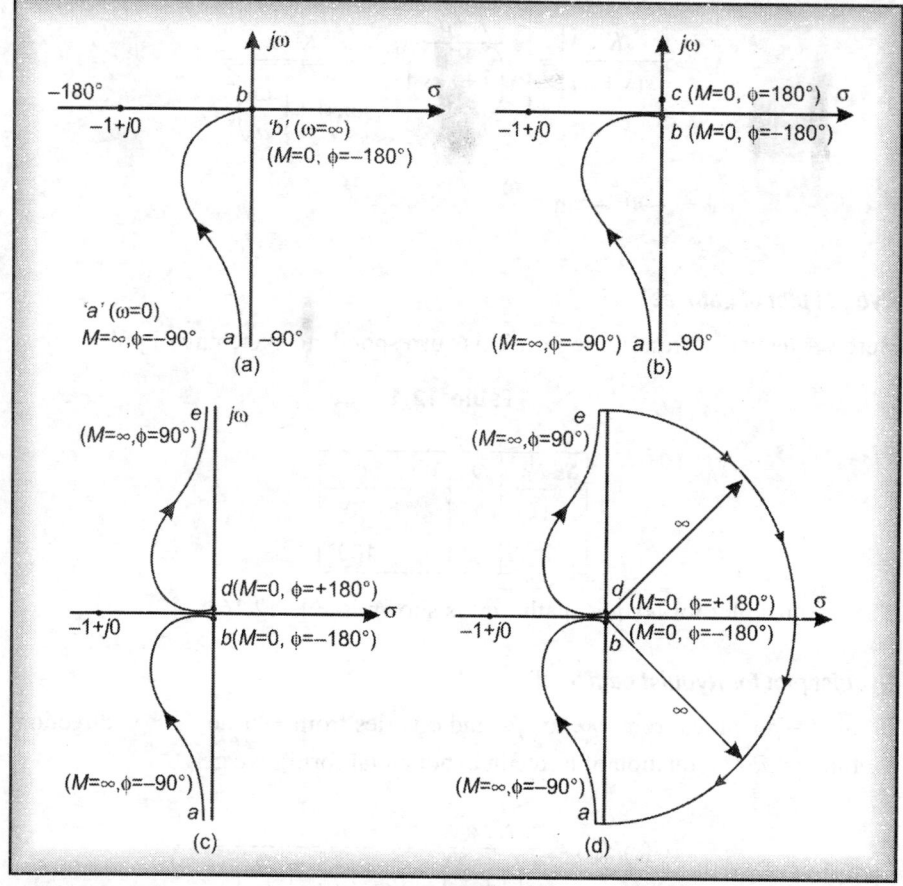

Fig. 12.16a to d

It implies that magnitude remains zero but the point 'b' which was at $-180°$ shifts to point 'd' at $+180°$ as shown in Fig. 12.16b.

Nyquist plot for Nyquist path 'de'

Here $s = j\omega$ where s varies from $j\infty^-$ to $j0^-$. This corresponds to mirror image of path 'ab' and is shown in Fig. 12.16c.

Nyquist plot for Nyquist path 'efa'

Here $s = j\omega$ varies from $j0^-$ to $j0^+$ and θ varies from $-90°$ to $+90°$ in CCW direction.

Putting $\qquad\qquad\qquad s = re^{j\theta}$ $\qquad$ (*equation of circle in exponential form*)

$$G(s)H(s) = \frac{K/a}{s\left(1+\dfrac{s}{a}\right)} = \frac{K/a}{re^{j\theta}\left(1+\dfrac{re^{j\theta}}{a}\right)}$$

Applying the limit $s \to 0$

$$\lim_{s \to 0} G(j\omega)H(j\omega) = \lim_{r \to 0} \frac{K/a}{re^{j\theta}\left(1 + \dfrac{re^{j\theta}}{a}\right)} = \lim_{r \to 0} \frac{K}{re^{j\theta}}$$

$$= \infty \angle e^{-j\theta} \qquad (\theta \text{ varies from } -90° \text{ to } +90°)$$

$$= \infty \angle -[-90° \text{ to } 90°]$$

$$= \infty \, [90° \text{ to } -90°]$$

$$= \infty \qquad [\text{point 'e' to point 'a' in CW direction}]$$

The Nyquist plot is shown in Fig. 12.16(d).

Stability analysis

$$N = Z - P$$

$$P = 0 \qquad (\text{as no pole is lying on the RHS of } s\text{-plane})$$

$$N = 0 \qquad (\text{as Nyquist plot has nil encirclement of critical point } -1 + j0)$$

Therefore,

$$0 = Z - 0$$

or $$Z = 0$$

In view of the above

(a) Closed-loop system is stable as $z = 0$

(b) Also, as $P = 0$, open-loop stability also exists.

Problem 12.2 $G(s) = \dfrac{1 + 4s}{s^2(s+1)(2s+1)}$.

Solution
The expression for magnitude is:

$$M = \frac{\sqrt{1 + 16\omega^2}}{\omega^2 \sqrt{1 + \omega^2}\,\sqrt{1 + 4\omega^2}}$$

The expression for 'ϕ' is:

$$\phi = \tan^{-1} 4\omega - 180° - \tan^{-1} \omega - \tan^{-1} 2\omega$$

Angle ϕ for different values of ω is tabulated in Table 12.2a.

Table 12.2a

ω	$\tan^{-1}4\omega$	$-180°$	$-\tan^{-1}\omega$	$-\tan^{-1}2\omega$	ϕ
0	0	$-180°$	0	0	$-180°$
∞	90°	$-180°$	$-90°$	$-90°$	$-270°$
0.1	21.8°	$-180°$	$-5.7°$	$-11.3°$	$-175°$
1	75.9°	$-180°$	$-45°$	$-63.4°$	$-212°$

$$\text{Poles} = 0, 0, \frac{-1}{2}, -1$$

$$\text{Zeros} = \frac{-1}{4}$$

Nyquist contour is shown on the s-plane in Fig. 12.17.

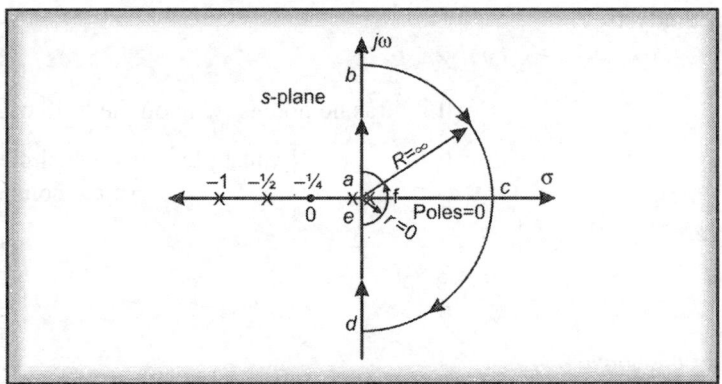

Fig. 12.17

The poles ($\times$) and zeros ($\cdot$) are also marked on the s-plane.

(a) Nyquist contour bypasses the two poles at origin with a small circle of radius r with ($r \to 0$). It is depicted by path 'efa'.

(b) Thereafter, it follows the path of positive imaginary axis from $\omega = j0^+$ to $j\infty^+$ depicted by path 'ab'.

(c) Path 'bcd' is a semicircle of radius R with ($R \to \infty$) and encloses all poles lying on the right hand side of s-plane, which in this case is zero, i.e. $P = 0$.

(d) Path 'de' is the path depicted on the negative imaginary axis where ω varies from $j\infty^-$ to $j0^-$.

Mapping of the Nyquist plot is done step by step.

Path 'ab' (corresponds to polar plot)

Put $s = j\omega$ for ω varying from 0^+ to ∞^+

$$G(j\omega) = \frac{4j\omega + 1}{(j\omega)^2 (1 + j\omega)(1 + 2j\omega)}$$

$$M\big|_{\omega=0} = |G(j\omega)|_{\omega=0} = \lim_{\omega \to 0} G(j\omega) = \lim_{\omega \to 0} \frac{1}{(j\omega)^2} = \infty \angle{-180°}$$

$$M\big|_{\omega=\infty} = \big|G(j\omega)\big|_{\omega=\infty} = \lim_{\omega\to\infty} G(j\omega) = \lim_{\omega\to\infty} \frac{j\omega\left(4+\dfrac{1}{j\omega}\right)}{(j\omega)^4\left(1+\dfrac{1}{j\omega}\right)\left(2+\dfrac{1}{j\omega}\right)}$$

$$= \lim_{\omega\to\infty} \frac{4j\omega}{2(j\omega)^4} = \lim_{\omega\to\infty} \frac{2}{(j\omega)^3} = 0 \angle -270°$$

$$\phi = \tan^{-1} 4\omega - 180° - \tan^{-1}\omega - \tan^{-1} 2\omega$$

$$\phi\big|_{\omega=0} = 0 - 180° - 0 - 0 = -180°$$

$$\phi\big|_{\omega=\infty} = 90° - 180° - 90° - 90° = -270°$$

Therefore,

$$G(j\omega)\big|_{\omega=0} = \infty \angle -180°$$

$$G(j\omega)\big|_{\omega=\infty} = 0 \angle -270°$$

To check, if there is any intersection of the polar plot on negative real axis, we separate $G(j\omega)$ into real and imaginary parts.

$$G(j\omega) = \frac{(1+ j4\omega)(1- j\omega)(1-2j\omega)}{-\omega^2(1+ j\omega)(1- j\omega)(1+2j\omega)(1-2j\omega)}$$

$$= \frac{(1+ j4\omega)[1-2j\omega - j\omega - 2\omega^2]}{-\omega^2[1+\omega^2][1+4\omega^2]}$$

$$= \frac{(1+ j4\omega)(1-3j\omega - 2\omega^2)}{-\omega^2(1+4\omega^2+\omega^2+4\omega^4)}$$

$$= \frac{1-3j\omega - 2\omega^2 + j4\omega + 12\omega^2 - 8j\omega^3}{\omega^2(1+4\omega^4+5\omega^2)}$$

$$= -\frac{1+10\omega^2 + j\omega - 8j\omega^3}{\omega^2(1+4\omega^4-4\omega^2+9\omega^2)}$$

$$= -\frac{1+10\omega^2 + j\omega - 8j\omega^3}{\omega^2[(1-2\omega^2)^2+9\omega^2]}$$

$$= \boxed{\frac{-(1+10\omega^2)}{\omega^2[(1-2\omega^2)^2+9\omega^2]} + j\frac{\omega - 8\omega^3}{-\omega^2[(1-2\omega^2)^2+9\omega^2]}}$$

$$= \frac{-(1+10\omega^2)}{\omega^2[(1-2\omega^2)^2+9\omega^2]} + j\frac{4\omega - 3\omega - 8\omega^3}{-\omega^2[(1-2\omega^2)^2+9\omega^2]}$$

$$= \frac{-(1+10\omega^2)}{\omega^2[(1-2\omega^2)^2+9\omega^2]} + j\frac{4\omega-8\omega^3-3\omega}{-\omega^2[(1-2\omega^2)^2+9\omega^2]}$$

$$= -\frac{1+10\omega^2}{\omega^2[(1-2\omega^2)^2+9\omega^2]} + j\frac{4\omega(1-2\omega^2)-3\omega}{-\omega^2[(1-2\omega^2)^2+9\omega^2]}$$

Equating imaginary part to zero, we get

$$4\omega(1-2\omega^2) - 3\omega = 0$$

or $$4(1-2\omega^2) - 3 = 0$$

or $$4 - 3 = 8\ \omega^2$$

or $$\omega = \sqrt{\frac{1}{8}} = \frac{1}{2\sqrt{2}} = \frac{1}{2.83} = 0.35 \Rightarrow \omega_\phi$$

Putting the value of $\omega = 0.35$ in the read part gives

$$-\frac{1+10\omega^2}{\omega^2[(1-2\omega^2)^2+9\omega^2]} = \frac{1+10\times0.35^2}{0.35^2[(1-2\times0.35^2)^2+9\times0.35^2]} = -10.85$$

Therefore, the polar plot will make an intercept of 10.85 on negative real axis as shown in Fig. 12.18a.

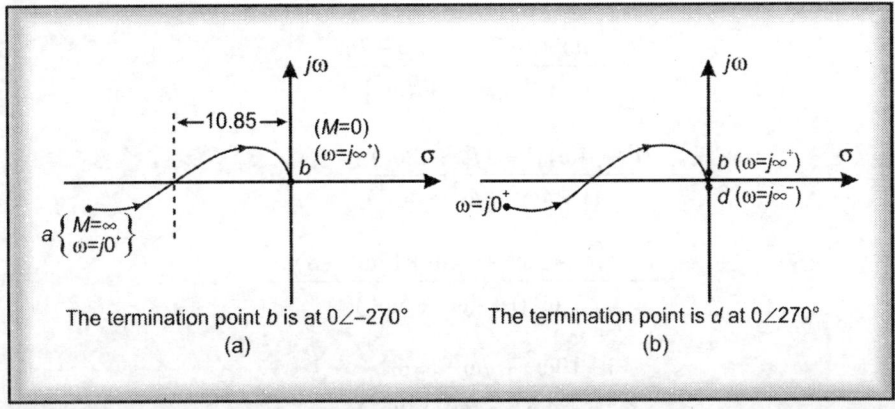

The termination point b is at $0\angle-270°$ The termination point is d at $0\angle270°$

(a)   (b)

Fig. 12.18a and b

Path 'bcd' (refer to Fig. 12.17)

Put $s = \lim\limits_{R\to\infty} Re^{j\theta}$ where, θ varies from $\dfrac{\pi}{2}$ to $\dfrac{-\pi}{2}$

Therefore,

$$G(j\omega) = \lim_{R\to\infty} \frac{4Re^{j\theta}+1}{(Re^{j\theta})^2\,(Re^{j\theta}+1)(2Re^{j\theta}+1)}$$

As R approaches infinity, $G(j\omega)$ can be written as:

$$= \lim_{R \to \infty} \frac{4Re^{j\theta}}{R^2 e^{j2\theta} \; Re^{j\theta} \; 2Re^{j\theta}}$$

$$= \lim_{R \to \infty} \frac{2Re^{j\theta}}{R^4 e^{j4\theta}}$$

$$= \lim_{R \to \infty} \left(\frac{2}{R^3}\right)\left(\frac{e^{-j3\theta}}{1}\right)$$

Application of limit gives

$$= 0 \angle - 3\theta \text{ where, } \theta \text{ varies from } \frac{\pi}{2} \text{ to } \frac{-\pi}{2}$$

$$= 0 \angle - 3\left(\frac{\pi}{2}\right) \text{ to } -3\left(\frac{-\pi}{2}\right)$$

$$= 0 \angle -270° \text{ to } 270°$$

The start point of path '*bcd*' is $0 \angle -270°$ which is also the termination point of path '*ab*' and also the termination point is point '*d*', i.e. $0\angle270°$. The path '*bcd*' is superimposed on part '*ab*' (Fig. 12.18a) and mapped as shown in Fig. 12.18b.

Path '*de*' (*refer* to Fig. 12.17)

The path '*de*' has to be mapped for $\omega = j\infty^-$ to $j0^-$. It is the reverse or mirror image of path '*ab*'. Since, it is a mirror image, it is shown by dashed line in Fig. 12.18c.

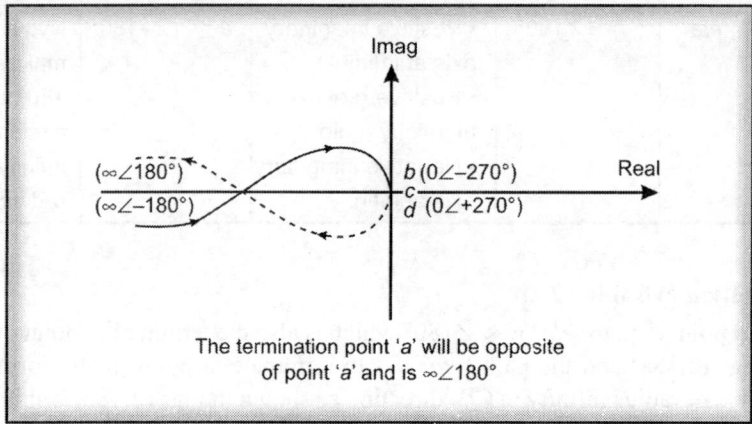

The termination point '*a*' will be opposite
of point '*a*' and is $\infty\angle180°$

Fig. 12.18c

Path '*efa*' (*refer* to Fig. 12.17)

Put $s = \lim_{r \to 0} re^{j\phi}$ where, ϕ varies from $\frac{-\pi}{2}$ to $\frac{\pi}{2}$

Therefore,

$$G(j\omega) = \lim_{r \to 0} \frac{4re^{j\phi} + 1}{(re^{j\phi})^2 \, (re^{j\phi} + 1) \, (2re^{j\phi} + 1)}$$

$$= \lim_{r \to 0} \frac{1}{r^2 e^{j2\phi}}$$

Application of limit gives

$$G(j\omega) = \infty \; \angle -2 \; \phi \text{ where, } \phi \text{ varies from } \frac{-\pi}{2} \text{ to } \frac{\pi}{2}$$

$$= \infty \angle - \left(2 \times \frac{-\pi}{2}\right) \text{ to } -\left(2 \times \frac{\pi}{2}\right) = \infty \angle 180° \text{ to } \infty \angle -180°$$

$$= \infty \angle \pi \text{ to } \infty \angle -\pi \; [\infty \angle 180° \text{ to } \infty \angle -180°]$$

Table 12.2b

S. No.	Path	Starting point	Intersection	Termination point	Remarks
1.	ab	$a = \infty \angle -180°$	$-10.85 + j0$ (negative real axis)	$b = 0 \; \angle -270°$	Polar plot
2.	bcd	$b = 0 \; \angle -270°$	—	$d = 0 \; \angle 270°$	
3.	de	$d = 0 \; \angle 270°$	$-10.85 + j0$ (negative real axis)	$e = \infty \angle 180°$	Mirror image of polar plot
4.	efa	$e = \infty \angle 180°$	• Positive imaginary axis at infinity • Positive real axis at infinity (point f) • Negative imaginary axis at infinity	$a = \infty \angle -180°$	Nyquist plot makes two 180° semi-circles of infinity radius

Explanation of Table 12.2b

The start point of path 'efa' is $\infty \angle 180°$ which is also the termination point of path 'de' (Fig. 12.18a) and the path 'efa' has to terminate at point 'a' by forming a semicircle of radius infinity in CW direction as shown in Fig. 12.18d. It should be noted that the termination point is 'a' which is at $\infty \angle -\pi$ and hence two semicircles (180° to 0 and 0 to –180°) traverse is required to reach termination point 'a'. Two

semicircles traverse also correspond to two integrates $\left(\dfrac{1}{s^2}\right)$ given in the expression

for $G(s)$ and also relate to two traverse of 180° in CW direction. The entire mapping of $G(s)$ is shown in Fig. 12.18d.

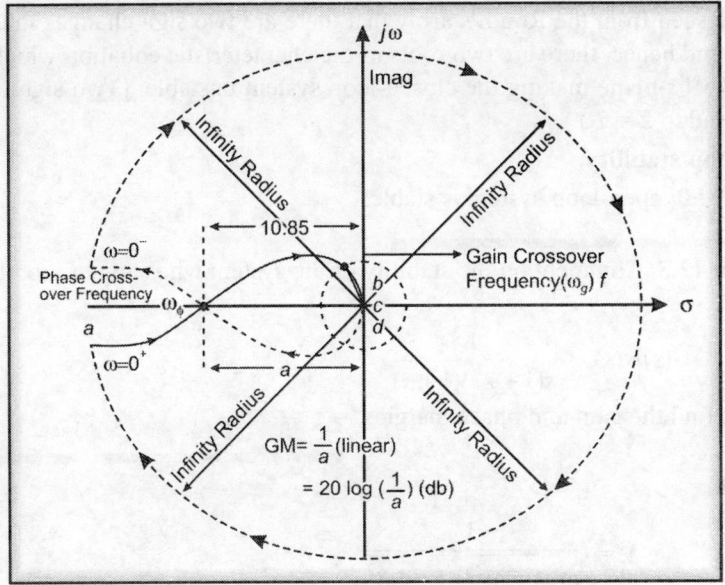

Fig. 12.18d

Encirclements of critical point $(-1 + j0)$

It can be seen from the Nyquist plot (Fig. 12.18d) that critical point $(-1 + j0)$ is encircled twice in clockwise direction. Therefore, $N = 2$. Since, $P = 0$ as $G(s)$ has no poles on the RHS of s-plane, therefore

$$N = Z - P$$
$$2 = Z - 0$$
$$Z = 2$$

It implies that closed-loop system is unstable as two roots of characteristic equation lie on the RHS of s-plane.

Verify

The characteristic equation is:

$$1 + G(s)H(s) = 1 + \left(\frac{4s+1}{s^2(1+s)(2s+1)} \right) \times 1 = 0 = \frac{s^2(1+s)(2s+1) + 4s + 1}{s^2(1+s)(2s+1)} = 0$$

or $s^2(1 + s)(2s + 1) + 4s + 1 = 0$

or $2s^4 + 3s^3 + s^2 + 4s + 1 = 0$

Routh's array is

$$
\begin{array}{c|ccc}
s^4 & 2 & 1 & 1 \\
s^3 & 3 & 4 & \times \\
s^2 & \dfrac{-5}{3} & 1 & \times \\
s^1 & \dfrac{29}{5} & \times & \times \\
s^0 & 1 & & \\
\end{array}
$$

It can be seen from the Routh's array that there are two sign changes in the first column and hence, there are two roots of the characteristic equation which lie on the RHS of s-plane making the closed-loop system unstable. (Two sign changes correspond to $Z = 2$.)

Open-loop stability

Since, $P = 0$, open-loop system is stable.

Problem 12.3 Comment on the stability of the system whose open-loop transfer function

$$G(s)H(s) = \frac{1}{s(1+2s)(1+s)}$$

Also, find the gain and phase margins.

Solution

$$M = \frac{1}{\omega\sqrt{1+\omega^2}\ \sqrt{1+4\omega^2}}$$

$$\phi = -90° - \tan^{-1}\omega - \tan^{-1}2\omega$$

Table 12.3

S. No.	ω rad/sec	M	ϕ	Remarks
1.	0	∞	−90°	It shows that the plot crosses
2.	0.5	1.03	−161.56	the negative real axis between
3.	1	0.32	−198.43	$\omega = 0.5$ and $\omega = 1$
4.	∞	0	−270°	

Putting $s = j\omega$ and rationalising, we get

$$G(j\omega)H(j\omega) = \frac{1}{j\omega(1+2j\omega)(1+j\omega)}$$

$$= \underbrace{\frac{-3\omega}{\omega[(1-2\omega^2)^2+9\omega^2]}}_{\text{Real part}} - j\underbrace{\frac{(1-2\omega^2)}{\omega[(1-2\omega^2)^2+9\omega^2)]}}_{\text{Imaginary part}}$$

Equating imaginary part to zero, we get the value of 'ω' at which, plot crosses −ve real axis.

$$(1-2\omega^2) = 0$$

$$\therefore \qquad \omega = \frac{1}{\sqrt{2}} = 0.707$$

Alternatively

$$\phi = -180°$$

$$-90° - \tan^{-1}\omega - \tan^{-1}2\omega = -180°$$

Table 12.4

Sr. No.	Path (Nyquist Contour)	M∠θ	Intersection	Starting Point	Terminating Point	Remarks
1.	ab	*ω varies from j0⁺ to j∞⁺* $M\big|_{\omega=0} = \lim_{\omega\to 0} \dfrac{1}{j\omega(1+2j\omega)(1+j\omega)} = \infty$ $\phi = -90° - 0° - 0° = -90°$ $M\big|_{\omega=0} = \lim_{\omega\to 0} \dfrac{1}{j\omega^3\left(1+\dfrac{2}{j\omega}\right)\left(1+\dfrac{1}{j\omega}\right)} = 0$ $\phi = -90° - 90° - 90° = -270°$	*With negative real axis* $\phi = -180°$ $-90° - \tan^{-1}\infty - \tan^{-1}2\omega = -180°$ $\tan[\tan^{-1}\omega + \tan^{-1}2\omega] = 90°$ $\dfrac{\omega+2\omega}{1-2\omega^2} = \infty$ or $1 - 2\omega^2 = 0$ $\omega = 0.707$ $OX = M\big|_{\omega=0.707} = \dfrac{1}{0.707\sqrt{1+\dfrac{1}{2}}\sqrt{1+2}}$ $= 0.66$	$a = \infty \angle{-90°}$	$b = 0\angle{-270°}$	Polar plot (Fig. 12.20) shows by the path 'ab'
2.	bcd	$s = \lim_{R\to\infty} Re^{j\theta}$ $\left(\theta \text{ varies from } \dfrac{\pi}{2} \text{ to } \dfrac{-\pi}{2}\right)$ $G(j\omega) = \lim_{R\to\infty} \dfrac{1}{Re^{j\theta}(1+2Re^{j\theta})(1+Re^{j\theta})}$ $= \lim_{R\to\infty} \dfrac{1}{Re^{j\theta} \times 2Re^{j\theta} \times Re^{j\theta}}$	—	$b = 0\angle{-270°}$	$d = 0\angle{270°}$	Fig. 12.20 shown by points 'b' and 'd'

(contd.)

Table 12.4 (contd.)

Sr. No.	Path (Nyquist Contour)	$M\angle\theta$	Intersection	Starting Point	Terminating Point	Remarks
		$= \lim\limits_{R\to\infty} \dfrac{1}{2R^3}\, e^{-j3\theta}$ $= 0\angle{-3\theta}$ $= 0\angle\left(\dfrac{-3\pi}{2}\right) \text{ to } \left(\dfrac{3\pi}{2}\right)$ $= 0\angle{-270°} \text{ to } 270°$				
3.	de	ω varies from $j\infty^-$ to $j0^-$ It is the reverse or mirror image of path 'ab'	Intersection on negative real axis will be at $\omega = -0.707$	$d = 0\angle270°$	$e = \infty\angle90°$	Fig. 11.20
4.	efa	$s = \lim\limits_{r\to 0} re^{j\phi} \left(\begin{array}{c}\phi \text{ varies from}\\ -\dfrac{\pi}{2} \text{ to } +\dfrac{\pi}{2}\end{array}\right)$ $G(j\omega) = \lim\limits_{r\to 0} \dfrac{1}{re^{j\phi}(1+2re^{j\phi})(1+re^{j\phi})}$ $= \lim\limits_{r\to 0} \dfrac{1}{re^{j\phi}}$ $= \infty\angle{-\phi}$ $= \angle{-\left(-\dfrac{\pi}{2} \text{ to } \dfrac{\pi}{2}\right)}$ $= \infty\angle90° \text{ to } -90°$	—	$e = \infty\angle90°$	$a = \infty\angle{-90°}$	Fig. 11.20 shows path 'e' for having radius $r\to\infty$

$$\tan^{-1}\omega + \tan^{-1} 2\omega = 90°$$
$$\tan[\tan^{-1}\omega + \tan^{-1} 2\omega] = \tan 90°$$

$$\frac{\omega + 2\omega}{1 - 2\omega^2} = \infty$$

$$\therefore \qquad 1 - 2\omega^2 = 0$$

$$\omega = \frac{1}{\sqrt{2}} = 0.707$$

Putting the value of 'ω' so obtained, we get magnitude

$$M\Big|_{\omega = 0.707} = \frac{1}{0.707\sqrt{1 + \dfrac{1}{2}}\sqrt{1 + 2}} = 0.66 = OX$$

Nyquist contour is shown in Fig. 11.19

$$\text{Poles} = 0, \frac{-1}{2}, -1$$

$$\text{Zeros} = \text{Nil}$$

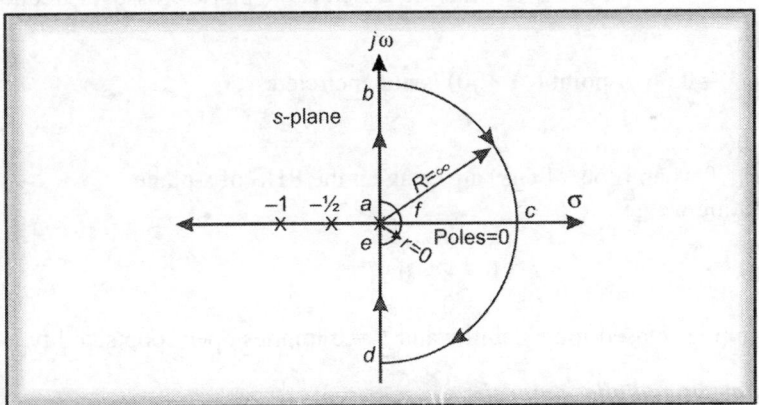

Fig. 12.19

(a) It by-passes the pole at origin with a small semicircle of radius 'r' ($r \to 0$). It is shown by path 'efa'.

(b) Thereafter, it follows the path of positive imaginary axis from $j0^+$ to $j\infty^+$ depicted by path 'ab'.

(c) Path 'bcd' is a semicircle of radius $R(R \to \infty)$ and encloses all poles lying on the RHS of s-plane, which in this case is zero, i.e. $P = 0$.

(d) Path 'de' is the path depicted on the negative imaginary axis where, ω varies from $j\infty^-$ to $j0^-$.

Mapping of the Nyquist plot is done step by step and is illustrated in Table 12.4. The Nyquist plot is shown in Fig. 12.20.

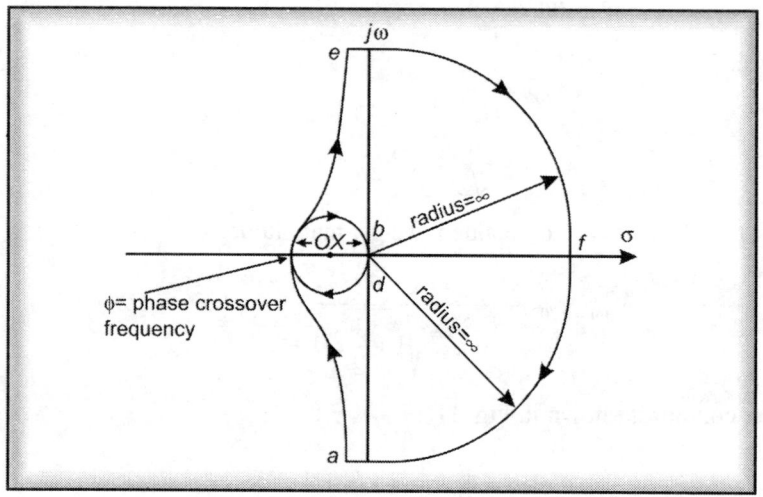

Fig. 12.20

Now = OX = 0.66 at ω = 0.707 rad/sec which is the phase crossover frequency.

Stability

Since OX = 0.66 $\therefore$ point $(-1 + j0)$ is not encircled.

$\therefore N = 0$

Also, $P = 0$ as no roots of $G(s)$ are lying on the RHS of s-plane.

Substituting we get,

$$N = Z - P$$
$$0 = Z - 0$$
$$\therefore \qquad Z = 0$$

$Z = 0$, implies closed-loop stability and $P = 0$ implies open-loop stability.

Gain margin and phase margin

$$\text{Gain margin} = 20 \log \frac{1}{OX} = 20 \log \frac{1}{0.66} = 3.61 \text{ db}$$

Frequency at which magnitude will become unity is given by:

$$\omega \sqrt{1 + \omega^2} \sqrt{1 + 4\omega^2} = 1$$
$$\omega^2 (1 + \omega^2)(1 + 4\omega^2) = 1$$
$$x(1 + x)(1 + 4x) = 1$$

By hit and trial $\qquad x = 0.33$

$$\therefore \qquad \omega = \sqrt{0.33} = 0.574 \text{ rad/sec}$$

$$\text{Phase margin} = -90° - \tan^{-1} \omega - \tan^{-1} 2\omega + 180°$$
$$= -90° - \tan^{-1} 0.57 - \tan^{-1} 2 \times 0.57 + 180° = 11.57°$$

Problem 12.4 The open-loop transfer function of a unity feedback system is:

$$G(s)H(s) = \frac{(s+2)}{(s+1)(s-1)}$$

Comment on the stability.

Solution

$$G(s)H(s) = \frac{(s+2)}{(s+1)(s-1)} = \frac{2(1+0.5\,s)}{(1+s)(-1+s)}$$

$$M = \frac{2\sqrt{1+(0.5\,\omega)^2}}{\sqrt{1+\omega^2}\,\sqrt{1+\omega^2}}$$

$$\phi = \tan^{-1}0.5\omega - \tan^{-1}\omega - \tan^{-1}\frac{\omega}{-1}$$

$$G(j\omega)H(j\omega) = \frac{2(1+0.5j\omega)}{(1+j\omega)(-1+j\omega)} = -\left[\frac{2}{1+\omega^2} + j\frac{\omega}{1+\omega^2}\right]$$

The Nyquist contour is shown in Fig. 12.21

$$\text{Poles} = 1, -1$$
$$\text{Zeros} = -2$$

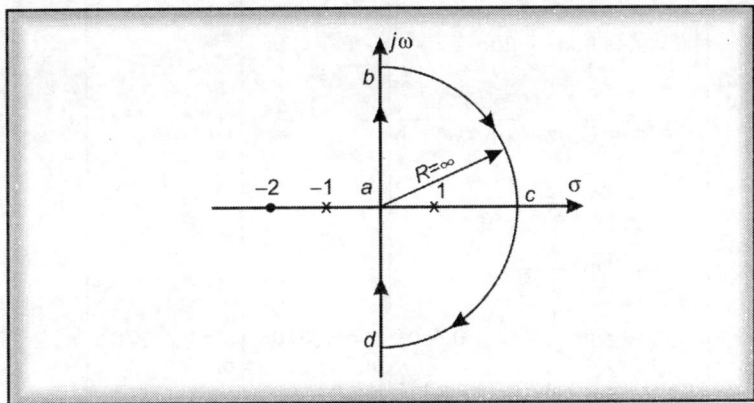

Fig. 12.21

Path 'ab': It follows the path of positive imaginary axis from $j0$ to $j\infty^+$ depicted by path ab.

Path 'bcd': Path '*bcd*' is a semicircle of radius $R(R \to \infty)$ and encloses all poles lying on the RHS of s-plane. In this case $s = 1$ is the pole which lies on the RHS and hence $P = 1$.

Path 'da': Path traced on the negative imaginary axis as 'ω' varies from $j\infty^-$ to $j0$. Table below shows, value of M and ϕ for varying 'ω'.

Table 12.5

Sr. No.	ω (rad/sec)	M	φ (degrees)	Remarks
1.	0	–2	–180°	• The magnitude is reducing to zero as 'ω' increases
2.	1	1.11	206.57°	• The angle remains in third qua-
3.	2	0.56	225°	drant and finally becomes 270°
4.	1000	0.00000316	269.94°	• Hence, the plot is traced in
5.	∞	0	270°	third quadrant

The Nyquist plot is traced as per the data given in the Table 12.6.

Table 12.6

Sr. No.	Path	$M\angle\theta$	Starting Point	Termination Point	Remarks
1.	ab	It is shown in Table 11.5 for 'ω' varying from $j0$ to $j\infty^+$	On negative real axis at point $M = -2$	Origin with magnitude equal to zero and angle equal to 270° (CW)	It is polar plot, the polar plot remains in third quadrant
2.	bcd	$s = \lim_{R\to\infty} Re^{j\theta}$ $\left(\theta \text{ varies from } \dfrac{\pi}{2} \text{ to } \dfrac{-\pi}{2}\right)$ $G(j\omega) = \lim_{R\to\infty} \dfrac{(Re^{j\theta}+2)}{(Re^{j\theta}+1)(Re^{j\theta}-1)}$ $= \lim_{R\to\infty} \dfrac{Re^{j\theta}}{Re^{j\theta}\,Re^{j\theta}}$ $= \lim_{R\to\infty} \dfrac{1}{Re^{j\theta}}$ $= \lim_{R\to\infty} \dfrac{1}{R}e^{-j\theta} = 0\angle-\theta$ $= 0\angle\left(\dfrac{-\pi}{2} \text{ to } -\left(\dfrac{-\pi}{2}\right)\right)$ $= 0\angle\left(\dfrac{-\pi}{2} \text{ to } \dfrac{\pi}{2}\right)$ or $= 0\angle\left(\dfrac{3\pi}{2} \text{ to } \dfrac{-3\pi}{2}\right)$ $= 0\angle(270° \text{ to } -270°)$	$b = 0\angle270°$ or $b = 0\angle-90°$	$d = 0\angle-270°$ or $d = 0\angle+90°$	
3.	da	'ω' varies from $j\infty^-$ to $j0$. It is reverse or mirror image of path ab.	$d = 0\angle-270°$ or $d = 0\angle+90°$	$a = -2\angle+180°$	

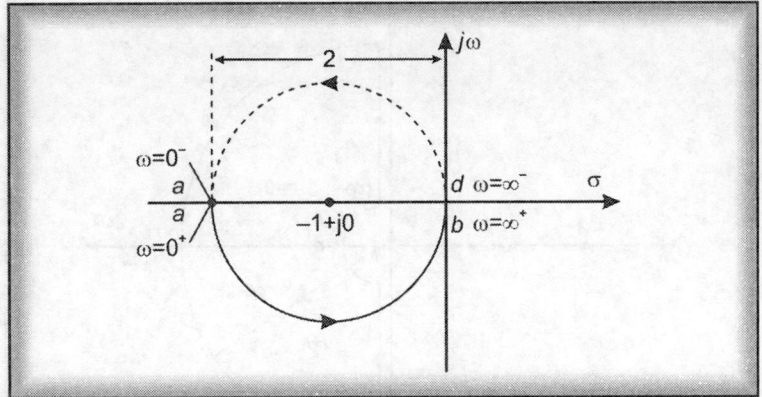

Fig. 12.22

Stability

$$N = Z - P$$

- Here, $P = 1$ as one pole of $G(s)$ lies on the RHS of s-plane.
- $N = -1$ as the Nyquist path encircles point $-1 + j0$ once in CCW direction.

Putting the values, we get

$$-1 = Z - 1$$

or $$Z = -1 + 1 = 0$$

It means number of closed-loop poles lying on the RHS of s-plane is zero. Hence, closed-loop system is stable. However, since $P = 1$, open-loop system is unstable.

Problem 12.5 The open-loop transfer function of a unity feedback system is:

$$G(s)H(s) = \frac{K(s+3)}{s(s-1)} \cdot \text{Comment on the stability.}$$

Solution

Let $$K = 1$$

$$\therefore \qquad G(s)H(s) = \frac{(s+3)}{s(s-1)} = \frac{3(1+0.33s)}{s(-1+s)}$$

$$M = \frac{3\sqrt{1+(0.33\,\omega)^2}}{\omega\sqrt{1+\omega^2}}$$

$$\phi = \tan^{-1}0.33\omega - 90° - \tan^{-1}\frac{\omega}{-1}$$

Nyquist path

$$\text{Poles} = 0, 1$$
$$\text{Zeros} = -3$$

The Nyquist path/contour is shown in Fig. 12.23.

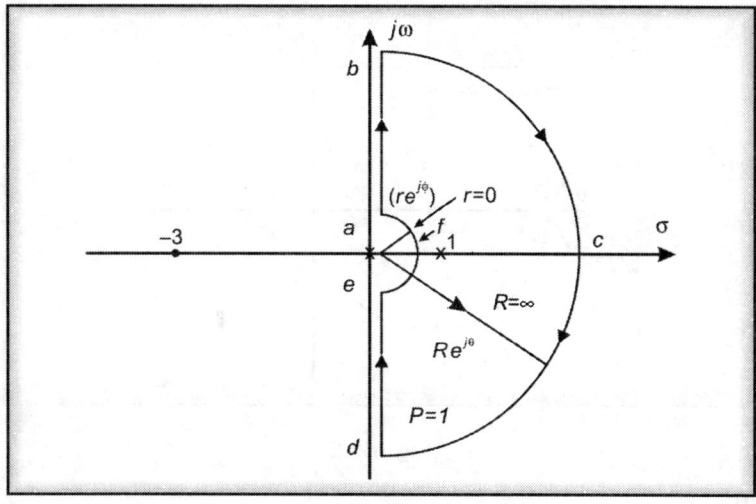

Fig. 12.23

Nyquist path 'ab'

Here $s = j\omega$ varies from $j0^+$ (point 'a') to $j\infty^+$ (point 'b'). This corresponds to sketching of the polar plot.

Table 12.7

Sr. No.	ω (rad/sec)	$M = \dfrac{3\sqrt{1 + (0.33\,\omega)^2}}{\omega\sqrt{1 + \omega^2}}$	$\phi = \tan^{-1} 0.33\,\omega$ $-90° - \tan^{-1}\omega/-1$	Remarks
1.	0	∞	90° (−270°)	As 'ω' varies from 0 to
2.	0.1	30	97.59° (−262.41°)	∞, angle varies from 90°
3.	10	0.102	245.85° (−114.15°)	and increases to 270°.
4.	100	0.0099	267.69° (−92.3°)	Hence, the plot crosses
5.	∞	0	270° (−90°)	the −ve real axis at
				some value of 'ω'

The polar plot is shown in Fig. 12.24a.

Nyquist path 'bcd'

Here, $s = Re^{j\theta}$, where $R \to \infty$ and θ varies from +90° (point 'b') to −90° (point 'd'). Application of limit gives

$$= \lim_{R \to \infty} \frac{(Re^{j\theta} + 3)}{Re^{j\theta}(Re^{j\theta} - 1)} = \lim_{R \to \infty} \frac{Re^{j\theta}}{Re^{j\theta}(Re^{j\theta} - 1)}$$

$$= \lim_{R \to \infty} \frac{1}{R} e^{-j\theta}$$

$$= 0\angle{-\theta} = 0\angle{-90°} \text{ to } +90°$$

The Nyquist plot for Nyquist path 'bcd' is shown in Fig. 12.24b.

Nyquist path 'de'

Here, $s = j\omega$ where, it varies from $j\infty^-$ to $j0^-$. This corresponds to mirror image of polar plot. $j\infty^-$ corresponds to point '*d*' and $j0^-$ corresponds to point '*e*'.

Nyquist path 'efa'

Here, $s = re^{j\phi}$ where, $r \to 0$ and ϕ varies from $-90°$ (point '*e*') to $+90°$ (point '*a*')

Application of limit gives

$$= \lim_{r\to 0} \frac{re^{j\phi} + 3}{re^{j\theta}(re^{j\theta} - 1)} = \lim_{r\to 0} \frac{-1}{re^{j\theta}} = \lim_{r\to 0} \frac{-1}{r} e^{j\theta}$$

$$= \infty\angle 180° - \phi = \infty\angle[180° - (-90°)] \text{ to } (180° - 90°)$$

$$= \infty\angle 270° \text{ (point '}e\text{') to } 90° \text{ (point '}a\text{')}$$

when the Nyquist path '*efa*' is traced, the mapping on $G(s)H(s)$ plane is traced from $270°$ to $90°$ in CW direction (Fig. 12.24d)

The Nyquist plot is shown in Fig. 12.24.

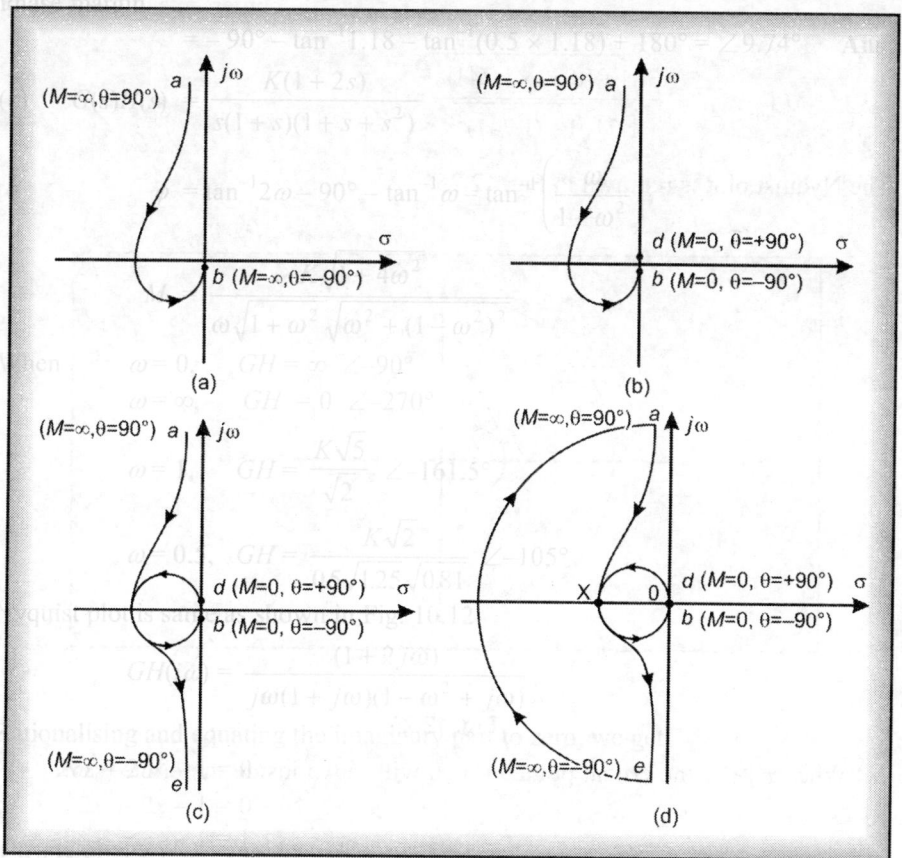

Fig. 12.24a to d $G(s)H(s)$ plane

Stability

We will first find out encirclements of critical point $(-1 + j0)$.

Putting $s = j\omega$, we get

$$G(j\omega)H(j\omega) = \frac{3(1 + 0.33\,j\omega)}{j\omega(-1 + j\omega)}$$

$$= \frac{-3j(1 + 0.33\,j\omega)}{\omega(-1 + j\omega)}$$

$$= -3\left[\frac{1.33\omega - j(1 - 0.33\omega^2)}{\omega(1 + \omega^2)}\right]$$

Equating imaginary part to zero, we get

$$1 - 0.33\,\omega^2 = 0$$

as $\qquad \omega = 1.741$ rad/sec. [This is the value at which Nyquist plot intersects −ve real axis]

$$M\big|_{\omega=1.741} = \frac{3\sqrt{1 + (0.33 \times 1.741)^2}}{1.71\sqrt{1 + (1.741)^2}} = 1$$

The Nyquist plot is redrawn in Fig. 12.25.

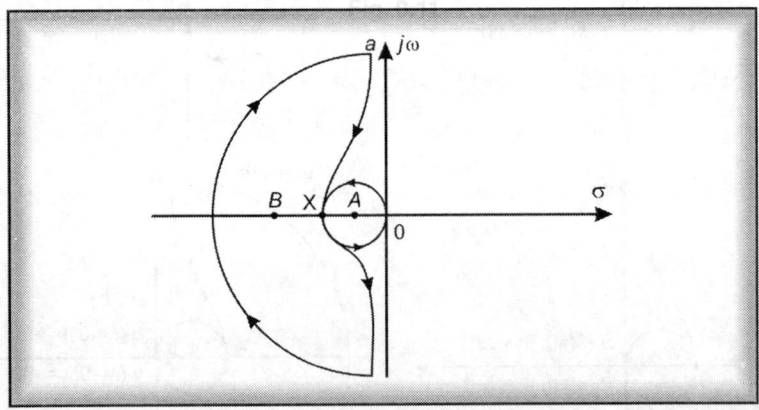

Fig. 12.25

(*a*) When $K > 1$, the crtical point $-1 + j0$ will move inside the circle (say at A). Then,

$$N = -1 \qquad \text{(since, one CCW encirclement)}$$

$$P = 1 \qquad \text{(as } G(s)H(s) \text{ has one pole on RHS of } s\text{-plane)}$$

Therefore, $N = Z - P$

$-1 = Z - 1$

or $Z = 0$ (hence closed-loop system is stable)

(b) If $K < 1$, then the critical point will move to left say at B. Then

$N = 1$ (one CW encirclement of critical point $-1 + j0$)

$P = 1$ (as one pole of $G(s)H(s)$ lie on the RHS side of s-plane)

Therefore, $N = Z - P$

$1 = Z - 1$

or $Z = 2$

Therefore, the characteristic equation has two roots on the RHS of s-plane. Therefore, the system is unstable.

(c) For $K = 1$, the critical point $(-1 + j0)$ lies at point X. For this value of K, the system is on the verge of instability.

Problem 12.6 The open-loop transfer function of a feedback control system is:

$$G(s)H(s) = \frac{-1}{2s(1 - 20\ s)} \cdot \text{Comment on the stability.}$$

Solution

$$G(s)H(s) = \frac{-1}{2s(1 - 20s)}$$

Poles are at $s = 0, \dfrac{1}{20}$

There is one pole which lies on the RHS of s-plane. Therefore,

$P = 1$

The Nyquist path/contour is shown in Fig. 12.26.

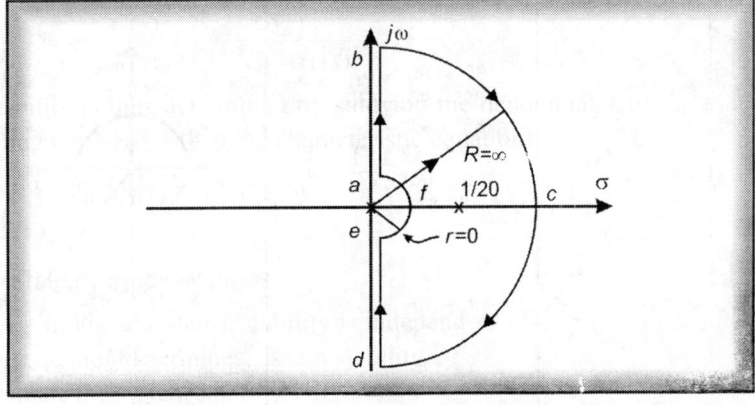

Fig. 12.26

Nyquist plot for nyquist path 'ab'

Here $s = j\omega$, where it varies from $j0^+$ (point 'a') to $j\infty^+$ (point 'b'). This corresponds to sketching of polar plot.

$$G(s)H(s) = \frac{-1}{2s(1 - 20s)}$$

$$G(j\omega)H(j\omega) = \frac{-1}{2j\omega(1 - 20j\omega)}$$

$$\phi = +180° - 90° - \tan^{-1}\frac{-20\omega}{1}$$

Note: $-1 = e^{j\pi} = \cos\pi + j\sin\pi = -1 + 0 = -1$

Thus, angle contributed by the term (-1) is $180°$.

Table 12.8

Sr. No.	ω (rad/sec)	$M = \dfrac{1}{2\omega\sqrt{1 + (20\,\omega)^2}}$	Angle $= 180° - 90°$ $- \tan^{-1}\dfrac{-20\omega}{1}$	Remarks
1.	0	∞	$\phi = 180° - 90° - (\phi_1)$ $= 180° - 90° - (360°)$ $= -270°$	Since, imaginary part is zero as $\omega = 0$, the angle ϕ_1 is $360°$ as shown above
2.	1	—	$\phi = 180° - 90° - \phi_1$ $= 180° - 90° - (360° - \phi_2)$ $= 180° - 90° -$ $(360° - 87.13°)$ $= 180° - 90° - 272.87°)$ $= -182.87°$	$\phi = \tan^{-1}\dfrac{-20 \times 1}{1} = -87.13°$ $\phi_1 = 360° - 87.13°$ $= 272.87°$

In row 1 remarks: $\phi = \tan\dfrac{-20\omega}{1} = \tan\dfrac{-20 \times 0}{1} = \tan\dfrac{-0}{1}$

In row 2 remarks: $\phi_2 = \dfrac{\tan^{-1}20\omega}{1} = \dfrac{\tan^{-1}20 \times 1}{1} = 87.13°$

(Contd.)

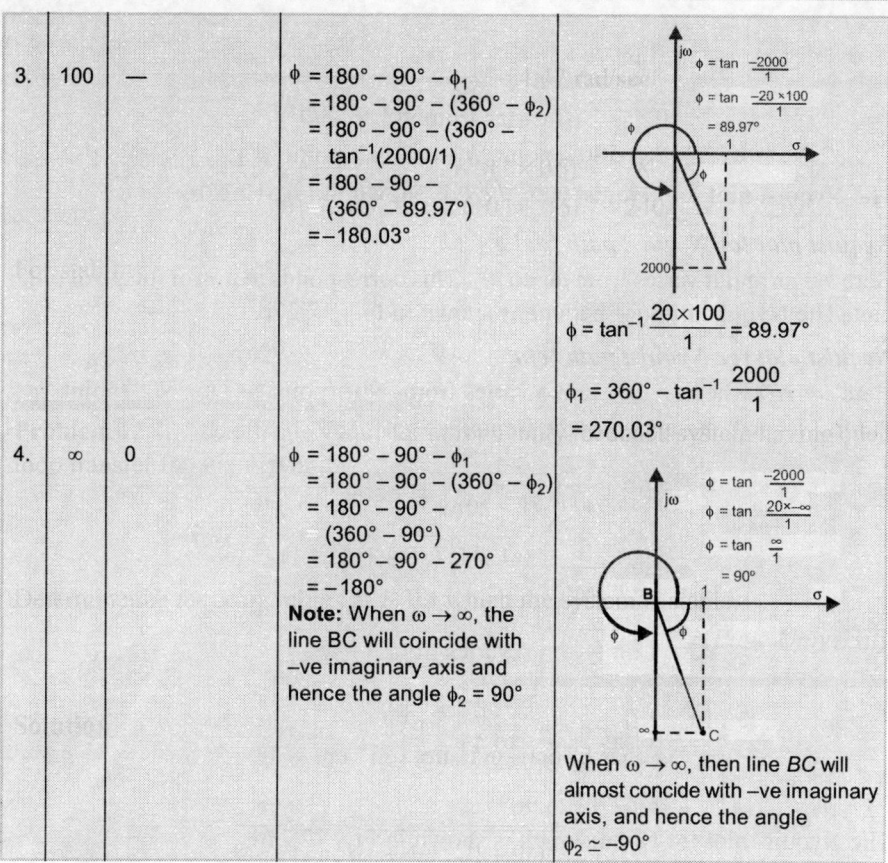

3.	100	—	$\phi = 180° - 90° - \phi_1$
			$= 180° - 90° - (360° - \phi_2)$
			$= 180° - 90° - (360° -$
			$\tan^{-1}(2000/1)$
			$= 180° - 90° -$
			$(360° - 89.97°)$
			$= -180.03°$

$$\phi = \tan^{-1}\frac{20 \times 100}{1} = 89.97°$$

$$\phi_1 = 360° - \tan^{-1}\frac{2000}{1}$$
$$= 270.03°$$

4.	∞	0	$\phi = 180° - 90° - \phi_1$
			$= 180° - 90° - (360° - \phi_2)$
			$= 180° - 90° -$
			$(360° - 90°)$
			$= 180° - 90° - 270°$
			$= -180°$

Note: When $\omega \to \infty$, the line BC will coincide with −ve imaginary axis and hence the angle $\phi_2 = 90°$

When $\omega \to \infty$, then line *BC* will almost concide with −ve imaginary axis, and hence the angle $\phi_2 \simeq -90°$

The Nyquist plot for Nyquist path '*ab*' is shown in Fig. 12.27a.

Nyquist Plot for Nyquist Path 'bcd'

Here, $s = Re^{j\theta}$ where, $R \to \infty$ and θ varies from $+90°$ (point '*b*') to $-90°$ (point '*d*'). Application of Limits give

$$= \lim_{R \to \infty} \frac{-1}{2Re^{j\theta}(1 - 20Re^{j\theta})}$$

$$= \lim_{R \to \infty} \frac{-1}{2Re^{j\theta}20Re^{j\theta}\left(\dfrac{1}{20Re^{j\theta}} - 1\right)}$$

$$= \lim_{R \to \infty} \frac{-1}{40R^2e^{j2\theta}\left(\dfrac{1}{20Re^{j\theta}} - 1\right)} = \lim_{R \to \infty} \frac{-1}{-40R^2e^{j2\theta}}\left(\dfrac{1}{20 \times \infty e^{j\theta}} - 1\right)$$

$$= \lim_{R \to \infty} \frac{-1}{40R^2e^{j2\theta}(0 - 1)} = \lim_{R \to \infty} \frac{-1}{-40R^2e^{j2\theta}}$$

$$= \lim_{R \to \infty} \frac{1}{40R^2 e^{j2\theta}}$$

$$= 0\angle e^{-j2\theta} \ (\theta \text{ varies from } 90° \text{ to } -90°)$$

$$= 0\angle -180° \ (\text{point } `b') \text{ to } 180° \ (\text{point } `d')$$

The Nyquist plot for Nyquist path 'bcd' is shown in Fig. 12.27b.

Nyquist plot for Nyquist path 'de'

Here $s = j\omega$ and it varies from $j\omega^-$ to $j0^-$. This corresponds to mirror image of polar plot. The Nyquist plot for path de is shown in Fig. 12.27c.

Nyquist plot for Nyquist path 'efa'

Here $s = re^{j\phi}$ where, $r \to 0$ and ϕ varies from $-90°$ (point 'e') to $+90°$ (point 'a'). Putting $s = re^{j\phi}$ and applying the limits, we get

$$= \lim_{r \to 0} \frac{-1}{2re^{j\phi} (1 - 20re^{j\phi})} = \lim_{r \to 0} \frac{-1}{2re^{j\phi}}$$

$$= \lim_{r \to 0} \frac{1}{2re^{j\theta}}$$

$$= \infty \angle \pi - \phi \quad \left(\phi \text{ varies from } \frac{-\pi}{2} \text{ to } \frac{\pi}{2} \right)$$

$$= \infty \left(\pi - \left(\frac{-\pi}{2} \right) \text{ to } \pi - \left(\frac{\pi}{2} \right) \right)$$

$$= \infty \angle 270° \text{ to } 90°$$

The Nyquist plot for Nyquist path is shown in Fig. 12.27d.

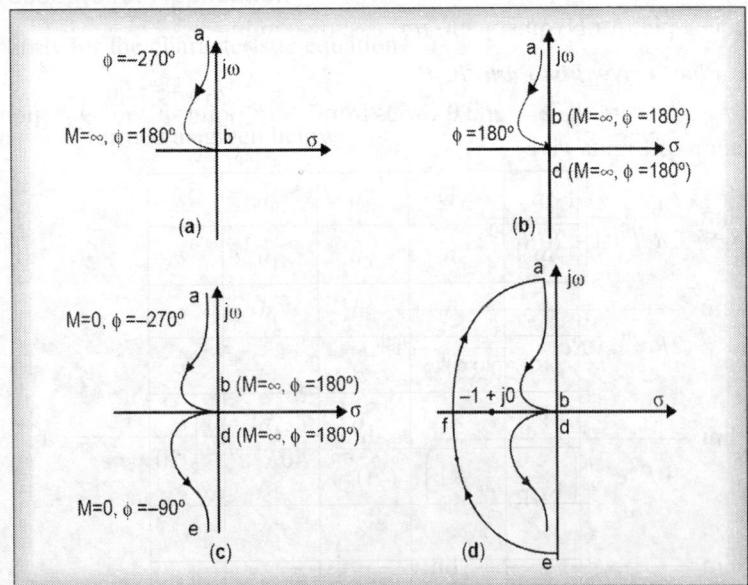

Fig. 12.27a to d

Stability

The Nyquist plot (Fig. 12.27d) encircles the critical point $(-1 + j0)$ once in CW direction. Hence,

$$N = 1$$

Also, $\quad\quad\quad P = 1$ (since, one pole of $G(s)H(s)$ lies on RHS of s-plane)

$$N = Z - P$$

Substituting the values of N and P, we get

$$1 = Z - 1$$

or $\quad\quad\quad Z = 2$

It means closed-loop system is unstable. Also, since $P = 1$, the open-loop system is also unstable.

Problem 12.7 The open-loop transfer function of a feedback system is

$$G(s)H(s) = \frac{K(1 + s)}{(1 - s)} \cdot \text{Comment on stability.}$$

Solution

$$G(s)H(s) = \frac{K(1 + s)}{(1 - s)}$$

$$\phi = \tan^{-1} \omega - \tan^{-1}\left(\frac{-\omega}{1}\right)$$

$$M = \frac{K\sqrt{1 + \omega^2}}{\sqrt{1 + \omega^2}} = K$$

Nyquist path/contour

From $G(s)H(s)$, we get

Pole $s = 1$. It means one poles of $G(s)H(s)$ is lying on the RHS of s-plane, i.e.

$$P = 1$$

The Nyquist path is shown in Fig. 12.28

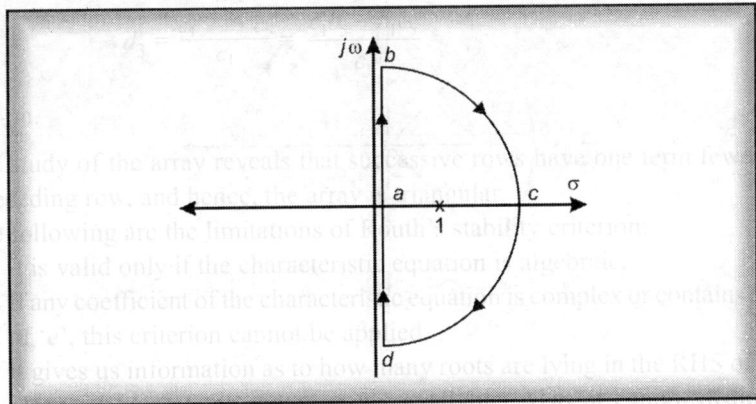

Fig. 12.28 Nyquist path

Nyquist path 'ab'

Here $s = j\omega$ where, ω varies from $j0$ (point 'a') to $j^{\infty+}$ (point 'b'). This corresponds to sketching of polar plot.

Table 12.9

Sr. No.	ω	M	$\phi = \tan^{-1}\omega - \tan^{-1}\left(\dfrac{-\omega}{1}\right)$	Remarks
1.	0	K	0	The magnitude remains constant for all values of 'ω'
2.	1	K	$45° - (-45°) = 90°$	
3.	2	K	$63.5° - (-63.5°) = 127°$	
4.	∞	K	$90° - (-90°) = 180°$	

Nyquist path 'bcd'

Here, $s = Re^{j\theta}$ where, $R \to \theta$ and θ varies from 90° (point 'b') to –90° (point 'd'). Substituting $s = Re^{j\theta}$ and applying the limits, we get

$$= \lim_{R \to \infty} \frac{K(1 + Re^{j\theta})}{(1 - Re^{j\theta})} = \lim_{R \to \infty} \frac{K\,Re^{j\theta}}{-Re^{j\theta}} = \frac{K}{-1} = \frac{K}{e^{j\pi}}$$

$$= Ke^{-j\pi} = K\angle{-180°} \text{ (point '}d\text{').}$$

The Nyquist plot for Nyquist path is shown in Fig. 12.29b

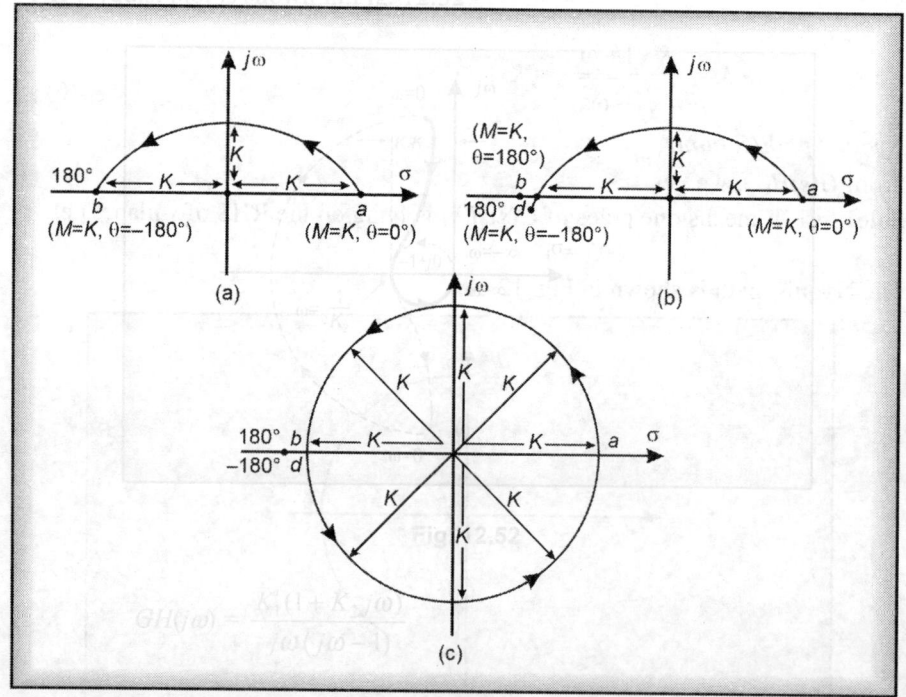

(a)

(b)

(c)

Fig. 12.29a to c

Nyquist path 'da'

Here $s = j\omega$ which varies from $j\infty^-$ (point 'd') to $j0$ (point 'a'). This corresponds to sketching mirror image of polar plot which was sketched for Nyquist path 'ab'. The Nyquist plot for Nyquist path 'da' is shown in Fig. 12.29c.

Stability

We have already found out that

$$P = 1$$

If K > 1

Then the Nyquist plot will encircle the critical point $(-1 + j0)$ once in CCW direction, which means

$$N = -1$$

Therefore, $\qquad N = Z - P$

or $\qquad -1 = Z - 1$

or $\qquad Z = 0$

The closed loop system is stable.

If K < 1

Then the Nyquist plot will not encircle the critical point $(-1 + j0)$. Hence,

$$N = 0$$

Therefore, $\qquad N = Z - P$

or $\qquad 0 = Z - 1$

or $\qquad Z = 1$

This means the closed loop system is unstable.

If K = 1

The system is on the verge of instability as the Nyquist plot passes through the critical point $(-1 + j0)$. The system is marginally stable.

Problem 12.8 The open-loop transfer function of a unity feedback control system is given by

$$G(s)H(s) = \frac{K(s + 5)(s + 40)}{s^3(s + 200)(s + 1000)}$$

Discuss the stability of closed-loop system as a function of K. Determine values of K which will cause sustained oscillations in the closed-loop system. What are the frequencies of oscillations. Use Nyquist approach.

Solution

$$G(s)H(s) = \frac{K(s + 5)(s + 40)}{s^3(s + 200)(s + 1000)}$$

$$= \frac{0.001K(1 + 0.2s)(1 + 0.025s)}{s^3(1 + 0.005s)(1 + 0.001s)}$$

$$M = \frac{0.001K\sqrt{1+(0.2\omega)^2}\;\sqrt{1+(0.025\omega)^2}}{\omega^3\sqrt{1+(0.005\omega)^2}\;\sqrt{1+(0.001\omega)^2}}$$

$$\phi = -270° - \tan^{-1} 0.005\,\omega - \tan^{-1} 0.001\,\omega + \tan^{-1} 0.2\,\omega + \tan^{-1} 0.025\,\omega$$

Poles = 0, 0, 0, – 200, –1000

These is no pole lying on the RHS of s-plane and hence

$$P = 0$$

Nyquist path/contour

The Nyquist path is modelled in Fig. 12.30.

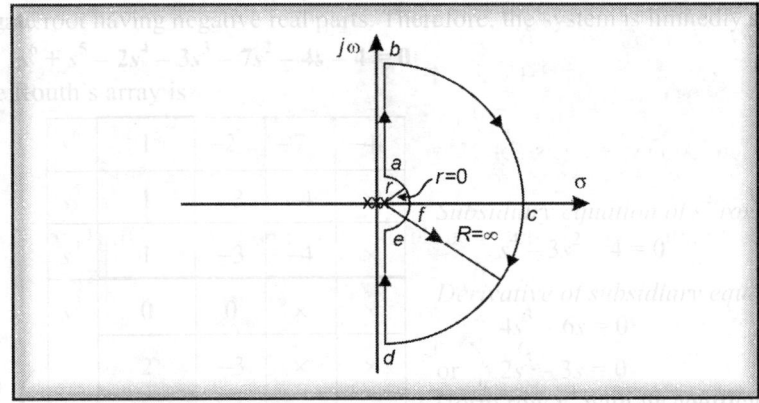

Fig. 12.30

Nyquist plot for Nyquist path 'ab'

Here $s = j\omega$, which varies from $j0^+$ (point 'a') to $j\infty^+$ (point 'b'). This corresponds to sketching of the polar plot.

Table 12.10

Sr. No.	ω	M	ϕ	Remarks
1.	0	∞	–270°	Careful study of the
2.	5		–219.6°	data reveals that
3.	16	$8.8 \times 10^{-7}\,K$	–180°	phase crossover of
4.	40		–155.7°	–180° occurs twice,
5.	200		–159°	i.e. between
6.	380	$0.6 \times 10^{-8}\,K$	–180°	(a) $\omega = 5$ and $\omega = 40$
7.	1000		–216°	(b) $\omega = 200$ and $\omega = 100$
8.	∞	0	–270°	

The Nyquist plot for Nyquist path 'ab' is shown in Fig. 12.31a.

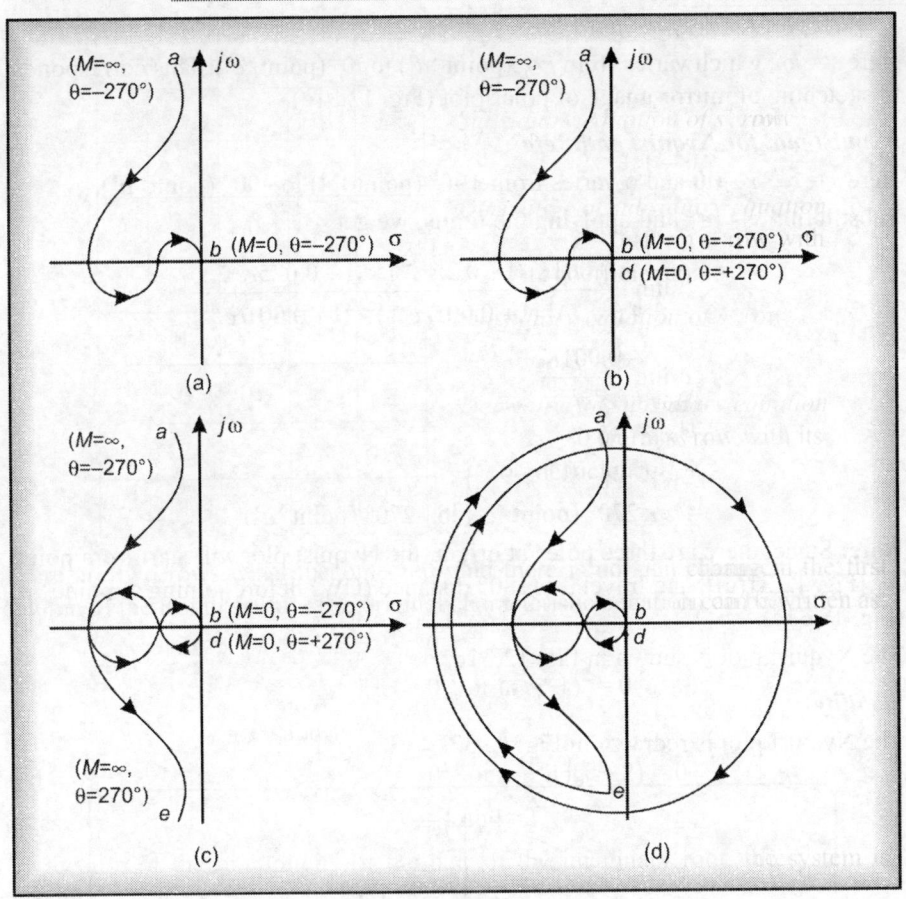

Fig. 12.31a to d

Nyquist plot for Nyquist path 'bcd'

Here $s = Re^{j\theta}$ where, $R \to \infty$ and θ varies from $+90°$ (point 'b') to $-90°$ (point 'd').

Substituting $s = Re^{j\theta}$ and application of limits gives

$$= \lim_{R \to \infty} \frac{0.001K\,(1 + 0.2Re^{j\theta})\,(1 + 0.025Re^{j\theta})}{(Re^{j\theta})^3\,(1 + 0.005Re^{j\theta})\,(1 + 0.001Re^{j\theta})}$$

$$= \lim_{R \to \infty} \frac{0.001K \times 0.2Re^{j\theta} \times 0.025Re^{j\theta}}{(Re^{j\theta})^3\,(0.005Re^{j\theta})\,(0.001Re^{j\theta})}$$

$$= \lim_{R \to \infty} \frac{0.001K \times 0.2 \times 0.025}{0.005 \times 0.001R^3 e^{j3\theta}}$$

$$= 0\angle e^{-j3\theta} = 0\angle{-3\theta} = 0\angle{-3}\ [+90° \text{ to } -90°]$$

$$= 0\angle{-270°}\ (\text{point '}b\text{'}) \text{ to } +270°\ (\text{point '}d\text{'})$$

The Nyquist plot for the Nyquist path 'bcd' is shown in Fig. 12.31b.

Nyquist plot for Nyquist path 'de'

Here $s = j\omega$, which varies from $j\infty^-$ (point 'd') to $j0^-$ (point 'e'). This corresponds to sketching of mirror image of polar plot (Fig. 12.31c).

Nyquist plot for Nyquist path 'efa'

Here $s = re^{j\phi}$ $r \to 0$ and ϕ varies from $+90°$ (point 'e') to $-90°$ (point 'a'). Substituting $s = re^{j\phi}$ and applying the limits, we get

$$= \lim_{r \to 0} \frac{0.001K \, (1 + 0.2re^{j\phi}) \times (1 + 0.025re^{j\phi})}{(re^{j\phi})^3 \, (1 + 0.005re^{j\phi}) \times (1 + 0.001re^{j\theta})}$$

$$= \lim_{r \to 0} \frac{0.001K}{r^3 e^{j3\phi}}$$

$$= \infty \angle -3\phi$$

$$= \infty \angle -3[-90° \text{ to } 90°]$$

$$= \infty \angle 270° \text{ (point '}e\text{') to } -270° \text{ (point '}a\text{')}$$

Note: Since, there are three poles at origin, the Nyquist plot will start from point 'e' [Fig. 12.31(c)], and travel $3 \times 180°$ distance (CW) before joining at point 'a' (Fig. 12.31c).

The Nyquist plot is shown in Fig. 12.31d.

Stability

The Nyquist plot is redrawn in Fig. 12.32.

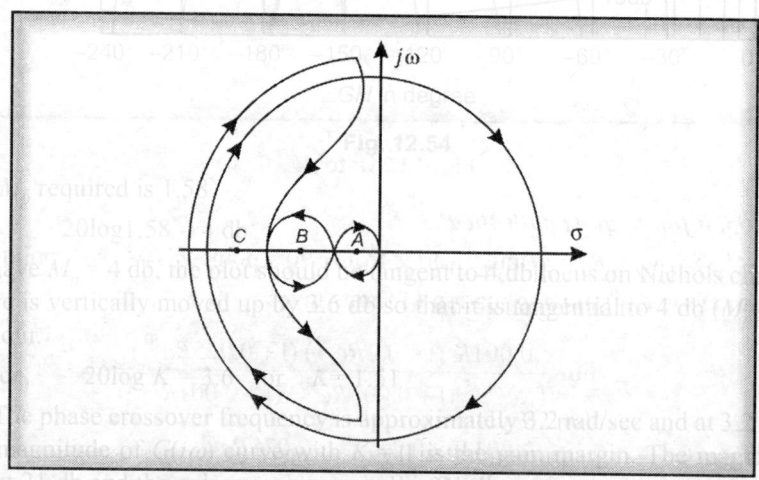

Fig. 12.32

When critical point $(-1 + j0)$ is at 'A' (Fig. 12.32)

$N = 2$ and $P = 0$ (as $G(s)H(s)$ has no poles lying on RHS of s-plane)

Therefore, $\qquad N = Z - P$

$$2 = Z - 0$$

or $\quad Z = 2$ (closed-loop system is unstable)

Also,

$\quad 0.6 \times 10^{-8} \, K > 1$ (Refer Table 12.10)

or $\qquad K > \dfrac{1}{0.6 \times 10^{-8}}$

or $\qquad K > 1.67 \times 10^{-8}$

When critical point $(-1 + j0)$ is at 'C' (Fig. 12.32)

$\qquad N = 2 \quad \text{and} \quad P = 0$

Therefore,

$\qquad N = Z - P$

$\qquad 2 = Z - 0$

or $\qquad Z = 2$ (closed-loop system is unstable)

Also,

$\quad 8.8 \times 10^{-17} \, K < 1$

or $\qquad K < 0.113 \times 10^7$

When critical point $(-1 + j0)$ is at 'B' (Fig. 12.32)

$\qquad N = 0$ (as there are two encirclements, one in CW direction another in CCW direction)

$\qquad P = 0$

Therefore,

$\qquad N = Z - P$

$\qquad 0 = Z - 0$

or $\qquad Z = 0$ (closed-loop system is stable)

This will happen when $K > 0.113 \times 10^7$

Condition for stability

Hence, the condition for stability is

$\qquad 0.113 \times 10^{-7} < K < 1.7 \times 10^{-8}$

Frequency of oscillations

$\qquad \omega = 16 \text{ rad/sec and } 380 \text{ rad/sec}$

Verification for condition for stability

This can be done by forming the characteristic equation and then forming the Routh's array.

Characteristic equation of the system under consideration is

$\qquad s^3(s + 200)(s + 1000) + K(s + 5)(s + 40) = 0$

or $\quad s^5 + 1200s^4 + 200000s^3 + Ks^2 + 45Ks + 200K = 0$

Routh's array is

s^5	1	200000	45K
s^4	1200	K	200K
s^3	$2 \times 10^5 - \dfrac{K}{1200}$	$45K - \dfrac{K}{6}$	×
s^2	$K - \dfrac{1200(45K - K/6)}{2 \times 10^5 - \dfrac{K}{1200}}$	200K	×
s^1	$\left(45K - \dfrac{K}{6}\right) - \dfrac{200K\left(2 \times 10^5 - \dfrac{K}{1200}\right)}{K - \dfrac{1200\,(45K - K/6)}{2 \times 10^5 - \dfrac{K}{1200}}}$	×	×
s^0	200K	×	×

For stability

(a) $$2 \times 10^5 - \frac{K}{1200} > 0 \text{ or } K < 24 \times 10^7$$

(b) $$K - \frac{1200\,(45K - K/6)}{2 \times 10^5 - K/1200} > 0 \text{ or } K < 1.75 \times 10^8$$

(c) $$45\,K - \frac{K}{6} - \frac{200K\left(2 \times 10^5 - \dfrac{K}{1200}\right)}{K - \dfrac{1200(45K - K/6)}{2 \times 10^5 - K/1200}} > 0 \text{ or } K > 0.113 \times 10^7$$

(d) $$200\,K > 0 \text{ or } K > 0.$$

∴   Condition for stability is $0.113 \times 10^7 < K < 1.75 \times 10^8$

when $K = 0.113 \times 10^7$, s^1 row is zero. Auxiliary equation is

$$0.73s^2 + 200 = 0$$

$$s^2 = \frac{-200}{0.73}$$

or $$s = \pm j16.55 \text{ rad/sec.}$$

Problem 12.9 Construct the Nyquist plot for a system whose open-loop transfer function is given by

$$G(s)H(s) = \frac{K(1+s)^2}{s^3}$$

Find the range of K for stability. *(Pune University)*

Solution

$$G(s)H(s) = \frac{K(1+s)^2}{s^3} = \frac{K(1+j\omega)^2}{(j\omega)^3}$$

$$M = \frac{K\sqrt{1+\omega^2}\sqrt{1+\omega^2}}{\omega^3}$$

$$\phi = -270° + \tan^{-1}\omega + \tan^{-1}\omega$$

Poles = 0, 0, and 0 [There is no pole lying on the RHS of s-plane]

Hence, $P = 0$

Nyquist path/contour

The Nyquist path is shown in Fig. 12.33

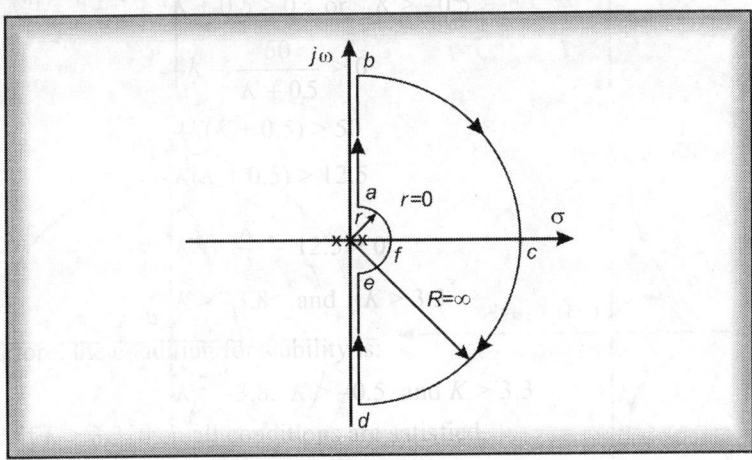

Fig. 12.33

Nyquist plot for Nyquist path 'ab'

Here $s = j\omega$ which varies from $j0^+$ to $j\infty^+$. This corresponds to sketching of the polar plot.

Table 12.11

Sr. No.	ω	$M = \dfrac{K\sqrt{1+\omega^2}\,\sqrt{1+\omega^2}}{\omega^3}$	$\phi = -270° + \tan^{-1}\omega + \tan^{-1}\omega$	Remarks
1.	0	∞	$-270°$	
2.	0.5	12.5 K	$-216.9°$	
3.	1	2 K	$-180°$	
4.	2	5/8 K	$-143°$	
5.	3	10 K/27 K	$-127°$	
6.	10	101 K/1000	$-101°$	
7.	∞	0	$-90°$	

The Nyquist plot for Nyquist path is shown in Fig. 12.34a.

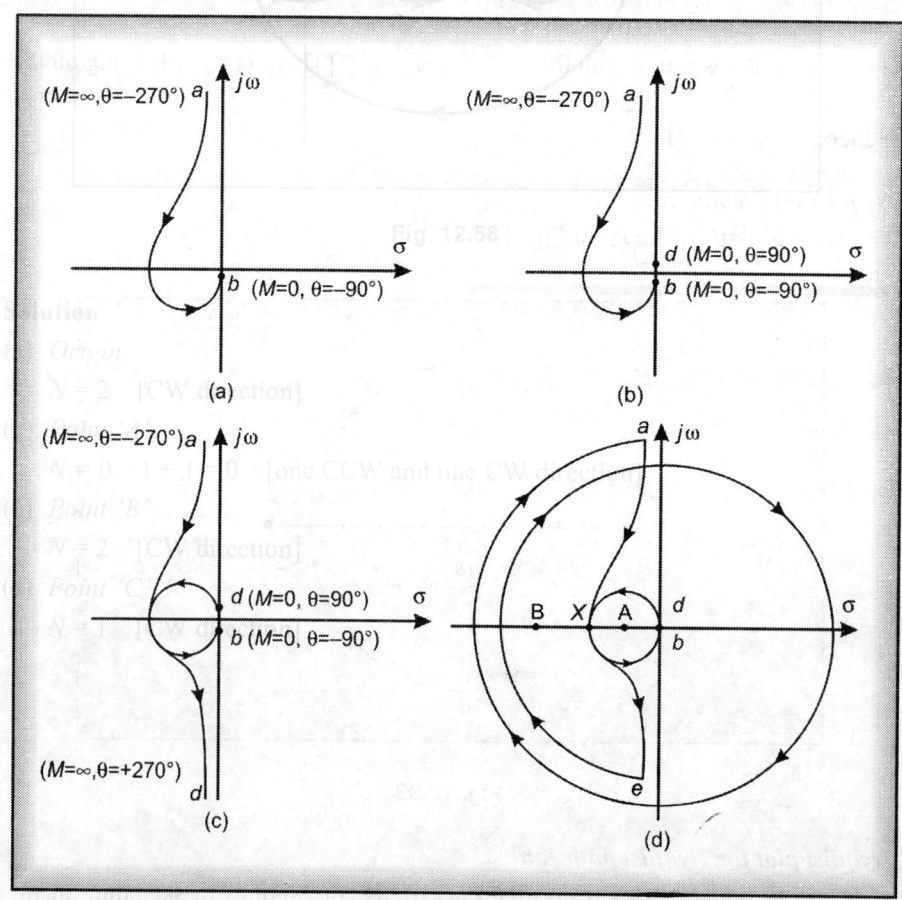

Fig. 12.34a to d

Nyquist plot for Nyquist path 'bcd'

Here $s = Re^{h\theta}$ where, $R \rightarrow \infty$ and θ varies from $+90°$ to $-90°$. Substituting $s = Re^{j\theta}$ and application of limits gives

$$= \lim_{R \rightarrow \infty} \frac{K[1 + (Re^{j\theta})^2]}{(Re^{j\theta})^3}$$

$$= \lim_{R \rightarrow \infty} \frac{K}{R} e^{-j\theta}$$

$$= \infty \angle -\theta$$

$$= \infty \angle -[90° \text{ to } -90°]$$

$$= \infty \angle -90° \text{ (point 'b') to } +90° \text{ (point 'd')}$$

The Nyquist plot for Nyquist path 'bcd' is shown in Fig. 12.34b.

Nyquist plot for Nyquist path 'de'

Here $s = j\omega$, which varies from $j\infty^-$ (point 'd') to $j0^+$ (point 'e'). This corresponds to drawing mirror image of polar plot (Fig. 12.34c).

Nyquist plot for Nyquist path 'efa'

Here $s = re^{j\phi}$, where $r \rightarrow \infty$ and ϕ varies from $-90°$ (point 'e') to $+90°$ (point 'a') substituting $s = re^{j\phi}$ and applying the limits, we get

$$= \lim_{r \rightarrow 0} \frac{K[1 + (re^{j\phi})^2]}{(re^{j\phi})^3}$$

$$= \lim_{r \rightarrow 0} \frac{K}{r^3 e^{j3\phi}}$$

$$= \infty \angle -3\phi = \infty \angle -[-270° \text{ to } +270°]$$

$$= \infty \angle +270° \text{ (point 'e') to } -270° \text{ (point 'a')}$$

The Nyquist plot for Nyquist path 'efa' is shown in Fig. 12.34d.

Note: Since, three poles at origin the Nyquist plot travels a $3 \times 180°$ starting from point 'e' in CW direction before joining point 'a'.

Stability

From the Table 12.11, it seen that when $\omega = 1$ rad/sec, then $\phi = -180°$ and the magnitude is $2K$. This is the magnitude, when the Nyquist plot crosses $-ve$ real axis point x marked on Fig. 12.34d. It can als be found below:

$$-270° + \tan^{-1} \omega + \tan^{-1} \omega = -180°$$

or

$$2 \tan^{-1} \omega = 90°$$

or

$$\tan^{-1} \omega = 45°$$

or

$$\omega = \tan 45° = 1 \text{ rad/sec}$$

At

$$\omega = 1 \text{ rad/sec}$$

$$|G(j\omega)| = \left| \frac{K(1 + \omega)^2}{\omega^3} \right|_{\omega=1} = 2K$$

This shows that

$$OX \text{ [Fig. 11.32(d)]} = -2K$$

and point

$$X = (-2K, 0)$$

Table 12.12

Sr. No.	K	n = 2K	Remarks
1.	0.4	0.8	
2.	0.5	1	
3.	0.6	1.2	
4.	0.7	1.4	

(a) When critical point (–1 + j0) is at 'B' (Fig. 12.34d)

Here $2K < 1$ or $K < 0.5$

Then $N = 2$ (these are two CW encirclements)

                  $P = 0$ (as no pole of $G(s)H(s)$ is lying on the RHS of s-plane)

Therefore,

$$N = Z - P$$
$$2 = Z - 0$$

or $Z = 2$ (closed-loop system is unstable)

(b) When critical point (–1 + j0) is at 'A' (Fig. 12.34d)

            $2K > 1$ or $K > 0.5$

From the table it can be seen that for all values of $K > 0.5$, the Nyquist plot will cross the –ve real axis at a point greater than the value of critical point $(-1 + j0)$. Hence, the critical point $(-1 + j0)$ will move inside the smaller circle (say at point 'A'). Then

                  $N = 0$ (These are two encirclements of point A represented by critical point $(-1 + j0)$, one CW and the other CCW.)

Hence, $N = +1 - 1 = 0$

                  $P = 0$

Therefore, $N = Z - P$

                  $0 = Z - 0$

or $Z = 0$ (Hence, closed-loop system is stable)

When critical point (–1 + j0) is at point 'C'

Then $2K = 1$ or $K = 0.5$. At this value of K, the Nyquist plot passes through the critical point $(-1 + j0)$. Therefore, the closed loop system is on the verge of instability.

Open loop system stability

Since $P = 0$, no poles of $G(s)H(s)$ are lying on the RHS of s-plane. Hence, open-loop system is stable.

Problem 12.10 Construct Nyquist plot for a feedback control system whose open-loop transfer function is given by

$$G(s)H(s) = \frac{5}{s\,(1-s)}$$

Comment on the stability of open-loop and closed-loop systems.

(*Pune University*)

Solution

$$G(s)H(s) = \frac{5}{s(1-s)}$$

$$G(j\omega)H(j\omega) = \frac{5}{j\omega(1-j\omega)}$$

$$M = \frac{5}{\omega\sqrt{1+\omega^2}}$$

$$\phi = -90° - \tan^{-1}\frac{-\omega}{1}$$

Nyquist path/contour

Poles = 0, 1

Since, one pole is lying on the RHS Hence,

$$P = 1$$

The Nyquist path is shown in Fig. 12.35

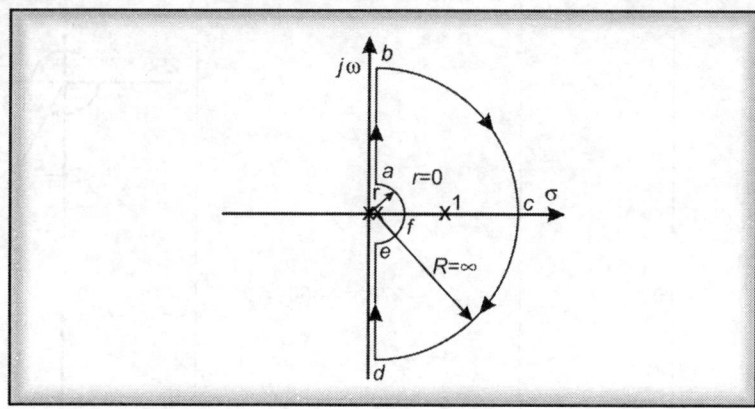

Fig. 12.35

Nyquist plot for Nyquist path 'ab'

Here $s = j\omega$ which varies from $j0^+$ (point '*a*') to $j\infty^+$ (point '*b*'). This corresponds to sketching of polar plot.

Table 12.13

Sr. No.	ω	$M = \dfrac{5}{\omega\sqrt{1+\omega^2}}$	$\phi = -90° - \tan^{-1}\dfrac{-\omega}{1}$	Remarks
1.	0	∞	$\phi = -90° - 360° = -450°$ or $-90°$	
2.	1	3.54	$\phi = -90° - 315° = -405°$ or $-45°$	
3.	2	1.12	$\phi = -90° - 296.57° = -386.57°$ or $-26.5°$	
4.	∞	0	$\phi = -90° - 270° = -360° = 0°$	

The Nyquist plot for Nyquist path '*ab*' is shown in Fig. 12.36a.

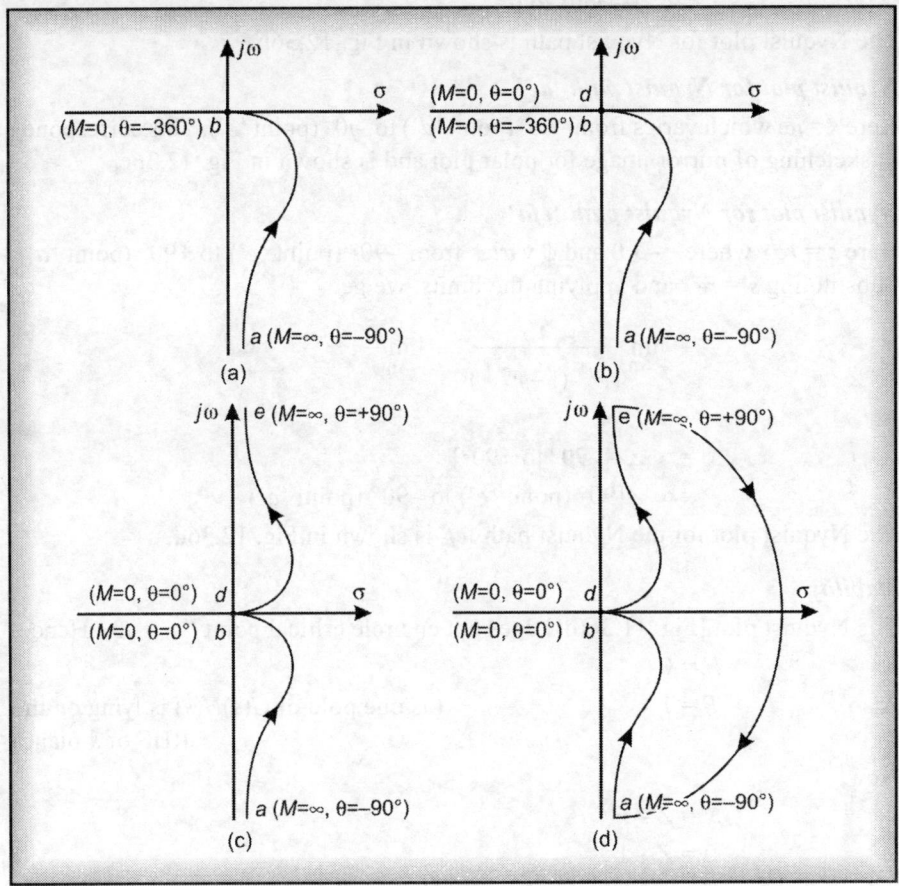

Fig. 12.36a to d

Nyquist plot for Nyquist path 'bcd'

Here, $s = Re^{j\theta}$, where $R \to \infty$ and θ varies from $+90°$ (point '*b*') to $-90°$ (point '*d*').

Substituting $s = Re^{j\theta}$ and application of limits gives

$$= \lim_{R \to \infty} \frac{5}{Re^{j\theta}(1 - Re^{j\theta})} = \lim_{R \to \infty} \frac{5}{-R^2 e^{j2\theta}}$$

$$= \lim_{R \to \infty} \frac{5}{R^2 e^{j\pi} e^{j2\theta}}$$

$$= \lim_{R \to \infty} \frac{5}{R^2} e^{-j\pi} e^{j2\theta}$$

$$= \infty \angle -\pi - 2\theta$$

$$= \infty \angle -180° - 2\theta$$

$$= \infty \ \angle(-180° - 2 \times 90°) \text{ to } (-180° - 2 \times -90°)$$
$$= \infty \ \angle-360° \text{ to } 0°$$

The Nyquist plot for Nyquist path is shown in Fig. 12.36b.

Nyquist plot for Nyquist path 'dc'

Here $s = j\omega$ which varies from $-j\omega^-$ (point 'd') to $-j0^-$ (point 'e'). This corresponds to sketching of mirror image for polar plot and is shown in Fig. 12.36c.

Nyquist plot for Nyquist path 'efa'

Here $s = re^{j\phi}$ where $r \to 0$ and ϕ varies from $-90°$ (point 'e') to $+90°$ (point 'a'). Substituting $s = re^{j\phi}$ and applying the limits, we get

$$= \lim_{r \to 0} \frac{5}{re^{j\phi}(1 - re^{j\phi})} = \lim_{r \to 0} \frac{5}{r} e^{-j\phi}$$

$$= \infty \ \angle-\phi$$
$$= \infty \ \angle-[-90° \text{ to } +90°]$$
$$= \infty \ \angle[90° \text{ (point 'e') to } -90° \text{ (point 'a')}]$$

The Nyquist plot for the Nyquist path 'ef' is shown in Fig. 12.36d.

Stability

The Nyquist plot [Fig. 11.36(d)] does not encircle critical point $(-1 + j0)$. Hence,

$$N = 0$$

Also, $\qquad P = 1$ $\qquad$ (as one pole of $G(s)H(s)$ is lying on the RHS of s-plane)

Now,

$$N = Z - P$$
$$0 = Z - 1$$

or $\qquad Z = 1$ $\qquad$ (Hence, closed-loop system is unstable)

Also, open-loop system is unstable as $P = 1$.

Check

$$1 + G(s)H(s) = 1 + \frac{5}{s(1-s)}$$

or $\qquad s - s^2 + 5 = 0$

or $\qquad s^2 - s - 5 = 0$

sign change

s^2	1	-5
s^1	-1	×
s^0	-5	×

There is one sign change in the first column of Routh's array, and hence, one root lies on the RHS of the s-plane. Therefore, the closed-loop system is unstable.

Problem 12.11 Sketch Nyquist plot for the system described by

$$G(s)H(s) = \frac{1.25(s+1)}{(s+0.5)(s-2)}$$

Solution

$$G(s)H(s) = \frac{1.25(s+1)}{(s+0.5)(s-2)}$$

Putting $s = j\omega$, we get

$$G(j\omega)H(j\omega) = \frac{1.25\,(1+j\omega)}{(0.5+j\omega)\,(-2+j\omega)}$$

$$M = \frac{1.25\sqrt{1+\omega^2}}{\sqrt{0.5^2+\omega^2}\,\sqrt{4+\omega^2}}$$

$$\phi = \tan^{-1}\omega - \tan^{-2}\frac{\omega}{0.5} - \tan^{-1}\left(\frac{\omega}{-2}\right)$$

Nyquist path/contour

Poles $P = -0.5, +2$

Zeros $Z = -1$

These is no pole lying on the imaginary axis and at origin and since, there is one pole lying the RHS of s-plane, hence

$$P = 1$$

The Nyquist path/contour is shown in Fig. 12.37.

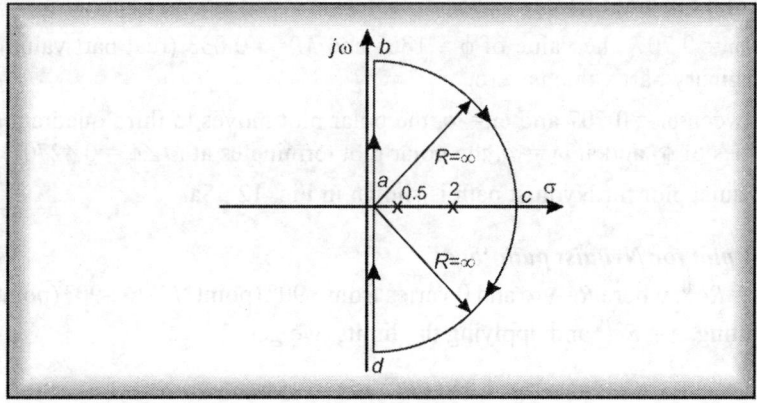

Fig. 12.37

Nyquist plot for Nyquist path 'ab'

Here $s = j\omega$, which varies from $j0$, (point 'a') to $j^{\infty+}$ (point 'b'). This corresponds to sketching of the polar plot.

Table 12.14

Sr. No.	ω	M	ϕ	Remarks
1.	0	−1.25	$0 - 0 - (-180°) = 180°$	**Intersection with negative real axis**
2.	0.5	0.825	—	$G(j\omega)H(j\omega)$
3.	0.707	−0.833	180°	$= \dfrac{1.25\,(1+j\omega)}{(0.5+j\omega)(j\omega-2)}$
4.	10	—	—	
5.	∞	0	$90° - 90° - (-270°)$ $= 270°$	$= \dfrac{1.25(1+j\omega)(0.5-j\omega)(-2-j\omega)}{(0.5+j\omega)(0.5-j\omega)(-2+j\omega)(-2-j\omega)}$

$$= \frac{1.25(1+j\omega)(-1+1.5j\omega-\omega^2)}{(0.25+\omega^2)(4+\omega^2)}$$

$$= \frac{-1.25(1+2.5\omega^2)}{(0.25+\omega^2)(4+\omega^2)}$$

$$+ \frac{1.25\,j\omega(0.5-\omega^2)}{(0.25+\omega^2)(4+\omega^2)}$$

Equating imaginary part to zero gives
$$0.5 - \omega^2 = 0 \quad \text{or} \quad \omega = 0.707$$
Putting the value of ω in the real part gives
$$a = \frac{-1.25(1+2.5\times0.5)}{(0.25+0.5)(4+0.5)} = -0.833$$

Note:

(a) As seen from the table, the polar plot between $\omega = 0$ and 0.707, remains in the second quadrant [Refer values of 'ϕ'].

(b) At $\omega = 0.707$, the value of $\phi = 180°$ and $M = -0.833$ (real part value). The imaginary part value is zero.

(c) Between $\omega = 0.707$ and $\omega = \infty$, the polar plot moves to third quadrant (refer values of ϕ) and at $\omega = \infty$, the polar plot terminates at $M\angle\phi = 0\angle270°$.

The Nyquist plot for Nyquist path is shown in Fig. 12.38a.

Nyquist plot for Nyquist path 'bcd'

Here, $s = Re^{j\theta}$, where $R \to \infty$ and θ varies from +90° (point 'b') to −90° (point 'd').

Substituting, $s = Re^{j\phi}$ and applying the limits, we get

$$= \lim_{R \to \infty} \frac{1.25\,(Re^{j\theta}+1)}{(Re^{j\theta}+0.5)\,(Re^{j\theta}-2)}$$

$$= \lim_{R \to \infty} \frac{1.25}{R}e^{-j\theta}$$

$$= 0\angle e^{-j\theta}$$

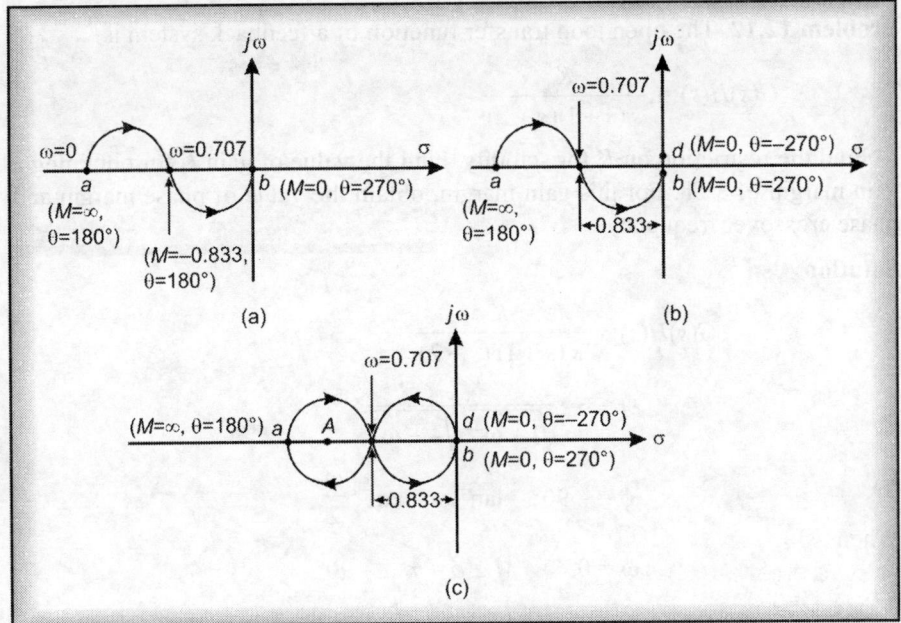

Fig. 12.38a to c

$$= 0\angle-\theta$$

$$= 0\angle-[90° \text{ to } -90°]$$

$$= 0\angle-90° \text{ (270°, } i.e. \text{ point '}b\text{') to } +90° (-270°, i.e. \text{ point '}d\text{')}$$

Nyquist plot for Nyquist path is shown in Fig. 12.38b.

Nyquist plot for Nyquist path 'da'

Here $s = j\omega$, which varies from $j\infty^-$ (point 'd') to $j0$ (point 'e'). This corresponds to sketching of mirror image of polar plot and is shown in Fig. 12.38c.

Stability

It can be seen that the Nyquist plot makes one clockwise encirclement of point 'A' $(-1 + j0)$ and hence $N = 1$. Also, from $G(s)H(s)$, we get $P = 1$, Substituting the values in

$$N = Z - P$$

we get $$1 = Z - P$$

or $$Z = 2$$

which means two roots of characteristic equation lying on the RHS of the s-plane. Therefore, the closed-loop system is unstable. The open-loop system is also unstable as $P = 1$.

Problem 12.12 The open loop transfer function of a feedback system is

$$G(s)H(s) = \frac{K}{s\,(s+1)(s+2)}$$

Find the restriction on K for stability. Find the value of gain K for obtaining a gain margin of 3 db. For this gain margin, obtain the value of phase margin and phase crossover frequency.

Solution

$$G(s)H(s) = \frac{K}{s\,(s+1)\,(s+2)}$$

$$M = \frac{K}{\omega\,\sqrt{1+\omega^2}\,\sqrt{4+\omega^2}}$$

$$\phi = -90° - \tan^{-1}\omega - \tan^{-1}\frac{\omega}{2}$$

When

$$\omega = 0, \quad M \angle\phi = \infty \,\angle{-90°}$$

$$\omega = 1, \quad M \angle\phi = \frac{K}{\sqrt{10}} \,\angle{-162°}$$

$$\omega = 2, \quad M \angle\phi = \frac{K}{4\sqrt{10}} \,\angle{-198.4°}$$

$$\omega = \infty, \quad M \angle\phi = 0 \,\angle{-270°}$$

The polar plot is shown in Fig. 12.39

$$\omega_c = \frac{1}{\sqrt{T_1 T_2}} = \frac{1}{\sqrt{1 \times 0.5}} = \sqrt{2}$$

$$a = \frac{K}{2}\left(\frac{T_1 T_2}{T_1 + T_2}\right) = \frac{K}{2}\left(\frac{1 \times 0.5}{1 + 0.5}\right) = 0.16667\,K = \frac{K}{6}$$

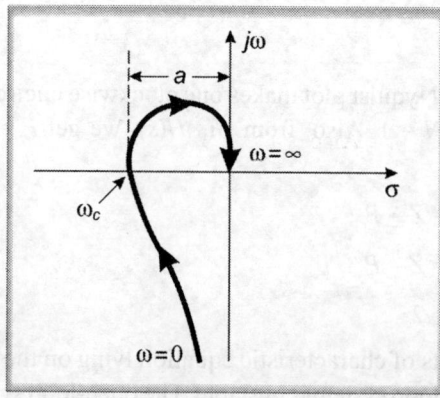

Fig. 12.39

For stability $\qquad N = Z - P.$

Since, $P = 0$, $Z = N$. But, for stability Z should be zero, *i.e.* N should be zero; which means that the point $(-1 + j0)$ should not be encircled. This will only happen, if

$$a < 1 \quad \text{or} \quad K/6 < 1 \quad \text{or} \quad K < 6$$

$$\text{Gain margin} = 20 \log \frac{1}{a} = 3$$

i.e. $\qquad 20 \log \dfrac{6}{K} = 3$

$$\frac{6}{K} = 1.41 \quad \text{or} \quad K = \frac{6}{1.41} = 4.25$$

Phase crossover frequency

$$M = 1, \text{ when gain is } 0 \ db$$

$\therefore \qquad M = \dfrac{K}{\omega \sqrt{1 + \omega^2} \ \sqrt{4 + \omega^2}} = 1$

or $\qquad (4.25)^2 = \omega^2(1 + \omega^2)(4 + \omega^2)$

By hit and trial method

$$\omega = 1.183 \text{ is the value which satisfies the equation.}$$

Phase margin

$$P.M. = \angle G(j\omega)H(j\omega) \mid_{\omega = 1.183} + 180°$$

$$= \left\{ \left(-90° - \tan^{-1} \frac{\omega}{1} - \tan^{-1} \frac{\omega}{2} \right) \Big|_{\omega = 1.183} \right\} + 180°$$

$$= -90° - \tan^{-1} \frac{1.183}{1} - \tan^{-1} \frac{1.183}{2} + 180° = 9.6°$$

Problem 12.13 The open-loop transfer function of a feedback control system is

$$G(s)H(s) = \frac{K(1 + 2s)}{s(1 + s)(1 + s + s^2)}$$

Find the restriction on K for stability. Find the value of K for the system to have a gain margin of 3 db. With this value of K, find the phase crossover frequency and phase margin.

Solution

$$G(s)H(s) = \frac{K(1 + 2s)}{s(1 + s)(1 + s + s^2)}$$

$$M = \frac{K\sqrt{1 + 4\omega^2}}{\omega\sqrt{1 + \omega^2} \ \sqrt{(1 - \omega^2)^2 + \omega^2}}$$

$$\phi = \tan^{-1}2\omega - 90° - \tan^{-1}\omega - \tan^{-1}\frac{\omega}{1-\omega^2}$$

When $\qquad \omega = 0, \qquad\qquad M\angle\phi = \infty \angle -90°,$

$$\omega = 1, \qquad\qquad M\angle\phi = \sqrt{\frac{5}{2}} \angle 161.5°,$$

$$\omega = 0.5, \qquad\qquad M\angle\phi = 2.8 \angle -105.5°,$$

$$\omega = \infty, \qquad\qquad M\angle\phi = 0 \angle -270°$$

$$G(j\omega)H(j\omega) = \frac{K(1+j2\omega)}{j\omega(1+j\omega)\{(1-\omega^2)+j\omega\}}$$

Separating into real and imaginary parts and equating the imaginary part to zero, *i.e.*

$$1 + 2\omega^2 - 2\omega^4 = 0$$

We get, by hit and trial method $\omega = 1.17$.

$$M\big|_{\omega = 1.17} = \frac{K\sqrt{1+(1.17 \times 2)^2}}{1.17\sqrt{1+1.17^2}\sqrt{(1-1.17^2)^2+1.17^2}} = 1.153\,K$$

$$\phi\big|_{\omega = 1.17} = \tan^{-1}(2 \times 1.17) - 90° - \tan^{-1}1.17 - \tan^{-1}\frac{1.17}{1-1.17^2} = -180°$$

$\therefore \quad G(j\omega)H(j\omega) = M\angle\phi = 1.153$ and $K\angle -180°$

The polar plot is shown in Fig. 12.40.

If $K = 1$, then point $(-1 + j0)$ will be encircled and the system will be unstable.

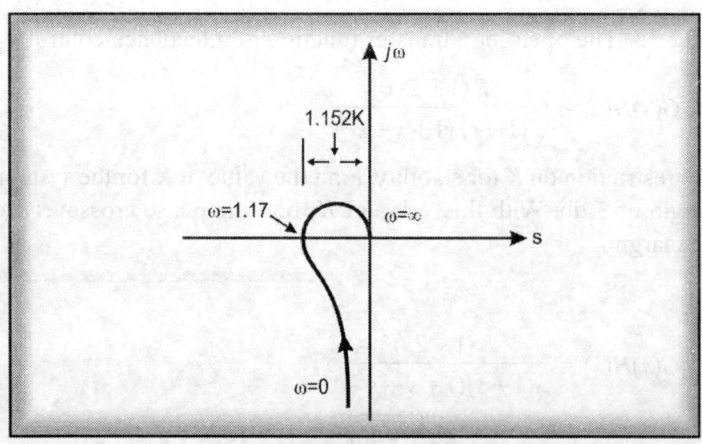

Fig. 12.40

If $\qquad K < \dfrac{1}{1.153}$

or $\qquad K < 0.87$; the system will be stable.

Phase crossover frequency

$$\frac{K\sqrt{1+4\omega^2}}{\omega\sqrt{1+\omega^2}\sqrt{\omega^2+(1-\omega^2)^2}} = 1$$

or

$$\frac{(1+4\omega^2)\times K}{\omega^2(1+\omega^2)\{\omega^2+(1-\omega^2)^2\}} = 1$$

$$\text{Gain margin} = 20\log\frac{1}{a} = 3$$

$$20\log\frac{1}{1.153K} = 3 \quad \text{or} \quad \frac{1}{1.153K} = 1.413$$

which gives $\qquad K = \dfrac{1}{1.153\times 1.413} = 0.614$

Putting the value of K in the equation

$$\frac{0.614(1+4\omega^2)}{\omega^2(1+\omega^2)\{\omega^2+(1-\omega^2)\}} = 1$$

By hit and trial method, $\omega = 0.97$ is the phase crossover frequency.

$$\text{P.M.} = \tan^{-1}(2\times 0.97) - 90° - \tan^{-1}0.97 - \tan^{-1}\frac{0.97}{1-0.97^2} + 180° = 22.5°$$

Problem 12.14 Draw Nyquist plot and examine the stability of the closed-loop system having the open-loop transfer function

$$G(s)H(s) = \frac{K}{s(1+T_1s)(1+T_2s)}$$

Solution

$$G(s)H(s) = \frac{K}{s(1+T_1s)(1+T_2s)}$$

$$M = \frac{K}{\omega\sqrt{1+\omega^2T_1^2}\sqrt{1+\omega^2T_2^2}}$$

$$\phi = -90° - \tan^{-1}\frac{\omega T_1}{1} - \tan^{-1}\frac{\omega T_2}{1}$$

When $\qquad \omega = 0, \qquad M\angle\phi = \infty\angle-90°$

$\qquad\qquad\qquad \omega = \infty, \qquad M\angle\phi = 0\angle-270°$

The Nyquist plot is shown in Fig. 12.41.

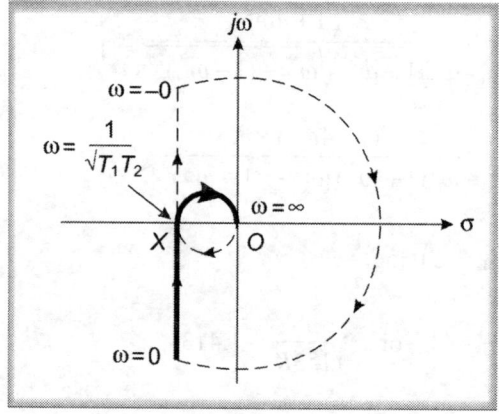

Fig. 12.41

The value of 'ω' when it crosses the negative real axis is $= \dfrac{1}{\sqrt{T_1T_2}}$

The value of 'M' at $\omega = \dfrac{1}{\sqrt{T_1T_2}}$ is obtained from the expression for 'M' and

is $\dfrac{K(T_1T_2)}{T_1 + T_2}$

The open-loop transfer function does not have any open-loop pole on the RHS of the s-plane. Therefore, for stability, the point $(-1 + j0)$ should not be encircled. This is obtained from the basic equation

$\qquad\qquad N = Z - P$

Since $\qquad\qquad P = 0$

$\qquad\qquad Z = N$

For closed-loop stability, Z should be zero. This can only happen if $N = 0$, *i.e.* point $(-1 + j0)$ must not be encircled. For this, the condition to be satisfied is

$$\frac{K(T_1T_2)}{T_1 + T_2} < 1 \quad \text{or} \quad K < \frac{T_1 + T_2}{T_1T_2}$$

Also, K should be greater than zero.

Problem 12.15 For each of the following transfer functions, sketch the Nyquist plots and ascertain the positive values of K for which the closed-loop operation is stable.

(a) $G(s)H(s) = \dfrac{K(1+s)}{(1-s)}$

(b) $G(s)H(s) = \dfrac{K}{s(1+s)(2+s)}$

(c) $G(s)H(s) = \dfrac{K(1+2s)}{s(1+s)(1+s+s^2)}$

Specify, in each case the gain K for obtaining a gain margin of 3 db. For this gain margin, find the phase margin and the gain and phase crossover frequencies.

Solution

(a) If K is positive then

$$\phi = \tan^{-1}\omega - \tan^{-1}\frac{-\omega}{1}$$

When, $\omega = 0$, $\phi = 0°$

$\omega = 1$, $\phi = 90°$

$\omega = 2$, $\phi = 126.9°$

$\omega = \infty$, $\phi = 180°$

$$M = \frac{K\sqrt{1+\omega^2}}{\sqrt{1+\omega^2}} = K$$

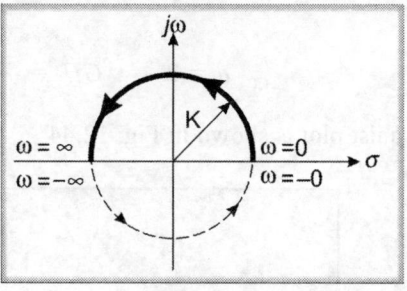

Fig. 12.42

Nyquist plot for positive value of K is shown in the Fig. 12.42.

(b) The above is true if K is positive. However, if K is negative, then

$$\phi = 180° + \tan^{-1}\omega - \tan^{-1}\left(\frac{\omega}{1}\right)$$

When $\omega = 0$, $\phi = 180°$

$\omega = 1$, $\phi = 270°$

$\omega = 2$, $\phi = 306.97°$

$\omega = \infty$, $\phi = 0$

Nyquist plot for negative values of K is shown in Fig. 12.43

When $|K| < 1$, then $N = 0$,

Since $N = Z - P$

or $0 = Z - 1$ or $Z = 1$

Closed-loop system is unstable

When $|K| > 1$, then $N = 1$,

Since $N = Z - P$

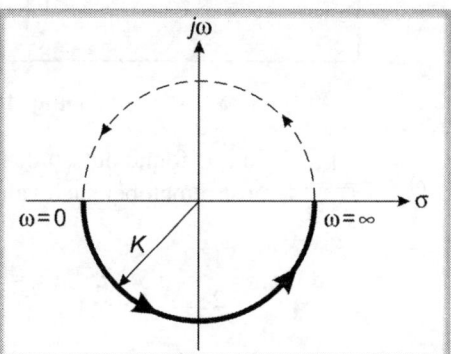

Fig. 12.43

or $-1 = Z - 1$ or $Z = 0$

Closed-loop system is stable.

System is unstable if

$$-1 < K < + 1, \quad i.e.\ K < -1 \text{ (unstable) and } K > 1 \text{ (stable)}.$$

(b) $G(s)H(s) = \dfrac{K}{s(1 + s)(2 + s)}$

$$\phi = -90° - \tan^{-1}\omega - \tan^{-1}0.5\ \omega$$

$$M = \dfrac{K}{2\omega\sqrt{1 + \omega^2}\ \sqrt{1 + 0.5\omega^2}}$$

When $\omega = 0,$ $GH = \infty\angle - 90°$

$\omega = \infty,$ $GH = 0\angle -270°$

$\omega = 1,$ $GH = \dfrac{K}{\sqrt{10}} \angle - 162°$

$\omega = 2,$ $GH = \dfrac{K}{2\sqrt{5}\sqrt{10}} \angle - 198°$

Nyquist plot is shown in Fig. 12.44

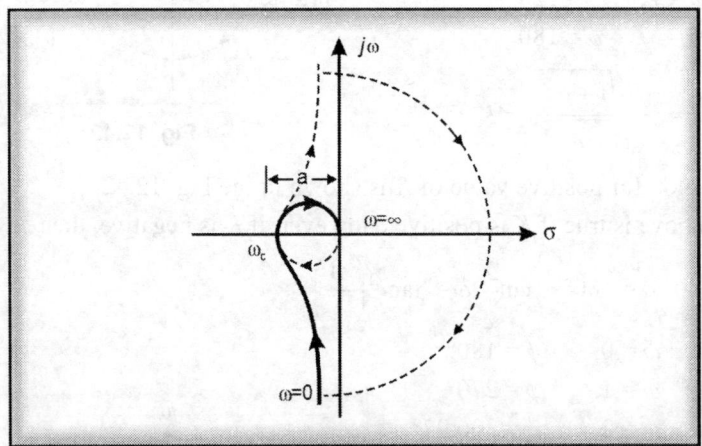

Fig. 12.44

$\omega_c = \dfrac{1}{\sqrt{\tau_1\tau_2}} \left(\begin{array}{l}\text{Can be found out, students}\\ \text{must remember such values}\end{array}\right)$

$\tau_1 = 1$ and $\tau_2 = \dfrac{1}{2} = 0.5$ $\therefore$ $\omega_c = \dfrac{1}{\sqrt{1 \times 0.5}} = \sqrt{2}$ **Ans**

$a = K\left(\dfrac{\tau_1\tau_2}{\tau_1 + \tau_2}\right) = K\left(\dfrac{1 \times 0.5}{1 + 0.5}\right) = \dfrac{K}{6}$

$\therefore$ For stability $\dfrac{K}{6} < 1$ or $K < 6$ **Ans**

Gain margin$= 20 \log \dfrac{1}{a}$

i.e. $3 = 20 \log \dfrac{6}{K}$ or $K = 4.25$

Since $GH = \dfrac{4.25}{s(1+s)(2+s)}$ $\therefore$ $M = \dfrac{4.25}{\omega \sqrt{1+\omega^2}\,\sqrt{4+\omega^2}}$

Frequency at which magnitude will become unity is given by

$$\omega \sqrt{1+\omega^2}\,\sqrt{4+\omega^2} = 4.25$$

or $\omega^2(1+\omega^2)(1+4\omega^2) = 18.1$

putting $x = \omega^2,$

we get, $x(1+x)(x+4) = 18.1$

By hit and trial method $x = 1.4$ $\therefore$ $\omega = \sqrt{1.4} = 1.18$ **Ans**

Phase margin

$$= -90° - \tan^{-1}1.18 - \tan^{-1}(0.5 \times 1.18) + 180° = \angle 9.74° \quad \textbf{Ans}$$

(c) $\quad G(s)H(s) = \dfrac{K(1+2s)}{s(1+s)(1+s+s^2)}$

$$\phi = \tan^{-1}2\omega - 90° - \tan^{-1}\omega - \tan^{-1}\left(\dfrac{\omega}{1-\omega^2}\right)$$

$$M = \dfrac{K\sqrt{1+4\omega^2}}{\omega\sqrt{1+\omega^2}\,\sqrt{\omega^2+(1-\omega^2)^2}}$$

When $\omega = 0,$ $GH = \infty\ \angle{-90°}$

$\omega = \infty,$ $GH = 0\ \angle{-270°}$

$\omega = 1,$ $GH = \dfrac{K\sqrt{5}}{\sqrt{2}}\ \angle{-161.5°}$

$\omega = 0.5,$ $GH = \dfrac{K\sqrt{2}}{0.5\sqrt{1.25}\sqrt{0.81}}\ \angle{-105°}$

Nyquist plot is same as shown in Fig. 10.12

$$GH(j\omega) = \dfrac{(1+2j\omega)}{j\omega(1+j\omega)(1-\omega^2+j\omega)}$$

Rationalising and equating the imaginary part to zero, we get

$2\omega^4 - 2\omega^2 - 1 = 0$ $\qquad\qquad \{\because\ x = \omega^2\}$

or $\quad 2x^2 - 2x - 1 = 0$

$\therefore \qquad\qquad x = 1.36$

or $\qquad\qquad \omega = \sqrt{1.36} = 1.17$ rad/sec

Check at $\omega = 1.17$; ϕ should be $-180°$

i.e. $\phi = 66.8° - 90° - 49.5° - 107.5° \simeq -180°$ (checked)

$$M\Big|_{\omega = 1.17} = a = \frac{K\sqrt{6.47}}{1.17\sqrt{2.37}\sqrt{1.5}} = 1.15K$$

For stability

$$1.15K < 1 \quad \text{or} \quad K < \frac{1}{1.15} = 0.87$$

Gain margin $= 20 \log 1/a$

$$3 = 20 \log \frac{1}{1.15K} \quad \text{or} \quad K = 0.615 \qquad\qquad \textbf{Ans}$$

$\therefore$ The transfer function $GH(s) = \dfrac{0.615(1 + 2s)}{s(1 + s)(1 + s + s^2)}$

$$M = \frac{0.615\sqrt{1 + 4\omega^2}}{\omega\sqrt{1 + \omega^2}\sqrt{\omega^2 + (1 - \omega^2)^2}}$$

Frequency at which $M = 1$ is found out by hit and trial method and is $= 0.97$

Phase Margin $= 63° - 90° - 44° - 86.5° + 180° = 22.5°$ $\qquad\qquad$ **Ans**

The results obtained are depicted on the polar plot shown in Fig. 12.45

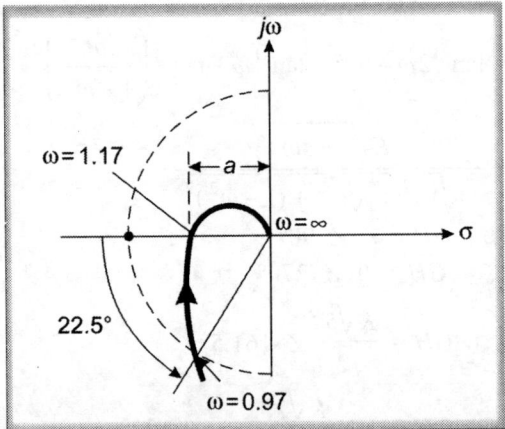

Fig. 12.45

Problem 12.16 Consider a feedback system having the characteristic equation

$$1 + \frac{K}{s(1.5 + s)(2 + s)} = 0$$

It is desired that all roots of the characteristic equation have real parts less than -1. Extend the Nyquist stability criterion to find the largest value of K satisfying this condition.

Solution

$$GH(s) = \frac{K}{s(1.5 + s)(2 + s)}$$

Put
$$s = -1 + j\omega$$

$$GH(-1 + j\omega) = \frac{K}{j\omega(0.5 + j\omega)(1 + j\omega)}$$

$$= \frac{2K}{j\omega(1 + 2j\omega)(1 + j\omega)}$$

$$M = \frac{2K}{\omega\sqrt{1 + 4\omega^2}\sqrt{1 + \omega^2}}$$

$$\phi = -90° - \tan^{-1} 2\omega - \tan^{-1}\omega$$

When $\omega = 0$, $\qquad\qquad M = \infty$ $\qquad$ and $\qquad$ $\phi = -90°$

$\qquad\omega = \infty$, $\qquad\qquad M = 0$ $\qquad$ and $\qquad$ $\phi = -270°$

$\qquad\omega = 0.5$, $\qquad\qquad\qquad\qquad\qquad\qquad\qquad$ $\phi = -161.56°$

$\qquad\omega = 0.1$, $\qquad\qquad\qquad\qquad\qquad\qquad\qquad$ $\phi = -107°$

$\qquad\omega = 1$, $\qquad\qquad\qquad\qquad\qquad\qquad\qquad$ $\phi = -198°$

$\qquad\omega = 10$, $\qquad\qquad\qquad\qquad\qquad\qquad\qquad$ $\phi = -261°$

Nyquist plot is shown in Fig. 12.46.

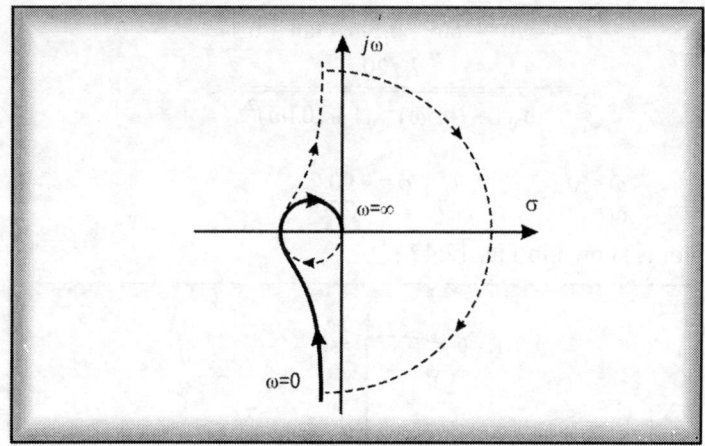

Fig. 12.46

Phase crossover frequency

$$-180° = 90° - \tan^{-1}2\omega - \tan^{-1}\omega$$

or $\qquad\qquad \tan^{-1}2\omega + \tan^{-1}\omega = 90°$

or $\qquad\qquad \dfrac{\omega + 2\omega}{1 - 2\omega^2} = \infty$

or $\qquad\qquad 1 - 2\omega^2 = 0$

or $\qquad\qquad \omega = 0.707$

At this frequency

$$M = \frac{2K}{0.707 \sqrt{1 + 4 \times 0.707^2} \sqrt{1 + 0.707^2}} = 1.33K$$

For stability

$$1.33K < 1$$

or

$$K < \frac{1}{1.33} \quad \text{or} \quad K < 0.75$$

Hence, $K = 0.75$ is the largest value of K.

Problem 12.17 Draw the complete Nyquist plot for a system whose open-loop transfer function is

$$G(s)H(s) = \frac{K}{s(s + 2)(s + 10)}$$

Determine the range of K for which closed-loop system is stable.

(Pune University)

Solution

$$G(s)H(s) = \frac{K}{s(s + 2)(s + 10)} = \frac{K/20}{s(1 + 0.5s)(1 + 0.1s)}$$

which gives

$$\phi = -90° - \tan^{-1} 0.5\omega - \tan^{-1} 0.1\omega$$

$$M = \frac{K/20}{\omega \sqrt{1 + (0.5\omega)^2} \sqrt{1 + (0.1\omega)^2}}$$

When,

$$\omega = 0 \qquad \phi = -90°$$
$$\omega = \infty \qquad \phi = -270°$$

Nyquist plot is shown in Fig. 12.47.

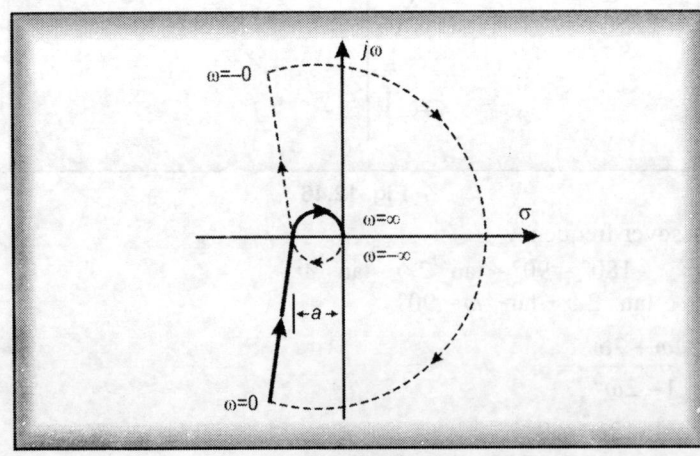

Fig. 12.47

$$\omega_\tau = \frac{1}{\sqrt{T_1 T_2}} = \frac{1}{\sqrt{0.5 \times 0.1}} = 4.47 \text{ rad/sec}$$

$$a = \frac{K}{20}\left(\frac{T_1 T_2}{T_1 + T_2}\right) = \frac{K(0.1 \times 0.5)}{20(0.1 + 0.5)} = \frac{K}{240}$$

For stability

$$\frac{K}{240} < 1 \quad \text{or} \quad K < 240$$

Problem 12.18 Sketch the Nyquist plot for a unity feedback system having open-loop transfer function given by

$$G(s) = \frac{K}{s(1+s)(1+2s)(1+3s)}$$

Determine the range of values of K for which the system is stable.

(Pune University)

Solution

$$\phi = -90° - \tan^{-1}\omega - \tan^{-1}2\omega - \tan^{-1}3\omega$$

$$M = \frac{1}{\omega\sqrt{1+\omega^2}\sqrt{1+4\omega^2}\sqrt{1+9\omega^2}}$$

When, $\omega = 0$ $\phi = -90°$

$\omega = \infty$ $\phi = -360°$

Nyquist plot is shown in Fig. 12.48

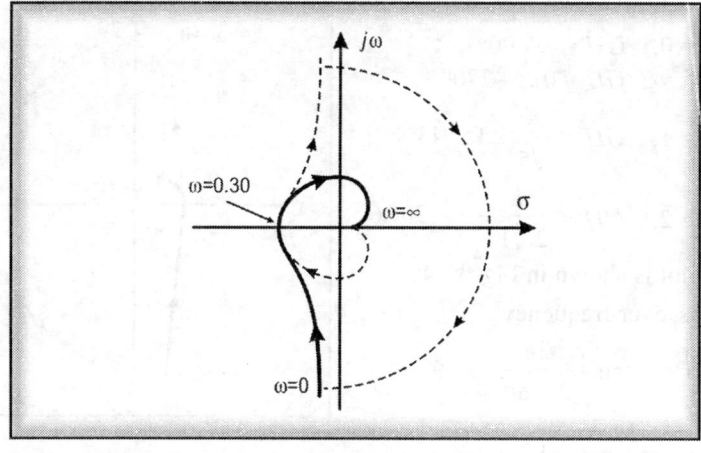

Fig. 12.48

$$G(j\omega) = \frac{K}{j\omega(1+j\omega)(1+2j\omega)(1+3j\omega)}$$

Rationalising and equating the imaginary part to zero, we get

$$11\omega^2 = 1$$

or $\qquad \omega = 0.30$ rad/sec

$$M\big|_{\omega=0.30} = \frac{K}{0.30\sqrt{1+0.30^2}\,\sqrt{1+4\times0.30^2}\,\sqrt{1+9\times0.30^2}} = \frac{K}{0.50}$$

for stability $\qquad 1 \le \dfrac{K}{0.50}$

$$\frac{K}{0.50} < 1 \quad \text{or} \quad K < 0.50$$

Problem 12.19 Construct the complete Nyquist plot for a unity feedback control system whose open-loop transfer function is

$$G(s)H(s) = \frac{K}{s(s^2+2s+2)}$$

Find maximum value of K for which the system is stable. (*Pune University*)

Solution

$$GH(j\omega) = \frac{K}{j\omega((j\omega)^2+2j\omega+2)} = \frac{K}{j\omega(2-\omega^2+2j\omega)}$$

which gives $\phi = -\tan^{-1}\dfrac{2\omega}{2-\omega^2} - 90°$

and $\quad M = \dfrac{K}{\omega\sqrt{(2-\omega^2)^2+4\omega^2}}$

when $\omega = 0$, $\quad GH = \infty\,{-}90°$

$\qquad \omega = \infty$, $\quad GH = 0\,\angle{-}270°$

$\qquad \omega = 1$, $\quad GH = \dfrac{K}{\sqrt{5}}\,\angle153.43°$

$\qquad \omega = 2$, $\quad GH = \dfrac{K}{2\sqrt{18}}\,\angle{-}206.6°$

Nyquist plot is shown in Fig. 12.49

Phase crossover frequency

$$-180° = -\tan^{-1}\frac{2\omega}{2-\omega^2} - 90°$$

or $\qquad \tan^{-1}\dfrac{2\omega}{2-\omega^2} = 90°$

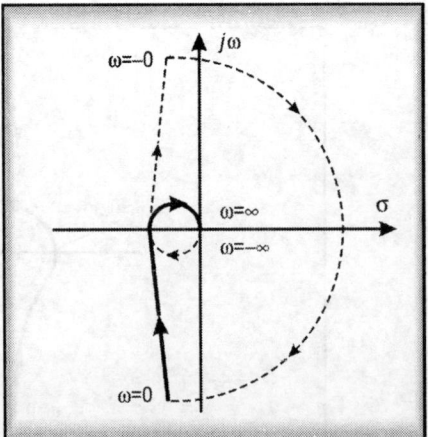

Fig. 12.49

or $\qquad 2 - \omega^2 = 0$

or $\qquad \omega = \sqrt{2}$

$$M\big|_{\omega = \sqrt{2}} = \frac{K}{\sqrt{2}\sqrt{(2-2)^2 + 4 \times 2}} = \frac{K}{4}$$

For stability $\qquad \dfrac{K}{4} < 1$ or $K < 4$ **Ans**

Problem 12.20 Determine the stability of the control system shown in Fig. 12.50 without and with derivative feedback (*i.e.* with $K_2 = 0$ and $K_2 > 0$)

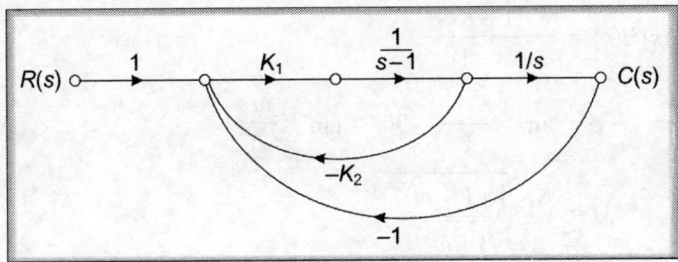

Fig. 12.50

Solution

(a) *Without derivative feedback* $(K_2 = 0)$

$$G(s)H(s) = \frac{K_1}{s(s-1)}$$

which gives $\qquad \phi = -90° - \tan^{-1}\dfrac{\omega}{-1}$ and $M = \dfrac{K_1}{\omega\sqrt{1+\omega^2}}$

When, $\qquad \omega = 0; \quad \phi = -90° - (-180°) = 90°$

$\qquad\qquad \omega = \infty; \quad \phi = 0 - (-270°) = 270°$

The Nyquist plot is shown in Fig. 12.51

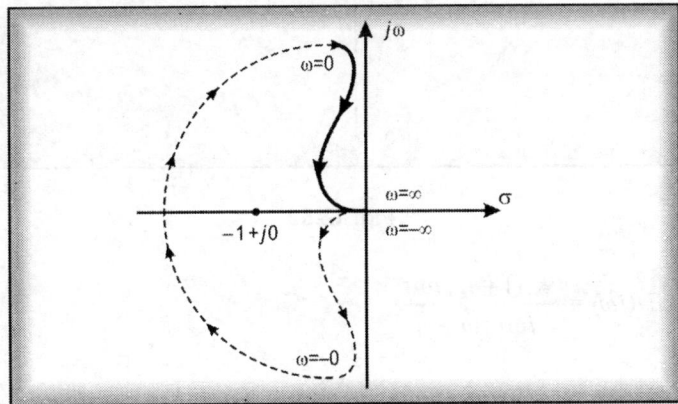

Fig. 12.51

Since, there is one pole of $G(s)H(s)$ on the RHS of s-plane; $P = 1$ which implies that open loop system is unstable. Since, there is one encirclement of the point $-1 + j0$, Hence, $N = -1$ and we have

$$N = Z - P$$

$$1 = Z - 1$$

or $\qquad Z = 2.$

This implies closed-cloop system is unstable,

(b) With derivative feedback

$$G(s)H(s) = \frac{K_1(1 + K_2 s)}{s(s - 1)}$$

which gives $\qquad \phi = \tan^{-1}\frac{K_2\omega}{1} - 90° - \tan^{-1}\frac{\omega}{-1}$

and $\qquad M = \dfrac{K_1\sqrt{1 + (K_2\omega)^2}}{\omega\sqrt{1 + \omega^2}}$

When, $\qquad \omega = 0; \quad \phi = 0° - 90° - (-180°) = 90°$

$\qquad\qquad \omega = \infty; \quad \phi = 90° - 90° - (-270°) = 270°$

The Nyquist plot is shown in Fig. 12.52

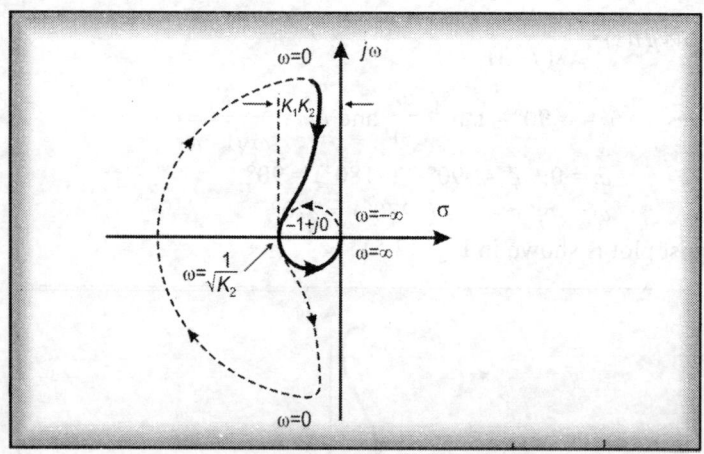

Fig. 12.52

$$GH(j\omega) = \frac{K_1(1 + K_2 j\omega)}{j\omega(j\omega - 1)}$$

$$= \frac{-K_1(\omega^2 + \omega^2 K_2) + j(\omega - K_2\omega^3)K_1}{\omega^2 + \omega^4}$$

Equating the imaginary part to zero, we get

$$\omega - K_2\omega^3 = 0$$

or $\omega = \dfrac{1}{\sqrt{K_2}}$; and the value of real part at this value of ω is

$$= \left.\frac{-\omega^2 K_1(1+K_2)}{\omega^2 + \omega^4}\right|_{\omega^2 = \frac{1}{K_2}} = -K_1K_2.$$

If $K_1K_2 > 1$, the Nyquist plot encircles the point $-1 + j0$ once anticlockwise, hence,
$N = -1$

$$N = Z - P$$
$$-1 = Z - 1$$

or $$Z = 1 - 1 = 0$$

Therefore, the system is stable if $K_1K_2 > 1$

Problem 12.21 Draw the Nyquist plot for the control system having

$$G(s)H(s) = \frac{s + 0.25}{s^2(s+1)(s+0.5)}$$

Solution

$$G(s)H(s) = \frac{s + 0.25}{s^2(s+1)(s+0.5)}$$

which gives

$$\phi = -180° - \tan^{-1}\frac{\omega}{1} - \tan^{-1}\frac{\omega}{0.5} + \tan^{-1}\frac{\omega}{0.25}$$

and $$M = \frac{\sqrt{0.25^2 + \omega^2}}{\omega^2\sqrt{1+\omega^2}\sqrt{0.5^2 + \omega^2}}$$

When,

$$\omega = 0; \quad M = \infty \angle -180°$$
$$\omega = \infty; \quad M = 0 \angle -270°$$

Intersection with the real axis

Put $s = j\omega$

$$\therefore \qquad G(j\omega)H(j\omega) = \frac{(0.25 + j\omega)}{(j\omega)^2 (1 + j\omega)(0.5 + j\omega)}$$

$$= \frac{(0.25 + j\omega)(1 - j\omega)(0.5 - j\omega)}{(j\omega)^2 (1 + j\omega)(1 - j\omega)(0.5 + j\omega)(0.5 - j\omega)}$$

$$= \frac{(0.25 + j\omega)(0.5 - 1.5j\omega - \omega^2)}{-\omega^2 (1 + \omega^2)(0.25 + \omega^2)}$$

$$= -\frac{0.125 + 1.25\omega^2}{-\omega^2 (1 + \omega^2)(0.25 + \omega^2)} + \frac{j\omega(0.125 - \omega^2)}{-\omega^2 (1 + \omega^2)(0.25 + \omega^2)}$$

Equating imaginary part to zero gives

$$\omega^2 = 0.125$$

or $\qquad \omega = \sqrt{0.125}$

Putting the value of $\omega = \sqrt{0.125}$ in the real part gives the intersection point

$$a = \frac{0.125 + 1.25 \times 0.125}{-(0.125)(1 + 0.125)(0.25 + 0.125)} = -5.4$$

The Nyquist plot is shown in Fig. 12.53

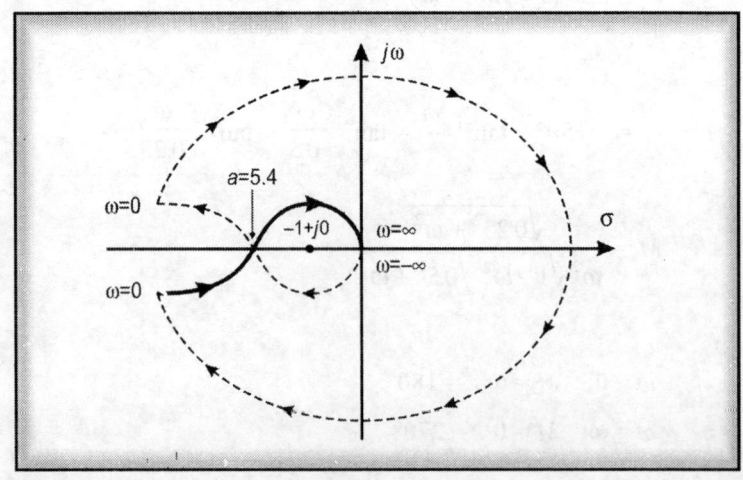

Fig. 12.53

From Fig. 12.53, we infer that

$N = 2$ as Nyquist plot makes two clockwise encirclements of the point $-1 + j0$ and since $P = 0$,

we get $\qquad 2 = Z - 0$

or $\qquad Z = 2$; Hence, two roots of the characteristic equation are lying on the RHS of the s-plane, therefore, closed-loop system is unstable. The open-loop system is stable as $P = 0$

Problem 12.22 The open-loop transfer function of a unity feedback system is

$$G(s) = \frac{K}{s\,(1 + s)\,(1 + 0.1\,s)}$$

Determine the value of K for the following cases:

(a) Resonance peak is required to be equal to 1.58.

(b) Gain margin of the system is 21 db.

(c) Phase margin of the system is 40°.

Solution

Assume $\qquad K = 1$

$$M = \frac{1}{\omega\sqrt{1 + \omega^2}\,\sqrt{1 + (0.1\,\omega)^2}}$$

$$\phi = -90° - \tan^{-1}\omega - \tan^{-1}0.1\omega$$

Sr. No.	ω (rad/sec)	Magnitude in db	ϕ (degree)
1	0.1	20	−96
2	0.2	13.8	−102
3	0.5	5	−119
4	0.7	1.3	−129
5	0.9	−1.7	−137
6	1.0	−3	−141
7	2.0	−13.2	−165
8	3.0	−20	−178
9	5.0	−29	−195
10	7.0	−35.6	−207
11	10	−43	−219
12	20	−59	−240
13	50	−82	−249

Nichols plot is drawn with the above data (Fig. 12.54)

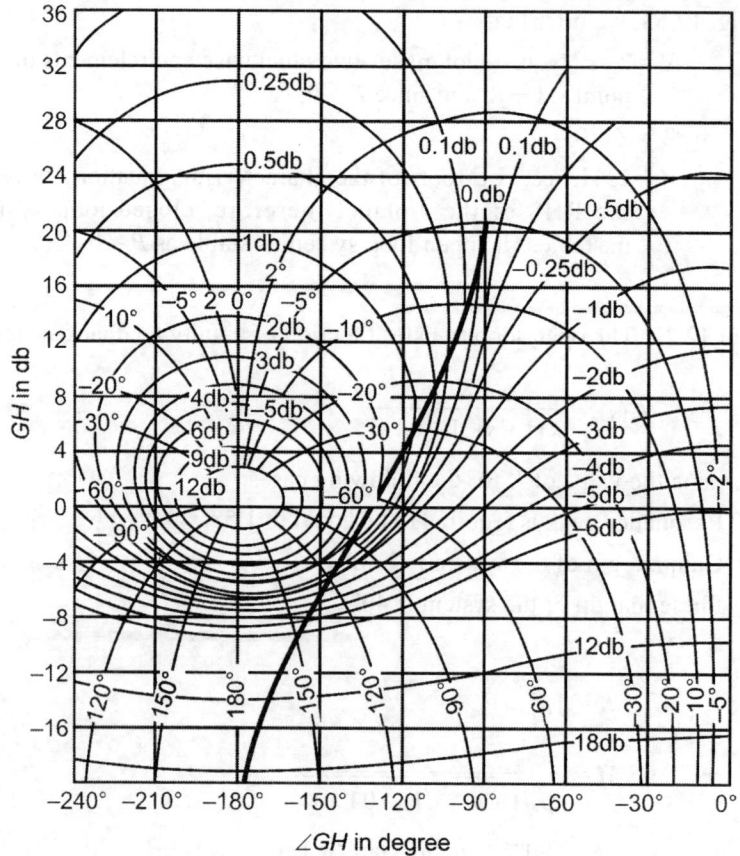

Fig. 12.54

(a) M_p required is 1.58

$20\log 1.58 = 4$ db

To have $M_p = 4$ db, the plot should be tangent to 4 db locus on Nichols chart. The curve is vertically moved up by 3.6 db so that it is tangential to 4 db ($M_p = 1.58$) contour.

Hence, $20\log K = 3.6$ or $K = 1.51$

(b) The phase crossover frequency is approximately 3.2 rad/sec and at 3.2 rad/sec the magnitude of $G(j\omega)$ curve with $K = 1$ is the gain margin. The magnitude is about 21 db and the gain margin required is 21 db.

$\therefore$ $20\log K = 21$ or $K = 11.2$

(c) Phase margin required is 40°. Therefore, the phase margin is $(-180° + 40°)$ $= -140°$. The point on the curve at $-140°$ has to be lifted up to coincide with 0 db line. A lift of 4 db is required.

$20\log K = 4$

or $K = 1.58$

Problem 12.23 From the Nichols chart find the value of peak resonance, resonant frequency and bandwidth for a unity feedback system whose open-loop transfer function is

$$G(s) = \frac{2}{s(2+s)(1+s)}$$

Also, determine gain and phase margin.

Solution

$$G(j\omega) = \frac{2}{j\omega\,(2j\omega)\,(1+j\omega)} = \frac{1}{j\omega\,(1+j\omega)\,(1+0.5j\omega)}$$

$$M = \frac{1}{\omega\,\sqrt{1+\omega^2}\,\sqrt{1+0.5\,\omega^2}}$$

$$\phi = 90° - \tan^{-1}\omega - \tan^{-1}0.5\omega$$

Values of magnitude in db and angle in degree are tabulated below:

Sr. No.	ω (rad/sec)	M (db)	ϕ (degree)
1	0.1	20	−98.6
2	0.2	13.8	−107
3	0.4	7	−123
4	0.6	2.7	−138
5	0.8	−0.85	−150.5
6	1.0	−4	161.5
7	1.2	−6.8	171.2
8	1.5	−10.6	183.2

Nichols plot is shown in Fig. 12.55.

Peak resonance. The plot is tangential to 5 db contour. Hence, peak value of the closed-loop frequency response is 5 db.

Resonant frequency. The plot is tangential to 5 db contour at ω = 0.8 rad/sec. Hence, resonant frequency is 0.8 rad/sec.

Bandwidth. Bandwidth is found at the frequency that corresponds to the intersection of the plotted curve with − 3 db contour. Frequency is 1.2 rad/sec.

Gain margin. It is the open-loop gain when the phase angle is 180°. It is read as 10.5 db.

Phase margin. It is the difference between −180° and actual phase angle when the open-loop gain is 0 db. Angle is read as −148°. Hence, the phase margin is

$$180 − 148° = 32°$$

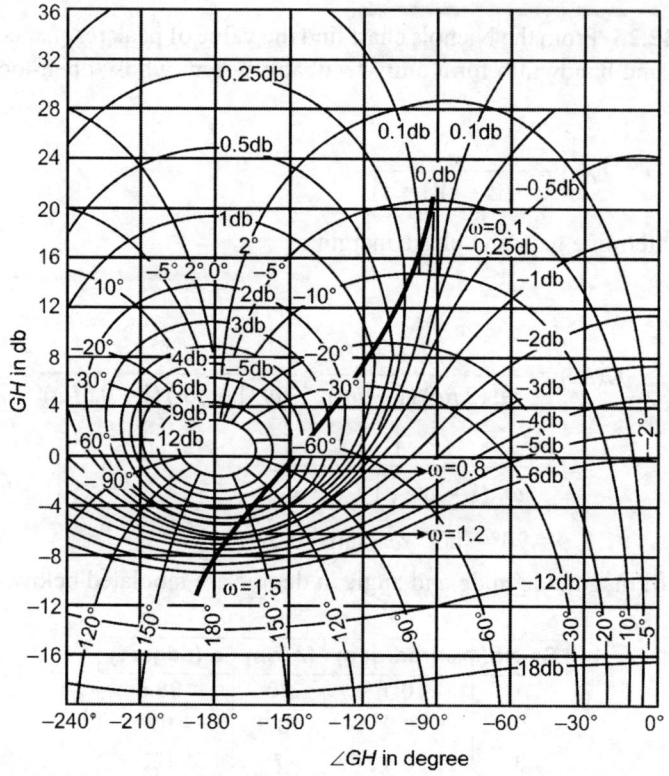

Fig. 12.55

Problem 12.24 If the transfer function is given as $G(j\omega) = s + 10$, the corresponding Nyquist plot is

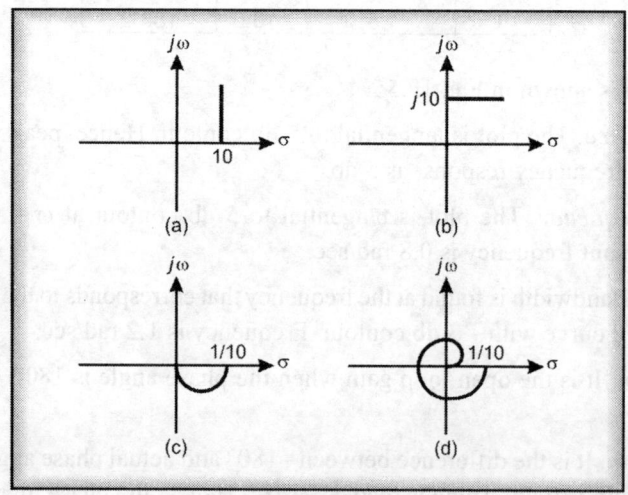

Fig. 12.56

Solution

$$G(j\omega) = j\omega + 10$$

$$M\angle\phi = \sqrt{\omega^2 + 100} \ \angle\tan^{-1}\frac{\omega}{10}$$

At $\omega = 0$, $M\angle\phi = 10\angle 0 = |j0 + 10| \angle 0$

  $\omega = \infty$, $M\angle\phi = \infty\angle 90 = |j\infty + 10| \angle 90°$

It shows that real part remains finite at 10 and the imaginary part varies from 0 to ∞.

Hence, the Nyquist plot for positive frequency is as given in Fig. 12.57a.

Problem 12.25 Find the encirclements of the critical point when located at a point marked 'A'.

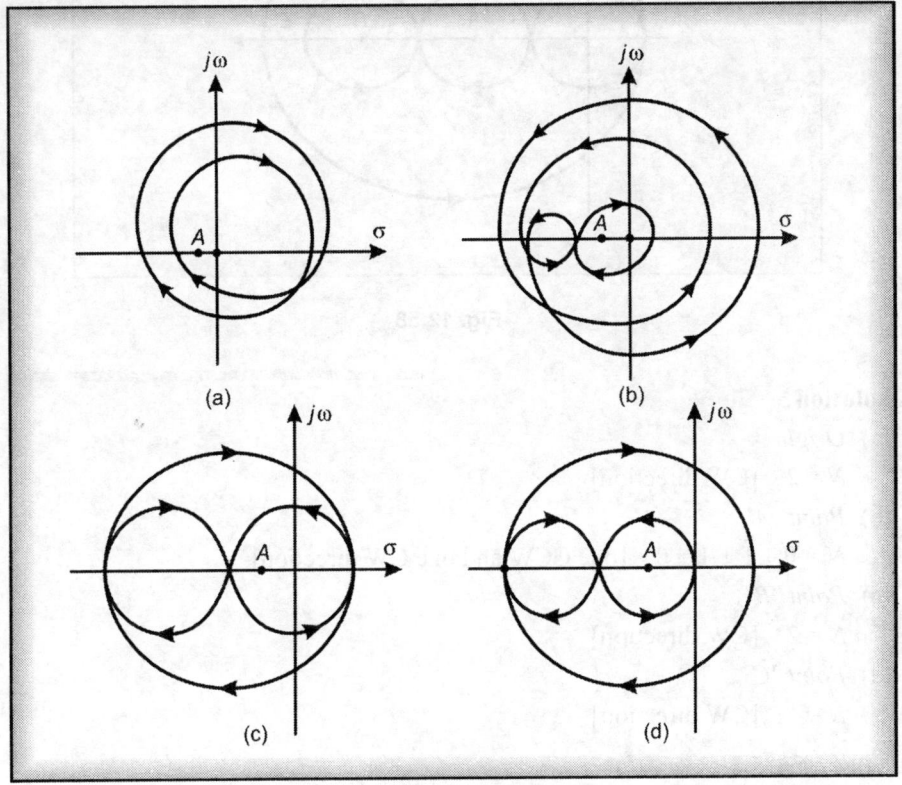

Fig. 11.57a to d

Solution

(a) $N = 2$ (CW encirclements)

(b) $N = 1 + (-2)$ (one CW and two CCW)

           $= -1$

(c) $N = -1 + (1) = 0$ (one CCW and one CW)

(d) $N = 1 + (-1) = 0$ (one CW and one CCW)

Problem 12.26 Find the number of encirclements of the

  (a) Origin   (b) Point 'A'

  (c) Point 'B'   (d) Point 'C'

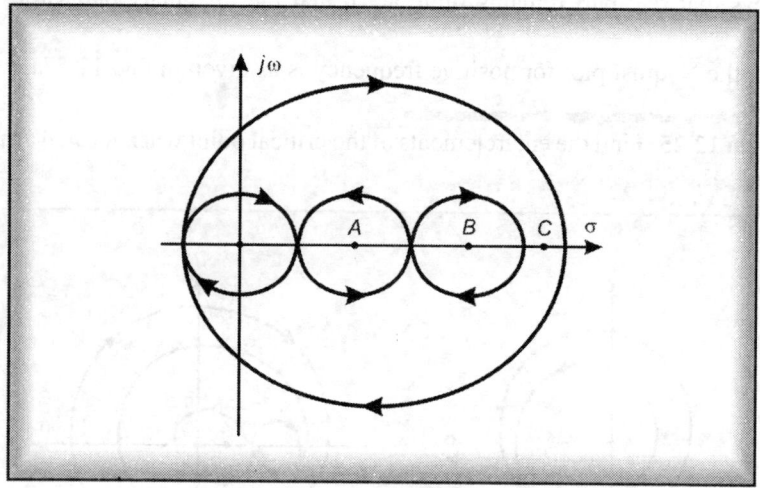

Fig. 12.58

Solution

(a) *Origin*

    $N = 2$ [CW direction]

(b) *Point 'A'*

    $N = 0 - 1 + 1 = 0$ [one CCW and one CW direction]

(c) *Point 'B'*

    $N = 2$ [CW direction]

(d) *Point 'C'*

    $N = 1$ [CW direction]

13

Bode Plot

13.1 Introduction

Sinusoidal transfer functions are commonly represented by Bode plot. It is a plot of magnitude and angle of transfer function against frequency. It is a logarithmic plot and consists of two plots:

- Plot of logarithm of magnitude against frequency on a logarithmic scale.

- Plot of phase angle against frequency on a logarithmic scale.

Logarithmic magnitude is represented as $20\log|G(j\omega)|$. The base of logarithm is 10 and the unit is 'decibel', usually abbreviated as 'db'.

The following are the advantages of Bode plot:

- Plotting of Bode plot is relatively easier as compared to other methods as the locus of $(1+sT)$ and $K/(1+sT)$ can be represented by straight line asymptotes.

- Multiplication is converted into addition. Therefore, if

$$G(s) = \frac{K}{s\,(1+sT)}, \text{ then putting } s = j\omega, \text{ we get}$$

$$20 \log |G(j\omega)| = 20 \log K - 20 \log |j\omega| - 20 \log |1 + j\omega T|$$

- Low and high frequency characteristics can be represented on a single diagram.

- Study of relative stability is easier, parameters of analysis of relative stability, *i.e.* gain and phase margins, are visible on the plots.

- If modification of an existing system is to be studied and analysed, it can be easily done on a Bode plot.

13.2 Method of Plotting

Basic terms that appear in a transfer function and method of plotting them is described below:

- **Constant term 'K':** It gives a constant magnitude of 20 log K. It does not reflect any phase shift. It is represented by a line parallel to 0 db line: (slope 0 db/dec) and starts from a point having a magnitude 20 log K.

- $(s)^{\pm 1}$: Its magnitude is $20\log\omega$. It is a straight line having a slope of ± 20 db/dec and passes through $\omega = 1$ rad/sec, 2 where its magnitude is 0 db.

- $(1 + sT)^{\pm n}$: Magnitude is given by $n \times 20\log|1 + j\omega T|$ having a slope of $\pm n \times 20$ db/dec. Asymptotes are approximated by

  (a) $\omega T \ll 1$: Magnitude is $n \times 20\log 1$, *i.e.* 0 db.

  (b) $\omega T \gg 1$: Magnitude is $\pm n \ 20\log\omega T$. It has a slope of $\pm n \times 20$ db/dec. Asymptotes meet at a point where,

    $20\log\omega T = 0$, *i.e.* $\omega T = 1$ or $\omega = 1/T$ which is called the *corner* frequency. Due to approximation, maximum error of 3db occurs at the corner frequency. Exact plot can be drawn by finding the magnitude, one octave to the left and right of the corner frequency.

- $s^2 + 2\xi\omega_n s + \omega_n^2$: This is written as

  (a) If $\xi > 1$, it can be reduced to two first order factors

  (b) If $\xi < 1$, the magnitude is

$$= \pm 20 \log \left| 1 - \left(\frac{\omega}{\omega_n} \right)^2 + j2\xi \frac{\omega}{\omega_n} \right|$$

$$= \pm 20 \log \left| \left\{ 1 - \left(\frac{\omega}{\omega_n} \right)^2 \right\}^2 + 4\xi^2 \left(\frac{\omega}{\omega_n} \right)^2 \right|^{1/2}$$

This term is approximated by drawing two straight lines. Corner frequency is ω_n. Below ω_n, the line follows 0 db line and beyond ω_n, it has a slope of: ± 40 db/dec. Error due to approximation depends upon the value of ξ. If ξ is small the error is large.

- **Phase Plot:** If the transfer function is as given below

$$G(s) = \frac{K(1 + sT_1)}{s(1 + sT_2)(s^2 + 2\xi\omega_n + \omega_n^2)}$$

then
$$G(j) = \phi = -90° + \tan^{-1}\omega T_1 - \tan^{-1}\omega T_2 - \tan^{-1}\frac{2\xi\omega/\omega_n}{1 - (\omega/\omega_n)^2}$$

SOLVED PROBLEMS

Problem 13.1 Sketch the Bode plot and determine the gain cross-over and phase cross over frequencies.

$$G(s) = \frac{10}{s(1 + 0.5s)(1 + 0.1s)}$$

(Pune University)

Solution

Corner frequency: The corner frequencies are 2 and 10.

Magnitude plot

Sr. No.	Factor	Corner Frequency (rad/sec)	Asymptotic Log-magnitude Characteristic
1	$1/s$	None	Straight line of constant slope (–20 db/dec) passing through $\omega = 1$
2	$1/(1 + 0.5s)$	$\omega_1 = 2$	Straight line of constant slope (–20 db/dec) originating from $\omega_1 = 2$
3	$1/(1 + 0.1s)$	$\omega_2 = 10$	Straight line of constant slope (–20 db/dec) originating from $\omega_2 = 10$.
4	10	None	Straight line of constant slope of 0 db/dec starting from 20log10 = 20 db point.

Magnitude plot for individual factors are shown by dotted line. Resultant line is shown by a firm line (Fig. 13.1).

Phase plot

$$\phi = -90° - \tan^{-1}0.5\omega - \tan^{-1}0.1\omega$$

Sr. No.	ω (rad/sec)	ϕ (degree)
1	0	– 90
2	0.1	– 93.43
3	1	– 122.3
4	2	– 146.31
5	5	– 184.76
6	10	– 213.7
7	15	– 228.7

Magnitude and phase plots are shown in Fig.13.1.

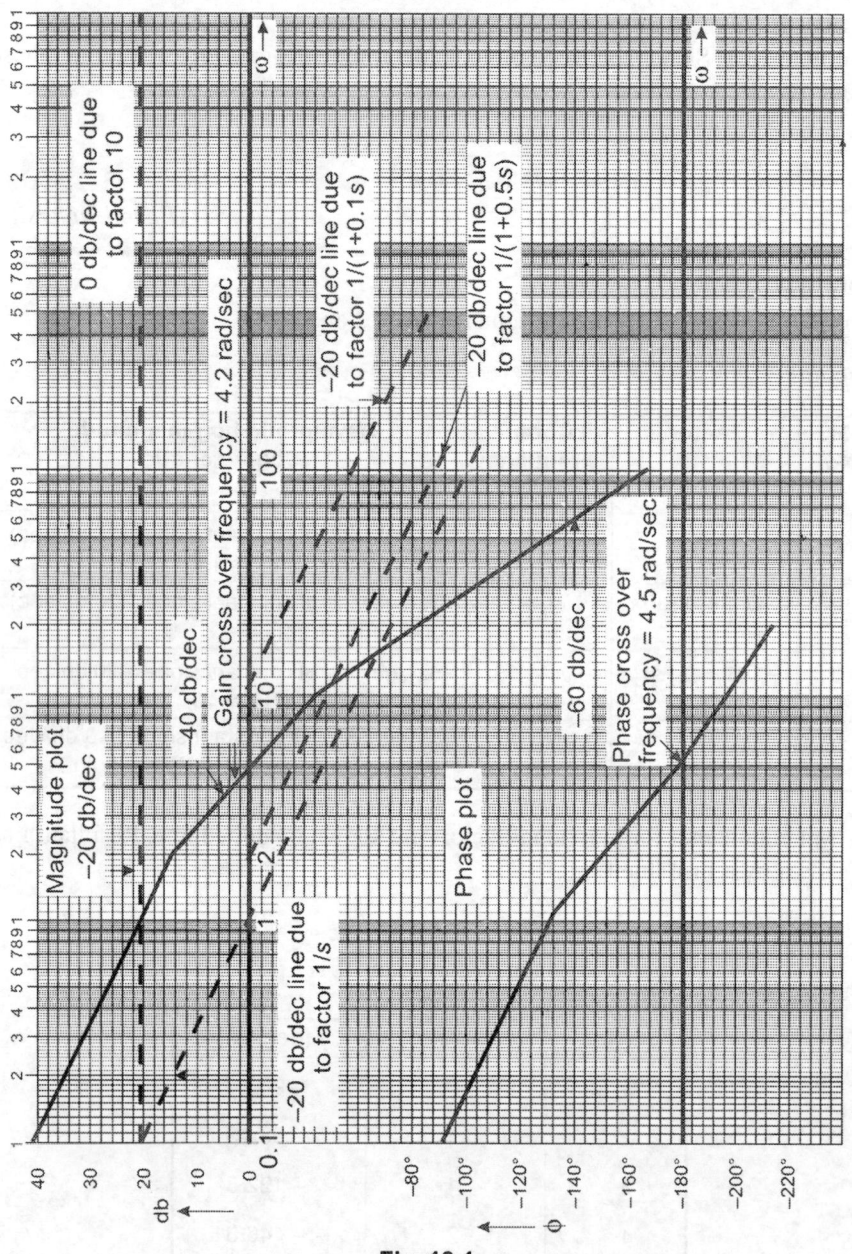

Fig. 13.1

From the plots

1. Gain crossover frequency = 4.2 rad/sec
2. Phase crossover frequency = 4.5 rad/sec.

Problem 13.2 Sketch the Bode plot for the transfer function

$$G(s) = \frac{Ks^2}{(1+0.2s)(1+0.02s)}$$

Determine the system gain K for the gain crossover frequency to be 5 rad/sec.

Solution

Let, $K = 1$, then

$$G(s) = \frac{s^2}{(1+0.2s)(1+0.02s)}$$

Corner frequency: The corner frequencies are 5 and 50 rad/sec.

Magnitude plot

Sr. No.	Factor	Corner Frequency (rad/sec)	Asymptotic Log-magnitude Characteristic
1	s^2	None	Straight line of constant slope (40 db/dec) passing through $\omega = 1$
2	$1/(1 + 0.2s)$	$\omega_1 = 5$	Straight line of constant slope (−20 db/dec) originating from $\omega = 5$
3	$1/(1 + 0.02s)$	$\omega_2 = 50$	Straight line of constant slope (−20 db/dec) originating from $\omega = 50$

The magnitude plot is shown in Fig.13.2. If the gain crossover frequency is required to be 5 rad/sec, then the magnitude plot must cross the 0 db line at 5 rad/sec. For this the plot has to be brought down by 28 db.

Hence $\quad$ 20 log $K = -28$

$\therefore \qquad\qquad K = 0.04$ $\hfill$ **Ans**

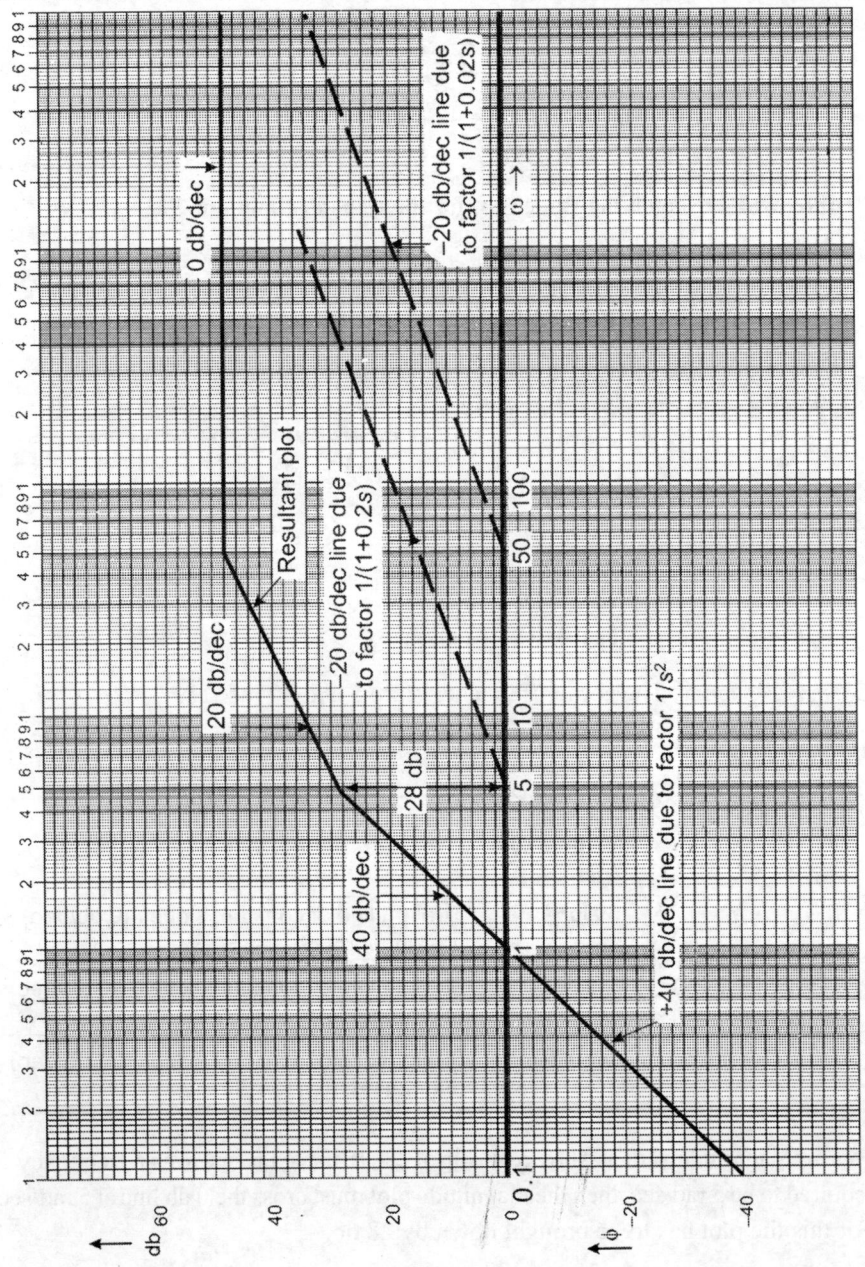

Fig. 13.2

Problem 13.3 Sketch the Bode plot for a unity feedback system characterized by the open-loop transfer function

$$G(s) = \frac{K(1 + 0.2s)(1 + 0.025s)}{s^3(1 + 0.001s)(1 + 0.005s)}$$

Show that the system is conditionally stable. Find the range of values of K for which the system is stable.

Solution

Let $K = 1$

Magnitude plot

Sr. No.	Factor	Corner Frequency (rad/sec)	Asymptotic Log-magnitude Characteristics
1	$1/s^3$	None	Straight line of constant slope of (–60 db/dec) passing through $\omega = 1$
2	$(1 + 0.2s)$	5	Straight line of constant slope of 20 db/dec originating from $\omega = 5$.
3	$(1 + 0.025 s)$	40	Straight line of constant slope of 20 db/dec originating from $\omega = 40$
4	$1/(1 + 0.005 s)$	200	Straight line of constant slope of (–20 db/dec) originating from $\omega = 200$
5	$1/(1 + 0.001s)$	1000	Straight line of constant slope of (–20 db/dec) originating from $\omega = 1000$

Phase plot
$$\phi = -270° + \tan^{-1}0.2\omega + \tan^{-1}0.025\omega - \tan^{-1}0.005\omega - \tan^{-1}0.01\omega$$

Sr. No.	ω (rad/sec)	ϕ (degree)
1	0.1	– 268
2	0.8	– 260
3	1	– 257
4	3	– 236
5	10	– 198
6	15	– 183
7	30	– 163
8	60	– 148
9	100	– 147
10	300	– 172
11	400	– 182

Resultant magnitude and phase plots are shown in Fig. 13.3

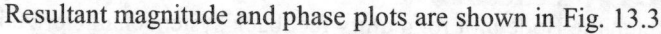

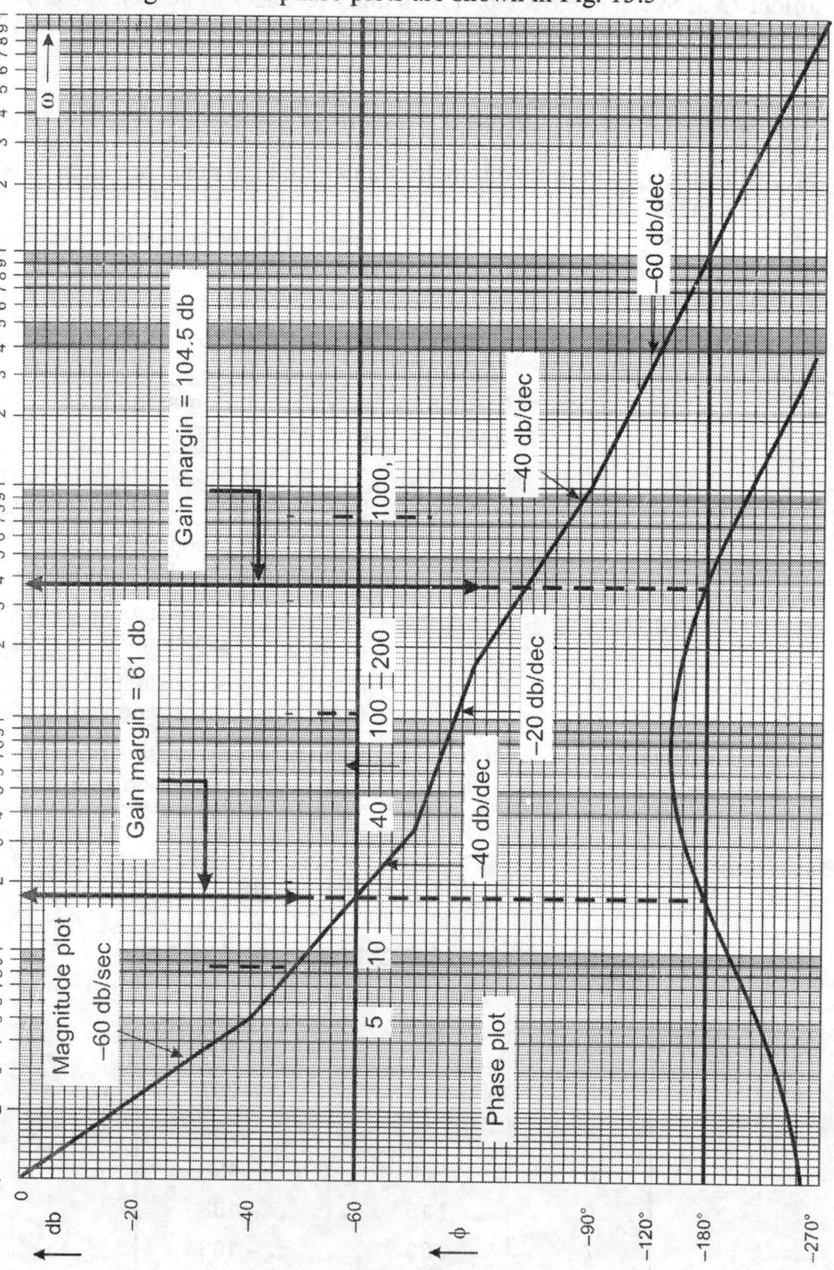

Fig. 13.3

The phase plot crosses the (– 180°) line twice indicating that the system under consideration is conditionally stable.

Gain margins are 61 db and 104.5 db

Now $20\log K = 61$ $\therefore$ $K = 1122$
Also $20\log K = 104.5$ $\therefore$ $K = 167880$
$\therefore$ Condition for stability is $1122 < K < 167880$ **Ans**

Problem 13.4 Draw the Bode plot for a system having

$$G(s)H(s) = \frac{100}{s(s+1)\,(s+2)}$$

Find
   (a) Gain margin   (b) Phase margin   (c) Gain crossover frequency
   (d) Phase crossover frequency   *(Pune University)*

Solution

$$G(s)H(s) = \frac{50}{s\,(1+s)\,(1+0.5s)}$$

Magnitude plot

Sr. No.	Factor	Corner Frequency (rad/sec)	Asymptotic log magnitude Characteristics
1	50	None	Straight line of slope (0 db/dec) starting from point 20log50 = 34 db
2	$1/s$	None	Straight line of slope (–20 db/dec) passing through $\omega = 1$
3	$1/(1 + s)$	1	Straight line of slope (–20 db/dec) originating from $\omega = 1$
4	$1/(1 + 0.5s)$	2	Straight line of slope (–20 db/dec) originating from $\omega = 2$

Phase plot

$$\phi = -90° - \tan^{-1}\omega - \tan^{-1}0.5\omega$$

Sr. No.	ω (rad/sec)	ϕ (degree)
1	0	– 90
2	0.1	– 98.6
3	0.2	– 107
4	0.5	– 130.6
5	1	– 161.6
6	1.3	– 175.5
7	1.4	– 179.5
8	1.5	– 183.2
9	2	– 198.4
10	4.45	– 233

Magnitude and phase plots are shown in Fig. 13.4

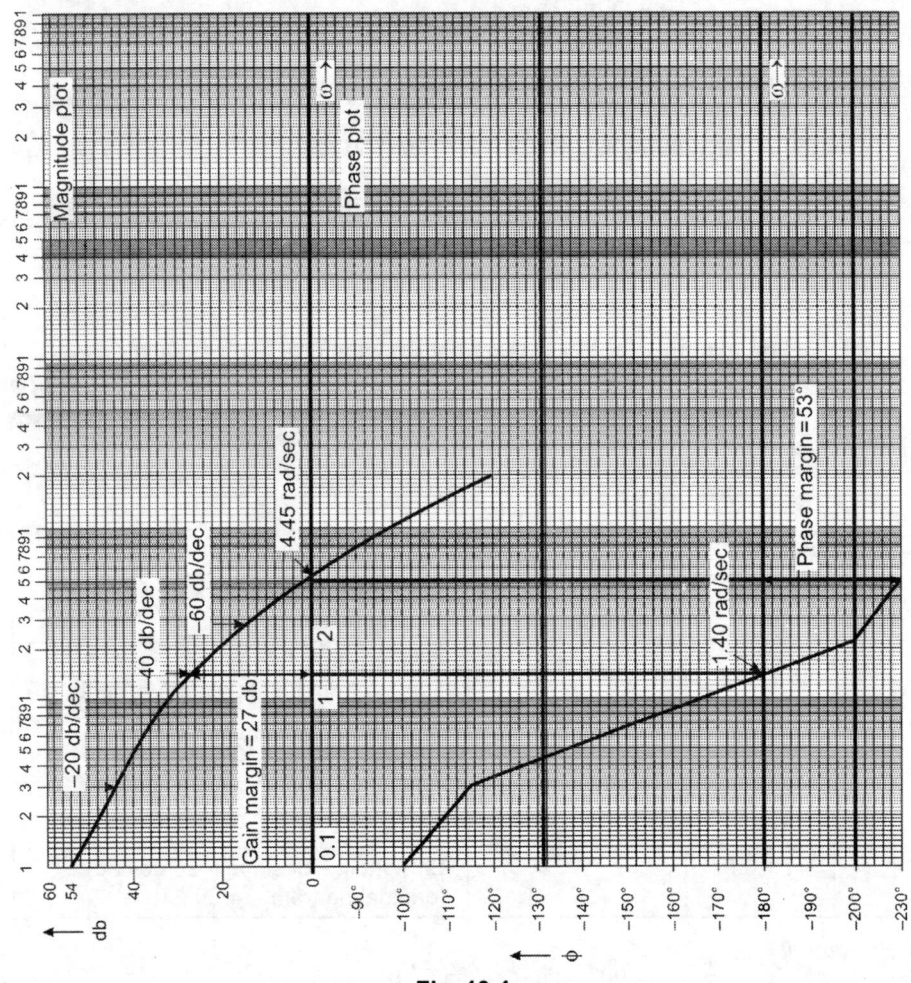

Fig. 13.4

Result:

1. Gain crossover frequency : 4.45 rad/sec
2. Phase crossover frequency : 1.40 rad/sec
3. Gain margin : 27 db
4. Phase margin : 53°

Problem 13.5 The open-loop transfer function of a unity feedback system is

$$G(s) = \frac{K}{s(s+2)(s+20)}$$

Construct Bode plot and determine.

    (a) Limiting value of K for system to be stable

    (b) Value of K for gain margin to be 10 db

    (c) Value of K for phase margin to be 50°   *(Pune Universty)*

Solution $G(s) = \dfrac{0.025K}{s(1+0.5s)(1+0.05s)}$

Let $0.025\,K = 1$, then $G(s) = \dfrac{1}{s(1+0.5s)(1+0.05s)}$

Magnitude plot

Sr. No.	Factor	Corner Frequency (rad/sec)	Asymptotic log magnitude Characteristics
1	$1/s$	None	Straight line of –20 db/dec passing through $\omega = 1$
2	$1/(1 + 0.5s)$	2	Straight line of (–20 db/dec) originating from $\omega = 2$
3	$1/(1 + 0.05s)$	20	Straight line of (–20 db/dec) originating from $\omega = 20$

Phase plot

$$\phi = -90° - \tan^{-1} 0.5\omega - \tan^{-1} 0.05\omega$$

Sr. No.	ω (rad/sec)	ϕ (degree)
1	0	– 90
2	1	– 119
3	2	– 141
4	2.5	– 148.5
5	3	– 155
6	4	– 165
7	4.5	– 168.7
8	5	– 172
9	6	– 178.3
10	6.5	– 181
11	10	– 195

Bode plot is shown in Fig. 13.5

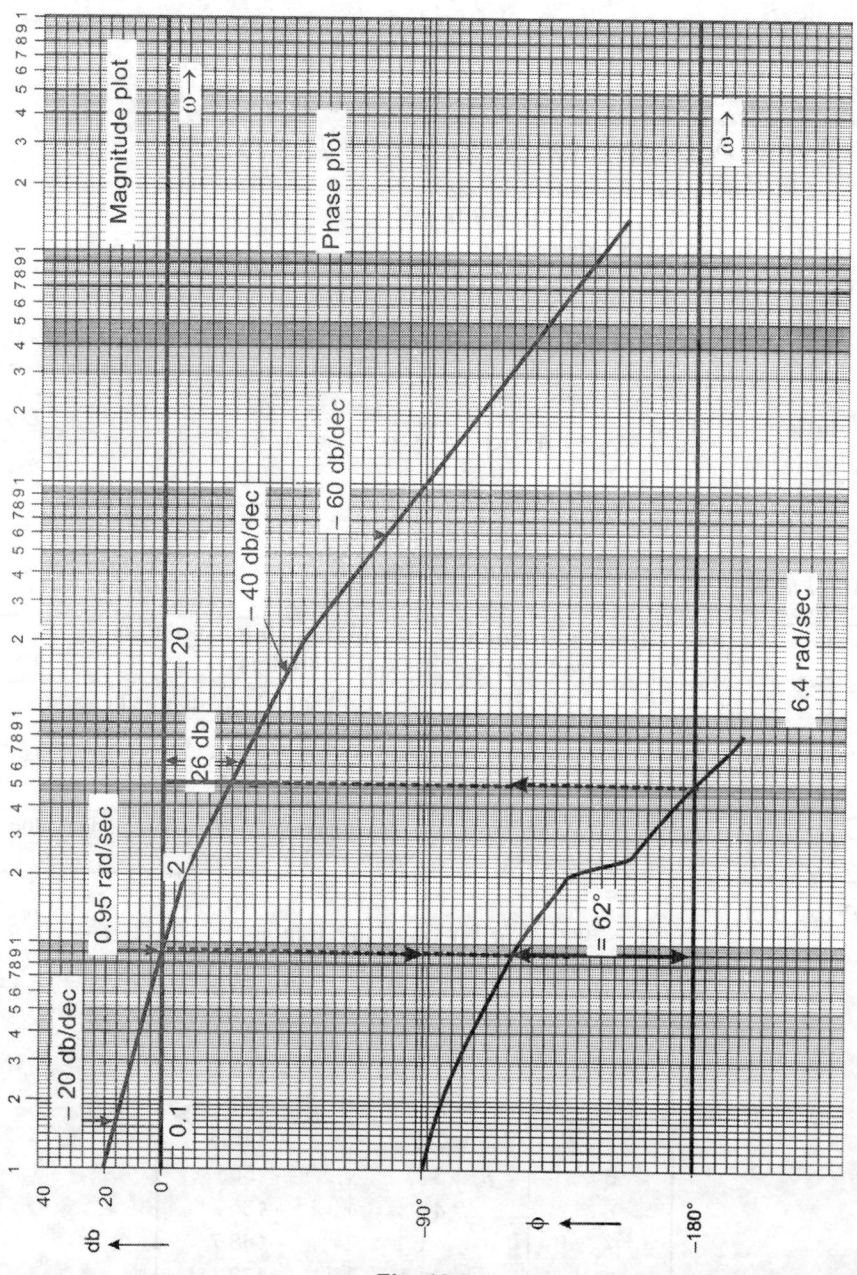

Fig. 13.5

(a) From the curves, the gain margin is 26 db

$\therefore$ $\qquad$ $20 \log K_1 = 26$

or $\qquad$ $K_1 = 19.95$

or $\qquad$ $0.025K = 19.95$

or $\qquad$ $K = \dfrac{19.95}{0.025} = 798$

(b) For the gain margin to be 10 db, the graph has to be lifted up by

$\qquad$ $26 - 10 = 16$ db

$\therefore$ $\qquad$ $20 \log K_1 = 16$

or $\qquad$ $K_1 = 6.3$

or $\qquad$ $0.025K = 6.3$

or $\qquad$ $K = 252$

(c) For the phase margin to be 50° the value of ω at $-180° + 50° = -130°$ is 1.9 rad/sec. Gain margin at 1.9 rad/sec is 5.5 db. Therefore, to have phase margin of 50°, magnitude plot has to be lifted up by 5.5 db so that gain cross-over frequency is 1.9 rad/sec

$\qquad$ $20 \log K_1 = 5.5$

or $\qquad$ $K_1 = 1.88$

or $\qquad$ $0.025K = 1.88$

or $\qquad$ $K = \dfrac{1.88}{0.025} = 75.35$ $\qquad$ **Ans**

Problem 13.6 Construct the Bode plots on a semilog graph paper for a unity feedback system whose open-loop transfer function is given by

$$G(s) = \frac{10}{s(1+s)(1+0.02s)}$$

From the Bode-plot determine
(a) Gain and phase crossover frequencies
(b) Gain and phase margin, and
(c) Stability of the closed-loop system $\qquad$ (*Pune University*)

Solution

Magnitude plot

Sr. No.	Factor	Corner Frequency (rad/sec)	Asymptotic Log-magnitude Characteristics
1	10	None	Straight line of 0 db/dec from point $20\log 10 = 20$ db
2	$1/s$	None	Straight line of $(-20$ db/dec) passing through $\omega = 1$
3	$1/(1 + s)$	1	Straight line of $(-20$ db/dec) originating from $\omega = 1$
4	$1/(1 + 0.02 s)$	50	Straight line of $(-20$ db/dec) originating from $\omega = 50$

Phase plot

$$\phi = -90° - \tan^{-1}\omega - \tan^{-1} 0.02\omega$$

Sr. No.	ω (rad/sec)	ϕ (degree)
1	0	-90
2	0.2	-101.5
3	0.5	-117
4	1	-136
5	10	-185.6
6	50	-224

Bode plot is shown in Fig. 13.6

Result

1. Gain crossover frequency = 3 rad/sec
2. Phase crossover frequency = 7.2 rad/sec
3. Gain margin = 15 db
4. Phase margin = 16°

Stability

Since phase margin and gain margin are positive, the system is absolutely stable.

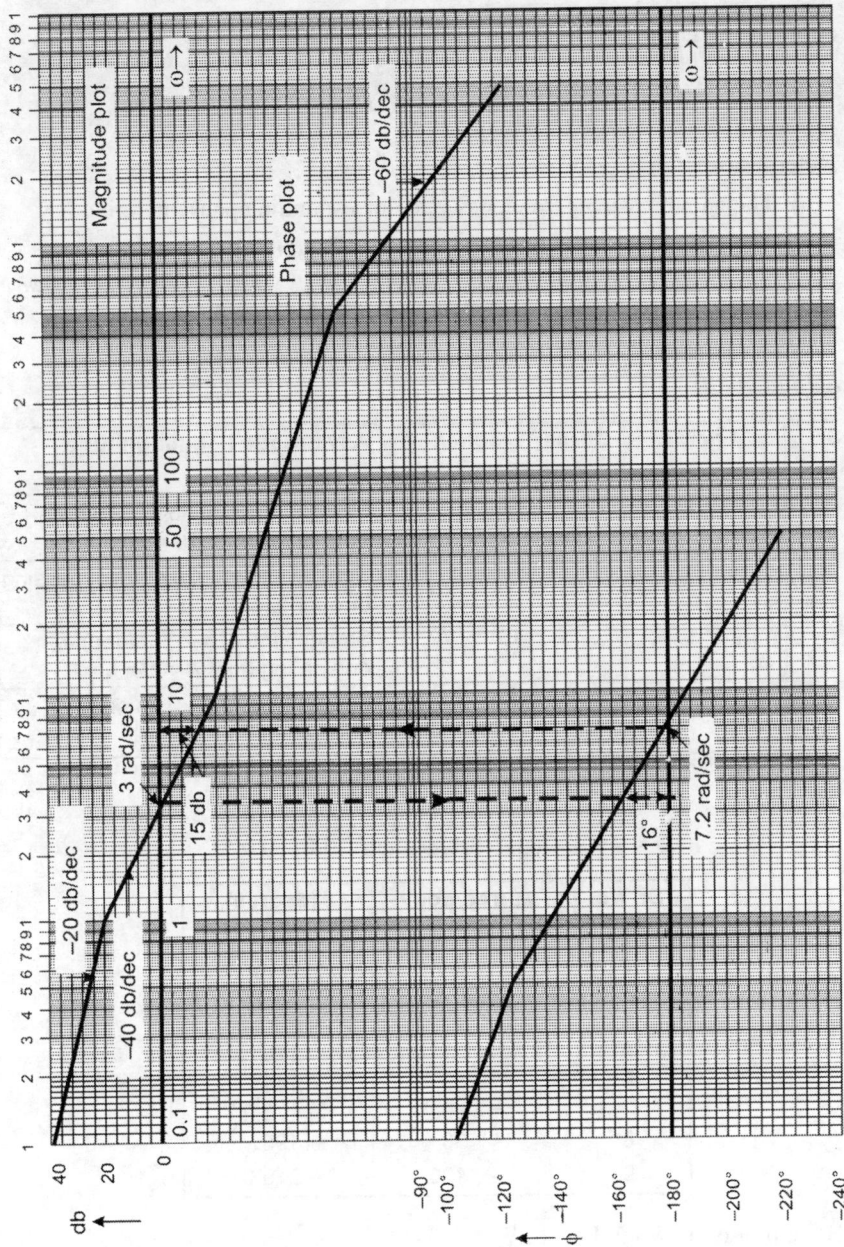

Fig. 13.6

Problem 13.7 Draw the Bode plot for the system having

$$G(s) = \frac{10}{s(1 + 0.01s)(1 + 0.1s)} \; ; H(s) = 1$$

Determine

(a) The gain crossover frequency and corresponding phase margin

(b) The phase crossover frequency and corresponding gain margin

(Pune University)

Solution

Magnitude plot

Sr. No.	Factor	Corner Frequency (rad/sec)	Asymptotic Log-magnitude Characteristics
1	10	None	Straight line of 0 db/dec starting from point 20log10 = 20 db
2	$1/s$	None	Straight line of (–20 db/dec) passing through $\omega = 1$
3	$1/(1 + 0.1s)$	10	Straight line of (–20 db/dec) originating from $\omega = 10$
4	$1/(1 + 0.01s)$	100	Straight line of (–20 db/dec) originating from $\omega = 100$

Phase plot

$$\phi = -90° - \tan^{-1} 0.1\,\omega - \tan^{-1} 0.01\,\omega$$

Sr. No.	ω (rad/sec)	ϕ (degree)
1	0	– 90
2	0.1	– 90.6
3	0.5	– 93.15
4	1	– 96.3
5	5	– 119
6	10	– 141
7	30	– 178.3
8	50	– 195
9	100	– 219.3

Bode plot is shown in Fig. 13.7

Result

1. Gain crossover frequency : 10 rad/sec

2. Phase crossover frequency : 31.5 rad/sec

3. Gain margin : 20 db

4. Phase margin : 39°

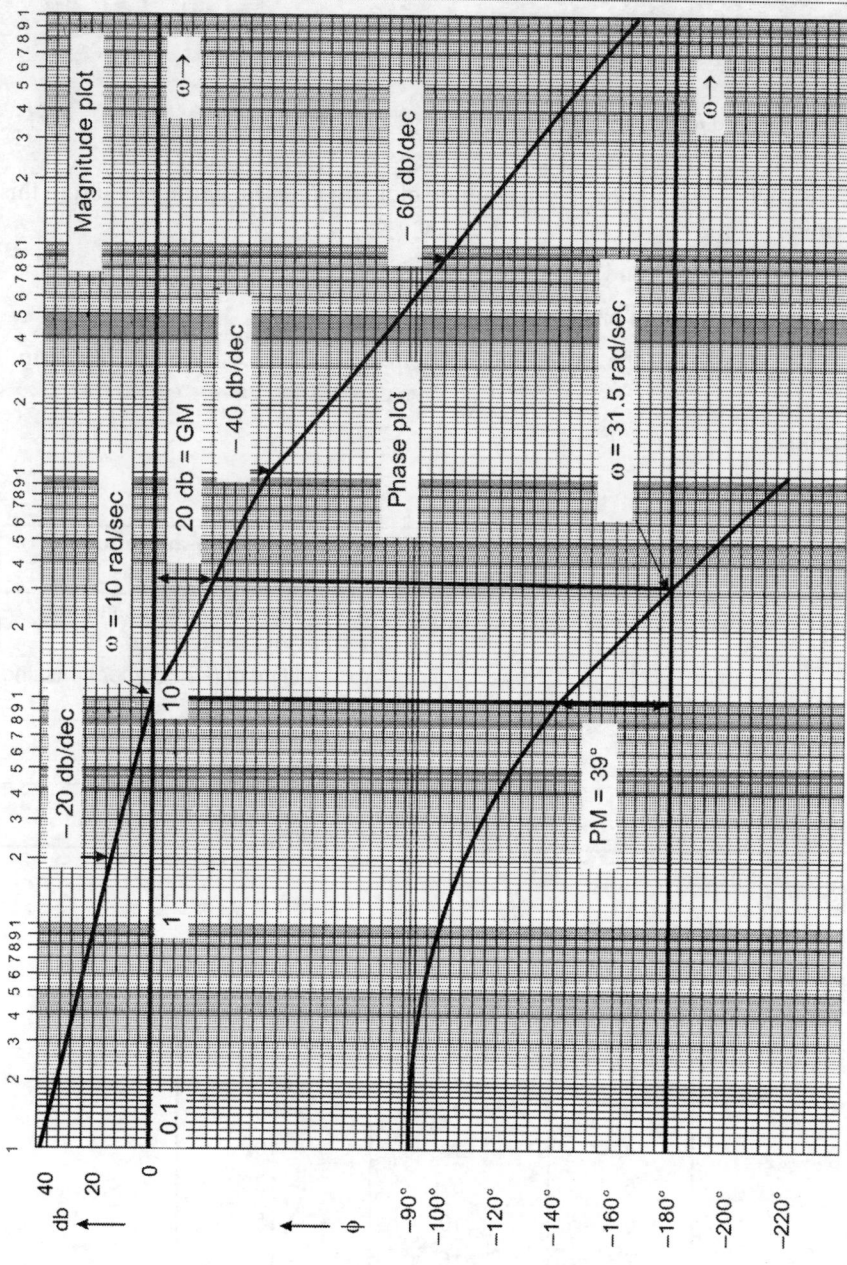

Fig. 13.7

Problem 13.8 A feedback control system is described by

$$G(s) = \frac{10}{s(1 + 0.2s)(1 + 0.01s)}; \quad H(s) = 1$$

Construct as asymptotic log-magnitude plot and an exact phase plot. From this determine:

(a) Gain crossover and phase crossover frequencies

(b) Gain margin and phase margin

(c) The stability of the closed-loop system. *(Pune University)*

Solution

Magnitude plot

Sr. No.	Factor	Corner Frequency (rad/sec)	Asymptotic Log-magnitude Characteristics
1	10	None	Straight line of 0 db/dec starting from point 20log 10 = 20 db
2	1/s	None	Straight line of (–20 db/dec) passing through $\omega = 1$
3	1/(1 + 0.2s)	5	Straight line of (–20 db/dec) originating from $\omega = 5$
4	1/(1 + 0.01s)	100	Straight line of (–20 db/dec) originating from $\omega = 100$

Phase plot

$$\phi = -90° - \tan^{-1}0.2\omega - \tan^{-1}0.01\omega$$

Sr. No.	ω (rad/sec)	ϕ (degree)
1	0	– 90
2	0.1	– 91.2
3	0.5	– 96
4	1	– 102
5	5	– 138
6	10	– 159
7	15	– 170
8	20	– 177
9	25	– 183
10	50	– 200.8
11	100	– 222

Bode plot is shown in Fig. 13.8

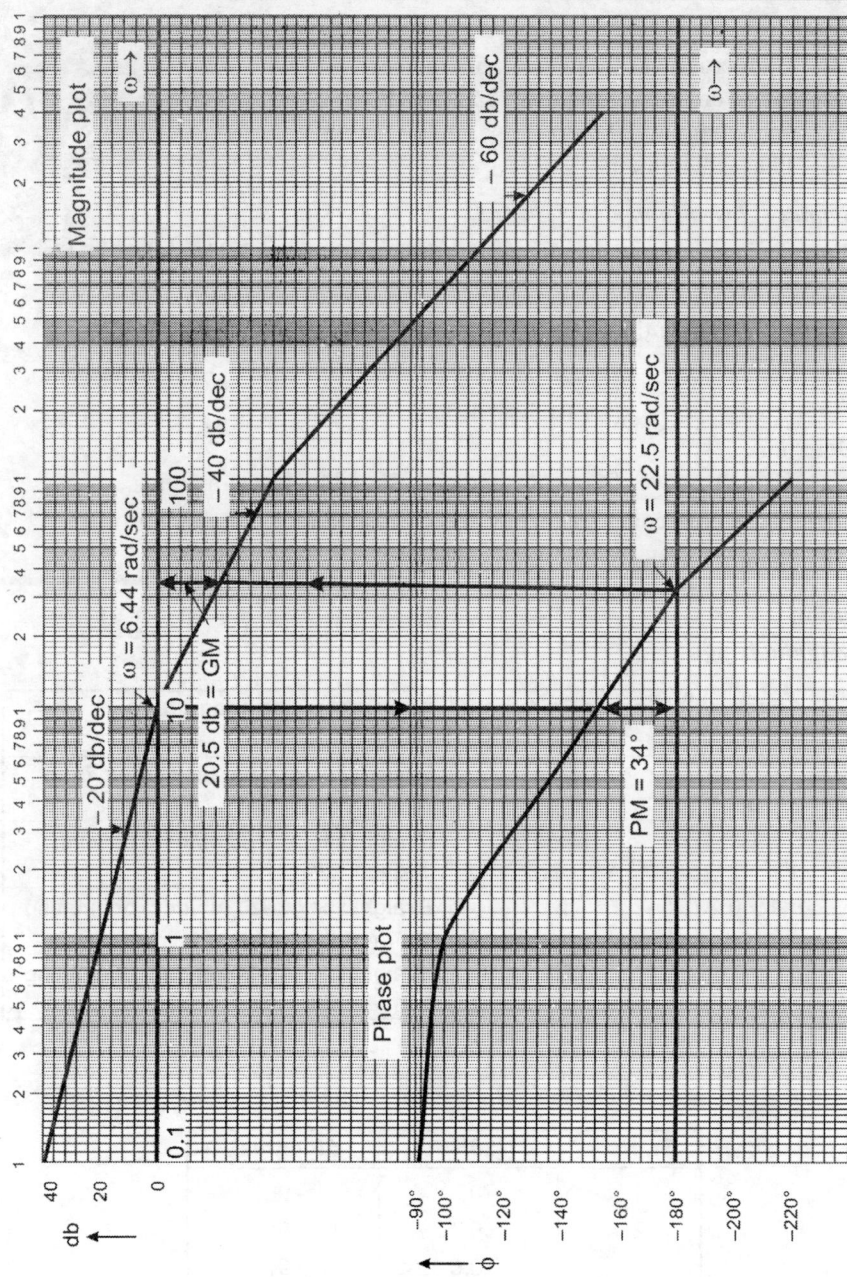

Fig. 13.8

Result
1. Gain crossover frequency : 6.4 rad/sec
2. Phase crossover frequency : 22.5 rad/sec
3. Gain margin : 20.5 db
4. Phase margin : 34°

Stability closed-loop

Since, gain and phase margins are positive, the system is absolutely stable.

Problem 13.9 Draw the Bode plot for a system having

$$G(s) = \frac{3}{s(1 + 0.05s)(1 + 0.2s)}; \quad H(s) = 1$$

Determine
(a) Gain crossover frequency and corresponding phase-margin.
(b) Phase crosscover frequency and corresponding gain-margin
(c) Stability of the closed-loop system.

Solution

Magnitude plot

Sr. No.	Factor	Corner Frequency (rad/sec)	Asymptotic Log-magnitude Characteristics
1	3	None	Straight line of 0 db/dec starting from 20 log 3 = 9.5 db
2	1/s	None	Straight line of (–20 db/dec) passing through $\omega = 1$
3	1/(1 + 0.2s)	5	Straight line of –20 (db/dec) originating from $\omega = 5$
4	1/(1 + 0.05s)	20	Straight line of (–20 db/dec) originating from $\omega = 20$

Phase plot

$$\phi = -90° - \tan^{-1} 0.2\omega - \tan^{-1} 0.05\omega$$

Sr. No.	ω (rad/sec)	ϕ (degree)
1	0	– 90
2	0.1	– 91.4
3	0.5	– 97
4	1	– 104.2
5	5	– 149
6	10	– 180
7	20	– 211

Bode plot is shown in Fig. 13.9

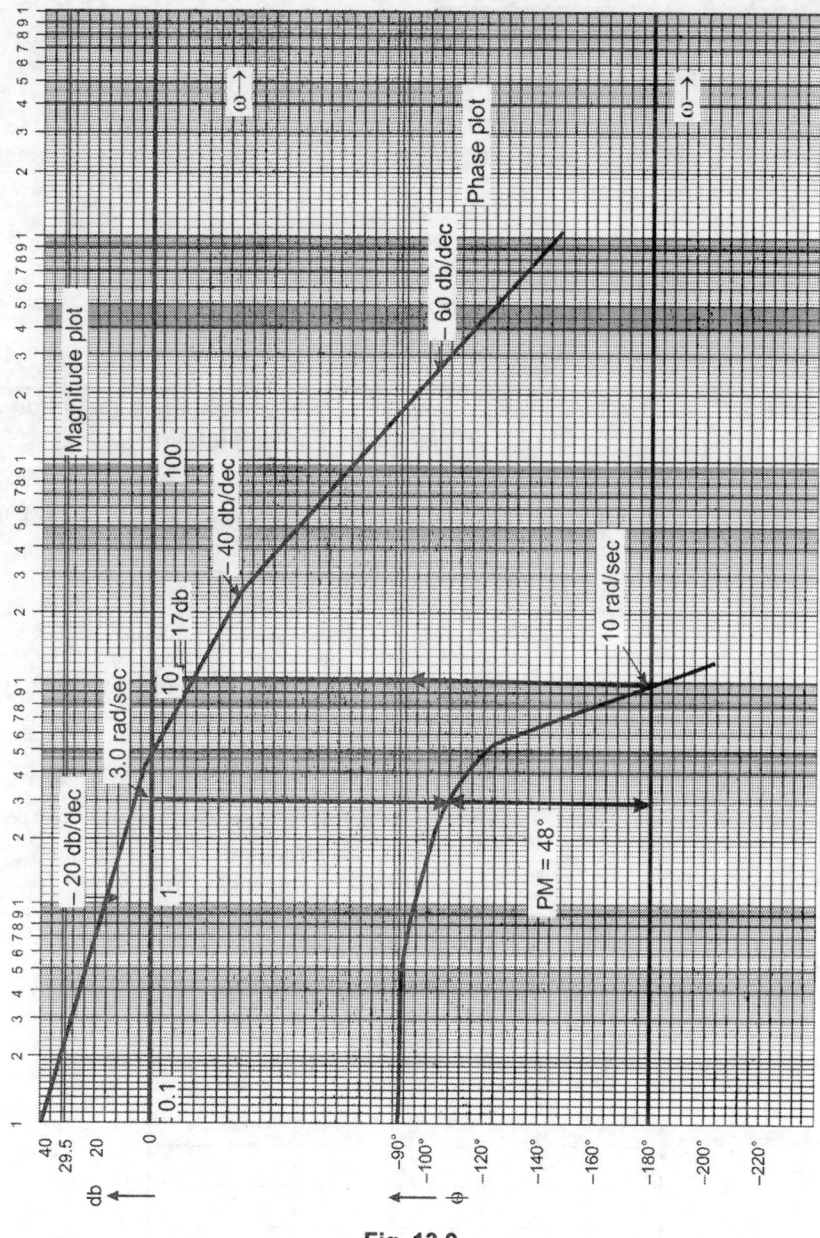

Fig. 13.9

Result

1. Gain crossover frequency = 3.0 rad/sec
2. Phase crossover frequency = 10 rad/sec
3. Gain margin = 17 db
4. Phase margin = 48°

Stability

Since, gain and phase margins are positive, the system is absolutely stable.

Problem 13.10 Construct the Bode plot for a unity feedback control system having

$$G(s) = \frac{10(s+10)}{s(s+2)(s+5)}$$

From the plot obtain the gain margin and phase margin. Comment on the stability of the system. *(Pune University)*

Solution $G(s) = \dfrac{10(1+0.1s)}{s(1+0.5s)(1+0.2s)}$

Magnitude plot

Sr. No.	Factor	Corner Frequency (rad/sec)	Asymptotic Log-magnitude Characteristics
1	10	None	Straight line of 0 db/dec starting from 20 log10 = 20 db point.
2	1/s	None	Straight line of (– 20 db/dec) passing through $\omega = 1$
3	1/(1 + 0.5s)	2	Straight line of (– 20 db/dec) originating from $\omega = 2$
4	1/(1 + 0.2s)	5	Straight line of (– 20 db/dec) originating from $\omega = 5$
5	(1 + 0.1s)	10	Straight line of (+ 20 db/dec) originating from $\omega = 10$

Phase plot

$$\phi = -90° + \tan^{-1} 0.1\omega - \tan^{-1} 0.5\omega - \tan^{-1} 0.2\omega$$

Sr. No.	ω (rad/sec)	ϕ (degrees)
1	0	– 90
2	0.1	– 93
3	0.5	– 107
4	1	– 122
5	2	– 145.5
6	5	– 176.6
7	8	– 185.3
8	10	– 187
9	20	– 205

Bode plot is shown in Fig. 13.10

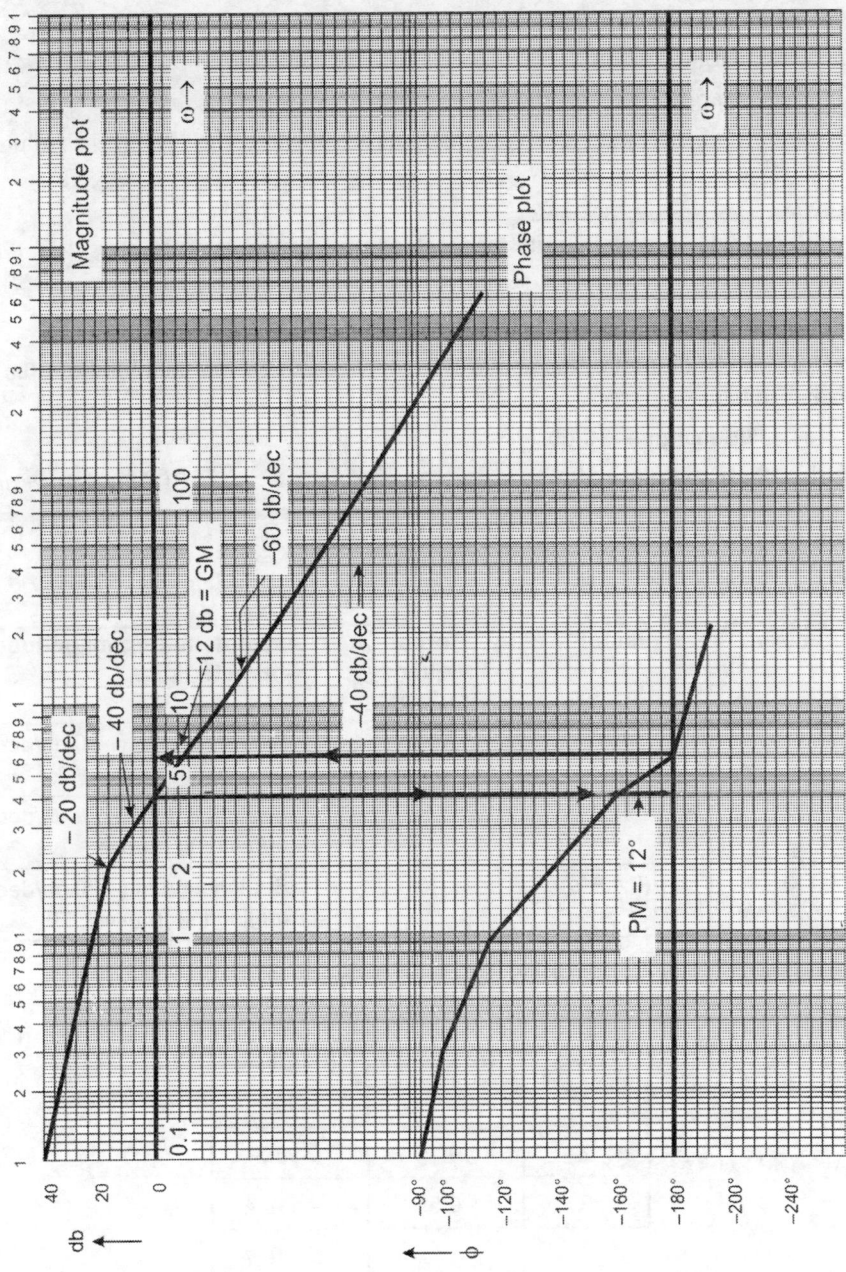

Fig. 13.10

Result

1. Gain margin : 12 db
2. Phase margin : 12°
3. *Stability:* Since, gain and phase margins are positive, the system is absolutely stable.

Problem 13.11 A certain unity feedback control system is given by

$$G(s) = \frac{K}{s\,(1+s)\,(1+0.1s)}$$

Draw the Bode-plot of the above system. Determine from the plot the value of K so as to have

(a) Gain margin = 10 db
(b) Phase margin = 50°

(*Pune University*)

Solution Let $K = 1$

Magnitude plot

Sr. No.	Factor	Corner Frequency (rad/sec)	Asymptotic Log-magnitude Characteristics
1	1/s	None	Straight line of (– 20 db/dec) passing through $\omega = 1$
2	1/(1 + s)	1	Straight line of (– 20 db/dec) originating from $\omega = 1$
3	1/(0.1 + s)	10	Straight line of (– 20 db/dec) originating from $\omega = 10$.

Phase plot $\phi = -90° - \tan^{-1}\omega - \tan^{-1}0.1\omega$

Sr. No.	ω (rad/sec)	ϕ (degree)
1	0	– 90
2	0.1	– 96.3
3	0.5	– 119.4
4	1	– 140.7
5	3	– 178.3
6	5	– 195
7	10	– 219

Bode plot is shown in Fig. 13.11

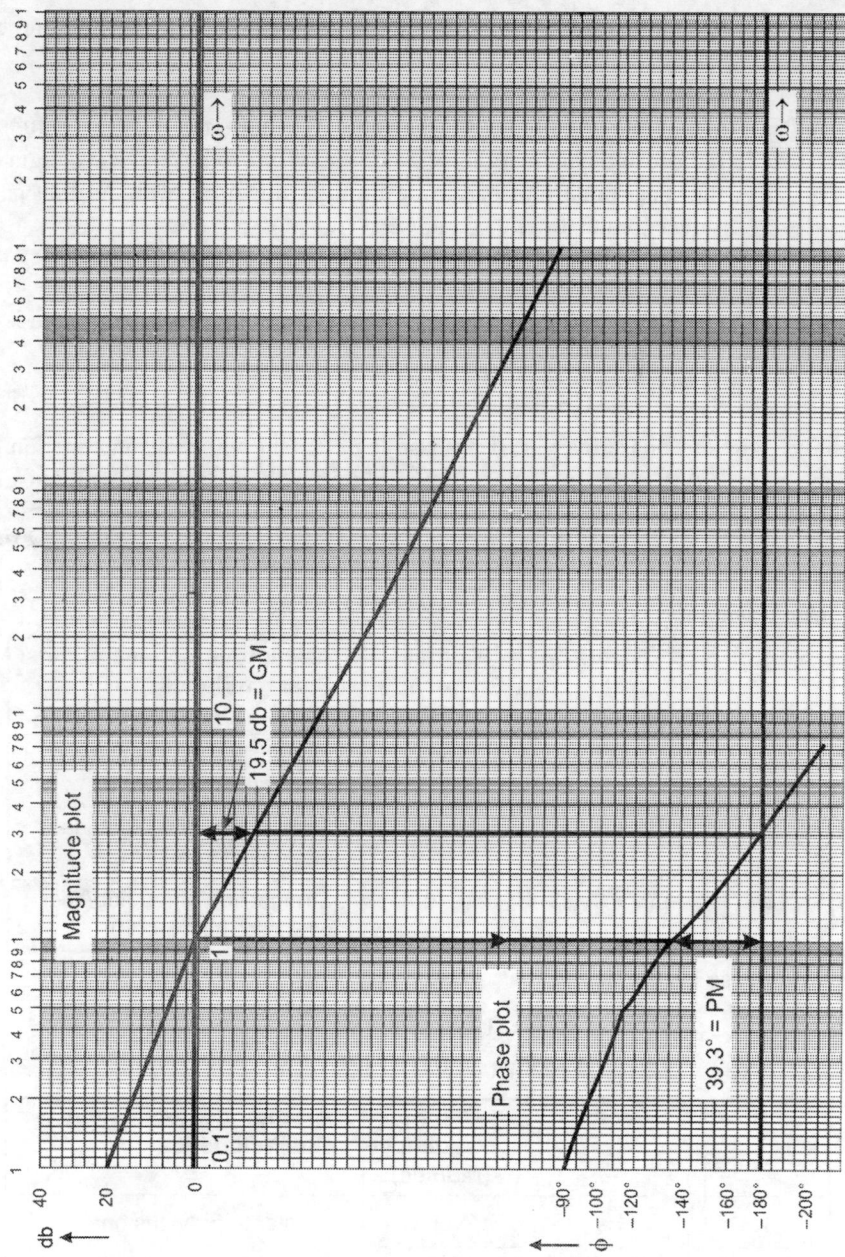

Fig. 13.11

(a) For the gain margin to be 10 db the magnitude plot has to be lifted up by
    $19.5 - 10 = 9.5$ db
    i.e. $20 \log K = 9.5$ or $K = 2.98$

(b) For the phase margin to be 50° the value of ω at $(-180° + 50° = -130°)$ is
    0.73 rad/sec and gain margin is –5 db. Therefore, to have phase margin of
    50°, the magnitude plot has to be brought down by 5 db, so that gain
    cross-over frequency is 0.73 rad/sec.
    $20 \log K = -5$ or $K = 0.56$

Problem 13.12 The open-loop transfer function of a unity feedback system is

$$G(s) = \frac{1}{s\,(1 + 0.5s)\,(1 + 0.1s)}$$

Find gain and phase margin. If a phase-lag element with transfer function of
$(1 + 2s)/(1 + 5s)$ is added in the forward path, find by how much the gain must be
changed to keep the margin same.

Solution

Magnitude plot

Sr. No.	Factor	Corner Frequency (rad/sec)	Asymptotic Log-magnitude Characteristics
1	$1/s$	None	Straight line of (–20 db/dec) passing through $\omega = 1$
2	$1/(1 + 0.5s)$	2	Straight line of (–20 db/dec) originating from $\omega = 2$
3	$1/(1 + 0.1s)$	10	Straight line of (–20 db/dec) originating from $\omega = 10$

With phase lag element

$$G(s) = \frac{(1 + 2s)}{s(1 + 0.5s)(s + 0.1s)(1 + 5s)}$$

Sr. No.	Factor	Corner Frequency (rad/sec)	Asymptotic Log-magnitude Characteristics
1	$1/s$	None	Similar to previous one.
2	$1/(1 + 5s)$	0.2	Straight line of (–20 db/dec) originating from $\omega = 0.2$
3	$(1 + 2s)$	0.5	Straight line of 20 db/dec originating from $\omega = 0.5$.
4	$1/(1 + 0.5s)$	2	Similar to previous one
5	$1/(1 + 0.1s)$	10	—do—

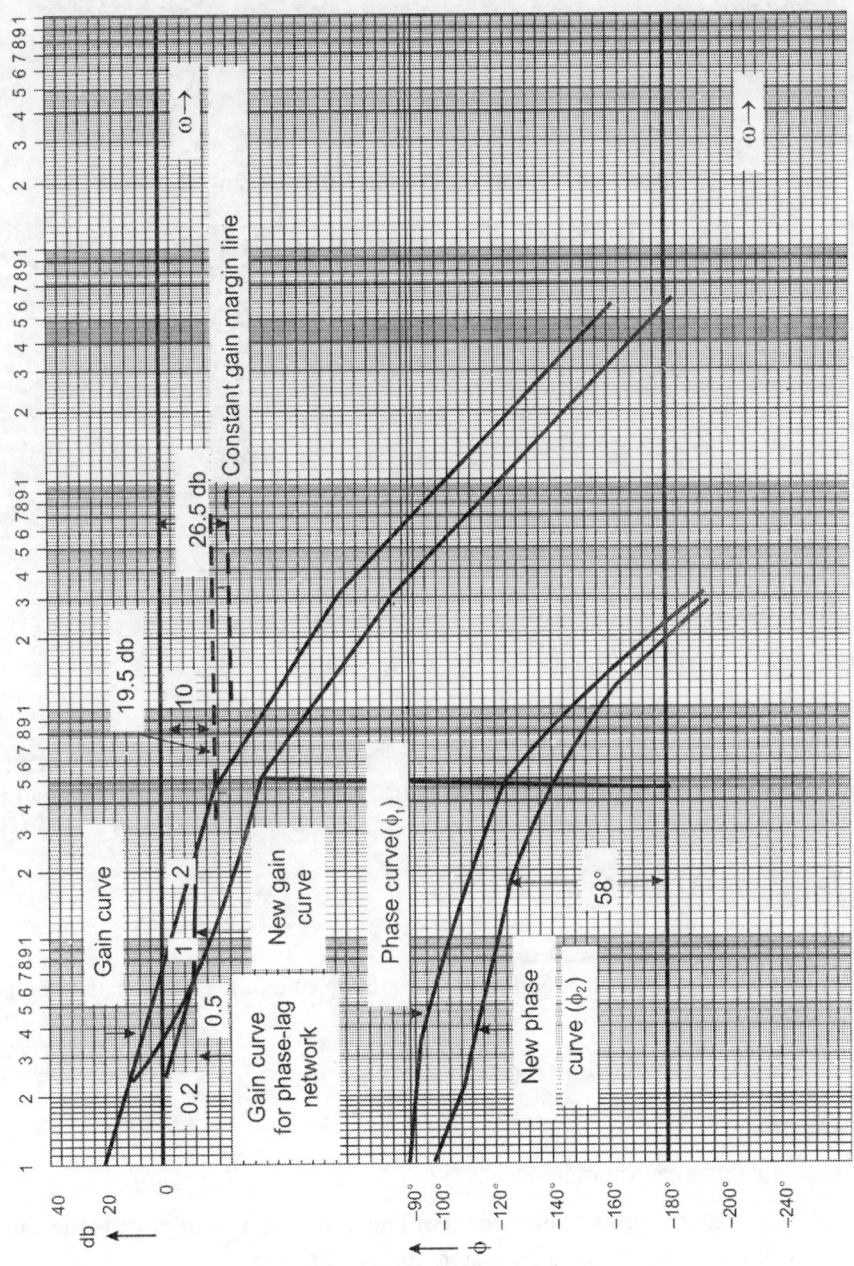

Fig. 13.12

Phase plot

(a) *Without phase lag*

$$\phi_1 = -90° - \tan^{-1}0.5\omega - \tan^{-1}0.1\omega$$

(b) *With phase lag*

$$\phi_2 = -90° - \tan^{-1}0.5\omega - \tan^{-1}0.1\omega + \tan^{-1}2\omega - \tan^{-1}5\omega$$

Sr. No.	ω (rad/sec)	ϕ_1 (degrees)	ϕ_2 (degree)
1	0.1	– 93.4	– 105
2	0.2	– 96.8	– 120
3	0.5	– 107	– 130
4	1	– 122.3	– 137.5
5	2	– 146.31	– 154.6
6	5	– 184.76	– 188
7	10	– 228.7	– 230.4

Bode plot is shown in Fig. 13.12

(a)   Gain margin = 19.5 db

(b)   Phase margin = 58°

With phase-lag network introduced the gain margin is 26.5 db. Since, gain margin is required to be kept at constant level of 19.5 db the gain must be changed by

26.5 – 19.5 = 7 db

$$20 \log K = 7 \quad \text{or} \quad K = 2.2$$

Problem 13.13   The open-loop transfer function of a unity feedback system is

$$G(s) = \frac{50}{s(s+10)(s+5)(s+1)}$$

Determine

(a) Gain and phase margin

(b) The value of steady-state error coefficient for a gain of 10 db and the value which will make the closed-loop system marginally stable.

Solution

$$G(s) = \frac{1}{s(1+s)(1+0.2s)(1+0.1s)}$$

Magnitude plot

Sr. No.	Factor	Corner Frequency (rad/sec)	Asymptotic Log-magnitude Characteristics
1	$1/s$	None	Straight line of (-20 db/dec) passing through $\omega = 1$
2	$1/(1 + s)$	1	Straight line of (-20 db/dec) originating from $\omega = 1$
3	$1/(1 + 0.2s)$	5	Straight line of (-20 db/dec) originating from $\omega = 5$
4	$1/(1 + 0.1s)$	10	Straight line of (-20 db/sec) originating from $\omega = 10$

Phase plot

$$\phi = -90° - \tan^{-1}\omega - \tan^{-1}0.1\omega - \tan^{-1}0.2\omega$$

Sr. No.	ω (rad/sec)	ϕ (degree)
1	0	-90
2	0.1	-97
3	0.2	-105
4	0.5	-125
5	1	-152
6	2	-187
7	3	-209
8	4	-227

Bode plot is shown in Fig. 13.13.

(a) (i) Gain margin = 10 db

    (ii) Phase margin = 28°

(b) Initial slope of the line is -20 db/sec and intersects 0 db line at $\omega = 1$ rad/sec. Therefore $K_v = 1$

Steady-state error coefficient for the system to be marginally stable is found by lifting the magnitude curve by 10 db (GM). The new magnitude curve is drawn and initial line having a slope of -20 db/sec is extended till it intersects 0 db line. Intersection occurs at $\omega = 3.2$ rad/sec. $\therefore K_v = 3.2$

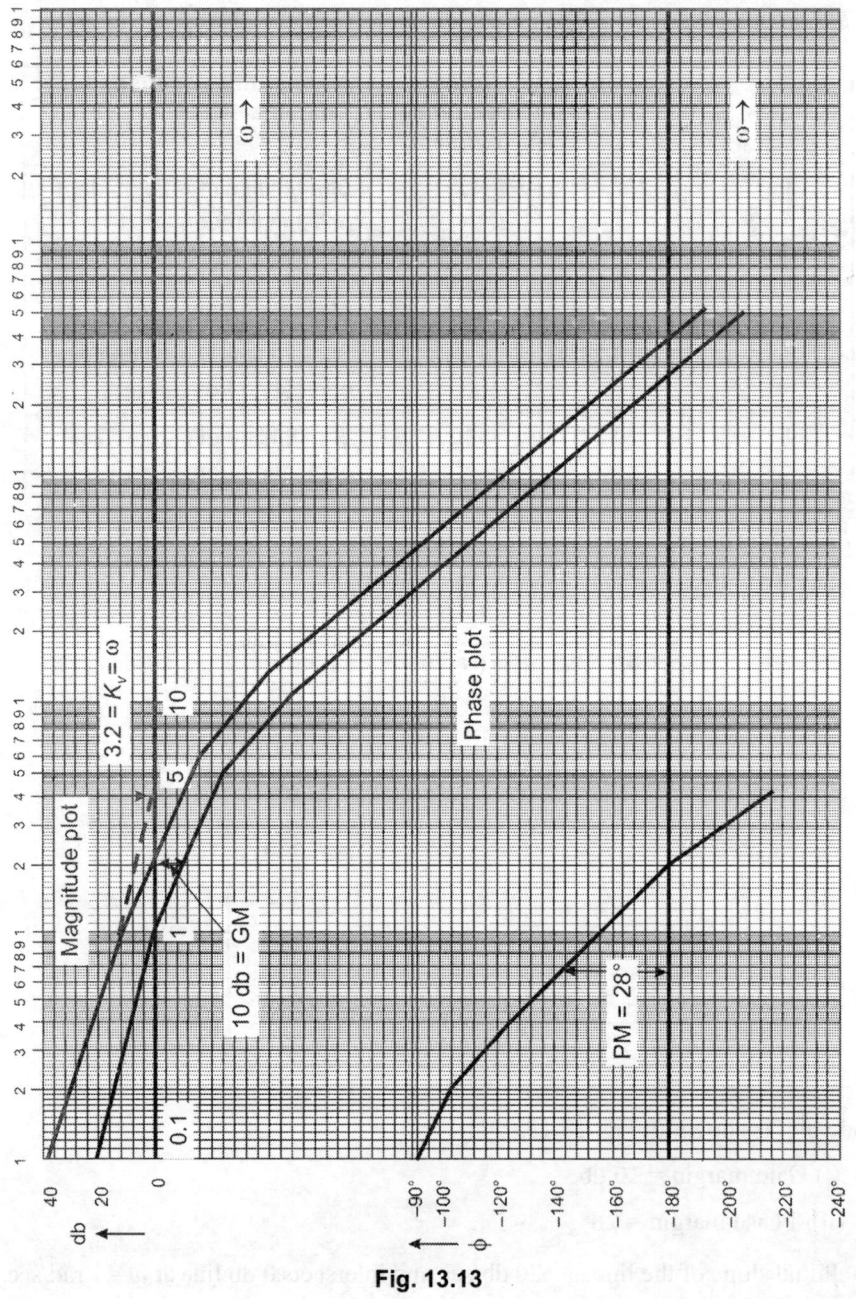

Fig. 13.13

Problem 13.14 Transfer function of a phase advance circuit is

$$\frac{1+0.2s}{1+0.02s}.$$

Find the maximum phase lag.

Solution

Magnitude plot

Sr. No.	Factor	Corner Frequency (rad/sec)	Asymptotic Log-magnitude Characteristics
1	(1 + 0.2s)	5	Straight line of (20 db/dec) originating from $\omega = 5$
2	(1 + 0.02s)	50	Straight line of (−20 db/dec) originating from $\omega = 50$

Phase plot

$$\phi = \tan^{-1} 0.2\, \omega - \tan^{-1} 0.02\, \omega$$

Sr. No.	ω (rad/sec)	ϕ (degrees)
1	0.1	− 1
2	0.2	− 2.1
3	0.4	− 4.1
4	0.5	− 5.1
5	1	− 10.2
6	5	− 39.3
7	10	− 52
8	15	− 54.8
9	16	− 54.9
10	17	− 54.8
11	20	− 54.2
12	50	− 39.3

Bode plot is shown in Fig. 13.14.

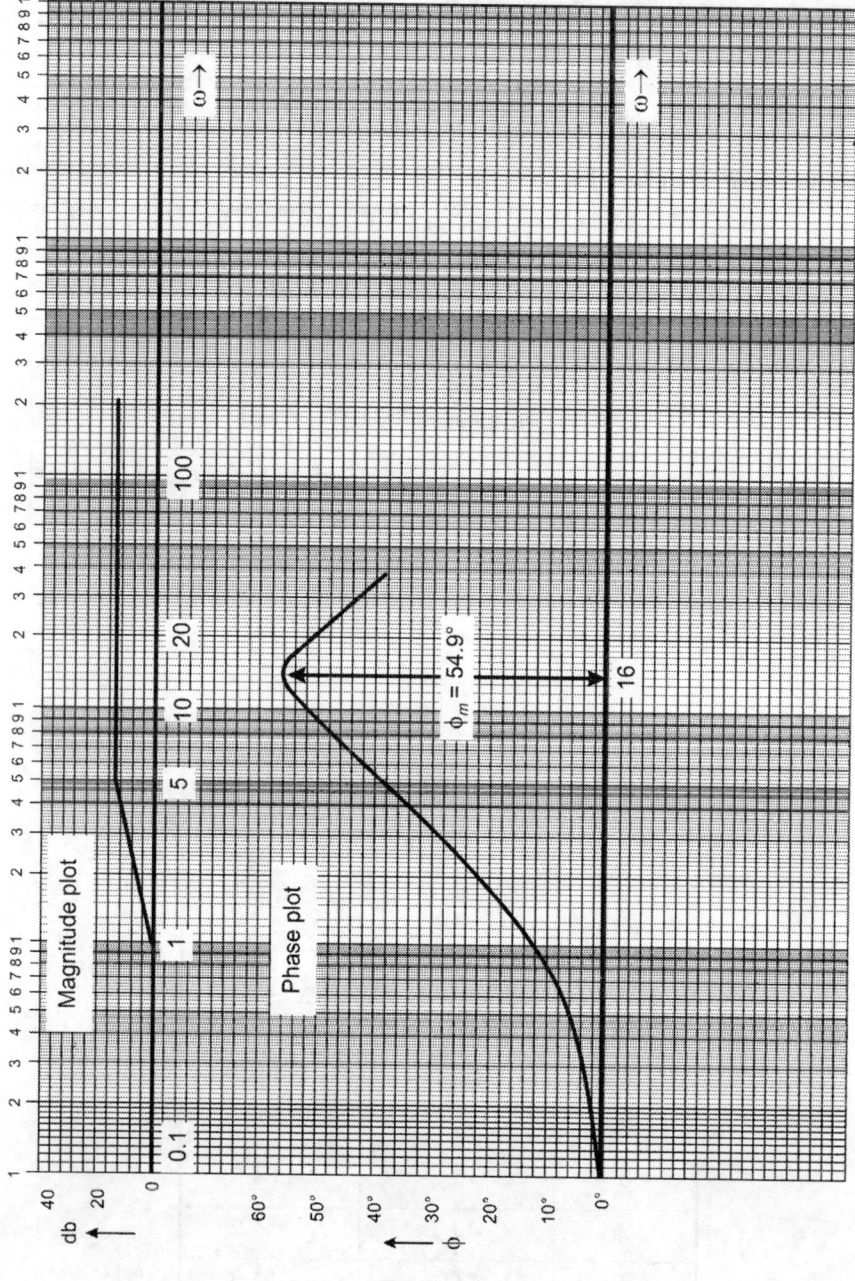

Fig. 13.14

From the phase-plot, maximum phase advance $\phi_m = 54.9°$

Problem 13.15 Sketch the Bode plot for the following transfer function and determine the system gain K for the gain crossover frequency to be 5 rad/sec

$$G(s) = \frac{Ke^{-0.1s}}{s(1+s)(1+0.1s)}$$

Solution

$M = 20 \log |G(s)|_{s=j\omega}$

$$= 20 \log K + 20 \log |e^{-0.1j\omega}| + 20 \log \left|\frac{1}{j\omega}\right| + 20 \log \left|\frac{1}{1+j\omega}\right| + 20 \log \left|\frac{1}{1+0.1j\omega}\right|$$

Now $\qquad |e^{-0.1j\omega}| = |\cos 0.1\omega - j\sin 0.1\omega| = 1$

$\therefore \quad 20 \log |e^{-0.1j\omega}| = 0 \text{ db}$

and $\qquad \angle e^{-j0.1\omega} = -57.3 \times 0.1\omega = -5.73\omega \text{ (degree)}$

Assuming $K = 1$, the magnitude and phase plots are drawn as given below

Magnitude plot

Sr. No.	Factor	Corner Frequency (rad/sec)	Asymptotic Log-magnitude Characteristics
1	$1/s$	None	Straight line of (–20 db/dec) originating from $\omega = 1$
2	$e^{-0.1s}$	None	Coincides with 0 db line
3	$1/1 + s$	1	Straight line of (–20 db/dec) originating from $\omega = 1$
4	$1/1 + 0.1s$	10	Straight line of (–20 db/dec) originating from $\omega = 10$

Phase plot $\qquad \phi = -90° - 5.73\omega - \tan^{-1}\omega - \tan^{-1}0.1\omega$

Sr. No.	ω (rad/sec)	ϕ (degree)
1	0	– 90
2	0.1	– 97
3	0.3	– 110
4	0.5	– 122
5	1	– 146
6	2.0	– 176
7	2.5	– 186.5
8	3	– 190
9	5	– 223
10	10	– 276.6

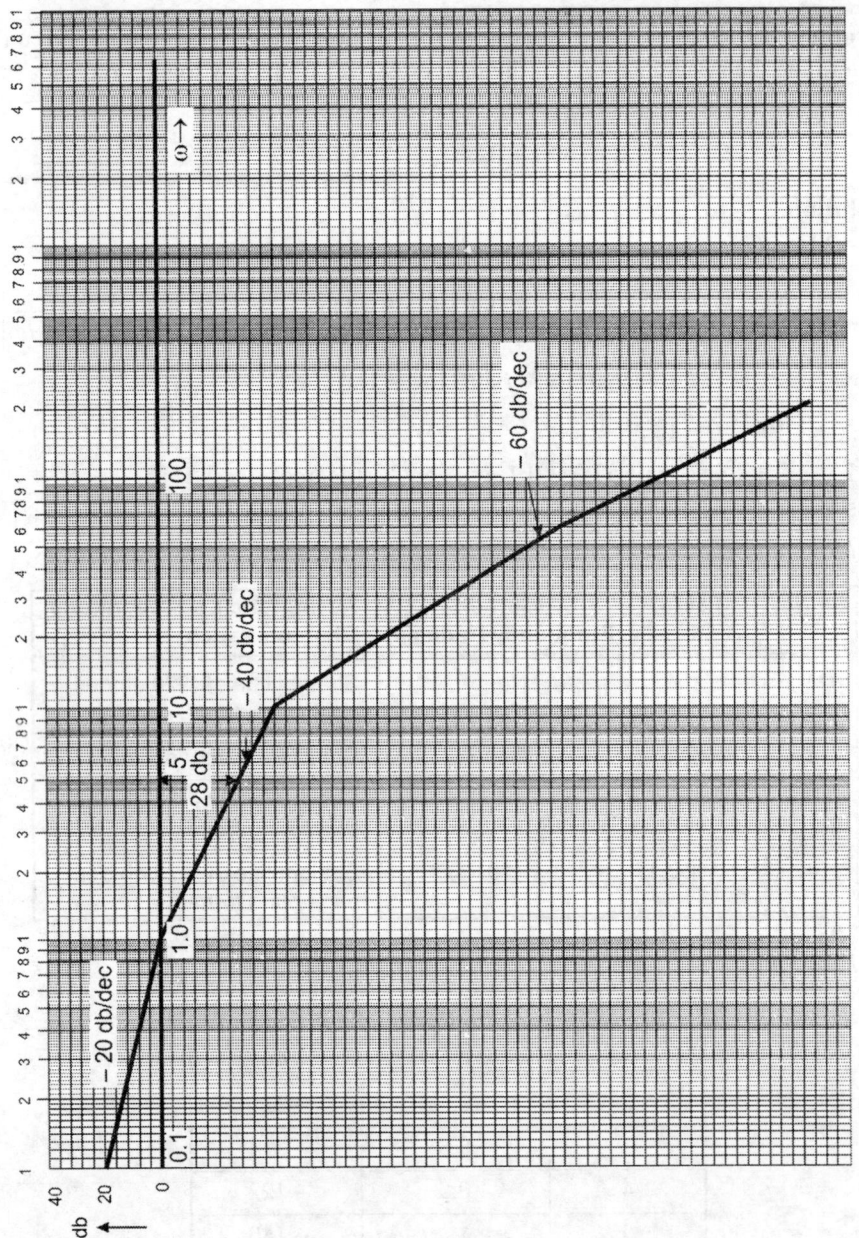

Fig. 13.15

Bode plot is shown in Fig. 13.15

If the gain crossover frequency is required to be 5 rad/sec then the magnitude plot is to be lifted up by 28 db, i.e.

$$20 \log K = 28 \quad \text{or} \quad K = 25.1$$

Problem 13.16   Consider the system shown in Fig. 13.16. Design lead compensator of this system to meet the following specification.

Damping ratio $\xi = 0.7$

Setting time $t_s = 1.4$ sec.

Velocity error constant $K_v = 2$ sec^{-1}

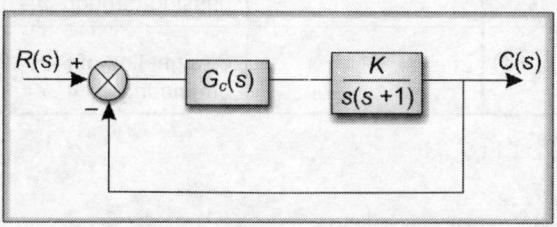

Fig. 13.16

Solution

$$\phi_m = \tan^{-1}[2\xi/[(1 + 4\xi^4)^{1/2} - 2\xi^2)]^{1/2}$$

$$= \tan^{-1}\{2 \times 0.7/[(1 + 4 \times 0.7^4)^{1/2} - 2 \times 0.7^2]^{1/2}\} = 65°$$

For 5% tolerance band

$$t_s = \frac{3}{\xi\omega_n}$$

or

$$\omega_n = \frac{3}{\xi t_n} = \frac{3}{0.7 \times 1.4} = 3.06 \text{ rad/sec}$$

$$\omega_n = \omega_n[1 - 2\xi^2 + \sqrt{2 - 4\xi^2 + 4\xi^4}\,]^{1/2}$$

$$= 3.06\,[1 - 2 \times 0.7^2 + \sqrt{2 - 4 \times 0.7^2 + 4 \times 0.7^4}\,]^{1/2}$$

$$= 3.1 \text{ rad/sec}$$

$$K_v = 2 \text{ sec}^{-1}$$

Open-loop transfer function of an uncompensated system is

$$G(s) = \frac{K}{s(s + 1)} = \frac{2}{s(1 + s)}$$

Magnitude plot

Sr. No.	Factor	Corner Frequency (rad/sec)	Asymptotic Log-magnitude Characteristics
1	2	None	Straight line of (–0 db/dec) starting from point 20 log 2 = 6 db
2	$1/s$	None	Straight line of (–20 db/sec) passing through $\omega = 1$
3	$1/(1 + s)$	1	Straight line of (–20 db/dec) originating from $\omega = 1$

Phase plot

$$\phi = -90° - \tan^{-1}\omega$$

Sr. No.	ω (rad/sec)	ϕ (degrees)
1	0.1	– 96
2	0.2	– 101
3	0.3	– 107
4	0.5	– 116.5
5	1	– 135
6	2	– 153
7	3	– 161.5
8	5	– 169
9	10	– 174
10	12	– 175
11	14	– 176
12	20	– 177
13	∞	– 180

Bode plot is shown in Fig. 13.17

From the plots, the phase and gain margins are 35.5° and ∞ db, respectively. But, the phase margin required is 65°.

Additional phase lead necessary to satisfy the relative stability requirement is 65° – 35.5° = 29.5° ≃ 30° (say). Addition of a lead compensator modifies the magnitude curve. Gain crossover frequency will shift. Therefore, we may assume that the maximum phase lead is

$$30° + 10° = 40°$$

∴ $$\alpha = \frac{1 - \sin 40°}{1 + \sin 40°} \simeq 0.2$$

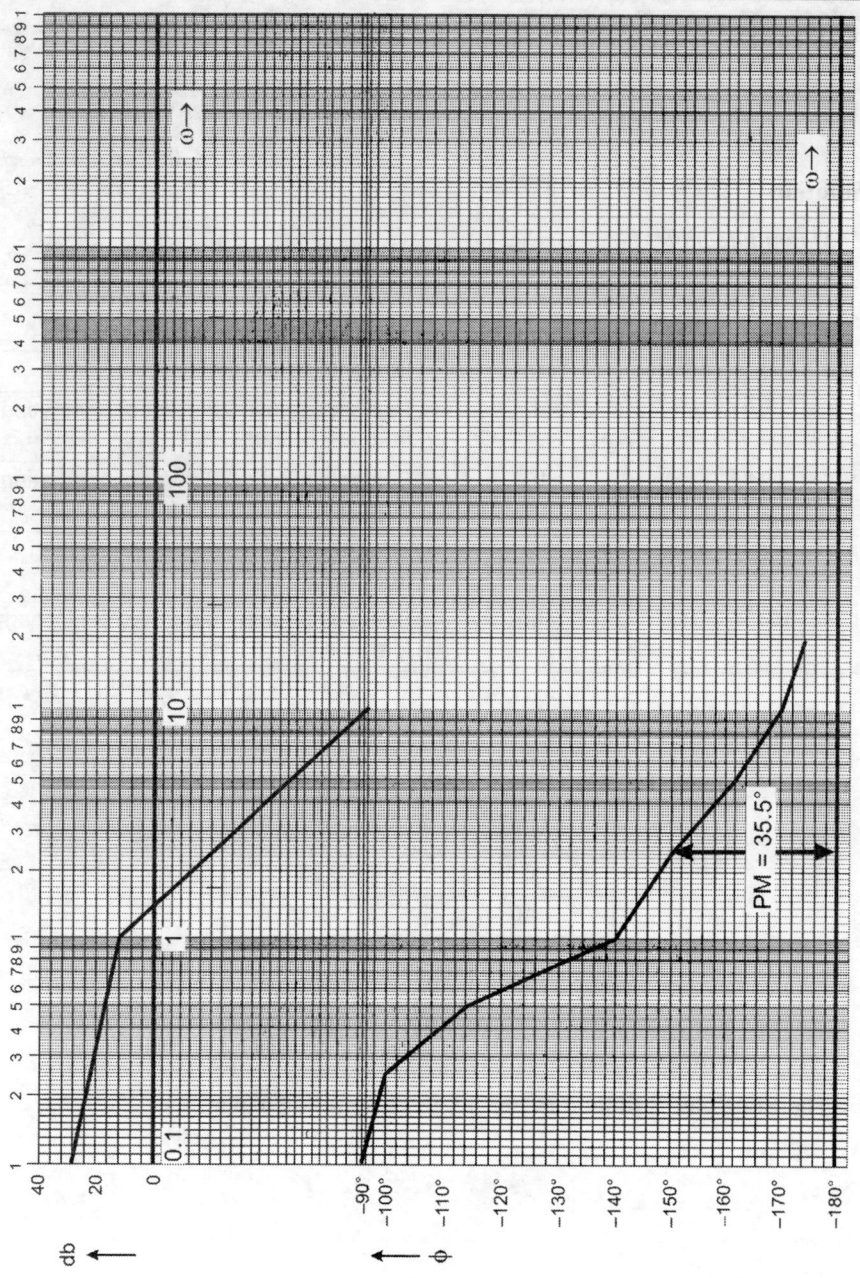

Fig. 13.17

Problem 13.17 Sketch the Bode plot for a control system having

$$G(s) = \frac{100(1 + 0.025s)}{(1 + s)(1 + 0.1s)(1 + 0.01s)^2}$$

Solution

Magnitude plot

Sr. No.	Factor	Corner Frequency (rad/sec)	Asymptotic Log-magnitude Characteristics
1	$K = 100$	None	Straight line of 40 db/dec.
2	$1/(1 + s)$	1	Straight line of (-20 db/dec) originating from $\omega = 50$ rad/sec
3	$1/(1 + 0.1s)$	10	Straight line of (-20 db/dec) originating from $\omega = 10$ rad/sec
4	$(1 + 0.02s)$	50	Straight line of 20 db/dec originating from $\omega = 50$ rad/sec
5	$1/(1 + 0.01s)^2$	100	Straight line of (-40 db/dec) originating form $\omega = 100$ rad/sec

Phase plot

$$\phi = \tan^{-1} 0.025\omega - \tan^{-1}\omega - \tan^{-1} 0.1\omega - 2\tan^{-1} 0.01\omega$$

ω (rad/sec)	ϕ (degree)
0.1	-6.28
10	-129.4
50	-175.6
100	-200.2
∞	-180

The Bode plot is plotted in Fig. 13.18 on semi-log paper.

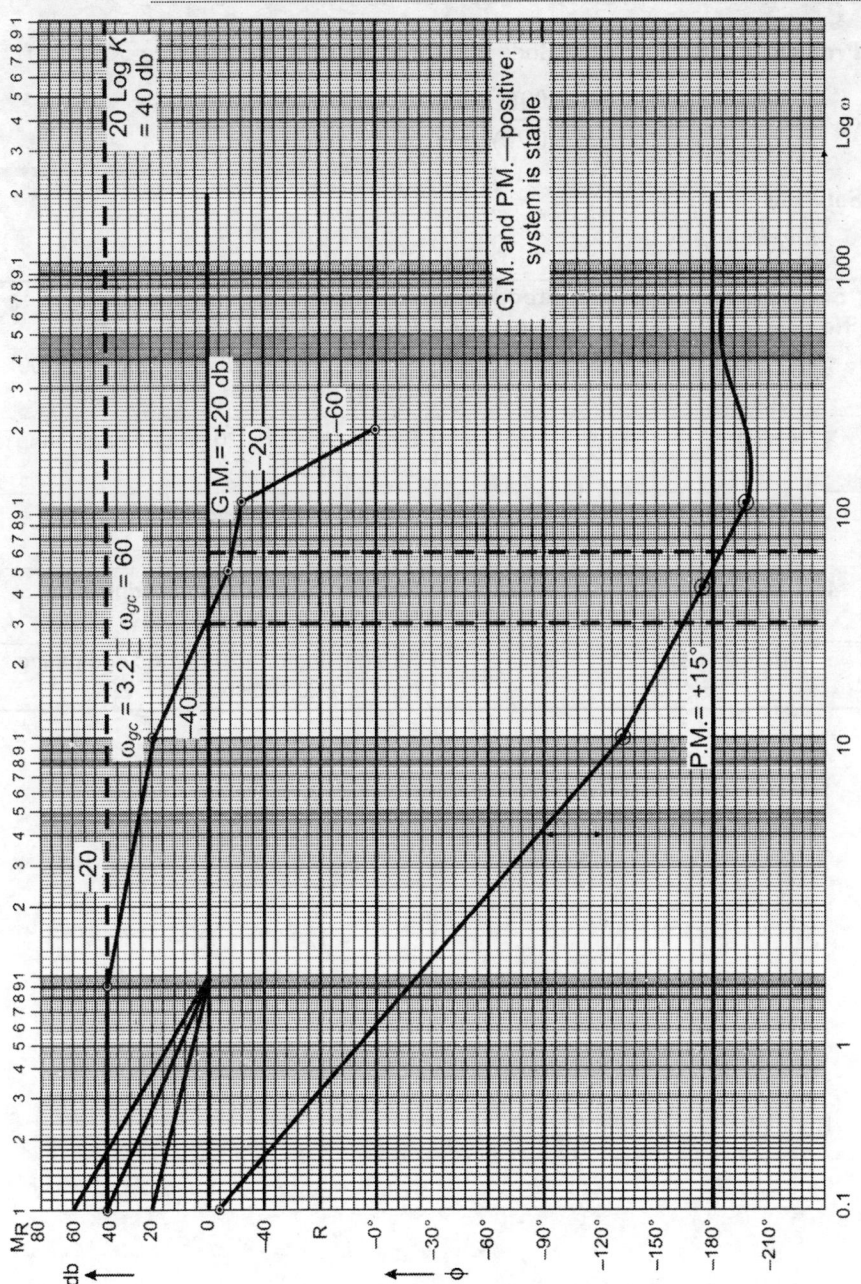

Fig. 13.18

Problem 13.18 Sketch the Bode plot for a control system having

$$G(s) = \frac{200(1+0.1s)}{s(1+0.2s)(1+0.05s)}$$

Solution

Magnitude plot

Sr. No.	Factor	Corner Frequency (rad/sec)	Asymptotic Log-magnitude Characteristics
1	200	None	Straight line of 0 db/dec from 20log200 = 46.02 db.
2	$1/s$	1	Straight line of – 20 db/dec originating ω = 1 rad/sec
3	$1/(1 + 0.2s)$	5	Straight of – 20 db/dec originating from ω = 5 rad/sec
4	$(1 + 0.2s)$	10	Straight line of 20 db/dec originating from ω = 10 rad/sec
5	$1/(1+0.05s)$	20	Straight line of – 20 db/dec originating from ω = 20 rad/sec

Phase plot

$$\phi = -90° + \tan^{-1} 0.100 - \tan^{-1} 0.200 - \tan^{-1} 0.0500$$

Sr. No.	ϕ (degree)
0.5	– 94.28
5	– 122.27
20	– 147.53
100	– 171.54
∞	– 180

Bode plot is shown in Fig. 13.19.

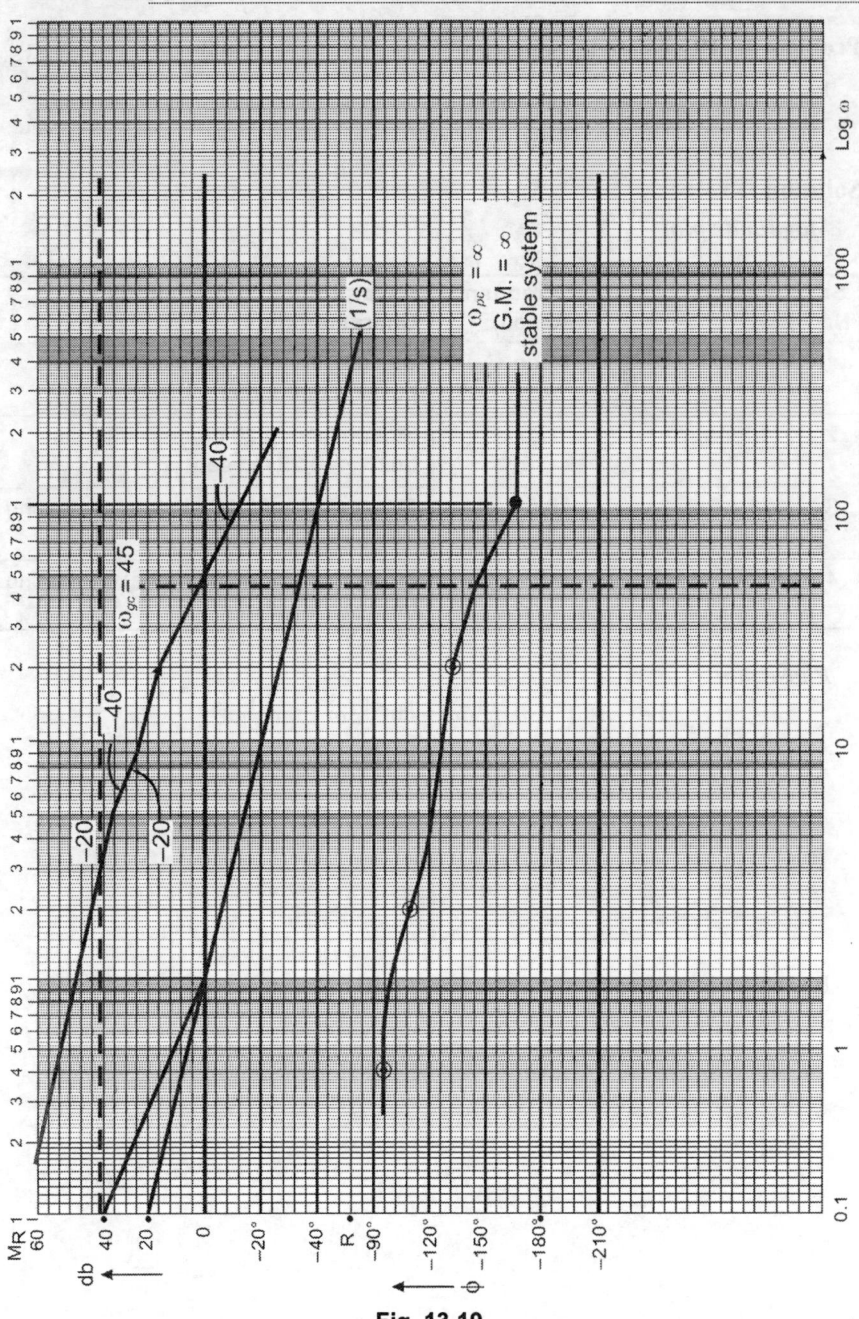

Fig. 13.19

Problem 13.19 Sketch the Bode plot for a control system having

$$G(s) = \frac{(1+4s)}{s^2(1+s)(1+2s)}$$

Solution

Magnitude plot

Sr. No.	Factor	Corner Frequency (rad/sec)	Asymptotic Log-magnitude Characteristics
1	$1/s^2$	None	Straight line of (–40 db/dec) passing through ω = 1 rad/sec
2	$(1 + 4s)$	0.25	Straight line of 20 db/dec originating from ω = 0.25 rad/sec
3	$1/(1 + 2s)$	0.5	Straight line of (–20 db/dec) originating from ω = 0.5 rad/sec
4	$1/(1 + s)$	1	Straight line of (–20 db/dec) originating from ω = 1 rad/sec.

Phase plot

$$\phi = -180° + \tan^{-1}4\omega - \tan^{-1}2\omega - \tan^{-1}\omega$$

ω (rad/sec)	ϕ (degree)
0.1	– 96.27
1	– 140.7
5	– 195.25
∞	– 270

Bode plot is plotted as shown in Fig. 13.20 on semi-log paper.

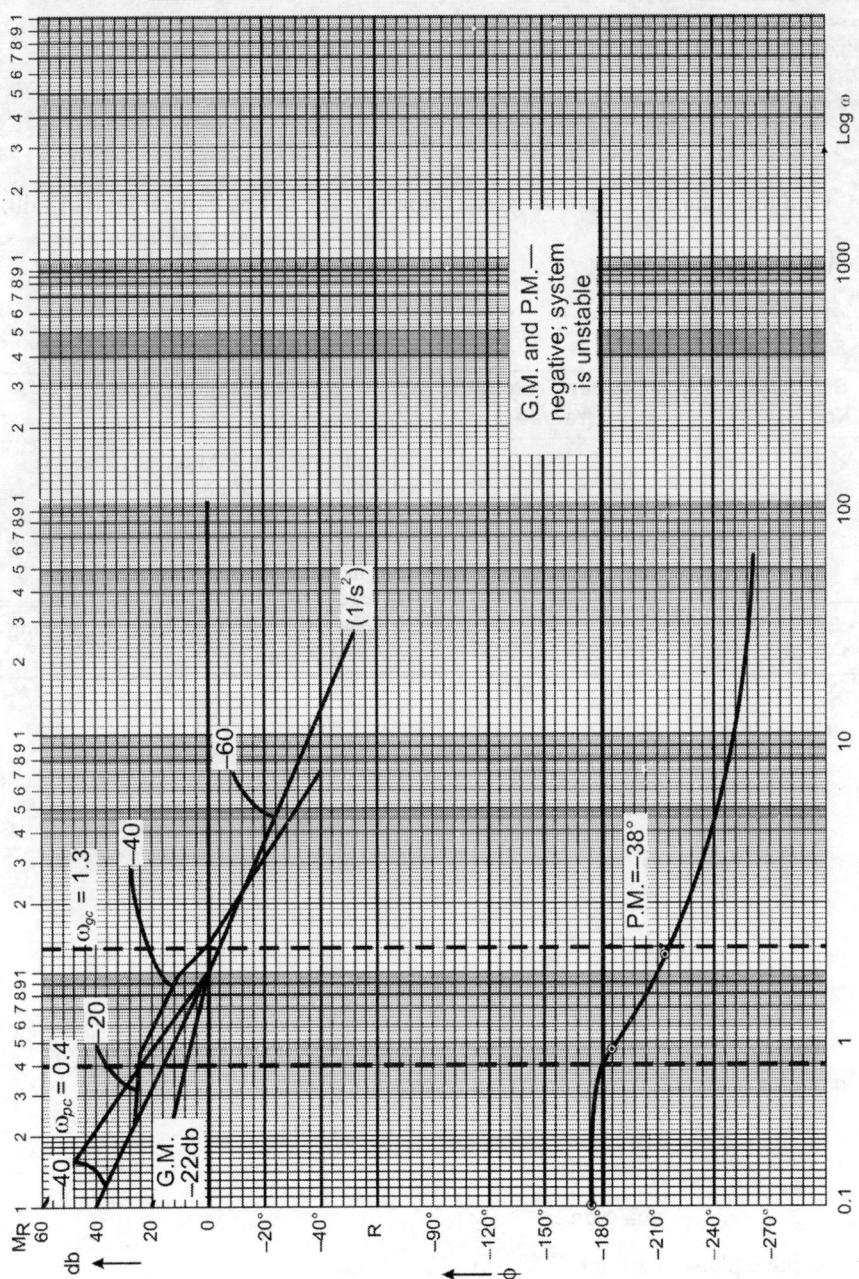

Fig. 13.20

Problem 13.20 Draw the Bode plot for the control system having

$$G(s) = \frac{K}{s(1 + 0.1s)(1 + 0.05s)}$$

Find the value of K to have gain margin of 7 db and phase margin of 18 db.

Solution

Magnitude plot

Sr. No.	Factor	Corner Frequency (rad/sec)	Asymptotic Log-magnitude Characteristics
1	$1/s$	None	Straight line of (–20 db/dec) passing through $\omega = 1$ rad/sec
2	$1/(1 + 0.1s)$	10	Straight line of (–20 db/dec) originating from $\omega = 10$ rad/sec
3	$1/(1 + 0.05s)$	20	Straight line of (–20 db/dec) originating from $\omega = 20$ rad/sec

Phase plot

$$\phi = -90° \tan^{-1} 0.1\omega - \tan^{-1} 0.05\omega$$

ω (rad/sec)	ϕ (degree)
0.1	–90.85
10	–161.56
20	–198.43
∞	–270.00

Bode plot is shown in Fig. 13.21 on a semi-log paper.

To have gain margin of 20 db, AB is shifted so that

$$20 \log K = 7 \text{ db}$$

or $K = 2.23$

To have phase margin of 38°, shift CD so that

$$20 \log K = 18 \text{ db}$$

or $K = 7.94$

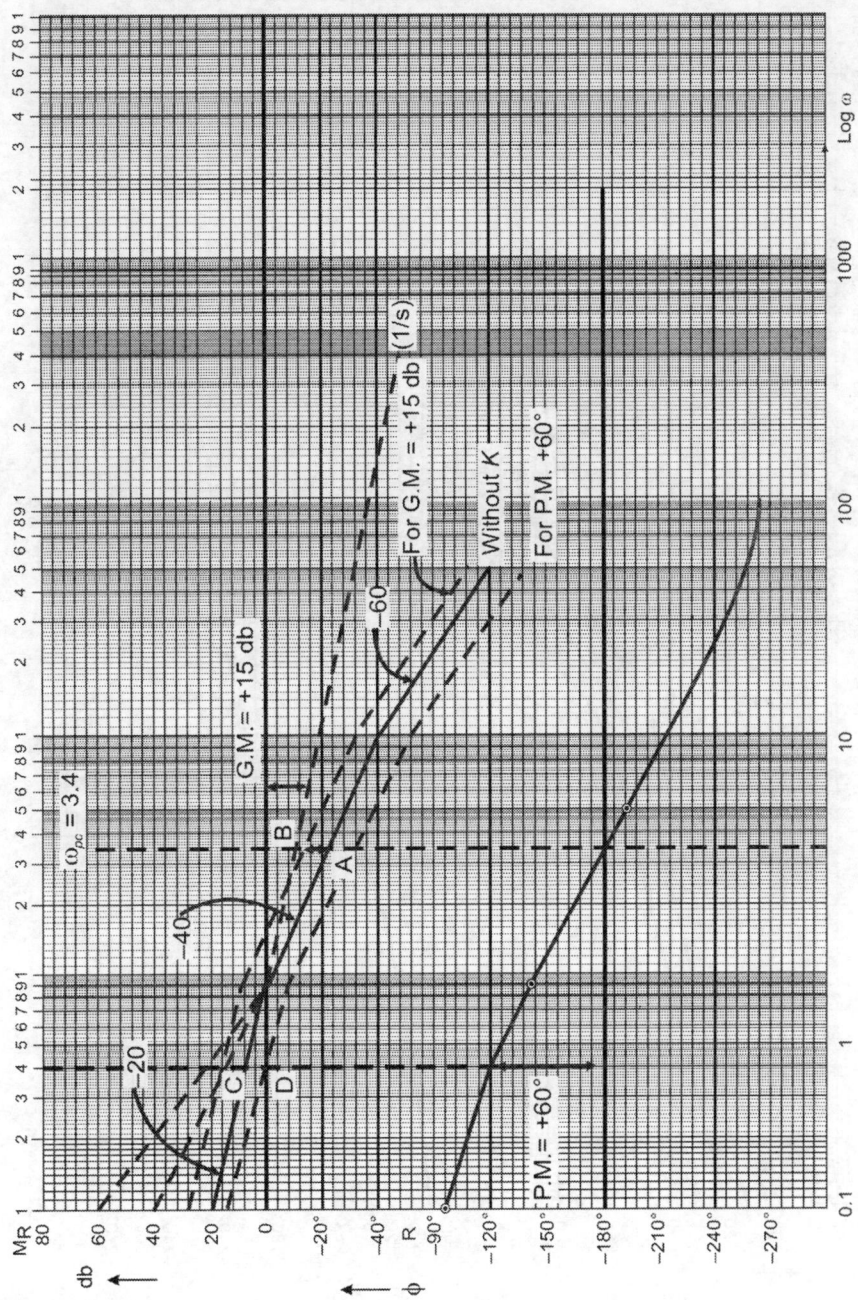

Fig. 13.21

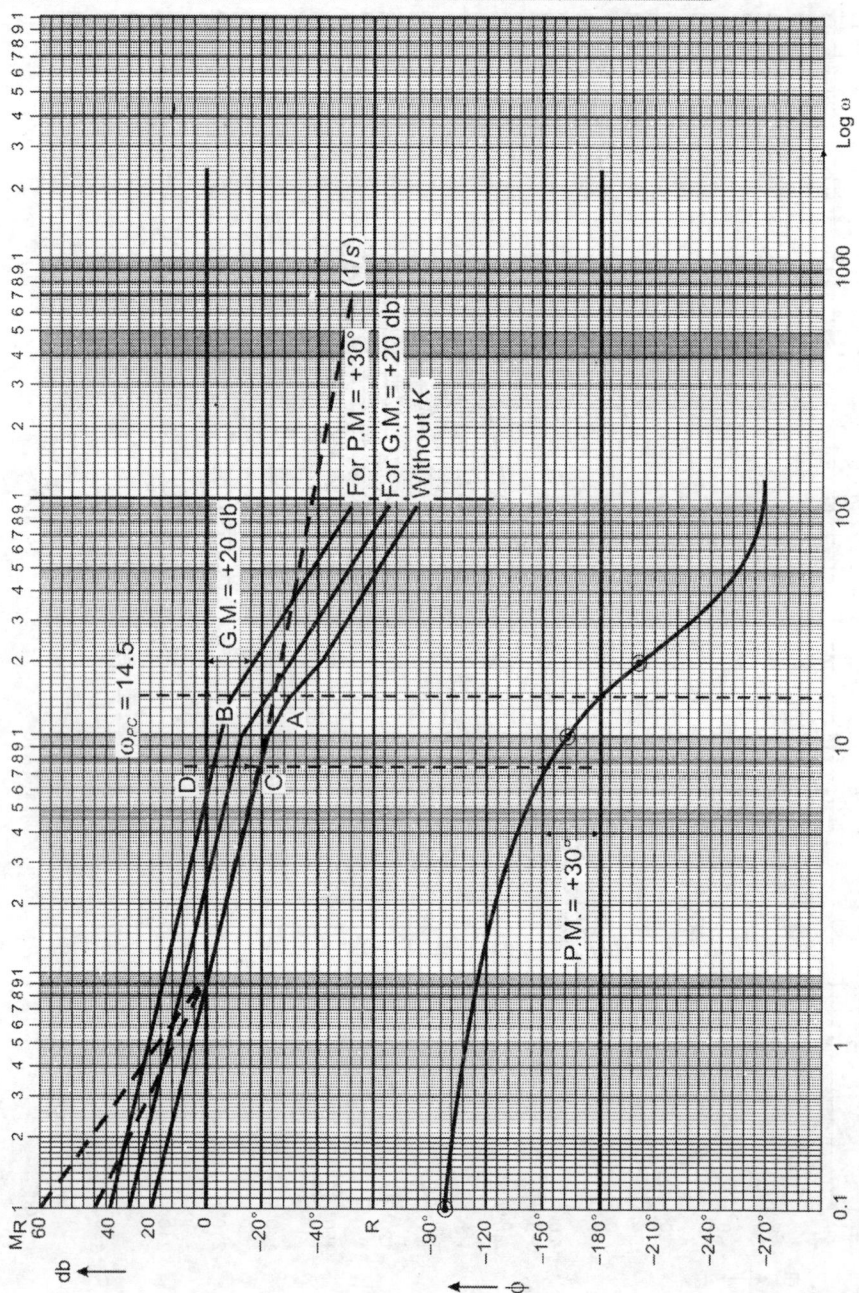

Fig. 13.22

Problem 13.21 Find the open-loop transfer function of a system whose approximate plot is shown in Fig.13.23.

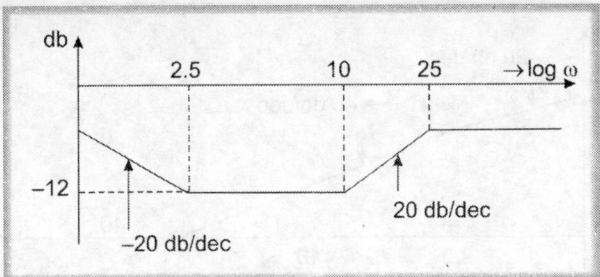

Fig. 13.23

Solution

Corner frequencies are

$$\omega_1 = 2.5; \quad \omega_2 = 10; \quad \omega_3 = 25$$

Change in magnitude in db

$$= \text{Slope} \times \text{Number of decades between two frequencies}$$

$$= -20\,(\log 2.5 - \log 1) = -20\,\log 2.5 = -7.95\ db$$

Magnitude $\quad = -12 + 7.95\ db = -4.05\ db$

$$20 \log K = -4.05 \quad \text{or} \quad K = 0.63$$

Since first line has a slope of -20 db/dec and starts from a point -4.05 db at $\omega = 1$ rad/sec, the factor contributing is

$$= \frac{K}{s} = \frac{0.63}{s}$$

Plot between $\omega = 2.5$ and $\omega = 10$ is having a slope of 0 db/dec. At $\omega = 2.5$ the slope has changed from -20 db/dec to 0 db/dec and this can only happen due to a factor in the numerator and is

$$= \left(\frac{s}{2.5} + 1\right) = (1 + 0.4s)$$

At $\omega = 10$, the slope has changed from 0 db/dec to $+20$ db/dec and is due to a factor in the numerator and is

$$= \left(\frac{s}{10} + 1\right) = (1 + 0.1s)$$

At $\omega = 25$, the slope has changed from $+20$ db/dec to 0 db/dec and is due to a factor in the denominator and is

$$= \left(\frac{s}{25} + 1\right) = (1 + 0.04s)$$

The open-loop transfer function is thus

$$G(s) = \frac{0.63(1 + 0.4s)(1 + 0.1s)}{s\,(1 + 0.4s)}$$

Problem 13.22 Determine the transfer function whose approximate plot is shown in Fig. 13.24.

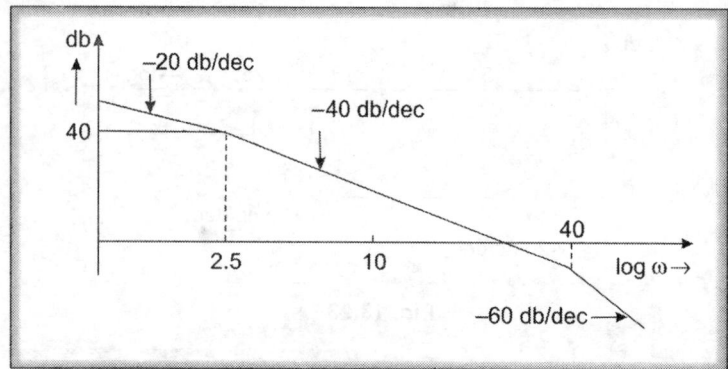

Fig. 13.24

Solution

Corner frequencies are 2.5 and 40 rad/sec

$$20 \log K = 40 + 20 \log 2.5 = 47.95 \quad \text{or} \quad K = 250$$

At $\omega = 2.5$ rad/sec slope changes from -20 db/dec to -40 db/dec due to a factor $1/(1 + s/2.5)$. At $\omega = 40$ rad/sec, slope changes from -40 db/dec to -60 db/dec due

to a factor $\dfrac{1}{\left(1 + \dfrac{s}{40}\right)}$. Since initial slope is -20 db/dec, it is due to factor $1/s$.

Therefore open-loop transfer function is

$$G(s) = \frac{250}{s\left(1 + \dfrac{s}{2.5}\right)\left(1 + \dfrac{s}{40}\right)} = \frac{250}{s(1 + 0.4s)(1 + 0.025s)}$$

Problem 13.23 Determine the open-loop transfer function of a system whose approximate plot is shown in Fig. 13.25.

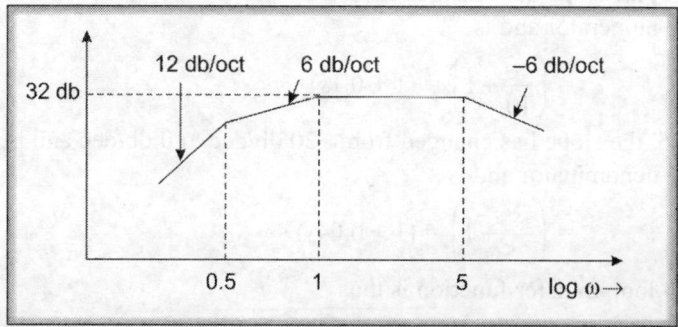

Fig. 13.25

Solution

First line is having a slope of 12 db/oct (40 db/dec). Therefore, there is a s^2 term in the numerator. At $\omega = 0.5$ rad/sec, slope changes to 6 db/oct (20 db/dec) due to a term in the denominator equal to

$$\left(1 + \frac{s}{0.5}\right)$$

At $\omega = 1$ rad/sec, slope becomes 0 db/dec due to a term in the denominator equal to $(1 + s)$.

At $\omega = 5$ rad/sec slope becomes –6 db/oct (–20 db/dec) due to a term in the denominator equal to $(1 + s/5)$

$$\therefore \qquad G(s) = \frac{Ks^2}{\left(1 + \dfrac{s}{0.5}\right)(1 + s)\left(1 + \dfrac{s}{5}\right)}$$

Calculation of 'K'

See Fig. 13.26.

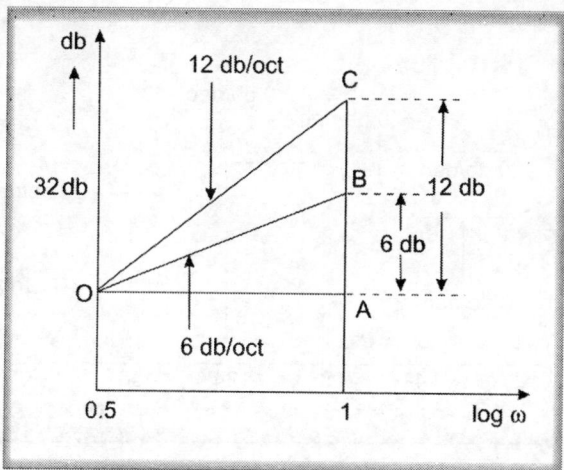

Fig. 13.26

Difference between $\omega = 0.5$ and $\omega = 1$ rad/sec is one octave, *i.e.* $AB = 6$ db since the slope of line OB is 6 db/oct. OC is the extended line having a slope 12 db/oct.

$$\therefore \qquad AC = 12 \text{ db}$$

$$BC = AC - AB = 12 - 6 = 6 \text{ db}$$

$$\therefore \qquad 20 \log K = (32 + 6) \text{ db} \quad \text{or} \quad K = 79.4$$

The open-loop transfer function is thus

$$G(s) = \frac{79.4s^2}{(1 + 2s)(1 + s)(1 + 0.2s)} \qquad\qquad \textbf{Ans}$$

Problem 13.24 From the asymptotic magnitude (in db) versus frequency (log scale) plot of Fig. 13.27, find the associated transfer function. Assume no right half plane poles or zeros present. (*Pune University*)

Solution

- Slope of the first line is −20 db/dec indicating a term $1/s$
- At $\omega = 3$ rad/sec slope changes to a 0 db/dec indicating a term $(1 + s/2)$ or $(1 + 0.5s)$ in the numerator.
- At $\omega = 4$ rad/sec slope changes to +20 db/dec indicating a term $(1 + s/4)$ or $(1 + 25s)$ in the numerator.
- At $\omega = 8$ rad/sec slope changes to 0 db/dec indicating a term $(1 + s/8)$ or $(1 + 0.125s)$ in the denominator.
- At $\omega = 24$ rad/sec slope changes to −20 db/dec indicating a term $(1 + s/24)$ or $(1 + 0.042s)$ in the denominator.
- At $\omega = 36$ rad/sec slope changes to −40 db/dec indicating a term $(1 + s/40)$ or $(1 + 0.028s)$ in the denominator.

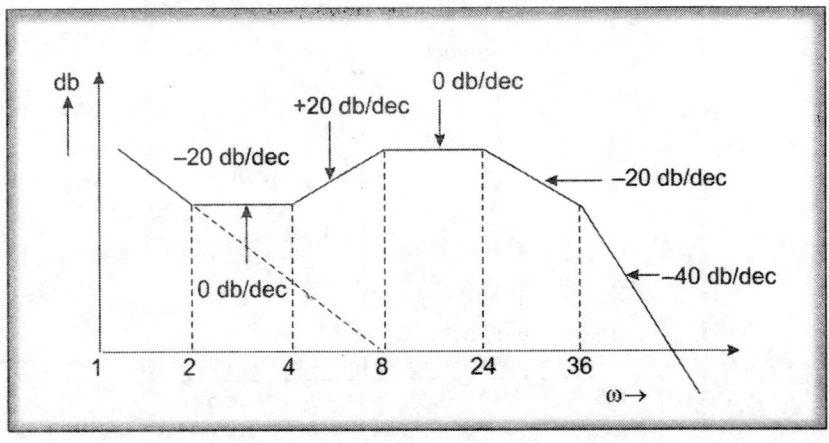

Fig. 13.27

Transfer function is thus

$$\frac{K(1 + 0.5s)(1 + 0.25s)}{s(1 + 0.125s)(1 + 0.042s)(1 + 0.028s)}$$

Calculation of 'K'

$$20\log K = 20\log 8 \quad \text{or} \quad K = 8$$

$$\therefore \quad G(s) = \frac{8(1 + 0.5s)(1 + 0.25s)}{s(1 + 0.125s)(1 + 0.042s)(1 + 0.028s)}$$

Problem 13.25 Derive the transfer function of the system from the data given on the Bode diagram shown in Fig. 13.28 below: (*AMIE*)

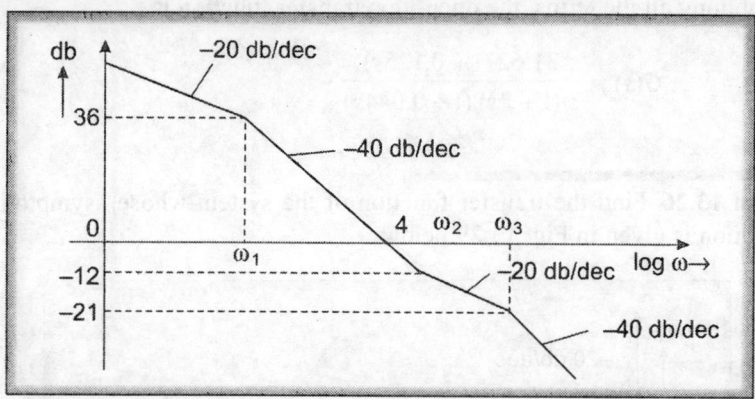

Fig. 13.28

Solution

Between ω_1 and $\omega = 4$ rad/sec, there is a decrease of 36 db

$$36 = -40 \log 4 - 40 \log \omega_1$$

$$36 = +40 \log 4 - 40$$

or $\omega_1 = 0.5036 \simeq 0.5$ rad/sec

Calculation of 'K'

$$20 \log K = 36 + 20 \log 0.5$$

or $K = 31.62$

Calculation of 'ω_2'

$$-12 = -40 (\log \omega_2 - \log 4)$$

or $\omega_2 = 8$ rad/sec.

Calculation of 'ω_3'

$$-21 + 12 = -20 (\log \omega_3 - \log 8)$$

or $\omega_3 = 22.5$ rad/sec

First line has a slope of –20 db/dec indicating a term $1/s$ and since it is not passing through $\omega = 1$ rad/sec, the term is K/s or $31.62/s$.

At $\omega_1 = 0.5$ rad/sec, slope changes to – 40 db/dec indicating a term $1/(1 + s/0.5)$ or $1/(1 + 2s)$.

At $\omega_2 = 8$ rad/sec, slope changes to –20 db/dec indicating a term $(1 + s/8)$ or $(1 + 0.125s)$.

At $\omega_3 = 22.5$ rad/sec, slope changes to -40 db/dec indicating a term $1/(1 + s/22.5)$ or $1/(1 + 0.044s)$.

Combining all the terms, the open-loop transfer function is

$$G(s) = \frac{31.62 \, (1 + 0.125s)}{s \, (1 + 2s) \, (1 + 0.044s)} .$$

Problem 13.26 Find the transfer function of the system whose asymptotic approximation is given in Fig. 13.29 below:

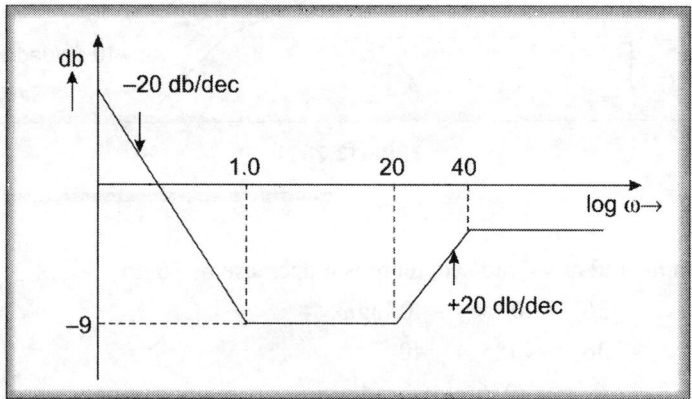

Fig. 13.29

Solution

First line has a slope of -20 db/dec and is not passing through $\omega = 1$ rad/sec. Therefore, it indicates a term K/s

$$20 \log K = -9$$

or $\qquad\qquad K = 0.35$

$\therefore$ the term is $0.35/s$

At $\omega = 1$ rad/sec, slope changes to 0 db/dec indicating a term $(1 + s)$.

At $\omega = 20$ rad/sec, slope changes to $+20$ db/dec indicating a term $(1 + s/20)$ or $(1 + 0.05s)$.

At $\omega = 40$ rad/sec, slope changes to 0 db/dec indicating a term $1/(1 + s/40)$ or $1/(1 + 0.025s)$. Combining all the terms, we get

$$G(s) = \frac{0.35 \, (1 + s) \, (1 + 0.05s)}{s \, (1 + 0.025s)}$$

Problem 13.27 Obtain the expression for open-loop transfer function for a system with unity feedback whose log-magnitude plot is shown in Fig. 13.30 below:

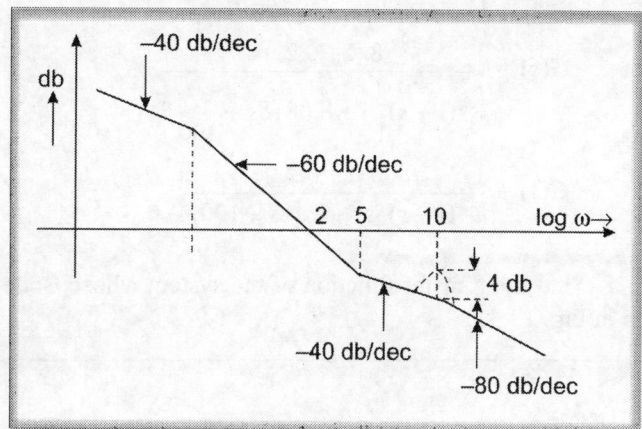

Fig. 13.30

Solution

First line is having a slope of -40 db/dec and since it is not passing through $\omega = 1$ rad/sec it indicates a term K/s^2.

Magnitude at $\omega = 1$ rad/sec of initial part $= 60 \log 2 = 18.06$

$20 \log K = 18.06$ or $K = 8$

At $\omega = 1$ rad/sec, the slope changes from -40 db/dec to -60 db/dec indicating a term $1/(1 + s)$

At $\omega = 5$ rad/sec, the slope changes from -60 db/dec to -40 db/dec indicating a term $(1 + s/5)$ or $(1 + 0.2s)$.

At $\omega = 10$ there is a term of the form

$$\left\{ \left(1 + \frac{2\xi s}{\omega_n} + \frac{s^2}{\omega_n^2} \right) \right\}^{-1}$$

because the slope changes from -40 db/dec to -80 db/dec and also a peak of 4 db is shown $\omega_n = 10$ rad/sec

Value of
$$\left\{ 1 + \frac{2\xi s}{\omega_n} + \frac{s^2}{\omega_n^2} \right\}^{-1} \text{ at } \omega = \omega_n$$

$$= \sqrt{\left\{ \left(1 - \left(\frac{10}{10} \right)^2 \right)^2 + \left(\frac{2 \times \xi \, 10}{10} \right)^2 \right\}^{-1}} = \frac{1}{2\xi}$$

$\therefore$ log magnitude $= 20 \log \dfrac{1}{2\xi} = 4$, or $\dfrac{1}{2\xi} = e^{1/5}$, or $\xi = 0.409$

$\therefore$ the term is $= \left(1 + \dfrac{2 \times 0.409 s}{10} + \dfrac{s^2}{100}\right)^{-1}$

$\therefore$ $G(s) = \dfrac{8\,(1 + 0.2s)}{s^2\,(1+s)\left(1 + 0.0818s + \dfrac{s^2}{100}\right)}$

or $G(s) = \dfrac{800\,(1 + 0.2s)}{s^2\,(1+s)(s^2 + 8.18s + 100)}$

Problem 13.28 Find the transfer function of the system whose Bode magnitude plot is shown in Fig. 13.31.

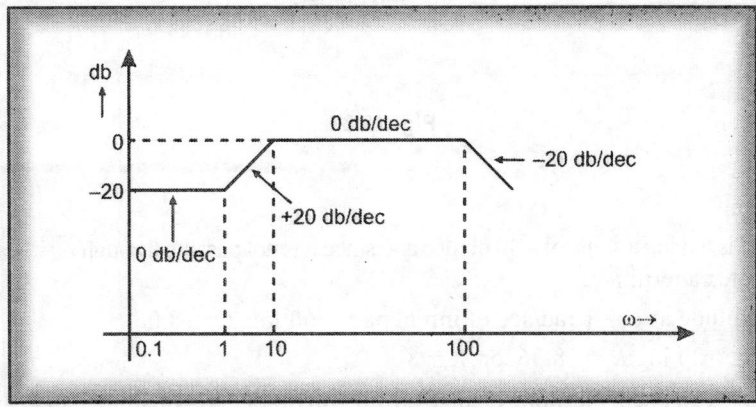

Fig. 13.31

Solution

The Fig. 13.31 can be redrawn as Fig. 13.32

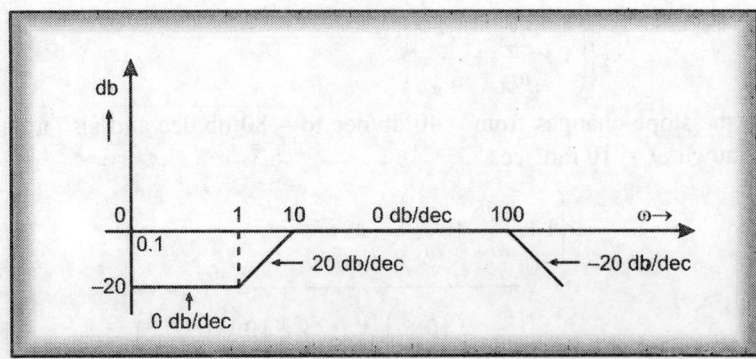

Fig. 13.32

It is seen that the initial slope is 0 db/dec and intercept is –20 db. This is only possible due to factor K. Since, the intercept is minus, the value of K will be less then 1 (one).

$$20 \log K = -20$$

or $$K = 0.1$$

The transfer function is calculated as shown below:

Sr. No.	Slope of the Line	Corner Frequency (rad/sec)	Factor
1.	0 db/dec (Initial slope of resultant magnitude plot)	—	The intercept on negative axis is –20 db. Hence, $20 \log K = -20$ $\therefore K = \dfrac{1}{10} = 0.1$
2.	+20 db/dec (Slope changes from 0 db/dec to +20 db/dec)	1	Slope is +20 db/dec Hence, there is a factor in the numerator $\left(1 + \dfrac{s}{1}\right) = (1 + s)$
3.	0 db/dec (Slope changes from 20 db/dec to 0 db/dec)	10	Resultant slope has become 0 db/dec due to line of slope –20 db/dec at $\omega = 10$ rad/sec Hence, there is a factor in the denominator $\left(1 + \dfrac{s}{10}\right) = (1 + 0.1s)$
4.	–20 db/dec (Slope changes from 0 db/dec to –20 db/dec)	100	Resultant slope has become –20 db/dec due to line of slope –20 db/dec at $\omega = 100$ rad/sec. Hence, there is a factor in the denominator $\left(1 + \dfrac{s}{100}\right) = (1 + 0.01s)$

Therefore, the transfer factor is

$$= \frac{0.1(1+s)}{(1+0.1s)(1+0.01s)} = \frac{\dfrac{1}{10}(1+s)}{\left(1 + \dfrac{s}{10}\right)\left(1 + \dfrac{s}{100}\right)}$$

$$= \frac{100(s+1)}{(s+10)(s+100)}.$$

Problem 13.29 Find the transfer function of the Bode magnitude plot shown in Fig. 13.33.

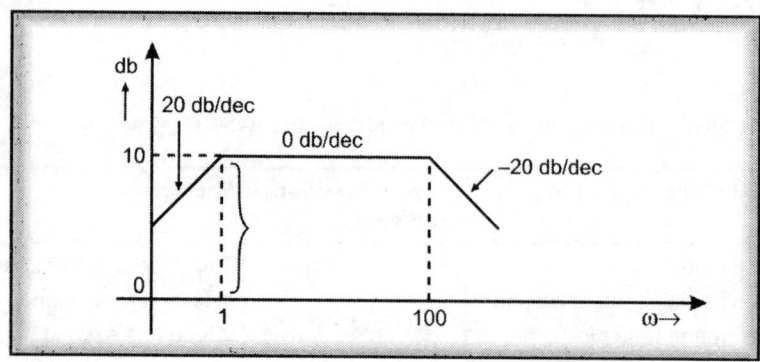

Fig. 13.33

Solution

Transfer function is calculated as below:

Sr. No.	Slope of the Line	Corner Frequency (rad/sec)	Factor
1.	20 db/dec	NIL	The line is not passing through $\omega = 1$ rad/sec. Hence, there is a factor K in the numerator. Had K been equal to 1, the 20 db/dec line would have passed through $\omega = 1$ rad/sec. From the figure above, it is clear that due to factor K, 20 db/dec line has got lifted by 10 db. Therefore, $\qquad 20 \log K = 10$ db or $\qquad K = 3.16$

(*contd.*)

Sr. No.	Slope of Line	Corner Frequency (rad/sec)	Factor
2.	0 db/dec	1	At ω = 1 rad/sec, the slope of the plot is 0 db/dec and is due to the presence of pole at 1 (one). To the original slope of the line of 20 db/dec, another line of slope −20 db/dec is added, to make the resultant slope 0 db/dec. Therefore, the factor in the denominator is $$\left(1+\frac{s}{1}\right) = (1 + s)$$
3.	−20 db/dec	100	At ω = 100 rad/sec, the slope changes from 0 db/dec to −20 db/dec and is due to presence of a pole in the transfer function. The factor in the denominator is $$\left(1+\frac{s}{100}\right) = (1 + 0.01s)$$

$$TF = \frac{K}{\left(1+\frac{s}{1}\right)\left(1+\frac{s}{100}\right)}$$

$$= \frac{K}{(1+s)(1+0.01s)}$$

Problem 13.30 Find the transfer function of the system whose Bode magnitude plot is shown in Fig. 13.34. What is the type of the system.

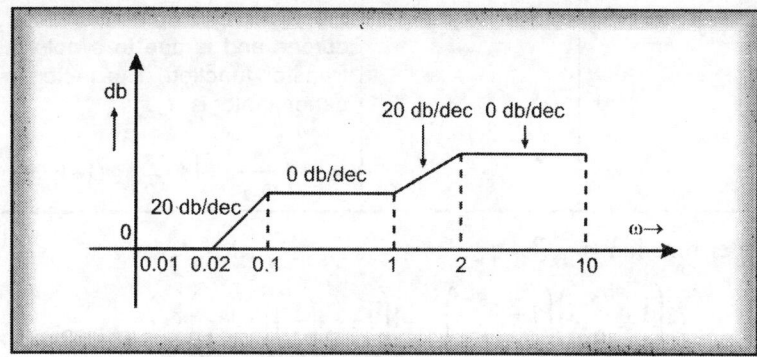

Fig. 13.34

Solution

Type of the system

Initial slope of the Bode magnitude plot is 0 db/dec. Hence, type of the system is '0'.

Transfer function

It is calculated as shown below:

Sr. No.	Slope of Line	Corner Frequency (rad/sec)	Factor
1.	0 db/dec	—	$20 \log K = 0$ Therefore, $K = 1$
2.	20 db/dec	$\omega_{c1} = 0.02$	The slope changes from 0 db/dec to +20 db/dec due to presence of a zero in the transfer function. The factor in the numerator is $= \left(1 + \dfrac{s}{\omega_{c1}}\right) = \left(1 + \dfrac{s}{0.02}\right) = (1 + 50s)$
3.	0 db/dec	$\omega_{c2} = 0.1$	Slope changes from 20 db/dec to 0 db/dec and is due to a pole in the transfer function. The factor in the denominator is $= \left(1 + \dfrac{s}{\omega_{c2}}\right) = \left(1 + \dfrac{s}{0.1}\right) = (1 + 10s)$
4.	20 db/dec	$\omega_{c3} = 1$	Slope changes fom 0 db/dec to +20 db/dec and is due a zero in the transfer function. The factor in the numerator is $= \left(1 + \dfrac{s}{\omega_{c3}}\right) = \left(1 + \dfrac{s}{1}\right) = (1 + s)$
5.	0 db/dec	$\omega_{c4} = 2$	Slope changes from +20 db/dec to 0 db/dec and is due to a pole in the transfer function. The factor in the denominator is $= \left(1 + \dfrac{s}{\omega_{c4}}\right) = \left(1 + \dfrac{s}{2}\right) = (1 + 0.5s)$

Transfer function is

$$= \frac{K\left(1 + \dfrac{s}{\omega_{c1}}\right)\left(1 + \dfrac{s}{\omega_{c3}}\right)}{\left(1 + \dfrac{s}{\omega_{c2}}\right)\left(1 + \dfrac{s}{\omega_{c4}}\right)} = \frac{1\left(1 + \dfrac{s}{0.02}\right)\left(1 + \dfrac{s}{1}\right)}{\left(1 + \dfrac{s}{0.1}\right)\left(1 + \dfrac{s}{2}\right)} = \frac{(1 + 50s)(1 + s)}{(1 + 10s)(1 + 0.5s)}$$

Problem 13.31 Find the transfer function of the system whose Bode magnitude plot is shown in Fig. 13.35.

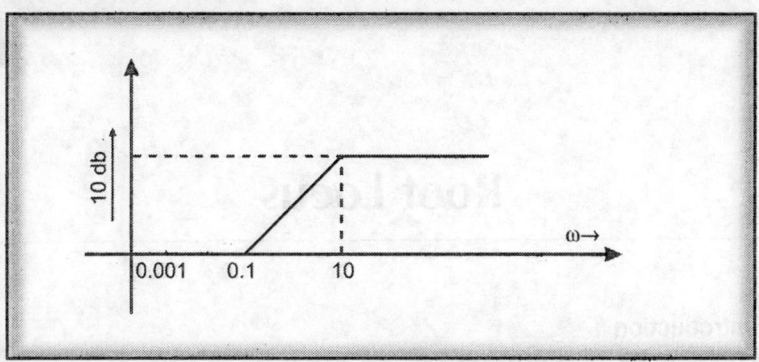

Fig. 13.35

Solution

Since, the initial slope is 0 db/dec, the type of the system is zero. Transfer function is calculated as shown below:

Sr. No.	Slope of Line	Corner Frequency (rad/sec)	Factor
1.	0 db/dec	—	$20 \log K = 0$ Therefore $K = 1$
2.	20 db/dec	$\omega_{c1} = 0.1$	The slope changes from 0 db/dec to +20 db/dec due to presence of a zero in the transfer function. The factor in the numerator is $= \left(1+\dfrac{s}{\omega_{c1}}\right) = \left(1+\dfrac{s}{0.1}\right) = (1+10s)$
3.	0 db/dec	$\omega_{c2} = 10$	The slope of the line changes from +20 db/dec to 0 db/dec due to presence of a pole in the transfer function. The factor in the denominator is $= \left(1+\dfrac{s}{\omega_{c2}}\right) = \left(1+\dfrac{s}{10}\right) = (1+0.1s)$

Transfer function

$$= \frac{K\left(1+\dfrac{s}{\omega_{c1}}\right)}{\left(1+\dfrac{s}{\omega_{c2}}\right)} = \frac{1\left(1+\dfrac{s}{0.1}\right)}{\left(1+\dfrac{s}{10}\right)} = \frac{(1+10s)}{(1+0.1s)}$$

Root Locus

14.1 Introduction

The transfer function of a second order control system is given as

$$\frac{C(s)}{R(s)} = \frac{\omega_n^2}{s^2 + 2\xi\omega_n s + \omega_n^2} \tag{14.1}$$

The characteristic equation is $s^2 + 2\xi\omega_n s + \omega_n^2 = 0$. Changing its coefficients will change the root locus, thus giving a different type of performance. The roots of the characteristic equation are the poles of the closed-loop transfer function. Therefore, a change in the amplifier gain will change the values of the poles of a transfer function. Adjusting gain of the amplifier improves system performance. A root locus plot gives a system's performance for any value of amplifier gain.

Consider a closed-loop system shown in Fig. 14.1. It's closed-loop transfer function is

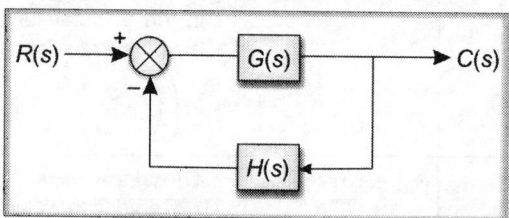

Fig. 14.1

$$T = \frac{C(s)}{R(s)} = \frac{G(s)}{1 + G(s)H(s)} \tag{14.2}$$

Now, $1 + G(s)H(s) = 0$; and is satisfied when $\tag{14.3}$

$$|G(s)H(s)| = 1 \tag{14.4}$$

and $\qquad |G(s)H(s)| = 180° \tag{14.5}$

The process of finding points on the s-plane which satisfy the above conditions, is the basis of the root locus method. Root locus is a graphical method used to find the position of the roots of the characteristic equation or the poles of closed-loop transfer function.

14.2 Simple Control System

Consider a simple control system shown in Fig. 14.2a

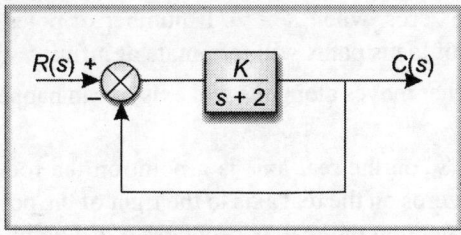

Fig. 14.2a

$$G(s)H(s) = \frac{K}{(s+2)} \text{ and has a pole at } s = -2$$

$$\frac{C(s)}{R(s)} = \frac{G(s)}{1 + G(s)H(s)} = \frac{K}{(s+K+2)} \qquad (14.6)$$

The closed-loop pole is $s = -(K+2)$

If K is varied from zero to infinity, the closed-loop pole moves to the left along the real axis away from its open-loop position as shown in Fig. 14.2b

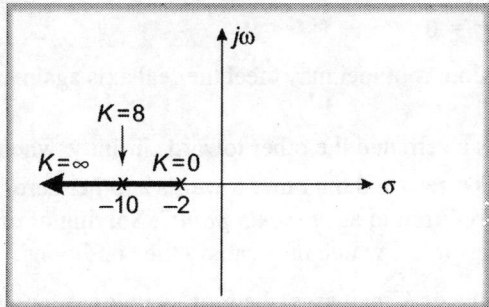

Fig. 14.2b Path of pole of $\dfrac{C(s)}{R(s)} = \dfrac{K}{s+K+2}$

Thus, we see that any point on the path shown in Fig. 14.2(b) satisfies the angle condition of Eq. (5), *i.e.* any point on the path (shown by dark line as in Fig. 14.2b) subtends an angle of $180°$ when seen from the open-loop at $s = -2$. A change in the gain constant 'K' gives a different position.

14.3 General Rules

Rule 1: Obtain the open-loop poles and zeros from the open-loop transfer function. The poles are obtained by first factorising the denominator of the open-loop transfer function, if already not existing and then equating it to zero. Similarly, the zeros are obtained by equating the numerator of the open-loop transfer function to zero.

Rule 2: The number of poles or the order of characteristic equation gives the number of root loci.

Rule 3: The starting point of the root loci are the location of poles, when $K = 0$ and terminates at the zeros, when $K = \infty$. If number of poles is greater than the zeros, the balance root locus paths will terminate at infinity, when $K = \infty$.

Rule 4: Root loci either moves along the real axis or can happen to be as complex conjugate pairs.

Rule 5: A value of 's' on the real axis is a point on the root-locus. If the total number of poles and zeros on the real axis to the right of the point is odd, *i.e.* while constructing the root-loci on the real axis, choose a test point on it. If the sum of poles and zeros to the right of this point is odd, then this point is a part of the root-loci. If it is even, then it is not part of root-loci.

Rule 6: On the root locus between two adjacent open-loop poles, the root loci originating from the two open-loop poles move towards each other, when $K = 0$ and meets at a point when K is maximum for that part of root loci. Thereafter, the root loci breaks away in two parts. The point at which the root loci breaks away is called the *breakway point* on the real axis. This is obtained by differentiating the open-loop transfer function gain with respect to 's' and equating it to zero and thereafter solving for 's'

$$\frac{dK}{ds} = 0$$

In a similar fashion, root loci may meet the real axis again and tend to move in two directions:

  (i) One towards a zero and the other towards infinity, when $K = \infty$, or

  (ii) One towards a zero and the other towards another zero, when $K = \infty$.

Such a point is referred to as "*breakin point*". Solving of dK/ds, in such cases will give two values of 's', which may satisfy the conditions.

Rule 7: As the value of K increases, the root loci moves far away from the poles and zeros. The path of the root loci then is approximated with the help of "asymptotes" or "asymptotic lines".

  (a) Number of asymptotes = number of poles – number of zeros

  (b) Angle with real axis of the asymptotes is given by $\alpha = \dfrac{(2K + 1) \times 180°}{P - Z}$,

where, K is equal to number of asymptotes starting with zero, *e.g.* if number of asymptotes is three, then $K = 0, 1, 2$ $(P - Z - 1)$....

  (c) Intersection of the asymptotes with the real axis is obtained as

$$s = \frac{\Sigma \text{ Real part of poles} - \Sigma \text{ Real part of poles zeros}}{P - Z}$$

Here, $s = $ *Centroid* of the open-loop pole zero configuration.

Rule 8: Intersection of the root loci with the imaginary axis is obtained by use of Routh's array or by putting $s = j\omega$ in open-loop transfer function and equating its real part to zero.

Rule 9: The angle of departure from complex pole or a complex zero helps in sketching the root loci with great accuracy. This is obtained by subtracting from 180°, the sum of all the angles subtended by all the other poles and zeros with appropriate summation sign.

In general $\quad \phi_d = 180° - (\phi_P - \phi_Z)$

ϕ_P = the sum of all the angles subtended by remaining poles ; and

ϕ_Z = the sum of all the angles subtended by the zeros.

Similarly, the angle of arrival at complex zero is given by

$\quad \phi_a = 180° - (\phi_Z - \phi_P)$

where

ϕ_Z = the sum of all angles subtended by remaining zeros, and

ϕ_P = the sum of all the angles subtended by the poles.

Rule 10: Value of K

(a) To find value of K at any point on the root loci, we have to first find out the coordinates of the point. Let the point be $s = -a + jb$. Then it is substituted in open-loop transfer function to find the value of K.

$$G(s)H(s)\big|_{s = a - jb} = -1$$

(b) **Value of K for given value of damping ratio**

  (i) Let us assume the value of damping ratio is 0.707. Find

$$\theta = \cos^{-1} 0.707 = 45°$$

In the case under consideration θ is 45°

  (ii) Draw a line from origin at angle $\theta = 45°$ intersecting the root loci point A (say) as shown in Fig. 14.3.

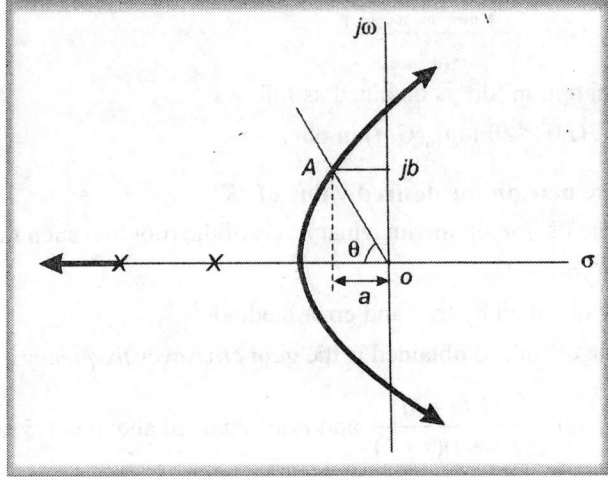

Fig. 14.3

(iii) Let the coordinates of point A is $(-a + jb)$.

(iv) Repeat steps under Rule 10(a) to find the value of K.

(c) **Value of 'K' for marginal stability.** This is obtained by the help of Routh's criterion:

   (i) Form the characteristic equation which is obtained by the relation

$$1 + G(s)H(s) = 0$$

  (ii) Construct Routh's array.

 (iii) Find the value of K for stability by equating the relevant row of the Routh's array to zero. This is the point from where any further increase in value of K will shift the roots to lie in the R.H.S. of s-plane and makes the system unstable.

Rule 11: Frequency of sustained oscillations. This occurs at a point where root loci crosses the imaginary axis. This point also gives the value of K for marginal stability (Rule 10 (c)). The frequency at this point is obtained as given below:

 (a) Find value of K for marginal stability (Rule 10(c)).

 (b) Formulate auxillary equation with the help of Routh's array from the row just above the row which gave value of K for marginal stability.

 (c) Solve the equation for the value of 's', after putting value of 'K' already found out in para 'a' (Rule 10) above.

 (d) The value of 's' will give frequency of sustained oscillations in rad/sec.

Rule 12: Gain margin for desired value of 'K'.

 (a) Find value of 'K' for marginal stability

 (b) Find gain margin (GM) as given below

$$GM = \frac{K_{(\text{marginal stability})}}{K_{(\text{desired})}}$$

 (c) Gain margin in 'db' is obtained as follows

$$GM = 20 \log_{10}(GM) \text{ in db}$$

Rule 13: Phase margin for desired value of 'K'

(a) Find value of '$j\omega$' on the imaginary axis of the root loci such that

$$G(j\omega)H(j\omega) = 1$$

This may be obtained by trial and error method.

(b) The value of 'ω', so obtained is the *gain crossover frequency*.

(c) Let $G(s)H(s) = \dfrac{K(s+3)}{s(s+1)(s+2)}$ and ω as obtained above is 1.5 rad/sec, then

phase margin $= 180° + G(j\omega)H(j\omega)$, where $\omega = 1.5$ rad/sec.

Therefore,

$$\text{phase margin} = 180° + \left(-90° - \tan^{-1}\frac{1.5}{1} - \tan^{-1}\frac{1.5}{2} + \tan^{-1}\frac{1.5}{3} \right)$$

SOLVED PROBLEMS

Problem 14.1 Plot the root-locus pattern of a system whose forward path transfer function is

$$G(s) = \frac{200}{s + 20}$$

Solution

See Fig. 14.4.

Since, there is one pole at $s = -20$. The plot starts from $s = -20$ on the real axis and the only root-locus terminates at a non-finite zero. If we select a point P_1 on the root-locus shown by the dark line, the sum of poles and zeros on the R.H.S. of this point are odd.

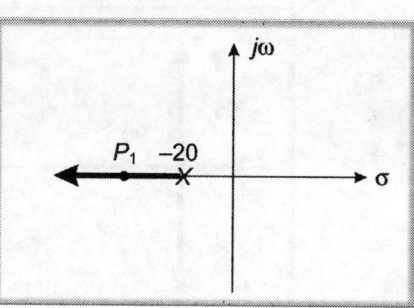

Fig. 14.4

Problem 14.2 Plot the root-locus pattern of a system whose forward path transfer function is

$$G(s) = \frac{K}{(s + 3)(s + 4)}$$

Solution

Refer Fig. 14.5. There are two poles at $s = -3$ and $s = -4$ and no zeros. These poles are shown in Fig. 14.5. Since, there are two poles, there will be two root-loci both terminating at non-finite zeros. Select two points as shown on Fig. 14.5.

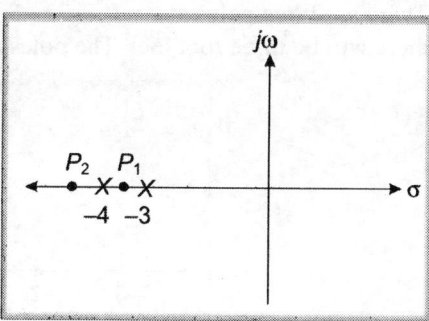

Fig. 14.5

Point P_1: It lies on the root-locus because the sum of poles and zeros to the right of it is odd.

Point P_2: It does not lie on the root-locus because the sum of poles and zeros to the right of it is even.

Breakaway point: Since, there are two root-loci and hence loci on the real axis must breakaway at a point.

$$\frac{d}{ds}K = 2s + 7 = 0 \quad \text{or} \quad s = -3.5$$

Therefore, the breakaway point is -3.5

Asymptotes to the root-loci at infinity

$$\alpha_0 = \pm\frac{180}{2-0} = \pm 90°$$

$$\alpha_1 = \pm\frac{3 \times 180}{2} = \pm 270° \text{ (which is same as above)}$$

The complete root-locus is shown in the Fig. 14.6 below by dark line. The asymptotes are part of the root-loci.

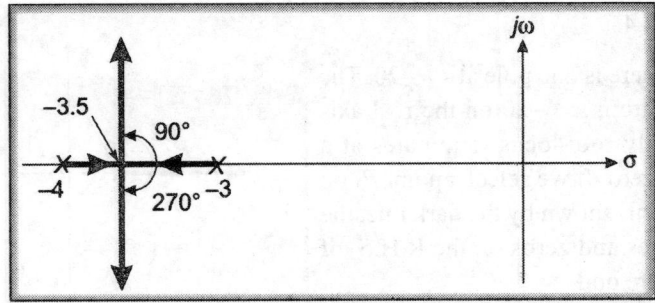

Fig. 14.6

Problem 14.3 Plot the root locus pattern of a system whose forward path transfer function is

$$G(s) = \frac{K}{s(s+2)(s+3)}$$

Solution

There are three poles at $s = 0, -2, -3$ and no zeros. Since there are three poles; there will be three root loci. The poles are plotted on Fig. 14.7.

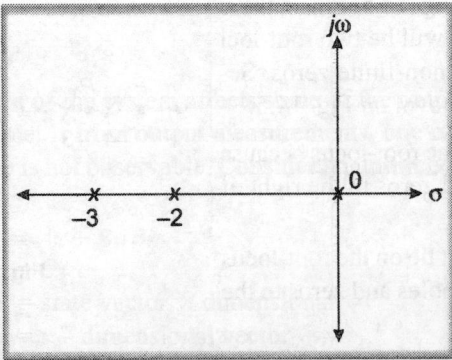

Fig. 14.7

Select three points P_1, P_2, P_3 as shown. Points P_1, and P_3 are on the root-loci because of Rule 5; whereas P_2 is not on the root-loci (Rule 5). See Fig. 14.8.

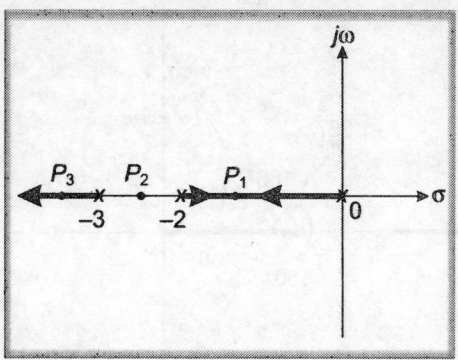

Fig. 14.8

Breakaway from the real axis

$$\frac{dK}{ds} = 3s^2 + 10s + 6 = 0 \quad \text{or} \quad s = -0.784 \quad \text{and} \quad -2.549$$

$s = -2.549$ cannot be a breakaway point because it does not lie on the root-locus. Therefore, the breakaway point is -0.784.

Asymptotes to the root-loci

$$\alpha_0 = \pm \frac{180}{3} = \pm 60°$$

$$\alpha_1 = \pm \frac{3 \times 180}{3} = \pm 180°$$

Intersection of the asymptotes

$$s = \frac{\Sigma \text{ poles} - \Sigma \text{ zeros}}{P - Z} = \frac{(0 - 2 - 3) - (0)}{3 - 0} = -1.667$$

The three asymptotes with centroids $(-1.667, j0)$ making angles $\pm 60°$ and $\pm 180°$ are shown in Fig. 14.9.

Intersection with imaginary axis

$$G(j\omega) = \frac{K}{j\omega(j\omega + 2)(j\omega + 3)}$$

or
$$G(j\omega) = K\left[\frac{-5\omega^2}{25\omega^4 + (6\omega - \omega^3)^2} - j\frac{(6\omega - \omega^3)}{25\omega^4 + (6\omega - \omega^3)^2}\right]$$

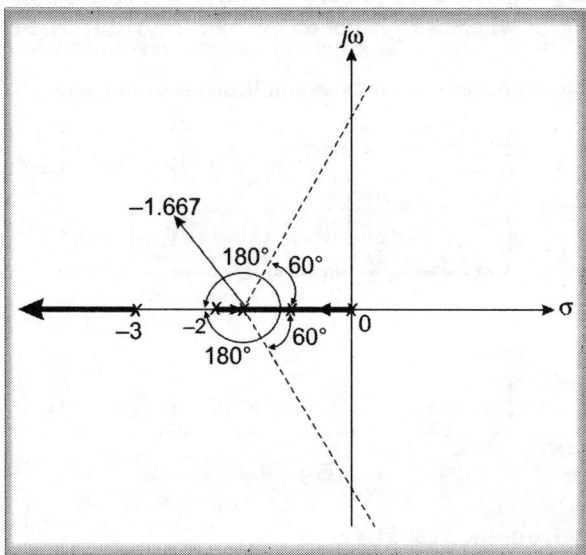

Fig. 14.9

Equating the real part to zero, we get $\omega = \pm 2.5$ rad/sec.

Therefore, the root loci intersects the imaginary axis at

$$\omega = \pm 2.5 \text{ rad/sec}$$

The complete root locus pattern is shown in Fig. 14.10.

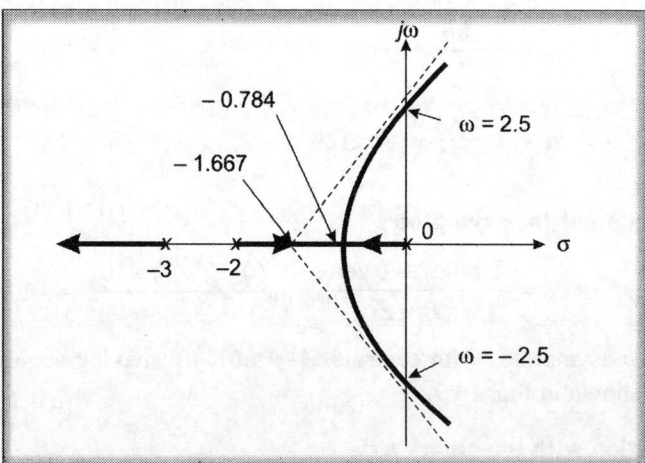

Fig. 14.10

Problem 14.4 Plot the root-locus pattern for a system whose forward path trans-

fer function is $\dfrac{K(s+1)}{s(s+2)(s^2+2s+5)}$

Solution
$$G(s) = \frac{K(s+1)}{s(s+2)(s+1+j2)(s+1-j2)}$$

The poles are $s = -2, (-1 - j2), (-1 + j2), 0$

The order of the characteristic equation is four and hence number of root-loci are four.

(a) All points located between $s = 0$ and $s = -1$ will be on the root locus path as the total number of poles and zeros to the right of them is odd.

(b) Same is true for the points lying on negative real axis beyond $s = -2$.

(c) However, there will be no root locus path between $s = -1$ and $s = -2$, as total poles and zeros to the right of the points located between $s = -1$ and $s = -2$ is even. Root loci along the real axis is shown in Fig. 14.11.

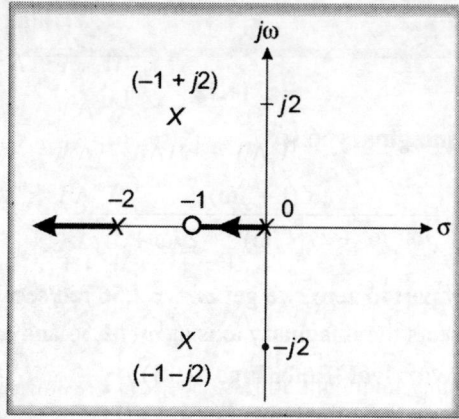

Fig. 14.11

Asymptotes
The number of asymptotes is $= P - Z = 4 - 1 = 3$.

Angles
$$\alpha_1 = \pm \frac{180}{4-1} = \pm 60°$$

$$\alpha_3 = \pm \frac{3 \times 180}{4-1} = \pm 180°$$

$$\alpha_5 = \pm \frac{5 \times 180}{3} = \pm 300° \text{ and so on.}$$

This can also be found by the formula $= \dfrac{(2K+1) \times 180°}{P-Z}$, where $K = 0, 1, 2$.

Intersection of asymptotes with real axis
$$s = \frac{(-2-1+j2-1-j2)-(-1)}{4-1} = -1$$

The asymptotes are shown in Fig. 14.12.

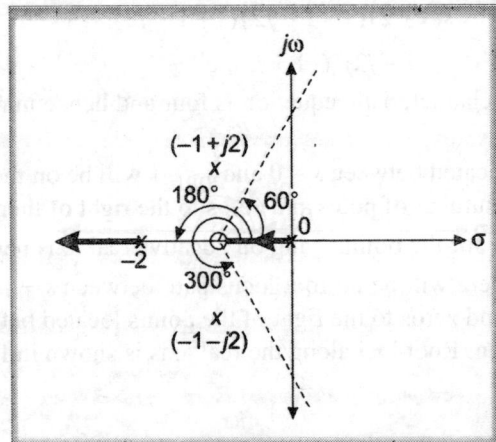

Fig. 14.12

Intersection with imaginary axis

$$G(j\omega) = \frac{K(1 + j\omega)}{j\omega(j\omega + 2)\{(j\omega)^2 + 2j\omega + 5\}}$$

Equating imaginary part to zero, we get $\omega = \pm 2.56$ rad/sec.

The root loci intersects the imaginary axis at $\omega = 2.56$ and at $\omega = -2.56$.

Angle of departure/arrival (Refer Fig. 14.13).

The angle of departure from the pole at $(-1 + j2)$ is

$$= 180° - 116.57° - 90° + 90° - 63.43° = 0$$

The angles from poles and zeros are obtained with the help of a protractor or

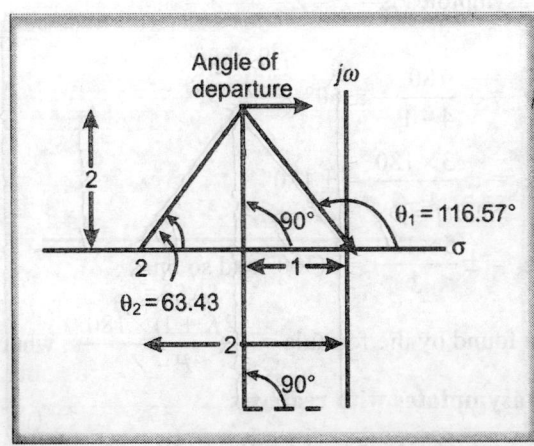

Fig. 14.13

$$\theta_2 = \tan^{-1}\frac{2}{1} = 63.43°$$

$$\theta_1 = 180° - \tan^{-1}\frac{2}{1} = 116.57$$

The angle of departure of other conjugate pole will be 0° due to symmetry. The root-locus pattern is shown in Fig. 14.14.

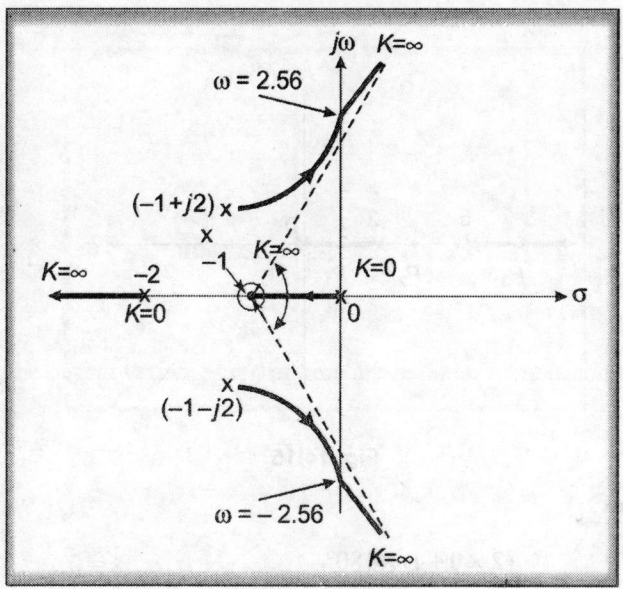

Fig. 14.14

Problem 14.5 A unity feedback control system has

$$G(s) = \frac{K}{s(s+2)(s+5)}$$

Sketch the root locus and show

(a) Breakaway point (b) Line for $\xi = 0.5$ and value of K for this damping ratio (c) The frequency at which the root locus crosses the imaginary axis and the corresponding value of K. (*AMIE, Pune University*)

Solution

Open-loop poles $s = 0, -2, -5$.

Since there are three poles, number of root loci are three, each branch describing the variation of a particular closed-loop pole with the variable K. When $K = 0$ the open and closed-loop poles concide. The pole-zero pattern is shown in

Fig. 14.15. Select points P_1, P_2 and P_3. By applying Rule (5), it is seen that P_1 and P_3 lie on the root locus and P_2 does not lie on the locus. Thus, there is a branch of the locus along the real axis between the origin and -2 and another branch between -5 and infinity. Thus, there will be root locus path between $s = 0$ and $s = -2$. These paths will move towards each other. There will be a breakaway point between $s = 0$ and $s = -2$. Since, there are no zeros, the three root locus paths will terminate at infinity. The root loci originating from pole $s = -5$, will extend towards negative real axis till infinity.

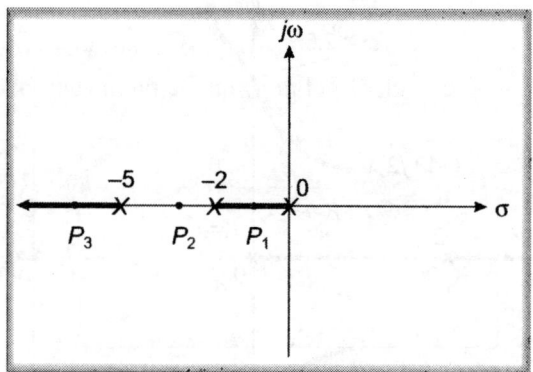

Fig. 14.15

Asymptotes

$$\alpha_1 = \frac{(2 \times 0 + 1) \times 180°}{3 - 0} = 60°$$

$$\alpha_2 = \frac{(2 \times 1 + 1) \times 180°}{3 - 0} = 180°$$

$$\alpha_3 = \frac{(2 \times 2 + 1) \times 180°}{3 - 0} = 300°$$

Point of intersection of asymptotes and real axis

$$s = \frac{0 - 2 - 5}{3} = -2.33$$

The asymptotes are shown in Fig. 14.16.

Breakaway point

$$\frac{dK}{ds} = 3s^2 + 14s + 10 = 0 \text{ gives } s = -3.8 \text{ and } -0.88.$$

$s = -3.8$ cannot be a breakaway point as it does not lie on the root-locus.

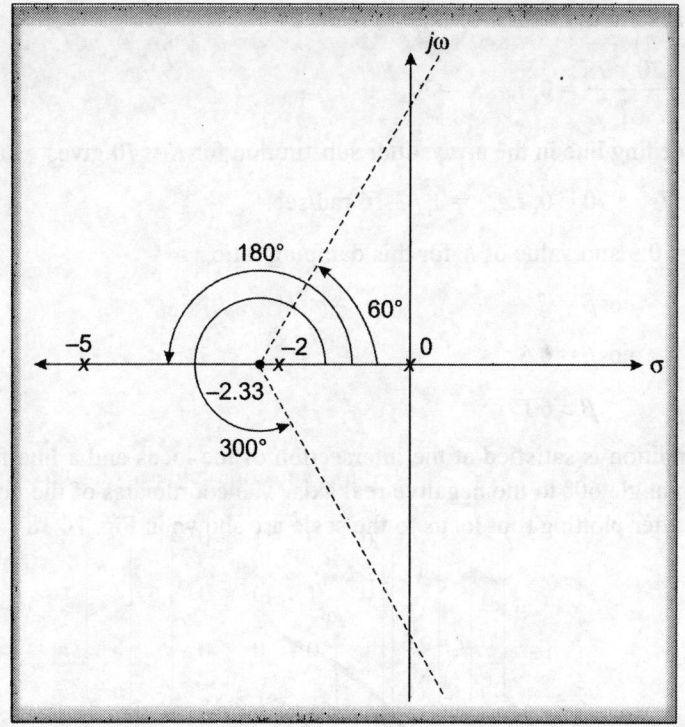

Fig. 14.16

Now, $\quad 1 + G(s) = 0$

$$G(s) = -1$$

$$\frac{K}{s(s+2)(s+5)} = -1$$

$$K = -s(s+2)(s+5)$$

For $\quad s = -0.88$

or $\quad K = 0.88(-0.88 + 2)(-0.88 + 5)$

or $\quad K = 4$

Intersection of locus with imaginary axis

The characteristic equation is

$$s^3 + 7s^2 + 10s + K = 0$$

s^3	1	10
s^2	7	K
s^1	$\left(\dfrac{70-K}{7}\right)$	0
s^0	K	0

For the system to be on the verge of stability

$$\frac{70 - K}{7} = 0, \text{ } i.e. \text{ } K = 70$$

The preceding line in the array, after substitution for $K = 70$ gives

$$7s^2 + 70 = 0, \text{ } i.e. \text{ } s = \pm j \text{ } 3.16 \text{ rad/sec.}$$

Line for $\xi = 0.5$ and value of K for this damping ratio

$$\cos\beta = \xi$$

or $$\cos\beta = 0.5$$

$$\therefore \qquad \beta = 60°$$

This condition is satisfied at the intersection of the locus and a line from the origin at an angle 60° to the negative real axis. The coordinates of the point s on Fig. 14.17 after plotting root locus to the scale are shown in Fig. 14.18.

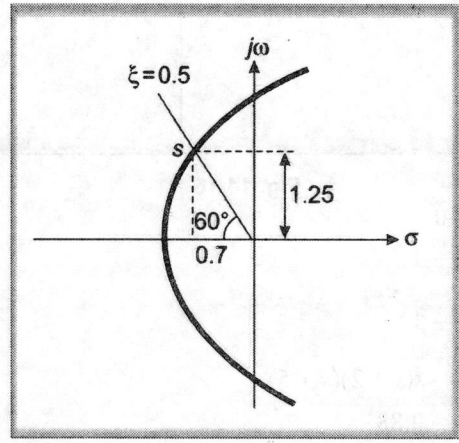

Fig. 14.17

$$s = 0.7 + j1.25$$

$$K = |s \text{ } (s + 2)(s + 5)|_s = -0.7 + j1.25$$

$$K = |(-0.7 + j1.25) \text{ } (1.3 + j1.25) \text{ } (4.3 + j1.25)|$$

$$= \left| \left(\sqrt{0.7^2 + 1.25} \right) \left(\sqrt{1.3^2 + 1.25^2} \right) \left(\sqrt{4.3^2 + 1.25^2} \right) \right|$$

$$= |1.45 \times 1.8 \times 4.45| = 11.5$$

The complete root locus is shown in Fig. 14.18

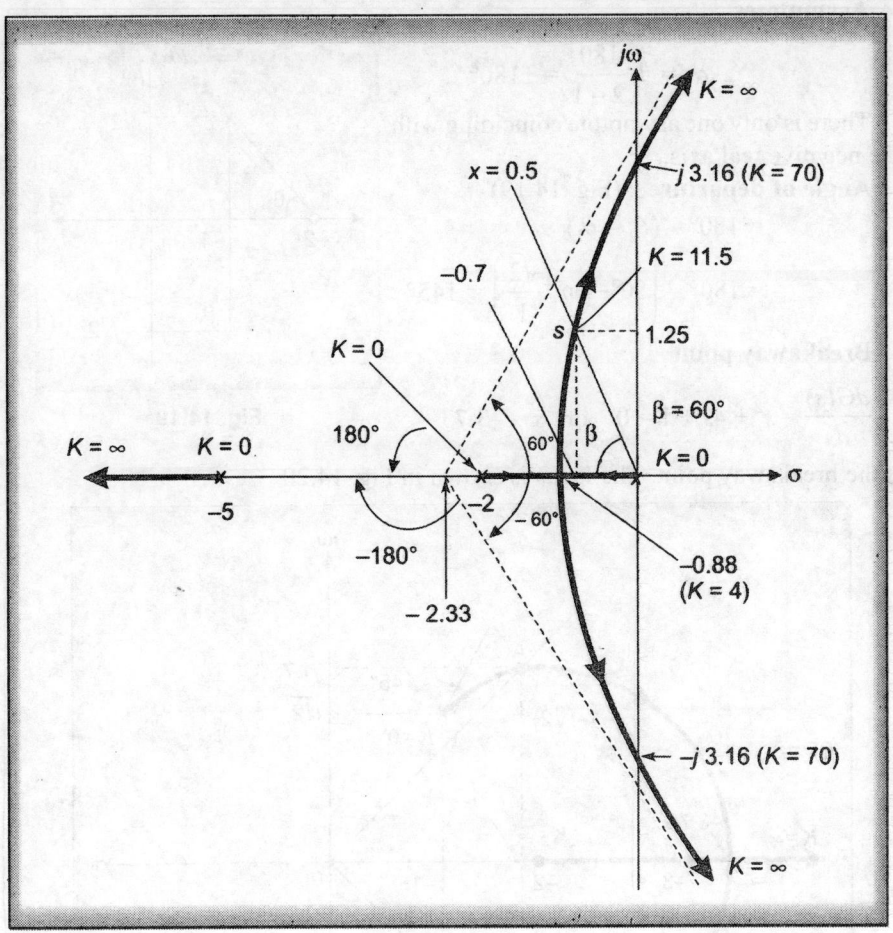

Fig. 14.18

Problem 14.6 Sketch the root-locus plot and determine the approximate damping ratio for a value of $K = 1.33$ for a control system having a forward transfer function.

$$G(s) = \frac{K(s+2)}{s^2 + 2s + 3}$$

Solution

Poles $= -1 \pm j\sqrt{2}$ and zeroes $= -2$

The characteristic equation is

$$s^2 + (2 + K)s + 2K + 3 = 0$$

Since, the order of the characteristic equation is two, the number of root-loci are two. Each root-locus starts from one of the two complex conjugate poles and breaks in on the negative real axis between -2 and $-\infty$. One of the root locus terminates at infinity and the other at $s = -2$.

Asymptotes

$$\alpha_1 = \pm \frac{180}{2-1} = \pm 180°$$

There is only one asymptote coinciding with the negative real axis.

Angle of departure   (Fig. 14.19)

$$= 180° - (\theta_2 - \theta_1)$$

$$= 180° - \left(90° - \tan^{-1}\frac{\sqrt{2}}{1}\right) = 145°$$

Breakaway point

$$\frac{dG(s)}{ds} = s^2 + 4s + 1 = 0 \quad \text{or} \quad s = -3.73$$

is the breakaway point. The locus is shown in Fig. 14.20

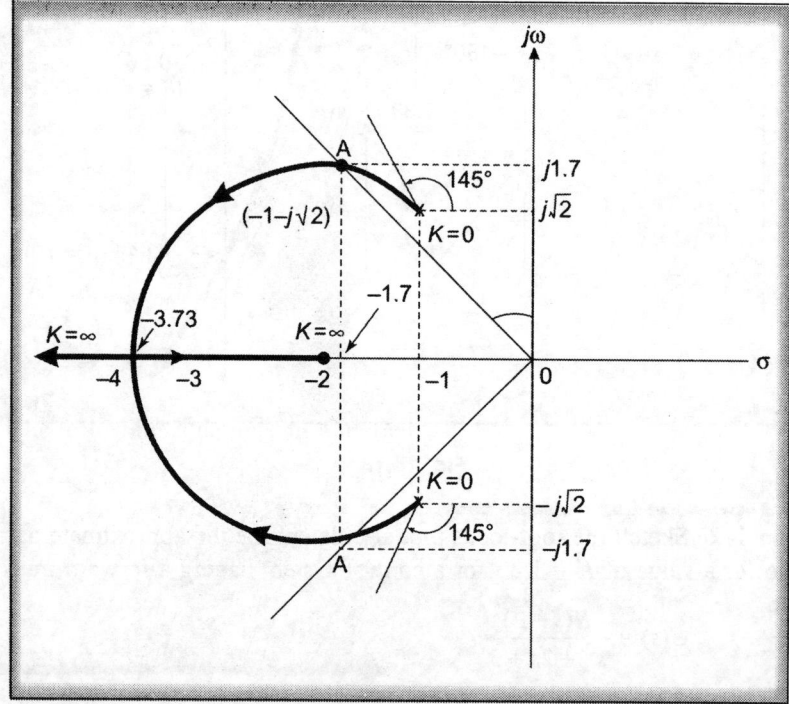

Fig. 14.20

Point on root locus where, $K = 1.33$

$$\frac{K(s+2)}{s^2 + 2s + 3} = -1 \text{ or } \frac{1.33(s+2)}{s^2 + 2s + 3} = -1 \text{ or } s^2 + 3.33s + 5.66 = 0$$

i.e. $s = -1.7 \pm j1.7$

$$s = 2.4\angle \pm 45° \text{ (This point is shown as point 'A').}$$

Problem 14.7 Sketch the root-locus of $G(s) = \dfrac{K}{s^2 + 10s + 100}$

Solution

Poles

There are two poles at $s = -5 \pm j8.66$. The characteristic equation is

$$s^2 + 10s + 100 = 0.$$

Since, the order of the characteristic equation is two, the number of root-locus are two. Since, there is no zero, the root locus paths originating from the two poles will terminate at infinity.

Asymptotes

$$\alpha_0 = \pm \frac{180°}{2-0} = \pm 90°$$

$$\alpha_1 = \pm \frac{3 \times 180°}{2-0} = \pm 270°$$

$$\alpha_2 = \pm \frac{5 \times 180°}{4-1} = \pm 300°$$

Intersection with real axis

$$s = \frac{(-5 + j8.66) + (-5 - j8.66)}{2} = -5$$

Angle of departure $= 180° - 90° = 90°$

The root locus is shown in Fig. 14.21.

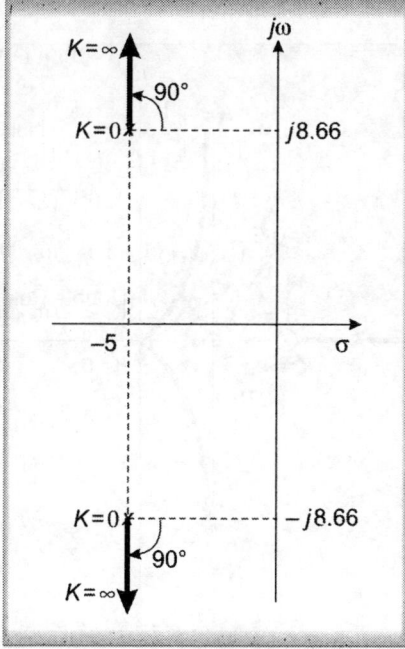

Fig. 14.21

Problem 14.8 Sketch the root-locus of $G(s) = \dfrac{K(s+1)}{s^2(s+2)}$

Solution

$$\text{Poles} = 0, 0, -2 \text{ and zeros} = -1$$

The characteristic equation is

$$s^3 + 2s^2 = 0$$

Since, the order of the characteristic equation is three, the number of root locus are three. Therefore, three root loci will originate from three poles. Out of the three, one root locus path will terminate at the zero $s = -1$ and other two root loci will terminate at infinity.

Asymptotes

$$\alpha_1 = \pm \frac{180°}{3-1} = \pm 90°$$

$$\alpha_3 = \pm \frac{3 \times 180°}{3-1} = \pm 270°$$

Intersection of the asymptotes with real axis $s = \dfrac{-2-(-1)}{3-1} = -0.5$

The root-locus is shown in Fig. 14.22.

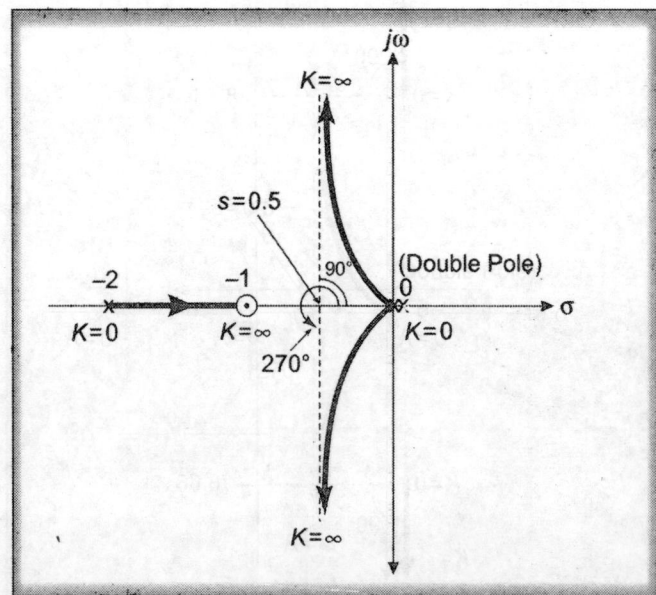

Fig. 14.22

Problem 14.9 Sketch the root locus for $G(s) = \dfrac{K(s+1)}{s^2(s+3)(s+5)}$

Solution

Poles $s = 0, 0, -3$ and -5 and zeros $= 1$.

Since, there are four poles, four root loci will originate from these four poles and since there is one zero, one of the four root locus paths will terminate at this zero and the other three root loci will terminate at infinity.

(*a*) The poles and zeros are shown in Fig. 14.23a

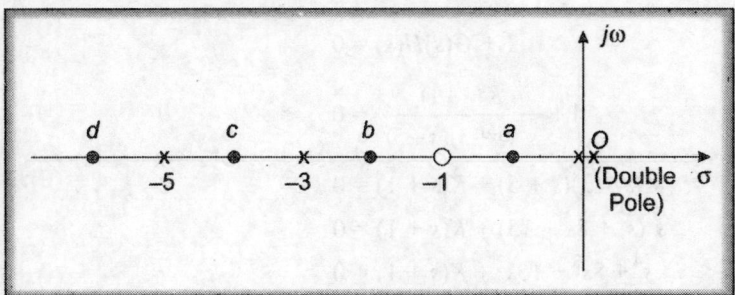

Fig. 14.23a

(i) **Points '*a*' and '*c*'.** These points will not lie on the root loci, as the total number of poles and zeros to the right of it, when considered pointwise is even. Hence, there will not be any root locus path between $s = 0$ and $s = -1$ and also between $s = -3$ and $s = -5$.

(ii) **Point '*b*'.** This point and also all the points between $s = -1$ and $s = -3$ will lie on the root loci as the total number of poles and zeros to the right of them is odd. Hence, there will be a root locus between path $s = -1$ and $s = -3$. It will originate from the pole $s = -3$ and terminate at $s = -1$ which is the zero.

(iii) **Point '*d*'.** This point and all the points lying on the negative real axis from $s = -5$ and beyond will lie on the root locus path as the total number of poles and zeros to the right of them is odd. Hence, there will be a root locus path originating from $s = -5$ (pole) and will proceed along the negative real axis and terminate at infinity.

(*b*) **Asymptotes.** The number of asymtotes are $P - Z$, *i.e.* $4 - 1 = 3$. The angle of asymtotes are

$$\frac{(2K+1)180°}{P-Z} \quad \text{where,} \quad K = 0, 1, 2$$

$$\alpha_1 = \frac{(2 \times 0 + 1) \times 180°}{4-1} = 60°$$

$$\alpha_2 = \frac{(2 \times 1 + 1) \times 180°}{4-1} = 180°$$

$$\alpha_3 = \frac{(2 \times 2 + 1) \times 180°}{4-1} = 300°$$

Intersection of the asymptotes with the real axis

$$= \frac{\Sigma P - \Sigma Z}{P - Z}$$

$$= \frac{(0+0-3-5)-(-1)}{4-1} = \frac{-7}{3} = -2.33$$

(c) Intersection of imaginary axis by root locus

This will be obtained with the help of Routh's array. The characteristic equation is

$$1 + G(s)H(s) = 0$$

or $$1 + \frac{K(s+1)}{s^2(s+3)(s+5)} = 0$$

or $$s^2(s+3)(s+5) + K(s+1) = 0$$

or $$s^2(s^2+8s+15) + K(s+1) = 0$$

or $$s^4 + 8s^3 + 15s^2 + K(s+1) = 0$$

or $$s^4 + 8s^3 + 15s^2 + Ks + K = 0$$

The Routh's array is

s^4	1	15	K
s^3	8	K	0
s^2	$\dfrac{120-K}{8}$	K	0
s^1	$\dfrac{\left(\dfrac{120-K}{8}\right) \times K - 8K}{\dfrac{120-K}{8}}$	0	0
s^0	K	0	0

For marginal stability

$$\frac{\left(\dfrac{120-K}{8}\right)K - 8K}{\dfrac{120-K}{8}} = 0$$

or $$(120 - K)K - 64K = 0$$

$$120 - K - 64 = 0$$

or $$K = 56$$

The auxilliary equation of s^3 row is

$$8s^3 + Ks = 0$$

or

$$s(8s^2 + K) = 0$$

which gives $s = 0$;

and

$$8s^2 + K = 0$$

Putting the value of K, we get

$$8s^2 + 56 = 0$$

or

$$s^2 + 7 = 0$$

or

$$s = \pm j\sqrt{7}$$

Therefore, the root locus is going to cross the imaginary axis at $s = \pm\sqrt{7}$

(d) **Root locus**

The root locus is shown in Fig. 14.23b

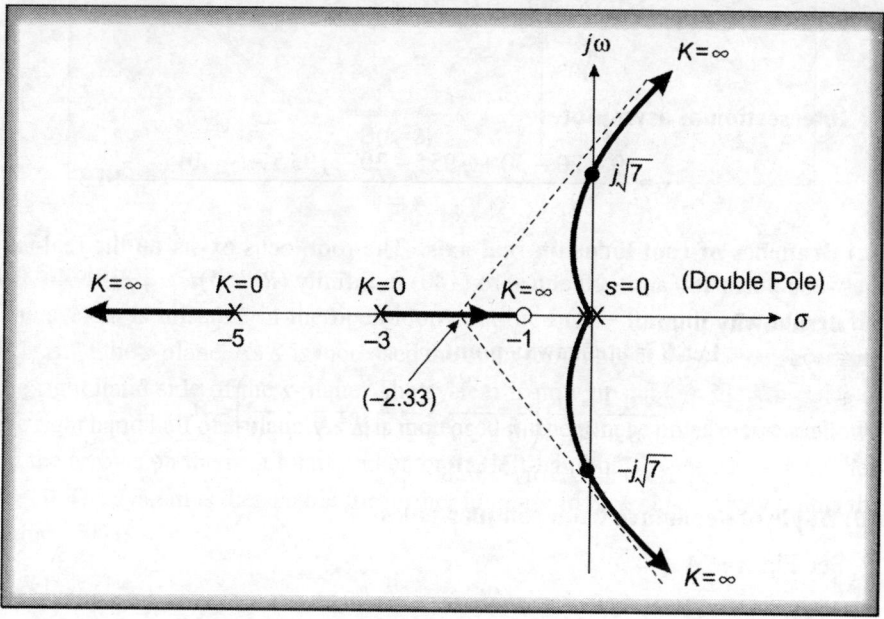

Fig. 14.23b

Problem 14.10 Sketch the root-locus of the transfer function

$$G(s) = \frac{s + 40}{s(s + 20)(s^2 + 60s + 100^2)}$$ (AMIE)

Solution

(a) **Starting and terminating points**

Poles are $s = 0, -20, -30 \pm j95.5$ and   zeros $s = -40$

Since, there are four poles, four root locus paths will originate from these poles and one of the four will terminate at the zero; $s = -40$ and the other three will terminate at infinity. Use *Rule 5* (Section 14.3) to find out which part of the real axis is part of the root locus.

(b) Asymptote
Directions and angles

Numbers $= P - Z = 4 - 1 = 3$

Angles

$$\alpha_0 = \pm \frac{180°}{4 - 1} = \pm 60°$$

$$\alpha_1 = \pm \frac{3 \times 180°}{4 - 1} = \pm 180°$$

$$\alpha_1 = \pm \frac{5 \times 180°}{4 - 1}$$
$$= \pm 300°$$

Intersection of asymptotes

$$s = \frac{0 - 20 - 30 + j95.5 - 30 - j95.5 - (-40)}{4 - 1} = -14.3$$

(c) Branches of root locus on real axis.
The root locus exists on the real-axis between poles at 0 and -20 and zero (-40) to infinity (*Rule 5*).

Breakaway points

Let B = breakaway point,

$$\frac{1}{B + 40} - \left(\frac{1}{B + 20} + \frac{1}{B + 0} + \frac{2}{B + 30} \right) = 0.$$
$$B = -11.72 \text{ or } -68.28$$

(d) Angle of departure from complex poles

See Fig. 14.24

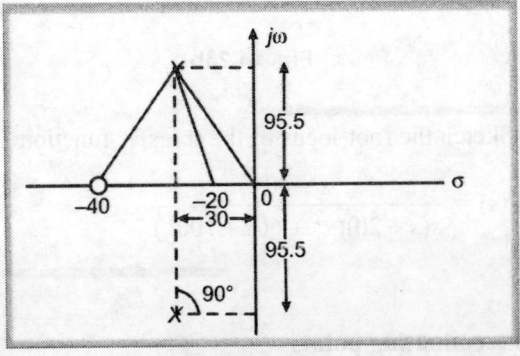

Fig. 14.24

$$= 180° - \left(180° - \tan^{-1}\frac{95.5}{30}\right) - \left(180° - \tan^{-1}\frac{95.5}{10}\right) - \left(90° + \tan^{-1}\frac{95.5}{10}\right)$$

$$= 180° - 107.4° - 96° - 90° + 84° = -29.5°$$

The complete locus is shown in Fig. 14.25.

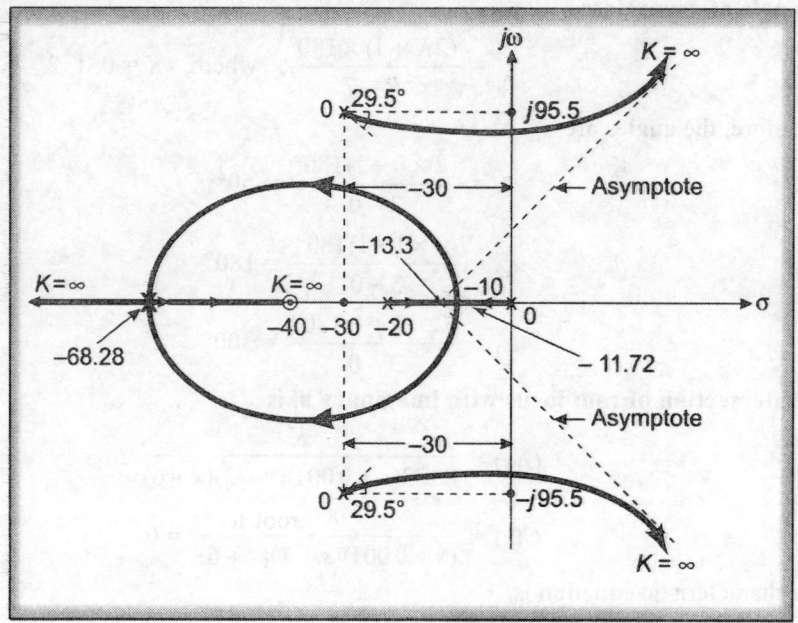

Fig. 14.25

Problem 14.11 Sketch the root locus of a control system having open-loop transfer function

$$G(s) = \frac{K}{83.33(s + 0.001)(s + 2)(s + 6)}$$

(a) Starting and terminating points

Poles $s = -0.001, -2, -6$

(b) Breakaway points

Let B is the breakaway point

$$\frac{1}{B + 0.001} + \frac{1}{B + 2} + \frac{1}{B + 6} = 0 \quad \text{or} \quad B = -0.9 \quad \text{or} \quad -4.45$$

When *Rule 5* is applied as given in Section 14.3, it is seen that the root locus exists between poles -2 and -0.001, -6 and infinity on the real axis. Therefore, $B = -4.45$ cannot be a breakaway point.

(c) Asymptotes intersection with real axis

$$s = \frac{(-0.001 - 2 - 6) - (0)}{3 - 0} = -2.67$$

Number of asymptotes

$$= P - Z = 3 - 0 = 3$$

Angle of asymptotes

$$= \frac{(2K + 1) \times 180°}{P - Z} \quad \text{where,} \quad K = 0, 1, 2$$

Therefore, the angles are

$$\alpha_1 = \frac{(2 \times 0 + 1)180°}{3 - 0} = 60°$$

$$\alpha_2 = \frac{(2 \times 1 + 1)180°}{3 - 0} = 180°$$

$$\alpha_3 = \frac{(2 \times 2 + 1)180°}{3 - 0} = 300°$$

(d) Intersection of root locus with imaginary axis

$$G(s) = \frac{K}{83.33(s + 0.001)(s + 2)(s + 6)}$$

or

$$G(s) = \frac{K'}{(s + 0.001)(s + 2)(s + 6)} = 0$$

The characteristic equation is

$$1 + G(s)H(s) = 0$$

$$1 + \frac{K'}{(s + 0.001)(s + 2)(s + 6)} = 0$$

$$s^3 + 8.001s^2 + 12.008s + 0.012 + K' = 0$$

$$
\begin{array}{c|ccc}
s^3 & 1 & 12.008 & \\
s^2 & 8.001 & 0.012 & K' \\
s^1 & \dfrac{(8.001 \times 12.008) - K' + 0.012}{8.001} & & \\
s^0 & K + 0.012 & &
\end{array}
$$

Putting s^1 row equal to zero, we get

$$\frac{(8.001 \times 12.008) - (K' + 0.012)}{8.001} = 0.$$

This gives

$$K' = 96.6$$

or

$$\frac{K}{83.33} = 96.6 \quad \text{or} \quad K = 96.6 \times 83.3 = 8050$$

Forming auxiliary equations for s^2 row, gives

$$8.001s^2 + K' + 0.012 = 0$$

or $\qquad 8.001s^2 + 96.6 + 0.012 = 0$

or $\qquad s^2 = \dfrac{-96.6 - 012}{8.001}$

or $\qquad s^2 = -12.074$

or $s = \pm j3.5$. This is the point where root locus crosses the imaginary axis.

Value of gain at the point where the root axis crosses has already been found out and is equal to K which is 8050.

(e) Gain at breakaway point

The breakaway point found out in part 'b' of the solution is $s = -0.9$. The gain at the point is

$$\left| \frac{K}{83.33(s + 0.001)(s + 2)(s + 6)} \right|_{s=-0.9} = -1$$

or $\qquad \dfrac{K}{83.33(-0.9 + 0.001)(-0.9 + 2)(-0.9 + 6)} = -1$

or $\qquad\qquad K = 420$ $\qquad\qquad\qquad\qquad\qquad\qquad\qquad$ **Ans**

(f) The root locus is shown in Fig. 14.26

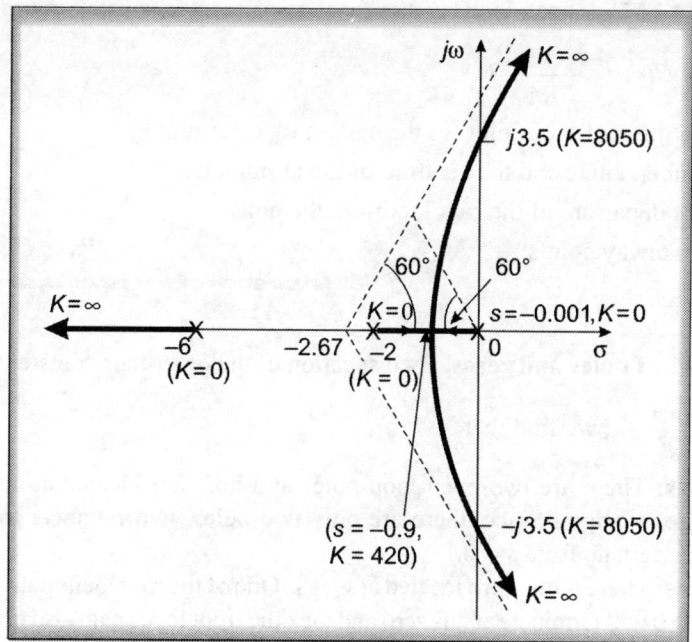

Fig. 14.26

(g) Gain at damping factor 0.5

Given $\qquad \xi = 0.5$

$$\cos\xi = 0.5$$

$\therefore \qquad\qquad \angle\xi = 60°$

Draw a line OA from the origin at an angle 60° with the real axis meeting the root locus at A as shown in Fig. 14.26. The approximate coordinate of the point 'A' as measured from the root locus is $-0.6 + j$.

The gain at this point will be

$$KG(s)|_{s = -0.6 + j|} = -1$$

$$\frac{K}{83.33(-0.599 + j)(1.4 + j)(5.4 + j)} = -1$$

$$K = 83.33 \times 1.16 \times 1.72 \times 5.49 = 913 \qquad\qquad \textbf{Ans}$$

(g) Time to reach 98% steady-state:
This is the time for the output to reach 2% tolerance band

$$T = \frac{4}{\xi\omega_n} = \frac{4}{0.6} = 6.67 \, \text{sec} \qquad\qquad \textbf{Ans}$$

$\xi\omega_n$ = intercept on the negative real axis made by $\xi = 0.5$ line as given in part (f) of the solution above.

Problem 14.12 The open-loop transfer function of a unity feedback control system

is given by $G(s) = \dfrac{K(s+1)}{s^2}$

Sketch the root-locus plot for the system by determining

(a) the number, angle and the centroid of the asymptotes

(b) angle of departure of the root loci from the poles

(c) the breakaway points $\qquad\qquad\qquad$ (*Pune University*)

Solution

(a) **Number of poles and zeros.** By inspection of the open-loop transfer function

$G(s) = \dfrac{K(s+1)}{s^2}$, we find that

(1) **Poles:** There are two open-loop poles and both are located at origin, *i.e.* double pole at origin. Since, there are only two poles, two numbers root locus paths will originate from $s = 0$.

(2) **Zeros:** There is one zero located at $s = -1$. One of the root locus path originating from $s = 0$ will terminate at this zero and the other root locus path also originating from $s = 0$ will terminate at infinity.

The poles and zeros are plotted in Fig. 14.27a.

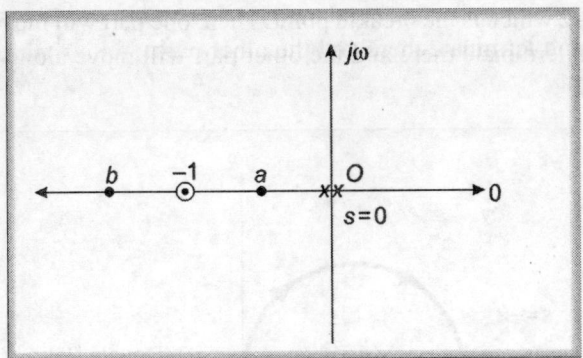

Fig. 14.27a

Select two points 'a' and 'b' as shown.

(i) **Point 'a':** This point and all the points located between $s = 0$ and $s = -1$ will not lie on the root locus path as total number of poles and zeros to the right of them are even. In other words, there will not be any root locus path between $s = 0$ and $s = -1$ on the negative part of the real axis.

(ii) **Point 'b':** This point and all the points located beyond $s = -1$ will lie on the root locus path as total number of poles and zeros to the right of them is odd. In other words, all these points will constitute root loci and extend till infinity. However, one of the root loci originating from $s = 0$ will breakin towards $s = -1$ (zero).

(iii) Since poles > zeros, *i.e.* $2 > 1$

$\therefore$ $N = P$ and hence, number of root locus paths will be two in number.

Asymptotes

Number of asymptotes $= 2 - 1 = 1$

$$\alpha_1 = \frac{(2K + 1)180°}{P - Z} \quad \text{where, } K = 0$$

$$= \frac{(2 \times 0 + 1)}{2 - 1} \times 180° = 180°$$

Centroid $\qquad$'s' $= \dfrac{0 + 0 - (-1)}{2 - 1} = 1$

Breakaway point

$$\frac{dK}{ds} = 0$$

i.e. $\qquad s\,(s + 2) = 0$

$\therefore$ $s = -2$ is the breakaway point. In this case it is a breakin point.

Angle of departure

$$= 180° - 0 - 0 = 180°$$

Root locus: The root-locus is shown in Fig. 14.27b. The root locus paths originating from $s = 0$ will take a circular path. The circular path will intersect the real axis at $s = -2$ which is the breakin point. Then, one part will move towards the zero $s = -1$ and terminate there and the other part will move along negative real axis till infinity.

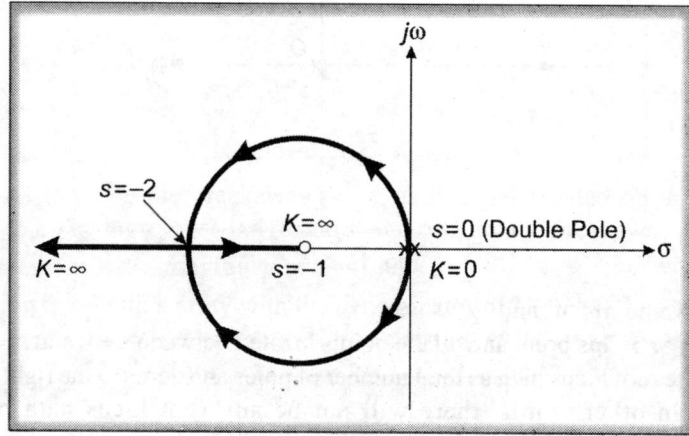

Fig. 14.27b

Problem 14.13 Draw a root locus plot for an unity feedback system whose forward path transfer function is given by $G(s) = \dfrac{K(s+1)}{s^2(s+5)}$.

Find the system gain corresponding to maximum value of damping ratio.

Solution

(a) **Poles and zeros:** By inspection of $G(s) = \dfrac{K(s+1)}{s^2(s+5)}$, we get

(1) **Poles:** Three poles at $s = 0, 0, -5$. Since, there are three poles, three root loci will originate from three poles.

(2) **Zeros:** There is one zero at $s = -1$. One root locus path will terminate at this point and the other two will terminate at infinity. The poles and zeros are plotted in Fig. 14.28a

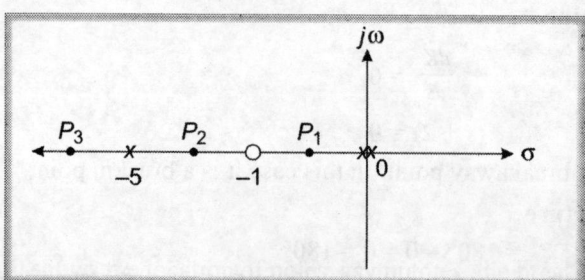

Fig. 14.28a

(i) **Point 'P_1':** This point and all the points located between $s = 0$ and $s = -1$ will not lie on the root loci as total number of poles and zeros to the right of them is even. Hence, between $s = 0$ and $s = -1$ on the negative real axis, there will not be any root locus path.

(ii) **Point 'P_3':** Similar argument as given for point 'P_1' concludes that there will not be any root locus path on negative real axis beyond $s = -5$.

(iii) **Point 'P_2':** This point and all the points located between $s = -1$ and $s = -5$ will lie on the root loci as total number of poles and zeros to the right of them is odd. Hence, there will be a root locus path on the negative real axis between $s = -1$ and $s = -5$.

The root locus path based on the above findings is shown in Fig. 14.28b.

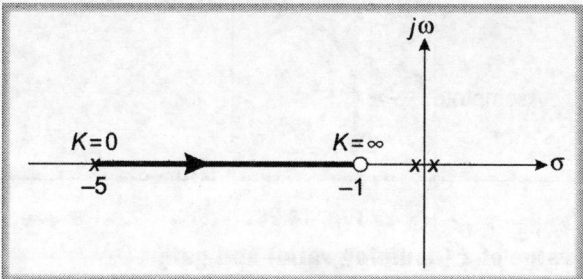

Fig. 14.28b

(b) Asymptotes

(i) Number $= P - Z = 3 - 1 = 2$

(ii) Angle $= \dfrac{(2K + 1)180°}{P - Z}$ where, $K = 0, 1$

$$\alpha_1 = \frac{(2 \times 0 + 1)180°}{3 - 1} = 90°$$

$$\alpha_2 = \frac{(2 \times 1 + 1)180°}{3 - 1} = 270°$$

(iii) Intersections of asymptotes with real axis

$$= \frac{\Sigma P - \Sigma Z}{P - Z} = \frac{(0 + 0 - 5) - (-1)}{3 - 1} = \frac{-4}{2} = -2$$

(c) Breakaway point

$$1 + G(s)H(s) = 0 \quad \text{or} \quad 1 + \frac{K(s + 1)}{s^2 (s + 5)} = 0 \quad \text{or} \quad K = \frac{-s^2(s + 5)}{s + 1}$$

$$\frac{dK}{ds} = -\left[\frac{(s + 1)(3s^2 + 10s) - s^2 (s + 5)}{(s + 1)^2} \right] = 0$$

or $\quad s(s^2 + 4s + 5) = 0$

which gives $s = 0$ and $s = -2 + j$. Since, the breakaway has to be on the real axis, $s = 0$ satisfies the condition.

(d) Root locus is shown in Fig. 14.28c.

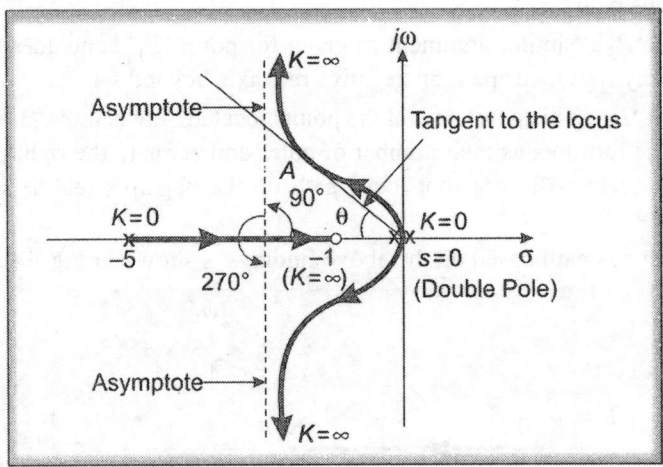

Fig. 14.28c

(e) **Maximum value of ξ (damping ratio) and gain**

   (i) Draw a line tangent to the locus originating from $s = 0$. Let this meet the locus at point A.

  (ii) Measure angle θ.

 (iii) Calculate damping ratio $= \cos\theta$

 (iv) Find the coordinate of point 'A'. Let it be $-\alpha + j\beta$

  (v) Put the values of coordinate in the following equation to find K.

$$KG(s)\big|_{s = -\alpha + j\beta} = -1$$

i.e.

$$\frac{K(s+1)}{s^2(s+5)}\bigg|_{s = -\alpha + j\beta} = -1$$

Note: Students are advised to plot the root locus on a graph paper. The value of K so obtained should be approximately 10.

Problem 14.14 The open loop control system transfer function is $G(s) = \dfrac{K}{s(s^2 + 4s + 8)}$. Sketch the root loci of the system touching the following points.

  (a) Number of root loci (b) Number of asymptotes (c) Angle of asymptotes (d) Point of intersection with real axis (e) Angle of departure (f) Imaginary axis intercepts (g) Real axis part of root locus.

(a) **Number of root loci**

Let us first find out number of poles and zeros. There is no zero. However, the poles are

(i) $s = 0$, and

(ii) The balance are obtained by solving $s^2 + 4s + 8 = 0$. On solving, we obtain
$s = -2 \pm j2$

Since, there are three poles, *i.e.* $s = 0, -2 + j2$ and $-2 - j2$, there are three root loci. The root loci originate at poles and terminate at infinity (as zeros are nil).

(b) **Number of asymptotes** = No. of poles – No. of zeros = $3 - 0 = 3$

(c) **Angle of asymptotes** = $\dfrac{(2K + 1)180°}{P - Z}$ where, $K = 0, 1$ and 2

The three asymptotes are at

$$\alpha_1 = \frac{(2 \times 0 + 1)180°}{3 - 0} = 60°, \quad \alpha_2 = \frac{(2 \times 1 + 1)180°}{3 - 0} = 180°$$

$$\alpha_3 = \frac{(2 \times 2 + 1)}{3 - 0} \times 180° = 300° = -60°$$

(d) **Intersection with real axis of the asymptotes**

$$x = \frac{\Sigma\, \text{Poles} - \Sigma\, \text{Zero}}{P - Z} = \frac{(0 - 2 + j2 - 2 - j2) - 0}{3 - 0} = \frac{-4}{3} = -1.33$$

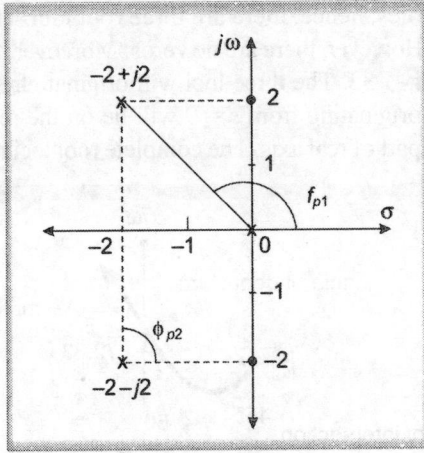

Fig. 14.29

(e) **Angle of departure**

$$\phi_{(-2+j2)} = 180° - (\phi_{p1} + \phi_{p2}) = 180° - (135° + 90°) = -45°$$

and the angle of departure from the complex pole $(-2-j2)$ is $+45°$

(f) **Intersection with imaginary axis**

we will use analytical approach to find the same.

$$1 + G(s)H(s) = 1 + \frac{K}{s(s^2 + 4s + 8)} \times 1 = 0 \quad \text{or} \quad s^3 + 4s^2 + 8s + K = 0$$

Constructing the Routh's array

s^3	1	8
s^2	4	K
s^1	$\dfrac{32-K}{4}$	0
s^0	K	0

for stability

(i) $\qquad\qquad K > 0$

(ii) $\qquad\dfrac{32-K}{4} > 0 \quad$ or $\quad K = 32$

The auxilliary equation of s^2 row is formed by taking into consideration the marginally stable condition and is

$$4s^2 + K = 0; \quad \text{putting} \quad K = 32$$
$$4s^2 + 32 = 0 \quad \text{or} \quad s^2 + 8 = 0 \quad i.e. \quad s = \sqrt{-8} = \pm j2.83$$

Therefore, the points at which the root loci are going to intersect imaginary axis are $\pm j2.83$.

(g) Real axis part of root locus

As there are three poles, hence, there are three root loci. The loci start from the poles and end at zeros. However, there are no zeros. Moreover, the intersection of the asymptotes is at a point -1.33. The three loci will originate from $s = 0, -2 + j2$ and $-2 - j2$. The root loci originating from $s = 0$ will lie on the real axis and extend to infinity on the negative part of real axis. The complete root loci is shown in Fig. 14.30

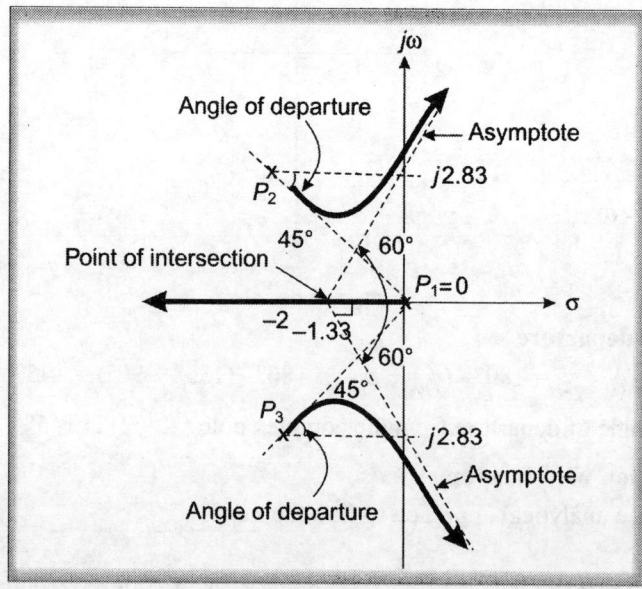

Fig. 14.30

Problem 14.15 The loop transfer function of a feedback control system is given by

$$G(s)H(s) = \frac{K(s+6)}{s(s+4)}$$

(a) Sketch the root locus plot with K as a variable parameter and show that loci of complex roots are part of a circle.

(b) Determine the breakaway/breakin points, if any.

(c) Determine the range of K for which the system is underdamped.

(d) Determine the value of K for critical damping.

(e) Determine the minimum value of damping ratio.

Solution

(a) **Number of poles and zeros.** The given transfer function has

(i) Poles at $s = 0$ and $s = -4$; and

(ii) Zeros at $s = -6$.

These are marked on Fig. 14.31 as P_1, P_2 and Z_1.

Since there are two poles, there are two root loci starting from two poles. Since there is one zero, one root loci will terminate at it. Select two points a and b as shown in Fig. 14.31

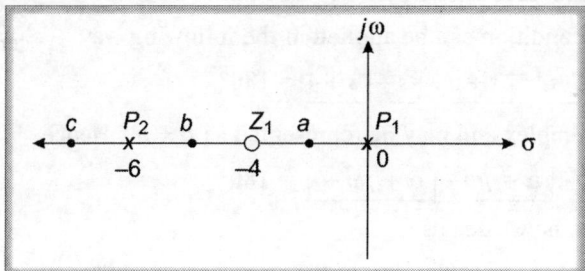

Fig. 14.31

Point a: It lies on the root loci as sum of poles and zeros to the right of it is odd, *i.e.* pole at $s = 0$; totalling one.

Point b: It does not lie on the root loci as sum of poles and zeros to the right of it is even, *i.e.* one pole at $s = -4$ and other of $s = 0$ totalling two. Therefore, as there are two root loci, there has to be a breakaway point.

Point c: It lies on the root locus as the number of poles and zeros to the right of it is odd, i.e. two poles and one zero totalling three.

(b) **Breakaway point.** The characteristic equation is

$$1 + GH = 0$$

$$1 + \frac{K(s+6)}{s(s+4)} = 0$$

or $\quad s^2 + 4s + K(s+6) = 0$

or
$$K = -\frac{s^2 + 4s}{s + 6}$$

Differentiating K with respect to s and equating it to zero, we get,

$$\frac{dK}{ds} = -\frac{(s + 6)(2s + 4) - (s^2 + 4s)}{(s + 6)^2} = 0$$

or
$$s^2 + 12s + 24 = 0$$

Solving the above equation, we get,
$$s = -2.5 \quad \text{and} \quad s = -9.5$$

The breakaway point is thus -2.5 as it lies between the two poles having values 0 and -4 and $s = -9.5$ is breakin point.

(c) Number of asymptotes
$$= \text{No. of poles} - \text{No. of zeros} = 2 - 1 = 1$$

(d) Angle of asymptotes
$$= \frac{(2K + 1)180°}{P - Z} \quad \text{where,} \quad K = 0$$

$$= \frac{(2 \times 0 + 1)180°}{2 - 1} = 180°$$

(e) To show that loci of complex roots are part of circle

The angle condition can be applied in the following way

$$\underline{/G(s)} = \underline{/s + 6} - \underline{/s} - \underline{/s + 4} = 180°$$

Now s is complex and may be represented as $\alpha + j\omega$. Hence

$$\underline{/\alpha + j\omega + 6} - \underline{/\alpha + j\omega} - \underline{/\alpha + j\omega + 4} = 180°$$

and this can be written as

$$\tan^{-1}\frac{\omega}{\alpha + 6} - \tan^{-1}\frac{\omega}{\alpha} = 180° + \tan^{-1}\frac{\omega}{\alpha + 4}$$

Taking tangents of both the sides, we get

$$\tan\left[\tan^{-1}\frac{\omega}{\alpha + 6} - \tan^{-1}\frac{\omega}{\alpha}\right] = \tan\left[180° + \tan + \frac{\omega}{\alpha + 4}\right]$$

$$\frac{\dfrac{\omega}{\alpha + 6} - \dfrac{\omega}{\alpha}}{1 + \left(\dfrac{\omega}{\alpha + 6}\right)\dfrac{\omega}{\alpha}} = \frac{0 + \dfrac{\omega}{\alpha + 4}}{1 - 0 \times \dfrac{\omega}{\alpha + 4}}$$

$$\frac{-6\omega}{\alpha(\alpha + 6) + \omega^2} = \frac{\omega}{\alpha + 4}$$

$$-6(\alpha + 4) = \alpha^2 + 6\alpha + \omega^2$$

$$-6\alpha - 24 = \alpha^2 + 6\alpha + \omega^2$$

$$\alpha^2 + 12\alpha + \omega^2 = -24$$
$$\alpha^2 + 2 \times 6\alpha + 36 + \omega^2 = -24 + 36$$
$$(\alpha + 6)^2 + \omega^2 = 12$$

$$(\alpha + 6)^2 + \omega^2 = \left(\sqrt{12}\right)^2$$

The above equation represents a circle with centre at $\alpha = -6$ and $\omega = 0$ and with radius $\sqrt{12}$ and the centre of the circle is the zero of the open-loop transfer function which is -6.

The root locus is shown in Fig. 14.32

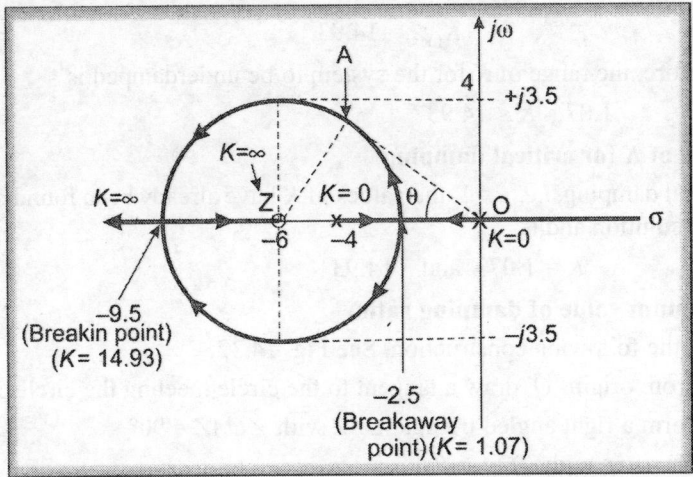

Fig. 14.32

The following points are noteworthy based on which the root loci has been drawn.

(i) There are two root loci starting from the poles at $s = 0$ and $s = -4$. Since, there is one zero at $s = -6$, one of the root loci will terminate at $s = -6$ and other at infinity.

(ii) Any point between poles $s = -4$ and zero at $s = -6$ does not lie on the root locus. Also, all the points beyond $s = -6$ on the negative part of real axis lie on the root loci.

(iii) There is a breakaway point at $s = -2.5$ between the poles $s = 0$ and $s = -4$ and breakin point at $s = -9.5$ to meet the above conditions.

(iv) The root locus also has a circular path having radius equal to $\sqrt{12}$, i.e. 3.5 and with centre at the zero, i.e. $s = -6$.

(f) Range of K for the system to be underdamped

At breakaway points, i.e. at $s = -2.5$ and $s = -9.5$, the value of ξ is 1.0. Therefore at $s = -2.5$, the value of gain K will be minimum and at $s = -9.5$, the value of K will be maximum.

$K_{min}(s = -2.5)$: This is obtained by the following condition:

$$K_{min} \, G(s)|_{s=-2.5} = -1$$

$$K_{min} \frac{(s+6)}{s(s+4)}\bigg|_{s=-2.5} = -1 \quad \text{or} \quad \frac{K_{min}(-2.5+6)}{-2.5(-2.5+4)} = -1$$

or
$$K_{min} = 1.07$$

$K_{max}(s = -9.5)$

$$K_{max} G(s)|_{s=-9.5} = -1$$

$$\frac{K_{max}(s+6)}{s(s+4)}\bigg|_{s=-9.5} = -1 \quad \text{or} \quad \frac{K_{max}(-9.5+6)}{-9.5(-9.5+4)} = -1$$

or
$$K_{max} = 14.93$$

Therefore, the range of K for the system to be underdamped is

$$1.07 < K < 14.93 \qquad \qquad \textbf{Ans}$$

(g) Value of K for critical damping

For critical damping, *i.e.* $\xi = 1$, the values of K have already been found out in part (*f*) of the solution and is

$$K = 1.07 \quad \text{and} \quad 14.93 \qquad \qquad \textbf{Ans}$$

(h) Minimum value of damping ratio

Carryout the following construction. See Fig. 14.32.

  (i)   From origin 'O' draw a tangent to the circle meeting the circle at A.

 (ii)   Form a right angled triangle OAZ with $\angle OAZ = 90°$

(iii)   Measure angle $ZOA = \theta$

The angle measured is 35°. Therefore, damping ratio

$$\xi_{min} = \cos 35° = 0.82 \qquad \qquad \textbf{Ans}$$

It can also be found out as given below:

Draw a triangle OAZ having right angle at angle A as OA is tangent to the circle (Fig. 14.33)

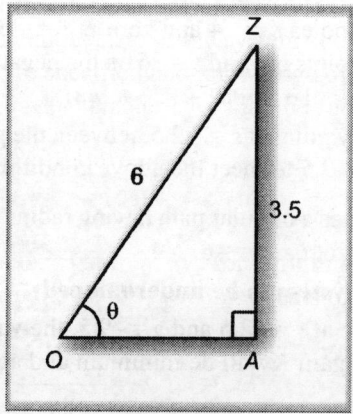

Fig. 14.33

Now

(i) $OZ = -6$ as this is the location of zero, $s = -6$ as explained in part (a) of the solution.

(ii) $AZ = \sqrt{12} = 3.5$ as it is the radius of the root loci which is a circle as explained in part (e) of the solution.

Therefore,

$$\theta = \sin^{-1}\frac{3.5}{6} = 35.68°; \text{ and hence}$$

damping ratio $\xi_{min} = \cos \theta = \cos 35.68° = 0.812$

Note: The value of ξ_{min} is same as found earlier in this part of the solution.

Problem 14.16 Sketch the root locus plot of unity feedback system with an open-loop transfer function

$$G(s) = \frac{K}{s(s+2)(s+4)}$$

Find the range of values of K for which the system has damped oscillatory response. What is the greatest value of K which can be used before continuous oscillations occur. Also, determine the frequency of continuous oscillations.

Solution

This problem is similar to the problem 14.3, as far as drawing of root loci is considered.

(a) **Poles and zeros**

By inspection of the transfer function, it is seen that

(i) **Poles:** There are three poles at

$$s = 0, s = -2 \quad \text{and} \quad s = -4$$

(ii) **Zeroes:** There are no zeros.

Since, there are three poles, there will be three root loci as well. As there is no open-loop zero, the root loci will terminate at infinity ($K = \infty$).

since $\Sigma P > \Sigma Z$, *i.e.* $3 > 0$ (three poles and no zeros)

Number of branches of root locus are

$$N = P,$$

i.e. $N = 3$

The poles are plotted in Fig. 14.34

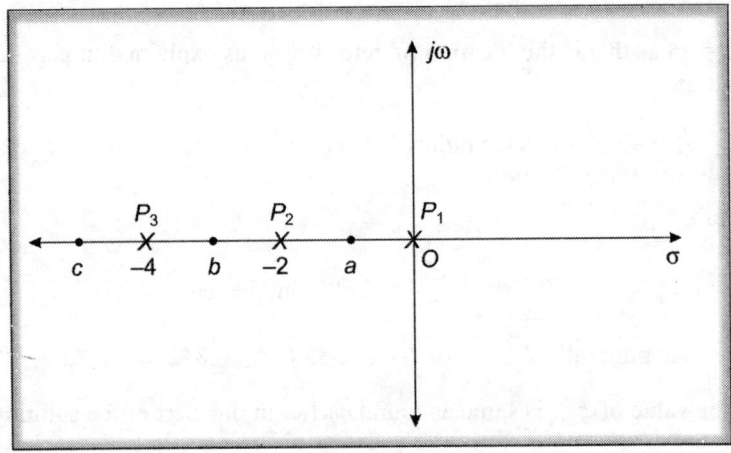

Fig. 14.34

Let us select three points a, b and c as shown in the Fig. 14.34

Point 'a': It lies on the root loci as sum of poles and zeros on the right of it is odd.

Point 'b': It does not lie on the root loci as sum of poles and zeros on the right of it is even.

Point 'c': It lies on the root loci as sum of poles and zeros on the right of it is odd.

Note: This has been explained in detail in problem 14.15 also.

(b) Breakaway point

From the above, we conclude that between the open-loop poles $s = 0$ and $s = -2$, there are root loci and move towards each other. This is because point 'b' does not lie on the root loci and as explained in part (a) above. Root loci originate from poles, *i.e.* $s = 0$ and $s = -2$ and also from $s = -4$. Therefore there has to be a breakaway point on the root loci between $s = 0$ and $s = -2$. The breakaway point is found by differentiating K with respect to term 's' in the characteristic equation and equating it to zero. The characteristic equation is given by:

$$1 + GH = 0$$

or $$1 + \frac{K}{s(s+2)(s+4)} = 0$$

or $$s\,(s+2)(s+4) + K = 0$$

or $$s^3 + 6s^2 + 8s + K = 0$$

$$K = -(s^3 + 6s^2 + 8s)$$

$$\frac{dK}{ds} = -(3s^2 + 12s + 8) = 0$$

Solving for 's', we get

$$s = -3.15 \text{ and } -0.85$$

Since, the breakaway point has to lie between the poles $s = 0$ and $s = -2$, the value $s = -0.85$ satisfies the condition and hence, is the break away point.

(c) **Number and angle of asymptotes and intersection with real axis**

The root loci and its branches are approximated with the help of asymptotic lines and the angle which it makes.

(i) Number of asymptotes $= P - Z = 3 - 0 = 3$

(ii) Angle of asymptotes $= \dfrac{(2K + 1) \times 180°}{P - Z}$

where, $K = 0, 1, 2$, as there are three asymptotes, then the angles are

$$\alpha_1 = \frac{(2 \times 0 + 1)180°}{3 - 0} = 60°$$

$$\alpha_2 = \frac{(2 \times 1 + 1)180°}{3} = 180°$$

$$\alpha_3 = \frac{(2 \times 2 + 1)180°}{3} = 300°$$

(iii) **Intersection of three asymtotes with the real axis**

$$x = \frac{\Sigma \text{ poles} - \Sigma \text{ zeros}}{P - Z} = \frac{(0 - 2 - 4) - 0}{3 - 0} = -2$$

(d) **Intersection of root loci with imaginary axis**

We will use analytical approach to find the points of intersection. The characteristic equation is

$$1 + GH = 0$$

or $$1 + \frac{K}{s(s + 2)(s + 4)} = 0$$

or $$s^3 + 6s^2 + 8s + K = 0$$

The Routh's array is

s^3	1	8
s^2	6	K
s^1	$\dfrac{48 - K}{6}$	0
s^0	K	0

The value of K at which root loci is going to intersect imaginary axis is obtained by equating s^1 row equal to zero.

Therefore,

$$\frac{48 - K}{6} = 0 \quad \text{or} \quad K = 48$$

Now we will form the auxilliary equation with the help of 's^2' row for taking into consideration marginally stable condition and solve for 's' to find the points of intersection of root loci with imaginary axis.

$$6s^2 + K = 0$$

Putting the value of $K = 48$, the above equation becomes

$$6s^2 + 48 = 0$$

or $s^2 + 8 = 0$

or $s^2 = -8$

or $s = \pm j\sqrt{8} = \pm j2.83$

(e) Sketching of root loci

With the help of data collected above, the root loci is shown in Fig. 14.35.

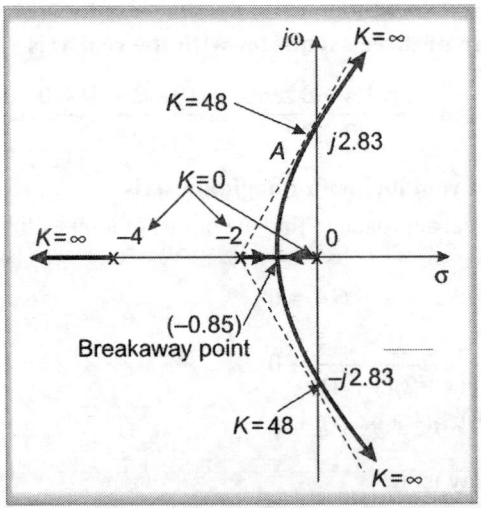

Fig. 14.35

(f) Range of values of K for which the system has damped oscillatory response

At the breakaway point $s = -0.85$ and at the point $s = -j2.83$, the point at which the system is marginally stable, gives the range for K for damped oscillatory response.

(i) **$K(s = -0.85)$.** This is obtained by the following condition

$$KG_{(s)}\Big|_{s = -0.85} = -1$$

or
$$\left.\frac{K \times 1}{s(s+2)(s+4)}\right|_{s=-0.85} = -1$$

or
$$\frac{K}{-0.85(-0.85+2)(-0.85+4)} = -1$$

or
$$K = 3.079; \text{ and}$$

(ii) $K(s = -j2.83)$

$$\left.\frac{K \times 1}{s(s+2)(s+4)}\right|_{s=-j2.83}$$

The value has already been found in part (d) of the solution and is
$$K = 48$$

Therefore, the range of K is
$$3.079 < K < 48 \qquad \qquad \textbf{Ans.}$$

(g) Greatest value of K which can be used before continuous oscillations

This will occur at a values of 's' which indicate the points at which root loci will cross the imaginary axis. This has already been found out in part (d) of the solution. The value of s is $s = \pm j2.83$, and at this point the value of $K = 48$ (already found in part (d) of the solution).

(h) Frequency of continuous oscillations

This is obtained by the length of the intercept from origin to a point where root loci crosses the imaginary axis and is given by OA as shown in Fig. 14.35 and is equal to 2.83. Therefore, frequency at which continuous oscillation will occur is
$$\omega = 2.83 \text{ rad/sec.} \qquad \qquad \textbf{Ans}$$

Problem 14.17 For a control system in Fig. 14.36 below, draw the root locus diagram neatly on a graph paper showing all the relevant calculations. From the diagram comment on how the value of K will make the system underdamped or overdamped.

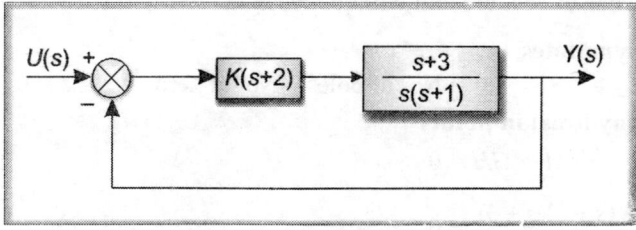

Fig. 14.36

Solution

The loop transfer function is

$$G(s) = \frac{K(s+2)(s+3)}{s(s+1)}$$

(a) **Number of poles and zeros.** By mere inspection $G(s)$ has

   (i) **Poles:** Two poles at $s = 0$ and $s = -1$

   (ii) **Zeros:** Two zeros at $s = -2$ and $s = -3$.

This shows that there will be two number root loci starting from two poles $s = 0$ and $s = -1$ and will terminate at the two zeros at $K = \infty$. The poles and zeros are plotted in Fig. 14.37. Select points a, b, c and d as shown in Fig. 14.37.

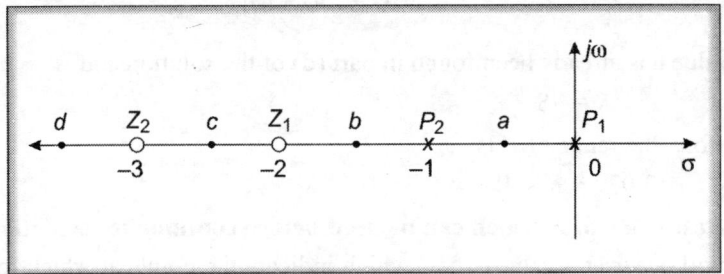

Fig. 14.37

  (i) **Points 'a' and 'c':** These points will lie on the root loci as the total number of poles and zeros on the right side of point 'a' is odd and that of 'c' is also odd. Since, point 'a' is between two poles $s = 0$ and $s = -1$, the root loci will originate from these points and move towards each other. Also, there will be a breakaway point.

  (ii) **Point 'b'.** This point will not lie on root loci as the total number of poles and zeros to the right of it is even. Infact all points between $s = -1$ and $s = -2$ will not lie on the root loci as these will also satisfy the condition of total number of poles and zeros to the right of these points is even.

  (iii) **Point 'd'.** This point or infact all points beyond $s = -3$ will not lie on the root loci as the total number of poles and zeros to the right of it is even. Since the root loci will originate from $s = 0$ and $s = -1$ with a breakaway point between them, will terminate at the two zeros, $s = -2$ and $s = -3$, there has to be a breakin point between the two zeros.

(b) **No. of asymptotes**

$$= \text{No. of poles} - \text{No. of zeros} = 2 - 2 = 0$$

(c) **Breakaway/breakin points**

$$1 + GH = 0$$

$$1 + \frac{K(s+2)(s+3)}{s(s+1)} = 0$$

$$s^2 + s + K(s^2 + 5s + 6) = 0$$

$$K = \frac{-(s^2 + s)}{s^2 + 5s + 6}$$

$$\frac{dK}{ds} = -\frac{(s^2 + 5s + 6)(2s + 1) - (s^2 + s)(2s + 5)}{s^2 + 5s + 6}$$

or $\qquad s^2 + 3s + 15 = 0$

Solving, we get $\quad s = -2.37$ and $- 0.634$

Therefore,

breakaway point $= -0.634$ (as it is between two poles $s = 0$ and $s = -1$) and

breakin point $= - 2.37$ (between the zeros at $s = -2$ and $s = -3$)

The breakaway point is at a distance of $1 - 0.634 = 0.366$ from the pole $s = -1$ towards the origin and the breakin point is a distance of $2.37 - 2 = 0.37$ from the zero, $s = -2$. This is shown in Fig. 14.38.

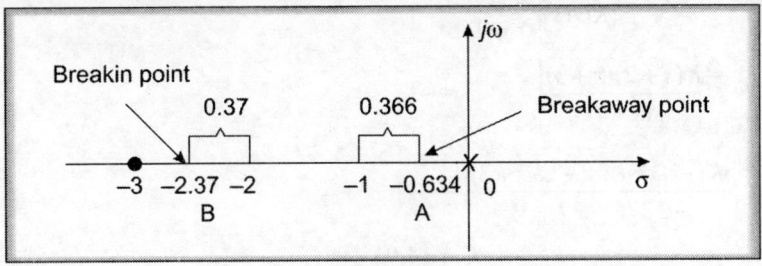

Fig. 14.38

The root loci will originate from $s = 0$ and $s = -1$ and move towards each other till $s = -0.634$; from where it will breakaway in a circular path with diameter equal to AB at point $s = -2.37$. There will be a breakin point between $s = -2$ and $s = -3$ which are zeros. The root locus is shown in Fig. 14.39.

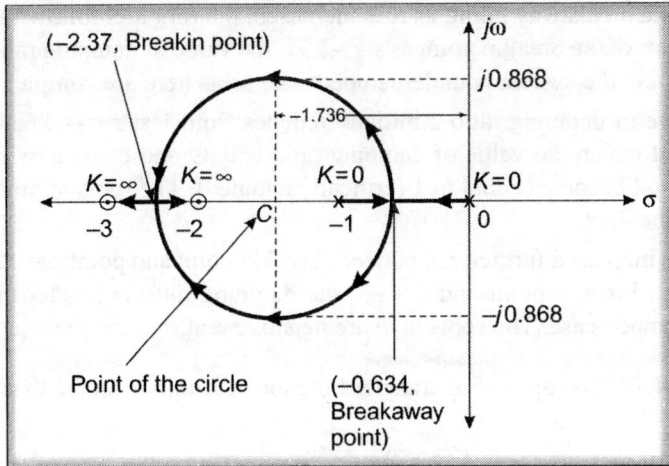

Fig. 14.39

(e) **Comments on how the value of K will make the system underdamped or overdamped**

(i) At the breakaway and breakin point value of $\xi = 1$, i.e. the system will be critically damped.

$$KG(s)\big|_{s=-0.634} = -1$$

or $$\left.\frac{K(s+2)(s+3)}{s(s+1)}\right|_{s=-0.634} = -1$$

or $$\frac{K(-0.634+2)(-0.634+3)}{-0.634(-0.634+1)} = -1$$

or $$K = 0.072;$$

and

$$KG(s)\big|_{s=-2.37} = -1$$

or $$\left.\frac{K(s+2)(s+3)}{s(s+1)}\right|_{s=-2.37} = -1$$

or $$\frac{K(-2.37+2)(-2.37+3)}{-2.37(-2.37+1)} = -1$$

or $$K = 14.93$$

Comments

(a) The value of K at $s = -0.634$, the breakaway point marked as A in Fig. 14.39 is 0.072 and the value of damping ratio is 1, *i.e.* critically damped. The roots here are coincident. The value of K is zero at poles $s = 0$, i.e. origin and at $s = -1$.

(b) After the breakaway point, as K is increased, the root loci follow the circular path and till the breakin point as $s = -2.37$, the value of damping ratio ξ is less than 1, *i.e.* the system is underdamped. The roots here are complex.

(c) The value of damping ratio ξ initially reduces from 1 at $s = -0.634$ to a low value at which the value of damping ratio is 0.89 and then starts increasing again and becomes equal to 1 (critically damped). The roots at breakin point are coincident.

(d) As K is increased further, *i.e.* between breakin point and point $s = -2$ and also between breakin point and $s = -3$, the damping ratio is greater than 1, *i.e.* overdamped case. The roots here are negative real.

Problem 14.18 The open-loop transfer function of a unity gain feedback is given by

$$G(s) = \frac{K(s+2)}{s^4 + 3s^3 + 4s^2 + 2s}, \quad K \geq 0$$

(a) Determine all the poles and zeros of $G(s)$

(b) Draw the root locus covering the following points which should be distinctly marked
   (i) loci start and end points
  (ii) real axis part of the locus
 (iii) angle of asymptotes and their real axis intercept
  (iv) breakaway and breakin points
   (v) imaginary axis crossing points
  (vi) four typical values of K on root locus.

Solution

$$G(s) = \frac{K(s+2)}{s^4 + 3s^3 + 4s^2 + 2s}$$

This can be written as

$$G(s) = \frac{K(s+2)}{s(s+1)(s^2+2s+2)}$$

or $$G(s) = \frac{K(s+2)}{s(s+1)(s+1+j)(s+1-j)}$$

(a) **Poles and zeros.** By inspection of open-loop transfer function G(s), we see that

(i) **Poles:** These are $s = 0$, $s = -1$, $s = -1 - j$, and $s = -1 + j$. There are four poles. These points are the start points of root loci.

(ii) **Zeros:** There is only one zero at $s = -2$. Therefore, the end points of root loci are -2, ∞, ∞, ∞, where ∞ = infinity.

Since, poles > zeros, i.e. $4 > 1$, the number of branches of root loci are

$$N = P, \text{ i.e. } N = 4$$

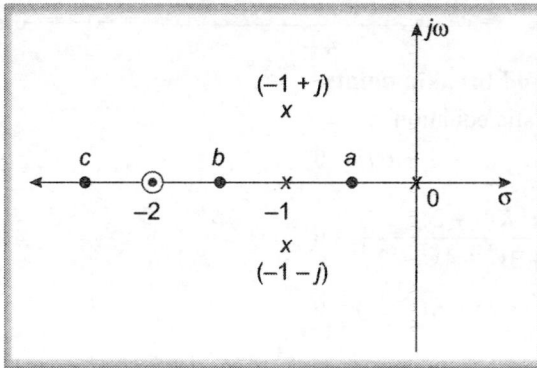

Fig. 14.40

The poles and zero have been shown in Fig. 14.40. Select points a, b and c.

Point 'a': This point and all points between poles $s = 0$ and $s = -1$ will lie on the root loci as total number of poles and zeros to the right of them are odd. Moreover, the root loci originating from poles $s = 0$ and $s = -1$ move towards each other and will have a breakaway point between $s = 0$ and $s = -1$.

Point 'b': This point and all points between $s = -1$ and $s = -2$ will not lie on the root loci as total number of poles and zeros to the right of them are even. This means there will not be any root locus path between points $s = 1$ and $s = -2$.

Point 'c': This point and all points beyond $s = -2$, towards the negative axis will lie on the root loci as number of poles and zeros to the right of them are odd. Since, one root locus will terminate at the zero, $s = -2$, there will be a breakin point also, on this part of the root loci.

(b) **Number of asymptotes**

$$= \text{No. of poles} - \text{No. of zeros} = 4 - 1 = 3$$

(c) **Angle of asymptotes**

$$= \frac{(2K + 1)180°}{P - Z} \quad \text{where, } K = 0, 1, 2$$

$$\alpha_1 = \frac{(2 \times 0 + 1)180°}{4 - 1} = 60°$$

$$\alpha_2 = \frac{(2 \times 1 + 1)180°}{4 - 1} = 180°$$

$$\alpha_3 = \frac{(2 \times 2 + 1)180°}{4 - 1} = 300°$$

(d) **Intersection of asymptotes with real axis**

$$x = \frac{\Sigma \text{ poles} - \Sigma \text{ zeros}}{P - Z}$$

$$= \frac{(0 - 1 - 1 + j - 1 - j) - (-2)}{4 - 1} = -\frac{1}{3} = -0.33$$

(e) **Breakaway and breakin points**

The characteristic equation is

$$1 + GH = 0$$

or $$1 + \frac{K(s + 2)}{s^4 + 3s^3 + 4s^2 + 2s} = 0$$

or $$s^4 + 3s^3 + 4s^2 + 2s + K(s + 2) = 0$$

or $$K = -\frac{s^4 + 3s^3 + 4s^2 + 2s}{s + 2}$$

$$\frac{dK}{ds} = -\frac{(s+2)(4s^3 + 9s^2 + 8s + 2) - (s^4 + 3s^3 + 4s^2 + 2s)}{(s+2)^2}$$

or $3s^4 + 14s^3 + 22s^2 + 16s + 4 = 0$

The above equation will have to solve by trial and error. The breakaway point has to lie between $s = 0$ and $s = -1$ and the breakin point beyond $s = -2$.

Breakaway point by trial and error $= -0.5$

Breakin point by trial and error $= -2.5$

(f) Angle of departure

$$\phi_{(-1+j)} = 180° - (\theta_1 + \theta_2 + \theta_3 - \theta_4)$$

From Fig. 14.41

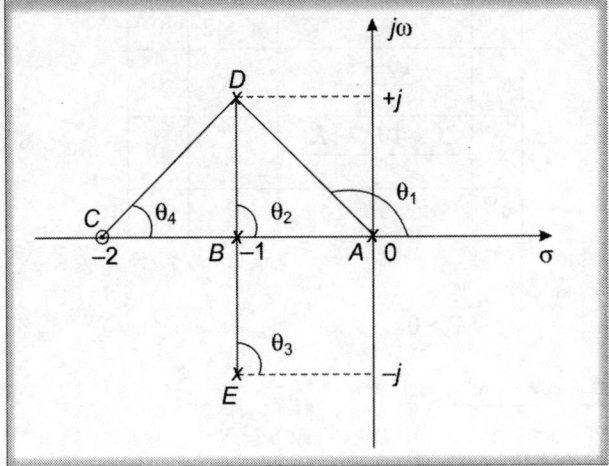

Fig. 14.41

$$\theta_1 = 180° - \tan^{-1}\frac{BD}{AB}$$

$$= 180° - \tan^{-1}\frac{1}{1}$$

$$= 180° - 45° = 135°$$

$\theta_2 = 90°$ (by inspection)

$\theta_3 = 90°$ (by inspection)

$$\theta_4 = \tan^{-1}\frac{BD}{BC} = \tan^{-1}\frac{1}{1} = 45°$$

Therefore,

$$\phi_{(-1+j)} = 180° - (\theta_1 + \theta_2 + \theta_3 - \theta_4)$$
$$= 180° - (135° + 90° + 90° - 45°) = -90°$$

Similarly,

$$\phi(-1-j) \quad \text{will be} \quad +90°$$

(g) Intersection of imaginary axis by root loci

This will be obtained by the help of Routh's array. The characteristic equation as found out in part 'e' of the solution is

$$s^4 + 3s^3 + 4s^2 + 2s + K(s + 2) = 0$$

or

$$s^4 + 3s^3 + 4s^2 (K + 2)s + 2K = 0$$

s^4	1	4	$2K$
s^3	3	$K+2$	0
s^2	$\dfrac{10-K}{3}$	$2K$	0
s^1	$\dfrac{20-10K-K^2}{3}$	0	0
s^0	$2K$	0	0

For stability

(i) $$K > 0$$

(ii) $$\frac{20 - 10K - K^2}{3} > 0$$

or $$K^2 + 10K - 20 = 0$$

$$K = \frac{-10 \pm \sqrt{100 + 80}}{2} = \frac{-10 \pm 13.42}{2}$$

Therefore,

$K = 1.71$ and -11.71. We will take only the positive value of K. Therefore, $K = 1.71$

The auxiliary equation of s^2 row is $\left(\dfrac{10-K}{3}\right)s^2 + 2K = 0$

Putting $K = 1.71$, we get

$$\frac{(10 - 1.71)s^2}{3} + 2 \times 1.71 = 0$$

$$2.76s^2 + 3.42 = 0$$

or

$$s = \sqrt{-\frac{3.42}{2.76}} = \pm j1.11$$

The complete root locus is shown in Fig. 14.42

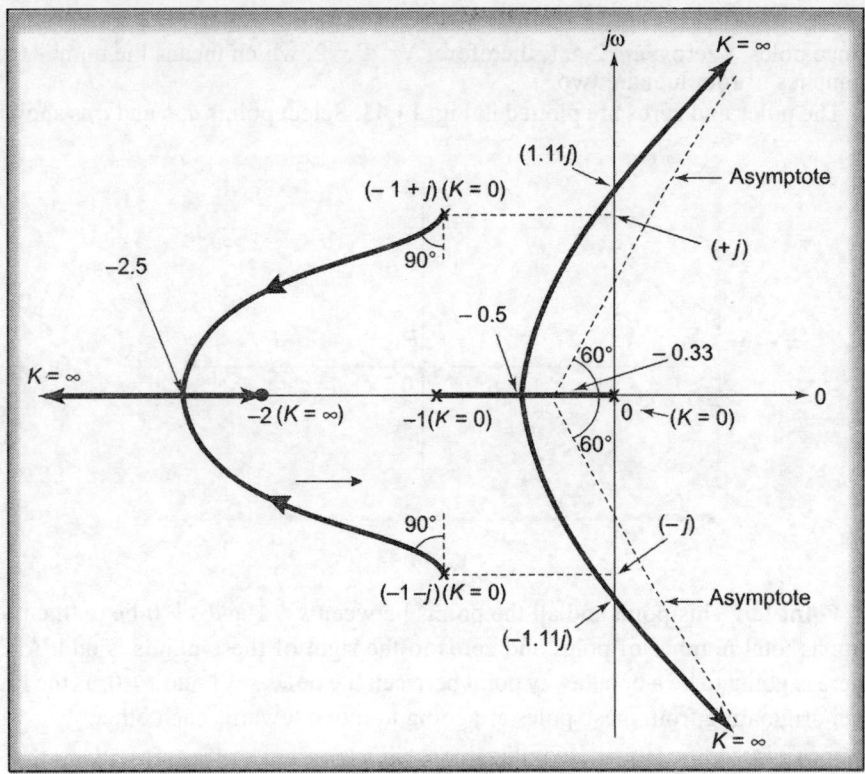

Fig. 14.42

Problem 14.19 A unity feedback system has an open-loop transfer function

$$G(s) = \frac{K(s+1)}{s(s-1)}$$

(a) Show that loci of centre and radius = $\sqrt{2}$

(b) Sketch the root locus with K as variable parameter

(c) Is the system stable for all values of K

(d) From the root locus plot, determine the value of K such that the system has settling time of 4 seconds. What are the corresponding closed-loop poles.

Solution

(a) **Poles and zeros.** By inspection of open-loop transfer function

$$G(s) = \frac{K(s+1)}{s(s-1)}, \text{ we get}$$

(i) **Poles:** There are two poles at $s = 0$ and $s = +1$. These are the start points of the root loci.

(ii) **Zeros:** There is only one zero at $s = -1$. One of the root loci is going to terminate at $s = -1$ and other at infinity.

Since poles > zeros, *i.e.* $2 > 1$, therefore, $N = P = 2$, which means the number of branches of root loci are two.

The poles and zeros are plotted in Fig. 14.43. Select points a, b and c as shown.

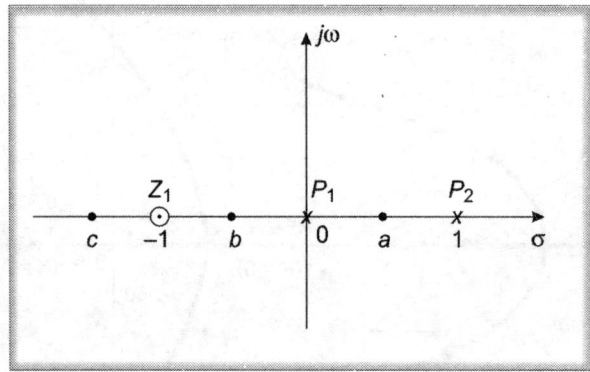

Fig. 14.43

Point 'a'. This point and all the points between $s = 1$ and $s = 0$ lie on the root loci as total number of poles and zeros to the right of these points is odd. Also, there is going to be a breakaway point between the poles $s = 1$ and $s = 0$, as the root loci originating from these poles are going to move towards each other.

Point 'b'. This point and all the points between $s = 0$ and $s = -1$ will not lie on the root loci as total number of poles and zeros to the right of these points is even. Hence, there will not be any root locus path between $s = 0$ and $s = -1$.

Point 'c'. This point and all the points beyond $s = -1$ will be on the root loci as total number of poles and zeros to the right of them is odd. Since, one of the root loci is going to terminate at $s = -1$ and other at infinity, there will be a breakin point beyond $s = -1$.

(b) Breakaway and breakin points

$$1 + G(s) = 0 \quad \text{or} \quad 1 + \frac{K(s+1)}{s(s-1)} = 0 \quad \text{or} \quad K = -\frac{s^2 - s}{s+1}$$

$$\frac{dK}{ds} = -\frac{(s+1)(2s-1) - (s^2 - s)}{(s+1)^2} \quad \text{or} \quad 2s^2 - s + 2s - 1 - s^2 + s = 0$$

or $\quad s^2 + 2s - 1 = 0$

$$s = \frac{-2 \pm \sqrt{4+4}}{2}$$

or $\quad\quad s = 0.414 \quad \text{or} \quad -2.414$

(i) **Breakaway point:** Since, breakaway point to lie between $s = 0$ and $s = 1$, the point $s = 0.414$ satisfies this condition and hence, is the breakaway point.

(ii) **Breakin point:** Since, breakin point is to lie beyond $s = -1$, the point $s = -2.414$ satisfies this condition and hence, is the breakin point.

(c) **Number of asymptotes**

$$= \text{No. of poles} - \text{No. of zeros} = 2 - 1 = 1$$

(d) **Angle of asymptotes**

$$= \frac{(2K + 1)180°}{P - Z} \quad \text{where, } K = 0$$

$$= \frac{(2 \times 0 + 1)180°}{2 - 1} = 180°$$

(e) **Intersection of the root locus with imaginary axis:** The characteristic equation is

$$1 + G(s)H(s) = 0 \quad \text{or} \quad 1 + \frac{K(s + 1)}{s(s - 1)} \times 1 = 0$$

or $\quad s^2 - s + Ks + 1 = 0$

or $\quad s^2 + s(K - 1) + 1 = 0$

The Routh's array is

s^2	1	1
s^1	$K - 1$	0
s^0	1	0

For marginal stability, put

$$K - 1 = 0 \quad \text{or} \quad K = 1$$

Forming auxiliary equation of s^2 row, we get

$$s^2 + 1 = 0$$

or $\quad s = \sqrt{-1} \quad \text{or} \quad s = \pm j$

(f) **Show the loci as circle with radius** $= \sqrt{2}$

$$G(s) = \frac{K(s + 1)}{s(s - 1)}$$

The angle condition can be applied in the following way

$$\underline{|G(s)} = \underline{|s+1} - \underline{|s} - \underline{|s-1} = 180°$$

Now s is complex and may be represented as $a + j\omega$. Hence,

$$\underline{|a+j\omega+1} - \underline{|a+j\omega} - \underline{|a+j\omega-1} = 180°$$

and this can be written as

or

$$\tan^{-1}\frac{\omega}{a+1} - \tan^{-1}\frac{\omega}{a} - \tan^{-1}\frac{\omega}{a-1} = 180°$$

or

$$\tan^{-1}\frac{\omega}{a+1} - \tan^{-1}\frac{\omega}{a} = 180° + \tan^{-1}\frac{\omega}{a-1}$$

Now taking tangents of both sides, we get

$$\tan\left[\tan^{-1}\frac{\omega}{a+1} - \tan^{-1}\frac{\omega}{a}\right] = \tan\left[180° + \tan^{-1}\frac{\omega}{a-1}\right]$$

or

$$\frac{\dfrac{\omega}{a+1} - \dfrac{\omega}{a}}{1 + \dfrac{\omega \times \omega}{a(a+1)}} = \frac{0 + \dfrac{\omega}{a-1}}{1 - 0 \times \dfrac{\omega}{a-1}}$$

or

$$\frac{-\omega}{a(a+1) + \omega^2} = \frac{\omega}{a-1}$$

or

$$\frac{-1}{a^2 + a + \omega^2} = \frac{1}{a-1}$$

or

$$-(a-1) = a^2 + a + \omega^2$$

or

$$1 = a^2 + 2a + \omega^2$$

or

$$a^2 + 2a + \omega^2 = 1$$

or

$$a^2 + 2a + 1 + \omega^2 = 1 + 1$$

or

$$(a+1)^2 + \omega^2 = 2$$

or

$$(a+1)^2 + \omega^2 = \left(\sqrt{2}\right)^2$$

The above equation represents a circle with centre at $a = -1$ and $\omega = 0$ and with radius equal to $\sqrt{2}$

(g) **Sketching of root locus:** The root locus is shown in Fig. 14.44.

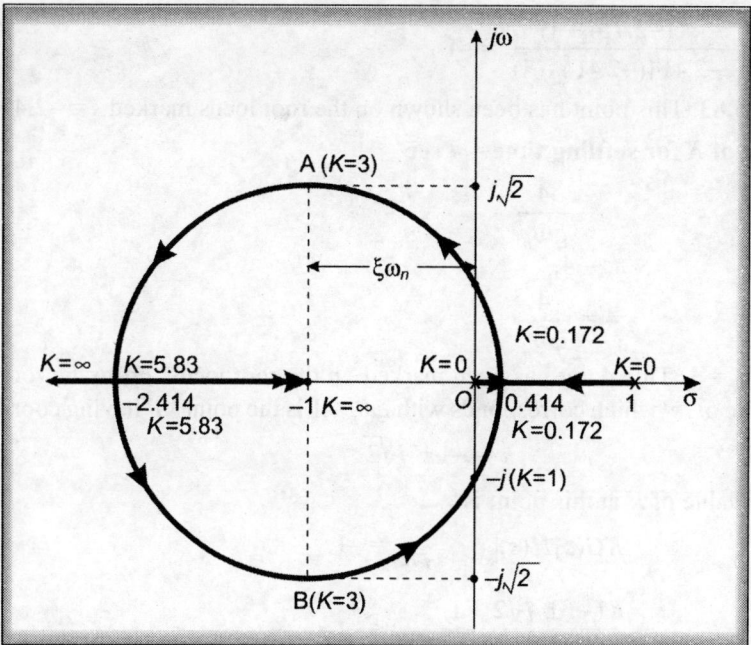

Fig. 14.44

(h) Stability

The system is unstable in the open-loop condition ($K = 0$) with one pole in the R.H.S. of the s-plane. As K is increased the open-loop pole at $s = 0$ also moves into the right hand side of the s-plane. The system is now unstable with two poles in the right hand half of s-plane. As K is increased further, these poles move as shown by the arrows on the root locus and enter the left-hand half plane at $s = \pm j$, when $K = 1$. The system is then stable for further increase in K. At breakaway points the value of K is

- $K|_{s = 0.414}$

$$|G(s)H(s)| = -1$$

or $$\left.\frac{K(s+1)}{s(s-1)}\right|_{s=0.414} = -1$$

or $$\frac{K(0.414+1)}{0.414(0.414-1)} = -1$$

or $K = 0.172$. This point has been shown on the root locus marked $s = 0.414$

- $K|_{s = -2.414}$

$$\frac{K(-2.414+1)}{-2.414(-2.414-1)} = -1$$

or $K = 5.83$. This point has been shown on the root locus marked $s = -2.414$

• **Value of K for settling time = 4 sec**

$$t_s = \frac{4}{\xi\omega_n}$$

putting 　　　　$t_s = 4$,

we get, 　　　　$4 = \frac{4}{\xi\omega_n}$

or 　$\xi\omega_n = 1$. This point has been marked on the root locus. From the root locus the value of 's' which corresponds with $\xi\omega_n = 1$ is the point 'A' having coordinates

$$s = -1 \pm j\sqrt{2} \qquad\qquad \textbf{Ans}$$

The value of K at this point is

$$KG(s)H(s)\big|_{s=-1\pm j\sqrt{2}} = -1$$

or 　　$\dfrac{K(-1\pm j\sqrt{2}+1)}{-1\pm j\sqrt{2}(-1\pm j\sqrt{2}-1)} = -1$

or 　　$\dfrac{K(\pm j\sqrt{2})}{-1\pm j\sqrt{2}(-1\pm j\sqrt{2}-1)} = -1$

Considering the positive value, we get

$$\frac{K(+j\sqrt{2})}{\left(-1+j\sqrt{2}\right)\left(-2+j\sqrt{2}\right)} = -1$$

or 　　$K = -\dfrac{(-1+j\sqrt{2})(-2+j\sqrt{2})}{+j\sqrt{2}} = \dfrac{\sqrt{3}\times\sqrt{6}}{\sqrt{2}} = 3$ 　　**Ans**

Problem 14.20 Consider the system shown in Fig. 14.45 below. Draw the locus of the poles of the overall system as K is varied from zero to infinity.

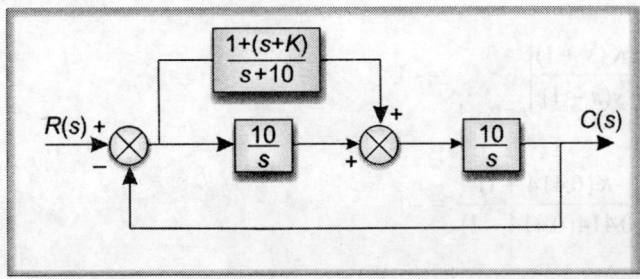

Fig. 14.45

Solution

The forward path transfer function for $H(s) = 1$ is given by

$$G(s) = \left[\frac{10}{s} + \frac{10(s + K)}{s + 10K}\right]\frac{10}{s}$$

$$G(s) = \left[\frac{10(s + 10K) + 10s(s + K)}{s(s + 10K)}\right] \times \frac{10}{s}$$

$$G(s) = \frac{100s + 1000K + 100s^2 + 100Ks}{s^3 + 10Ks^2}$$

$$= \frac{100s^2 + 100s + 100Ks + 1000K}{s^3 + 10Ks^2}$$

The block diagram of Fig. 14.45 reduces to as shown in Fig. 14.46.

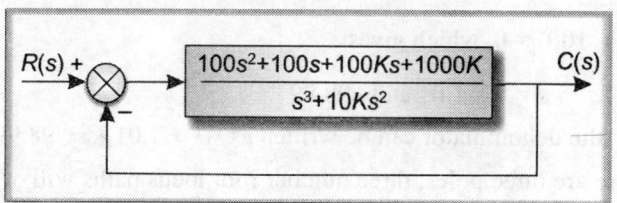

Fig. 14.46

The characteristic equation is

$$1 + G(s)H(s) = 0$$

or substituting $G(s) = \dfrac{100s^2 + 100s + 100Ks + 1000K}{s^3 + 10Ks^2}$ and $H(s) = 1$

or $1 + G(s)H(s) = 1 + \left(\dfrac{100s^2 + 100s + 100Ks + 1000K}{s^3 + 10Ks^2}\right) \times 1 = 0$

or $s^3 + 10Ks^2 + 100s^2 + 100s + 100Ks + 1000K = 0$

Rearranging, we get

$$s^3 + 100s^2 + 100s + K(10s^2 + 100s + 1000) = 0$$

or $s^3 + 100s^2 + 100s = -K(10s^2 + 100s + 1000)$

or $1 = -\dfrac{K(10s^2 + 100s + 1000)}{s^3 + 100s^2 + 100s}$

or
$$1 + \frac{K(10s^2 + 100s + 1000)}{s^3 + 100s^2 + 100s} = 0$$

Comparing the above with

    $1 + G(s)H(s) = 0$, we get

$$G(s) = \frac{K(10s^2 + 100s + 1000)}{s^3 + 100s^2 + 100s}$$

or
$$G(s) = \frac{K(10s^2 + 100s + 1000)}{s(s^2 + 100s + 100)}$$

(a) Poles and zeros

(i) **Poles:** The open-loop poles are obtained by equating the denominator equal to zero, i.e.

$s(s^2 + 100s + 100) = 0$, which gives

$$s = 0, -1.01 \text{ and } -98.99$$

Therefore, the denominator can be written as $s(s + 1.01)(s + 98.99)$

Since, there are three poles, three number root locus paths will originate from these three poles respectively.

(ii) **Zeros:** The zeros are obtained by equating the numerator equal to zero, i.e.
$10s^2 + 100s + 1000 = 0$, which gives

$$s = \frac{-100 \pm \sqrt{100^2 - 4 \times 10 \times 1000}}{2 \times 10}$$

$$= -5 \pm j5\sqrt{3}$$

$$= -5 \pm j8.66$$

Hence, there are two zeros at

$$s = -5 + j8.66 \text{ and } s = -5 - j8.66$$

Since, there are two zeros, two number root locus paths originating from two poles are going to terminate at these two zeros and the other root locus path originating from the third pole will terminate at infinity.

The poles and zeros have been plotted in Fig. 14.47.

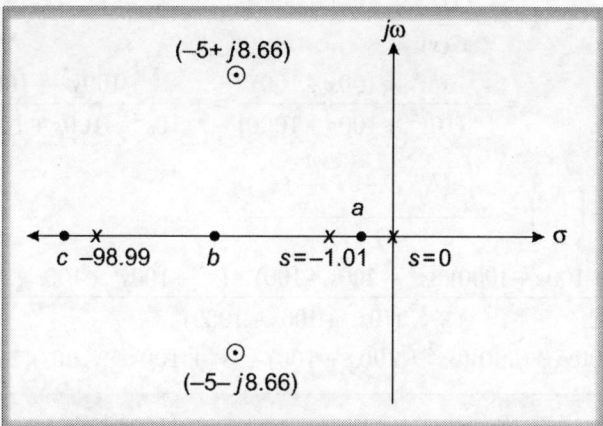

Fig. 14.47

Consider three points *a, b* and *c* as marked in Fig. 14.47

(i) **Point 'a':** This point and all the points located between $s = 0$ and $s = -1.01$ will lie on the root loci as total number of poles and zeros to the right of them are odd. Hence, the real axis between $s = 0$ and $s = -1.01$ will be the root locus path. The locus originating from $s = 0$ and 1.01 will move towards each other and hence there will be a breakaway point between $s = 0$ and $s = -1.01$ located on the real axis.

(ii) **Point 'b':** This point and all the points located between $s = -1.01$ and $s = -98.99$ are not going to lie on the root locus as the total number of poles and zeros to the right of it is even. In other words, there will not be any root locus path between $s = -1.01$ and $s = -98.99$.

(iii) **Point 'c':** This point and all the points located beyond $s = -98.99$ will lie on the root locus path as the total number of poles and zeros to the right of them is odd. This means, that the root loci originating from $s = -98.99$ will move towards left on the negative real axis and extend to infinity.

(b) **Asymptotes**

(i) **Number of asymptotes** = No. of poles – No. of zeros = $3 - 2 = 1$

(ii) **Angle of asymptotes** $= \dfrac{(2K+1) \times 180°}{P - Z}$

(where, $K = 0$, as number of asymptotes is equal to one number only)

$$= \frac{(2 \times 0 + 1) \times 180°}{3 - 2} = 180°$$

(c) **Breakaway point:** There will be a breakaway point between $s = 0$ and $s = -1.01$ (refer explanation given in part 'a' of the solution against point 'a').

$$1 + KG(s) = 1$$
$$KG(s) = 1 - 1 = 0$$

or
$$K = -\frac{1}{G(s)}$$

or
$$K = -\frac{s(s^2 + 100s + 100)}{(10s^2 + 100s + 1000)} = -\frac{s^3 + 100s^2 + 100s}{10s^2 + 100s + 1000}$$

$$\frac{d}{dx}\left(\frac{N}{D}\right) = \frac{\frac{d}{dx}(N)D - N\frac{d}{dx}(s^2)}{D^2}$$

$$\frac{dK}{ds} = \frac{(10s^2 + 100s + 1000)(3s^2 + 200s + 100) - (s^3 + 100s^2 + 100s)(20s + 100)}{(10s^2 + 100s + 1000)^2} = 0$$

$$(100s^2 + 100s + 1000)(3s^2 + 200s + 100) - (s^3 + 100s^2 + 100s)(20s + 100) = 0$$

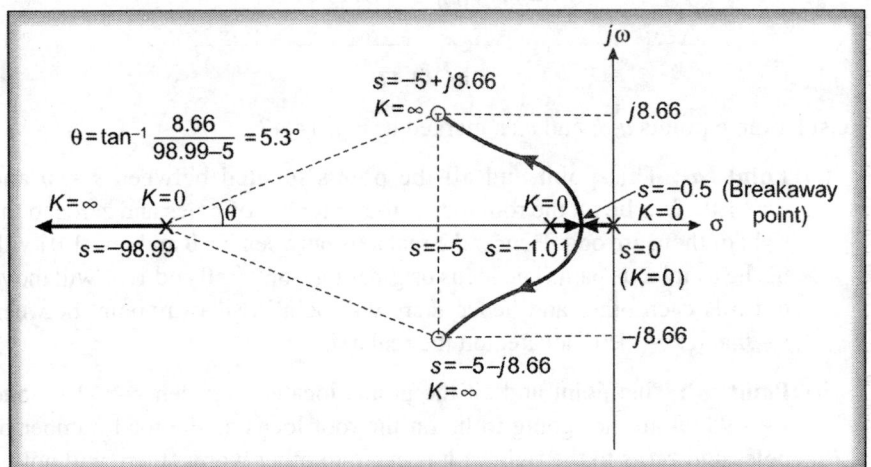

Fig. 14.48

On trial and error, we find that $s = -0.5$ satisfies the condition of the equation hence, breakaway point $= -0.5$.

(d) The root locus is shown in Fig. 14.48.

Problem 14.21 Sketch the root locus plot for a system having

$$G(s) = \frac{K}{s(s+1)(s+3)(s+4)}$$

Find the range of K that yields a stable system.

Solution

(a) **Poles/zeros:** There are four poles at $s = 0, -1, -3$ and -4, and no zeros. Since, there are four poles, there will be four root loci originating from four poles. Select from P_1, P_2, P_3 and P_4 as shown in Fig. 14.49. Applying *Rule 5*, we find that points P_1 and P_3 are on the root loci, as sum of poles and zeros to the right of these points is odd. Root loci are shown with dark lines in Fig. 14.49.

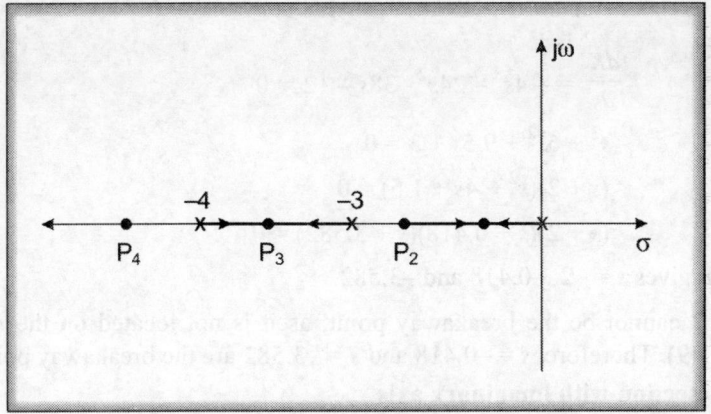

Fig. 14.49

(b) Asymptotes

(i) Number of asymptotes = No. of poles – No. of zeros = 4 – 0 = 4.

(ii) Angle of asymptotes:

$$\alpha = \frac{(2K+1) \times 180°}{P-Z}, \text{ which gives}$$

$$\alpha_1 = \frac{(2 \times 0 + 1) \times 180°}{4} = \frac{180}{4} = 45°$$

$$\alpha_2 = \frac{(2 \times 1 + 1) \times 180°}{4} = \frac{3 \times 180°}{4} = 135°$$

$$\alpha_3 = \frac{(2 \times 2 + 1) \times 180°}{4} = \frac{5 \times 180°}{4} = 225°$$

$$\alpha_4 = \frac{(2 \times 3 + 1) \times 180°}{4} = \frac{7 \times 180°}{4} = 315°$$

(iii) Centroid $= s = \dfrac{\Sigma \text{ poles} - \Sigma \text{ zeros}}{P-Z} = \dfrac{0-1-3-4}{4} = -2$

(c) Breakaway point

$$1 + G(s)H(s) = 0$$

or

$$1 + \frac{K}{s(s+1)(s+3)(s+4)} = 0$$

or

$$s(s+1)(s^2+7s+12) + K = 0$$

or $\qquad K = -s^4 - 8s^3 - 19s^2 - 12s$

or $\qquad \dfrac{dK}{ds} = -4s^3 - 24s^2 - 38s - 12 = 0$

or $\qquad s^3 + 6s^2 + 9.5s + 3 = 0$

or $\qquad (s + 2)(s^2 + 4s + 1.5) = 0$

or $\qquad (s + 2)(s + 0.418)(s + 3.582) = 0$

which gives $s = -2, -0.418$ and -3.582.

$s = -2$ cannot be the breakaway point, as it is not located on the root loci (Fig. 14.49). Therefore, $s = -0.418$ and $s = -3.582$ are the breakaway points.

(d) **Intersection with imaginary axis**

Forming the Routh's array with the denominator of $G(s)H(s)$, i.e.

$$s^4 + 8s^3 + 19s^2 + 12s + K^6$$

s^4	1	19	K
s^3	8	12	0
s^2	17.5	K	0
s^1	$\dfrac{210-8K}{17.5}$	0	0
s^0	K	0	0

A complete row of zero gives the possibility of existence of root lying on the imaginary axis. For positive values of gain s^1 row can only become zero row. Thus,

$$\dfrac{210 - 8K}{17.5} = 0$$

$$K = \dfrac{210}{8} = 26.25$$

Forming the even polynomial with $K = 26.25$ by utilising the s^2 row, we get

$$17.5s^2 + K = 0$$

$$17.5s^2 + 26.25 = 0$$

or $\qquad s^2 = -\dfrac{26.25}{17.5}$

or $\qquad s = +j1.2247$

Figure 14.50 shows the location of poles, asymptotes, the breakaway point and the points at which the root loci is going to cross imaginary axis.

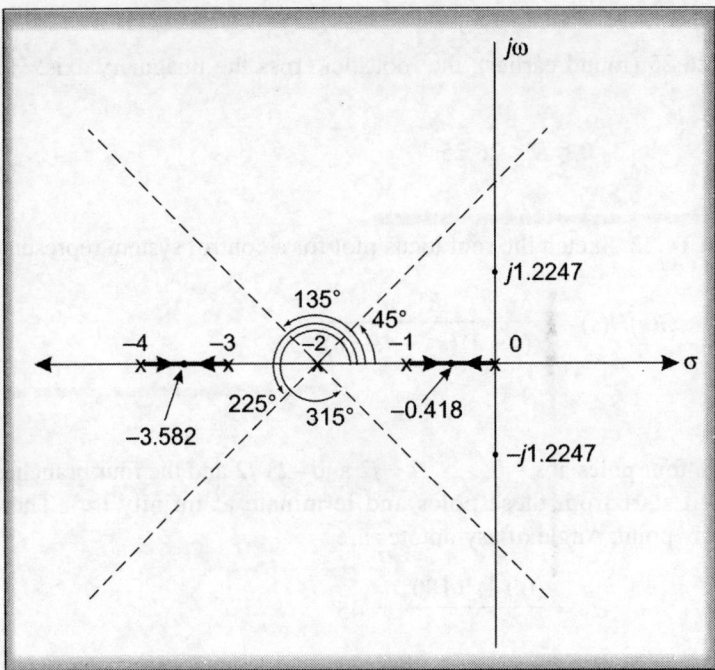

Fig. 14.50

The complete root locus plot is shown in Fig. 14.51.

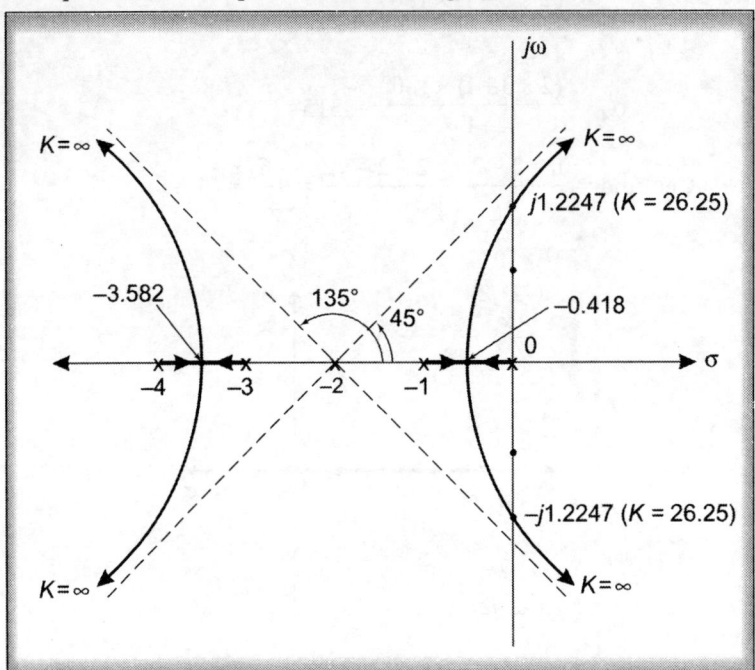

Fig. 14.51

Range of K

At $K = 26.25$ (found earlier), the root loci cross the imaginary axis. Hence, for stability

$$0 < K < 26.25$$

Problem 14.22 Sketch the root locus plot for a control system represented by

$$G(s)H(s) = \frac{K}{s(s+2)(s^2+4s+8)}$$

Solution

There are four poles at $s = 0, -2, -2+j2$ and $-2-j2$ and the four branches of root locus will start from these poles and terminate at infinity (∞). There is one breakaway point. Angle of asymptotes are

$$\alpha_1 = \frac{(0+1) \times 180°}{4} = 45°$$

$$\alpha_2 = \frac{(2 \times 1+1) \times 180°}{4} = 135°$$

$$\alpha_3 = \frac{(2 \times 2+1) \times 180°}{4} = 225°$$

$$\alpha_4 = \frac{(2 \times 3+1) \times 180°}{4} = 315°$$

$$\text{Centroid} = \frac{0-2-2+j2-2-j2}{4} = \frac{-6}{4} = -1.5$$

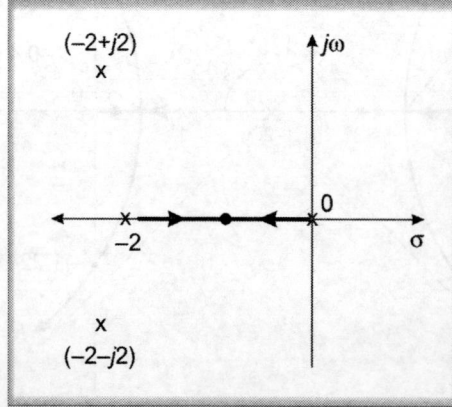

Fig. 14.52

Breakaway point

$$1 + G(s)H(s) = 0$$

or $$1 + \frac{K}{s(s+2)(s^2+4s+8)} = 0$$

or $$s(s^3 + 6s^2 + 16s + 16) + K = 0$$

or $$K = -s^4 - 6s^3 - 16s^2 - 163$$

or $$\frac{dK}{ds} = -4s^3 - 18s^2 - 32s - 16 = 0$$

or $$s^3 + 4.5s^2 + 8s + 4 = 0$$

The equation has three roots. By hit and trial, we find $s = -0.8$ is one of the roots. The other two roots are imaginary and hence, cannot be valid breakaway points. Therefore, $s = -0.8$ is the breakaway point.

Intersection with imaginary axis

The characteristic equation is $1 + G(s)H(s) = 0$, which gives

$$s^4 + 6s^3 + 16s^2 + 16s + K = 0$$

Forming the Routh's array

s^4	1	16	K
s^3	6	16	0
s^2	13.33	K	
s^1	$\dfrac{213.33 - 6K}{13.33}$	0	
s^0	K		

The s^1 row can be a zero row and gives existence of roots lying on the imaginary axis

$$\frac{213.33 - 6K}{13.33} = 0$$

or $$K = \frac{213.33}{6} = 35.55$$

Forming polynomial of s^2 row, we get

$$14.33s^2 + K = 0$$

or $\quad 14.33s^2 + 35.5 = 0$

or $\qquad\qquad s = +j1.633.$

Angles of departure at complex poles

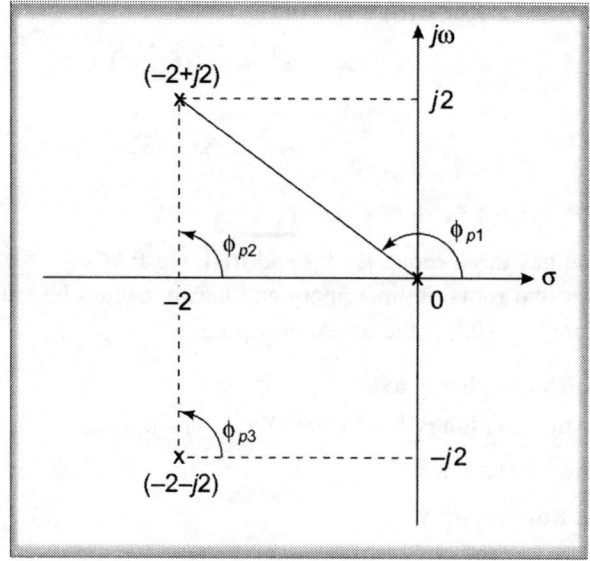

Fig. 14.53

$$\phi = \Sigma\phi_p - \Sigma\phi_z$$

$$\Sigma\phi_p = \phi_{p1} + \phi_{p2} + \phi_{p3}$$

or $\qquad \Sigma\phi_p = \left(180° - \tan^{-1}\dfrac{2}{2}\right) + 90° + 90° = 315°$

Therefore,

$$\phi_d = 180° - \Sigma\,\phi_p$$

$$= 180° - 315°$$

$$= -135° \text{ at } -2 + j2$$

∴ $\qquad\qquad \phi_d = \pm135° \text{ at } -2 - j2$

The complete root locus is shown in Fig. 14.54

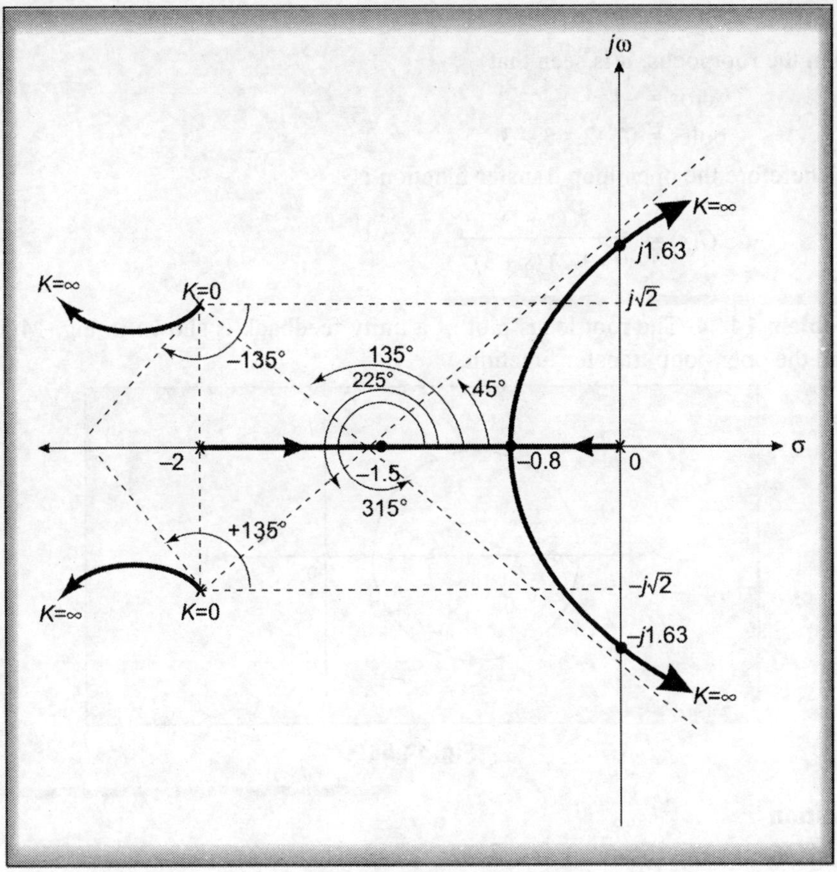

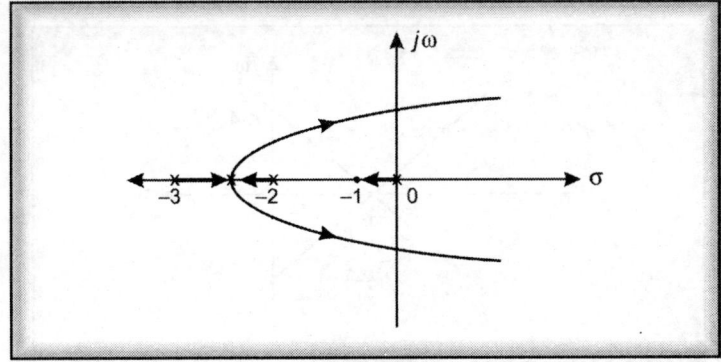

Fig. 14.54

Problem 14.23 The root locus plot for a system is given in Fig. 14.55. Find the open-loop transfer function.

Fig. 14.55

Solution

From the root locus, it is seen that

$$\text{zeros} = -1$$
$$\text{poles} = 0, -2, -3, -3$$

Therefore the open loop transfer function is

$$G(s) = \frac{K(s+1)}{s(s+2)(s+3)^2}$$

Problem 14.24 The root locus plot of a unity feedback is shown in Fig. 14.56. Find the open-loop transfer function.

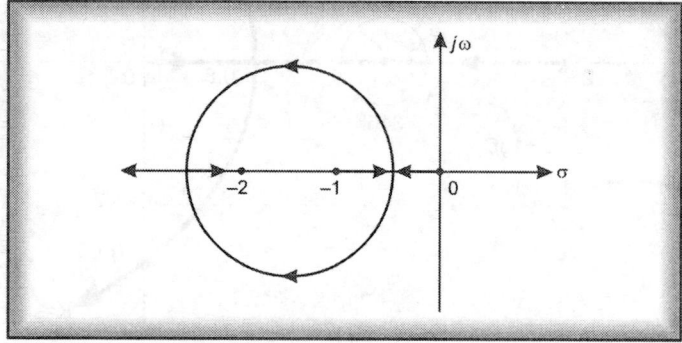

Fig. 14.56

Solution

$$\text{Zeros are at } s = -2$$
$$\text{Poles are at } s = 0, -1$$
$$G(s)H(s) = \frac{K(s+2)}{s(s+1)}$$

Problem 14.25 Find the transfer function of the root loci shown in Fig. 14.57.

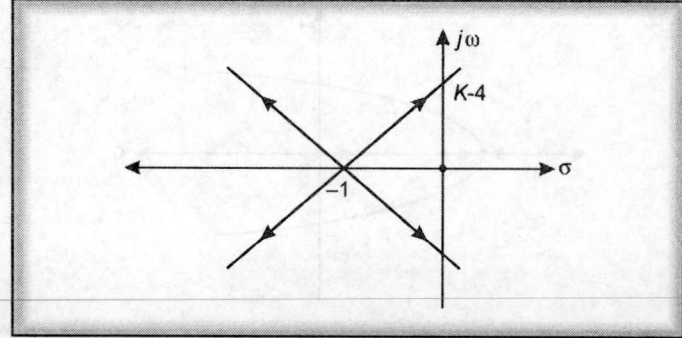

Fig. 14.57

Solution

From the root locus it is seen that four root loci originate from the pole at $s = -1$. Therefore,

$$\text{poles are at } s = -1, -1, -1 \text{ and } -1$$

and since $K = 4$

$$G(s)H(s) = \frac{4}{(s+1)(s+1)(s+1)(s+1)} = \frac{4}{(s+1)^2}$$

Problem 14.26 The root locus plot of unity feedback control system is shown below. Find the closed-loop transfer function.

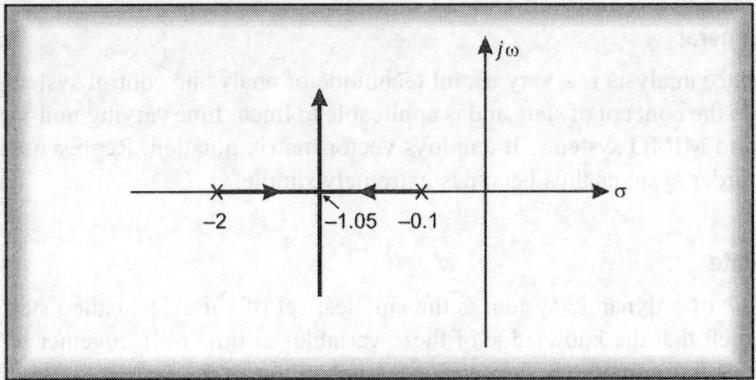

Fig. 14.58

Solution

Poles are at $s = -0.1$ and -2

Therefore,

$$G(s)H(s) = \frac{K}{(s+0.1)(s+2)}$$

Since $H(s) = 1$

$$G(s) = \frac{K}{(s+0.1)(s+2)} = \frac{K}{(0.5s+1)(10s+1)}$$

$$\frac{C(s)}{R(s)} = \frac{G(s)}{1+G(s)} = \frac{\dfrac{K}{(0.5s+1)(10s+1)}}{1+\dfrac{K}{(0.5s+1)(10s+1)}}$$

$$= \frac{K}{(0.5s+1)(10s+1)+K}$$

State Space Analysis

15.1 General

State space analysis is a very useful technique of analyzing control systems. It is based on the concept of state and is applicable to linear time varying/non-varying-linear and MIMO systems. It employs vector matrix notation. Representation of higher order systems thus becomes extremely simple.

15.2 State

The *state* of a dynamic system is the smallest set of variables, called state variables, such that the knowledge of these variables at time $t = t_0$ together with the input for $t \geq t_0$ completely determines the behaviour of the system for any time t. The state of a dynamic system is uniquely found out by state of the dynamic system at time $t = t_0$ and the input for $t \geq t_0$.

15.3 State Variables

The state variables of a dynamic system are the minimal or the smallest set of variables which describe the dynamics of a linear system. Let $x_1(t), x_2(t) \ldots, x_n(t)$ are the state variable chosen to describe the dynamics of a linear system. $x_1(t_0)$, $x_2(t_0) \ldots, x_n(t_0)$ define the initial state of the system at time $t = t_0$. However, once the inputs for time $t \geq t_0$ are specified and knowledge of initial state is known set of variables $x_1, x_2 \ldots, x_n$ have characteristics that the future behaviour of the system can be determined.

15.4 State Vector

The 'n' set of state variables chosen above for defining the dynamics of a linear system can be contemplated to be the 'n' components of a state vector $x(t)$.

15.5 State Space

It is the n dimensional space whose co-ordinate axes consist of x_1 axis, x_2 axis $\ldots$, x_n axis. Any state can be uniquely represented by a point in the state space.

15.6 Controllability and Observability

The concepts of controllability and observability were first put forward by Kalman. If a control function $u(t)$ can transform the initial state x_0 of a system to some required final state $x(t)$ in a finite period of time, then the system is *controllable* and if from measurements of output carried out over a finite period of time, state of a system can be determined, then the system is *observable*. Concepts of controllability and observability are closely related to the cancellation of a Pole and a Zero in the system transfer function.

15.6.1 Controllability

Consider a linear time invariant described by the state equation

$$\dot{x} = Ax + Bu$$

where

x = state vector
u = input vector
$A = n \times n$ matrix
$B = n \times 1$ matrix

The solution of the equation is

$$x(t) = \phi(t - t_0)x(t_0) + \int_{t_0}^{t_f} \phi(t - \tau)Bu(\tau)d\tau$$

or

$$x(t) = e^{A(t - t_0)}x(t_0) + \int_{t_0}^{t_f} e^{A(t - \tau)}Bu(\tau)dt \qquad (t_0 = 0)$$

If the system is completely controllable, then any state can be transferred to zero state in a finite interval of time, then

$$0 = x(0) + \int_0^{t_f} e^{-A\tau}Bu(\tau)d\tau$$

or

$$x(0) = -\int_0^{t_f} e^{-A\tau}Bu(\tau)d\tau$$

The state transition matrix as obtained from Cayley-Hamilton theorem

$$e^{At} = \phi(t) = \sum_{k=o}^{n-1} \alpha_k(t)A^k$$

Here

$x = (n \times 1)$ matrix
$A = (n \times n)$ matrix
$u = (p \times 1)$ input vector
$B = (n \times p)$ matrix

$\alpha_\kappa(t)$ = is a scaler function of t.

Substitution yields

$$x(t_0) = -\int_{t_0}^{t_f} \sum_{k=0}^{n-1} \alpha_k(t_0 - \tau) A^K Bu(\tau) d\tau$$

since the matrices A and B are not functions of τ

$$x(t_0) = -\sum_{k=0}^{n-1} A^k B \int_{t_0}^{t_f} \alpha_k(t_0 - t_f) u(\tau) d\tau$$

which can be written as

$$= -[B \quad AB \quad A^2 B \cdots A^{n-1} B] \begin{bmatrix} P_0 \\ P_1 \\ \vdots \\ P_{n-1} \end{bmatrix}$$

where
$$P_k = \int_{t_0}^{t_f} \alpha_k(t_0 - \tau) u(\tau) d\tau$$

$$Q_c = [B \quad AB \quad A^2 B \quad ... \quad A^{n-1} B]$$

$$P = [P_0 \quad P_1 \quad P_2 \quad ... \quad P_{n-1}]'$$

If the system is completely state controllable

(a) if the rank of $n \times n$ matrix
$$Q_c = [\, B : AB : A^2 B : ... : A^{n-1} B\,] \text{ be } n\,;$$

or

(b) if the vector $B, AB, ..., A^{n-1} B$ are linearly independent.

15.6.2 Observability

If every state variable of the system affects some of the output, then the system is completely observable. If from output measurements, one of states cannot be observed, then the state is not observable. Consider an unforced system described by equations

$$\dot{x} = Ax + Bu$$
$$y = Cx$$

x = state vector, n dimensional
$u = r -$ dimensional vector
$A = n \times n$ constant matrix

$$B = n \times r \text{ Constant matrix}$$

Taking Laplace transform of the above equation, we get

$$sX(s) - x(0) = AX(s) + BU(s)$$

$$X(s) = (sI - A)^{-1}x(0) + (sI - A)^{-1}BU(s)$$

where $\qquad x(0) = $ initial state vector at $t = 0$

$$I = \text{unity matrix}$$

$(sI - A)^{-1} = $ matrix inverse of $(sI - A)$ and $(sI - A)$ is assumed to be nonsingular which means that its inverse matrix exists.

Now,

$$(sI - A)^{-1} = \frac{I}{s} + \frac{A}{s^2} + \frac{A^2}{s^3} + \dots$$

Inverse Laplace transform is given by

$$\mathcal{L}^{-1}[sI - A]^{-1} = I + At + \frac{A^2 t^2}{2!} + \dots \frac{A^n t^n}{n!} + \dots \text{ for } t \geq 0$$

By definition

$$\mathcal{L}^{-1}\left[(sI - A)^{-1}\right] = e^{At} = \phi(t) \text{ for } t \geq 0 \qquad (1)$$

where,

$\phi(t)$ is termed *state transition matrix*.

By solving Eq. (1), by making use of convolution integral, we get

$$x(t) = e^{At}x(0) + \int_0^t e^{A(t-\tau)} Bu(\tau)d\tau$$

$$x(t) = \phi(t)x(0) + \int_0^t \phi(t - \tau) Bu(\tau)d\tau \qquad (2)$$

Let

initial time $= t_0$

initial state $= x(t_0)$

$$\therefore \quad x(t_0) = \phi(t) x(0) + \int_0^{t_0} \phi(t_0 - \tau) Bu(\tau)d\tau \qquad (3)$$

Solving for $x(0)$ gives

$$x(0) = \phi^{-1}(t_0)x(t_0) - \phi^{-1}(t_0) \int_0^{t_0} \phi(t_0 - \tau) Bu(\tau)d\tau \qquad (4)$$

Since $\phi^{-1}(t) = \phi(-t)$, we get

$$x(0) = \phi(-t_0)x(t_0) - \phi(-t_0) \int_0^{t_0} \phi(t_0 - \tau) Bu(\tau)d\tau \qquad (5)$$

Substituting Eq. (5) in Eq. (2), we get

$$x(t) = \phi(t)\phi(-t_0) x (t_0) - \phi(t)\phi(-t_0) \int_0^{t_0} \phi(t_0 - \tau) Bu(\tau)d\tau + \int_0^t \phi(t - \tau) Bu(\tau)d\tau$$

or $x(t) = \phi(t - t_0) x(t_0) + \int_{t_0}^{t} \phi(t - \tau) Bu(\tau)d\tau$ \hfill (6)

Note: The significance of state transition matrix is

(1) It gives the solution of linear homogeneous and non-homogeneous equations.

(2) It gives free response of the system.

$$\boxed{\textbf{SOLVED PROBLEMS}}$$

Problem 15.1 Prove the following

(1) $\qquad\qquad \phi(0) = I$

(2) $\qquad\qquad \phi^{-1}(t) = \phi(-t)$

(3) $\quad \phi(t_2 - t_1) \phi(t_1 - t_0) = \phi(t_2 - t_0)$

(4) $\qquad\qquad [\phi(t)]^k = \phi(kt)$ \hfill *(ME Pune University)*

Solution

(1) $\qquad\qquad \phi(t) = I + At + \dfrac{A^2 t^2}{2!} + \dfrac{A^3 t^3}{3!} + ...$

Put $\quad t = 0$, we get $\phi(t) = I$

(2) $\qquad\qquad \phi(t) = e^{At}$

$\qquad\qquad\quad \phi(t)e^{-At} = e^{At} e^{-At}$

or $\qquad\qquad \phi(t)e^{-At} = I$

or $\qquad \phi^{-1}(t)\phi(t)e^{-At} = \phi^{-1}(t) I$

or $\qquad\qquad\quad e^{-At} = \phi^{-1}(t)$

or $\qquad\qquad\quad \phi^{-1}(t) = e^{-At}$

or $\qquad\qquad\quad \phi^{-1}(t) = \phi(-t)$

(3) $\quad \phi(t_2 - t_1) \phi(t_1 - t_0) = e^{A(t_2 - t_1)} e^{A(t_1 - t_0)}$

$\qquad\qquad\qquad\qquad\quad = e^{A(t_2 - t_0)} = \phi(t_2 - t_0)$

(4) $\qquad\quad [\phi(t)]^K = e^{At} . e^{At} ... e^{At}$ (*K* terms)

$\qquad\qquad\qquad\quad = e^{KAt} = \phi(Kt)$

Problem 15.2 Explain various methods of evaluation of state transition matrix.

\hfill *(ME Pune University)*

Solution

Various methods available for finding out state transition matrix are:

(1) **By numerical computation** We know that

$$\phi(t) = I + At + \frac{A^2 t^2}{2!} + \frac{A^3 t^3}{3!} + \dots$$

Put $\qquad At = M$

$\therefore \qquad \phi(t) = I + M + \frac{M(M)}{2!} + \frac{M}{3}\left(\frac{M^2}{2!}\right) + \dots$

Careful study of the above equation points that each term in the parenthesis is equal to the entire preceding term. This property is useful in computer computations. If the above expression consists of exponential terms, the convergence occurs rapidly. For higher order systems this method is the easiest. There is no requirement of finding matrix A.

(2) **By use of Λ matrix**

Let $\qquad \Lambda = P^{-1}AP$

or $\qquad A = P\Lambda P^{-1}$

or $\qquad A^2 = P\Lambda P^{-1}P\Lambda P^{-1} = P\Lambda^2 P^{-1}$

or $\qquad A^m = P\Lambda^m P^{-1}$

$$e^{At} = I + At + \frac{A^2 t^2}{2!} + \dots$$

$$= I + P\Lambda P^{-1}t + \frac{P\Lambda^2 P^{-1}t^2}{2!} + \dots = Pe^{\Lambda t}P^{-1}$$

- Find *eigen values*. Let $\lambda_1, \lambda_2 \dots \lambda_n$ be the eigen values
- Find Λ matrix

$$\Lambda = \begin{bmatrix} \lambda_1 & \dots & \dots \\ \dots & \lambda_2 & \dots \\ \dots & \dots & \lambda_n \end{bmatrix}$$

- Find $e^{\Lambda t}$

$$e^{\Lambda t} = \begin{bmatrix} e^{\lambda_1 t} & \dots & \dots \\ \dots & e^{\lambda_2 t} & \dots \\ \dots & \dots & e^{\lambda_3 t} \end{bmatrix}$$

- Find e^{At} where

$$e^{At} = Pe^{\Lambda t}P^{-1} \text{ where } P = \text{modal matrix}$$

(3) **By Laplace transform approach**

We know that

$$\phi(t) = \mathcal{L}^{-1}(sI - A)^{-1}$$

$\therefore \qquad = \mathcal{L}^{-1}\dfrac{adj\,(sI - A)}{|sI - A|}$

Problem 15.3 Prove that a system described by state model of the form

$$\dot{x}(t) = Ax(t) + Bu(t)$$
$$y(t) = Cx(t) + Du(t)$$

where A, B, C, D are constant matrices is linear and time invariant.

Solution

A system is linear if the dependent variable x and u and their derivatives have a linear effect on the system output. A system is said to be time invariant or constant, if the independent variable 't' does not appear explicitly.

$$\dot{x}(t) = f[x(t), u(t)]$$

$$f(\cdot) = \begin{bmatrix} f_1(\cdot) \\ f_2(\cdot) \\ f_3(\cdot) \\ \vdots \\ f_n(\cdot) \end{bmatrix} ; \text{ and}$$

$$y(t) = g[x(t), u(t)]$$

Let us consider

$u_1, u_2, \ldots, u_m$ as the input variables

$y_1, y_2, \ldots, y_p$ as the output variables

$x_1, x_2, \ldots, x_n$ as the state variables.

In a linear time invariant system derivatives of the state variables, i.e. $x_1, x_2, \ldots,$ x_n are the linear combination of system states and inputs.

$$\dot{x}_1 = a_{11}x_1 + x_{12}x_2 + a_{13}x_3 + \ldots + a_{1n}x_n + b_{11}u_1 + b_{12}u_2 + b_{13}u_3 + \ldots + b_{1m}u_m$$
$$\dot{x}_2 = a_{21}x_1 + a_{22}x_2 + a_{23}x_3 + \ldots + a_{2n}x_n + b_{21}u_1 + b_{22}u_2 + b_{23}u_3 + \ldots + b_{2m}u_m$$

$$\ldots\ldots\ldots\ldots\ldots\ldots\ldots\ldots\ldots\ldots\ldots\ldots\ldots\ldots\ldots\ldots\ldots\ldots\ldots\ldots$$

$$\dot{x}_n = a_{n1}x_1 + a_{n2}x_2 + \ldots + a_{nn}x_n + \ldots + bn_1u_1 + b_{n2}u_2 + \ldots + b_{nm}u_m$$

This can be written as

$$\dot{x}(t) = Ax(t) + Bu(t) ; \text{ where}$$

$$x(t) = n \times 1 \text{ state vector}$$

$$u(t) = m \times 1 \text{ input vector}$$

$$A = \begin{bmatrix} a_{11} & a_{12} & \dots & a_{1n} \\ a_{21} & a_{22} & \dots & a_{2n} \\ a_{31} & a_{32} & \dots & a_{3n} \\ \dots & \dots & \dots & \dots \\ a_{n1} & a_{n2} & \dots & a_{nn} \end{bmatrix}$$
$(n \times n)$
matrix

$$A = \begin{bmatrix} b_{11} & b_{12} & \dots & b_{1m} \\ b_{21} & b_{22} & \dots & b_{2m} \\ b_{31} & b_{32} & \dots & b_{3m} \\ \dots & \dots & \dots & \dots \\ b_{m1} & b_{m2} & \dots & b_{mm} \end{bmatrix}$$
$(n \times m)$
matrix

Similarly, the output variables $y(t)$ at time 't' are linear combination of state $x(t)$ and input $u(t)$.

$$y_1(t) = c_{11}x_1(t) + c_{12}x_2(t) + \dots + c_{1n} x_n(t)$$
$$+ d_{11}u_1(t) + d_{12}u_2(t) + \dots + d_m u_m(t)$$
$$\dots \qquad \dots \qquad \dots$$
$$y_p(t) = c_{p1}x_1(t) + c_{p2}x_2(t) + \dots + c_{pn}x_n(t)$$
$$+ d_{p1}u_1(t) + d_{p2}u_2(t) + \dots + d_{pm}u_m(t)$$

where $y(t) = p \times 1$ output vector

$$C = \begin{bmatrix} c_{11} & c_{12} & \dots & c_{1n} \\ c_{21} & c_{22} & \dots & c_{2n} \\ \dots & \dots & \dots & \dots \\ \dots & \dots & \dots & \dots \\ c_{p1} & c_{p2} & \dots & c_{pn} \end{bmatrix}$$
$(p \times n)$
output
matrix

$$D = \begin{bmatrix} d_{11} & d_{12} & \dots & d_{1m} \\ d_{21} & d_{22} & \dots & d_{2m} \\ \dots & \dots & \dots & \dots \\ d_{p1} & d_{p2} & \dots & d_{pm} \end{bmatrix}$$
$(p \times n)$
matrix

Therefore, the state model for a linear time invariant system is

$$\dot{x}(t) = Ax(t) + Bu(t)$$
$$y(t) = Cx(t) + Du(t)$$

Problem 15.4 Discuss the merits and demerits of representing a state model into (a) physical variable form (b) phase variable form (c) canonical variable form.

(ME Pune University)

Solution

(a) *Physical variable form*

Let us consider the circuit as shown in Fig.15.1.

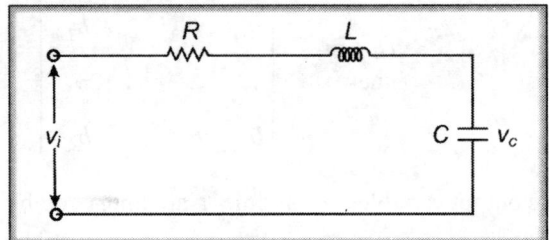

Fig. 15.1

The following equations are satisfied by the circuit:

$$\frac{Ldi}{dt} + Ri + v_c = v_i$$

$$\frac{Cdv_c}{dt} = i$$

∴
$$\frac{di}{dt} = -\frac{R}{L}i - \frac{v_c}{L} + \frac{v_i}{L} ; \text{ and}$$

$$\frac{dv_c}{dt} = \frac{i}{C}$$

The state variables choosen are i and v_c. Therefore the state model is

$$\begin{bmatrix} \dot{i} \\ \dot{v_c} \end{bmatrix} = \begin{bmatrix} -\dfrac{R}{L} & -\dfrac{1}{L} \\ \dfrac{1}{C} & 0 \end{bmatrix} \begin{bmatrix} i \\ v_c \end{bmatrix} + \begin{bmatrix} \dfrac{1}{L} \\ 0 \end{bmatrix} v_i$$

1. State variables choosen are physical quantities which can be measured.
2. In addition to the output, the other state variables could be used for feedback and hence, the implementation of design by selecting physical variables as state variables is straight forward.

3. The solution of the state equations give time variance of state variable. It has a direct relevance to the physical system.

4. The solution of state equation is difficult.

(b) *Phase variable representation*

Phase variables are state variable obtained, from one of the system variable and its derivatives. System output is taken as the state variable and the other state variables are its derivatives. The merits/demerits of such a representation are

(1) State model can be easily obtained, if the system model is readily available in differential equation or transfer function form, e.g.

Let $\qquad T(s) = \dfrac{Y(s)}{U(s)} = \dfrac{2}{s^4 + 3s^3 + 3s^2 + 3s + 2}$

then the phase variable representation is

$$\begin{bmatrix} \dot{x}_1 \\ \dot{x}_2 \\ \dot{x}_3 \\ \dot{x}_4 \end{bmatrix} = \begin{bmatrix} 0 & 1 & 0 & 0 \\ 0 & 0 & 1 & 0 \\ 0 & 0 & 0 & 1 \\ -2 & -3 & -3 & -3 \end{bmatrix} \begin{bmatrix} x_1 \\ x_2 \\ x_3 \\ x_4 \end{bmatrix} + \begin{bmatrix} 0 \\ 0 \\ 0 \\ 2 \end{bmatrix} u$$

$$y = \begin{bmatrix} 1 & 0 & 0 & 0 \end{bmatrix} \begin{bmatrix} x_1 \\ x_2 \\ x_3 \\ x_4 \end{bmatrix}$$

Matrix 'A' so formulated is termed 'Bush' or 'Companion' form.

(2) When the transfer function has a zero, the formulation of state model is difficult. Help of the Mason's gain formula is taken to form the state equations and then the state model is obtained.

(3) Phase variables are not practical set of variables from measurement, analysis and control point of view.

(4) Phase variables provide to the control engineer, a link between the time domain and transfer function design approach.

(c) *Canonical variables*

Matrix A in such a representation is in diagonal form consisting of eigenvalues. State model is

$$\begin{bmatrix} \dot{x}_1 \\ \dot{x}_2 \\ \vdots \\ \dot{x}_n \end{bmatrix} = \begin{bmatrix} \lambda_1 & 0 & 0 \ldots 0 \\ 0 & \lambda_2 & 0 \ldots 0 \\ \ldots & \ldots & \ldots \ldots \\ 0 & 0 & 0 \ldots \lambda_n \end{bmatrix} \begin{bmatrix} x_1 \\ x_2 \\ \vdots \\ x_n \end{bmatrix} + \begin{bmatrix} 1 \\ 1 \\ \vdots \\ 1 \end{bmatrix} u$$

$$y = [c_1 \quad c_2 \ldots c_n] \begin{bmatrix} x_1 \\ x_2 \\ \vdots \\ x_n \end{bmatrix}$$

or

$$\begin{bmatrix} \dot{x}_1 \\ \dot{x}_2 \\ \vdots \\ \dot{x}_n \end{bmatrix} = \begin{bmatrix} \lambda_1 & 0 & 0\ldots0 \\ 0 & \lambda_2 & 0\ldots0 \\ 0 & 0 & 0\ldots0 \\ \cdots & \cdots & \cdots \\ 0 & 0 & 0\ldots\lambda_n \end{bmatrix} \begin{bmatrix} x_1 \\ x_2 \\ \vdots \\ x_n \end{bmatrix} + \begin{bmatrix} c_1 \\ c_2 \\ \vdots \\ c_n \end{bmatrix} u$$

$$y = [1 \quad 1\ldots1] \begin{bmatrix} x_1 \\ x_2 \\ \vdots \\ x_n \end{bmatrix}$$

The decoupled nature of this is useful in analysis of the control systems. Canonical variables are not physical variables which can be measured.

Problem 15.5 Derive the expression for the transfer function from the state model

$$\dot{x} = Ax + Bu$$

$$y = Cx + Du$$

(ME Pune University)

Solution

$$\dot{x} = Ax + Bu$$

$$sX(s) - x(0) = AX(s) + BU(s)$$

$$(sI - A) X(s) = x(0) + BU(s)$$

$$X(s) = (sI - A)^{-1} x(0) + (sI - A)^{-1} BU(s) \tag{1}$$

$$y = Cx + Du$$

$$Y(s) = CX(s) + DU(s) \tag{2}$$

Substituting the value of $X(s)$ from Eq. (1) in Eqn. (2), we get
$$Y(s) = C(sI - A)^{-1} x(0) + C(sI - A)^{-1} BU(s) + DU(s)$$

Assuming zero initial conditions
$$Y(s) = C(sI - A)^{-1} BU(s) + DU(s)$$

$$\frac{Y(s)}{U(s)} = T(s) = C(sI - A)^{-1}B + D$$

Problem 15.6 Obtain state variable representation of

(1) an armature controlled *DC* motor,

(2) field controlled *DC* motor.

Solution

(1) *Armature controlled DC motor*. An armature controlled DC motor is shown in Fig. 15.2

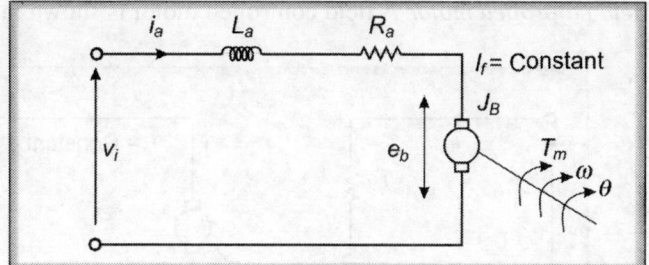

Fig. 15.2

Equations satisfied are

(a)
$$v_i = i_a R_a + L_a \frac{di_a}{dt} + e_b$$

or
$$\frac{di_a}{dt} = -\frac{R_a}{L_a} i_a - \frac{e_b}{L_a} + \frac{v_i}{L_a} \qquad (1)$$

(b)
$$e_b = K_b \frac{d\theta}{dt} = K_b \omega \qquad (2)$$

Note: Substitute the value of e_b in Eqn. (1)

(c)
$$J\frac{d^2\theta}{dt^2} + B\frac{d\theta}{dt} = T_m = K_T i_a$$

or
$$J\frac{d\omega}{dt} + B\omega = K_T i_a$$

or
$$\frac{d\omega}{dt} = -\frac{B}{J}\omega + \frac{K_T}{J} i_a \qquad (3)$$

(d)
$$\frac{d\theta}{dt} = \omega \qquad (4)$$

The state equations are

$$
\begin{bmatrix} \dfrac{di_a}{dt} \\[2mm] \dfrac{d\omega}{dt} \\[2mm] \dfrac{d\theta}{dt} \end{bmatrix} = \begin{bmatrix} -\dfrac{R_a}{L_a} & -\dfrac{K_b}{L_a} & 0 \\[2mm] \dfrac{K_T}{J} & -\dfrac{B}{J} & 0 \\[2mm] 0 & 1 & 0 \end{bmatrix} \begin{bmatrix} i_a \\ \omega \\ \theta \end{bmatrix} + \begin{bmatrix} \dfrac{1}{L_a} \\[2mm] 0 \\[2mm] 0 \end{bmatrix} v_i
$$

(2) *Field controlled motor* A field controlled motor is shown in Fig.15.3

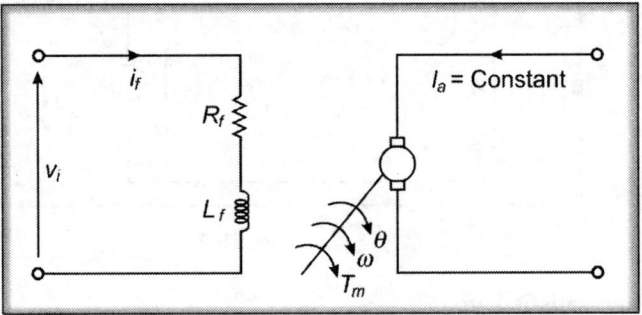

Fig. 15.3

The following equations are satisfied

(a) $$v_i = R_f i_f + L_f \dfrac{d i_f}{dt}$$

$$\dfrac{d i_f}{dt} = \dfrac{-R_f}{L_f} i_f + \dfrac{v_i}{L_f}$$

(b) $$J \dfrac{d\omega}{dt} + B d\omega = T_m = K_T i_f$$

$$\dfrac{d\omega}{dt} = \dfrac{K_T}{J} i_f - \dfrac{B}{J} \omega$$

(c) $$\dfrac{d\theta}{dt} = \omega$$

$$
\begin{bmatrix} \dfrac{di_f}{dt} \\[2ex] \dfrac{d\omega}{dt} \\[2ex] \dfrac{d\theta}{dt} \end{bmatrix} = \begin{bmatrix} \dfrac{-R_f}{L_f} & 0 & 0 \\[2ex] \dfrac{K_T}{J} & \dfrac{-B}{J} & 0 \\[2ex] 0 & 1 & 0 \end{bmatrix} \begin{bmatrix} i_f \\ \omega \\ \theta \end{bmatrix} + \begin{bmatrix} \dfrac{1}{L_f} \\[2ex] 0 \\[2ex] 0 \end{bmatrix} v_1
$$

Problem 15.7 Given

$$
\dot{x}(t) = \begin{bmatrix} 0 & 1 \\ -2 & 3 \end{bmatrix} x(t) = Ax(t)
$$

Find eigenvalues, eigenvectors and response when

(a) $\quad x(0) = \begin{bmatrix} 1 \\ 1 \end{bmatrix}$ and

(b) $\quad x(0) = \begin{bmatrix} 1 \\ 2 \end{bmatrix}$ *(ME Pune University)*

Solution

(1) *Eigenvalues and eigenvectors*

$$
|\lambda I - A| = \det \begin{vmatrix} \lambda & -1 \\ +2 & \lambda - 3 \end{vmatrix}
$$

i.e., $\quad \lambda(\lambda - 3) - (-1 \times 2) = 0$

or $\quad \lambda^2 - 3\lambda + 2 = 0 \quad$ or $\quad (\lambda - 1)(\lambda - 2) = 0$

Therefore eigenvalues are $\lambda = 1, 2$

Eigenvectors
$$\lambda = 1.$$

$\therefore \quad \begin{bmatrix} 1 & -1 \\ 2 & -2 \end{bmatrix} \begin{bmatrix} x_1 \\ x_2 \end{bmatrix} = \begin{bmatrix} 0 \\ 0 \end{bmatrix}$

or $\quad x_1 - x_2 = 0$

and $\quad 2x_1 - 2x_2 = 0$

i.e. $\quad x_2 = x_1$

and $\quad x_2 = x_1$

Therefore, eigenvector is $\begin{bmatrix} 1 \\ 1 \end{bmatrix}$

$$\lambda = 2.$$

$$\therefore \quad \begin{bmatrix} 2 & -1 \\ 2 & -1 \end{bmatrix} \begin{bmatrix} x_1 \\ x_2 \end{bmatrix} = \begin{bmatrix} 0 \\ 0 \end{bmatrix}$$

or $\qquad 2x_1 - x_2 = 0 \text{ or } x_2 = 2x_1$
$$2x_1 - x_2 = 0 \text{ or } x_2 = 2x_1$$

$\therefore$ eigenvector is $\begin{bmatrix} 1 \\ 2 \end{bmatrix}$

Check: Correctness of eigenvectors is checked as given below

$$M^{-1}AM = \begin{bmatrix} 1 & 0 \\ 0 & 2 \end{bmatrix}$$

where $\qquad M = \begin{bmatrix} 1 & 1 \\ 1 & 2 \end{bmatrix}$

$$\therefore \qquad M^{-1} = \begin{bmatrix} 2 & -1 \\ -1 & 1 \end{bmatrix}$$

$$\therefore \qquad M^{-1}AM = \begin{bmatrix} 2 & -1 \\ -1 & 1 \end{bmatrix} \begin{bmatrix} 0 & 1 \\ -2 & 3 \end{bmatrix} \begin{bmatrix} 1 & 1 \\ 1 & 2 \end{bmatrix}$$

$$= \begin{bmatrix} 2 & -1 \\ -1 & 1 \end{bmatrix} \begin{bmatrix} 1 & 2 \\ 1 & 4 \end{bmatrix} = \begin{bmatrix} 1 & 0 \\ 0 & 2 \end{bmatrix}$$

$\therefore$ Correct

Response

$$\phi(t) = \pounds^{-1}(sI - A)^{-1}$$

$$(sI - A) = \begin{bmatrix} s & 0 \\ 0 & s \end{bmatrix} - \begin{bmatrix} 0 & 1 \\ -2 & 3 \end{bmatrix} = \begin{bmatrix} s & -1 \\ 2 & s-3 \end{bmatrix}$$

$$(sI - A)^{-1} = \frac{adj[sI - A]}{det \, |sI - A|}$$

$$= \frac{\begin{bmatrix} s-3 & 1 \\ -2 & s \end{bmatrix}}{s(s-3) - (-1 \times 2)}$$

$$= \frac{1}{s^2 - 3s + 2}\begin{bmatrix} s-3 & 1 \\ -2 & s \end{bmatrix} = \frac{1}{(s-1)(s-2)}\begin{bmatrix} s-3 & 1 \\ -2 & s \end{bmatrix}$$

$$= \begin{bmatrix} \dfrac{s-3}{(s-1)(s-2)} & \dfrac{1}{(s-1)(s-2)} \\ \dfrac{-2}{(s-1)(s-2)} & \dfrac{s}{(s-2)(s-2)} \end{bmatrix} = \begin{bmatrix} \dfrac{2}{s-1} + \dfrac{-1}{s-2} & \dfrac{-1}{s-1} + \dfrac{1}{s-2} \\ \dfrac{2}{s-1} + \dfrac{-2}{s-1} & \dfrac{-1}{s-2} + \dfrac{1}{s-2} \end{bmatrix}$$

$$\phi(t) = \mathcal{L}^{-1}(sI - A)^{-1} = \begin{bmatrix} 2e^t - e^{2t} & -e^t + e^{2t} \\ 2e^t - 2e^{2t} & -e^t + e^{2t} \end{bmatrix}$$

Now $x(t) = \phi(t)\, x(0)$

(a) $x(0) = \begin{bmatrix} 1 \\ 1 \end{bmatrix}$

∴ $x(t) = \begin{bmatrix} 2e^t - e^{2t} & -e^t + e^{2t} \\ 2e^t - 2e^{2t} & -e^t + e^{2t} \end{bmatrix}\begin{bmatrix} 1 \\ 1 \end{bmatrix}$

$x_1(t) = 2e^t - e^{2t} - e^t + e^{2t} = e^t$

$x_2(t) = 2e^t - 2e^{2t} - e^t + 2e^{2t} = e^t - e^{2t}$

(b) $x(0) = \begin{bmatrix} 1 \\ 2 \end{bmatrix}$

$$x(t) = \phi(t)\, x(0) = \begin{bmatrix} 2e^t - e^{2t} & -e^t + e^{2t} \\ 2e^t - 2e^{2t} & -e^t + 2^{2t} \end{bmatrix}\begin{bmatrix} 1 \\ 2 \end{bmatrix}$$

$x_1(t) = 2e^t - e^{2t} - 2e^t + 2e^{2t} = e^{2t}$

$x_2(t) = 2e^t - 2e^{2t} - 2e^t + 2e^{2t} = 0$

Problem 15.8 Given

$$\begin{bmatrix} \dot{x}_1 \\ \dot{x}_2 \end{bmatrix} = \begin{bmatrix} 1 & 0 \\ 1 & 1 \end{bmatrix}\begin{bmatrix} x_1 \\ x_2 \end{bmatrix} + \begin{bmatrix} 1 \\ 1 \end{bmatrix} u(t)$$

$x^T(0) = [1 \quad 0]$

Find the response for a step input

Solution

$$x(t) = e^{At} x(0) + \int_0^t e^{A(t-\tau)}\, Bu(\tau)d\tau$$

Computation of $e^{At}(\phi(t))$

$$A = \begin{bmatrix} 1 & 0 \\ 1 & 1 \end{bmatrix}$$

$$= [sI - A]^{-1} \begin{bmatrix} s & 0 \\ 0 & s \end{bmatrix} - \begin{bmatrix} 1 & 0 \\ 1 & 1 \end{bmatrix} = \begin{bmatrix} s-1 & 0 \\ -1 & s-1 \end{bmatrix}$$

$$[sI - A] = \frac{1}{(s-1)(s-1)} \begin{bmatrix} s-1 & 0 \\ 1 & s-1 \end{bmatrix} = \begin{bmatrix} \dfrac{1}{s-1} & 0 \\ \dfrac{1}{(s-1)^2} & \dfrac{1}{(s-1)} \end{bmatrix}$$

$$e^{At} = \mathcal{L}^{-1}[sI - A]^{-1} = \begin{bmatrix} e^t & 0 \\ te^t & e^t \end{bmatrix}$$

Alternatively

$$e^{At} = I + At + \frac{A^2 t^2}{2!} + \dots$$

$$A = \begin{bmatrix} 1 & 0 \\ 1 & 1 \end{bmatrix}$$

$$A^2 = \begin{bmatrix} 1 & 0 \\ 1 & 1 \end{bmatrix}\begin{bmatrix} 1 & 0 \\ 1 & 1 \end{bmatrix} = \begin{bmatrix} 1 & 0 \\ 2 & 1 \end{bmatrix}$$

$$e^{At} = \begin{bmatrix} 1 & 0 \\ 0 & 1 \end{bmatrix} + \begin{bmatrix} 1 & 0 \\ 1 & 1 \end{bmatrix}t + \frac{1}{2}\begin{bmatrix} 1 & 0 \\ 2 & 1 \end{bmatrix}t^2$$

or $\quad e^{At} = \begin{bmatrix} 1+t+\dfrac{t^2}{2} & 0 \\ t+t^2 & t+t+\dfrac{t^2}{2} \end{bmatrix} = \begin{bmatrix} 1+t+\dfrac{t^2}{2} & 0 \\ t(1+t) & 1+t+\dfrac{t^2}{2} \end{bmatrix} = \begin{bmatrix} e^t & 0 \\ te^t & e^t \end{bmatrix}$

$$x(t) = \phi(t)\,x(0) + \int_0^t \phi(t-\tau)\,Bu(\tau)d\tau = \phi(t)\left[x(0) + \int_0^t \phi(-\tau)\,Bu(\tau)d\tau\right]$$

Computation of $\phi(-\tau)\,Bu(\tau)$

$$\phi(t) = \begin{bmatrix} e^t & 0 \\ te^t & e^t \end{bmatrix}$$

$$\phi(-\tau) = \begin{bmatrix} e^{-\tau} & 0 \\ -\tau e^{-\tau} & e^{-\tau} \end{bmatrix}$$

$$B = \begin{bmatrix} 1 \\ 1 \end{bmatrix}$$

$$u = 1$$

$\therefore \qquad \phi(-\tau)Bu(\tau) = \begin{bmatrix} e^{-\tau} & 0 \\ -\tau e^{-\tau} & e^{-\tau} \end{bmatrix} \begin{bmatrix} 1 \\ 1 \end{bmatrix} 1 = \begin{bmatrix} e^{-\tau} \\ -\tau e^{-\tau} + e^{-\tau} \end{bmatrix}$

Therefore,

$$\int_0^t \phi(-\tau)Bu(\tau)d\tau = \int_0^t \begin{bmatrix} e^{-\tau} \\ -\tau e^{-\tau} + e^{-\tau} \end{bmatrix} d\tau = \begin{bmatrix} 1 - e^{-t} \\ te^{-t} \end{bmatrix}$$

Therefore $\qquad \dot{x}(t) = \begin{bmatrix} e^t & 0 \\ te^t & e^t \end{bmatrix} \left\{ \begin{bmatrix} 1 \\ 0 \end{bmatrix} + \begin{bmatrix} 1 - e^{-t} \\ te^{-t} \end{bmatrix} \right\}$

or $\qquad \dot{x}(t) = \begin{bmatrix} e^t & 0 \\ te^t & e^t \end{bmatrix} \begin{bmatrix} 2 - e^{-t} \\ te^{-t} \end{bmatrix} = \begin{bmatrix} 2e^t - 1 \\ 2te^t \end{bmatrix}$

Problem 15.9 Find the state transition matrix for

$$\dot{x} = \begin{bmatrix} -2 & 1 & 0 \\ 0 & -2 & 1 \\ 0 & 0 & -2 \end{bmatrix} x$$

Solution

$$A = \begin{bmatrix} -2 & 1 & 0 \\ 0 & -2 & 1 \\ 0 & 0 & -2 \end{bmatrix}$$

$$[sI - A] = \begin{bmatrix} s+2 & -1 & 0 \\ 0 & s+2 & -1 \\ 0 & 0 & s+2 \end{bmatrix}$$

$$[sI - A]^{-1} = \frac{adj[sI - A]}{det\ |sI - A|} = \frac{\begin{bmatrix} (s+2)(s+2) & (s+2) & 1 \\ 0 & (s+2)(s+2) & (s+2) \\ 0 & 0 & (s+2)(s+2) \end{bmatrix}}{(s+2)(s+2)(s+2)}$$

$$= \begin{bmatrix} \dfrac{1}{(s+2)} & \dfrac{1}{(s+2)(s+2)} & \dfrac{1}{(s+2)(s+2)(s+2)} \\ 0 & \dfrac{1}{(s+2)} & \dfrac{1}{(s+2)(s+2)} \\ 0 & 0 & \dfrac{1}{(s+2)} \end{bmatrix}$$

$$\phi(t) = e^{At} = \mathcal{L}^{-1}(sI - A)^{-1} = \begin{bmatrix} e^{-2t} & te^{-2t} & \dfrac{t^2}{2}e^{-2t} \\ 0 & e^{-2t} & te^{-2t} \\ 0 & 0 & e^{-2t} \end{bmatrix}$$

Problem 15.10 Obtain eigenvalues, eigenvectors and the state model in canonical form for a system described by

$$\dot{x}(t) = \begin{bmatrix} 0 & 1 & 0 \\ 3 & 0 & 2 \\ -12 & -7 & -6 \end{bmatrix} x(t) + \begin{bmatrix} 1 \\ 0 \\ 2 \end{bmatrix} u(t)$$

$$y(t) = [1\ \ 0\ \ 0]\, x(t)$$

Solution

(1) *Eigenvalues*

$$[\lambda I - A] = \begin{bmatrix} \lambda & 0 & 0 \\ 0 & \lambda & 0 \\ 0 & 0 & \lambda \end{bmatrix} - \begin{bmatrix} 0 & 1 & 0 \\ 3 & 0 & 2 \\ -12 & -7 & -6 \end{bmatrix} = \begin{bmatrix} \lambda & -1 & 0 \\ -3 & \lambda & -2 \\ 12 & 7 & \lambda+6 \end{bmatrix}$$

Characteristic equation is

$$\lambda\,\{\lambda(\lambda + 6) - (-2)\,7\} - (-1)\,\{-3(\lambda + 6) - (-2)(12)\} = 0$$

or

$$(\lambda + 1)(\lambda + 2)(\lambda + 3) = 0$$

∴ eigenvalues are

$$\lambda = -1, -2, -3.$$

(2) *Eigenvectors*

(a) Corresponding to $\lambda = -1$, the eigenvector is

$$\begin{bmatrix} -1 & -1 & 0 \\ -3 & -1 & -2 \\ 12 & 7 & 5 \end{bmatrix} \begin{bmatrix} x_1 \\ x_2 \\ x_3 \end{bmatrix} = 0$$

i.e.
$$x_1 + x_2 = 0 \text{ or } x_2 = -x_1$$
$$3x_1 + x_2 + 2x_3 = 0 \text{ or } x_3 = -x_1$$
$$12x_1 + 7x_2 + 5x_3 = 0$$

If $x_1 = 1$, then x_2 and x_3 are -1. Therefore, eigenvector is $\begin{bmatrix} 1 \\ -1 \\ -1 \end{bmatrix}$

(b) Corresponding to $\lambda = -2$, the eigenvector is

$$\begin{bmatrix} -2 & -1 & 0 \\ -3 & -2 & -2 \\ 17 & 7 & 4 \end{bmatrix} \begin{bmatrix} x_1 \\ x_2 \\ x_3 \end{bmatrix} = 0$$

i.e.,
$$2x_1 + x_2 = 0 \text{ or } x_2 = -2x_1$$
$$3x_1 + 2x_2 + 2x_3 = 0 \text{ or } x_3 = \frac{1}{2}x_1$$
$$17x_1 + 7x_2 + 4x_3 = 0$$

If
$$x_1 = 2$$
then
$$x_2 = -4$$
and
$$x_3 = 1$$

therefore the eigenvector is $\begin{bmatrix} 2 \\ -4 \\ 1 \end{bmatrix}$

(c) Corresponding to $\lambda = -3$, the eigenvector is

$$\begin{bmatrix} -3 & -1 & 0 \\ -3 & -3 & -2 \\ 12 & 7 & 3 \end{bmatrix} \begin{bmatrix} x_1 \\ x_2 \\ x_3 \end{bmatrix} = 0$$

$$3x_1 + x_2 \quad = 0 \text{ or } x_2 = -3x_1$$
$$3x_1 + 3x_2 + 2x_3 = 0 \text{ or } x_3 = 3x_1$$
$$12x_1 + 7x_2 + 3x_3 = 0$$

If $\qquad x_1 = 1,$

then $\qquad x_2 = -3$

and $\qquad x_3 = 3$

therefore, the eigenvector is $= \begin{bmatrix} 1 \\ -3 \\ 3 \end{bmatrix}$

Modal matrix is $= \begin{bmatrix} 1 & 2 & 1 \\ -1 & -4 & -3 \\ -1 & 1 & 3 \end{bmatrix}$

$$adj[M] = \begin{bmatrix} +(-12+3) & -(6-1) & +(-6+4) \\ -(-3-3) & +(3+1) & -(-3+1) \\ +(-1-4) & -(1+2) & +(-4+2) \end{bmatrix} = \begin{bmatrix} -9 & -5 & -2 \\ 6 & 4 & 2 \\ -5 & -3 & -2 \end{bmatrix}$$

$$|M| = \{(-4-3) - (-3 \times 1)\} - 2\{(-1 \times 3) - (-1 \times -3)\} + 1\{(-1 \times -1) - (-4 \times -1)\} = -2$$

$$\therefore \qquad M^{-1} = -\frac{1}{2} \begin{bmatrix} -9 & -5 & -2 \\ 6 & 4 & 2 \\ -5 & -3 & -2 \end{bmatrix}$$

$$M^{-1}AM = -\frac{1}{2} \begin{bmatrix} -9 & -5 & -2 \\ 6 & 4 & 2 \\ -5 & -3 & -2 \end{bmatrix} \begin{bmatrix} 0 & 1 & 0 \\ 3 & 0 & 2 \\ -12 & -7 & -6 \end{bmatrix} \begin{bmatrix} 1 & 2 & 1 \\ -1 & -4 & -3 \\ -1 & 1 & 3 \end{bmatrix}$$

$$= \begin{bmatrix} -1 & 0 & 0 \\ 0 & -2 & 0 \\ 0 & 0 & -3 \end{bmatrix}$$

$$B = \begin{bmatrix} 1 \\ 0 \\ 2 \end{bmatrix}$$

$$M^{-1}B = -\frac{1}{2} \begin{bmatrix} -9 & -5 & -2 \\ 6 & 4 & 2 \\ -5 & -3 & -2 \end{bmatrix} \begin{bmatrix} 1 \\ 0 \\ 2 \end{bmatrix} = -\frac{1}{2} \begin{bmatrix} -13 \\ -10 \\ -9 \end{bmatrix} = \begin{bmatrix} 13/2 \\ 5 \\ 9/2 \end{bmatrix}$$

Therefore, canonical representation is

$$\begin{bmatrix} \dot{z}_1 \\ \dot{z}_2 \\ \dot{z}_3 \end{bmatrix} = \begin{bmatrix} -1 & 0 & 0 \\ 0 & -2 & 0 \\ 0 & 0 & -3 \end{bmatrix} \begin{bmatrix} z_1 \\ z_2 \\ z_3 \end{bmatrix} + \begin{bmatrix} 13/2 \\ 5 \\ 9/2 \end{bmatrix}$$

$$y = \begin{bmatrix} 1 & 1 & 1 \end{bmatrix} \begin{bmatrix} z_1 \\ z_2 \\ z_3 \end{bmatrix}$$

Problem 15.11 Obtain phase variable representation for a system whose transfer function is given by

$$\frac{Y(s)}{U(s)} = \frac{6s^3 + 4s^2 + 3s + 10}{s^3 + 8s^2 + 4s + 20}$$

Solution

$$\frac{Y(s)}{U(s)} = \frac{6 + 4/s + 3/s^2 + 10/s^3}{1 - (-8/s - 4/s^2 - 20/s^3)} \tag{1}$$

The signal flow graph is shown in Fig.15.4. Comparing, the above with the Mason's gain formula, the following information is obtained:

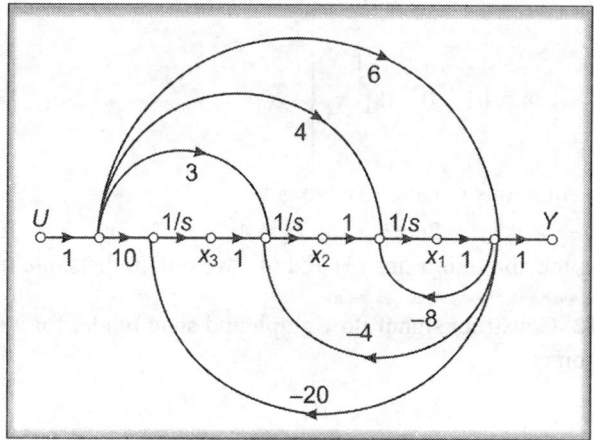

Fig. 15.4

(1) There are four forward paths having gain 6, $4/s$; $3/s^2$ and $10/s^3$.

(2) There are three feedback paths having gain $-8/s$, $-4/s^2$ and $-20/s^3$.

Above information helps in drawing the signal flow graph.

$$y = x_1 + 6u$$

$$x_1 = \frac{1}{s}(x_2 - 8y + 4u)$$

$$\dot{x}_1 = x_2 - 48u - 8x_1 + 4u = -8x_1 + x_2 - 44u$$

$$\dot{x}_2 = x_3 - 4y + 3u = -4x_1 + x_3 - 21u$$

$$\dot{x}_3 = -20y + 10u = -20x_1 - 120u + 10u = -20x_1 - 110u$$

Phase variable representation is

$$\begin{bmatrix} \dot{x}_1 \\ \dot{x}_2 \\ \dot{x}_3 \end{bmatrix} = \begin{bmatrix} -8 & 1 & 0 \\ -4 & 0 & 1 \\ -20 & 0 & 0 \end{bmatrix} \begin{bmatrix} x_1 \\ x_2 \\ x_3 \end{bmatrix} + \begin{bmatrix} -44 \\ -21 \\ -110 \end{bmatrix} u$$

$$y = \begin{bmatrix} 1 & 0 & 0 \end{bmatrix} \begin{bmatrix} x_1 \\ x_2 \\ x_3 \end{bmatrix} + 6u$$

Alternatively

$$T_{(s)} = \frac{Y(s)}{U(s)} = \frac{b_0 s^3 + b_1 s + b_2 s + b_3}{s^3 + a_1 s^2 + a_2 s + a_3} = \frac{b_0 + b_1/s + b_2/s^2 + b_3/s^3}{1 - (-a_1/s - a_2/s^2 - a_3/s)} \quad (2)$$

Phase variable representation of such a system is given as

$$\begin{bmatrix} \dot{x}_1 \\ \dot{x}_2 \\ \dot{x}_3 \end{bmatrix} = \begin{bmatrix} -a_1 & 1 & 0 \\ -a_2 & 0 & 1 \\ -a_3 & 0 & 0 \end{bmatrix} + \begin{bmatrix} b_1 - a_1 b_0 \\ b_2 - a_2 b_0 \\ b_3 - a_3 b_0 \end{bmatrix} u \quad (3)$$

$$y = \begin{bmatrix} 1 & 0 & 0 \end{bmatrix} \begin{bmatrix} x_1 \\ x_2 \\ x_3 \end{bmatrix} + b_0 u \quad (4)$$

Comparing equations (1) and (2), we get

$$a_1 = 8; \ a_2 = 4; \ a_3 = 26; \ b_0 = 6; \ b_1 = 4; \ b_2 = 3; \ \text{and} \ b_3 = 10$$

Substituting the above in Eqns (3) and (4), we obtain the same result.

Problem 15.12 Construct signal flow graph and state model for a system whose transfer function is

$$T(s) = \frac{s^2 + 3s + 3}{s^3 + 2s^2 + 3s + 1}$$

Solution

$$T(s) = \frac{1/s + 3/s^2 + 3/s^3}{1 - (-2/s - 3/s^2 - 1/s^3)}$$

Signal flow graph is shown in Fig. 15.5

$$y = x_1$$
$$\dot{x}_1 = x_2 - 2y + u = x_2 - 2x_1 + u = -2x_1 + x_2 + u$$
$$\dot{x}_2 = x_3 - 3y + 3u = x_3 - 3x_1 + 3u = -3x_1 + x_3 + 3u$$
$$\dot{x}_3 = 3u - y = 3y - x_1 = -x_1 + 3u$$

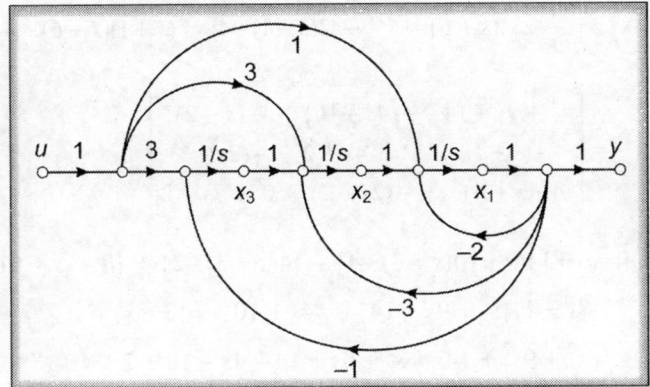

Fig. 15.5

Therefore, state model is

$$\begin{bmatrix} \dot{x}_1 \\ \dot{x}_2 \\ \dot{x}_3 \end{bmatrix} = \begin{bmatrix} -2 & 1 & 0 \\ -3 & 0 & 1 \\ -1 & 0 & 0 \end{bmatrix} \begin{bmatrix} x_1 \\ x_2 \\ x_3 \end{bmatrix} + \begin{bmatrix} 1 \\ 3 \\ 3 \end{bmatrix} u$$

and

$$y = \begin{bmatrix} 1 & 0 & 0 \end{bmatrix} \begin{bmatrix} x_1 \\ x_2 \\ x_3 \end{bmatrix}$$

Problem 15.13 A system is described by

$$\dot{x} = \begin{bmatrix} -1 & -4 & -1 \\ -1 & -6 & -2 \\ -1 & -2 & -3 \end{bmatrix} x + \begin{bmatrix} 0 \\ 1 \\ 1 \end{bmatrix} u$$

$$y = \begin{bmatrix} 1 & 1 & 1 \end{bmatrix} x$$

Find the transfer function and construct the signal flow graph.

Solution

$$G(s) = C[sI - A]^{-1}B$$

$$[sI - A] = \begin{bmatrix} s & 0 & 0 \\ 0 & s & 0 \\ 0 & 0 & s \end{bmatrix} - \begin{bmatrix} -1 & -4 & -1 \\ -1 & -6 & -2 \\ -1 & -2 & -3 \end{bmatrix} = \begin{bmatrix} s+1 & 4 & 1 \\ 1 & s+6 & 2 \\ 1 & 2 & s+3 \end{bmatrix}$$

$$Adj\,[sI - A] = \begin{bmatrix} (s+6)(s+3)-4 & -\{4(s+3)-2\} & 8-(s+6) \\ -\{(s+3)-2\} & (s+1)(s+3)-1 & -\{2(s+1)-1 \\ 2-(s+6) & -\{2(s+1)-4\} & (s+1)(s+6)-4 \end{bmatrix}$$

$$= \begin{bmatrix} s^2+9s+14 & -(4s+10) & -(s-2) \\ -(s+1) & s^2+4s+2 & -(2s+1) \\ -(s+4) & -(2s-2) & s^2+7s+2 \end{bmatrix}$$

$$|sI - A| = s+1\{(s+6)(s+3)-4\} - \{4(s+3)-2\} + \{8-(s+6)\}$$
$$= (s+1)\{s^2+9s+14\} - \{4s+10\} + \{2-s\}$$
$$= s^2+9s^2+14s+s^2+9s+14-4s-10+2-s$$
$$= s^3+10s^2+18s+6$$

$$G(s) = C[sI - A]^{-1}B$$

$$= [1 \quad 1 \quad 1]\frac{1}{s^3+10s^2+18s+6}$$

$$\times \begin{bmatrix} s^2+9s+14 & -(4s+10) & -(s-2) \\ -(s+1) & s^2+4s+2 & -(2s+1) \\ -(s+4) & -(2s-2) & s^2+7s+2 \end{bmatrix} \begin{bmatrix} 0 \\ 1 \\ 1 \end{bmatrix}$$

$$= [1 \quad 1 \quad 1]\frac{1}{s^3+10s^2+18s+6} \begin{bmatrix} -4s-10-s+2 \\ s^2+4s+2-2s-1 \\ -2s+2+s^2+7s+2 \end{bmatrix}$$

$$= [1 \quad 1 \quad 1]\frac{1}{s^3+10s^2+18s+6} \begin{bmatrix} -5s-8 \\ s^2+2s+1 \\ s^2+5s+4 \end{bmatrix}$$

$$= \frac{-5s - 8 + s^2 + 2s + 1 + s^2 + 5s + 4}{s^3 + 10s^2 + 18s + 6}$$

$$= \frac{2s^2 + 2s - 3}{s^2 + 10s^2 + 18s + 6} = \frac{\dfrac{2}{s} + \dfrac{2}{s^2} - \dfrac{3}{s^3}}{1 - \left(-\dfrac{10}{s} + \dfrac{18}{s^2} - \dfrac{6}{s^3}\right)}$$

Signal flow graph is shown in Fig. 15.6.

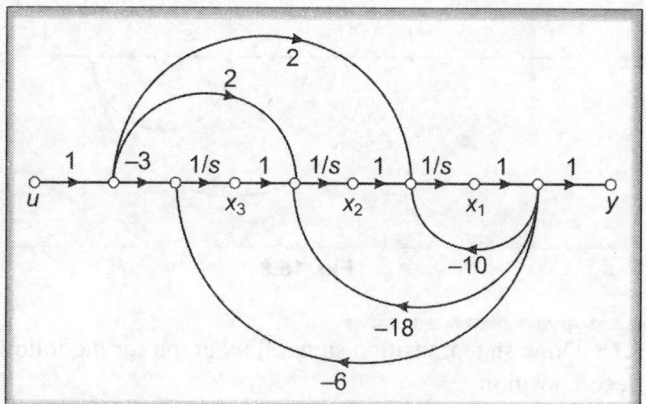

Fig. 15.6

Problem 15.14 Draw state transition signal flow graphs for the following transfer functions by means of direct decomposition.

(1) $\quad G(s) = \dfrac{10}{s^3 + 5s^2 + 4s + 10}$ (2) $G(s) = \dfrac{6(s+1)}{s(s+2)(s+3)}$

Solution

(1) $\qquad G(s) = \dfrac{10}{s^3 + 5s^2 + 4s + 10} = \dfrac{10/s^3}{1 - (-5/s - 4/s^2 - 10/s^3)}$

The signal flow graph is shown in Fig. 15.7.

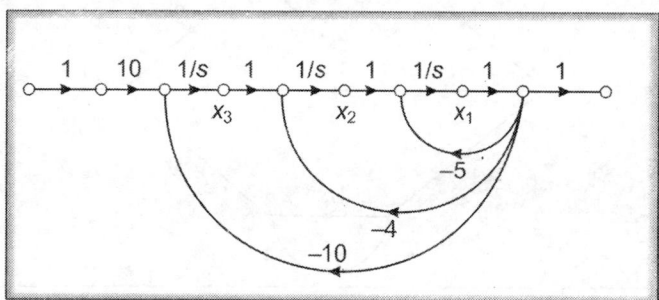

Fig. 15.7

(2) $$G(s) = \frac{6(s+1)}{s(s+2)(s+3)} = \frac{6s+6}{s^3+5s^2+6s} = \frac{6/s^3+6/s^2}{1-(-5/s-6/s^2)}$$

The signal flow graph is shown in Fig. 15.8.

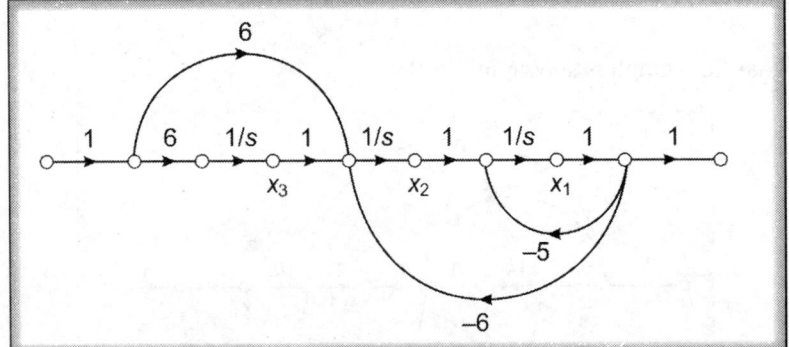

Fig. 15.8

Problem 15.15 Draw state transition signal flow graph for the following system by parallel decomposition.

$$G(s) = \frac{6(s+1)}{s(s+2)(s+3)}.$$

Solution

$$G(s) = \frac{6(s+1)}{s(s+2)(s+3)} = \frac{A}{s} + \frac{B}{s+2} + \frac{C}{s+3} = \frac{1}{s} + \frac{3}{s+2} + \frac{-4}{s+3}$$

Signal flow graph is shown in Fig. 15.9.

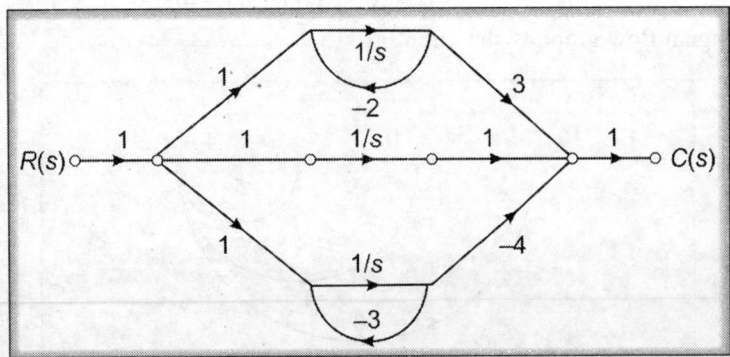

Fig. 15.9

Problem 15.16 Draw state transition signal flow graph for the following system by cascade decomposition.

$$G(s) = \frac{6(s+1)}{s(s+2)(s+3)}$$

Solution

$$G(s) = \frac{6(s+1)}{s(s+2)(s+3)} = 6 \times \frac{1}{s} \frac{(s+1)}{(s+2)} \times \frac{1}{(s+3)}$$

The signal flow graph is shown in Fig. 15.10

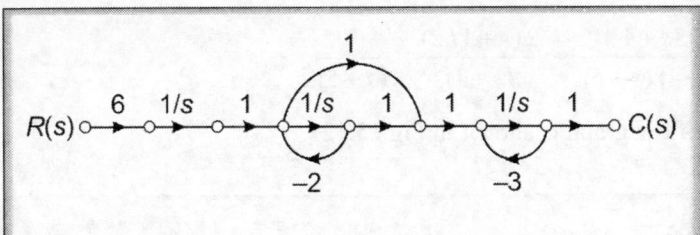

Fig. 15.10

Problem 15.17 Draw state transition signal flow graph for the following system

$$\frac{d^2c(t)}{dt^2} + \frac{6dc(t)}{dt} + 5c(t) = \frac{2dr(t)}{dt} + r(t)$$

by means of
(1) Parallel decomposition   (2) Cascade decomposition

Solution

The given system equation can be written as

$$(s^2 + 6s + 5)\,C(s) = (2s + 1)\,R(s)$$

$$\therefore \qquad \frac{C(s)}{R(s)} = \frac{2s+1}{s^2 + 6s + 5}$$

(1) *By parallel decomposition*

$$\frac{2s+1}{s^2 + 6s + 5} = \frac{2s+1}{(s+1)(s+5)} = \frac{A}{(s+1)} + \frac{B}{(s+5)} = \frac{-1/4}{(s+1)} + \frac{9/4}{(s+5)}$$

Signal flow graph is shown in Fig. 15.11

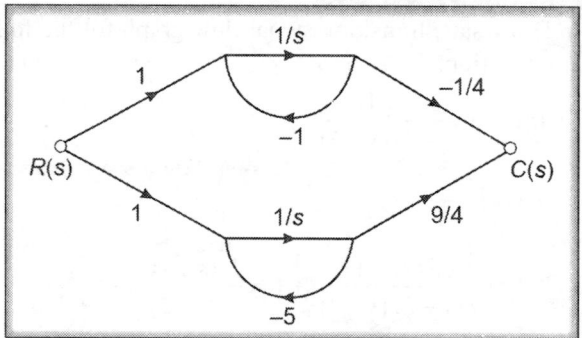

Fig. 15.11

(2) *By cascade decomposition*

$$\frac{2s+1}{(s+1)(s+5)} = \frac{2(s+1/2)}{(s+1)} \times \frac{1}{(s+5)}$$

Signal flow graph is shown in Fig.15.12

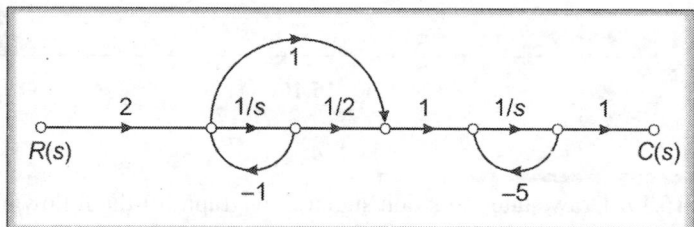

Fig. 15.12

Problem 15.18 A control system has a transfer function given by

$$G(s) = \frac{s+3}{(s+1)(s+2)^2}$$

Using the method of parallel decomposition, draw the state diagram showing minimum number of integrals. Also, obtain the state model.

Solution

$$G(s) = \frac{Y(s)}{U(s)} = \frac{s+3}{(s+1)(s+2)^2} = \frac{A}{(s+1)} + \frac{B}{(s+2)} + \frac{C}{(s+2)^2}$$

$$A = \frac{s+3}{(s+2)^2}\bigg|_{s=-1} = \frac{-1+3}{(-1+2)^2} = 2$$

$$C = \frac{s+3}{(s+1)}\Bigg|_{s=-2} = \frac{-2+3}{(-2+3)} = -1$$

$$B = \frac{d}{ds}[(s+2)^2\,G(s)]\Bigg|_{s=-2}$$

$$= \frac{d}{ds}\left[\frac{(s+2)^2\,(s+3)}{(s+1)(s+2)^2}\right]\Bigg|_{s=-2}$$

$$= \frac{d}{ds}\left[\frac{s+3}{s+1}\right]\Bigg|_{s=-2}$$

$$= \frac{(s+1)-(s+3)}{(s+1)^2}\Bigg|_{s=-2} = -2$$

$$G(s) = \frac{Y(s)}{U(s)} = \frac{2}{s+1} - \frac{2}{s+2} - \frac{1}{(s+2)^2}$$

Let $\qquad X_1(s) = \dfrac{U(s)}{s+1};\quad X_2(s) = \dfrac{U(s)}{s+2}\quad$ and $\quad X_3(s) = \dfrac{U(s)}{(s+2)^2}$

$$Y(s) = 2X_1(s) - 2X_2(s) - X_3(s)$$

This system can be depicted by four number of integrals as shown in Fig. 15.13.

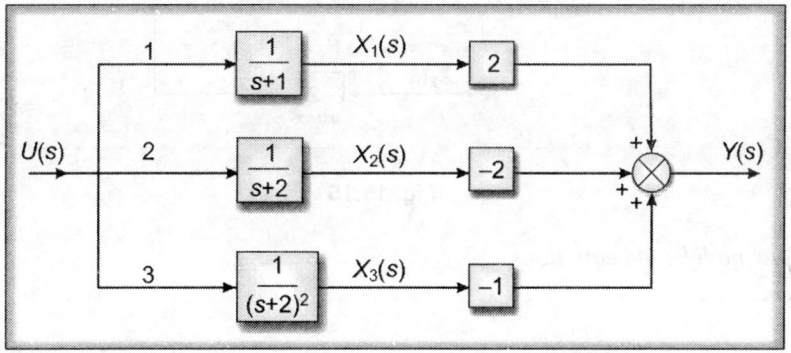

Fig. 15.13

The blocks in branch 1 and 2 can be depicted by one integrator each. However, the block in branch 3 has a squared term and hence require two integrators as shown in Fig. 15.14.

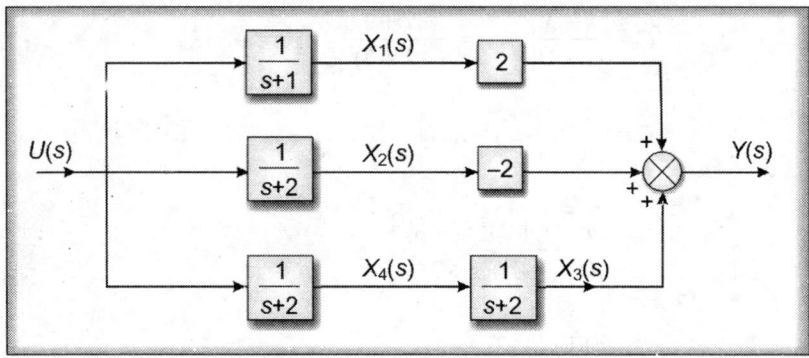

Fig. 15.14

where $\qquad X_3(s) = \dfrac{X_4(s)}{s+2}$ and $X_4(s) = \dfrac{U(s)}{s+2}$

Also $\qquad X_2(s) = \dfrac{U(s)}{s+2}$

Thus, it is seen that $X_2(s)$ and $X_4(s)$ are similar and hence we can get rid of one integrator and draw the state diagram with three integrators as shown in Fig. 15.15.

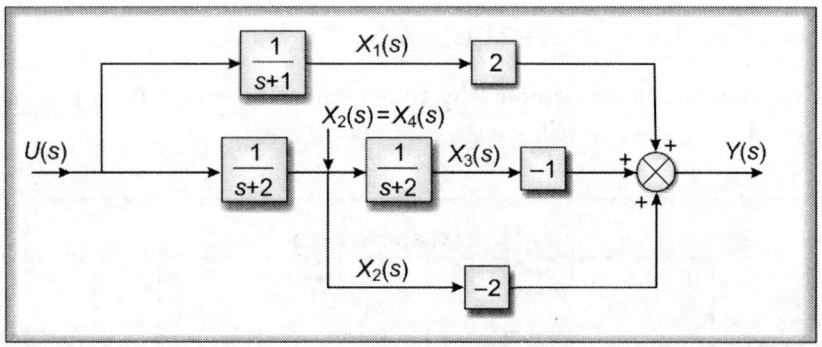

Fig. 15.15

State model/state equations

Now

$$X_1(s) = \dfrac{U(s)}{s+1}$$

i.e. $\quad sX_1(s) + X_1(s) = U(s)$

$$\dot{x}_1 + x_1 = u$$

or $\qquad\qquad \dot{x}_1 = -x_1 + u \qquad\qquad\qquad\qquad (1)$

Similarly

$$\dot{x}_2 = -2x_2 + u \qquad (2)$$

and

$$\dot{x}_3 = -2x_3 + x_2 \qquad (3)$$

Also

$$y = 2x_1 - 2x_2 - x_3 \qquad (4)$$

Hence, the state model in Jordon canonical form is

$$\begin{bmatrix} \dot{x}_1 \\ \dot{x}_2 \\ \dot{x}_3 \end{bmatrix} = \begin{bmatrix} -1 & 0 & 0 \\ 0 & -2 & 0 \\ 0 & 1 & -2 \end{bmatrix} \begin{bmatrix} x_1 \\ x_2 \\ x_3 \end{bmatrix} + \begin{bmatrix} 1 \\ 1 \\ 0 \end{bmatrix} u$$

$$y = \begin{bmatrix} 2 & -2 & -1 \end{bmatrix} \begin{bmatrix} x_1 \\ x_2 \\ x_3 \end{bmatrix}$$

The state signal flow graph is shown in Fig. 15.16.

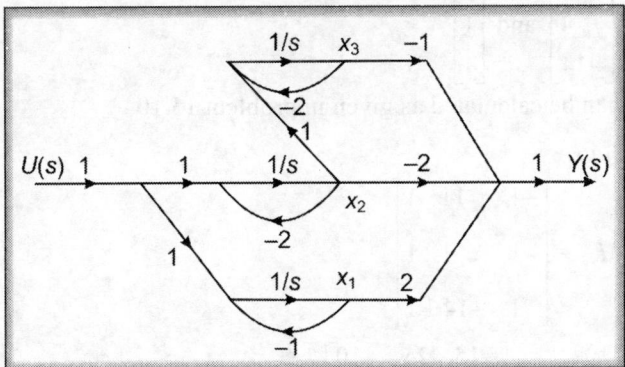

Fig. 15.16

Problem 15.19 Find the canonical form of representation and state transition matrix.

$$\begin{bmatrix} \dot{x}_1 \\ \dot{x}_2 \\ \dot{x}_3 \end{bmatrix} = \begin{bmatrix} 2 & -2 & 3 \\ 1 & 1 & 1 \\ 1 & 3 & -1 \end{bmatrix} \begin{bmatrix} x_1 \\ x_2 \\ x_3 \end{bmatrix} + \begin{bmatrix} 11 \\ 1 \\ -14 \end{bmatrix} u$$

$$y = \begin{bmatrix} -3 & 5 & -2 \end{bmatrix} \begin{bmatrix} x_1 \\ x_2 \\ x_3 \end{bmatrix}$$

Solution

 1. *State transition matrix*

$$|\lambda I - A| = \begin{bmatrix} \lambda - 2 & 2 & -3 \\ -1 & \lambda - 1 & -1 \\ -1 & -3 & \lambda + 1 \end{bmatrix}$$

or $\qquad \lambda^3 - 2\lambda^2 - 5\lambda + 6 = 0$

or $\qquad (\lambda - 1)(\lambda + 2)(\lambda - 3) = 0$

The eigenvalues are $\lambda = 1, -2, 3$

$$\Lambda = \begin{bmatrix} 1 & 0 & 0 \\ 0 & -2 & 0 \\ 0 & 0 & 3 \end{bmatrix} \qquad e^{\Lambda t} = \begin{bmatrix} e^t & 0 & 0 \\ 0 & e^{-2t} & 0 \\ 0 & 0 & e^{3t} \end{bmatrix}$$

Eigenvectors corresponding to eigenvalues of $1, -2$ and 3 are

$$\begin{bmatrix} -1 \\ 1 \\ 1 \end{bmatrix} \quad \begin{bmatrix} 11 \\ 1 \\ -14 \end{bmatrix} \text{ and } \begin{bmatrix} 1 \\ 1 \\ 1 \end{bmatrix}$$

Note: These can be calculated as given in problem 15.10

 2. *Modal matrix*

$$M = \begin{bmatrix} -1 & 11 & 1 \\ 1 & 1 & 1 \\ 1 & -14 & 1 \end{bmatrix}$$

$$M^{-1} = \frac{1}{30} \begin{bmatrix} -15 & 25 & -10 \\ 0 & 2 & -2 \\ 15 & 3 & 12 \end{bmatrix}$$

$$e^{At} = M e^{\Lambda t} M^{-1}$$

$$= \begin{bmatrix} -1 & 11 & 1 \\ 1 & 1 & 1 \\ 1 & -14 & 1 \end{bmatrix} \begin{bmatrix} e^t & 0 & 0 \\ 0 & e^{-2t} & 0 \\ 0 & 0 & e^{3t} \end{bmatrix} \times \begin{bmatrix} -15 & 25 & -10 \\ 0 & 2 & -2 \\ 15 & 3 & 12 \end{bmatrix}$$

$$= \frac{1}{30} \begin{bmatrix} 15e^t + 15e^{3t} & -25e^t + 22e^{-2t} + 3e^{3t} & 10e^t - 22e^{-2t} + 12e^{3t} \\ -15e^t + 15e^{3t} & 25e^t + 2e^{-2t} + 3e^{3t} & -10e^t - 2e^{-2t} + 12e^{3t} \\ -15e^t + 15e^{3t} & 25e^t - 28e^{-2t} + 3e^{3t} & -10e^t + 28e^{-2t} + 12e^{3t} \end{bmatrix}$$

Now
$$B = \begin{bmatrix} 11 \\ 1 \\ -14 \end{bmatrix}$$

$$M^{-1}B = \frac{1}{30} \begin{bmatrix} -15 & 25 & -10 \\ 0 & 2 & -2 \\ 15 & 3 & 12 \end{bmatrix} \begin{bmatrix} 11 \\ 1 \\ -14 \end{bmatrix} = \frac{1}{30} \begin{bmatrix} 0 \\ 30 \\ 0 \end{bmatrix} = \begin{bmatrix} 0 \\ 1 \\ 0 \end{bmatrix}$$

Therefore, canonical representation is

$$\begin{bmatrix} \dot{z}_1 \\ \dot{z}_2 \\ \dot{z}_3 \end{bmatrix} = \begin{bmatrix} 1 & 0 & 0 \\ 0 & -2 & 0 \\ 0 & 0 & 3 \end{bmatrix} \begin{bmatrix} z_1 \\ z_2 \\ z_3 \end{bmatrix} + \begin{bmatrix} 0 \\ 1 \\ 0 \end{bmatrix} u$$

$$y = \begin{bmatrix} 1 & 1 & 1 \end{bmatrix} \begin{bmatrix} z_1 \\ z_2 \\ z_3 \end{bmatrix}$$

Problem 15.20 The dynamics of linear system relating to its input u and output y is represented by

$$y^{(n)} + a_{n-1} y^{(n-1)} + \dots + a_1 \dot{y} + a_0 \, y = b_0 \, u.$$

Represent the dynamics of the system in a state model.   (*AMIE*)

Solution

Let the state variables $x_1 \, x_2 \, \dots$ are defined as
$$x_1 = y$$
$$x_2 = \dot{y}$$
$$\dots\dots\dots$$
$$\dots\dots\dots$$
$$x_n = y^{(n-1)}$$
Then
$$\dot{x}_1 = x_2$$
$$\dot{x}_2 = x_3$$
$$\dots\dots\dots$$
$$\dot{x}_{n-1} = x_n$$
$$\dot{x}_2 = -a_0 x_1 - a_1 x_2 - \dots - a_{n-1} x_n + b_0 u$$

The above results in the following state model

$$\begin{bmatrix} \dot{x}_1 \\ \dot{x}_2 \\ \vdots \\ \dot{x}_{n-1} \\ \dot{x}_n \end{bmatrix} = \begin{bmatrix} 0 & 1 & 0\ldots0 \\ 0 & 0 & 1\ldots0 \\ \vdots & \vdots & \vdots\ldots\vdots \\ 0 & 0 & 0\ldots1 \\ -a_0 & -a_1 & -a_1\cdots-a_{n-1} \end{bmatrix} \begin{bmatrix} x_1 \\ x_2 \\ \vdots \\ x_n \end{bmatrix} + \begin{bmatrix} 0 \\ 0 \\ \vdots \\ 0 \\ b_0 \end{bmatrix} u$$

Problem 15.21 For the given transfer function $(T(s))$

$$T(s) = \frac{b_0}{s^3 + a_2 s^2 + a_1 s + a_0}$$

draw the signal flow graph and obtain the state model.   (*AMIE*)

Solution

$$T(s) = \frac{b_0/s^3}{1 + \dfrac{a_2}{s} + \dfrac{a_1}{s^2} + \dfrac{a_0}{s^3}} = \frac{\dfrac{b_0}{s^3}}{1 - \left(-\dfrac{a_2}{s} - \dfrac{a_1}{s^2} - \dfrac{a_0}{s^3}\right)}$$

The signal flow graph consist of (i) three feedback loops (touching each other) with gains

$$\frac{-a_2}{s},\quad \frac{-a_1}{s^2}\quad \text{and}\quad \frac{-a_0}{s^3}$$

(ii) One forward path which touches the loops and have gain b_0/s^3. A signal flow graph which satisfies the above, is shown in Fig. 15.17

From the Fig. 15.17, we have

$$y = x_1$$

$$x_1 = \frac{1}{s}(x_2 - a_2 y) \tag{1}$$

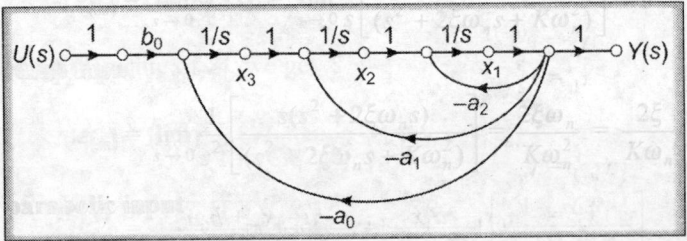

Fig. 15.17

$$\dot{x}_1 = (x_2 - a_2 y)$$
$$\dot{x}_1 = x_2 - a_2 x_1 \tag{2}$$

$$x_2 = \frac{1}{s}(x_3 - a_1 y)$$

$$\dot{x}_2 = x_3 - a_1 y$$
$$\dot{x}_2 = x_3 - a_1 x_1 \tag{3}$$

$$x_3 = \frac{1}{s}(b_0 u - a_0 y)$$

$$\dot{x}_3 = b_0 u - a_0 y$$
$$\dot{x}_3 = b_0 u - a_0 x_1$$

Therefore, the state model is

$$\begin{bmatrix} \dot{x}_1 \\ \dot{x}_2 \\ \dot{x}_3 \end{bmatrix} = \begin{bmatrix} -a_2 & 1 & 0 \\ -a_1 & 0 & 1 \\ -a_0 & 0 & 0 \end{bmatrix} \begin{bmatrix} x_1 \\ x_2 \\ x_3 \end{bmatrix} + \begin{bmatrix} 0 \\ 0 \\ b_0 \end{bmatrix} u$$

$$y = [1 \quad 0 \quad 0] \begin{bmatrix} x_1 \\ x_2 \\ x_3 \end{bmatrix}$$

Problem 15.22 State equation of a control system is given by

$$\begin{bmatrix} \dot{x}_1 \\ \dot{x}_2 \end{bmatrix} = \begin{bmatrix} 0 & 1 \\ -2 & -3 \end{bmatrix} \begin{bmatrix} x_1 \\ x_1 \end{bmatrix} + \begin{bmatrix} 0 \\ 1 \end{bmatrix} u$$

Obtain the state transition matrix and find the response. $u(t)$ is the unit step function occurring at $t = 0$. *(AMIE)*

Solution

$$A = \begin{bmatrix} 0 & 1 \\ -2 & -3 \end{bmatrix}$$

State transition matrix

$$\phi(t) = e^{At}$$
$$= \mathcal{L}^{-1}[(sI - A)^{-1}]$$

$$(sI - A) = \begin{bmatrix} s & 0 \\ 0 & s \end{bmatrix} - \begin{bmatrix} 0 & 1 \\ -2 & -3 \end{bmatrix} = \begin{bmatrix} s & -1 \\ 2 & s+3 \end{bmatrix}$$

$$(sI - A)^{-1} = \frac{1}{(s+1)(s+2)} \begin{bmatrix} s+3 & 1 \\ -2 & s \end{bmatrix} = \begin{bmatrix} \dfrac{(s+3)}{(s+1)(s+2)} & \dfrac{1}{(s+1)(s+2)} \\ \dfrac{-2}{(s+1)(s+2)} & \dfrac{s}{(s+1)(s+2)} \end{bmatrix}$$

$$\phi(t) = \mathcal{L}^{-1}[(sI-A)^{-1}] = \begin{bmatrix} 2e^{-t} - e^{-2t} & e^{-t} - e^{-2t} \\ -2e^{-t} + 2e^{-2t} & -e^{-t} + 2e^{-2t} \end{bmatrix}$$

The response to the unit step input is given below

$$x(t) = e^{At}x(0) + \int_0^t e^{A(t-\tau)}Bu(\tau)d\tau$$

$$= e^{At}x(0) + \int_0^t \begin{bmatrix} 2e^{-(t-\tau)} - e^{-2(t-\tau)} & e^{-(t-\tau)} - e^{-2(t-\tau)} \\ -2e^{-(t-\tau)} + 2e^{-2(t-\tau)} & -e^{-(t-\tau)} + 2e^{-2(t-\tau)} \end{bmatrix}\begin{bmatrix} 0 \\ 1 \end{bmatrix}[1]d\tau$$

$$\begin{bmatrix} x_1(t) \\ x_2(t) \end{bmatrix} = \begin{bmatrix} 2e^{-t} - e^{-2t}e^{-t} - e^{-2t} \\ -2e^{-t} + 2e^{-2t} - e^{-t} + 2e^{-2t} \end{bmatrix}\begin{bmatrix} x_1(0) \\ x_2(0) \end{bmatrix} + \begin{bmatrix} \dfrac{1}{2} - e^{-t} + \dfrac{1}{2}e^{-2t} \\ e^{-t} - e^{-2t} \end{bmatrix}$$

Problem 15.23 A linear time-invariant system is characterised by the homogeneous state equation

$$\begin{bmatrix} \dot{x}_1 \\ \dot{x}_2 \end{bmatrix} = \begin{bmatrix} 1 & 0 \\ 1 & 1 \end{bmatrix}\begin{bmatrix} x_1 \\ x_2 \end{bmatrix}$$

(a) Compute the solution of the homogeneous equation assuming that initial state vector

$$x(0) = \begin{bmatrix} 1 \\ 0 \end{bmatrix}$$

(b) Consider now that the system has a forcing function and is represented by the following non-homogeneous state equation

$$\begin{bmatrix} \dot{x}_1 \\ \dot{x}_2 \end{bmatrix} = \begin{bmatrix} 1 & 0 \\ 1 & 1 \end{bmatrix}\begin{bmatrix} x_1 \\ x_2 \end{bmatrix} + \begin{bmatrix} 0 \\ 1 \end{bmatrix}u$$

where u is a unit input state function. Compute the solution of this state equation assuming

$$x(0) = \begin{bmatrix} 1 \\ 0 \end{bmatrix}$$

Solution

$$x(t) = \phi(t)\, x(0)$$

$$\phi(t) = \begin{bmatrix} e^t & 0 \\ te^t & e^t \end{bmatrix}$$

(Refer to problem 15.8)

$\therefore$ $\quad x(t) = \begin{bmatrix} x_1(t) \\ x_2(t) \end{bmatrix} = \begin{bmatrix} e^t & 0 \\ te^t & e^t \end{bmatrix} \begin{bmatrix} 1 \\ 0 \end{bmatrix}$

$x_1(t) = e^t$

$x_2(t) = te^t$ or $x(t) = \begin{bmatrix} e^t \\ te^t \end{bmatrix}$

(b) $\quad x(t) = e^{At}x(0) + \int_0^t e^{A(t-\tau)}Bu(\tau)d\tau$

$= \phi(t)x(0) + \int_0^t \phi(t-\tau)Bu(\tau)d\tau = \phi(t)\left[x(0) + \int_0^t \phi(-\tau)Bu(\tau)d\tau \right]$

$\phi(t) = \begin{bmatrix} e^t & 0 \\ te^t & e^t \end{bmatrix}$

$\phi(-\tau) = \begin{bmatrix} e^{-\tau} & 0 \\ -\tau e^{-\tau} & e^{-\tau} \end{bmatrix}$

$B = \begin{bmatrix} 0 \\ 1 \end{bmatrix}$

$u = 1$

$\therefore$ $\quad \phi(-\tau)Bu = \begin{bmatrix} e^{-\tau} & 0 \\ -\tau e^{-\tau} & e^{-\tau} \end{bmatrix} \begin{bmatrix} 0 \\ 1 \end{bmatrix} 1 = \begin{bmatrix} 0 \\ e^{-\tau} \end{bmatrix}$

$\int_0^t \phi(-\tau)Bu(\tau)d\tau = \int_0^t \begin{bmatrix} 0 \\ e^{-\tau} \end{bmatrix} d\tau = \begin{bmatrix} 0 \\ 1-e^{-t} \end{bmatrix}$

Therefore,

$\dot{x}(t) = \begin{bmatrix} e^t & 0 \\ te^t & e^t \end{bmatrix} \left\{ \begin{bmatrix} 1 \\ 0 \end{bmatrix} + \begin{bmatrix} 0 \\ 1-e^{-t} \end{bmatrix} \right\}$

or $\quad \dot{x}(t) = \begin{bmatrix} e^t & 0 \\ te^t & e^t \end{bmatrix} \begin{bmatrix} 1 \\ 1-e^{-t} \end{bmatrix}$

or $\quad \dot{x}(t) = \begin{bmatrix} e^t \\ te^t + e^t - 1 \end{bmatrix} = \begin{bmatrix} e^t \\ e^t(t+1)-1 \end{bmatrix}$

Problem 15.24 Construct the state model for a system characterised by the differential equation

$$\frac{d^3y}{dt^3} + \frac{6d^2y}{dt^2} + 11\frac{dy}{dt} + 6y = u$$

Give the block diagram representation of the state model.

Solution

$$T(s) = \frac{Y(s)}{U(s)} = \frac{1}{s^3 + 6s^2 + 11s + 6} = \frac{1}{(s+1)(s+2)(s+3)}$$

$$= \frac{1/2}{(s+1)} + \frac{-1}{(s+2)} + \frac{1/2}{(s+3)}$$

Hence, $$Y(s) = \frac{1/2}{(s+1)}U(s) + \frac{-1}{(s+2)}U(s) + \frac{1/2}{(s+3)}U(s)$$

We define $$X_1(s) = \frac{1/2}{(s+1)}U(s)$$

$$X_2(s) = \frac{-1}{(s+2)}U(s)$$

$$X_3(s) = \frac{1/2}{(s+3)}U(s)$$

Taking inverse Laplace transforms of above three equations; we get

$$\dot{x}_1 = -x_1 + \frac{1}{2}u$$

$$\dot{x}_2 = -2x_2 - u$$

$$\dot{x}_3 = -3x_3 + \frac{1}{2}u$$

Since $$Y(s) = X_1(s) + X_2(s) + X_3(s)$$

We get $$y = x_1 + x_2 + x_3$$

In terms of vector-matrix notation

$$\begin{bmatrix} \dot{x}_1 \\ \dot{x}_2 \\ x_3 \end{bmatrix} = \begin{bmatrix} -1 & 0 & 0 \\ 0 & -2 & 0 \\ 0 & 0 & -3 \end{bmatrix} \begin{bmatrix} x_1 \\ x_2 \\ x_3 \end{bmatrix} + \begin{bmatrix} 1/2 \\ -1 \\ 1/2 \end{bmatrix} u$$

$$y = \begin{bmatrix} 1 & 1 & 1 \end{bmatrix} \begin{bmatrix} x_1 \\ x_2 \\ x_3 \end{bmatrix}$$

Fig. 15.18a shows a block diagram representation

Alternatively

$$\ddot{y} + 6\ddot{y} + 11\dot{y} + 6y = u$$

Let $$x_1 = y$$

$$x_2 = \dot{y} = \dot{x}_1$$

$$x_3 = \dot{x}_2 = \ddot{y} = \ddot{x}_1$$

or

$$\dot{x}_1 = x_2 = \dot{y}$$

$$\dot{x}_2 = x_3 = \ddot{y}$$

$$\dot{x}_3 = \ddot{y} = u - 6\ddot{y} - 11\,\dot{y} - 6y$$

$$= u - 6x_3 - 11x_2 - 6x_1$$

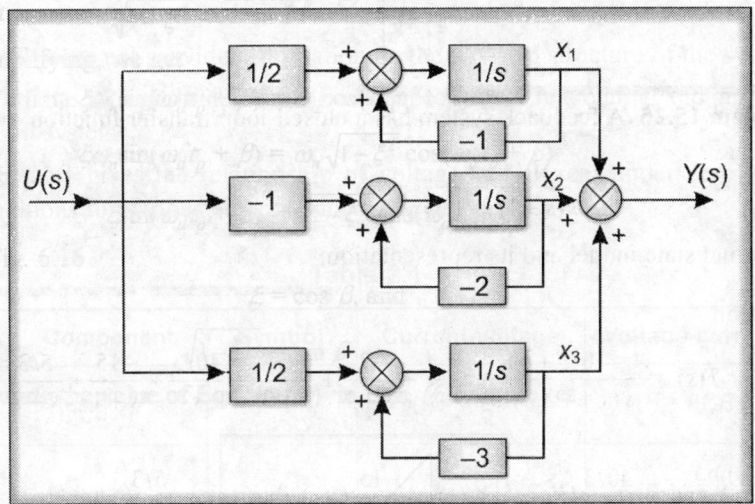

Fig. 15.18a

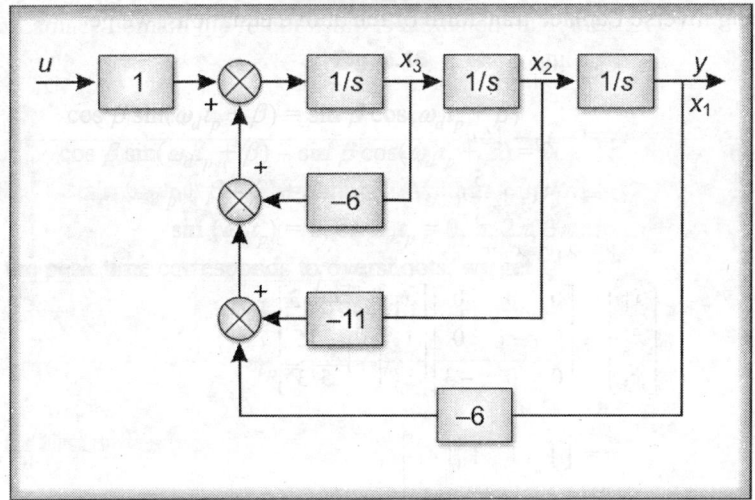

Fig. 15.18b

In vector matrix notation

$$
\begin{bmatrix} \dot{x}_1 \\ \dot{x}_2 \\ \dot{x}_3 \end{bmatrix} = \begin{bmatrix} 0 & 1 & 0 \\ 0 & 0 & 1 \\ -6 & -11 & -6 \end{bmatrix} \begin{bmatrix} x_1 \\ x_2 \\ x_3 \end{bmatrix} + \begin{bmatrix} 0 \\ 0 \\ 1 \end{bmatrix} u
$$

$$
y = \begin{bmatrix} 1 & 0 & 0 \end{bmatrix} \begin{bmatrix} x_1 \\ x_2 \\ x_3 \end{bmatrix}
$$

Problem 15.25 A feedback system has a closed-loop transfer function

$$
\frac{10(s+4)}{s(s+1)(s+3)}
$$

Construct state model and its representation.

Solution

$$
T(s) = \frac{10(s+4)}{s(s+1)(s+3)} = \frac{A}{s} + \frac{B}{s+1} + \frac{C}{s+3} = \frac{40/3}{s} + \frac{-15}{s+1} + \frac{5/3}{s+3}
$$

Let

$$
X_1(s) = \frac{40/3}{s} U(s), \quad X_2(s) = \frac{-15}{s+1} U(s), \quad X_3(s) = \frac{5/3}{s+3} U(s)
$$

$$
\therefore \qquad Y(s) = \frac{40}{3} X_1(s) - 15 X_2(s) + \frac{5}{3} X_3(s)
$$

Hence, $Y(s) = X_1(s) + X_2(s) + X_3(s)$

Taking inverse Laplace transform of the above equations, we get

$$
\dot{x}_1 = \frac{40}{3} u
$$

$$
\dot{x}_2 = -x_2 - 15u
$$

$$
\dot{x}_3 = -3x_3 + \frac{5}{3} u
$$

$$
y = x_1 + x_2 + x_3
$$

$$
\begin{bmatrix} \dot{x}_1 \\ \dot{x}_2 \\ \dot{x}_3 \end{bmatrix} = \begin{bmatrix} 0 & 0 & 0 \\ 0 & -1 & 0 \\ 0 & 0 & -3 \end{bmatrix} \begin{bmatrix} x_1 \\ x_2 \\ x_3 \end{bmatrix} + \begin{bmatrix} 40/3 \\ -15 \\ 5/3 \end{bmatrix} u
$$

$$
y = \begin{bmatrix} 1 & 1 & 1 \end{bmatrix} \begin{bmatrix} x_1 \\ x_2 \\ x_3 \end{bmatrix}
$$

The block diagram representation is shown in Fig. 15.19.

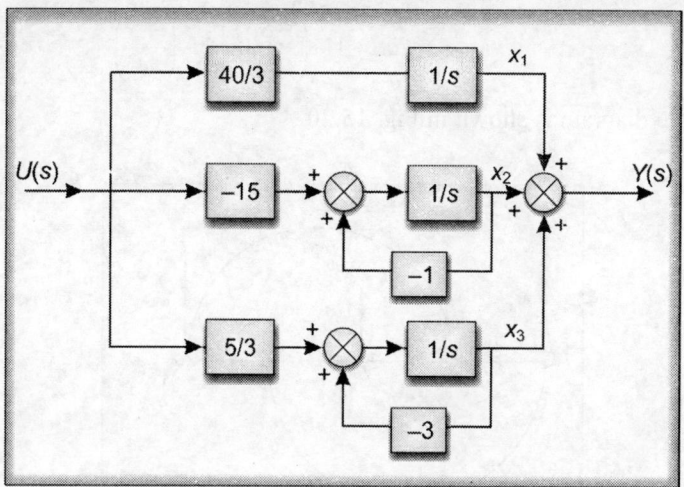

Fig. 15.19

Problem 15.26 A control system is described by the state equation

$$\dot{x} = \begin{bmatrix} 0 & 1 \\ -6 & -5 \end{bmatrix} x + \begin{bmatrix} 1 \\ 1 \end{bmatrix} u \qquad x(0) = \begin{bmatrix} 1 \\ 2 \end{bmatrix}$$

Construct signal flow graph and from it obtain the solution of state equation.

(ME Pune University)

Solution

$$\hat{x}(s) = \hat{G}(s)x^0 + \hat{H}(s)\hat{u}(s)$$

$$\hat{G}(s) = \begin{bmatrix} \hat{G}_{11} & \hat{G}_{12} \\ \hat{G}_{21} & \hat{G}_{22} \end{bmatrix} \qquad \hat{H}(s) = \begin{bmatrix} \hat{H}_1 \\ \hat{H}_2 \end{bmatrix}$$

Now

$$\dot{x} = \begin{bmatrix} 0 & 1 \\ -6 & -5 \end{bmatrix} \begin{bmatrix} x_1 \\ x_2 \end{bmatrix} + \begin{bmatrix} 1 \\ 1 \end{bmatrix} u$$

$$\dot{x}_1 = x_2 + u$$
$$\dot{x}_2 = -6x_1 - 5x_2 + u$$

or

$$x_1 = \frac{x_2}{s} + \frac{u}{s} \quad \text{and} \quad x_2 = \frac{-6x_1}{s} + \frac{-5x_2}{s} + \frac{u}{s}$$

or

$$x_2 = \frac{-6x_1}{s+5} + \frac{u}{s+5} = \frac{\dfrac{-6x_1}{s}}{1 - \left(\dfrac{-5}{s}\right)} + \frac{\dfrac{u}{s}}{1 - \left(\dfrac{-5}{s}\right)}$$

The state diagram is shown in Fig. 15.20

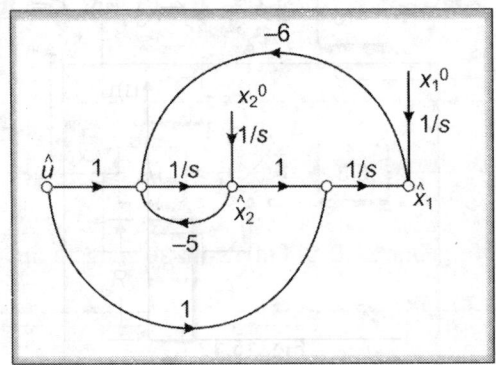

Fig. 15.20

Computation of $\hat{G}(s)$

$$\hat{G}_{11} = \text{system gain } \frac{\hat{x}_1}{x_1(0)} \ (\text{with } x_2^0 = \hat{u} = 0) = \frac{P_1 \Delta_1}{\Delta}$$

Where P_1, Δ_1 and Δ have the meanings as defined by Mason's gain formula

$$P_1 = \text{path gain of forward path} = \frac{1}{s}$$

$$\Delta = \text{determinant}$$

$$= 1 - \left(\frac{-5}{s} - \frac{6}{s^2}\right) = 1 + \frac{5}{s} + \frac{6}{s^2}$$

$$\Delta_1 = 1 - \left(\frac{-5}{s}\right) = 1 + \frac{5}{s}$$

$$\hat{G}_{11} = \frac{\dfrac{1}{s}\left(1 + \dfrac{5}{s}\right)}{1 + \dfrac{5}{s} + \dfrac{6}{s^2}} = \frac{s+5}{s^2 + 5s + 6} = \frac{s+5}{(s+3)(s+2)} = \frac{-2}{s+3} + \frac{3}{s+2}$$

$$\hat{G}_{12} = \text{system gain} \frac{\hat{x}_1}{x_2^0} \ (\text{with } x_1^0 = \hat{u} = 0) = \frac{P_1 \Delta_1}{\Delta}$$

$$P_1 = \frac{1}{s^2}$$

$$\Delta = 1 - \left(\frac{-5}{s} - \frac{6}{s^2}\right) = 1 + \frac{5}{s} + \frac{6}{s^2}$$

$$\Delta_1 = 1$$

$$\hat{G}_{12} = \frac{\dfrac{1}{s^2}}{1 + \dfrac{5}{s} + \dfrac{6}{s^2}} = \frac{1}{(s^2 + 5s + 6)} = \frac{-1}{s+3} + \frac{-1}{s+2}$$

$$\hat{G}_{21} = \text{system gain } \frac{\hat{x}_2}{x_1^0} \text{ (with } x_2^0 = \hat{u} = 0) = \frac{P_1 \Delta_1}{\Delta}$$

$$P_1 = \frac{-6}{s^2}$$

$$\Delta = 1 - \left(\frac{-5}{s} - \frac{6}{s^2}\right) = 1 + \frac{5}{s} + \frac{6}{s^2}$$

$$\Delta_1 = 1$$

$$\hat{G}_{21} = \frac{-6/s^2}{1 + \dfrac{5}{s} + \dfrac{6}{s^2}} = \frac{-6}{(s^2 + 5s + 6)} = \frac{6}{s+3} + \frac{-6}{s+2}$$

$$\hat{G}_{22} = \text{system gain } \frac{\hat{x}_2}{x_2^0} \text{ (with } x_1^0 = \hat{u} = 0) = \frac{P_1 \Delta_1}{\Delta}$$

$$P_1 = \frac{1}{s}$$

$$\Delta = 1 + \frac{5}{s} + \frac{6}{s^2}$$

$$\Delta_1 = 1$$

$$\hat{G}_{22} = \frac{\dfrac{1}{s}}{1 = \dfrac{5}{s} + \dfrac{6}{s^2}} = \frac{s}{s^2 + 5s + 6} = \frac{3}{s+3} + \frac{-2}{s+2}$$

Computation of $\hat{H}(s)$

$$\hat{H}_1 = \text{system gain } \frac{\hat{x}_1}{\hat{u}} \text{ (with } x_1^0 = x_2^0 = 0) = \frac{P_1 \Delta_1 + P_2 \Delta_2}{\Delta}$$

$$P_1 = \frac{1}{s^2}$$

$$P_2 = \frac{1}{s}$$

$$\Delta_1 = 1$$

$$\Delta_2 = 1 - \left(\frac{-5}{s}\right) = 1 + \frac{5}{s}$$

$$\Delta = 1 + \frac{5}{s} + \frac{6}{s^2}$$

$$\hat{H}_1 = \frac{\frac{1}{s^2} + \frac{1}{s}\left(1 + \frac{5}{s}\right)}{1 + \frac{5}{s} + \frac{6}{s^2}} = \frac{s+6}{s^2 + 5s + 6} = \frac{-3}{s+3} + \frac{4}{s+2}$$

$$\hat{H}_2 = \text{system gain } \frac{\hat{x}_2}{\hat{u}} \text{ (with } x_1^0 = x_2^0 = 0) = \frac{P_1 \Delta_1}{\Delta}$$

$$P_1 = \frac{1}{s}$$

$$\Delta = 1 + \frac{5}{s} + \frac{6}{s^2}$$

$$\Delta_1 = 1$$

$$\hat{H}_2 = \frac{\frac{1}{s}}{1 + \frac{5}{s} + \frac{6}{s^2}} = \frac{s}{s^2 + 5s + 6} = \frac{3}{s+3} + \frac{-2}{s+2}$$

$$\hat{x}(s) = \begin{bmatrix} \frac{-2}{s+3} + \frac{3}{s+2} & \frac{-1}{s+3} + \frac{1}{s+2} \\ \frac{6}{s+3} + \frac{-6}{s+2} & \frac{3}{s+3} + \frac{-2}{s+2} \end{bmatrix} \begin{bmatrix} x_1(0) \\ x_2(0) \end{bmatrix} + \begin{bmatrix} \frac{-3}{s+3} + \frac{4}{s+2} \\ \frac{3}{s+3} + \frac{-2}{s+2} \end{bmatrix} \hat{u}$$

If the input is zero, the response is

$$= e^{At}x(0)$$

(Laplace inverse of $\hat{G}(s)$ is e^{At})

$$e^{At} = \begin{bmatrix} -2e^{-3t} + 3e^{-2t} & -e^{-3t} + e^{-2t} \\ 6e^{-3t} - 6e^{-2t} & 3e^{-3t} - 2e^{-2t} \end{bmatrix}$$

$$x(0) = \begin{bmatrix} 1 \\ 2 \end{bmatrix}$$

$$x(t) = \begin{bmatrix} -2e^{-3t} + 3e^{-2t} & -e^{-3t} + e^{-2t} \\ 6e^{-3t} - 6e^{-2t} & 3e^{-2t} - 2e^{-2t} \end{bmatrix} \begin{bmatrix} 1 \\ 2 \end{bmatrix}$$

$$x(t) = \begin{bmatrix} -2e^{-3t} + 3e^{-2t} - 2e^{-3t} + 2e^{-2t} \\ 6e^{-3t} - 6e^{-2t} + 6e^{-3t} - 4e^{-2t} \end{bmatrix}$$

$$x(t) = \begin{bmatrix} -4e^{-3t} + 5e^{-2t} \\ 12e^{-3t} - 10e^{-2t} \end{bmatrix}$$

To obtain zero-state response, with a unit step input

$$\hat{u}(s) = \frac{1}{s}$$

$$x(s) = \hat{H}(s)\hat{u}(s)$$

$$= \begin{bmatrix} \dfrac{s+6}{(s^2+5s+6)} \\ \dfrac{s}{(s^2+5s+6)} \end{bmatrix} \dfrac{1}{s} = \begin{bmatrix} \dfrac{s+6}{s(s^2+5s+6)} \\ \dfrac{1}{(s^2+5s+6)} \end{bmatrix}$$

$$= \begin{bmatrix} \dfrac{s+6}{s(s+3)(s+2)} \\ \dfrac{1}{(s+3)(s+2)} \end{bmatrix} = \begin{bmatrix} \dfrac{1}{s} + \dfrac{1}{s+3} + \dfrac{-2}{s+3} \\ \dfrac{-1}{s+3} + \dfrac{1}{s+2} \end{bmatrix}$$

$$x(t) = \begin{bmatrix} 1 + e^{-3t} - 2e^{-2t} \\ -e^{-3t} + e^{-2t} \end{bmatrix}$$

e^{At} can be obtained by another method. Also to check your result

$$e^{At} = \mathcal{L}^{-1}[(sI-A)^{-1}]$$

$$[sI-A] = \begin{bmatrix} s & 0 \\ 0 & s \end{bmatrix} \begin{bmatrix} 0 & 1 \\ -6 & -5 \end{bmatrix} = \begin{bmatrix} s & -1 \\ 6 & s+5 \end{bmatrix}$$

$$[sI-A]^{-1} = \frac{1}{s^2+5s+6} \begin{bmatrix} s+5 & 1 \\ -6 & s \end{bmatrix}$$

$$= \begin{bmatrix} \dfrac{s+5}{(s+3)(s+2)} & \dfrac{1}{(s+3)(s+2)} \\ \dfrac{-6}{(s+3)(s+2)} & \dfrac{s}{(s+3)(s+2)} \end{bmatrix}$$

$$e^{At} = \mathcal{L}^{-1}-[(sI-A)^{-1}] = \begin{bmatrix} -2e^{-3t} + 3e^{-2t} & -e^{-3t} + e^{-2t} \\ 6e^{-3t} - 6e^{-2t} & 3e^{-3t} - 2e^{-2t} \end{bmatrix}$$

which is same as calculated earlier.

Problem 15.27 Obtain the transfer function if

$$\begin{bmatrix} \dot{x}_1 \\ \dot{x}_2 \end{bmatrix} = \begin{bmatrix} -5 & -1 \\ 3 & -1 \end{bmatrix} \begin{bmatrix} x_1 \\ x_2 \end{bmatrix} + \begin{bmatrix} 2 \\ 5 \end{bmatrix} u \qquad y = \begin{bmatrix} 1 & 2 \end{bmatrix} \begin{bmatrix} x_1 \\ x_2 \end{bmatrix}$$

Solution

$$G(s) = C(sI-A)^{-1}B$$
$$C = \begin{bmatrix} 1 & 2 \end{bmatrix}$$

$$B = \begin{bmatrix} 2 \\ 5 \end{bmatrix}$$

$$[sI - A] = \begin{bmatrix} s+5 & 1 \\ -3 & s+1 \end{bmatrix}$$

$$[sI - A]^{-1} = \frac{\begin{bmatrix} s+1 & -1 \\ 3 & s+5 \end{bmatrix}}{(s+5)(s+1) - (-3 \times 1)} = \frac{1}{(s+4)(s+2)} \begin{bmatrix} s+1 & -1 \\ 3 & s+5 \end{bmatrix}$$

$$G(s) = [1 \quad 2] \frac{1}{(s+4)(s+2)} \begin{bmatrix} s+1 & -1 \\ 3 & s+5 \end{bmatrix} \begin{bmatrix} 2 \\ 5 \end{bmatrix}$$

or
$$G(s) = [1 \quad 2] \frac{1}{(s+4)(s+2)} \begin{bmatrix} 2s-3 \\ 5s+31 \end{bmatrix}$$

$$= [1 \quad 2] \begin{bmatrix} \dfrac{2s-3}{(s+4)(s+2)} \\ \dfrac{5s+31}{(s+4)(s+2)} \end{bmatrix}$$

$$= \frac{2s-3+10s+62}{(s+4)(s+2)} = \frac{12s+59}{(s+4)(s+2)}$$

Problem 15.28   Find the state transition matrix for the following unforced system

$$\dot{x} = \begin{bmatrix} -2 & 1 & 0 \\ 0 & -2 & 1 \\ 0 & 0 & -2 \end{bmatrix}$$

(ME Pune University)

Solution

$$(sI - A) = \begin{bmatrix} s+2 & 1 & 0 \\ 0 & s+2 & -1 \\ 0 & 0 & s+2 \end{bmatrix}$$

$$[sI - A]^{-1} = \frac{1}{(s+2)^3} \begin{bmatrix} (s+2)(s+2) & 0 & 0 \\ (s+2) & (s+2)(s+2) & 0 \\ 1 & (s+2) & (s+2)(s+2) \end{bmatrix}$$

$$= \begin{bmatrix} \dfrac{1}{(s+2)} & 0 & 0 \\[3mm] \dfrac{1}{(s+2)^2} & \dfrac{1}{(s+2)} & 0 \\[3mm] \dfrac{1}{(s+2)^3} & \dfrac{1}{(s+2)^2} & \dfrac{1}{(s+2)} \end{bmatrix}$$

By making use of the formula

$$\frac{n!}{(s+a)^{n+1}} = t^n e^{-at}$$

$$e^{At} = L^{-1}[(sI - A)^{-1}] = \begin{bmatrix} e^{-2t} & 0 & 0 \\[2mm] te^{-2t} & e^{-2t} & 0 \\[2mm] \dfrac{1}{2}t^2 e^{-2t} & te^{-2t} & e^{-2t} \end{bmatrix}$$

Problem 15.29   A control system has a transfer function

$$G(s) = \frac{3(s+1)}{(s+5)(s+2)^2}$$

Draw state transition signal flow graph by method of parallel decomposition. State diagram should contain minimum number of integrators. Show states are decoupled.

(ME Pune University)

Solution

$$G(s) = \frac{3(s+1)}{(s+5)(s+2)^2} = \frac{A}{(s+5)} + \frac{B}{(s+2)} + \frac{C}{(s+2)^2}$$

$$A = \frac{3(s+1)}{(s+2)^2}\bigg|_{s=-5} = \frac{-4}{3} \quad \text{and} \quad C = \frac{3(s+1)}{(s+5)}\bigg|_{s=-2} = -1$$

$$B = \frac{d}{ds}[(s+2)^2 \, G(s)]\bigg|_{s=-2}$$

$$= \frac{d}{ds}\left[\frac{(s+2)^2 \times 3(s+1)}{(s+5)(s+2)^2}\right]\bigg|_{s=-2} = \frac{d}{ds}\left[\frac{3(s+1)}{(s+5)}\right]\bigg|_{s=-2}$$

$$= 3\left[\frac{(s+5)-(s+1)}{(s+5)^2}\right]\bigg|_{s=-2} = \frac{3 \times 4}{(s+5)^2}\bigg|_{s=-2} = \frac{3 \times 4}{3 \times 3} = \frac{4}{3}$$

$\therefore$ $$G(s) = \frac{-1}{(s+2)^2} + \frac{4/3}{(s+2)} + \frac{-4/3}{(s+5)}$$

State transition signal flow graph is shown in Fig. 15.21

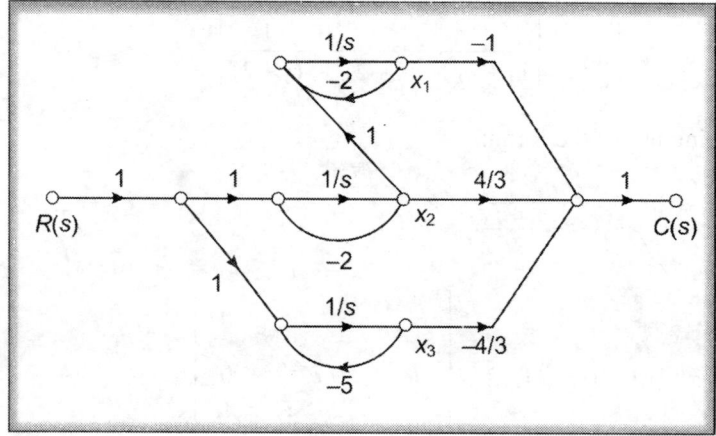

Fig. 15.21

State model is

$$\begin{bmatrix} \dot{x}_1 \\ \dot{x}_2 \\ \dot{x}_3 \end{bmatrix} = \begin{bmatrix} -2 & 1 & 0 \\ 0 & -2 & 0 \\ 0 & 0 & -5 \end{bmatrix} \begin{bmatrix} x_1 \\ x_2 \\ x_3 \end{bmatrix} + \begin{bmatrix} 0 \\ 1 \\ 1 \end{bmatrix} r$$

$$y = \begin{bmatrix} -1 & \dfrac{4}{3} & \dfrac{-4}{3} \end{bmatrix} \begin{bmatrix} x_1 \\ x_2 \\ x_3 \end{bmatrix}$$

Problem 15.30 A system is governed by the differential equation:

$$\frac{d^3 y}{dt^3} + \frac{6 d^2 y}{dt^2} + \frac{11 dy}{dt} + 10y = 8u(t)$$

where y is the output and u is the input of the system. Obtain a state space representation of the system.   *(Pune University)*

Solution

The given equation is $\dddot{y} + 6\ddot{y} + 11\dot{y} + 10y = 8u$

Let $x_1 = y$

$x_2 = \dot{y}$

$x_3 = \ddot{y}$

then $\dot{x}_1 = x_2$

$$\dot{x}_2 = x_3$$
$$\dot{x}_3 = -10x_1 - 11x_2 - 6x_3 + 8u$$

It can be represented as

$$\begin{bmatrix} \dot{x}_1 \\ \dot{x}_2 \\ \dot{x}_3 \end{bmatrix} = \begin{bmatrix} 0 & 1 & 0 \\ 0 & 0 & 1 \\ -10 & -11 & -6 \end{bmatrix} \begin{bmatrix} x_1 \\ x_2 \\ x_3 \end{bmatrix} + \begin{bmatrix} 0 \\ 0 \\ 8 \end{bmatrix} u$$

$$y = [1 \ 0 \ 0] \begin{bmatrix} x_1 \\ x_2 \\ x_3 \end{bmatrix}$$

State model representation is shown in Fig. 15.22

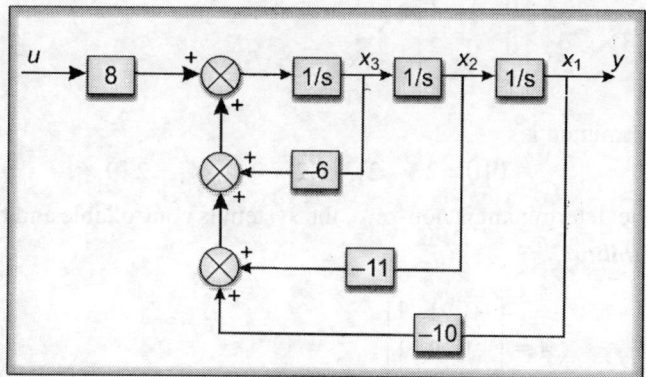

Fig. 15.22

Problem 15.31 Determine the state controllability and observability of the system described by

$$\dot{x} = \begin{bmatrix} -3 & 1 & 1 \\ -1 & 0 & 1 \\ 0 & 0 & 1 \end{bmatrix} x + \begin{bmatrix} 0 & 1 \\ 0 & 0 \\ 2 & 1 \end{bmatrix} u$$

$$y = \begin{bmatrix} 0 & 0 & 1 \\ 1 & 1 & 0 \end{bmatrix} x$$

(ME Pune Universty)

Solution

$$B = \begin{bmatrix} 0 & 1 \\ 0 & 0 \\ 2 & 1 \end{bmatrix}$$

$$AB = \begin{bmatrix} -3 & 1 & 1 \\ -1 & 0 & 1 \\ 0 & 0 & 1 \end{bmatrix} \cdot \begin{bmatrix} 0 & 1 \\ 0 & 0 \\ 2 & 1 \end{bmatrix} = \begin{bmatrix} 2 & -2 \\ 2 & 0 \\ 2 & 1 \end{bmatrix}$$

$$A^2B = \begin{bmatrix} -3 & 1 & 1 \\ -1 & 0 & 1 \\ 0 & 0 & 1 \end{bmatrix} \begin{bmatrix} 0 & -2 \\ 0 & 0 \\ 2 & 1 \end{bmatrix} = \begin{bmatrix} -2 & -5 \\ 0 & 3 \\ 0 & 1 \end{bmatrix}$$

$$Q_c = [B : AB : A^2B]$$

or

$$Q_c = \begin{bmatrix} 0 & 1 & 2 & -2 & -2 & -5 \\ 0 & 0 & 2 & 0 & 0 & 3 \\ 2 & 1 & 2 & 1 & 2 & 1 \end{bmatrix}$$

We should be able to collect three columns from Q_c to form a square matrix whose determinant is non-zero

Let it be

$$\begin{bmatrix} 0 & 1 & 2 \\ 0 & 0 & 2 \\ 2 & 1 & 2 \end{bmatrix}$$

and its determinant is

$$= 0[0 \times 2 \quad -2] - 1[2 \times 0 \quad -4] + 2[0 \times 1 \quad -0 \times 2] = +4$$

Since, the determinant is non-zero, the system is controllable and the rank is 3.

Observability

$$A = \begin{bmatrix} -3 & 1 & 1 \\ -1 & 0 & 1 \\ 0 & 0 & 1 \end{bmatrix}$$

$$C = \begin{bmatrix} 0 & 0 & 1 \\ 1 & 1 & 0 \end{bmatrix}$$

$$C^T = \begin{bmatrix} 0 & 1 \\ 0 & 1 \\ 1 & 0 \end{bmatrix}$$

$$A^T = \begin{bmatrix} -3 & -1 & 0 \\ 1 & 0 & 0 \\ 1 & 1 & 1 \end{bmatrix}$$

$$A^T C^T = \begin{bmatrix} -3 & -1 & 0 \\ 1 & 0 & 0 \\ 1 & 1 & 1 \end{bmatrix} \begin{bmatrix} 0 & 1 \\ 0 & 1 \\ 1 & 0 \end{bmatrix} = \begin{bmatrix} 0 & -4 \\ 0 & 1 \\ 1 & 2 \end{bmatrix}$$

$$A^T (A^T C^T) = \begin{bmatrix} -3 & -1 & 0 \\ 1 & 0 & 0 \\ 1 & 1 & 1 \end{bmatrix} \begin{bmatrix} 0 & -4 \\ 0 & 1 \\ 1 & 2 \end{bmatrix} = \begin{bmatrix} 0 & 11 \\ 0 & -4 \\ 1 & -1 \end{bmatrix}$$

$$T = [C^T : A^T C^T : A^T (A^T C^T)]$$

$$= \begin{bmatrix} 0 & 1 & 0 & -4 & 0 & 11 \\ 0 & 1 & 0 & 1 & 0 & -4 \\ 1 & 0 & 1 & 2 & 1 & -1 \end{bmatrix}$$

Let us choose

$$\begin{bmatrix} -4 & 0 & 11 \\ 1 & 0 & -4 \\ 2 & 1 & -1 \end{bmatrix}$$

$$det = -4\,[0+4] + 11\,[1-0] = -5$$

Since, determinant is not equal to zero, hence, rank is 3 and the system is completely observable.

Problem 15.32 Transfer function of a system is given by

$$\frac{Y(s)}{U(s)} = \frac{2}{s^3 + 6s^2 + 11s + 6}$$

Find controllability and observability.

Solution In matrix notation

$$\begin{bmatrix} \dot{x}_1 \\ \dot{x}_2 \\ \dot{x}_3 \end{bmatrix} = \begin{bmatrix} 0 & 1 & 0 \\ 0 & 0 & 1 \\ -6 & -11 & -6 \end{bmatrix} \begin{bmatrix} x_1 \\ x_2 \\ x_3 \end{bmatrix} + \begin{bmatrix} 0 \\ 0 \\ 2 \end{bmatrix} u$$

$$y = [1 \ 0 \ 0] \begin{bmatrix} x_1 \\ x_2 \\ x_3 \end{bmatrix}$$

Eigenvalues are

$$\lambda_1 = -1$$
$$\lambda_2 = -2$$
$$\lambda_3 = -3$$

Choosing Vandermonde matrix as modal matrix we get

$$M = \begin{bmatrix} 1 & 1 & 1 \\ \lambda_1 & \lambda_2 & \lambda_3 \\ \lambda_1^2 & \lambda_2^2 & \lambda_3^2 \end{bmatrix} = \begin{bmatrix} 1 & 1 & 1 \\ -1 & -2 & -3 \\ 1 & 4 & 9 \end{bmatrix}$$

$$M^{-1} = \begin{bmatrix} 3 & 5/2 & 1/2 \\ -3 & -4 & -1 \\ 1 & 3/2 & 1/2 \end{bmatrix}$$

$$B' = M^{-1} B = \begin{bmatrix} 3 & 5/2 & 1/2 \\ -3 & -4 & -1 \\ 1 & 3/2 & 1/2 \end{bmatrix} \begin{bmatrix} 0 \\ 0 \\ 2 \end{bmatrix} = \begin{bmatrix} 1 \\ -2 \\ 1 \end{bmatrix}$$

Since, none of rows in matrix B' is zero, the system is completely controllable.
Observability

$$X = CM$$

$$X = [1\ 0\ 0]\begin{bmatrix} 1 & 1 & 1 \\ -1 & -2 & -3 \\ 1 & 4 & 9 \end{bmatrix} = [1\ 1\ 1]$$

$$y = CX = CMZ$$

$$= [1\ 1\ 1]\begin{bmatrix} z_1 \\ z_2 \\ z_3 \end{bmatrix}$$

The system is completely observable as $X = CM = [1\ 1\ 1]$ is not having zero element.

Problem 15.33   A system is described by the equations

$$\begin{bmatrix} \dot{x}_1 \\ \dot{x}_2 \\ \dot{x}_4 \end{bmatrix} = \begin{bmatrix} 0 & 1 & 0 \\ 0 & 0 & 1 \\ -6 & -11 & -6 \end{bmatrix}\begin{bmatrix} x_1 \\ x_2 \\ x_3 \end{bmatrix} + \begin{bmatrix} 0 \\ 0 \\ 1 \end{bmatrix}u$$

$$y = [1\ 1\ 0]$$

Find if the system is completely observable. If not, find the mode which is not observable.

Solution

$$A = \begin{bmatrix} 0 & 1 & 0 \\ 0 & 0 & 1 \\ -6 & -11 & -6 \end{bmatrix}$$

$$A^T = \begin{bmatrix} 0 & 0 & -6 \\ 1 & 0 & -11 \\ 0 & 1 & -6 \end{bmatrix}$$

$$C^T = \begin{bmatrix} 1 \\ 1 \\ 0 \end{bmatrix}$$

$$A^T C^T = \begin{bmatrix} 0 \\ 1 \\ 1 \end{bmatrix}$$

$$A^T(A^T C^T) = \begin{bmatrix} -6 \\ -11 \\ -5 \end{bmatrix}$$

$$T = \begin{bmatrix} 1 & 0 & -6 \\ 1 & 1 & -11 \\ 0 & 1 & -5 \end{bmatrix}$$

$det\ |T| = 1\,[(1\times-5)-(1\times-11)-0+(-6)\,(1-0)] = -5+11-6 = 0$

Since determinant is zero, rank is not 3 and hence system is not completely observable.

Eigenvalues are $-1, -2$ and -3

$$M = \begin{bmatrix} 1 & 1 & 1 \\ -1 & -2 & -3 \\ 1 & 4 & 9 \end{bmatrix} = \text{Vandermonde matrix}$$

$$X = CM = \begin{bmatrix} 1 & 1 & 0 \end{bmatrix} \begin{bmatrix} 1 & 1 & 1 \\ -1 & -2 & -3 \\ 1 & 4 & 9 \end{bmatrix} = \begin{bmatrix} 0 & -1 & 1 \end{bmatrix}$$

$$y = CMZ = \begin{bmatrix} 0 & -1 & 1 \end{bmatrix} \begin{bmatrix} z_1 \\ z_2 \\ z_3 \end{bmatrix}$$

$z_1(t)$ is not observable, a mode due to eigenvalue $\lambda = -1$

Problem 15.34 Give a suitable state variable representation (state equation) for the system described by the following third-order differential equation. (*AMIE*)
$$[\dddot{y} + 2\ddot{y} + 5\dot{y} + y = 10]$$

Solution

Let
$$x_1 = y$$
$$x_2 = \dot{y}$$
$$x_3 = \ddot{y}$$

then
$$\dot{x}_1 = \dot{y}$$
$$\dot{x}_2 = \ddot{y}$$
$$\dot{x}_3 = \dddot{y}$$

or
$$\dot{x}_1 = x_2$$
$$\dot{x}_2 = x_3$$
$$\dot{x}_3 = \dddot{y} = 10 - 2\ddot{y} - 5\dot{y} - y = 10 - 2x_3 - 5x_2 - x_1$$
$$= -x_1 - 5x_2 - 2x_3 + 10$$

In vector-matrix notation

$$\begin{bmatrix} \dot{x}_1 \\ \dot{x}_2 \\ \dot{x}_3 \end{bmatrix} = \begin{bmatrix} 0 & 1 & 0 \\ 0 & 0 & 1 \\ -1 & -5 & -2 \end{bmatrix} \begin{bmatrix} x_1 \\ x_2 \\ x_3 \end{bmatrix} + \begin{bmatrix} 0 \\ 0 \\ 10 \end{bmatrix}$$

$$y = \begin{bmatrix} 1 & 0 & 0 \end{bmatrix} \begin{bmatrix} x_1 \\ x_2 \\ x_3 \end{bmatrix}$$

Problem 15.35 Consider the control system shown in Fig. 15.23. Obtain a state space model of the system. $\hspace{2cm}$ *(B.Sc.Engg. (Elect.)–1989)*

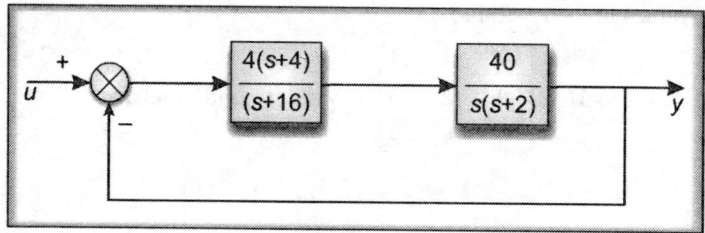

Fig. 15.23

Solution

$$\frac{y}{u} = \frac{\dfrac{40 \times 4(s+4)}{s(s+2)(s+16)}}{1 + \dfrac{40 \times 4(s+4)}{s(s+2)(s+16)}}$$

$$= \frac{160(s+4)}{s(s+2)(s+16) + 160(s+4)}$$

$$= \frac{160s + 640}{s(s^2 + 18s + 32) + 160s + 640}$$

$$= \frac{160s + 640}{s^3 + 18s^2 + 32s + 160s + 640}$$

$$= \frac{160s + 640}{s^3 + 18s^2 + 192s + 640}$$

$$= \frac{\dfrac{160}{s^2} + \dfrac{640}{s^3}}{1 - \left(-\dfrac{18}{s} - \dfrac{192}{s^2} - \dfrac{640}{s^3} \right)}$$

Signal flow graph is shown in Fig. 15.24.

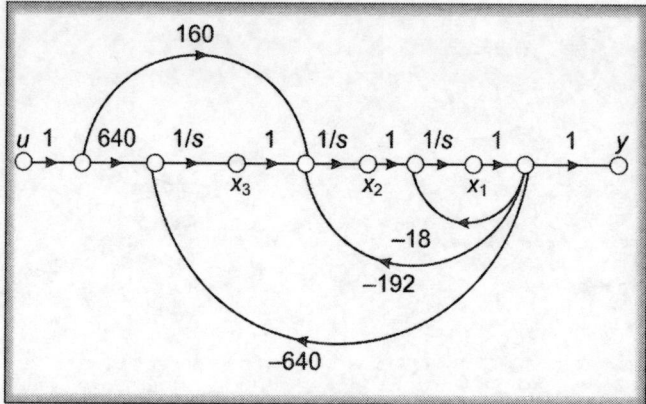

Fig. 15.24

From the signal flow graph, we have

$$y = x_1$$

$$x_1 = \frac{1}{s}(x^2 - 18y)$$

$$\dot{x}_1 = x_2 - 18x_1$$

$$x_2 = \frac{1}{s}(x_3 + 160u - 192y)$$

$$\dot{x}_2 = x_3 - 192x_1 + 160u$$

$$x_3 = \frac{1}{s}(-640y + 640u)$$

$$\dot{x}_3 = -640x_1 + 640u$$

Phase variable representation is thus

$$\begin{bmatrix} x_1 \\ x_2 \\ x_3 \end{bmatrix} = \begin{bmatrix} -18 & 1 & 0 \\ -192 & 0 & 1 \\ -640 & 0 & 0 \end{bmatrix} \begin{bmatrix} x_1 \\ x_2 \\ x_3 \end{bmatrix} + \begin{bmatrix} 0 \\ 160 \\ 640 \end{bmatrix} u$$

$$y = \begin{bmatrix} 1 & 0 & 0 \end{bmatrix} \begin{bmatrix} x_1 \\ x_2 \\ x_3 \end{bmatrix}$$